WORKING DRAWINGS METHODS
THREAD DETAILS AUXILIARIES
SECTIONS PICTORIALS VECTORS
DESCRIPTIVE GEOMETRY
ORTHOGRAPHIC VIEWS GRAPHS
EMPIRICAL EQUATIONS POINTS
LINES PLANES GEARS CAMS
COMPUTER GRAPHICS SKETCHES
ISOMETRICS OBLIQUES SPRINGS
CONTOURS STRIKE & DIP SLOPE
OUTCROP NOMOGRAPHY
EMPIRICAL EQUATIONS

Architectural drawings made with computer graphics
using MEGA CADD by MEGA CADD, Inc.

DRAFTING TECHNOLOGY
SECOND EDITION

DRAFTING TECHNOLOGY
SECOND EDITION

James H. Earle
TEXAS A&M UNIVERSITY

ADDISON-WESLEY PUBLISHING COMPANY
Reading, Massachusetts • Menlo Park, California
Don Mills, Ontario • Wokingham, England • Amsterdam
Sydney • Singapore • Tokyo • Madrid • Bogotá
Santiago • San Juan

Sponsoring Editor: Tom Robbins
Production Supervisor: Laura Skinger
Text Design: Rose Design Associates
Illustrator and Cover Design: James H. Earle
Production Coordinator: Ezra C. Holston
Layout Artist: Lorraine Hodsdon

Library of Congress Cataloging in Publication Data
Earle, James H.
  Drafting technology.
  Includes index.
  1. Mechanical drawing.   I. Title.
T353.E28   1986   604.2'4   85-7487
ISBN 0-201-10239-0

Dedicated to
Elizabeth and Susan Earle
from whom this book took many hours
that I would have liked to have given them.

The following supplements are compatible with the following textbooks from Addison-Wesley Publishing Co.:

Drafting Technology,

Engineering Design Graphics,

Design Drafting.

Graphics for Engineers,

Geometry for Engineers

DRAFTING TECHNOLOGY PROBLEMS is a problem book designed to cover basic graphics, descriptive geometry, and specialty drafting areas. It is designed to accompany *Drafting Technology,* that is available from Addison-Wesley Publ. Co., Reading, Mass. 01867. 131 pages.

BASIC DRAFTING (With Computer Graphics) is a problem book that covers the basics for a one-semester high school drafting course. 67 pages.

CREATIVE DRAFTING (With Computer Graphics) is a problem book that covers mechanical drawing and architectural drafting for the high school. 106 pages.

DRAFTING & DESIGN (With Computer Graphics) is a problem book for mechanical drawing for a high school or college course. 106 pages.

DRAFTING FUNDAMENTALS 1 is a problem book for a one-year high school course in mechanical drawing. 94 pages.

DRAFTING FUNDAMENTALS 2 is a second version of **DRAFTING FUNDAMENTALS 1** for the same level and content. 94 pages.

TECHNICAL ILLUSTRATION is a problem book for a course in pictorial drawing for the college or high school. 70 pages.

ARCHITECTURAL DRAFTING is a problem book for a first course in architectural drafting. 71 pages.

GRAPHICS FOR ENGINEERS 1 (With Computer Graphics) is a problem book for a first course in engineering graphics for the college student. 100 pages.

GRAPHICS FOR ENGINEERS 2 (With Computer Graphics) is a problem book for a first course in engineering graphics for the college student. 100 pages.

GRAPHICS FOR ENGINEERS 3 (With Computer Graphics) is a problem book for a first course in engineering graphics for the college student. 108 pages.

GEOMETRY FOR ENGINEERS 1 (With Computer Graphics) is a problem book for a college-level descriptive geometry course. 100 pages.

GEOMETRY FOR ENGINEERS 2 (With Computer Graphics) is a problem book for a college-level descriptive geometry course. 100 pages.

GEOMETRY FOR ENGINEERS 3 (With Computer Graphics) is a problem book for a college-level descriptive geometry course. 100 pages.

GRAPHICS & GEOMETRY 1 (With Computer Graphics) is a problem book for a college course in graphics and descriptive geometry. 119 pages.

GRAPHICS & GEOMETRY 2 (With Computer Graphics) is a problem book for a college course in graphics and descriptive geometry. 121 pages.

GRAPHICS & GEOMETRY 3 (With Computer Graphics) is a problem book for a college course in graphics and descriptive geometry. 138 pages.

Creative Publishing Co.
BOX 9292   COLLEGE STATION, TEXAS   77840
PHONE 409-775-6047

Preface

DRAFTING TECHNOLOGY has been written to cover the principles of drafting, engineering drawing, and graphical problem solving for a college-level course. This second edition is a major revision of the first, but it retains the same classroom-tested format and sequence of topics.

RATIONALE

The major areas presented in the text are communications, working drawings, descriptive geometry, computer graphics, and design concepts. Each of these areas is an important part of the background of the technician, technologist, and engineer.

Communications is covered as it relates to the presentation of ideas, three-dimensional concepts, data analysis, and pictorials. The methods of making working drawings are given in accordance with ANSI standards and include materials, tolerances, welding specifications, pipe drafting, and electrical/electronic drafting. The design process is covered in Chapter 17 to show the importance of engineering graphics in developing creative solutions to technical problems. Descriptive geometry is included in Chapters 19 through 24. Computer graphics (Chapter 30) has been presented in an entirely fresh format where specific examples have been given using AutoCAD software and microcomputers. Vectors, nomography, empirical equations, and graphical calculus are other specialty areas receiving emphasis.

OBJECTIVES

The objective of this book is to support courses in which the student learns the various standards and techniques of preparing engineering drawings and specifications, learns how to read and interpret drawings, learns how to supervise the preparation of drawings by others, and learns how to solve three-dimensional technical problems that require the application of descriptive geometry and graphical analysis. Above all, this textbook is designed to help students expand their creative talents and communicate their ideas in an effective manner.

FORMAT

DRAFTING TECHNOLOGY can be used as a self-instructional book enabling the student to work independently. Many problems are presented in a step-by-step sequence that shows the progressive steps in solving a problem and a second color is used to emphasize important points of construction. There are over 600 problems, with many new to this edition, at the ends of chapters. These problems offer a wide range of assignments that can be solved by students to aid them in grasping the principles covered in each chapter.

REVISION FEATURES

This revision entailed the modification of several hundred illustrations, the preparation of several hundred new drawings, the editing of all chapters, and the complete revision of several chapters. The chapter on computer graphics was entirely redone to offer an introduction to microcomputer graphics using AutoCAD software, the most widely used software on the market today. Also, the chapters on tolerancing and the design process are major revisions.

SUPPLEMENTARY PROBLEMS

Fifteen different problem books are available to be used with this textbook. A listing of these books and their source is given on the page preceeding the Preface. Eight of the manuals have computer graphics problems on the backsides of the problem sheets, that allow the coverage of the problems by both computer and traditional methods with the same problem books.

ACKNOWLEDGEMENTS

We are grateful for the assistance of many who have influenced the development of this volume.

We have been significantly aided by the use of many illustrations developed by the late William E. Street and Carl L. Svensen for their earlier publications. Many industries have furnished photographs and drawings that have been acknowledged in the legends where they are used. The Engineering Design Graphics staff of Texas A & M University has been helpful in making suggestions for the preparation of this book. Professor Tom Pollock provided valuable information on various metals and their designations.

We are appreciative of the many institutions who have thought enough of our publications to adopt them for classroom use. It is indeed an honor for one's work to be accepted by his colleagues. We are hopeful that this textbook will fill the needs of technology and engineering programs. As always, comments and suggestions for improvement and revision of this book will be appreciated.

College Station, Texas   J.H.E.

Contents

DRAFTING TECHNOLOGY
SECOND EDITION

CHAPTER 1

Introduction to Drafting Technology

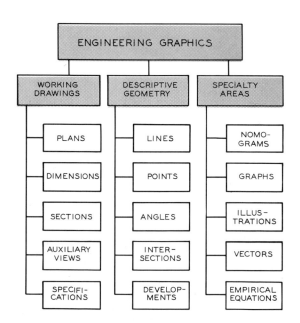

FIG. 1.1 The major divisions within the broad field of engineering graphics.

1.1
Engineering graphics

The term "engineering graphics" describes the broad field that uses drawings as a means of solving problems and presenting their solutions. It is also used to describe college courses required of most students in engineering and engineering technology programs. Engineering graphics consists of three major divisions—working drawings, descriptive geometry, and specialty areas—as illustrated in Fig. 1.1. A typical work station of a drafting technologist is shown in Fig. 1.2.

1.2
Working drawings

Working drawings are the plans that are drawn to explain the construction of something before it is built. A typical working drawing for an inspection hole

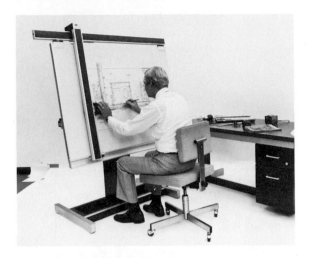

FIG. 1.2 A typical work station of a drafting technologist working in industry. (Courtesy of Keuffel and Esser Co., Morristown, N.J.)

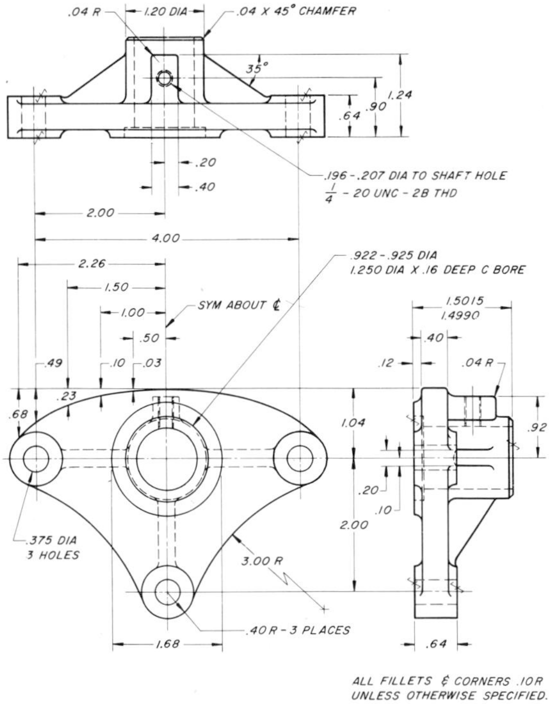

flange is illustrated in Fig. 1.3. Since working drawings must be understood by a great number of people, standard practices have been developed for their preparation. Without standardization, the designer and technologist would not have a common language for communicating their ideas.

Although the drafter is responsible for preparing working drawings, the engineer or the designer who originated the design is responsible for the accuracy and correctness of the plans. Consequently, the engineer and designer must have a clear understanding of working drawings and the methods used to prepare them.

1.3
Descriptive geometry

Descriptive geometry is a graphical method used for solving three-dimensional problems to determine geometric information. For example, the angle of a bend in a support rod can be determined by auxiliary views projected from the given orthographic views (Fig. 1.4).

FIG. 1.3 A detail or working drawing that gives the necessary details for the construction of a part. (Courtesy of General Motors Corp.)

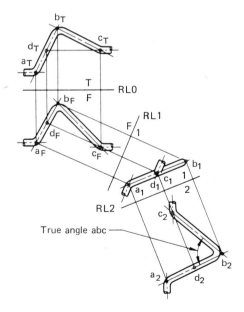

FIG. 1.4 Using descriptive geometry, the angle between two lines can be found by projecting auxiliary views from the given orthographic views. (Courtesy of General Motors Corp.)

Descriptive geometry was developed by a young French student, Gaspard Monge (Fig. 1.5), who used this method for designing fortifications. Descriptive geometry was found to be so much easier and more effective than the mathematical methods that it was kept a military secret for a number of years. Monge became a scientific aide to Napoleon, and was highly respected as a man of knowledge.

FIG. 1.5 Gaspard Monge (1746–1818) was the originator of descriptive geometry.

1.4
Specialty areas

A few of the many specialty areas of engineering graphics are described in the following sections.

NOMOGRAMS AND MATHEMATICS Mathematical relationships can be shown and solved by the application of engineering graphics. A nomogram that is used for the solution of the equation $A = BC$ is given in Fig. 1.6. It is possible to find many solutions to this equation by laying a straightedge across the two outside scales and reading the answer from the middle scale. Advanced mathematical problems involving calculus can be solved graphically as well.

GRAPHS Data, numbers, and lists of information may be difficult to understand unless shown in picture form. The picture may be a graph, a chart, or a schematic diagram. For example, the graph in Fig. 1.7 illustrates the relationship of clay tile strength to its absorption characteristics. Almost all technical reports use graphs and diagrams to explain technical data.

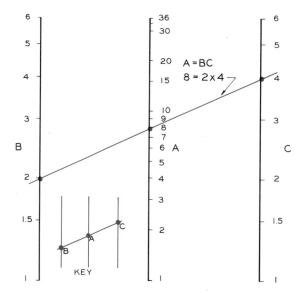

FIG. 1.6 This nomogram can be used for solving the equation of $A = BC$ by laying a straightedge across the outside scales and reading the answer on the middle scale.

TECHNICAL ILLUSTRATION The pictorial presentation of technical objects and products is called technical illustration. The technical illustrator's finished drawing is usually reproduced in a catalog or a publication that explains a particular project or product. The technical illustration of the automobile in Fig.

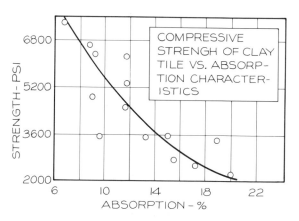

FIG. 1.7 This graph presents a comparison of the compressive strength of clay tile with its absorption characteristics.

FIG. 1.8 This technical illustration presents a view of the automobile that would be impossible by any other method. (Courtesy of Ford Motor Co.)

1.8 is an example of an illustration that would not have been possible without the skill of the technical illustrator.

Illustrators must be able to prepare and read working drawings since most of their illustrations are made from reference to detail drawings.

VECTOR ANALYSIS Graphical methods are used to analyze structural systems to determine the loads and forces in various members. The forces exerted by loads are represented by arrows, which are known as vectors and are drawn to scale. The forces in chains A and B in Fig. 1.9 are found by a vector polygon when the load in the lower chain is known.

Trusses and structural frames can be designed with the help of graphical vector analysis. Used in this manner, engineering graphics supplements the mathematical methods that are conventionally used in solving problems of this type.

EMPIRICAL DATA Data that are collected by experimentation to identify mathematical relationships are called empirical data. Graphical methods can be applied to determine the mathematical equations of data that have been plotted on a graph. Once the equation has been determined, it is possible to manipulate the data mathematically and to analyze them in detail.

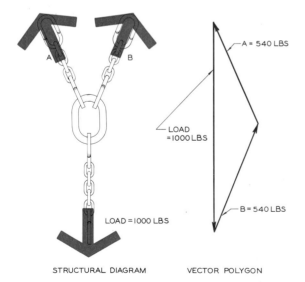

FIG. 1.9 The loads in this double sling chain can be determined by a graphical procedure known as vector analysis. The vector arrows are measured to find their loads.

1.5
Applications of drawings

Without the use of engineering graphics, engineers and technologists could not function effectively; consequently, most applications of drawings have been directed toward the area of engineering.

There are many other areas of specialization in which engineering graphics can be applied. A number of other fields using this discipline are covered in the following sections.

Designers

The designer develops new products and methods that improve our way of life. These products may be as simple as a bicycle (Fig. 1.10) or as complex as an entire manufacturing process (Fig. 1.11). The designer

FIG. 1.10 Bicycles are examples of mass-produced products developed by a designer. (Courtesy of Huffy Co.)

FIG. 1.12 Designers must master both mechanical drawing and freehand sketching as tools of their trade that are used daily. (Courtesy of Keuffel and Esser Co., Morristown, N.J.)

is often an engineer, but this is not a requirement. In general, the designer is a problem solver who uses imagination and creativity to develop a marketable product.

The designer must use engineering graphics techniques in addition to many freehand sketches in the development of a problem's solution (Fig. 1.12). It would be almost impossible for the designer to function without the ability to employ engineering graphics.

Architects and engineers

Engineers and architects deal with designs that cannot be explained without the use of drawings.

The architect designs and details buildings, homes, schools, shopping centers, and residential developments. He or she must prepare drawings that show floor plans, elevations, construction details, and pictorials that explain the appearance of the proposed design. An architect may even be involved in designing entire city blocks (Fig. 1.13).

The engineer assists the architect in designing structural systems and mechanical systems that are incorporated within buildings. He or she may design the air-conditioning system, supply the specifications for the foundations, or design the mechanical systems in a commercial building. The engineer must use engineering graphics to communicate ideas to contractors and construction workers who will build the structure that has been designed.

The engineer is responsible for the design of traffic systems (Fig. 1.14), aircraft, machines, utilities that serve entire communities, and superstructures of all kinds. The Buffalo Bill Dam in Fig. 1.15 is an example of an engineering project that was developed to conserve water and provide electrical power to the area of Cody, Wyoming.

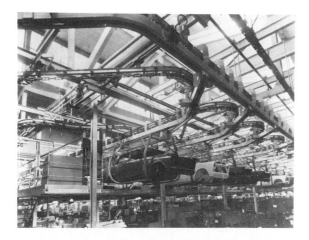

FIG. 1.11 Designers are involved in all products and systems, including the development of entire manufacturing processes.

FIG. 1.13 Architects may be responsible for designing an entire city block such as Houston Center, shown here. (Courtesy of Texas Eastern Transmission Co.)

FIG. 1.15 Engineers use engineering graphics to develop plans for projects such as this one, the Buffalo Bill Dam, outside of Cody, Wyoming. (Courtesy of the Bureau of Reclamation.)

FIG. 1.14 Engineers are responsible for designing the traffic systems of the city and countryside and the structures that support them.

Technologists and technicians

The technologist and technician are semiprofessional assistants who work with engineers and architects. The technologist usually has finished a four-year engineering technology program at the college level, while the technician is usually a graduate of a two-year technician program.

Either may be in charge of a drafting room, be a laboratory assistant, or work under the supervision of an architect or engineer. They might assist in the design and construction of complex structures such as the cement plant illustrated in Fig. 1.16.

Like the engineer, they must be able to prepare or supervise the preparation of working drawings.

FIG. 1.16 Technologists and technicians might assist with the design and construction of complex projects such as this cement plant, which needed extensive application of engineering graphics.

Average citizens

Most people do not make many drawings as part of their jobs. However, everyone should be able to sketch a map to give directions, communicate ideas to a repairman who is modifying a home, and use drawings to explain ideas when other methods fail. Almost everyone can interpret survey results that are illustrated by a graph (Fig. 1.17), interpret a weather report that is graphically presented, or read a complicated travel map.

Drawings illustrating objects that will be used by a nontechnical person are in most cases pictorials rather than working drawings. An example is a set of instructions that show how unassembled products are put together.

Engineering graphics is an important method of communicating and solving problems. It is more essential in our technological world now than ever before.

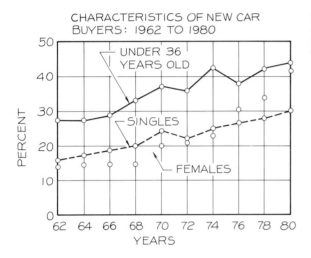

FIG. 1.17 The average citizen can read and interpret the information found by a survey when it is presented graphically.

CHAPTER

Drawing Instruments

2.1
Introduction

The preparation of technical drawings is possible only through knowledge of and skill in the use of a variety of drafting instruments.

You will find that your skill and productivity will increase with practice and as you become more familiar with using the tools that are available to you. People with little artistic ability can produce technical drawings of a professional quality when they learn to use drawing instruments and tools properly.

2.2
Pencils

Pencils may be either the conventional wood pencil or the lead holder, which is a mechanical pencil (Fig. 2.1).Both types are identified by a number and/or let-

FIG. 2.1 The mechanical pencil (lead holder) or the wood pencil can be used for mechanical drawing. The ends of the lead holder and the wood pencil are labeled to indicate the grade of the pencil lead.

ter at the end. Sharpen the end opposite these markings so you will not sharpen away the identity of the grade of lead.

Pencil grades are shown in Fig. 2.2, ranging from the hardest, 9H, to the softest, 7B.

The wood pencil can be sharpened with a small knife, which removes the wood and leaves approximately ⅜ inch of lead exposed (Fig. 2.3). The point can then be sharpened to a conical point with a sandpaper pad, (Fig. 2.4). The excess graphite is wiped from the point with a cloth or tissue.

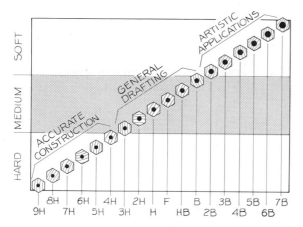

FIG. 2.2 The hardest pencil lead is 9H and the softest is 7B. Note that the diameter of the hard leads is smaller than the soft leads.

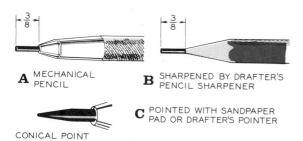

A MECHANICAL PENCIL

B SHARPENED BY DRAFTER'S PENCIL SHARPENER

C POINTED WITH SANDPAPER PAD OR DRAFTER'S POINTER

CONICAL POINT

FIG. 2.3 The drafting pencil should be sharpened to a tapered conical point (not a needle point) with a sandpaper pad or other type of sharpener.

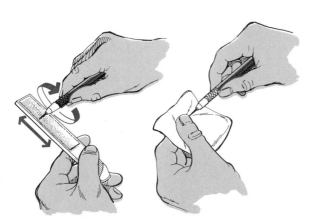

FIG. 2.4 The drafting pencil is revolved about its axis as it is stroked against the sandpaper pad to form a conical point. The graphite is wiped from the sharpened point with a tissue or a cloth.

Professional drafters often use a pencil pointer (Fig. 2.5) to sharpen either wood or mechanical pencils. Insert the pencil in the hole and revolve it to sharpen the lead to a conical point. Other types of small hand-held point sharpeners that work on the same principle are available.

FIG. 2.5 The professional drafter will often use a pencil pointer of this type to sharpen pencils.

2.3
Papers and drafting media

SIZES The surface on which a drawing is made must be carefully selected to yield the best results for a given application. This usually begins by selecting the sheet size. Sheet sizes are specified by letters such as Size A, Size B, and so forth.

Size A	8½″ × 11″	9″ × 12″
Size B	11″ × 17″	12″ × 18″
Size C	17″ × 22″	18″ × 24″
Size D	22″ × 34″	24″ × 36″
Size E	34″ × 44″	36″ × 48″

DETAIL PAPER An opaque paper, called **detail paper,** can be used as the drawing surface when drawings are not to be reproduced by the diazo process (which is a blue-line print).

The higher the rag content (cotton additive) of the paper, the better its quality and durability will be because cotton fiber is stronger than wood pulp alone.

Preliminary layouts can be drawn on detail paper and then traced onto the final tracing surface.

TRACING PAPER   A thin translucent paper that is used for making detail drawings is **tracing paper** or **tracing vellum.** These papers are translucent to permit the passage of light through them so drawings can be reproduced by the diazo process (blue-line process). The highest quality tracing papers are very translucent and yield the best reproductions.

Vellum is tracing paper that has been treated chemically to improve its translucency. Vellum does not retain its original quality as long as high-quality, untreated tracing paper does.

TRACING CLOTH   **Tracing cloth** is a permanent drafting medium that is available for both ink and pencil drawings. It is made of cotton fabric that has been covered with a compound of starch to provide a tough, erasable drafting surface that yields excellent blue-line reproductions. This material is more stable than paper, which means that it does not change its shape with variations in temperature and humidity as much as does tracing paper.

Erasures can be made on tracing cloth repeatedly without damaging the surface. This is especially important when drawing with ink.

POLYESTER FILM   An excellent drafting surface is polyester film. It is available under several trade names such as **Mylar film.** This material is highly transparent (much more so than paper and cloth), very stable, and is the toughest medium available. It is waterproof and is very difficult to tear.

Mylar film is used for both pencil and ink drawings. The drawing is made on the matte side of the film; the other side is glossy and will not take pencil or ink lines. Some films specify that a plastic-lead pencil be used; others adapt well to standard lead pencils. India ink and inks especially made for Mylar film can be used for ink drawings.

Ink lines will not wash off with water and will not erase with a dry eraser, but erasures can be made with a dampened hand-held eraser. An electric eraser is not recommended for use with this medium unless it is equipped with an eraser of the type recommended by the manufacturer of the film.

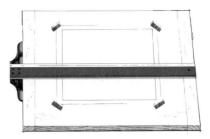

FIG. 2.6   The T-square and drafting board are the basic tools used by the student drafter. The drawing is taped to the board with drafting tape.

2.4
T-squares and drafting boards

The T-square and drafting board are the basic pieces of equipment used by the beginning drafter (Fig. 2.6). The T-square can be moved with its head in contact with the edge of the board for drawing parallel horizontal lines. The drawing paper should be attached with drafting tape to the drawing board parallel to the blade of the T-square (Fig. 2.7).

The drafting board is made of basswood, which is lightweight but strong. Standard sizes of boards are 12" × 14", 15" × 20", and 21" × 26". The working edge of a drawing board is the edge where the T-square head is held firmly in position when each horizontal line is drawn. Some drawing boards have built-in steel working edges for a higher degree of accuracy.

FIG. 2.7   The drafting tape placed at each corner of a drawing should be cut square prior to taping the drawing.

FIG. 2.8 The drafting machine is often used instead of the T-square in both industry and school laboratories. (Courtesy of Keuffel & Esser Co., Morristown, N.J.)

FIG. 2.10 The professional who uses a drafting station may work in an environment similar to the one shown here. (Courtesy of Martin Instrument Co.)

2.5
Drafting machines

Although the T-square is used in industry and in the classroom, most professional drafters prefer the mechanical **drafting machine,** which is attached to the drawing board or table top (Fig. 2.8). These machines have fingertip controls for drawing lines at any angle, and can be easily returned to their original position. These machines provide a considerable degree of convenience to the drafter, but the same drawing can be made with the T-square and triangle in the hands of a skilled drafter.

For large drawings such as those made by architects, a parallel blade is available (Fig. 2.9). The blade is attached to a cable at each end that keeps it parallel in any position. The angle of the blade can be changed by adjusting the cables.

The professional will use equipment that is more sophisticated and advanced than that used by the student. A modern, fully equipped drafting station is shown in Fig. 2.10. Also, in the future most offices will be equipped with computer graphics equipment to supplement manual equipment and techniques (Fig. 2.11).

FIG. 2.9 The parallel blade is advantageous to the drafter who makes very large drawings. It is guided by the cables attached to the board. (Courtesy of Keuffel & Esser Co., Morristown, N.J.)

FIG. 2.11 More and more professional work stations are being equipped with computer graphics equipment. (Courtesy of Bausch & Lomb.)

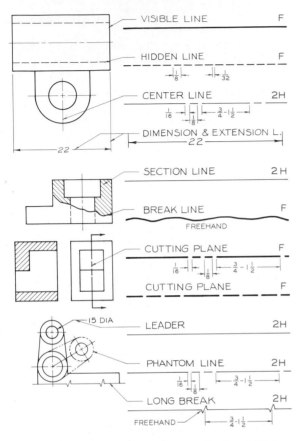

FIG. 2.12 The alphabet of lines varies in width to produce a finished mechanical drawing of an object. The full-size lines are shown in the right column along with the recommended pencil grades for drawing the lines.

2.6
Alphabet of lines

The type of line produced by a pencil depends upon the hardness of the lead, the drawing surface, and the technique of the drafter. Examples of the standard lines, or the **alphabet of lines,** are shown in Fig. 2.12, along with the recommended pencils for drawing the lines.

Guidelines are used to aid in lettering and in laying out a drawing. They are very light lines, just dark enough to be seen. A 4H pencil is recommended for drawing most guidelines.

Except for guidelines, all other lines should be drawn dark and black so they will reproduce well. The important characteristic of pencil lines is their relative widths, as shown in Fig. 2.12. When drawing these lines, assume that you are drawing them in ink where their blackness is uniform and the only variable is their widths.

2.7
Horizontal lines

A horizontal line is drawn using the upper edge of your horizontal straightedge, and for the right-handed person, the line is drawn from left to right (Fig. 2.13).

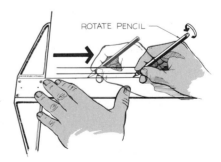

FIG. 2.13 Horizontal lines are drawn left to right along the upper edge of the T-square.

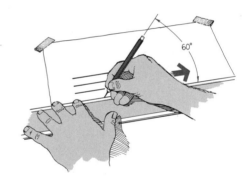

FIG. 2.14 As the horizontal lines are drawn, the pencil should be rotated about its axis so that the point will wear down evenly.

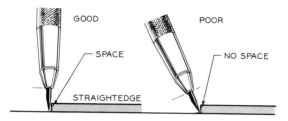

FIG. 2.15 The pencil point should be held in a vertical plane and inclined 60° to leave a space between the point and the straightedge being used.

As horizontal lines are drawn, the pencil should be rotated about its axis to allow its point to wear evenly (Fig. 2.14). If necessary, lines can be darkened by drawing over them one or more times. A small space should be left between the straightedge and the pencil point for the best line (Fig. 2.15).

2.8
Vertical lines

A triangle is used in conjunction with the blade of a T-square for drawing vertical lines since all drawing triangles have one 90° angle. While the T-square is held firmly with one hand, the triangle can be posi-

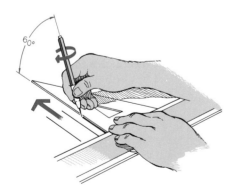

FIG. 2.16 Vertical lines are drawn along the left side of a triangle in an upward direction with the pencil held in a vertical plane at a 60° angle to the surface.

tioned where needed and the vertical lines drawn (Fig. 2.16).

Vertical lines are drawn upward along the left side of the triangle while holding the pencil in a vertical plane at a 60° angle to the drawing surface.

2.9
Drafting triangles

The two most often used triangles are the 45° triangle and the 30°–60° triangle. The 30°–60° triangle is specified by the longer of the two sides adjacent to the 90° angle (Fig. 2.17). Standard sizes of 30°–60° triangles range in 2-inch intervals from 4 in. to 24 in. The variety of lines that can be drawn with this triangle and straightedge is shown in Fig. 2.17D.

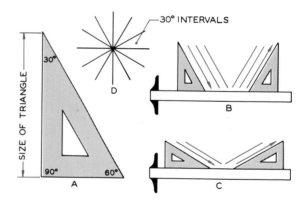

FIG. 2.17 The 30°-60° triangle can be used to construct lines at angles of 60° and 30° to the horizontal.

The 45° triangle is specified by the length of the sides adjacent to the 90° angle. These range in size from 4 in. to 24 in. at 2-inch intervals, but the 6-in. and 10-in. sizes are adequate for most classroom applications. The various angles that can be drawn with this triangle are shown in Fig. 2.18.

By using the 45° and 30°–60° triangles in combination, angles can be drawn at 15° intervals throughout 360° (Fig. 2.19).

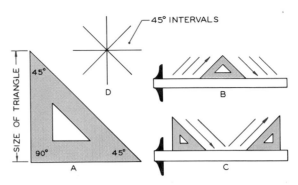

FIG. 2.18 The 45° triangle can be used to draw lines at 45° intervals throughout 360°.

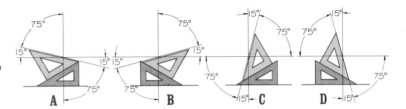

FIG. 2.19 By using the 30°–60° triangle in combination with the 45° triangle, angles can be drawn at 15° intervals without the use of a protractor.

2.10
The protractor

When lines must be drawn or measured at angles that are not multiples of 15°, a protractor is used (Fig. 2.20). Protractors are available as semicircles (180°) or as circles (360°).

An adjustable triangle serves as a protractor and a drawing edge at the same time (Fig. 2.20B). Most drafting machines have built-in protractors that are more convenient than traditional protractors.

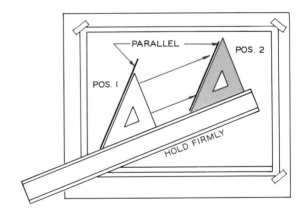

FIG. 2.21 A straightedge and a 45° triangle can be used for drawing a series of parallel lines. The T-square is held firmly in position and the triangle is moved from Position 1 to Position 2.

2.11
Parallel lines

A series of lines can be drawn parallel to a given line by using a triangle and a straightedge (Fig. 2.21).

The 45° triangle is placed parallel to a given line and is held in contact with the straightedge (which may be another triangle). By holding the straightedge

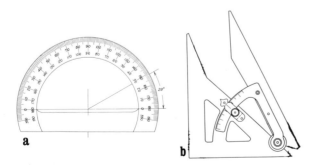

FIG. 2.20 The semicircular protractor can be used to measure angles. The adjustable triangle can be used as a drawing edge in addition to being used to measure angles.

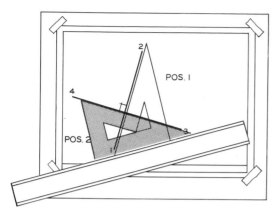

FIG. 2.22 A 30°–60° triangle and a straightedge can be used to construct a line perpendicular to Line 1–2. The triangle is aligned with 1–2 in Position 1 and is then rotated to Position 2 to construct line 3–4.

in one position, the triangle can be moved to various positions for drawing series of parallel lines.

2.12
Perpendicular lines

Perpendicular lines can be constructed by using either of the standard triangles. A 30°–60° triangle is used with a straightedge or another triangle to draw Line 3–4 perpendicular to Line 1–2 (Fig. 2.22). One edge of the triangle is placed parallel to Line 1–2 in Position 1 with the straightedge in contact with the triangle. The triangle is then rotated and moved to Position 2 to draw the perpendicular line.

2.13
Irregular curves

Curves that are not arcs must be drawn with an **irregular curve.** These plastic curves come in a variety of sizes and shapes, but the one shown in Fig. 2.23 is typical of those that are used.

The irregular curve shown in this figure is used to connect a series of points to form a smooth curve. Note that the plastic curve must be repositioned several times to draw the complete curve.

The **flexible spline** is used for drawing long irregular curves (Fig. 2.24). The spline is held in position by weights while the curve is drawn.

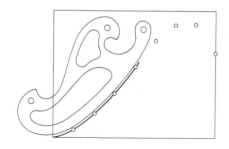  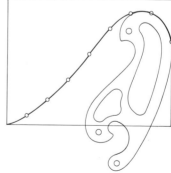

FIG. 2.23 Use of the irregular curve

Step 1 To connect points with a smooth curve, the irregular curve is positioned to pass through as many points as possible and a portion of the curve is drawn.

Step 2 The irregular curve is repositioned for drawing another portion of the curve.

Step 3 The last portion is drawn to complete the smooth curve. Most irregular curves must be drawn in separate steps, in this manner.

FIG. 2.24 A flexible spline can be used to draw large irregular curves. The spline is held in position by weights.

2.14
Erasing

Erasures should be made with the softest eraser that will serve the purpose. For example, ink erasers should not be used to erase pencil lines, because ink erasers are coarse and may damage the surface of the paper.

When it is necessary to erase in small areas, an erasing shield can be used to prevent accidental era-

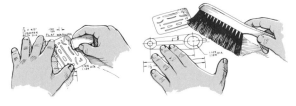

FIG. 2.25 The erasing shield is used for erasing in tight spots without removing the wrong lines. The dusting brush is used to remove the erased material when finished. Do not brush with the palm of your hand; this will smear the drawing.

FIG. 2.26 This cordless electric eraser is typical of those used by professional drafters.

sures of adjacent lines (Fig. 2.25). All erasing should be followed by brushing away the "crumbs" from the drawing with a brush. Do not use your hands for this purpose; it may smudge your drawing.

Electric erasers are available. A cordless model is shown in Fig. 2.26. This eraser is recharged by its desk stand that is left plugged into a wall outlet. The erasers that are used in these machines are available in several grades for erasing ink and pencil lines.

2.15
Scales

All engineering drawings require the use of scales to measure lengths, sizes, and other measurements. Triangular engineers' and architects' scales are shown in Fig. 2.27.

Most scales are either 6 in. or 12 in. long. The scales covered in this section are architects', engineers', mechanical engineers', and metric scales.

Architects' scale

The architects' scale is used to dimension and scale features encountered by the architect, such as cabinets, plumbing, and electrical layouts. Most indoor measurements on a drawing are made in feet and inches with an architects' scale.

The basic form of indicating on a drawing the scale that is being used is shown in Fig. 2.28. This form should be placed on the drawing in the title block or in some other prominent location.

Since the dimensions made with the architects' scale are in feet and inches, it is very difficult to handle the arithmetic associated with these dimensions. It is necessary to convert all dimensions to decimal equivalents (all feet or all inches) before the simplest arithmetic can be performed.

SCALE: FULL SIZE The 16 scale is used for measuring full-size lines (Fig. 2.29a). An inch on the 16 scale is divided into sixteenths to match the ruler used by the carpenter. The line shown in the figure is measured to be 3⅛ in. Note that when the measurement is less than one foot, a zero may be used to precede the inch measurements, and the inch marks are omitted in all cases.

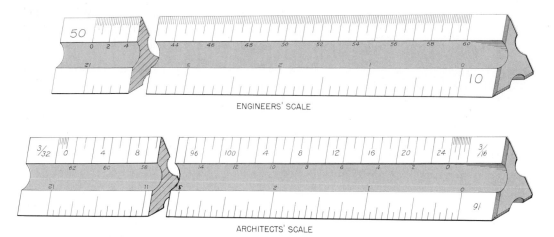

ENGINEERS' SCALE

ARCHITECTS' SCALE

FIG. 2.27 The architects' scale is used to measure in feet and inches, whereas the engineers' scale measures in decimal units.

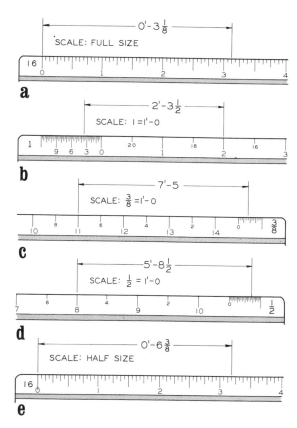

FIG. 2.29 Examples of lines measured using an architects' scale.

FIG. 2.28 The basic form for indicating the scale when the architects' scale is used, and the variety of scales that is available.

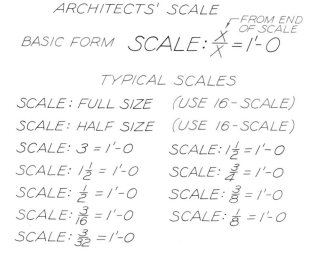

ARCHITECTS' SCALE

BASIC FORM SCALE: $\dfrac{X}{X} = 1'-0$ ┌─ FROM END OF SCALE

TYPICAL SCALES

SCALE: FULL SIZE (USE 16-SCALE)

SCALE: HALF SIZE (USE 16-SCALE)

SCALE: 3 = 1'-0   SCALE: $1\frac{1}{2}$ = 1'-0

SCALE: $1\frac{1}{2}$ = 1'-0   SCALE: $\frac{3}{4}$ = 1'-0

SCALE: $\frac{1}{2}$ = 1'-0   SCALE: $\frac{3}{8}$ = 1'-0

SCALE: $\frac{3}{16}$ = 1'-0   SCALE: $\frac{1}{8}$ = 1'-0

SCALE: $\frac{3}{32}$ = 1'-0

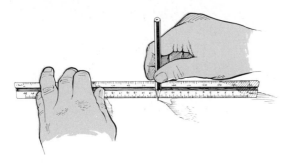

FIG. 2.30 When marking off measurements along a scale, hold your pencil vertically for the most accurate measurement.

SCALE: 1 = 1'–0 In Fig. 2.29b, a line is measured to its nearest whole foot (2 ft in this case) and the remainder is measured in inches at the end of the scale (3½ in.) for a total of 2'–3½. At the end of each architects' scale, a foot has been divided into inches for measuring dimensions that are less than a foot.

The scale 1 = 1'–0 is the same as saying 1 in. is equal to 12 in., or a ¹⁄₁₂th scale (inch marks omitted).

SCALE: ⅜ = 1'–0 When this scale is used, ⅜ in. is used to represent 12 in. on a drawing. The line in Fig. 2.29c is measured to be 7'–5 (inch marks omitted).

SCALE: ½ = 1'–0 A line is measured to be 5'–8½ in Fig. 2.29d (inch marks omitted).

SCALE: HALF SIZE The 16 scale is also used to measure or draw a line that is half size. This is sometimes specified as Scale: 6 = 12 (inch marks omitted). The line in Fig. 2.29e is measured to be 0'–6⅜.

When locating dimensions using any scale, hold your pencil in a vertical position for the greatest accuracy when marking measurements (Fig. 2.30).

OMIT INCH MARKS   ZERO HERE   ZERO OPTIONAL

FIG. 2.31 Inch marks are omitted, conforming to current standards, but foot marks are shown. A leading zero is used when the inch measurements are less than a whole inch. When representing feet, a zero is optional if the measurement is less than a foot.

When indicating dimensions in feet and inches, they should be in the form shown in Fig. 2.31. Notice that fractions are twice as tall as whole numerals.

Engineers' scale

The engineers' scale is a decimal scale on which each division is a multiple of 10 units. It is used for making drawings of engineering projects that are located outdoors, such as streets, structures, land measurements, and other large dimensions associated with topography. For this reason, it is sometimes called the civil engineers' scale.

Since the measurements are in decimal form, it is easy to perform arithmetic operations without the need to convert feet to inches as when the architects' scale is used.

The correct form of specifying scales on the engineers' scale is shown in Fig. 2.32, such as Scale: 1 = 10'. Each end of the scale is labeled 10, 20, 30, etc., indicating the number of units per inch on the scale. Many combinations may be obtained by moving the decimal places of a given scale, as indicated in Fig. 2.32.

ENGINEERS' SCALES

BASIC FORM   FROM END OF ENGR. SCALE   *SCALE: 1 = XX*

EXAMPLE SCALES

10	*SCALE: 1=10';*	*SCALE: 1 = 1,000'*
20	*SCALE: 1=200';*	*SCALE: 1 = 20 LB*
30	*SCALE: 1= 0.3;*	*SCALE: 1 =3,000'*
40	*SCALE: 1=4';*	*SCALE: 1 = 40'*
50	*SCALE: 1= 50';*	*SCALE: 1 = 500'*
60	*SCALE: 1= 6;*	*SCALE: 1 = 0.6'*

FIG. 2.32 The basic form for indicating the scale and the variety of scales that is available, when the engineers' scale is used.

10 SCALE In Fig. 2.33a, the 10 scale is used to measure a line at the scale of 1 = 10'. The line shown is 32.0 feet long.

20 SCALE In Fig. 2.33b, the 20 scale is used to measure a line drawn at a scale of 1 = 200.0'. The line shown is 540.0 feet long.

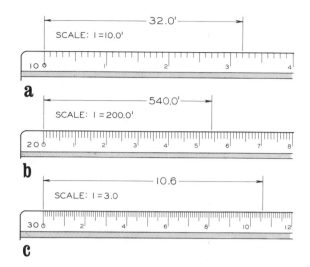

FIG. 2.33 Examples of lines measured with the engineers' scale.

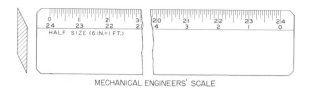

FIG. 2.35 The mechanical engineers' scale is used for measuring small parts at scales of half size, quarter size, and one-eighth size. These units are in inches with common fractions.

Mechanical engineers' scale

The mechanical engineers' scale is used to draw small parts (Fig. 2.35) in inches using common fractions. These scales are available in ratios of half size, one-quarter size, and one-eighth size. For example, on the half-size scale, 1 inch is used to represent 2 inches. On a quarter-size scale, 1 inch would represent 4 inches.

Metric system—SI units

The English system (Imperial system) of measurements has been used in the United States, Britain, and Canada since these countries were established. A movement is now underway to convert to the more universal metric system.

The English system was based on the arbitrary units the inch, foot, cubit, yard, and mile (Fig. 2.36). There is no common relationship between these units of measurement; consequently the system is cumbersome to use when simple arithmetic is performed. For example, finding the area of a rectangle that is 25 inches by 6¾ yards is a complex problem.

The metric system was proposed by France in the fifteenth century. In 1793, the French National Assembly agreed that the meter (m) would be one ten-millionth of the meridian quadrant of the earth (Fig.

FIG. 2.34 When using English units (inches), decimal fractions do not have leading zeros and inch marks are omitted. Be sure to provide adequate space for decimal points between the numbers. Foot marks are shown.

30 SCALE A line of 10.6 (inch marks omitted) is measured using the scale of 1 = 3.0 in Fig. 2.33c.

The format for indicating measurements in feet and inches is shown in Fig. 2.34. It is customary to omit zeros in front of decimal points when dimensioning an object using English (Imperial) units, and inch marks are always omitted if the dimensions are given in inches.

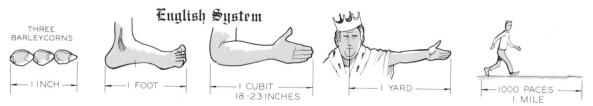

FIG. 2.36 The units of the English system (Imperial system) were based on arbitrary dimensions.

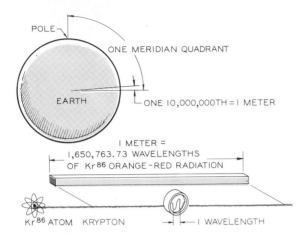

FIG. 2.37 The origin of the meter was based on the dimensions of the earth, but it has since been based on the wavelength of krypton-86. A meter is 39.37 inches.

2.37). Fractions of the meter were expressed as decimal fractions. Debate on the merits of each system continued until an international commission officially adopted the metric system in 1875. Since a slight error in the first measurement of the meter was found, the meter was later established as equal to 1,650,763.73 wavelengths of the orange-red light given off by krypton-86 (Fig. 2.37).

The international organization charged with the establishment and the promotion of the metric system is called the **International Standards Organization (ISO).** The system they have endorsed is called **Système International d'Unités** (International System of Units) and is abbreviated SI. The basic SI units are shown in Fig. 2.38 with their abbreviations. It is important that lowercase and uppercase abbreviations be used properly as shown.

Several practical units of measurement have been derived (Fig. 2.39) to make them easier to use in

SI UNITS			DERIVED UNITS		
LENGTH	METER	m	AREA	SQ METER	m²
MASS	KILOGRAM	kg	VOLUME	CU METER	m³
TIME	SECOND	s	DENSITY	KILOGRAM/CU MET	kg/m³
ELECTRICAL			PRESSURE	NEWTON/SQ MET	N/m²
CURRENT	AMPERE	A			
TEMPERATURE	KELVIN	K			
LUMINOUS					
INTENSITY	CANDELA	cd			

FIG. 2.38 The basic SI units and their abbreviations. The derived units are units that have come into common usage.

PARAMETER	PRACTICAL UNITS		SI EQUIVALENT
TEMPERATURE	DEGREES CELSIUS	°C	0°C = 273.15 K
LIQUID VOLUME	LITER	l	l = dm³
PRESSURE	BAR	BAR	BAR = 0.1 MPa
MASS WEIGHT	METRIC TON	t	t = 10³ kg
LAND MEASURE	HECTARE	ha	ha = 10⁴ m²
PLANE ANGLE	DEGREE	°	1° = π/180 RAD

FIG. 2.39 These practical metric units are a few of those that are widely used because they are easier to than the official SI units.

many applications. Note that degrees Celsius (centigrade) is recommended over the official temperature measurement, Kelvin. When using Kelvin, the freezing and boiling temperatures are 273.15°K and 373.15°K, respectively. Pressure is measured in bars, where one bar is equal to 0.1 megapascal, or 100,000 pascals.

Many SI units have prefixes to indicate placement of the decimal. The more common prefixes and their abbreviations are shown in Fig. 2.40.

VALUE		PREFIX	SYMBOL
1,000,000	= 10⁶ =	MEGA	M
1,000	= 10³ =	KILO	k
100	= 10² =	HECTO	h
10	= 10¹ =	DEKA	da
1	= 10⁰ =		
.1	= 10⁻¹ =	DECI	d
.01	= 10⁻² =	CENTI	c
.001	= 10⁻³ =	MILLI	m
.000 001	= 10⁻⁶ =	MICRO	μ

FIG. 2.40 The prefixes and abbreviations used to indicate the decimal placement for SI measurements.

Several comparisons of English and SI units are given in Fig. 2.41. Other conversion factors are given in Appendix 2.

2.16
Metric scales

The basic unit of measurement on an engineering drawing is the millimeter (mm), which is one-thousandth of a meter, or one-tenth of a centimeter. The width of the fingernail of your index finger can serve as a convenient gauge to approximate the dimension of a centimeter, or ten millimeters (Fig. 2.42).

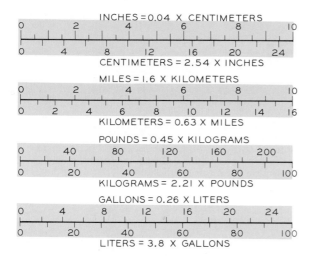

FIG. 2.41 A comparison of metric units with those used in the English system of measurement.

Metric scales are indicated on a drawing in the form shown in Fig. 2.43. A colon is placed between the numeral 1 and the ratio of the drawing size. The units are not specified since the millimeter is understood to be the unit of measurement.

Decimal fractions are unnecessary on most metrically dimensioned drawings; consequently, the dimensions are usually rounded off to whole numbers except for those measurements that are dimensioned with specified tolerances. For metric units less than 1, a leading zero is placed in front of the decimal (Fig.

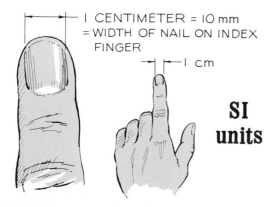

FIG. 2.42 The width of the nail on your index finger is approximately equal to one centimeter or ten millimeters.

METRIC SCALES FROM END OF SCALE

BASIC FORM SCALE: 1:2

TYPICAL SCALES

SCALE: 1:1 (1mm=1mm; 1cm=1cm: ETC)

SCALE: 1:20 (1mm=20 mm; 1mm=2cm)

SCALE: 1:300 (1mm=300 mm; 1mm=0.3m)

OTHERS: 1:125; 1:250; 1:500

FIG. 2.43 The basic form for indicating scales, and the variety of metric scales that is available, when the metric scale is used.

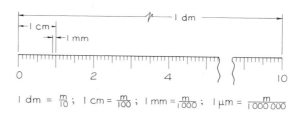

FIG. 2.44 When decimal fractions are shown in metric units, a zero is used to precede the decimal. Be sure to allow adequate space for the decimal point when numbers with decimals are lettered.

2.44). In the English system, the zero is omitted from inch measurements.

SCALE 1:1 The full-size metric scale (Fig. 2.45) shows the relationship between the metric units of the decameter, centimeter, millimeter, and the micrometer. There are 10 decameters in a meter; 100 centimeters in a meter; 1,000 millimeters in a meter; and 1,000,000 micrometers in a meter. A line of 59 mm is measured in Fig. 2.46a.

$$1\ dm = \frac{m}{10}; \quad 1\ cm = \frac{m}{100}; \quad 1\ mm = \frac{m}{1\,000}; \quad 1\ \mu m = \frac{m}{1\,000\,000}$$

FIG. 2.45 A decameter is one-tenth of a meter; a centimeter is one-hundredth of a meter; a millimeter is one-thousandth of a meter; and a micrometer is one-millionth of a meter.

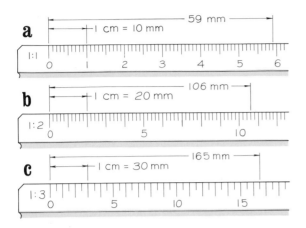

FIG. 2.46 Examples of lines measured with metric scales.

SCALE 1:2 This scale is used when 1 mm is equal to 2 mm, 20 mm, 200 mm, etc. The line in Fig. 2.46b is 106 mm long.

SCALE 1:3 A line of 165 mm is measured in Fig. 2.46c where 1 mm is used to represent 33 mm.

Other scales

Many other metric (SI) scales are used: 1:250, 1:400, 1:500, and so on. The scale ratios mean that the one unit on the left of the colon represents the number of units on the right of the colon. For example, 1:20 means that one millimeter equals 20 mm, or one centimeter equals 20 cm, or one meter represents 20 m.

Metric symbols

When drawings are made in metric units, it can be noted in the titleblock or elsewhere using the SI symbol shown in Fig. 2.47. The large SI indicates Systéme International. The two views of the partial cone are used to denote whether the orthographic views were drawn in the U.S. system (third-angle projection) or the European system (first-angle projection).

Scale conversion

Tables for converting inches to millimeters are given in Appendix 2; however, this conversion can be performed by multiplying decimal inches by 25.4 to ob-

a METRIC UNITS— 3RD ANGLE PROJ.   **b** METRIC UNITS— 1ST ANGLE PROJ.

FIG. 2.47 The large letters SI indicate that measurements are in metric units. The partial cones indicate that the views are arranged using third-angle projection (the U.S. system) or first-angle projection (the European system).

USE ZERO PRECEDING DECIMALS
0.72   NOT   .72
OMIT COMMAS & GROUP INTO THREES
2 000 000   NOT   2,000,000
USE RAISED DOT FOR MULTIPLICATION
N·M   NOT   NM
INDICATE DIVISION BY EITHER
kg/m   OR   kg·m⁻¹
INDICATE METRIC SCALE AS EITHER
SCALE: 1:2 SI OR SCALE: 1:3 METRIC

FIG. 2.48 General rules to be used with the SI system.

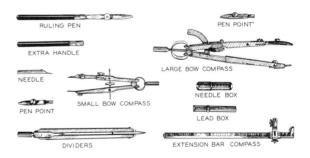

FIG. 2.49 The parts of a set of drafting instruments

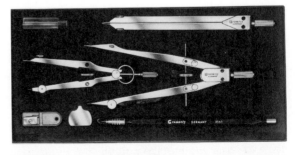

FIG. 2.50 Instruments usually come as a cased set. (Courtesy of Gramercy Guild.)

tain millimeters. For example, 1.5 inches would be 1.5 × 25.4 = 38.1 mm.

To convert an architects' scale to an approximate metric scale, the scale must be multiplied by 12. For example, Scale: ⅛ = 1′–0 is the same as ⅛ inch = 12 inches, or 1 inch = 96 inches. This scale closely approximates the metric scale of 1:100. The scale of 1 = 5′ converts exactly to the metric scale of 1:60.

Expression of metric units

The general rules for expressing SI units are given in Fig. 2.48. Commas are not used between sets of zeros; instead, a space is left between them.

2.17 ──────────
The instrument set

A basic set of drawing instruments and the name of each part are shown in Fig. 2.49. These can be purchased separately, but they are available assembled as a set in a case similar to the one shown in Fig. 2.50. A more elaborate set of instruments is shown in Fig. 2.51.

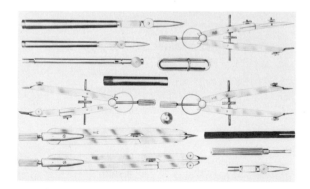

FIG. 2.51 A set of instruments with three bows for the advanced student. (Courtesy of Keuffel & Esser Co., Morristown, N.J.)

Compass

The compass is used to draw circles and arcs in ink and in pencil (Fig. 2.52). In order to obtain good results with the compass, its pencil point must be sharp-

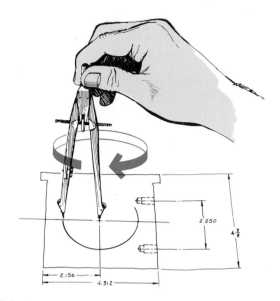

FIG. 2.52 The compass is used to draw circles.

ened on its outside with a sandpaper board (Fig. 2.53). When the compass point is set in the drawing surface, it should be inserted just enough for a firm set, not to the shoulder of the point.

When the table top has a hard covering, several sheets of paper should be placed under the drawing to provide a seat for the compass point. A center tack (the small circular part shown in Fig. 2.51) can be placed on the paper and used to set the compass point as well as to prevent enlargement of the center hole when it is used repeatedly.

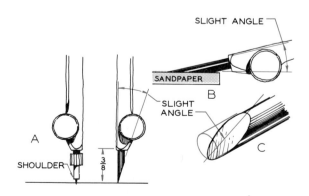

FIG. 2.53 The compass lead should be sharpened from the outside on a sandpaper pad at B, to a wedge point (C). The pencil point should be about the same length as the compass point at A.

Bow compasses are provided in some sets (Fig. 2.54) to draw small circles with ink and pencil. Extension bars are provided to extend the range of the large-bow compass for large circles. If the circles are to be much larger, a beam compass can be used (Fig. 2.55).

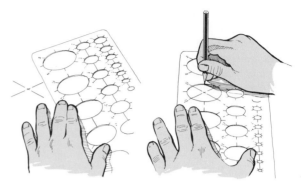

FIG. 2.56 The circle template can be used to draw circles without the use of a compass. The circle or ellipse template is aligned with the center lines.

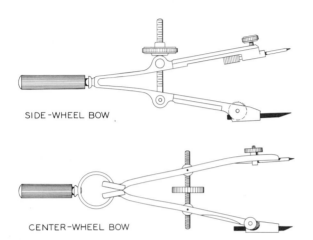

SIDE-WHEEL BOW

CENTER-WHEEL BOW

FIG. 2.54 Two types of small-bow compasses for drawing circles of about one-inch radius.

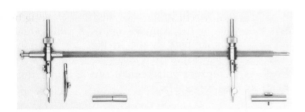

FIG. 2.55 Large circles can be drawn with the beam compass. Note that ink attachments are available also.

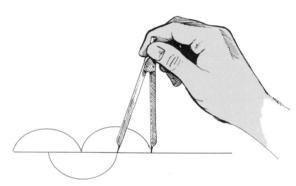

FIG. 2.57 The dividers are used to step off measurements on a drawing.

Small circles and ellipses can be effectively drawn with a circle template that is aligned with the perpendicular center lines of the circle. The circle or ellipse is drawn with a pencil to match the other lines of the drawing (Fig. 2.56).

Dividers

The dividers look much like a compass but are used for laying off dimensions onto a drawing. For example, equal divisions can be stepped off rapidly along a line

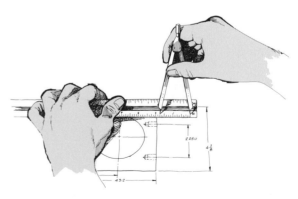

FIG. 2.58 Dividers are also used to transfer dimensions from a scale to a drawing.

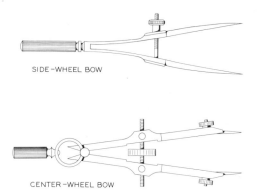

SIDE-WHEEL BOW

CENTER-WHEEL BOW

FIG. 2.59 Two types of bow dividers for transferring small dimensions, such as the spacing between guidelines for lettering.

(Fig. 2.57). A slight impression is made in the drawing surface with the points as each measurement is made.

Dividers can be used to transfer dimensions from a scale to a drawing (Fig. 2.58). Another use for the dividers is the division of a line into a number of equal parts. This is done by trial and error: estimating the spacing and stepping off the space until the correct spacing is found.

Small-bow dividers can be used for transferring smaller dimensions, such as the spacing between the guidelines for lettering. Two types of bow dividers are shown in Fig. 2.59.

Proportional dividers

Dimensions can be transferred from one scale to another by using the **proportional dividers.** The central pivot point can be moved from position to position to vary the ratio of the spacing at one end of the dividers to the ratio at the other end (Fig. 2.60). This instrument is very helpful in enlarging and reducing dimensions on a drawing.

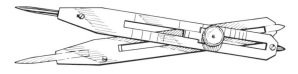

FIG. 2.60 Proportional dividers can be used for making measurements that are proportional to other dimensions. The pivot can be set to give the desired ratio.

2.18
Ink drawing

Although the majority of drafting and design work is done in pencil, inking is required for many applications, especially for drawings that will be reproduced. Pencil drawings have a tendency to lose their sharpness as instruments are moved about the drawing surface during preparation. Ink drawings remain dark and distinct without danger of losing their quality.

Materials for ink drawing

DRAWING SURFACE A good grade of tracing paper can be used for ink drawings, but erasing errors may result in holes in the paper and the loss of time. Therefore, film or tracing cloth that will withstand many erasures and corrections should be used.

When using **drafting film,** the drawing should be made on the matte surface in accordance with the manufacturer's directions. A cleaning solution is available to prepare the surface for ink and to remove spots that might not take the ink properly. These films can also be cleaned with a damp cloth.

Tracing cloths need to be prepared for inking by applying a coating of powder or pounce to absorb oily spots that will otherwise repel an ink line. These oily spots can be left on a drawing by fingerprints.

INK The drawing ink used for engineering drawings is called India ink and is available under numerous trade names. This dense black, carbon ink is much thicker and faster-drying than regular fountain pen ink. Some draftsmen prefer to "season" their ink by leaving the top of the bottle open for several days to thicken the ink.

Ink should be wiped from instruments before it dries to prevent clogging, which restricts an easy flow of ink from the pen to the paper. Ultrasonic pen cleaners are available to aid in cleaning pen points that have been used for inking. Dirty points are immersed in a cup of pen cleaner solution that is vibrated at about 80,000 cycles per second to free the dried particles of ink.

RULING PEN Two types of ruling pens are shown in Fig. 2.61. Both have set screws for varying the widths of the lines that are drawn.

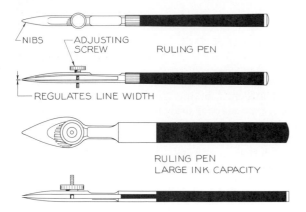

FIG. 2.61 The ruling pen has two nibs that are adjusted by a set screw to vary the width of the ink lines drawn with the pen.

FIG. 2.62 The pen is inked between the nibs with the spout on the ink bottle cap.

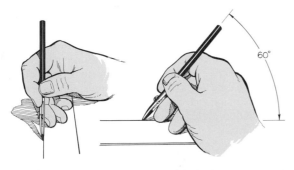

FIG. 2.63 The ruling pen is held in a vertical plane at a 60° angle to the drawing surface.

The ruling pen should be inked with the spout on the cap of the ink bottle (Fig. 2.62). Experiment with your particular pen to learn the proper amount of ink to apply to the nibs.

When ruling horizontal lines, the ruling pen is held in the same position as a pencil (Fig. 2.63), maintaining a space between the nibs and the straightedge. An extra margin of safety can be obtained by placing a triangle or template under the straightedge, as shown in Fig. 2.64.

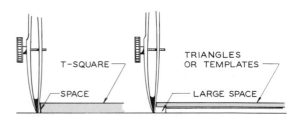

FIG. 2.64 The ruling pen should be held in a vertical plane so that there will be a space between the straightedge and the nibs. A triangle or template can be placed under the straightedge for a greater margin of safety.

FIG. 2.65 An India ink technical pen that can be used to make ink drawings.

Another pen that can be used is the technical ink fountain pen, illustrated in Fig. 2.65. These pens come in sets (Fig. 2.66) with various sized pen points that are used for the alphabet of lines (Fig. 2.67).

Examples of poorly drawn ink lines are shown in Fig. 2.68. These mistakes can be avoided by learning the technical faults that cause them.

INKING COMPASS The inking compass is usually the same compass used for circles drawn by pencil, with the inking attachment inserted in place of the

FIG. 2.66 A set of inking pens available in various line weights.

pencil attachment. The inking compass may have an elbow in its legs to allow the points to be approximately perpendicular to the drawing surface. The circle can be drawn with one continuous line, as shown in Fig. 2.69.

Two types of compasses for drawing smaller circles are shown in Fig. 2.70. These can be used to draw

Available in 18 "Kolor-Koded" line widths *

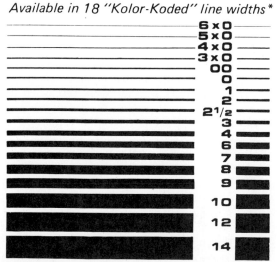

6 x 0
5 x 0
4 x 0
3 x 0
00
0
1
2
2½
3
4
6
7
8
9
10
12
14

*Approximate only. (Line widths will vary, depending on type of surface, type of ink, speed at which line is drawn, etc.)

FIG. 2.67 This chart shows the variety of technical pens that is available for drawing lines of graduated widths. (Courtesy of Koh-I-Noor Rapidograph, Inc.)

A. GOOD-EVEN LINE

B. POOR - INK RAN UNDER STRAIGHTEDGE

C. POOR-TOO MUCH INK; SLOW AT ENDS

D. POOR - NIBS TOUCH IMPROPERLY

FIG. 2.68 Examples of poor ink lines are shown at B, C, and D.

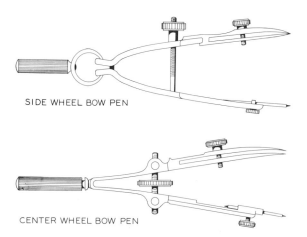

FIG. 2.69 An inking attachment can be used with a large-bow compass to draw circles and arcs in ink.

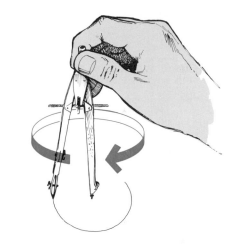

SIDE WHEEL BOW PEN

CENTER WHEEL BOW PEN

FIG. 2.70 Two types of small-bow compasses to draw arcs in ink up to a radius of about 1 inch.

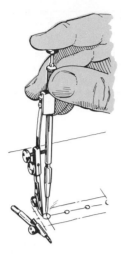

FIG. 2.71. The spring-bow compass is used to draw circles as small as ⅛ inch in diameter. A pencil attachment is available for pencil circles.

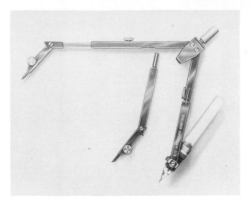

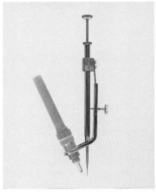

FIG. 2.73 Special compasses with adapters are available for using technical ink pens to draw arcs. (Courtesy of Koh-I-Noor Rapidograph, Inc.)

circles up to a radius of 1 inch. The spring-bow compass (Fig. 2.71) can be used to draw pencil and ink circles that are very small, usually about ⅛ inch in radius.

Larger circles can be drawn using the extension bar with the large compass (Fig. 2.72). Much larger circles can be drawn with the beam compass (as shown in Fig. 2.55) using the inking attachment. Special compasses are available for drawing circles with a technical fountain pen (Fig. 2.73).

TEMPLATES   A wide variety of templates is available for preparing drawings in both pencil and ink (Fig. 2.74). These are available for drawing nuts and bolts, circles and ellipses, architectural symbols, and many other applications. They work best when used

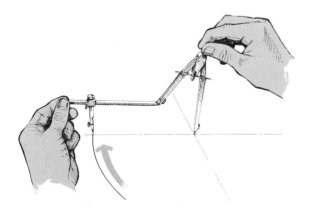

FIG. 2.72 An extension bar can be used with a bow compass to draw large arcs.

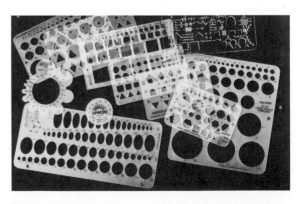

FIG. 2.74 Many types of templates are available to aid drafters in their work. (Courtesy of Rapidesign.)

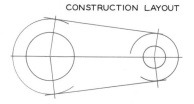
CONSTRUCTION LAYOUT

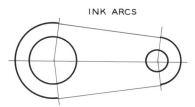
INK ARCS

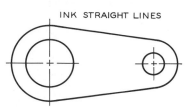
INK STRAIGHT LINES

FIG. 2.75 Order of inking

Step 1 The drawing is laid out with light pencil construction lines. All centers and tangent points are accurately located.

Step 2 The arcs and circles are always inked first. Arcs should stop at their points of tangency.

Step 3 Straight lines are drawn to match the ends of the arcs. Centerlines are shown to complete the drawing.

with technical fountain pens rather than with the traditional ruling pen.

Order of inking

When a drawing is to be inked, begin by locating the centerlines and tangent points associated with the arcs, as shown in Step 1 of Fig. 2.75. This construction should be laid out in pencil prior to inking.

Ink the arcs and circles first, being careful to stop all arcs at their points of tangency, Step 2. Connect the arcs with straight lines at the points of tangency to complete the drawing.

When a drawing is composed mostly of straight lines, begin at the top of the drawing and draw all of the horizontal lines as you move from the top of the sheet to the bottom. In this manner, you are moving away from the wet lines, which are left to dry as you progress. After allowing the last horizontal line to dry, the right-handed person should ink the vertical lines by beginning with the far left and moving across your drawing to the right, away from the wet lines. The direction should be reversed for the left-handed drafter.

2.19 Solutions of problems

The following formats are suggested for the layout of problem sheets. Most problems will be drawn on $8\frac{1}{2}'' \times 11''$ sheets as shown in Fig. 2.76. A title strip is suggested in this figure, with a border as shown. The

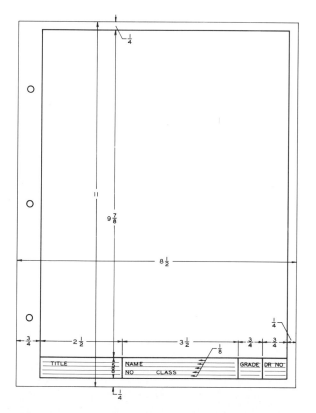

FIG. 2.76 The format and title strip for a Size A sheet ($8\frac{1}{2}'' \times 11''$) that is suggested for solving the problems at the end of each chapter. When the sheet is in the vertical format it will be called a Size AV sheet.

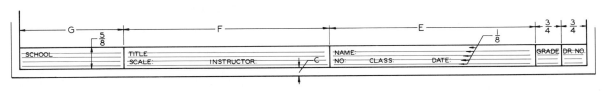

FIG. 2.77 The format for Size AH (an 8½″ × 11″ sheet in a horizontal position) and the sizes of other sheets. The dimensions under columns A though H give the various layouts.

				DIMENSIONS					
SIZE	A	B	C	D	E	F	G	H	
(A)	8½	11	¼	¾	5	4		¼	
(B)	11	17	¼	¼	6	8¼		1	
(C)	17	22	½	½	6	6	7	1	
(D)	22	34	½	½	6	6	6	1	
(E)	33	44	½	½	6	6	6	1	

FIG. 2.78 An alternative title strip that can be used on sheet Sizes B, C, D, and E instead of the one given in Fig. 2.77.

FIG. 2.79 A title block and a parts list that can be used on some problem sheets if needed.

8½″ × 11″ sheet in the vertical format is called Size AV throughout the remainder of this textbook. When this size sheet is in the horizontal format, as shown in Fig. 2.77, it will be called Size AH.

The standard sizes of sheets from Size A through Size E are shown in Fig. 2.77. An alternative title strip for Sizes B, C, D, and E is shown in Fig. 2.78. Guide-

lines should always be used for lettering title strips.

Another title block and a parts list are given in Fig. 2.79. These are placed in the lower right-hand corner of the sheet, against the borders. When both are used on the same drawing, the parts list is placed directly above and in contact with the title block or the title strip as the case may be.

Problems

The problems (Figs. 2.80–2.84) are to be constructed on Size AH (8½″ × 11″) paper, plain or with a printed grid, using the format shown in Fig. 2.76. Two problems can be constructed per sheet. Use pencil or ink as assigned by your instructor.

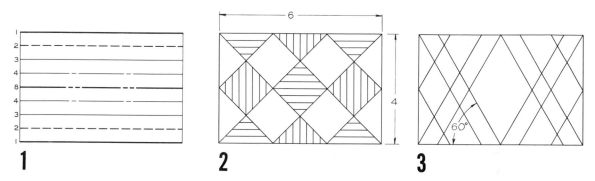

FIG. 2.80 Each of these problems is to be drawn on a Size AV sheet. Begin by lightly laying out a 4″ × 6″ rectangle. Problem 1: Construct the following lines: 1—visible line, 2—hidden line, 3—dimension line, 4—center line, and 8—cutting plane line. Problems 2 and 3: Construct the patterns shown using line weights equal to that of visible lines.

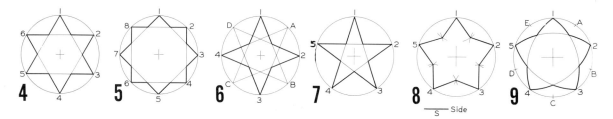

FIG. 2.81 Study the figures closely and draw one per sheet of Size AV paper. The circles are 4″ in diameter. The dimension S in Problem 8 is 1.2 inches.

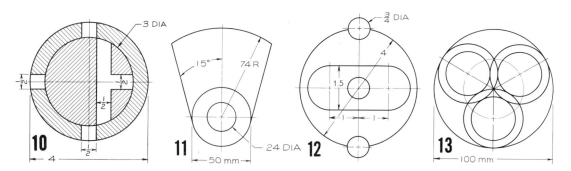

FIG. 2.82 Draw Problems 10–13 on a Size AV sheet, one per sheet, using the given dimensions and your instruments.

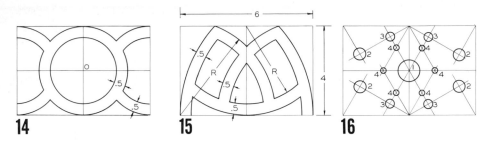

FIG. 2.83   Construct the problems inside of the 4″ × 6″ rectangles, using the given dimensions. The hole diameters in Problem 16 are as follows: No. 1—1″, No. 2—0.5″, No. 3—0.4″, and No. 4—0.25″.

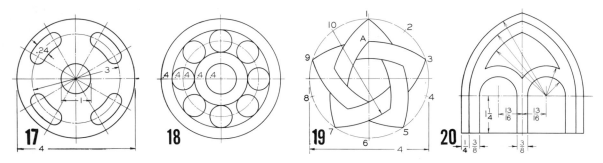

FIG. 2.84   Construct the drawings in Problems 17–20 using the given dimensions (the large circles are 4″ in diameter), one problem per sheet. Use a Size AV sheet.

CHAPTER

3

Lettering

3.1
Lettering

All drawings are supplemented with notes, dimensions, and specifications that must be lettered. Consequently, the ability to construct legible freehand letters is a very important skill to develop since it affects the usage and interpretation of a drawing.

3.2
Tools of lettering

Good lettering requires the use of proper instruments. The best grade pencil for lettering on most surfaces is a pencil in the H–HB grade range, with an F pencil being the most commonly used grade. Some papers and films are coarser than others and may require a harder pencil lead. The point of the pencil should be slightly rounded to give the desired line width (Fig. 3.1).

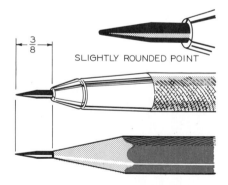

FIG. 3.1 Good lettering begins with a properly sharpened pencil point. The point should be slightly rounded, not a needle point. The F pencil is usually the best grade for lettering.

When lettering, the pencil should be revolved about its axis between your fingers as the strokes are being made so the lead will wear down evenly. Bear down firmly to make letters black and bright for good reproduction. To prevent smudging it is helpful to place a sheet of paper under your hand (Fig. 3.2).

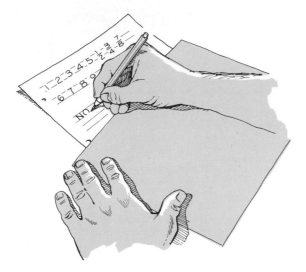

FIG. 3.2 When lettering a drawing, use a protective sheet under your hand to prevent smudges. Your lettering will be best when you are working from a comfortable position; you may wish to turn your paper for the most natural strokes.

An inking pen that is widely used is the India ink technical pen (Fig. 3.3). These pens have tubular points that are kept clear by gently shaking the pen up and down to activate the plunger inside the point. They are available in a range of tubular point sizes for making lines of varying widths.

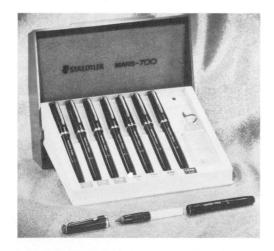

FIG.3.3 Ink fountain pens have been widely used by professionals who work with ink. The line widths vary with each pen, each of which has its own number. (Courtesy of J. S. Staedtler, Inc.)

3.3
Gothic lettering

The standard type of lettering that is recommended for engineering drawings is **single-stroke Gothic lettering.** It was given this name because the letters are made with a series of single strokes and the letter form is a variation of Gothic lettering.

Two general categories of Gothic lettering are **vertical** and **inclined** lettering (Fig. 3.4). Each is acceptable; however, these types should not be mixed on the same drawing.

FIG. 3.4 Two types of Gothic lettering recommended by engineering standards are vertical and inclined lettering.

3.4
Guidelines

The most important rule of lettering is: **Use guidelines at all times.** This applies whether you are lettering a paragraph or a single letter or numeral. The method of constructing and using guidelines can be seen in Fig. 3.5. Use a sharp pencil in the 3H–5H grade range and draw the lines very lightly, just dark enough for them to be seen.

Most lettering is done with the capital letters ⅛ in. (3 mm) high. The spacing between the lines of lettering should be no closer than half the height of the capital letters, 1/16 in. in this case.

Vertical guidelines should be drawn at random to serve as visual guides in addition to the horizontal lines. These guidelines will be slanted for inclined lettering.

Lettering guides

The two instruments used most often for drawing guidelines for lettering are the **Braddock-Rowe lettering triangle** and the **Ames lettering instrument.**

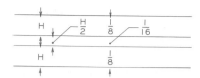

FIG. 3.5 Lettering Guidelines

Step 1 Letter heights, *H*, are laid off and light guidelines are drawn with a 4H pencil. The spacing between the lines should be no closer than *H/2*.

Step 2 Vertical guidelines are drawn as light, thin lines. These are randomly spaced to serve as visual guides for lettering.

Step 3 The letters are drawn with single strokes using a medium-grade pencil, H–HB. The guidelines need not be erased since they were drawn lightly.

The Braddock-Rowe triangle is pierced with sets of holes for spacing guidelines (Fig. 3.6). The numbers under each set of holes represent thirty-seconds of an inch. For example, the numeral 4 represents ⁴/₃₂ in., or guidelines that are placed ⅛ in. apart for making uppercase (capital) letters. Some triangles are marked for metric lettering in millimeters. Note in Fig. 3.6 that intermediate holes are provided for guidelines for lowercase letters, which are not as tall as the capital letters.

The Braddock-Rowe triangle is used in conjunction with a horizontal straightedge held firmly in position with the triangle placed against its edge. A sharp 4H pencil is placed in one hole of the desired set of holes to contact the drawing surface, and the pencil point is guided across the paper to draw the guideline while the triangle slides against the straightedge. This process is repeated as the pencil point is moved successively to each hole until the desired number of guidelines is drawn.

A slanted slot for drawing guidelines for inclined lettering is cut in the triangle. These slanting guidelines are spaced randomly by eye.

The Ames lettering guide is a very similar device with a circular dial for selecting the proper spacing of guidelines. Again, the numbers around the dial represent thirty-seconds of an inch. The number 8 represents ⁸/₃₂ in., or guidelines for drawing capital letters

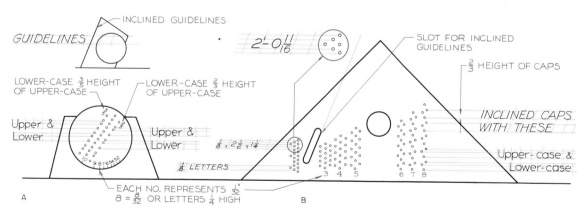

FIG. 3.6 A. The Ames lettering guide can be used to draw guidelines for uppercase and lowercase letters, vertical or inclined. The dial is set to the desired number of thirty-seconds of an inch for the height of uppercase letters. B. The Braddock-Rowe triangle can be used as a 45° triangle as well as an instrument for constructing guidelines. The numbers designating the guidelines represent thirty-seconds of an inch. For example, the number 4 represents ⁴/₃₂ or ⅛ inch for the height of uppercase letters.

that are ¼ in. tall. Metric guides are labeled in millimeters.

This instrument is used with a pencil and straightedge, as previously explained for using the Braddock-Rowe triangle.

> Be sure to keep guidelines very light so that they will not interfere with the legibility of the lettering.

3.5
Vertical letters

Vertical capital letters

The capital letters for the **single-stroke Gothic** alphabet are shown in Fig. 3.7. Each letter is drawn inside a square box of guidelines to help you learn their correct proportions. Each straight-line stroke should be drawn as a single stroke without stopping. For example, the letter A is drawn with three single strokes. Letters composed of curves can best be drawn in segments. The letter O can be drawn by joining two semicircles to form the full circle.

Memorize the shape of each letter given in this alphabet. Small wiggles in your strokes will not detract from your lettering if the letter forms are correct.

Examples of poor lettering are shown in Fig. 3.8. Observe the reason given for the lettering being poor in each example.

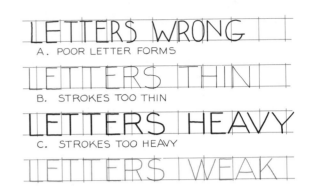

FIG. 3.8 There are many ways to letter poorly. A few of them, and the reasons why the lettering is inferior, are shown here. *Do not* make these mistakes.

Vertical lowercase letters

An alphabet of lowercase letters is shown in Fig. 3.9. Lowercase letters are either two-thirds or three-fifths as tall as the uppercase letters that they are used with. Both of these ratios are labeled on the Ames guide. Only the two-thirds ratio is available on the Braddock-Rowe triangle.

Some lowercase letters have ascenders that extend above the body of the letter, such as the letter b; and some have descenders that extend below the body of the letter, such as the letter y. The ascenders and the descenders should be the same length.

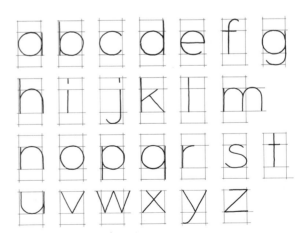

FIG. 3.9 The lowercase alphabet used in single-stroke Gothic lettering. The body of each letter is drawn inside a square to help you learn the proportions.

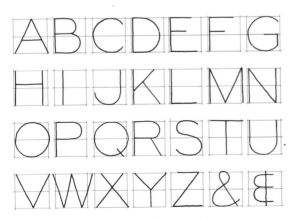

FIG. 3.7 The uppercase letters used in single-stroke Gothic lettering. Each is drawn inside a square to help you learn the proportions of each letter.

The guidelines in Fig. 3.9 form perfect squares about the body of each letter to illustrate the proportions.

Capital and lowercase letters are used together in Fig. 3.10 as in a title. You can see the difference between the lowercase letters that are two-thirds the height of capitals and those that are three-fifths the height of capitals.

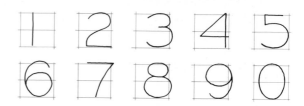

FIG. 3.11 The numerals for single-stroke Gothic lettering. Each is drawn inside a square to help you learn their proportions.

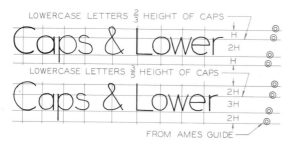

FIG. 3.10 The ratio of the lowercase letters to the uppercase letters will be either two-thirds or three-fifths. The Ames guide has both, and the Braddock-Rowe triangle has only the two-thirds ratio.

Vertical numerals

Vertical numerals are shown in Fig. 3.11, with each number enclosed in a square box of guidelines. As with lettering, you must learn the proportions of the numerals in order to use them properly. Each number is made the same height as the capital letters being used; usually ⅛ in. high. The numeral zero is an oval and the letter O is a perfect circle in vertical lettering.

3.6 ⸺
Inclined letters

Inclined capital letters

Inclined uppercase letters have the same heights and proportions as vertical letters; the only difference is their inclination of 68° to the horizontal. The inclined alphabet is shown in Fig. 3.12.

Inclined guidelines should be drawn using the Braddock-Rowe triangle or the Ames guide, as illustrated in Fig. 3.6.

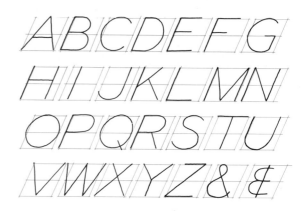

FIG. 3.12 The uppercase alphabet for single-stroke inclined Gothic lettering.

Lettering features that appear as circles in vertical lettering will appear as ellipses when inclined lettering is used.

Inclined lowercase letters

Lowercase inclined letters are drawn in the same manner as the vertical lowercase letters. This alphabet is shown in Fig. 3.13. Ovals (ellipses) are used instead of the circles used in vertical lettering. The angle of inclination is 68°, the same as that used for uppercase letters.

Inclined numerals

The inclined numerals that should be used in conjunction with inclined lettering are shown in Fig. 3.14. Except for the inclination of 68° to the horizontal, they are drawn the same as vertical numbers.

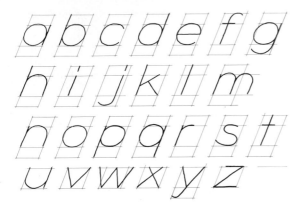

FIG. 3.13 The lowercase alphabet for single-stroke inclined Gothic lettering. The body of each letter is drawn inside a rhombus to help you learn their proportions.

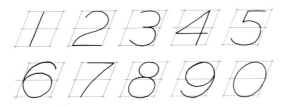

FIG. 3.14. The numerals for single-stroke inclined Gothic lettering. Each number is drawn inside a rhombus to help you learn their proportions.

The use of inclined letters and numbers in combination is seen in Fig. 3.15. The guidelines in this example were constructed using the Braddock-Rowe triangle (Fig. 3.6).

3.7 ——
Spacing numerals and letters

Common fractions are twice as tall as single numerals (Fig. 3.16). The fractions will be ¼ in. tall when they are used with ⅛ in. lettering. A separate set of holes for common fractions is given on the Braddock-Rowe triangle and on the Ames guide. These holes are equally spaced ¹⁄₁₆ in. apart, with the centerline being used for the fraction's crossbar. (Fig. 3.16 and Fig. 3.6).

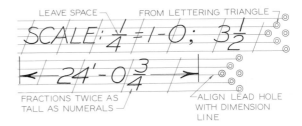

FIG 3.15 Inclined common fractions are twice as tall as single numerals. Inch marks are omitted when numerals are used to show dimensions.

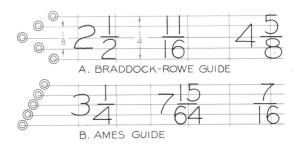

FIG 3.16 Common fractions are twice as tall as single numerals. Guidelines for these can be drawn by using the Ames or the Braddock-Rowe triangle.

When numbers are used with decimals, space should be provided for the decimal point (Fig. 3.17). The correct method of drawing common fractions is illustrated at D, and often-encountered errors are shown at E, F, and G.

When letters are grouped together to spell words, the areas between the letters should be approximately

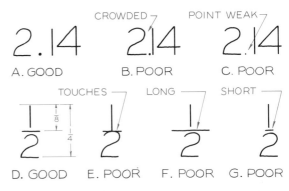

FIG. 3.17 Examples of poor spacing of numerals that result in poor lettering.

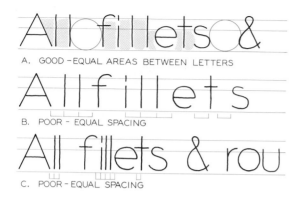

A. GOOD – EQUAL AREAS BETWEEN LETTERS

B. POOR – EQUAL SPACING

C. POOR – EQUAL SPACING

FIG. 3.18 Proper spacing of letters is necessary for good lettering. The areas between letters should be approximately equal.

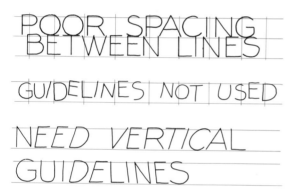

FIG. 3.19 Always leave space between lines of lettering. After constructing guidelines, *use them.* Use vertical guidelines to improve the angle of your vertical strokes.

equal (Fig. 3.18). The incorrect use of guidelines and other violations of good lettering practice are shown in Fig. 3.19.

3.8

Mechanical lettering

Drawings and illustrations that are to be reproduced by a printing process are usually drawn in India ink, and the lettering must also be in ink.

The Wrico lettering template (Fig. 3.20) can be placed against a fixed straightedge for aligning the let-

FIG. 3.20 A Wrico lettering template can be used for mechanical lettering. (Courtesy of Wood-Reagan Instrument Co.)

ters. You move the template from position to position while drawing each letter (with a lettering pen) through the raised portion of the template where the holes form the letters.

A slightly different template and pen are used by the Rapid-O-Graph system (Fig. 3.21). A variety of pen sizes is available from each manufacturer for drawing different sized letters and numbers with thin or bold lines.

Another system of mechanical lettering uses a grooved template in conjunction with a scriber. The scriber follows the grooves in the template and inks the letters on the drawing surface. A standard India

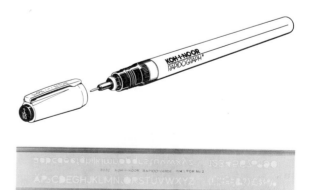

FIG. 3.21 A typical India ink fountain pen and template that can be used for mechanical lettering. (Courtesy of Koh-I-Noor Rapidograph, Inc.)

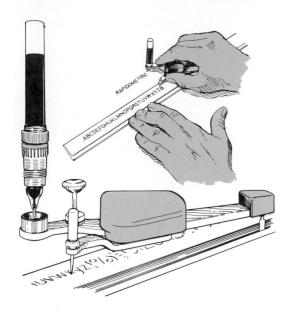

FIG. 3.22 Templates and scribers of this type are available for mechanical lettering.

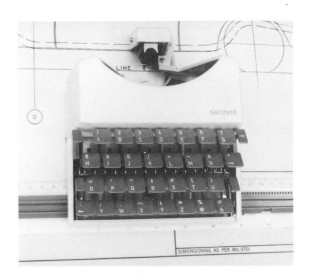

FIG. 3.23 This portable typewriter can be used to type notes and numerals on a drawing. (Courtesy of Gritzner, Inc.)

FIG. 3.24 Transfer lettering can be transferred from film to the drawing surface. Transfer lettering comes in many sizes and styles. (Courtesy of Artype, Inc.)

FIG. 3.25 A computer-based lettering machine that transcribes and stores lettering and symbols for engineering drawings. (Courtesy of AlphaMerics Corp.)

ink technical pen can be unscrewed from its barrel and attached to the scriber (Fig. 3.22).

3.9
Lettering by typing

Some drafting departments type many notes and specifications to reduce drafting time and to improve the readability of a drawing. Typewriters are available with long carriages that will accept large drawings; the notes are typed on the drawings in the conventional manner.

A portable typewriter without a carriage is available (Fig. 3.23) for typing directly on the drawing.

3.10
Transfer lettering

Transfer lettering comes printed on transparent sheets with adhesive backings on one side (Fig. 3.24). The letters are cut from the sheet, aligned with guidelines, and then burnished down permanently.

Some types of transfer lettering are burnished directly from the transparent sheet to the drawing surface. Each letter is rubbed firmly and evenly to make it transfer.

A multitude of letter forms and sizes is available from many manufacturers of transfer letters. Although many symbols and patterns are available in addition to lettering, it is possible to have custom transfer sheets produced for trademarks, title blocks, and other often-used applications.

3.11
Lettering by computer

Computer-based equipment may also be used for lettering on engineering drawings. The Datascribe IV (Fig. 3.25) is a portable machine that operates on a microprocessor with a typewriter keyboard. The text is then transcribed onto the drawing with the selected pen and the selected type font, which ranges in letter size from $\frac{1}{16}$ inch to $\frac{1}{2}$ inch. Often-repeated notes can be stored on cassette tapes for recall when needed.

Most computer graphics systems have programs for plotting lettering and notes on drawings. A typical alphabet and its operating program are shown in Fig. 3.26. Additional information on computer graphics lettering is given in Chapter 30.

```
100 REM ALPHABET OF CHARACTERS
110 REM INITIALIZE
120 IP% = 0: GOSUB 9000: GOSUB 9400
130 REM DEFINE CHARACTER STRING
140 REM DEFINE CHAR ORIGIN & HEIGHT
150 CHAR$ = "ABCDEFGHI"
160 REM DEFINE ORIGIN
170 CHAR.X = .5: CHAR.Y = 5
180 REM DEFINE ROTATION & HEIGHT
190 CHAR.A = 0: CHAR.H = .6
200 REM PLOT CHARACTER STRING
210 GOSUB 9300
220 REM PLOT 3 MORE STRINGS
230 CHAR.X = .5: CHAR.Y = 4
240 CHAR.A = 0: CHAR.H = .6
250 CHAR$ = "JKLMNOPQR"
260 GOSUB 9300
270 CHAR.X = .5: CHAR.Y = 3
280 CHAR.A = 0: CHAR.H = .6
290 CHAR$ ="STUVWXYZ"
300 GOSUB 9300
310 CHAR.X = .5: CHAR.Y = 2
320 CHAR.A = 0: CHAR.H = .6
330 CHAR$ = "0123456789"
340 GOSUB 9300
350 XP =0: YP = 0: IP%=3: GOSUB 9000
360 STOP
370 END
```

FIG. 3.26 Computer-drawn symbols, and the program that was written to plot them.

Problems

Lettering problems are to be presented on Size AV (8½'' × 11'') paper, plain or grid, using the format shown in Fig. 3.27.

1. Practice lettering the vertical uppercase alphabet shown in Fig. 3.27. Construct each letter four times: four A's, four B's, etc. Use a medium-weight pencil—H, F, or HB.

2. Practice lettering the vertical numerals and the lowercase alphabet shown in Fig. 3.28. Construct each letter and numeral three times: three 1's, three 2's, etc. Use a medium-weight pencil—H, F, or HB.

3. Practice lettering the inclined uppercase alphabet shown in Fig. 3.29. Construct each letter four times: four A's, four B's, etc. Use a medium-weight pencil—H, F, or HB.

4. Practice lettering the inclined numerals and the lowercase alphabet shown in Fig. 3.30. Construct

each letter three times: three 1's, three 2's, etc. Use a medium-weight pencil—H, F, or HB.

5. Construct guidelines for ⅛-in. capital letters starting ¼ in. from the top border. Each guideline should end ½ in. from the left and right borders. Using these guidelines and all vertical capitals, letter the first paragraph of the text of this chapter. Spacing between the lines should be ⅛ in.

6. Repeat Problem 5, but use all inclined capital letters. Use inclined guidelines to assist you in slanting your letters uniformly.

7. Repeat Problem 5 but use vertical capitals and lowercase letters in combination. Capitalize only those words that are capitalized in the text.

8. Repeat Problem 5 but use inclined capitals and lowercase letters in combination. Capitalize only the words that are capitalized in the text.

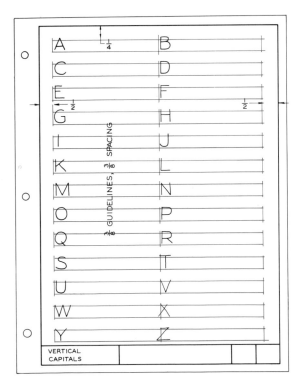

FIG. 3.27 Problem 1. Construct each vertical uppercase letter four times.

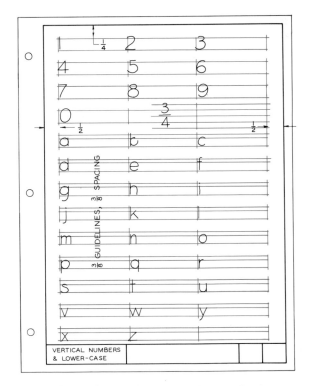

FIG. 3.28 Problem 2. Construct each vertical numeral and lowercase letter three times.

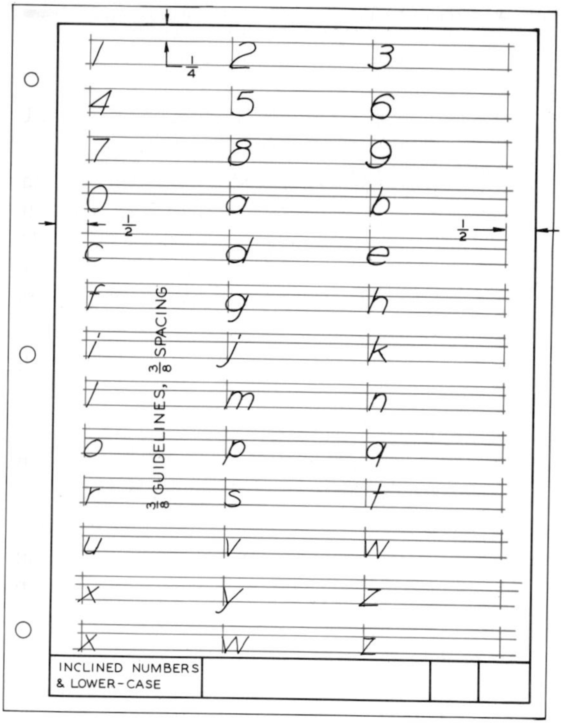

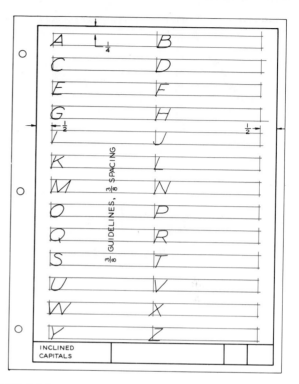

FIG. 3.29 Problem 3. Construct each inclined uppercase letter four times.

FIG. 3.30 Problem 4. Construct each inclined numeral and lowercase letter three times.

4

Geometric Construction

4.1
Introduction

Many technical problems can be solved only by the application of geometry and geometric construction. Mathematics was an outgrowth of graphic construction; consequently, there is a close relationship between the two areas. The proofs of many principles of plane geometry and trigonometry may be developed by using graphics. Graphic methods can be applied to algebra and arithmetic, and virtually all problems of analytical geometry can be solved graphically.

4.2
Angles

A fundamental requirement of geometric construction is the construction of lines that join at specified angles with each other. The definitions of various angles are given in Fig. 4.1.

> The unit of angular measurement is the degree, and a circle has 360 degrees. A degree (°) can be divided into 60 minutes ('), and a minute can be divided into 60 seconds ('').

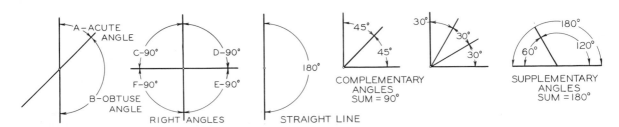

FIG. 4.1 Standard types of angles and their definitions

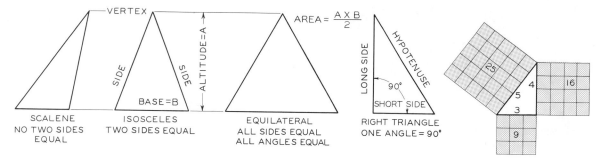

FIG. 4.2 Types of triangles and their definitions. The hypotenuse of the right triangle is equal to the square root of the sum of the squares of the other two sides. This is the Pythagorean theorem.

An angle of 15°32'14'' is an angle of 15 degrees, 32 minutes, and 14 seconds.

4.3
Triangles

The **triangle** is a three-sided figure (or polygon) that is named according to its shape. The four types of triangles are the **scalene, isosceles, equilateral,** and **right triangles** (Fig. 4.2). The sum of the angles inside a triangle will always be equal to 180°.

4.4
Quadrilaterals

A **quadrilateral** is a four-sided figure of any shape. The sum of the angles inside a quadrilateral is 360°. The various types of quadrilaterals and their respective names are shown in Fig. 4.3, along with the equations for their areas.

4.5
Polygons

A **polygon** is a multi-sided plane figure of any number of sides. (The triangle and quadrilateral are polygons.) If the sides of the polygon are equal in length, the polygon is a **regular polygon.** Four types of reg-

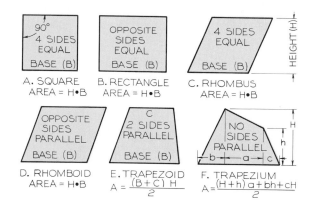

FIG. 4.3 Types of quadrilaterals (four-sided plane figures)

ular polygons are shown in Fig. 4.4. A regular polygon can be inscribed in a circle and all of the corner points will lie on the circle.

Other regular polygons not pictured are: the **heptagon** with seven sides, the **nonagon** with nine sides, the **decagon** with 10 sides, and the **dodecagon** with 12 sides.

The sum of the angles inside any polygon can be found by the equation: Sum = $(n - 2) \times 180°$, where n is equal to the number of sides of the polygon.

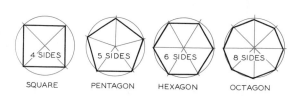

FIG. 4.4 Regular polygons inscribed in circles

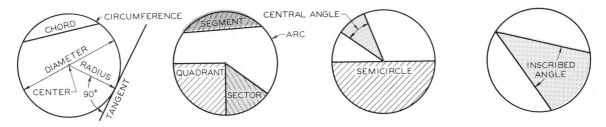

FIG. 4.5 Definitions of the elements of a circle

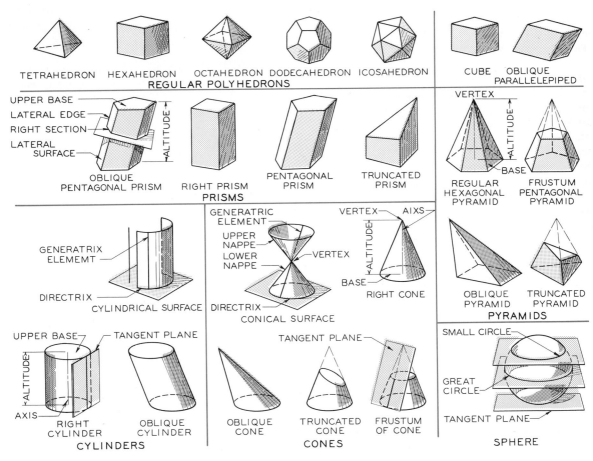

FIG. 4.6 Types of geometric solids and their elements and definitions

4.6
Elements of a circle

A circle can be divided into a number of parts, each of which has its own special name (Fig. 4.5). The equation for finding the area (A) of a circle is $A = \pi r^2$, where r is the radius and π is equal to 3.14 (pi). The equation for finding the circumference is $C = \pi d$; where d is the diameter.

4.7
Geometric solids

The various types of solid geometric shapes are shown in Fig. 4.6 along with their names and definitions.

POLYHEDRA   A multi-sided solid formed by intersecting planes is a **polyhedron.** If the faces of a polyhedron are regular polygons, it is called a **regular polyhedron.** The five regular polyhedra are the **tetrahedron** with four sides, the **hexahedron** with six sides, the **octahedron** with eight sides, the **dodecahedron** with 12 sides, and the **icosahedron** with 20 sides.

PRISMS   A **prism** is a solid that has two parallel bases that are equal in shape. The bases are connected by sides that are called **parallelograms.** The line from the center of one base to the other is the **axis.** If the axis is perpendicular to the bases, the axis is called the **altitude** and the prism is a **right prism.** If the axis is not perpendicular to the base, the prism is an **oblique prism.**

A prism that has been cut off to form a base that is not parallel to the other is called a **truncated prism.** A **parallelepiped** is a prism with a base that is either a rectangle or a parallelogram.

PYRAMIDS   The **pyramid** is a solid with a polygon as a base and triangular faces that converge at a point called the **vertex.** The line from the vertex to the center of the base is the **axis.** If the axis is perpendicular to the base, it is the **altitude** of the pyramid, and the pyramid is a **right pyramid.** If the axis is not perpendicular to the base, the pyramid is an **oblique pyramid.** A truncated pyramid is called a **frustum** of a pyramid.

CYLINDERS   The **cylinder** is formed by a line or element, called a **generatrix,** that moves about the circle while remaining parallel to its axis.

The axis of a cylinder connects the centers of each end of a cylinder. If the axis is perpendicular to the bases, it is the **altitude** of the cylinder, and the cylinder is a **right cylinder.** When the axis does not make a 90° angle with the base, the cylinder is an **oblique cylinder.**

CONES   A **cone** is formed by a line or an element, called a generatrix, with one end that moves about the curved base while the other end remains at a fixed point called the **vertex.** The line from the center of the base to the vertex is called the **axis.** If the axis is perpendicular to the base, it is called the **altitude,** and the cone is a **right cone.** A truncated cone is called a **frustum** of a cone.

SPHERES   The **sphere** is generated by the plane of a circle that is revolved about one of its diameters to form a solid. The ends of an axis through the center of the sphere are called **poles.**

The equations for find the volumes of geometric solids are given in Fig. 4.7.

4.8
Constructing a triangle

When three sides of a triangle are given, the triangle can be constructed by using a compass as shown in Fig. 4.8.

A right triangle can be constructed by inscribing it inside a semicircle, as shown in Fig. 4.9. Any triangle inscribed in a semicircle will always be a right triangle.

4.9
Constructing polygons

A regular polygon (having equal sides) can be inscribed in a circle or circumscribed about a circle. When inscribed, all the corner points will lie along the circle. This makes it possible to divide the circle into

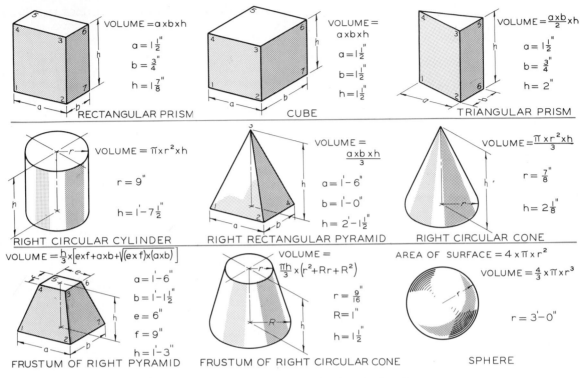

FIG. 4.7 Standard geometric solids and the equations for finding their volumes

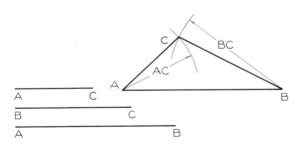

FIG. 4.8 A triangle can be drawn with a compass when three sides are given.

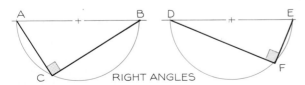

FIG. 4.9 Any angle inscribed in a semicircle will be a right triangle.

the desired number of sectors to locate the points (Fig. 4.10).

For example, a 12-sided polygon is constructed by dividing the circle into 12 sectors of 30° each and connecting the points to form the polygon.

4.10
Hexagon

Examples of inscribed and circumscribed hexagons are shown in Fig. 4.11. These are drawn with 30°–60° triangles either inside or outside the circles. Note that the circle represents the distance from corner to corner when inscribed, and from side to side when circumscribed.

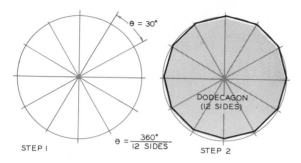

FIG. 4.10 The regular polygon

Step 1 To construct a regular polygon, divide the circumference of a circle into the same number of divisions as the polygon has sides, 12 in this case.

Step 2 Connect the divisions marked on the circumference of the circle with straight lines to form the sides of the polygon.

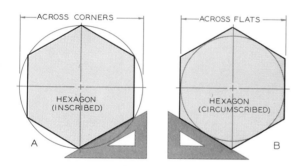

FIG. 4.11 A circle can be inscribed or circumscribed to form a hexagon with a 30°–60° triangle.

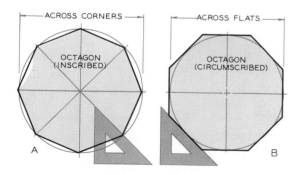

FIG. 4.12 A circle can be inscribed or circumscribed to form an octagon with a 45° triangle.

4.11
Octagon

The octagon, an eight-sided regular polygon, can be inscribed in or circumscribed about a circle (Fig. 4.12) by using a 45° triangle. A second method inscribes the octagon inside a square (Fig. 4.13).

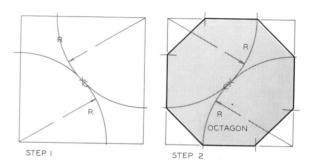

Fig. 4.13 Octagon in a square

Step 1 Construct a square and locate its center, C. Construct two arcs from opposite corners that will pass through C.

Step 2 Repeat this step using the other two corners. Connect the points marked on the square with straight lines to form the octagon.

4.12
Pentagon

Since the pentagon is a five-sided regular polygon, it can be inscribed in or circumscribed about a circle, as previously covered. Another method of constructing a pentagon is shown in Fig. 4.14.

4.13
Bisecting lines and angles

Finding the midpoint of a line, or the perpendicular bisector of a line, is a basic technique of geometric construction. Two methods are illustrated in Fig. 4.15.

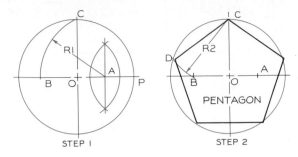

STEP 1 | STEP 2 | PENTAGON

FIG. 4.14 The pentagon

Step 1 Bisect radius *OP* to locate Point *A*. With *A* as the center and *AC* as the radius *R*1, locate Point *B* on the diameter.

Step 2 With Point *C* as the center and *BC* as the radius *R*2, locate Point *D* on the arc. Line *CD* is the chord that can be used to locate the corners of the pentagon.

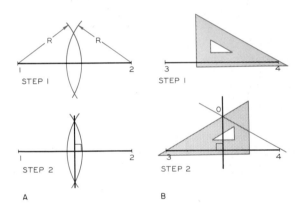

STEP 1 | STEP 1
STEP 2 | STEP 2
A | B

FIG. 4.15 Bisecting a line A line can be bisected by using a compass and any radius or a standard triangle and straightedge.

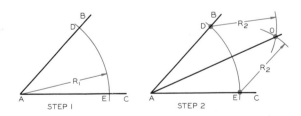

STEP 1 | STEP 2

FIG. 4.16 Bisecting an angle

Step 1 Swing an arc of any radius to locate Points *D* and *E*.

Step 2 Using the same radius, draw two arcs from *D* and *E* to locate Point *O*. Line *AO* is the bisector of the angle.

The first method involves the use of a compass to construct a perpendicular to a line. The second method uses a standard triangle.

The compass method can be used to find the midpoint of an arc as well as a straight line. The angle in Fig. 4.16 was bisected with a compass by drawing three arcs.

4.14
Revolution of a figure

Rotating a triangle about one of its points is demonstrated in Fig. 4.17. In this case, the triangle is rotated about Point 1 of Line 1–4. Point 4 is rotated to its desired position using a compass. Points 2 and 3 are found by triangulation, to complete the rotated view. This principle will work on any plane regardless of its shape.

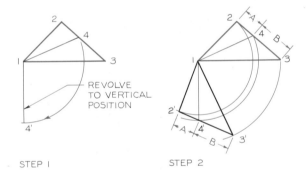

STEP 1 | STEP 2

FIG. 4.17 Rotation of a figure

Step 1 A plane figure can be rotated about any point. Line 1–4 is rotated about Point 1 to its desired position with a compass.

Step 2 Points 2' and 3' are located by measuring distances *A* and *B* from Point 4'.

4.15
Enlargement and reduction of a figure

In Fig. 4.18, the smaller figure is enlarged by using a series of radial lines from the lower left corner. The smaller figure is completed as a rectangle, and the

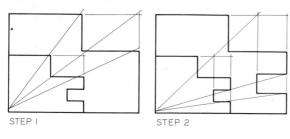

STEP 1   STEP 2

FIG. 4.18 Enlargement of a figure

Step 1 A proportional enlargement can be made by using a series of diagonals drawn through a single point, in this case the lower left corner.

Step 2 Additional diagonals are drawn to locate the other features of the object. This process can be reversed to reduce an object.

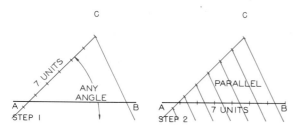

FIG. 4.19 Division of a line

Step 1 To divide Line *AB* into seven equal divisions construct a line through *A* and divide it into seven known units with your dividers. Connect Point *C* to Point *B*.

Step 2 Draw a series of lines parallel to *CB* to locate the divisions along Line *AB*.

larger rectangle is drawn proportional to the small one. The upper right notch is located using the same technique.

The smaller notch is found by using three construction lines projected from the lower left corner. This method is based on the principle of similar and proportional triangles, and could have been used to reduce the larger drawing to the smaller one.

4.16
Division of a line

It is often necessary to divide a line into a number of equal parts when a convenient scale is not available. For example, suppose a one-inch line is to be divided

into seven equal parts. No scale is available that divides an inch into sevenths, and mathematical units involve hard-to-measure decimals.

The method shown in Fig. 4.19 is an efficient way to solve this problem.

An application of the same principle is used to locate lines on a graph that are equally spaced (Fig. 4.20). A scale with the desired number of units (0 to 5) is laid across from left to right on the graph.

4.17
Circle through three points

An arc can be drawn through any three points by connecting the points with two lines (Fig. 4.21). Perpendicular bisectors are found for each line to locate the center at Point *C*. The radius is drawn, and the lines *AB* and *BD* become chords of the circle.

This system can be reversed to find the center of a given circle. Draw two chords that intersect at a point on the circumference and bisect them. The perpendicular bisectors will intersect at the center of the circle.

4.18
Parallel lines

A line can be drawn parallel to another by using either of the methods shown in Fig. 4.22.

The first method involves the use of a compass to draw two arcs to locate a parallel line that is the desired distance away.

The second method requires constructing a perpendicular from a given line and measuring the distance, *R*, to locate the parallel, which is drawn with a straightedge.

4.19
Points of tangency

A point of tangency is the theoretical point where a straight line joins an arc, or where two arcs join making a smooth transition. In Fig. 4.23, a line is tangent to an arc. The point of tangency is located by con-

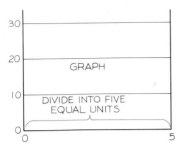 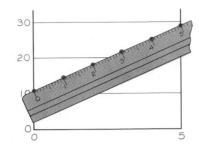 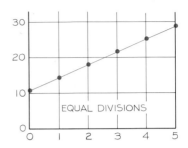

FIG. 4.20 Division of a space

Given It is desired to divide a graph into five equal divisions along the x-axis.

Step 1 A scale with five units of measurement that approximate the horizontal distance is laid across the graph and five divisions are marked.

Step 2 Construct vertical lines through the points found in Step 1. This method could have been used to calibrate the divisions along the y-axis.

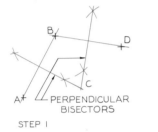

FIG. 4.21 An arc through three points

Step 1 Connect Points A, B, and D with two lines and find their perpendicular bisectors. The bisectors will intersect at the center, C.

Step 2 Using the center, C, and the distance to the points as the radius, construct the arc through the points.

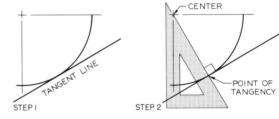

FIG. 4.23 Locating a tangent point

Step 1 Align your triangle with the tangent line while holding it firmly against a straightedge.

Step 2 To locate the point of tangency, hold the straightedge in position, rotate the triangle, and draw a line through the center of the arc that is perpendicular to the line.

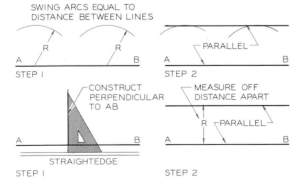

FIG. 4.22 Construction of parallel lines Either of these two methods can be used to construct one line parallel to another. The first method uses a compass and a straightedge; the second uses a triangle and straightedge.

structing a perpendicular to the line from the center of the arc. A thin line is drawn from the center through the point to mark the point of tangency.

The conventional methods of marking points of tangency are shown in Fig. 4.24. Thin lines are drawn through the points from the centers of the arcs.

4.20
Line tangent to an arc

Although you can approximate the point of tangency between a line and an arc, the method of finding the exact point of tangency is shown in Fig. 4.25. Point A

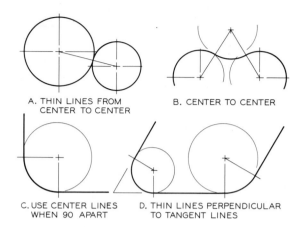

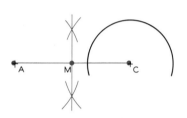

A. THIN LINES FROM
CENTER TO CENTER

B. CENTER TO CENTER

C. USE CENTER LINES
WHEN 90 APART

D. THIN LINES PERPENDICULAR
TO TANGENT LINES

FIG. 4.24 Thin lines that extend beyond the arcs from the centers are used to mark points of tangency.

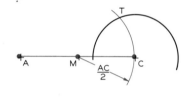

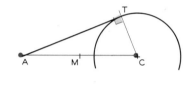

FIG. 4.25 Line from a point tangent to an arc

Step 1 Connect Point *A* with Center *C*. Locate Point *M* by bisecting *AC*.

Step 2 Using Point *M* as the center and *MC* as the radius, locate Point *T* on the arc.

Step 3 Draw the line from *A* to *T* that is tangent to the arc at Point *T*.

is connected to the center in Step 1, *AC* is bisected in Step 2, and *T,* the exact point of tangency, is located in Step 3.

The point of tangency can also be found by using a standard triangle, as shown in Fig. 4.26. One edge of the triangle is aligned with *TA* while the straight-

edge is held firmly. The triangle is rotated to construct a line through the center that is perpendicular to *AT,* locating the point of tangency, *T.*

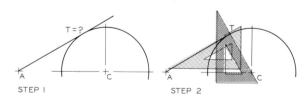

FIG. 4.26 Line from a point tangent to an arc

Step 1 A line can be drawn from Point *A* tangent to the arc by eye.

Step 2 By rotating your triangle, the point of tangency can be located at a 90° angle with Line *AT* and pass through the center of the arc.

4.21
Arc tangent to a line from a point

If an arc is to be constructed tangent to Line *CD* at *T* (Fig. 4.27) and pass through Point *P*, a perpendicular bisector of *TP* is drawn. A perpendicular to *CD* is drawn at *T* to locate the center at *O*.

A similar problem in Fig. 4.28 requires that you draw an arc of a given radius that will be tangent to Line *AB* and pass through Point *P*. In this case the point of tangency on the line is not known until the problem has been solved.

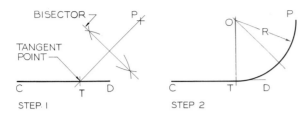

FIG. 4.27 Arc through two points

Step 1 If an arc must be tangent to a given line at a certain point and pass through *P*, find the perpendicular bisector of line *TP*.

Step 2 Construct a perpendicular to the line at *T* to intersect the bisector. The arc is drawn from Center *O* with Radius *OT*.

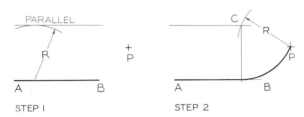

FIG. 4.28 Arc tangent to a line and a point

Step 1 When an arc of a given radius is to be drawn tangent to a line and through Point *P*, draw a line parallel to *AB* and distance *R* from it.

Step 2 Draw an arc from *P* with Radius *R* to locate the center, *C*. The arc is drawn with Radius *R* and Center *C*.

4.22
Arc tangent to two lines

An arc of a given radius can be constructed tangent to two nonparallel lines. This construction may be used to round a corner of a product or to design a curb at a traffic intersection. The method shown in Fig. 4.29 is used when two lines form an acute angle.

The same steps are used to find an arc that is tangent to two lines that form an obtuse angle (Fig. 4.30). In both cases the points of tangency are located with thin lines drawn from the centers through the points of tangency.

A different technique can be used to find an arc of a given radius that is tangent to perpendicular lines (Fig. 4.31). This method will work only for perpendicular lines.

4.23
Arc tangent to an arc and a line

When a radius is given, an arc can be drawn that is tangent to an arc and a line. These steps of construction are given in Fig. 4.32.

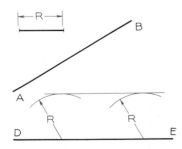
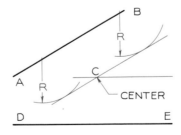
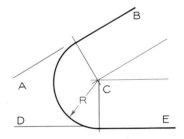

FIG. 4.29 Arc tangent to two lines

Step 1 Construct a line parallel to *DE* with the radius of the specified arc, *R*.

Step 2 Draw a second construction line parallel to *AB* to locate the center, *C*.

Step 3 Thin lines are drawn from *C* perpendicular to *AB* and *DE* to locate the points of tangency, and the tangent arc is drawn.

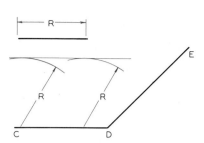

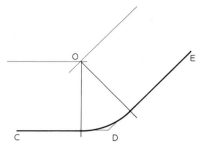

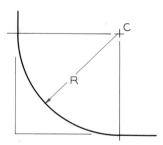

FIG. 4.30 Arc tangent to two lines

Step 1 Using the specified Radius *R*, construct a line parallel to *CD*.

Step 2 To locate center *O*, construct a line parallel to *DE* that is distance *R* from it.

Step 3 Construct thin lines from *O* perpendicular to *CD* and *DE* to locate the points of tangency. Draw the arc using Radius *R*.

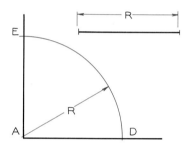

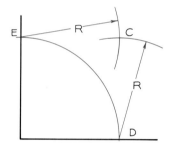

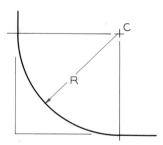

FIG. 4.31 Arc tangent to perpendicular lines

Step 1 Using the specified radius, *R*, locate Points *D* and *E* by using Center *A*.

Step 2 To locate Point *C*, swing two arcs using the radius, *R*, that was used in Step 1.

Step 3 Locate the tangent points with lines from *C*. Draw the arc with Radius *R*.

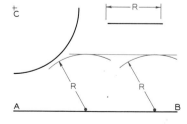

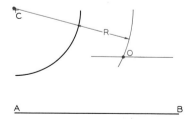

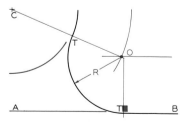

FIG. 4.32 Arc tangent to an arc and a line

Step 1 Construct a line parallel to *AB* that is distance *R* from it. Use thin construction lines.

Step 2 Add Radius *R* to the extended radius through Point *C*. Use this large radius to locate Point *O*.

Step 3 Draw Lines *OC* and *OT* to locate the tangency points. Draw the arc with Radius *R*.

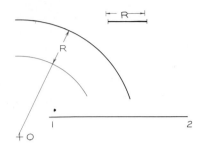

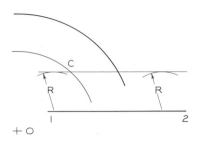

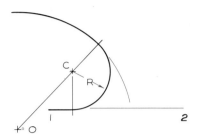

FIG. 4.33 Arc tangent to an arc and a line

Step 1 The specified arc, R, is subtracted from the extended radius. A concentric arc is drawn with the shortened radius.

Step 2 A line parallel to 1–2 is drawn a distance of R from it to locate the center, Point C.

Step 3 The tangent points are located with lines from O through C, and through C perpendicular to 1–2. Draw the tangent arc with Radius R.

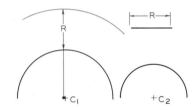

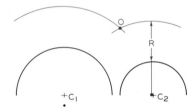

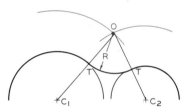

FIG. 4.34 Arc tangent to two arcs

Step 1 The radius of one circle is extended and Radius R is added to it. The extended radius is used to draw a concentric arc.

Step 2 The radius of the other circle is extended and R is added to it. The extended radius is used to construct an arc and to locate Center O.

Step 3 The centers are connected with Center O to locate the points of tangency. The arc is drawn tangent to the two arcs with Radius R.

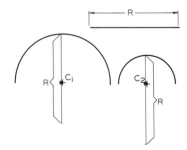

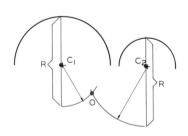

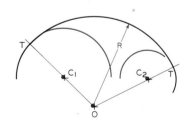

FIG. 4.35 Arc tangent to two arcs

Step 1 The radius of each arc is extended from the arc past its center and the specified radius, R, is laid off from the arcs along these lines.

Step 2 The distance from each center to the ends of the extended radii are used for two concentric arcs to locate the center, O.

Step 3 Thin lines from O through Centers C_1 and C_2 locate the points of tangency. The arc is drawn using Point O as the center.

A variation of this principle of construction is shown in Fig. 4.33, where the arc is drawn tangent to an arc and line with the arc in a reverse position.

4.24
Arc tangent to two arcs

An arc is drawn tangent to two given arcs in Fig. 4.34. Thin lines are drawn from the centers to locate the points of tangency. This tangent arc is concave from the top.

A convex arc can be drawn tangent to the given arcs if the radius of the arc is greater than the radius of either of the given arcs (Fig. 4.35).

A variation of this problem is shown in Fig. 4.36, where an arc of a given radius is drawn tangent to the top of one arc and the bottom of the other. A similar problem is shown in Fig. 4.37, where an arc is drawn tangent to a circle and a larger arc.

The points of tangency are located and marked by thin lines drawn from center to center of the tangent arcs.

4.25
Ogee curve

The ogee curve can be though of as double curve formed by tangent arcs. The ogee curve in Fig. 4.38

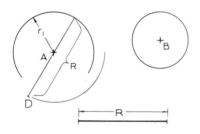
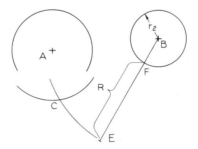
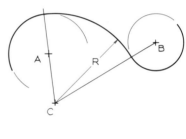

FIG. 4.36 Arc tangent to two circles

Step 1 The specified radius, *R*, is laid off from the arc along the extended radius to locate Point *D*. Radius *AD* is used to construct a concentric arc.

Step 2 The radius through Center *B* is extended and Radius *R* is added to it from Point *F*. Radius *BE* is used to locate the center, *C*.

Step 3 The tangent arc is drawn from Center *C* with Radius *R*. The points of tangency are located with thin lines from *O* to the given centers.

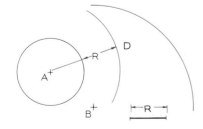
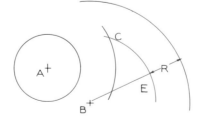
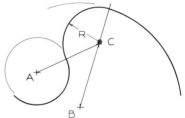

FIG. 4.37 Arc tangent to two arcs

Step 1 Radius *R* is added to the radius through Center *A*. Radius *AD* is used to draw a concentric arc with Center *A*.

Step 2 Radius *R* is subtracted from the radius through *B*. Radius *BE* is used to construct a concentric arc and locate Center *C*.

Step 3 The points of tangency are located with thin lines *BC* and *AC*. The tangent arc is drawn with Center *C*.

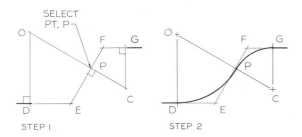

FIG. 4.38 Ogee curve

Step 1 To draw an ogee curve between two parallel lines, draw Line *EF* at any angle. Locate a point of your choosing along *EF*, *P* in this case. Find the tangent points by making *FG* equal to *FP* and *DE* equal to *EP*. Draw perpendiculars at *G* and *D* to intersect the perpendicular bisector of *EF*.

Step 2 Using Radii *CP* and *OP*, draw the two tangent arcs to complete the ogee curve.

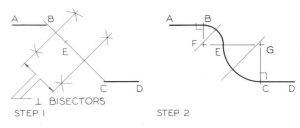

FIG. 4.40 Unequal ogee curve

Step 1 Two parallel lines are to be connected by an ogee curve that passes through *B* and *C*. Draw Line *BC* and select Point *E* on the line. Construct perpendicular bisectors of *BE* and *EC*.

Step 2 Construct perpendiculars at *B* and *C* to intersect the bisectors to locate Centers *F* and *G*. Locate the points of tangency and draw the ogee curve using Radii *FB* and *GC*.

was found by constructing two arcs tangent to three intersecting lines.

An ogee curve can be drawn between two parallel lines (Fig. 4.39) from Points *B* to *C* by geometric construction. An alternative method of drawing an ogee curve that passes through points *B*, *E*, and *C* is illustrated in Fig. 4.40. The method of drawing an ogee curve through two nonparallel lines is shown in Fig. 4.41.

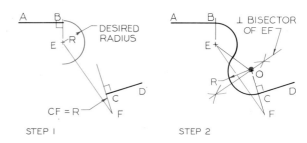

FIG. 4.41 Ogee curve between nonparallel lines

Step 1 Draw a perpendicular to *AB* at *B* and draw the arc with the desired radius from *E*. Draw a perpendicular to Line *CD* at *C* and make *CF* equal to Radius *R*. Connect *E* and *F*.

Step 2 Draw a perpendicular bisector of *EF* to locate Point *O*. Use Radius *OC* (which may be different from the first radius) to complete the curve. Mark points of tangency.

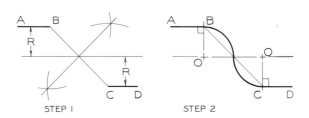

FIG. 4.39 An ogee curve

Step 1 To draw an ogee curve formed by two equal arcs passing through Points *B* and *C*, draw a line between the points. Bisect the Line *BC* and draw a line parallel to *AB* and *CD* to find the radius, *R*.

Step 2 Construct perpendiculars at *B* and *C* to locate the centers at both Points *O*. Draw the arcs to complete the ogee curve.

4.26
Curve of arcs

An irregular curve formed with tangent arcs can be constructed as shown in Fig. 4.42. In this case the radii of the arcs are selected to give the desired curve by moving from one set of points to the next.

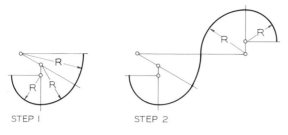

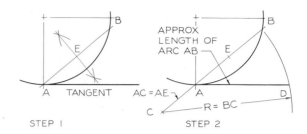

FIG. 4.42 Curve of arcs

Step 1 A series of arcs can be joined to form a smooth curve. Begin with the small arc, extend the radius through its center, draw the second arc, and then the third.

Step 2 The curve can be reversed by extending the radius in the opposite direction and repeating the same process.

FIG. 4.44 Rectifying an arc—compass method

Step 1 If you wish to rectify an arc from A to B, draw Chord AB and find its midpoint.

Step 2 Extend AB to make AC equal to AE. The length of Arc AB is approximated to be the distance from A to D found by swinging Arc BC.

4.27
Rectifying an arc

An arc is rectified when its true length is laid out along a straight line. One method of rectifying an arc is illustrated in Fig. 4.43.

A second method of rectifying an arc is given in Fig. 4.44, using a different form of geometric construction.

An arc can also be rectified by using the mathematical equation for finding the circumference of the circle. Since a circle has 360°, the arc of a 30° sector is $\frac{1}{12}$ of the full circumference. Therefore, if the circumference is 12 inches, the arc of 30° is equal to 1 inch.

4.28
Conic sections

Conic sections are plane figures that can be described graphically as well as mathematically. They are formed by passing imaginary cutting planes through a right cone (Fig. 4.45).

Ellipse

The **ellipse** is a conic section formed by passing a plane through a right cone at an angle (Fig. 4.45). The ellipse is mathematically defined as the path of a point that moves in such a way that the sum of the distances from two focal points is a constant.

The construction of an ellipse is found by revolving the edge view of a circle, as shown in Fig. 4.46. This ellipse could have been drawn using the ellipse template, shown in Fig. 4.47. The angle between the line of sight and the edge view of the circle is the angle of the ellipse template that should be used (or the one closest to this size).

The largest diameter of an ellipse is always the true length and is called the **major diameter.** The shortest diameter is perpendicular to the major diam-

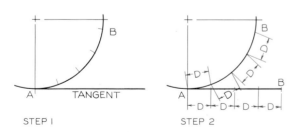

FIG. 4.43 Rectifying an arc

Step 1 An arc has been rectified when its length has been laid out along a straight line. Construct a line tangent to the arc and divide the arc into a series of equal divisions from A to B.

Step 2 The cordal distances, D, along the arc are laid out along the straight line until Point B is located.

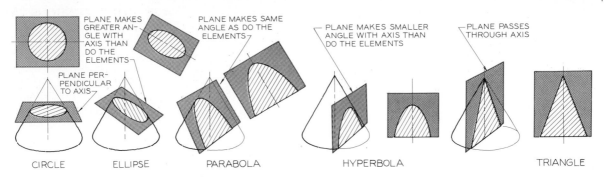

CIRCLE   ELLIPSE   PARABOLA   HYPERBOLA   TRIANGLE

FIG. 4.45 The conic sections are formed by passing cutting planes at various angles through right cones. The conic sections are the circle, ellipse, parabola, hyperbola, and triangle.

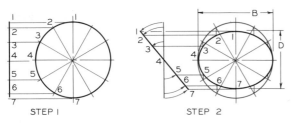

STEP I   STEP 2

FIG. 4.46 An ellipse by revolution

Step 1 When the edge view of a circle is perpendicular to the projectors between its adjacent view, the view will be a true circle. Mark equally spaced points along the arc and project them to the edge.

Step 2 Revolve the edge of the circle to the desired position and project the points to the circular view. Note that the points are projected vertically downward to their new positions, to form an ellipse.

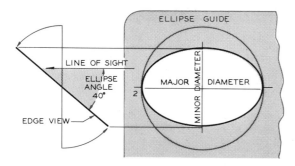

FIG. 4.47 The ellipse template When the line of sight is not perpendicular to the edge view of a circle, it will appear as an ellipse. The major diameter remains constant but the minor diameter will vary. The angle between the line of sight and the edge view of the circle is the angle of the ellipse template that should be used.

eter and is called the **minor diameter.** The crossing diameters are used to align the ellipse template.

Ellipse templates are available in intervals of 5° and in variations in size of the major diameter of about ⅛ inch (Fig. 4.48).

The mathematical equation of the ellipse is

$$\frac{x^2}{a^2} + \frac{y^2}{b^2} = 1, \qquad \text{where } a, b \neq 0.$$

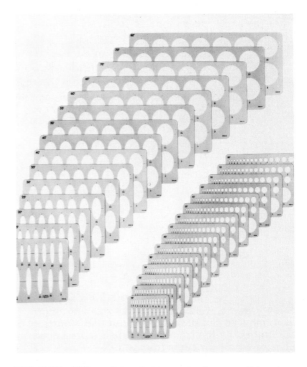

FIG. 4.48 Ellipse templates come in a variety of sizes. Most are calibrated at 5° intervals from 15° up to 60°. (Courtesy of Timely Products, Inc.)

60

Letters *a* and *b* are constants, and *x* and *y* are variables. This equation can be plotted on graph paper.

The ellipse can be constructed inside a rectangular box or a parallelogram, as illustrated in Fig. 4.49, with a series of points plotted to form an elliptical curve.

Two circles can be used to construct an ellipse by making the diameter of the large one equal to the major diameter and the diameter of the small one equal to the minor diameter (Fig. 4.50).

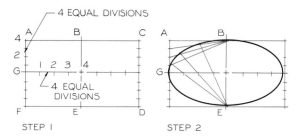

FIG. 4.49 Ellipse—parallelogram method

Step 1 An ellipse can be drawn inside of a rectangle or a parallelogram by dividing the horizontal center line into the same number of equal divisions as the shorter sides, *AF* and *CD*.

Step 2 The construction of the curve in one quadrant is shown by using sets of rays from *E* and *B* to plot the points.

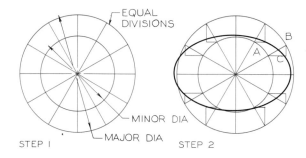

FIG. 4.50 Ellipse—circle method

Step 1 Two concentric circles are drawn with the large one equal to the major diameter and the small one equal to the minor diameter. Divide them into equal sectors.

Step 2 Plot points on the ellipse by projecting downward from the large curve to intersect horizontal construction lines drawn from the intersections on the small circle.

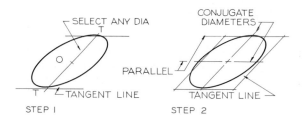

FIG. 4.51 Conjugate diameters

Step 1 Many sets of conjugate diameters can be constructed for an ellipse. A conjugate diameter is one that is parallel to the tangents at the ends of the other conjugate diameter. A diameter is selected and parallel tangents are drawn.

Step 2 The horizontal conjugate diameter is drawn parallel to the horizontal tangents, and the inclined tangents are drawn parallel to the conjugate diameter found in Step 1.

The **conjugate diameters** of an ellipse are diameters that are parallel to the tangents at the ends of each, as illustrated in Fig. 4.51. A single ellipse has an infinite number of sets of conjugate diameters.

When the ellipse and a pair of conjugate diameters are given, the major and minor diameters of the ellipse can be found by using the method illustrated in Fig. 4.52. If the conjugate diameters are not given, they can be constructed as shown in Fig. 4.51, and the major and minor diameters found. The major and minor diameters are necessary to use the ellipse template.

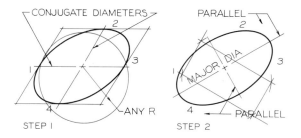

FIG. 4.52 Finding the axes of an ellipse

Step 1 When the conjugate diameters of an ellipse are given, you can find the major and minor diameters of the ellipse by drawing a circle of any radius from the intersection of the diameters.

Step 2 The circle cuts four points along the ellipse. The points are connected to form a rectangle. The major and minor diameters are parallel to the sides of the rectangle.

Parabola

The parabola is mathematically defined as a plane curve, each point of which is equidistant from a directrix (a straight line) and its focal point. The parabola is formed when the cutting plane makes the same angle with the base of a cone as do the elements of the cone.

The construction of a parabola by using its mathematical definition is shown in Fig. 4.53.

A parabolic curve can be constructed by geometric construction, as shown in Fig. 4.54, by dividing the two perpendicular lines into the same number of divisions.

A third method of construction is illustrated in Fig. 4.55, using the parallelogram method.

The mathematical equation of the parabola is

$$y = ax^2 + bx + c, \qquad \text{where } a \neq 0.$$

Letters a, b, and c are constants, and x and y are variables. This equation can be written by interchanging the y's and x's. The equation can be plotted on graph paper.

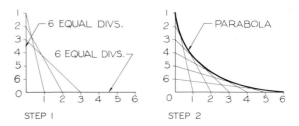

FIG. 4.54 Parabola—tangent method

Step 1 Construct two lines at a convenient angle and divide each of them into the same number of divisions. Connect the points with a series of diagonals.

Step 2 Construct the parabolic curve to be tangent to the diagonals.

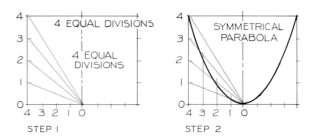

FIG. 4.55 Parabola—parallelogram method

Step 1 Construct a rectangle or a parallelogram to contain the parabola, and locate its axis parallel to the sides through 0. Divide the sides into equal divisions. Connect the divisions with Point 0.

Step 2 Construct lines parallel to the sides (vertical in this case) to locate the points along the rays from 0. Draw the parabola.

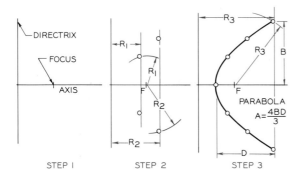

FIG. 4.53 Parabola—mathematical method

Step 1 Draw an axis perpendicular to a line (called a directrix). Choose a point for the focus, F.

Step 2. Locate points by using a series of selected radii to plot points on the curve. For example, draw a line parallel to the directrix and R_2 from it. Swing R_2 from F to intersect the line and plot the point.

Step 3 Continue the process with a series of arcs of varying radii until an adequate number of points have been found to complete the curve.

Hyperbola

The hyperbola is a two-part conic section that is mathematically defined as the path of a point that moves in such a way that the difference of its distances from two focal points is a constant. This definition is used to construct the hyperbola in Fig. 4.56.

A second method of construction is shown in Fig. 4.57. Two perpendicular lines are drawn through Point B as asymptotes. The hyperbolic curve becomes more nearly parallel and closer to the asymptotes as

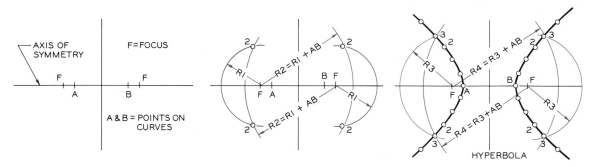

FIG. 4.56 Hyperbola

Step 1 A perpendicular is drawn through the axis of symmetry, and focal points F are located equidistant from it on both sides. Points on the curve, A and B, are located equidistant from the perpendicular at a location of your choice but between the focal points.

Step 2 Radius R_1 is selected to draw arcs using focal points F as the centers. R_1 is added to AB (the distance between the nearest points on the hyperbolas) to find R_2. Radius R_2 is used to draw arcs using the focal points as centers. The intersections of R_1 and R_2 establish Points 2 on the hyperbola.

Step 3 Other radii are selected and added to distance AB to locate additional points in the same manner as described in Step 2. A smooth curve is drawn through the points to form the hyperbolic curves.

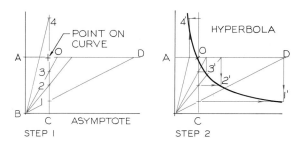

FIG. 4.57 Equilateral hyperbola

Step 1 Two perpendiculars are drawn through B, and any point O on the curve is located. Horizontal and vertical lines are drawn through O. Line CO is divided into equal divisions and rays from B are drawn through them to the horizontal line.

Step 2 Horizontal construction lines are drawn from the divisions along Line OC, and lines from AD are projected vertically to locate Points 1' through 4' on the curve.

the hyperbola is extended, but the curve never merges with the asymptotes.

4.29
Spiral of Archimedes

A spiral is a coil that begins at a point and becomes larger as it travels around the orign. A spiral lies in a single plane. The steps used to construct a spiral are shown in Fig. 4.58.

4.30
Helix

A helix is a curve that coils around a cylinder or a cone at a constant angle of inclination. Examples of helixes are corkscrews or threads on a screw. A helix is constructed about a cylinder in Fig. 4.59, and about a cone in Fig. 4.60.

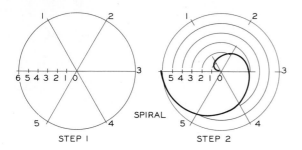

FIG. 4.58 Spiral

Step 1   Draw a circle and divide it into equal parts. The radius is divided into the same number of equal parts—six in this example.

Step 2   By beginning on the inside, draw Arc 0–1 to intersect Radius 0–1. Then swing Arc 0–2 to Radius 0–2 and continue until the last point is reached at 6, which lies on the original circle.

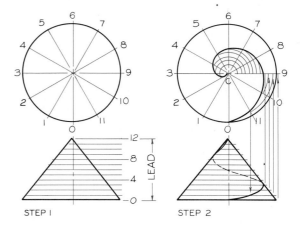

FIG. 4.60   Conical helix

Step 1   Divide the cone's base into equal parts. Pass a series of horizontal cutting planes through the front view of the cone. Use the same number as the divisions on the base, 12 in this case.

Step 2   Project all of the divisions along the front view of the cone to Line C9, and draw a series of arcs from Center C to their respective radii in the top view to plot the points. Project the points to their respective cutting planes in the front view.

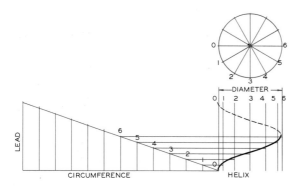

FIG. 4.59 Helix   Divide the top view of the cylinder into equal divisions and project them to the front view. Lay out the circumference and the height of the cylinder, which is the lead. Divide the circumference into the same number of equal parts by taking the measurements from the top view. Project the points along the inclined rise to their respective elements in the front view to find the helix.

4.31
Involute

The involute is the path of the end of a line as it unwinds from a line or a plane figure. In Fig. 4.61, an involute is formed by unwinding a line from a rectangle. Successively different radii that are equal in length to the sides are used to develop the involute.

4.32
Cycloid

The **cycloid** is a plane curve formed by a point on a circle as the circle rolls along a straight line.

In Fig. 4.62, the distance from 1 to 9 must be equal to the circumference of the circle. The circle is located at the center point 5, and it is rolled to the left to locate points A_4, A_3, A_2, and A_1 as the center moves

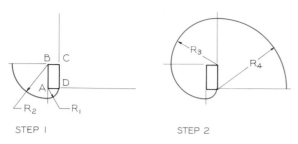

STEP 1   STEP 2

FIG. 4.61 Involute

Step 1 Side *AD* is used as a radius for drawing Arc *AD*. *AB* is added to *AD* to form Radius R_2. Draw a second arc using Center *B*.

Step 2 The two remaining sides are used to unwind the involute back to its point of origin.

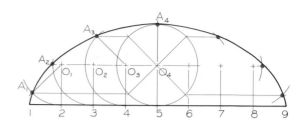

FIG. 4.62 Cycloid

Step 1 A circle is centered on a horizontal tangent line. It is divided into a number of equal divisions. The circumference of the circle is laid off along Line 1–9, which is divided into the same number of parts as the circle.

Step 2 The center of the arc is moved from O_4 to O_3 to locate point A_3, then to O_2. The points are connected to form the cycloidal curve.

from O_4 to O_1. These points are connected with a smooth curve to complete the left side of the cycloid.

The same construction is repeated as the circle moves to the right to complete the symmetrical curve.

4.33
Epicycloid and hypocycloid

The **epicycloid** is a curve formed by a point on a circle as the circle rolls along the convex side of a larger circle (Fig. 4.63). The circumference of the rolling cir-

cle is divided into equal divisions. These same units are measured along the circumference of the large arc. These points will be the contact points as the circle rolls along the arc. The plotted points are connected to form the epicycloid.

The **hypocycloid** is a curve formed by a point on a circle as the circle rolls along the concave side of a larger arc. As in the epicycloid, the circumference of the circle is divided into equal divisions that are laid off along the arc. The curve is plotted and drawn as a smooth curve as the circle is rolled left and right.

The elliptical-appearing curve in Step 3 of Fig. 4.63 is a curve formed by connecting the epicycloid and hypocycloid curves that are obtained by rolling a circle of the same size on each side of a given arc.

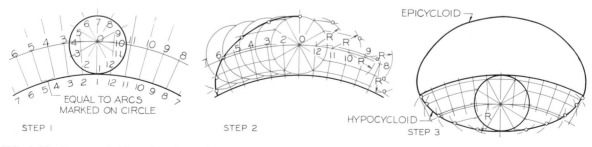

STEP 1   STEP 2   STEP 3

FIG. 4.63 Hypocycloid and epicycloid

Step 1 Divide the circle into a number of equal parts. Measure the same lengths along the large arc. Locate the positions of the Center *O* along an arc drawn through the Center *O*.

Step 2 Move the circle to Positions 1 through 6 to find the points along the epicycloid. Repeat this process at the right side and draw the curve.

Step 3 The hypocycloid is found in the same manner, but the circle rolls along the inside of the large arc. This figure shows both the epicycloid and the hypocycloid drawn together.

Problems

These problems are to be solved on Size AV paper similar to that shown in Fig. 4.64, where Problems 1–5 are laid out. Each inch on the grid is equal to 0.20 inches; therefore, use your engineers' 10 scale to lay out the problems. By equating each grid to 5 mm, you can use your full-size metric scale to lay out and solve the problems.

Show your construction and mark all points of tangency, as discussed in the chapter.

1. Draw triangle *ABC* using the given sides.

2–3. Inscribe an angle in the semicircles with the vertexes at Point *P*.

4. Inscribe a nine-sided regular polygon inside the circle.

5. Circumscribe a ten-sided regular polygon about the circle.

6. Circumscribe a hexagon about the circle.

7. Inscribe an octagon in the circle.

8. Circumscribe an octagon about the circle.

9. Construct a pentagon inside the circle using the compass method.

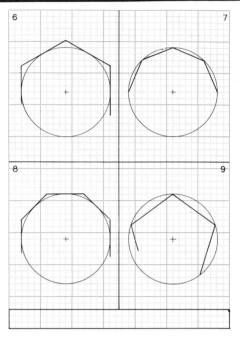

FIG. 4.65 Problems 6–9: Construction of regular polygons.

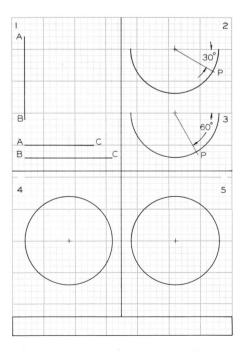

FIG. 4.64 Problems 1–5: Basic constructions.

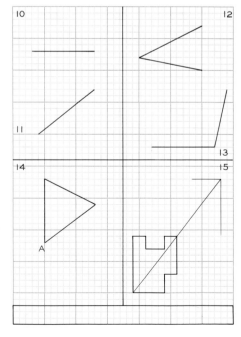

FIG. 4.66 Problems 10–15: Basic constructions.

10–11. Bisect the lines.

12–13. Bisect the angle.

14. Rotate the triangle 80° in a clockwise direction about Point *A*.

15. Enlarge the given shape to the size indicated by the diagonal.

16. Divide *AB* into seven equal parts. Draw the construction line through *B* for your construction.

17. Divide the two vertical lines into four equal divisions. Draw three equally spaced vertical lines at the divisions.

18. Construct an arc with a radius, *R*, that is tangent to the line at *J* and passes through Point *P*.

19. Construct an arc with a radius *R*, that is tangent to the line and passes through *P*.

20. Construct a line from *P* that is tangent to the semicircle. Locate the points of tangency. Use the compass method.

21–23. Construct arcs with the given radii tangent to the lines.

24–27. Construct arcs that are tangent to the arcs and/or lines. The radii are given for each problem.

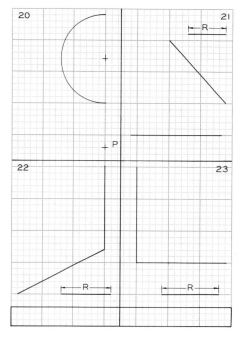

FIG. 4.68 Problems 20–23: Tangency construction.

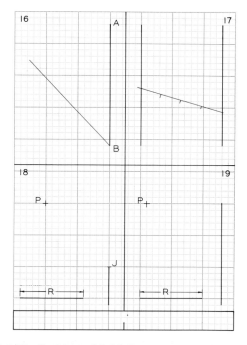

FIG 4.67 Problems 16–19: Tangency construction.

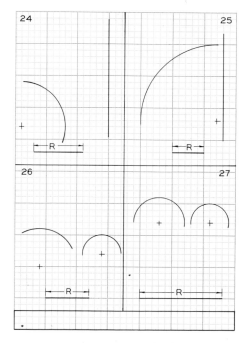

FIG. 4.69 Probems 24–27: Tangency construction.

28–31. Construct ogee curves that connect the ends of the given lines and pass through Points *P* where given. In Problem 31, the radii for the arcs are given.

32–33. Using the given radii, connect the given arcs with a tangent arc as indicated in the freehand sketches.

34. Rectify the arc along the given line by dividing the circumference into equal divisions and laying them off with your dividers.

35. Rectify the arc by using the compass method, as shown in Fig. 4.44.

36. Construct an ellipse inside the rectangular layout.

37. Construct an ellipse inside the large circle. The small circle represents the minor diameter.

38. Construct an ellipse inside the circle when the edge view has been rotated 45°, as shown.

39. Using the focal point, *F*, and the directrix, plot and draw the parabola formed by these elements.

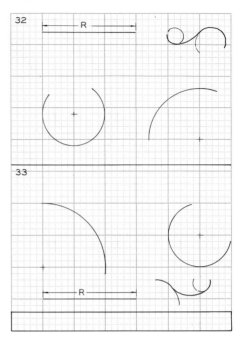

FIG. 4.71 Problems 32–33: Tangency construction.

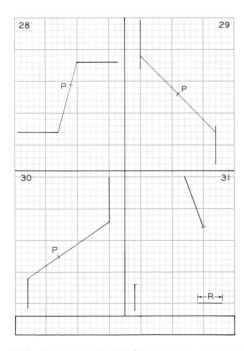

FIG. 4.70 Problems 28–31: Ogee curve construction.

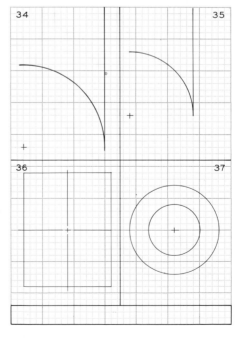

FIG. 4.72 Problems 34–37: Rectifying an arc, ellipse construction.

40. Construct a parabola using the perpendicular lines by either of the methods shown in Figs. 4.54 and 4.55.

41. Using the focal points, *F,* Points *A* and *B* on the curve, and the axis of symmetry, construct the hyperbolic curve.

42. Construct a hyperbola that passes through *O.* The perpendicular lines are asymptotes.

43. Construct a spiral of Archimedes by using the four divisions that are marked along the radius.

44–45. Construct a helix that has a rise equal to the heights of the cylinder and cone. Show construction and the curve in all views.

46–47. Construct involutes by unwinding the triangle and the square in a clockwise direction, beginning with Point *A* in each.

48. Construct a hypocycloid by rolling the circle along the curve whose center is at Point *C.*

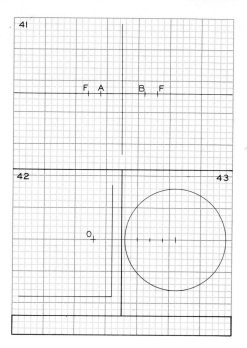

FIG. 4.74 Problems 41–43: Hyperbola and spiral construction.

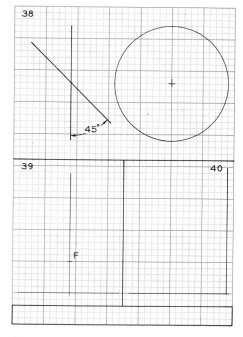

FIG. 4.73 Problems 38–40: Ellipse and parabola construction.

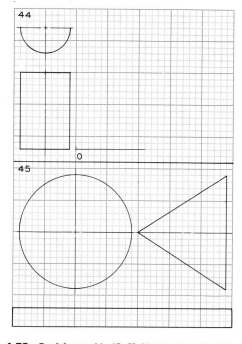

FIG. 4.75 Problems 44–45: Helix construction.

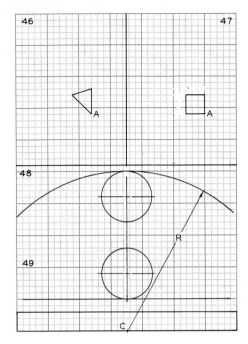

FIG. 4.76 Problems 46–49: Involute, hypocycloid, and cycloid construction.

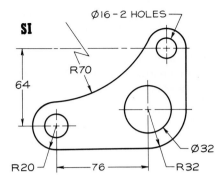

FIG. 4.78 Problem 51: Lever crank.

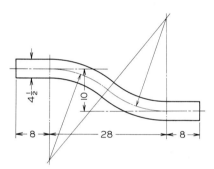

FIG. 4.79 Problem 52: Road tangency.

49. Construct a cycloid by rolling the circle about the horizontal line.

50–61. Construct these problems on Size A sheets, one problem per sheet. Select the scale that will best fit the problem to the sheet. Mark all points of tangency and strive for good line quality.

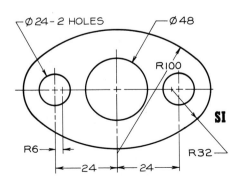

FIG. 4.77 Problem 50: Gasket.

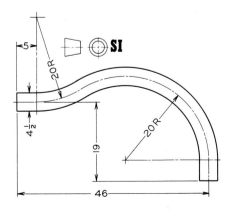

FIG. 4.80 Problem 53: Road tangency.

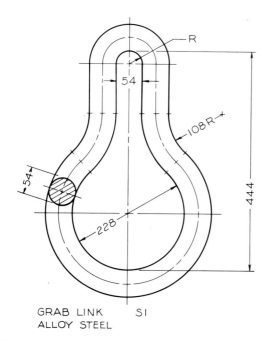

GRAB LINK   SI
ALLOY STEEL

FIG. 4.81   Problem 54: Grab link.

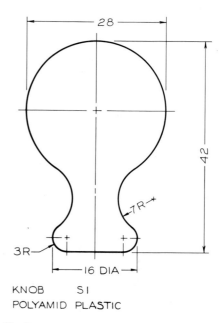

KNOB   SI
POLYAMID PLASTIC

FIG. 4.83   Problem 56: Knob.

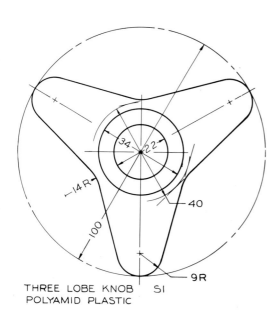

THREE LOBE KNOB   SI
POLYAMID PLASTIC

FIG. 4.82   Problem 55: Three-lobe knob.

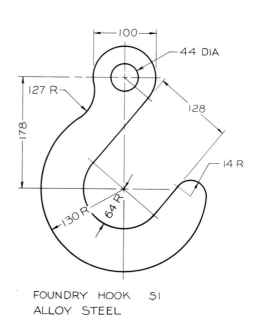

FOUNDRY HOOK   SI
ALLOY STEEL

FIG. 4.84   Problem 57: Foundry hook.

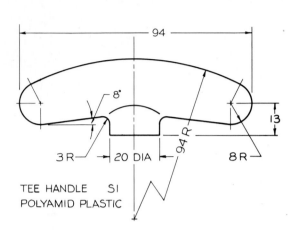

FIG. 4.85 Problem 58: Tee handle.

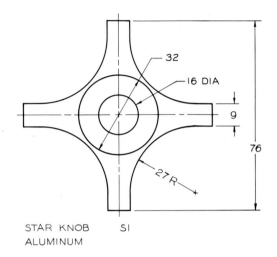

FIG. 4.87 Problem 60: Star knob.

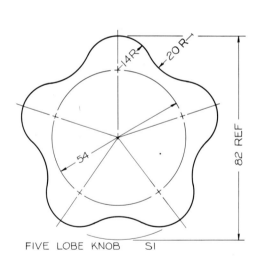

FIG. 4.86 Problem 59: Five-lobe knob.

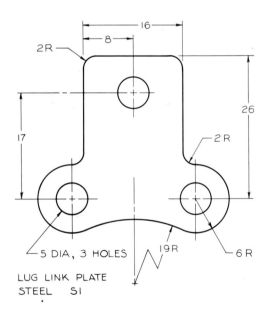

FIG. 4.88 Problem 61: Lug link plate.

CHAPTER **5**

Multiview Sketching

5.1

Purpose of sketching

Sketching is a thinking process as well as a technique of communication. Designers must develop their ideas by making many sketches, revising them, and finally arriving at the desired solution. Later, these sketches are converted into instrument drawings.

Sketching is not an effective technique unless it can be done rapidly. If you develop your sketching skills, you can assign drafting work to assistants who can then prepare the finished drawings by working from your sketches. If your sketches are not sufficiently clear to communicate your ideas to someone else, then it is likely that you have not thought out the solution well enough, even for your own understanding. The ability to communicate by any means is a great asset, and sketching is one of the more powerful techniques of communication.

5.2

Shape description

A pictorial of an object is shown in Fig. 5.1 with three arrows that indicate the directions of sight that will give top, front, and right-side views of it. Each view is

a two-dimensional view rather than a three-dimensional pictorial.

Presenting an object this way is called **orthographic projection** or **multiview projection.** In multiview projection, it is important that the views be located as shown in Fig. 5.1. The top view is placed over the front view, since both views share the dimension of width. The side view is placed to the right of the front view, since these views share the dimension of height. The distance between the views can vary, but they must be positioned so that the views project from each other as shown.

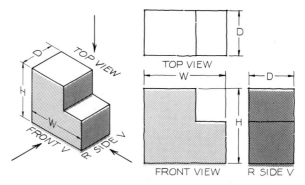

FIG. 5.1 Three views of an object can be found by looking at the object in this manner. The three views—top, front, and right side—describe the object.

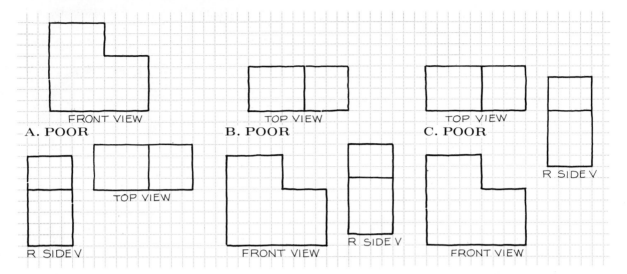

FIG. 5.2 Arrangement of views

A. These views are positioned incorrectly; they are scrambled.

B. These views are nearly correct, but they do not project from view to view.

C. The top and front views are correctly positioned, but the right-side view is incorrectly positioned.

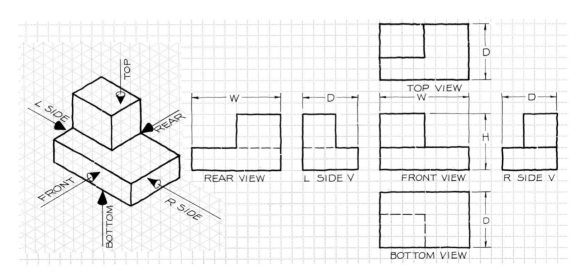

FIG. 5.3 Six principal views can be sketched by looking at the object in the directions indicated by the lines of sight. Note how the dimensions are placed on the views. Height (*H*) is shared by all four of the horizontally positioned views.

Several examples of poorly arranged views are shown in Fig. 5.2. You can see that although these views are correct, they are hard to interpret because the views are not placed in their standard positions.

5.3
Six-view drawings

Six principal views may be found for any object by using the rules of orthographic or multiview projection. Six is the maximum number of principal views.

The directions of sight for the six views are shown in Fig. 5.3, where the views are drawn in their standard positions. The width dimension is common to the

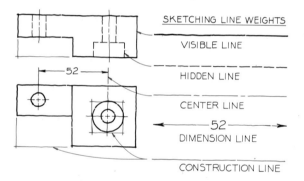

FIG. 5.4 These lines are examples of those that you should sketch with an F or an HB pencil when drawing views of an object. Note that some lines are thin and others are wider, but all are black except construction lines.

top, front, and bottom views. Height is common to the right side, front, left-side, and rear views.

Seldom will an object be so complex as to require six orthographic views, but if six views are needed, they should be arranged as shown in this example.

5.4
Sketching techniques

Sketching means freehand drawing without the use of instruments or straightedges. The best pencil grades for sketching are medium weight pencils such as H, F, or HB grades. The standard lines used in multiview drawing and their respective line weights are shown in Fig. 5.4. Using the correct line weight improves the readability of a drawing.

By sharpening the pencil point to match the desired line width, you will be able to use the same grade of pencil for all lines when you are sketching. The different point sizes are shown in Fig. 5.5. A line that is drawn freehand should have a freehand appearance; no attempt should be made to give the line the appearance of one drawn by instruments, since these two drawing techniques are completely different.

Sketching technique can be improved by using a printed grid on sketching paper, or by overlaying a printed grid with translucent tracing paper (Fig. 5.6) so the grid can be seen through the paper.

If you do not tape your drawing to the table top, you will be able to position the sheet for the most comfortable freehand strokes (left to right for the right-handed drafter) (Fig. 5.7).

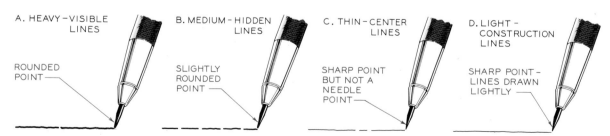

FIG. 5.5 The alphabet of lines that are sketched freehand and are made with the same pencil grade (F or HB). The variation in the lines is achieved by varying the sharpness of the pencil point.

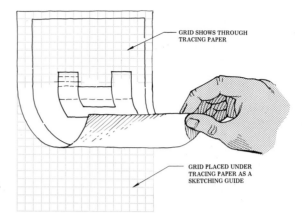

GRID SHOWS THROUGH
TRACING PAPER

GRID PLACED UNDER
TRACING PAPER AS A
SKETCHING GUIDE

FIG. 5.6 A grid can be placed under a sheet of tracing paper as an aid in freehand sketching. The grid can be used as guidelines for sketching.

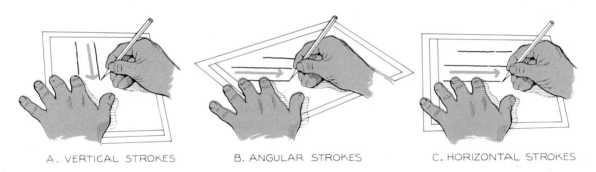

A. VERTICAL STROKES B. ANGULAR STROKES C. HORIZONTAL STROKES

FIG. 5.7 Freehand sketching techniques

A. Vertical lines should be sketched in a downward direction.

B. Angular strokes can be sketched left to right, if you rotate your sheet slightly.

C. Horizontal strokes are made in a left-to-right direction. Always sketch from a comfortable position, and turn your paper if necessary.

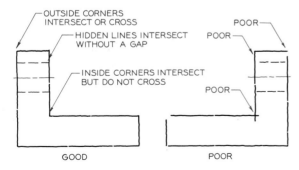

OUTSIDE CORNERS
INTERSECT OR CROSS

HIDDEN LINES INTERSECT
WITHOUT A GAP

INSIDE CORNERS INTERSECT
BUT DO NOT CROSS

POOR

POOR

POOR

GOOD   POOR

FIG. 5.8 For good sketches, follow these examples of good technique. Compare the good drawing with the drawing made when these rules were not followed.

The lines that are sketched to form the various views should intersect as indicated in Fig. 5.8 for the best effect.

5.5
Three-view sketch

The steps of drawing three orthographic views on a printed grid are shown in Fig. 5.9. The most commonly used combination of views consists of the front, top, and right-side views, as in this example. The overall dimensions of the object are sketched in Step 1.

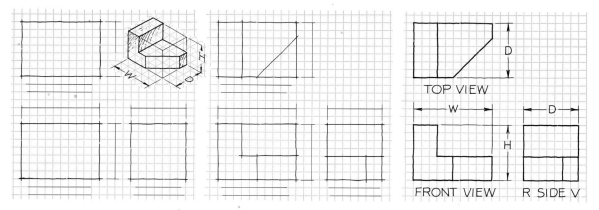

FIG. 5.9 Three-view sketching

Step 1 Block in the views by using the overall dimensions. Allow proper spacing for labeling and dimensioning the views.

Step 2 Remove the notches and project from view to view as shown.

Step 3 Check your layout for correctness; darken the lines and complete the labels and dimensions.

The slanted surface is drawn in the top view and projected to the other views. The final lines are darkened, the views labeled, and the overall dimensions of height, width, and depth are applied to the views.

Slanted surfaces will not appear true shape in the principal views of orthographic projection (Fig. 5.10). Surfaces that do not appear true size are either **foreshortened,** or they appear as **edges.** In Fig. 5.10C, two planes of the object are slanted, thus both of them appear foreshortened in the right-side view.

A good exercise for analyzing the given views is to find the missing view when two views are given (Fig. 5.11).

The right-side view is found in Fig. 5.12, where the top and front views are given. The right-side view has the depth dimension in common with the top

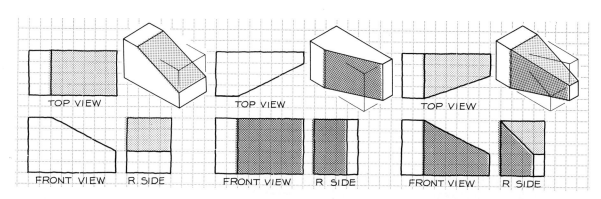

FIG. 5.10 Views of planes

A. The plane appears as an edge in the front view and as foreshortened in the top and side views.

B. The plane is an edge in the top view and is foreshortened in the front and side views.

C. These two planes appear foreshortened in the right-side view. Each appears as an edge in either the top or front views.

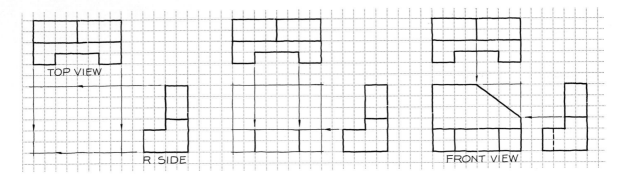

FIG. 5.11 Missing views

Step 1 When sketches of two views are given and the third is required, begin by projecting the overall dimensions from the top and right-side views.

Step 2 The various features of the object are sketched using construction lines.

Step 3 The features are completed, the view is checked for correctness, and the lines are darkened to the proper line quality.

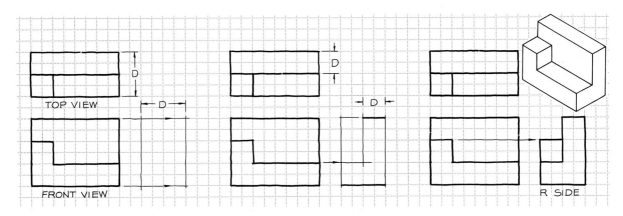

FIG. 5.12 Missing views

Step 1 To find the right-side view when the top and front views are given, block in the view with the overall dimensions.

Step 2 Develop the features of the part by analyzing the views together. Use light construction lines.

Step 3 Check the view for correctness and darken the lines to their proper line weight.

view, and height in common with the front view. Knowing this enables us to block in the side view in Step 1. The side view is developed in Step 2, and is completed in Step 3.

Another exercise is that of completing the views when some or all of them have missing lines (Fig. 5.13). Remember that depth is common to the top and side views, as shown in Step 2.

5.6
Circular features

The lines that are used with circles and cylinders to indicate that the features are true circles or cylinders are called **centerlines.** Examples of these are shown in Fig. 5.14.

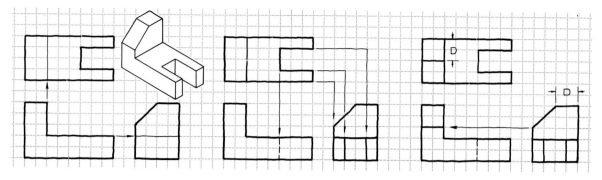

FIG. 5.13 Missing lines

Step 1 Lines may be missing in all views in this type of problem. The first missing line is found by projecting the edges of the planes from the front to the top and side views.

Step 2 The notch in the top view is projected to the front and side views. The line in the front view is a hidden line.

Step 3 The line formed by the beveled surface is found in the front view by projecting from the side view.

Centerlines cross in the circular views to indicate the center of the circle. Centerlines are thin lines with short dashes spaced at intervals of about one inch along the line. Refer to Fig. 5.4 for several examples of centerlines applied to a drawing.

If a centerline coincides with an object line—visible or hidden—the centerline should be omitted since the object lines are more important (Fig. 5.14C).

The application of centerlines is shown in Fig. 5.15, where they indicate whether or not circles and arcs are concentric (share the same centers). The cen-

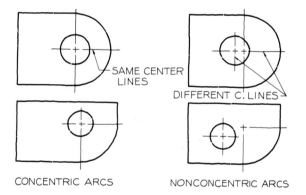

FIG. 5.15 The centerline should extend beyond the last arc that has the same center. When the arcs are not concentric, separate centerlines should be drawn.

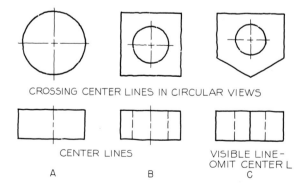

FIG. 5.14 Centerlines are used to indicate the centers of circles and the axes of cylinders. These are drawn as very thin lines. When they coincide with visible or hidden lines, the centerlines are omitted.

terline should extend beyond the arc by about one-eighth of an inch. The correct manner of applying centerlines is shown in Fig. 5.16.

Sketching circles

Circles can be sketched by using light guidelines in conjunction with centerlines (Fig. 5.17). The arcs are drawn using the guidelines and a series of short arcs. Arcs of less than a full circle are drawn by using light guidelines and centerlines, as shown in Fig. 5.18.

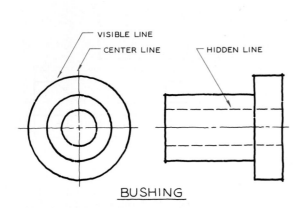

VISIBLE LINE

CENTER LINE

HIDDEN LINE

BUSHING

FIG. 5.16 Here you can see the application of centerlines of concentric cylinders, and the relative weight of hidden, visible, and centerlines.

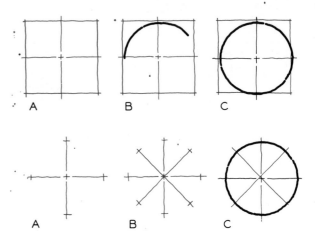

A   B   C

A   B   C

FIG. 5.17 Circles can be sketched using either of the construction methods shown here. The use of guidelines is essential to freehand sketching circles and arcs.

FIG. 5.18 Partial circles (arcs) should be drawn using guidelines.

RADIUS   RADIUS

STEP 1   STEP 2   STEP 3

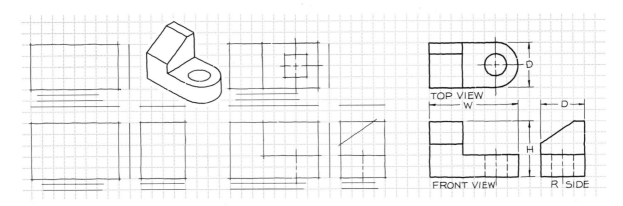

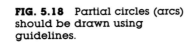

TOP VIEW
W

FRONT VIEW   R SIDE

D

H

D

FIG. 5.19 Circular features in orthographic views

Step 1 To draw orthographic views of the object shown in this pictorial; begin by blocking in the overall dimensions. Leave room for labels and dimensions.

Step 2 Construct the centerlines and the squares about the centerlines in which the circles will be drawn. Show the slanted surface in the side view.

Step 3 Sketch the arcs, and darken the final lines of the views. Label the views and show dimensions of *W*, *D*, and *H*.

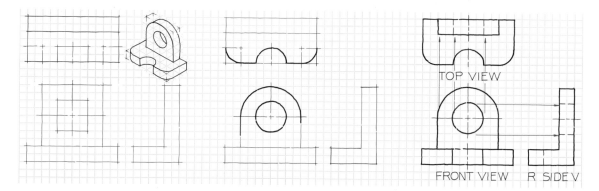

FIG. 5.20 Circular features in orthographic views

Step 1 When sketching orthographic views with circular features, begin by sketching the centerlines and guidelines first.

Step 2 Using the guidelines, sketch the circular features. Lines can be darkened as they are drawn if they will be final lines.

Step 3 The corresponding outlines of the circular features are found by projecting from the views found in Step 2. Darken all lines.

The construction of three orthographic views with circular features are located with centerlines and light guidelines are sketched in Fig. 5.19.

A similar example is given in Fig. 5.20, where the part is composed of circular features and arcs. The circles should be drawn first so that their corresponding rectangular views (such as the hidden hole in the top view) can be found by projecting from the circular view.

5.7
Isometric sketching

Another type of pictorial is the **isometric drawing,** which may be drawn on a specially printed grid. You will notice that the grid in Fig. 5.21 is composed of a series of lines making 120° angles with each other to form the axes for drawing isometrics.

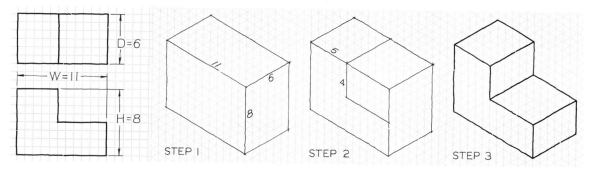

FIG. 5.21 Isometric pictorial sketching

Step 1 When orthographic views are given, an isometric pictorial can be sketched by using a printed isometric grid. Begin by constructing a box using the overall dimensions from the given views.

Step 2 The notch can be located by measuring over 5 squares as shown in the orthographic views. The notch is measured 4 squares downward by counting the units.

Step 3 The pictorial is completed and the lines are darkened. This is a three-dimensional pictorial, whereas the orthographic views are each two-dimensional views.

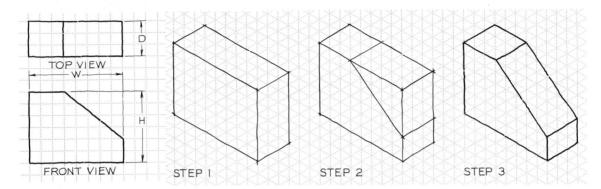

FIG. 5.22 Angles in isometric pictorials

Step 1 Begin by drawing a box using the overall dimensions given in the orthographic views. Count the squares and transfer them to the isometric grid.

Step 2 Angles cannot be measured with a protractor. Find each end of the angle by measuring along the axes.

Step 3 The ends of the angles are connected to give the pictorial view of the slanting surface. Dimensions can be measured only in directions parallel to the three axes.

The squares in the orthographic views can be laid off along the isometric grids as shown in Step 1. The notch is located in the same manner in Steps 2 and 3 to complete the isometric pictorial.

Angles cannot be measured with a protractor in isometric pictorials; they must be drawn by measuring coordinates along the three axes of the printed grid. In Fig. 5.22, the sloping surface is located by measuring in the direction of width and height to locate the ends of the angular slope.

When an object has two sloping planes that intersect (Fig. 5.23), it is necessary to draw the sloping planes one at a time to find Point B. The line from A to B is the line of intersection between the two planes.

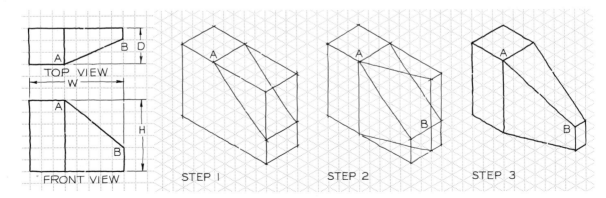

FIG. 5.23 Double angles in isometric pictorials

Step 1 When part of an object has a double angle, begin by constructing the overall box and finding one of the angles.

Step 2 Find the second angle that locates point B. Point A will connect to point B to give the intersection line.

Step 3 The final lines are darkened. Line AB is the line of intersection between the two sloping planes.

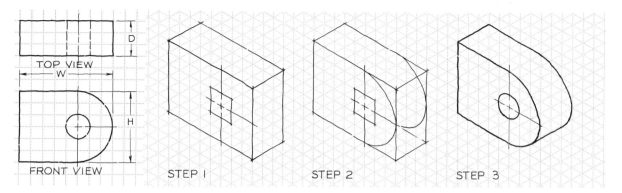

FIG. 5.24 Circles in isometric pictorials

Step 1 Begin by constructing a box using the overall dimensions and omit the arcs and circles. Draw the centerlines and a square of guidelines around the circular hole.

Step 2 Draw the pictorial views of the arcs tangent to the boxes formed by the guidelines. These arcs will appear as ellipses instead of circular arcs.

Step 3 Construct the small hole and darken the lines. Hidden lines are normally omitted in pictorial sketches.

Circles in isometric pictorials

Circles will appear as ellipses in isometric pictorials. These can be sketched by locating their centerlines, as shown in Step 1 of Fig. 5.24. The center must be located equidistant from the top, bottom, and end of the front view (Step 1).

The circles are blocked in with guidelines, which will appear as rhombuses. The approximate elliptical views of the circular features will pass through the points where the centerlines intersect the guidelines (Step 2). The rear of the object is found in the same manner to complete the pictorial in Step 3.

The steps of constructing elliptical views of circles by two methods are shown in Fig. 5.25. The first method is without the use of guidelines and the second method utilizes guidelines.

Techniques of sketching similar circular and cylindrical shapes are shown in Fig. 5.26. When drawing cylinders, the outside elements are drawn parallel to the axis of the cylinder and tangent to the elliptical ends of the cylinder.

A part that is composed of a number of cylindrical forms is shown in Fig. 5.27a. These three views are used as the basis for an isometric pictorial shown in the steps of construction. When an isometric grid is not used, the axes of the isometric sketch are positioned 120° apart. In other words, the height dimension is vertical, and the width and depth dimensions make 30° angles with the horizontal.

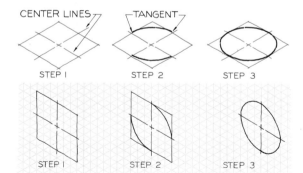

FIG. 5.25 Sketching ellipses

Two methods of sketching ellipses are shown: one without a grid, and one with a grid. When a grid is not used, the centerlines are drawn at a horizontal angle of 30° (for a horizontal circle). A rhombus is drawn about them and the ellipse is sketched inside the guidelines. When there is a grid, the same technique is used except that the lines of the grid become the guidelines.

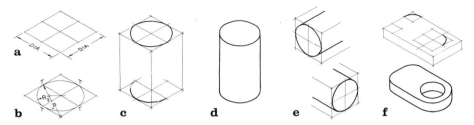

FIG. 5.26 Sketching circular features

A variety of examples of sketching circles and cylinders are shown here. Note that in all cases, guidelines are used to aid in sketching.

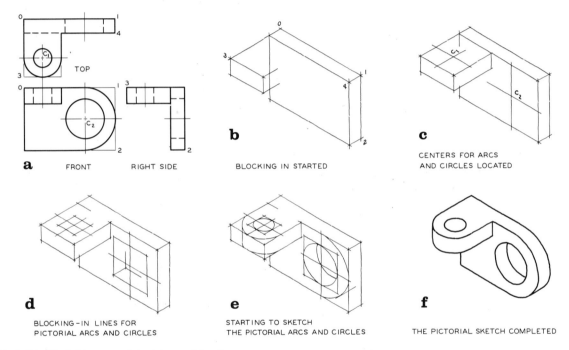

FIG. 5.27 Sketching an object isometrically

The five steps of constructing an isometric pictorial of a complex part are shown here. Since a grid is not given, the guidelines are drawn vertically and at 30° to the horizontal. Centerlines are used for locating the circular features at c. The object is developed at d and e and darkened at f.

Problems

These sketching problems should be drawn on Size AV (8½″ × 11″) paper with or without a printed grid. A typical format for this size sheet is shown in Fig. 5.28 where a one-quarter inch grid is given. (This grid can be converted to an approximate metric grid by equating each square to five millimeters.) All sketches and lettering should be neatly executed by applying the principles covered in this chapter. Figures 5.29–5.31 contain the problems and instructions.

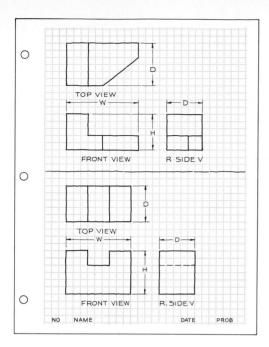

FIG. 5.28 The layout of a Size A sheet for sketching problems

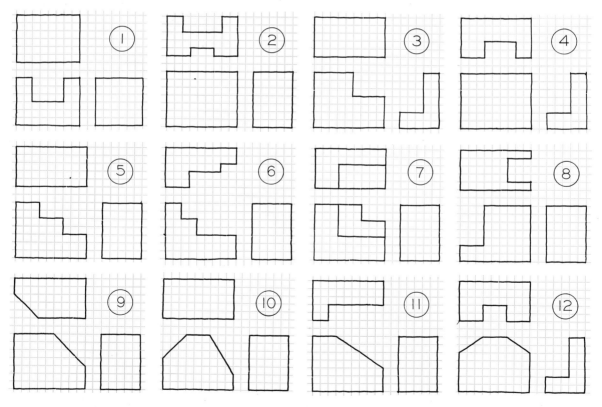

(Continued)

FIG. 5.29 On Size AV paper sketch top, front, and right-side views of the problems assigned. Two problems can be drawn on each sheet. Give the overall dimensions of *W*, *D*, and *H* and label each view.

Figure 5.29 (continued)

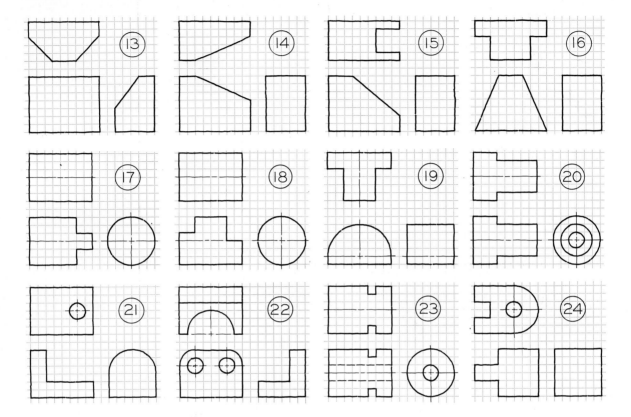

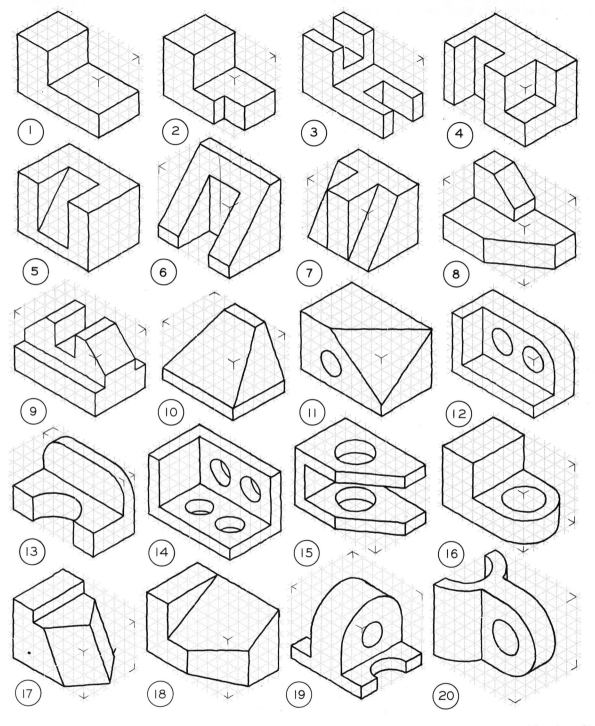

FIG. 5.30 Multiview problems and isometric sketching

On Size AV paper, sketch the top, front, and right-side views of the problems assigned. Supply the lines that may be missing from all views. Then sketch isometric pictorials of the object assigned, two per sheet.

(Continued)

Figure 5.30 (continued)

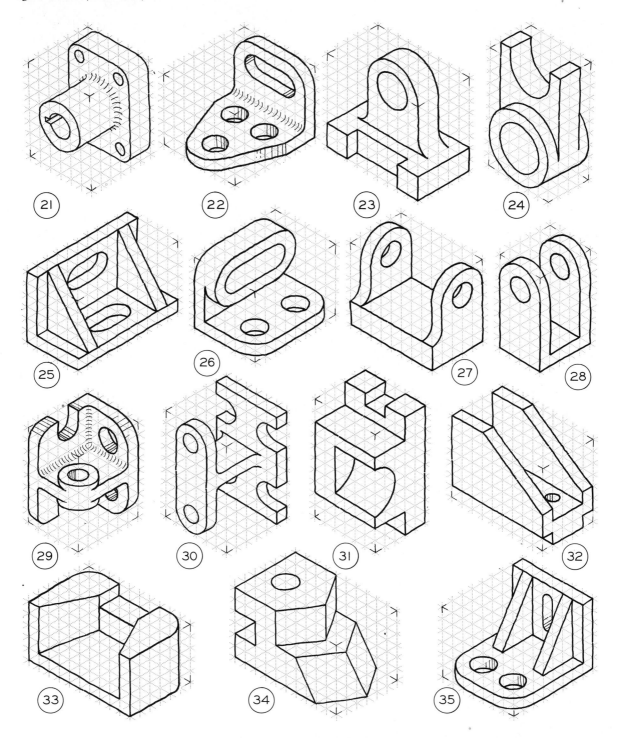

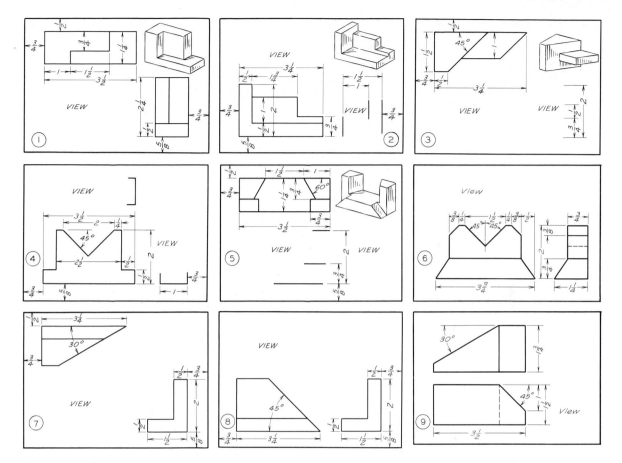

FIG. 5.31 Multiview problems and isometric pictorials

Sketch the problems assigned by your instructor, two per sheet on Size AV paper. Use the given dimensions to locate and draw the views. Draw the missing views. Then sketch isometric pictorials of the objects assigned, two per sheet on Size A paper.

6 CHAPTER

Multiview Drawing with Instruments

6.1
Introduction

Multiview drawing is the system of representing three-dimensional objects on a sheet of paper using separate views arranged in a manner standard to the engineering and technological fields. These drawings are usually executed with the instruments and drafting aids that were discussed in Chapter 2; consequently, these drawings are often referred to as **mechanical drawings.** When dimensions and notes have been added to complete the specifications of the parts that have been drawn, they are called **working drawings** or **detailed drawings.**

This system of constructing multiview drawings is called **orthographic projection.** It has evolved over the years into a system that, when the rules of the system are followed, is readily understood by the technical community. Orthographic drawing truly is the language of the engineer.

6.2
Orthographic projection

In the *orthographic projection system,* the views of an object are projected perpendicularly onto projection planes with parallel projectors. The process of finding one orthographic view, the front view, is illustrated in Fig. 6.1. The resulting front view is two dimensional; it has no depth and lies in a single plane described by two dimensions, width and height.

The top view of the same object is projected onto a horizontal projection plane that is perpendicular to the frontal projection plane in Fig. 6.2A. The right-side view is projected onto a vertical profile plane that is perpendicular to both the horizontal and frontal planes (Fig. 6.2B).

Imagine that the same object has been enclosed in a glass box composed of the frontal, horizontal, and profile projection planes. While in the glass box, the views are projected onto the projection planes (Fig.

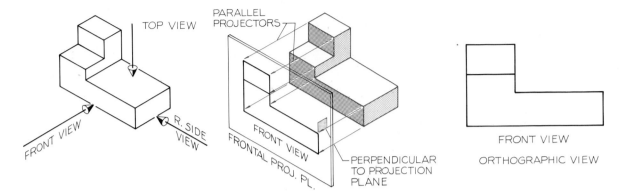

FIG. 6.1 Orthographic projection

Step 1 Three mutually perpendicular lines of sight are drawn to obtain three views of the object.

Step 2 The frontal plane is a vertical plane on which the front view is projected with parallel projectors that are perpendicular to the frontal plane.

Step 3 The resulting view is the front view of the object. This is a two-dimensional orthographic view.

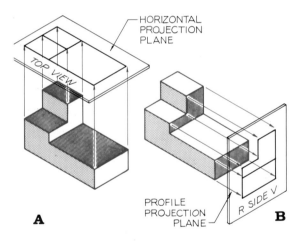

FIG. 6.2 The top view is projected onto a horizontal projection plane. The right-side view is projected onto a vertical profile plane that is perpendicular to the horizontal and frontal planes.

6.3), and then the box is opened into the plane of the drawing surface. This gives the standard positions for the three orthographic views.

A similar example of the same principle is the object in the projection box in Fig. 6.4. Again, the three views are positioned in the same manner: the top view over the front, and the right-side view to the right of the front view.

The three principal projection planes of orthographic projection are referred to as the **horizontal (H), frontal (F),** and **profile (P)** planes.

Any view that is projected onto one of these principal planes is called a **principal view.** The three dimensions of an object that are used to show its three-dimensional form are **height, width,** and **depth.**

6.3
Alphabet of lines

The use of proper line weights in a drawing will greatly improve its readability and appearance. All lines should be drawn dark and dense, as if they were drawn with ink, varying the lines only by their width. The only lines that are exceptions are guidelines and construction lines that are drawn very lightly for lettering and laying out a drawing.

The lines of an orthographic view are labeled in Fig. 6.5, along with the suggested pencil grades for drawing them. The lengths of dashes in hidden lines and centerlines are drawn longer as the size of a drawing increases. Additional specifications for these lines are given in Fig. 6.6.

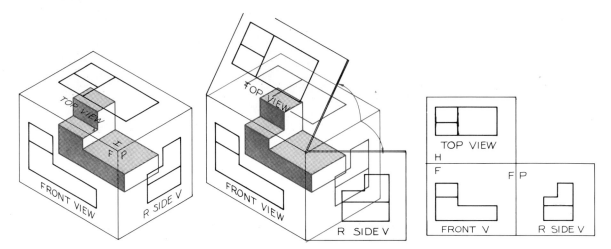

FIG. 6.3 Glass box theory

Step 1   Imagine that the object has been placed inside a box formed by the horizontal, frontal, and profile planes on which the top, front, and right-side views are projected.

Step 2   The three projection planes are then opened into the plane of the drawing surface.

Step 3   The three views are positioned with the top view over the front view and the right-side view to the right. The planes are labeled *H, F,* and *P* at the fold lines.

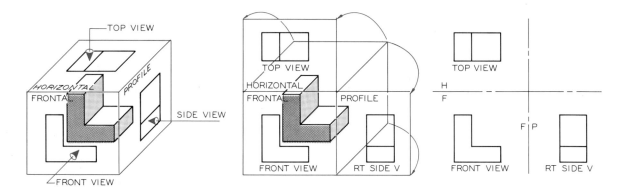

FIG. 6.4 The principal projection planes

A. The three principal projection planes of orthographic projection can be thought of as planes of a glass box.

B. The views of an object are projected onto the projection planes that are opened into the plane of the drawing surface.

C. The outlines of the planes are omitted. The fold lines are drawn and labeled.

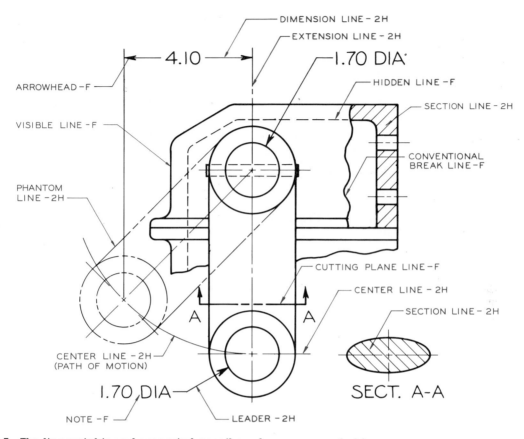

DIMENSION LINE – 2H
EXTENSION LINE – 2H
4.10
1.70 DIA
ARROWHEAD –F
HIDDEN LINE –F
SECTION LINE – 2H
VISIBLE LINE –F
CONVENTIONAL BREAK LINE –F
PHANTOM LINE – 2H
CUTTING PLANE LINE –F
CENTER LINE – 2H
SECTION LINE – 2H
A   A
CENTER LINE – 2H (PATH OF MOTION)
1.70 DIA
SECT. A-A
NOTE –F
LEADER – 2H

FIG. 6.5 The line weights and suggested pencil grades recommended for orthographic views.

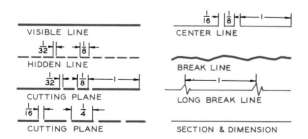

VISIBLE LINE
HIDDEN LINE
$\frac{1}{32}$ $\frac{1}{8}$
CUTTING PLANE
$\frac{1}{16}$ $\frac{1}{4}$
CUTTING PLANE
CENTER LINE
$\frac{1}{16}$ $\frac{1}{8}$
BREAK LINE
LONG BREAK LINE
SECTION & DIMENSION

FIG. 6.6 A comparison of the line weights for orthographic views. These dimensions will vary for different sizes of drawings and should be approximated by eye.

6.4
Six-view drawing

A view that is projected onto a principal plane—the horizontal, frontal, or profile plane—is called a **principal view.** Again, if you visualize an object placed inside a glass box, you will see that there are two horizontal planes, two frontal planes, and two profile planes (Fig. 6.7). Consequently, the maximum number of principal views that can be used to represent an object is six.

The top and bottom views are projected onto horizontal planes, the front and rear views are projected onto frontal planes, and the right- and left-side views are projected onto profile planes.

To draw these views on a sheet of paper, the glass box is imagined to be opened up into the plane of the

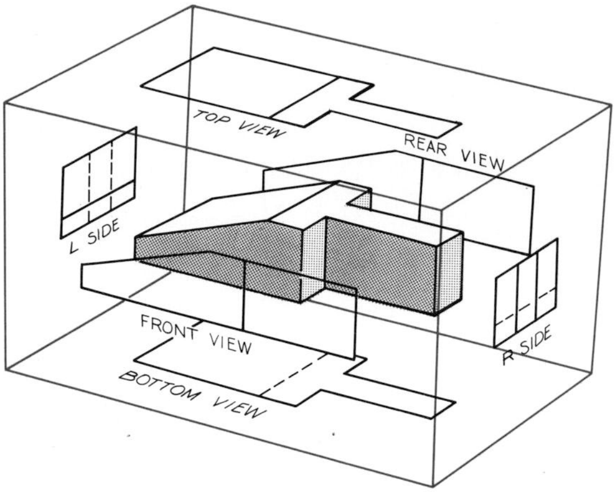

FIG. 6.7 Six principal views of an object can be drawn in orthographic projection. You can imagine that the object is in a glass box with the views projected onto the six planes of projection.

drawing paper (Fig. 6.8). When fully opened, the views will appear as shown in Fig. 6.9.

The top view is placed over the front view, the bottom view under the front view, the right-side view to the right, the left-side view to the left, and the rear view is placed to the left of the left-side view.

The three dimensions of an object that are necessary to give its size are height, width, and depth. The standard arrangement of the six views allows some of the views to share dimensions by projection. For example, the height dimension applies to the four views that are arranged horizontally, and this dimension is

shown only once between the front and right-side views. The width dimension is placed between the top and front views, but it also applies to the bottom view.

Note that in Fig. 6.9 the projectors align the views both horizontally and vertically about the front view. This is one of the reasons why this system of drawing is referred to as a system of projection.

Each side of the fold lines of the glass box is labeled with the letters *H*, *F*, and *P*. These letters identify the projection planes on a given side of the fold lines.

FIG. 6.8 The glass box is opened onto the plane of the drawing surface, which locates the views in their standard positions.

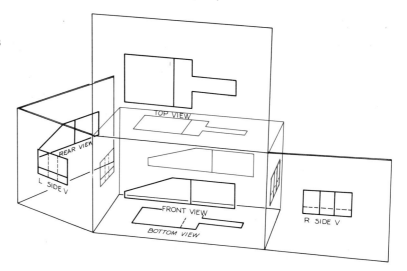

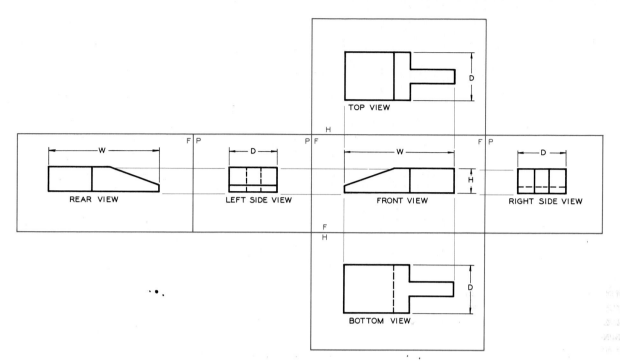

FIG. 6.9 Once the box is completely opened into a single plane, the six views are arranged to describe the object. The outlines of the planes are usually omitted. They are shown here to assist you in relating this figure to the previous one.

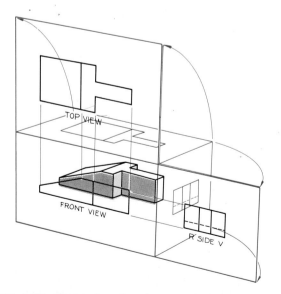

FIG. 6.10 Three-view drawings are commonly used for describing machine parts and small parts. The glass box is used to illustrate how the views are projected onto their projection planes.

6.5
Three-view drawing

The most commonly used orthographic arrangement is the three-view drawing composed of the front, top, and right-side views. This is true because three views are usually adequate to describe an object.

The same object used in the previous example is shown placed in a glass box in Fig. 6.10, which is opened onto the plane of the drawing surface. The resulting three-view arrangement is shown in Fig. 6.11, where the views are labeled and dimensioned.

Since the views are understood to be projections from one view to the next, the dimensions are placed between the adjacent views and are not repeated. A single dimension will apply to all views that project along a single direction.

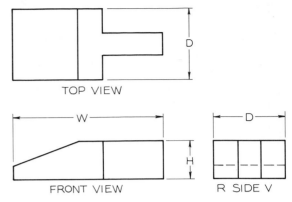

FIG. 6.11 The resulting three-view drawing of the object from the previous figure.

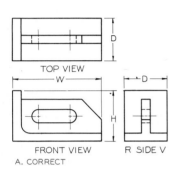

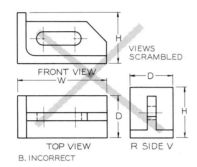

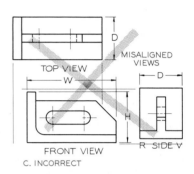

FIG. 6.12 Positioning orthographic views

A. This is a correct arrangement of views, labels, and dimensions. The views project from each other in proper alignment.

B. These views are scrambled into unconventional positions, making it hard to interpret them. The dimensions are unnecessarily repeated.

C. These views have been misaligned where they do not project from one to the other to give an incorrect arrangement.

FIG. 6.13 A. These views are properly aligned and the dimensions are correctly located. B. A number of errors are indicated in this incorrect arrangement.

6.6

Arrangement of views

The proper arrangement of orthographic views is essential to their understanding and interpretation. The standard positions for a three-view drawing consisting of the top, front, and right-side views are shown in Fig. 6.12A. The top and side views are projected directly from the front view. The views are properly labeled and dimensioned.

You can see the problems that you would cause by rearranging the views in a nonstandard sequence as shown in Fig. 6.12B. Similarly, views that do not project from view to view are improperly drawn, as in the example in Fig. 6.12C. These rules of arrangement are emphasized in Fig. 6.13.

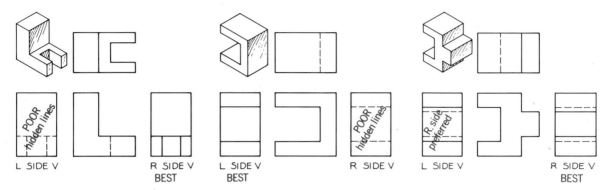

FIG. 6.14 Selection of views

A. In orthographic projection, you should select the sequence of views with the fewest hidden lines.

B. The left-side view has fewer hidden lines; therefore this view is selected over the right-side view.

C. When both views have an equal number of hidden lines, the right-side view is traditionally selected.

6.7
Selection of views

When drawing an object by orthographic projection, you should select the views with the **fewest hidden lines.** That is why in Fig. 6.14A the right-side view is preferred over the left-side view.

Although the three-view arrangment of top, front, and right-side views is the most commonly used, the arrangement of the top, front, and left-side views (Fig. 6.14B) is equally acceptable if this view has fewer hidden lines than the right-side view.

Some objects have standard views that are considered to be the front view, top view, and so forth. For example, a chair has front and top views that are recognized as such by everyone; therefore the accepted front view should be used as the orthographic front view.

6.8
Line techniques

As drawings become more complex, you will encounter more instances where lines will overlap and inter-

sect in a variety of ways similar to those shown in Fig. 6.15. This illustration shows the standard techniques of handling intersecting lines of most types.

The methods of constructing hidden lines composed of straight lines and curved segments is shown in Fig. 6.16.

You should become familiar with the order of precedence (priority) of lines (Fig. 6.17). The most important line that dominates all others is the visible object line. It will be shown regardless of any other type of line that lies behind it. The hidden object line is of next importance, and it takes precedence over the centerline.

6.9
Point numbering

It will be helpful to you in constructing orthographic views if you become familiar with the method of numbering points and lines of an object. The rules of point numbering are introduced in Fig. 6.18.

An object is shown in Fig. 6.19 that has been numbered to aid in the construction of the missing front view when the top and side views are given. By projecting a selected point from the top and side views, the front view of the object can be found.

This method is recommended when you are having difficulty in interpreting given views.

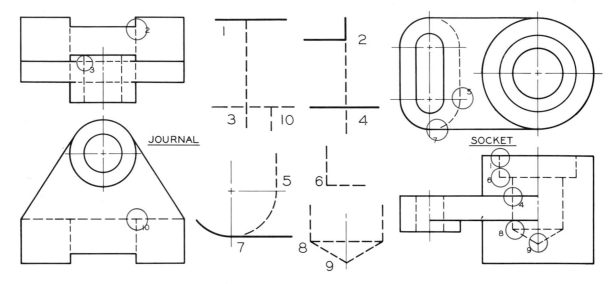

FIG. 6.15 When lines intersect in orthographic projection, they should intersect as shown here.

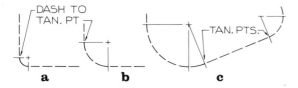

FIG. 6.16 Hidden lines in orthographic projection that are composed of curves should be drawn in this manner.

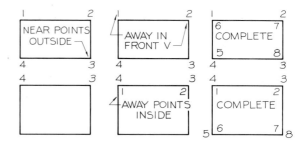

FIG. 6.18 When numbering points, the near points are labeled on the outside of the view, and away points are labeled inside the view.

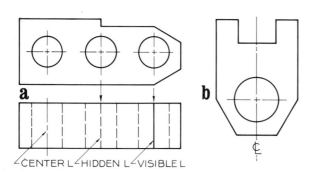

FIG. 6.17 The order of importance (priority) of lines is: visible, hidden, and centerlines, Part A. The symbol made of the letters *C* and *L* in Part B is used to label a centerline when this is needed on symmetrical parts.

6.10
Lines and planes

An orthographic view of a line can appear true length, foreshortened, or as a point (Fig. 6.20). When a line appears true length, it must be parallel to the reference line in the previous view.

A plane in orthographic projection can appear true size, foreshortened, or as an edge (Fig. 6.20).

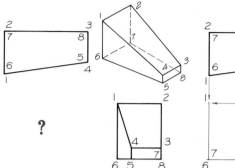
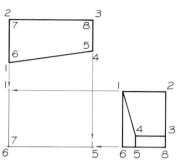
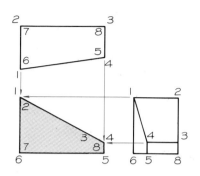

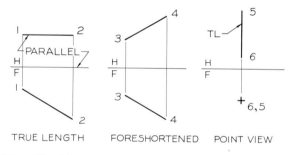

FIG. 6.19 Point numbering

Step 1 When a missing orthographic view is to be drawn, it is helpful to number the points in the given views.

Step 2 Points 1, 5, 6, and 7 are found by projecting from the given views of these points.

Step 3 The plotted points are connected to form the missing front view.

TRUE LENGTH   FORESHORTENED   POINT VIEW   TRUE SIZE   FORESHORTENED   EDGE VIEW

FIG. 6.20 Lines and planes

A line will project in orthographic projection as true length, foreshortened, or a point.

A plane in orthographic projection will appear as true size, foreshortened, or an edge.

6.11
Alternate arrangement of views

Although the right-side view is usually placed to the right of the front view (Fig. 6.21A), the side view can be projected from the top view as in Part B. This is advisable when the object has a large depth dimension as compared to its height. Figure 6.22 shows different methods of positioning and arranging views. All of these arrangements are acceptable.

6.12
Laying out the three-view drawing

The depth dimension applies to both the top and side views, but these views are usually positioned where this dimension does not project between them (Fig. 6.23). The depth dimension can be graphically transferred to the two views by using a 45° line, an arc, or a pair of dividers. It is preferable that you learn to use your dividers for this purpose.

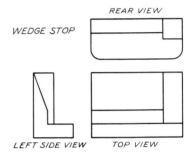

FIG. 6.21 When an object has a much larger depth than height, the right-side view can be projected from the top view and positioned as shown in Part B to save space.

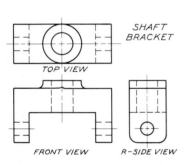

FIG. 6.22 Arrangement of views

A. This is the standard arrangement of views for a three-view drawing.

B. The right-side view is projected off the top view in this arrangement.

C. This is an unconventional but acceptable arrangement of three views.

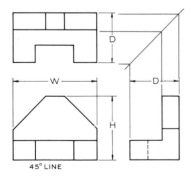

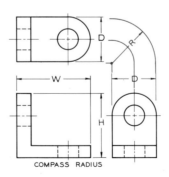

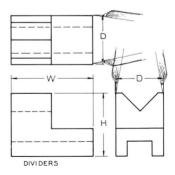

FIG. 6.23 Transferring depth dimensions

A. The depth dimension can be projected from the top view to the right-side view by constructing a 45° line positioned as shown.

B. The depth dimension can be projected from the top view to the side view using a compass and a center point.

C. The depth dimension can be transferred from the top view to the side view by using dividers.

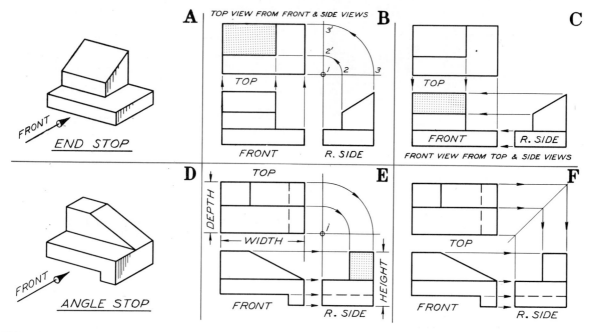

FIG. 6.24 A number of examples of methods of transferring dimensions from the views are shown here.

These and similar methods of transferring dimensions from view to view are illustrated in Fig. 6.24.

The method of laying out a three-view drawing in a 10″ × 15″ space is shown in Fig. 6.25. The views are blocked in with light construction lines using the overall dimensions of the views. Centerlines and tan-

gent points are drawn next, and then the straight lines. Once the layout has been checked, the lines are darkened to their proper weight.

When overall dimensions and labels are required, the three-view sketch should be positioned as shown in Figs. 6.26–6.30.

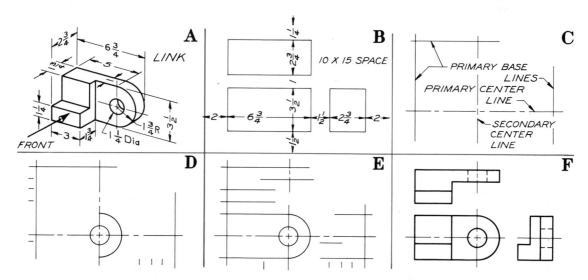

FIG. 6.25 Lay out a three-view drawing in this order: Center the views, draw the centerlines, draw the arcs, draw the straight lines, and darken your layout.

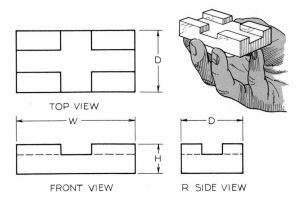

FIG. 6.26 A three-view drawing of an object

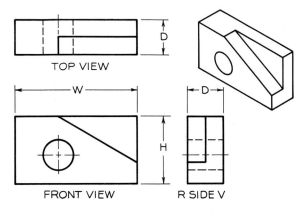

FIG. 6.29 A three-view drawing of an object

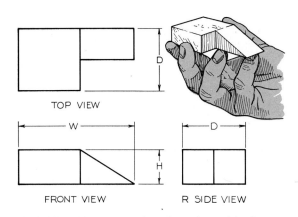

FIG. 6.27 A three-view drawing of an object

6.13
Two-view drawing

Some objects can be adequately explained in two views, and it is good economy of time and space to use only the views that are necessary to depict an object. Two parts that require only two views are shown in Fig. 6.31.

Cylindrical parts need only two views, as shown in Fig. 6.32. It is preferable to select the views with the fewest hidden lines; consequently the right-side view is the best view in this case.

The layout of a two-view drawing in a 10″ × 15″ space is shown in Fig. 6.33. The overall dimensions

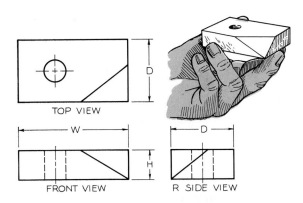

FIG. 6.28 A three-view drawing of an object

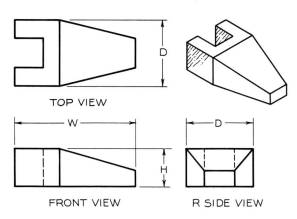

FIG. 6.30 A three-view drawing of an object

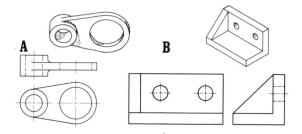

FIG. 6.31 These objects can be adequately described with two orthographic views.

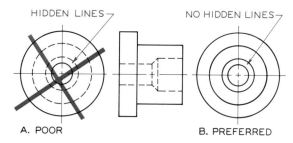

FIG. 6.32 Cylindrical objects can be depicted with two views. Always select views with the fewest hidden lines.

are used to block in and center the views. Second, the centerlines are located and drawn using light construction lines with a 2H–4H pencil. When the views have been completed, the lines are darkened to complete the two-view drawing.

Views that involve arcs and tangent lines should be laid out by locating the centers and centerlines of the arcs as the first step (Fig. 6.34). The tangent points are found using light construction lines. Begin by drawing the arcs and the lines are darkened to their proper weights.

6.14

One-view drawings

Cylindrical parts and those with a uniform thickness can be described in one view (Fig. 6.35).

In both cases, notes are used to explain the missing feature or dimension. The note DIA is placed after

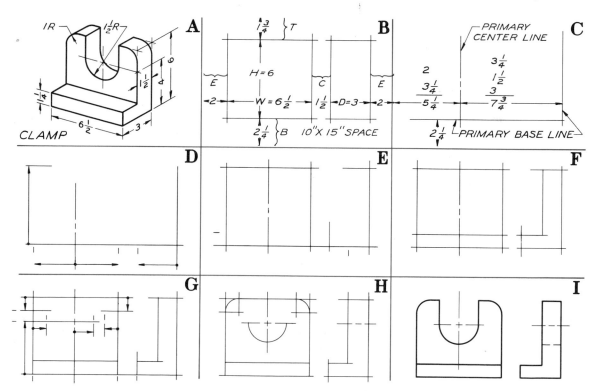

FIG. 6.33 Lay out two-view drawings in this order: Position the views, locate the centerlines, block in the views, locate the centers, draw the arcs, draw the straight lines, and darken the lines of the finished views.

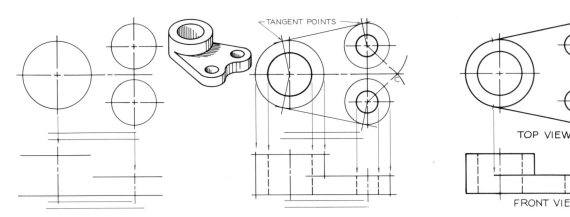

FIG. 6.34 Views with circular features

Step 1 Begin the layout by locating the centers and drawing the centerlines and circles. Draw these lines lightly.

Step 2 Locate the tangent points and draw the tangent lines. Block in the front view.

Step 3 After checking the views for correctness, darken the lines to their proper widths. Draw the arcs first, being careful to stopping them at the tangent points.

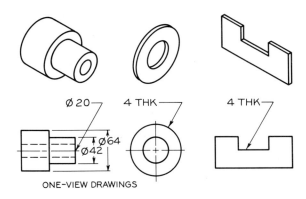

FIG. 6.35 Objects of this type can be described with only one orthographic view and supplementary notes.

the diameter dimension for the cylindrical part or 0 is placed before the diameter dimension. The thickness of the washer and the shim is specified by a note.

6.15
Incomplete and removed views

The right- and left-side views of the part in Fig. 6.36 would be very complex and hard to interpret if all hidden lines were shown as specified by the rules of or-

FIG. 6.36 Unnecessary and confusing hidden lines have been omitted in the side views to improve the clarity of the part.

thographic projection. Therefore, it is best to omit lines to make views more understandable. The two side views that are shown are easier to understand than views drawn in their entirety.

In many cases, it is difficult to show a feature because of its location. Standard views can be confusing when lines overlap from other features. The view in Fig. 6.37 is more clearly shown when the view indicated by the lines of sight is moved to an isolated position (called a removed view).

6.16
Curve plotting

An irregular curve can be drawn by following the rules of orthographic projection (Fig. 6.38).

The process of plotting points is begun by locating a series of points along the curve in two given views. By following the rules of projection, these points are

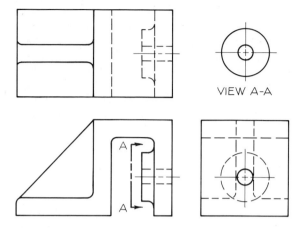

FIG. 6.37 A removed view, indicated by the directional arrows, can be used to draw views in hard-to-view locations.

projected to the top view, where each point is located and the points are connected by a smooth curve.

Figure 6.39 is a similar example, with an ellipse plotted in the top view by projecting from the front and side views. You will find it helpful to number points that are to be plotted when curves are being located by projection.

6.17
Partial views

A partial view can be used to save time and space when the parts are symmetrical or cylindrical. By omitting the rear of the circular top view in Fig. 6.40, space can be saved without sacrificing clarity.

A partial break may be used to make it more apparent that a portion of the view has been omitted. Either method is correct and acceptable.

6.18
Conventional revolutions

The readability of an orthographic view may be improved if the rules of projection are violated.

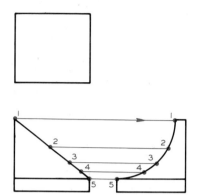

Step 1 Curves in orthographic projection can be plotted by locating and numbering the points in two views.

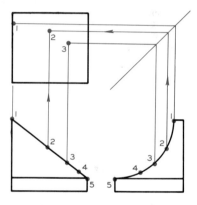

Step 2 Two views of each point are projected to the third view, where the projectors intersect.

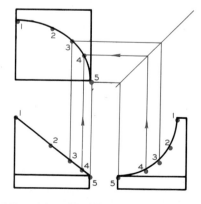

Step 3 All points are projected in this manner. The points are connected with an irregular curve.

FIG. 6.38 Plotting curved lines

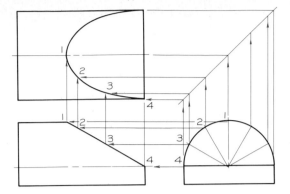

FIG. 6.39 The ellipse in the top view was found by numbering points in the front and side views and then projecting them to the top view.

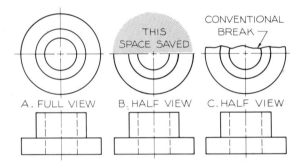

FIG. 6.40 To save space and drawing time, the top view of a cylindrical part can be drawn as a partial view using any of these methods.

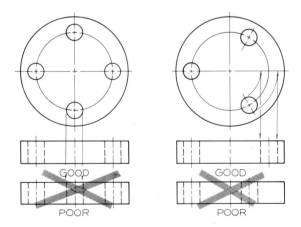

FIG. 6.41 Revolving holes

A. A true projection of equally spaced holes gives a misleading impression that the center hole passes through the center of the plate.

B. A conventional view is used to show the true radial distances of the holes from the center by revolution. The third hole is omitted.

Established violations of rules that are customarily made for the sake of clarity are called **conventional practices.**

When holes are symmetrically spaced in a circular plane, as shown in Fig. 6.41, it is conventional practice to shown them at their true radial distance from the center of the plane. This requires an imagined revolution of the holes in the top view.

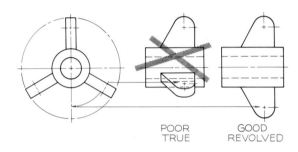

FIG. 6.42 Symmetrically positioned external features, such as these lugs, are revolved to their true-size positions for the best views.

This same principle of revolution applies to symmetrically positioned features such as the three lugs on the outside of the part in Fig. 6.42. The conventional view is better than the true orthographic projection.

The conventional and desired method of drawing holes and ribs in combination in the same view is shown in Fig. 6.43.

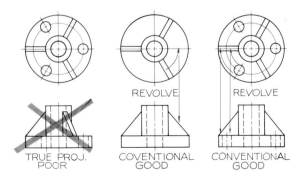

FIG. 6.43 The conventional methods of revolving holes and ribs in combination for improved clarity.

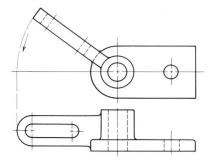

FIG. 6.44 It is conventional practice to imagine that features of this type have been revolved in order for them to appear true shape in the front view.

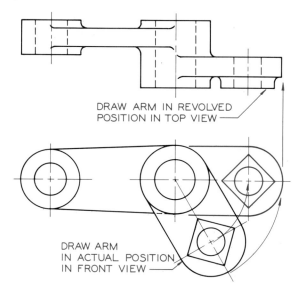

DRAW ARM IN REVOLVED POSITION IN TOP VIEW

DRAW ARM IN ACTUAL POSITION IN FRONT VIEW

FIG. 6.45 The arm in the front view is imagined to be revolved so its true length can be drawn in the top view. This is an accepted conventional practice.

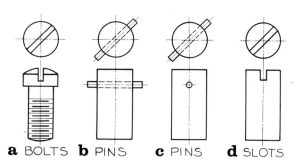

a BOLTS **b** PINS **c** PINS **d** SLOTS

FIG. 6.46 Features of this type are drawn at 45° angles in the top views, but they are imagined to be revolved to show their detail in the front view.

Another conventional practice is illustrated in Fig. 6.44, where an inclined feature is revolved to a frontal position in the top view, so it can be drawn as true size in the front view. The revolution of the part is not drawn since it is an imagined revolution. A similar object with a revolved feature is shown in Fig. 6.45.

Other parts whose views are improved by revolution are shown in Fig. 6.46. Unless there is a reason for them to be positioned otherwise, it is desirable to show the top view features at a 45° angle. The front views are drawn by imagining that the features have been revolved.

A closely related type of conventional view is the developed view, where a bent piece of material is drawn as if it were flattened out (Fig. 6.47).

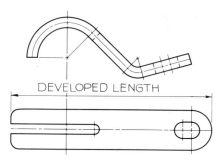

DEVELOPED LENGTH

FIG. 6.47 Objects that have been shaped by bending thin stock can be shown as developed views as if the views have been flattened out.

6.19
Intersections

In orthographic projection, an intersection between planes results in a line. In Fig. 6.48 examples of views are shown where lines may or may not be required.

The standard types of intersections between cylinders are shown in Fig. 6.49. Those at A and B are conventional intersections, which means that they are approximations, and they are drawn in this manner for ease of construction. The example at C is a true intersection where cylinders of equal diameters intersect. Similar intersections are shown in Fig. 6.50.

The types of intersections that are formed by holes in cylinders are shown in Fig. 6.51. These are conventional intersections that are sufficient for depicting these features.

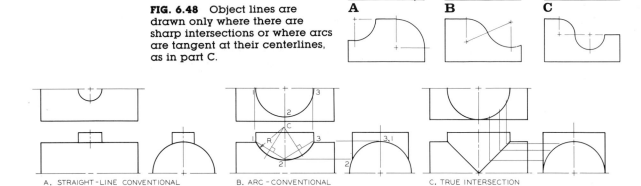

FIG. 6.48 Object lines are drawn only where there are sharp intersections or where arcs are tangent at their centerlines, as in part C.

A. STRAIGHT-LINE CONVENTIONAL   B. ARC-CONVENTIONAL   C. TRUE INTERSECTION

FIG. 6.49 The conventional methods of showing intersections between cylinders. The intersections at A and B are approximations.

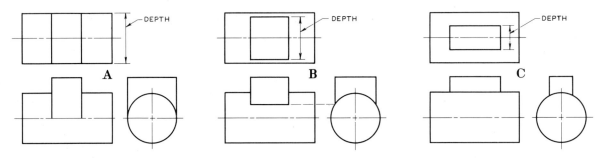

FIG. 6.50 Conventional intersections between cylinders and rectangular shapes. The intersection at C is an approximate intersection.

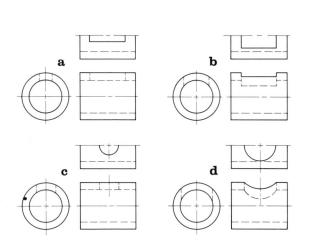

FIG. 6.51 Conventional intersections between cylinders and holes piercing them.

FIG. 6.52 The edges of this Collet Index Fixture are rounded to form fillets and rounds. Note that the surface of the casting is rough except where it has been machined. (Courtesy of Hardinge Brothers Inc.)

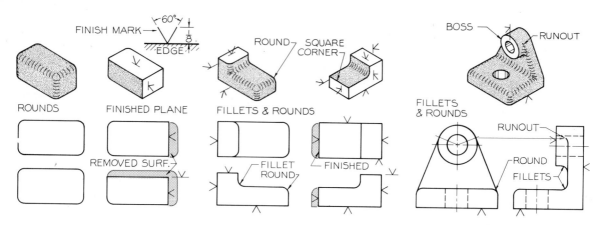

FIG. 6.53 Fillets and rounds

A. When a surface has been finished by machining, the rounds are removed and the corners are squared. The finish mark is a V that is placed on the edge view of finished surfaces.

B. A fillet is a rounded inside corner. The rounds are removed when the outside surfaces are finished. The fillets can be seen only in the front view.

C. The views of an object with fillets and rounds must be drawn using runouts as a way as to call attention to them.

6.20
Fillets and rounds

Fillets and **rounds** are rounded corners that are used on castings, such as the body of the Collet Index Fixture shown in Fig. 6.52. A **fillet** is an **inside rounding,** and a **round** is an **external rounding.** The radii of fillets and rounds are usually about ¼ in.; they are used on castings for added strength and improved appearance.

The orthographic views of a part with fillets and rounds must be drawn in such a manner that these detailed features can be seen (Fig. 6.53).

A casting will have square corners only when its surfaces have been finished, which is the process of machining away a portion of the surface to a smooth finish (Fig. 6.53B).

> The finished surface is indicated on a view by placing a **finish mark** (V) on the outside edge views of the surface that is to be finished in all views, whether the edges are **visible** or **hidden.**

You can see in Fig. 6.53 how fillets and rounds are shown, as well as the square corners without rounded corners. Note that a **boss** is a raised cylindri-

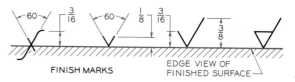

FIG. 6.54 Alternate types of finish marks that are applied to the edge views of a finished surface, whether hidden or visible.

cal feature that is thickened to receive a shaft or to be threaded. Types of finish marks are shown in Fig. 6.54. The techniques of showing fillets and rounds on orthographic views are given in Fig. 6.55.

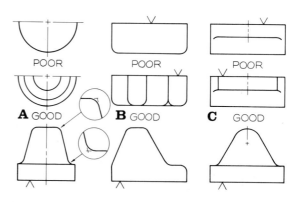

FIG. 6.55 Examples of conventionally drawn fillets and rounds

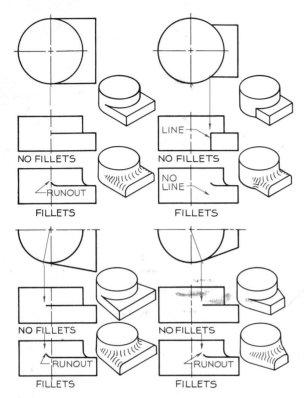

FIG. 6.56 Intersections between features of types of objects. The intersections with fillets have runouts.

The curve formed by a fillet at a point of tangency is called a **runout.** A comparison of intersections and runouts of parts with and without fillets and rounds is shown in Fig. 6.56. Large runouts are constructed with a compass as an eighth of a circle, as illustrated in Fig. 6.57. When the runouts are small, they can be drawn with a circle template.

When properly drawn, the runouts on orthographic views will tell much about the details of an object (Fig. 6.58). For example, the runouts in the top views tell us that the ribs at A have rounded corners and the rib at B is completely round.

Methods of showing intersections of other types are shown in Fig. 6.59.

6.21
Left-hand and right-hand views

Two parts are often required that are very similar to each other, but one part is actually a "mirror image" of the other (Fig. 6.60). Your first impression is that the parts are interchangeable, but the parts are actually as different as a pair of shoes.

The drafter can reduce drawing time by drawing views of only one of the parts and labeling these views as shown in Fig. 6.60. A note can be added to indicate that the other matching part has the same dimensions.

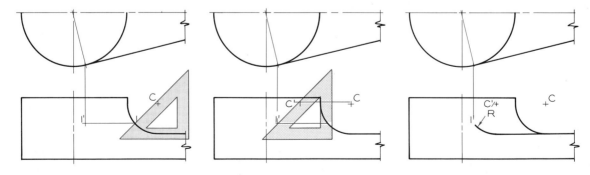

FIG. 6.57 Plotting runouts

Step 1 Find the point of tangency in the top view and project it to the front view. A 45° triangle is used to find Point 1, which is projected to Point 1'.

Step 2 A 45° triangle is used to locate Point C', which is on the horizontal projector from the center of the fillet, C.

Step 3 The radius of the fillet is used to draw the runout with C' as the center. The runout arc is equal to one-eighth of a circle.

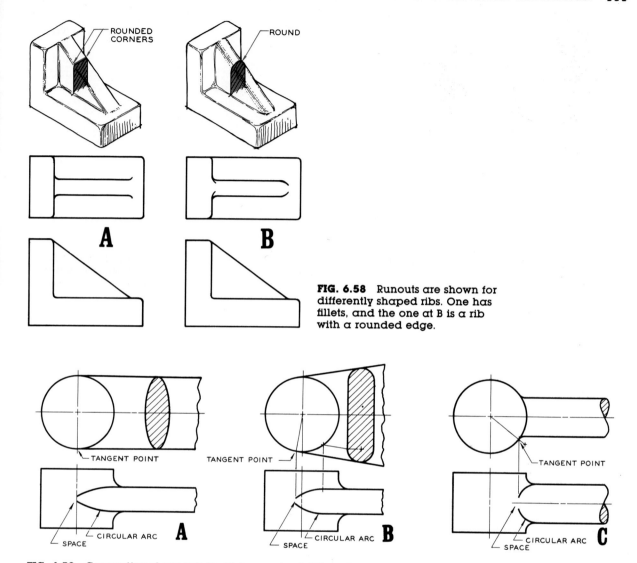

FIG. 6.58 Runouts are shown for differently shaped ribs. One has fillets, and the one at B is a rib with a rounded edge.

FIG. 6.59 Conventional runouts involving parts of different cross sections.

6.22
First-angle projection

The problems of this chapter have been presented as **third-angle projections,** where the top view is placed over the front view, and the right-side view to the right of the front view. This method is used extensively in America, Britain, and Canada. Most of the rest of the industrial world uses the **first-angle** of projection.

The first-angle system is illustrated in Fig. 6.61, where an object is placed above the horizontal plane and in front of the frontal plane. When these projection planes are opened onto the surface of the drawing paper, the front view is projected over the top view, and the left-side view is placed to the right of the front view.

To indicate the angle of projection of a drawing, a truncated cone is placed in or near the title block (Fig. 6.62). When metric units of measurements are used, the cone and the symbol SI are placed together on the drawing.

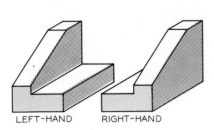

LEFT-HAND   RIGHT-HAND

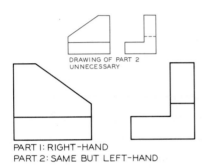

DRAWING OF PART 2
UNNECESSARY

PART 1: RIGHT-HAND
PART 2: SAME BUT LEFT-HAND

FIG. 6.60   Left-hand and right-hand parts

A. Some parts are required to be similar except that one is a left-hand part and the other is a right-hand part.

B. One of the parts can be drawn and labeled, the right-hand part in this case. The other part need not be drawn, but merely indicated by a note.

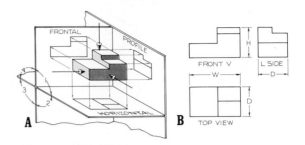

FIG. 6.61   First-angle projection

A. The first angle of projection is used by many of the countries that use the metric system. You imagine that the object is placed above the horizontal and in front of the frontal plane.

B. The views are drawn in this location, which is different from the third-angle of projection that is usually used in America.

U. S. SYSTEM—VISIBLE   EUROPEAN—VISIBLE

a METRIC UNITS—
3ᴿᴰ ANGLE PROJ

b METRIC UNITS—
1ˢᵀ ANGLE PROJ

FIG. 6.62   The angle of projection that is used to prepare a set of drawings is indicated by this truncated cone. It is placed in or near the title block of a drawing.

Problems

The following problems are to be drawn as orthographic views on Size A or Size B paper as assigned by your instructor. (Refer to Section 2.19 for instructions on laying out a sheet.)

1–15. (Figs. 6.63–6.77) Draw the given views using the dimensions provided, and then construct the missing view: either the top, front, or right-side view. Use Size A sheets and draw one or two problems per sheet as assigned by your instructor.

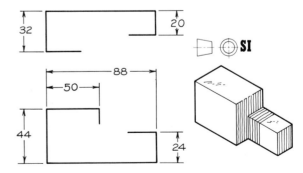

FIG. 6.63   Problem 1: Guide block.

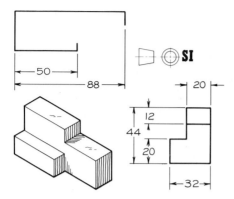

FIG. 6.64 Problem 2: Double step.

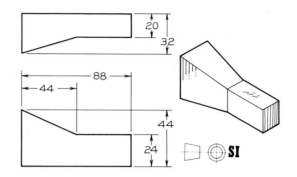

FIG. 6.67 Problem 5: Two-way adjuster.

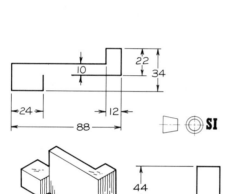

FIG. 6.65 Problem 3: Adjustable stop.

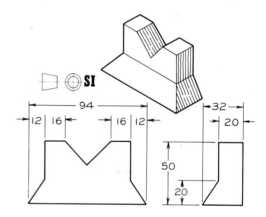

FIG. 6.68 Problem 6: Vee block.

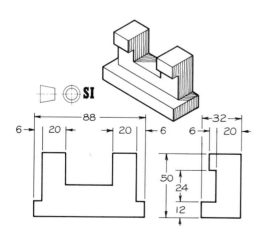

FIG. 6.66 Problem 4: Lock catch.

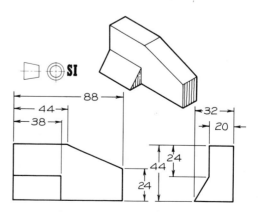

FIG. 6.69 Problem 7: Filler.

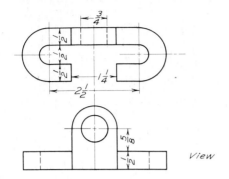

FIG. 6.70 Problem 8: Slide stop.

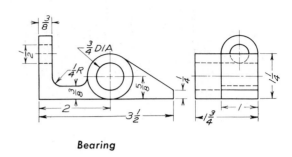

Bearing

FIG. 6.73 Problem 11: Bearing.

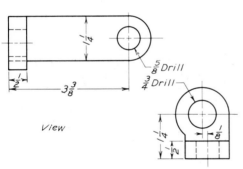

Angle brace

FIG. 6.71 Problem 9: Angle brace.

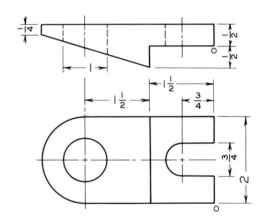

FIG. 6.74 Problem 12: Latch plate.

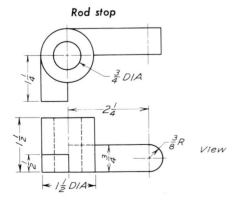

Rod stop

FIG. 6.72 Problem 10: Rod stop.

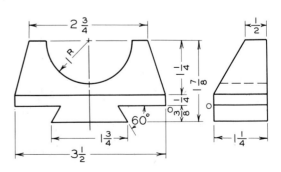

FIG. 6.75 Problem 13: Shaft support.

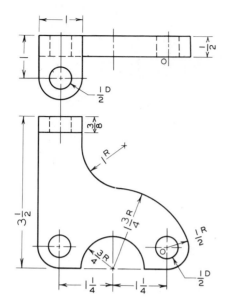

FIG. 6.76 Problem 14: Support brace.

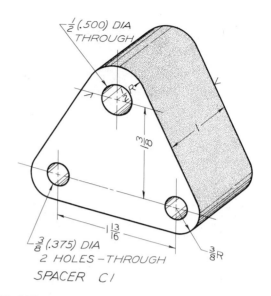

SPACER C1

FIG. 6.78 Problem 16: Spacer.

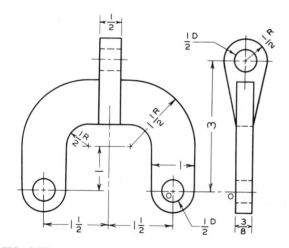

FIG. 6.77 Problem 15: Yoke.

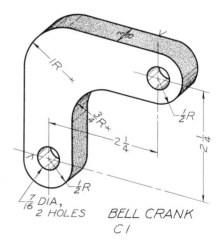

BELL CRANK
C1

FIG. 6.79 Problem 17: Bell crank.

16–54. Construct the necessary orthographic views to describe the objects in Figs. 6.78 to Fig. 6.116. Draw these on Size A or Size B sheets. Label the views and show the overall dimensions of *W*, *D*, and *H* on the appropriate views.

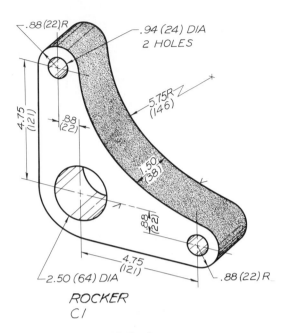

.88 (22) R
.94 (24) DIA
2 HOLES
5.75 R (146)
4.75 (121)
.88 (22)
1.50 (38)
.88 (22)
4.75 (121)
2.50 (64) DIA
.88 (22) R

ROCKER
C1

FIG. 6.80 Problem 18: Rocker.

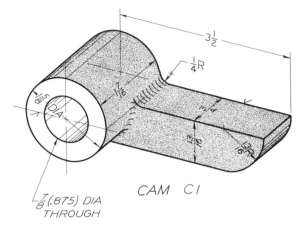

$3\frac{1}{2}$
$\frac{1}{4}$ R
$\frac{7}{16}$
$\frac{5}{16}$ DIA
$\frac{7}{8}$ (.875) DIA THROUGH

CAM C1

FIG. 6.82 Problem 20: Cam.

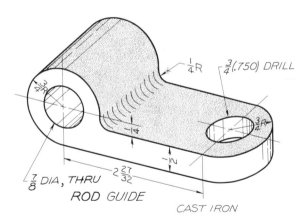

$\frac{1}{4}$ R
$\frac{3}{4}$ (.750) DRILL
$4\frac{3}{16}$
$\frac{3}{4}$
$\frac{1}{2}$
$\frac{3}{4}$
$2\frac{27}{32}$
$\frac{7}{8}$ DIA, THRU

ROD GUIDE
CAST IRON

FIG. 6.81 Problem 19: Rod guide.

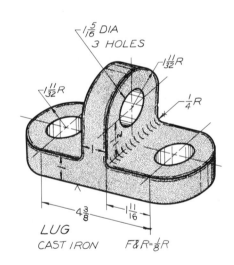

$1\frac{5}{16}$ DIA
3 HOLES
$1\frac{11}{32}$ R
$1\frac{11}{32}$ R
$\frac{1}{4}$ R
$4\frac{3}{8}$
$1\frac{11}{16}$

LUG
CAST IRON   F & R = $\frac{1}{8}$ R

FIG. 6.83 Problem 21: Lug.

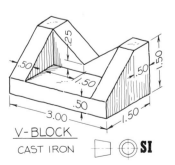

2.5
.50
1.50
.50
1.50
.50
3.00
.50
1.50

V-BLOCK
CAST IRON   **SI**

FIG. 6.84 Problem 22: V-block.

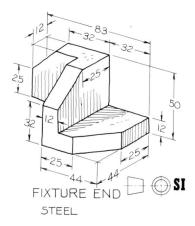

FIXTURE END
STEEL

FIG. 6.85 Problem 23: Fixture end.

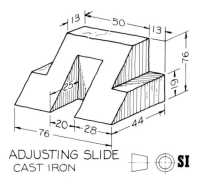

ADJUSTING SLIDE
CAST IRON

FIG. 6.86 Problem 24: Adjusting slide.

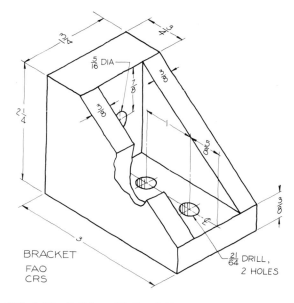

BRACKET
FAO
CRS

FIG. 6.87 Problem 25: Bracket.

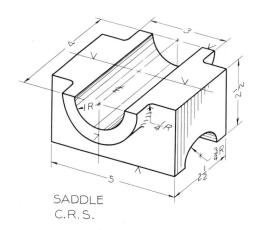

SADDLE
C.R.S.

FIG. 6.88 Problem 26: Saddle.

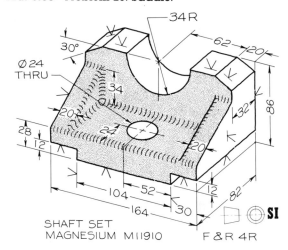

SHAFT SET
MAGNESIUM M11910

F & R 4R

FIG. 6.89 Problem 27: Shaft set.

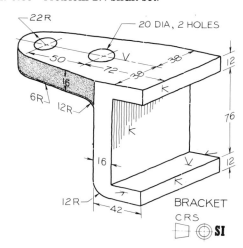

BRACKET
CRS

FIG. 6.90 Problem 28: Bracket.

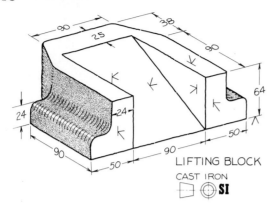

FIG. 6.91 Problem 29: Lifting block.

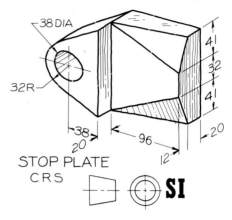

STOP PLATE
CRS

FIG. 6.94 Problem 32: Stop plate.

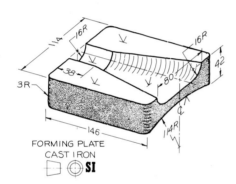

FORMING PLATE
CAST IRON

FIG. 6.92 Problem 30: Forming plate.

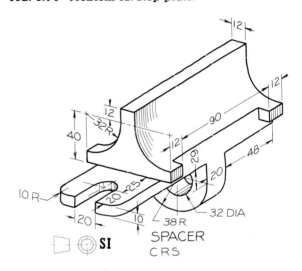

SPACER
CRS

FIG. 6.95 Problem 33: Spacer.

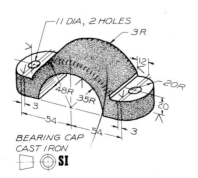

BEARING CAP
CAST IRON

FIG. 6.93 Problem 31: Bearing cap.

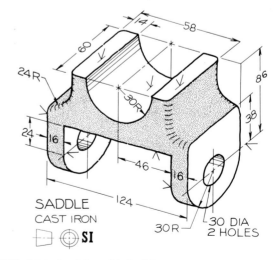

SADDLE
CAST IRON

FIG. 6.96 Problem 34: Saddle.

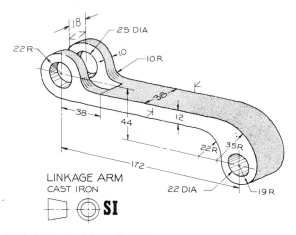

FIG. 6.97 Problem 35: Linkage arm.

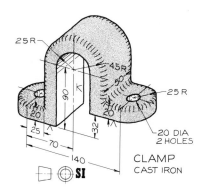

FIG. 6.100 Problem 38: Clamp.

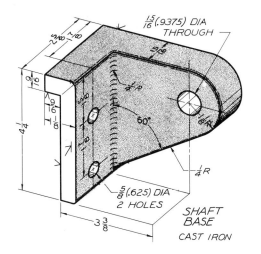

FIG. 6.98 Problem 36: Shaft base.

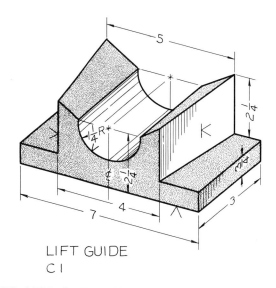

LIFT GUIDE
C 1

FIG. 6.101 Problem 39: Lift guide.

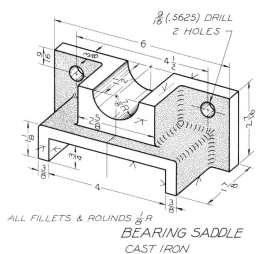

FIG. 6.99 Problem 37: Bearing saddle.

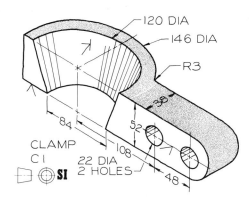

FIG. 6.102 Problem 40: Clamp.

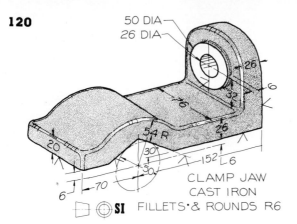

50 DIA
26 DIA

CLAMP JAW
CAST IRON
SI FILLETS·& ROUNDS R6

FIG. 6.103 Problem 41: Clamp jaw.

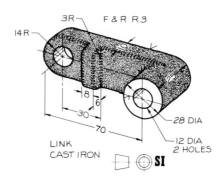

3R   F & R R3
14R

8
6
30
70

28 DIA
12 DIA
2 HOLES

LINK
CAST IRON
SI

FIG. 6.106 Problem 44: Link.

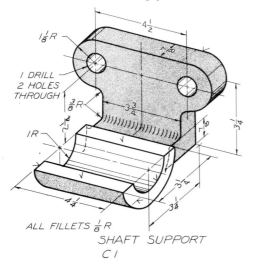

$4\frac{1}{2}$
$1\frac{1}{8}R$

I DRILL
2 HOLES
THROUGH
$\frac{3}{8}R$
IR

$3\frac{3}{4}$
$3\frac{1}{4}$
$3\frac{1}{4}$
$3\frac{1}{4}$
$4\frac{1}{4}$

ALL FILLETS $\frac{1}{8}R$

SHAFT SUPPORT
C I

FIG. 6.104 Problem 42: Shaft support.

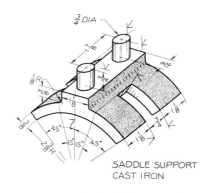

$\frac{3}{4}$ DIA

$\frac{1}{8}R$

45°
15° 15°
45°
$\frac{1}{8}$
$\frac{3}{4}$
$\frac{1}{8}$

SADDLE SUPPORT
CAST IRON

FIG. 6.107 Problem 45: Saddle support.

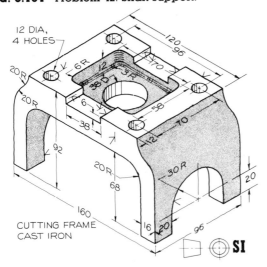

12 DIA,
4 HOLES
20R
6R   12   70
20R
3R
38   58
6
38
92   70
20R   -12
68   30R   20
160
16   20   96

CUTTING FRAME
CAST IRON
SI

FIG. 6.105 Problem 43: Cutting frame.

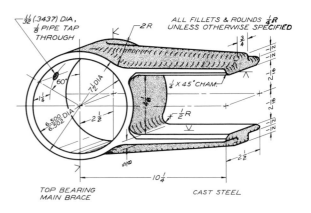

$\frac{11}{32}$ (.3437) DIA,
$\frac{1}{8}$ PIPE TAP
THROUGH

2R

ALL FILLETS & ROUNDS $\frac{1}{4}R$
UNLESS OTHERWISE SPECIFIED

$\frac{3}{4}$

60°
$7\frac{1}{2}$ DIA
$1\frac{1}{4}$
6.500 DIA
6.502 DIA
$2\frac{1}{2}$

$\frac{1}{4}$ X 45° CHAM.

$\frac{1}{2}R$
$2\frac{1}{2}$

$10\frac{1}{4}$

TOP BEARING
MAIN BRACE

CAST STEEL

FIG. 6.108 Problem 46: Top bearing.

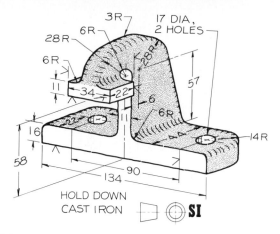

FIG. 6.109 Problem 47: Hold down.

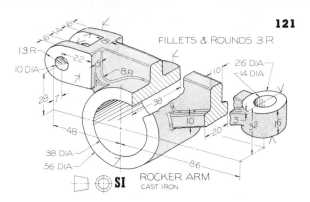

FIG. 6.112 Problem 50: Rocker arm.

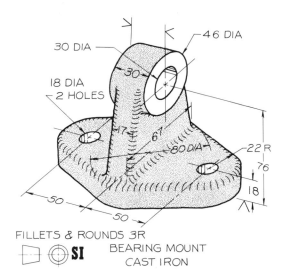

FIG. 6.110 Problem 48: Bearing mount.

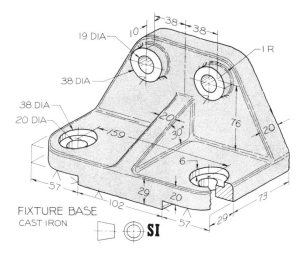

FIG. 6.113 Problem 51: Fixture base.

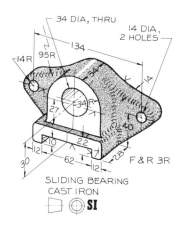

FIG. 6.111 Problem 49: Sliding bearing.

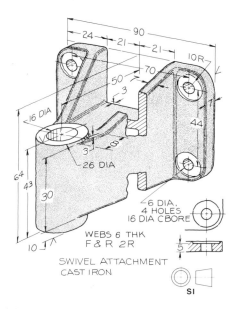

FIG. 6.114 Problem 52: Swivel attachment.

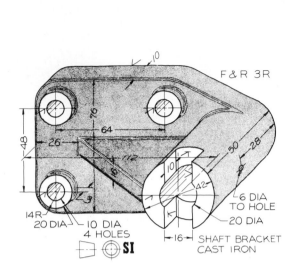

FIG. 6.115 Problem 53: Shaft bracket.

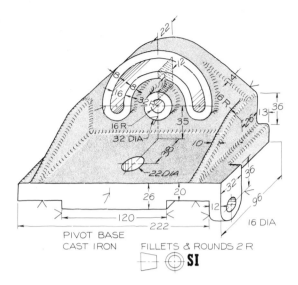

FIG. 6.116 Problem 54: Pivot base.

CHAPTER

7

Auxiliary Views

7.1
Introduction

The auxiliary view is a type of orthographic view, but it is not a **principal** view as previously covered. The principal views, such as the top, front, and side views, are usually drawn so that their surfaces are parallel to

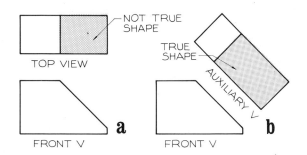

FIG. 7.1 When a surface appears as an inclined edge in a principal view, it can be found true size by an auxiliary view. The top view at a is foreshortened, but this plane is true size in an auxiliary view at b.

the principal projection planes, and therefore, appear true size.

However, if the surfaces of an object are not mutually perpendicular, all planes cannot be parallel to the projection planes. This means that an **auxiliary plane,** must be used on which to project a true-size view of a nonprincipal view.

An example of an auxiliary plane is shown in Fig. 7.1. The edge view of the surface is not horizontal in the front view; therefore it cannot appear true shape in the top view. This surface is true shape in the auxiliary view that is projected from the front view.

7.2
The folding-line approach

The three principal planes of orthographic projection are the **frontal, horizontal,** and **profile** planes. These are the names of the planes of an imaginary glass box in which an object is placed to draw its various views.

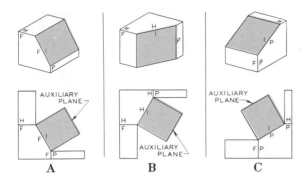

FIG. 7.2 A primary auxiliary plane can be folded from the frontal, horizontal, or profile planes. The fold lines are labeled F-1, H-1, and P-1, respectively.

A primary auxiliary plane is a plane that is perpendicular to one of the principal planes and oblique to the other two planes. In other words, a **primary auxiliary view** is an auxiliary view that is projected from a **primary orthographic view.**

The auxiliary planes can be thought of as planes that fold from the principal planes, as shown in Fig. 7.2. The plane at A folds from the frontal plane to make a 90° angle with it. The fold line between the two planes is labeled F-1. The F is an abbreviation for frontal and the numeral 1 represents **first** or **primary** auxiliary plane.

7.3
Auxiliaries projected from the top view

The inclined plane in Fig. 7.3 is an edge in the top view, and it is perpendicular to the top projection plane, the horizontal plane. An auxiliary plane can be drawn that is parallel to the inclined surface of the object, and the view projected onto it will be a true-size view of the inclined plane.

It is necessary that a surface appear as an edge in a principal view before it can be found as true size in a primary auxiliary view.

Fold line H-1 is drawn parallel to the edge view of the inclined plane in Step 1. The line of sight is drawn perpendicular to the edge view. Each corner of the inclined plane is projected perpendicularly to the auxiliary plane and is located by using the dimension of height (*H*) from the side or front view.

A similar example is the object shown in the glass box in Fig. 7.4. Since this object has an inclined surface that appears as an edge in the top view, it can be found true size in a primary auxiliary view. The fold line between the horizontal and auxiliary plane is labeled H-1. The height dimension (*H*) is transferred

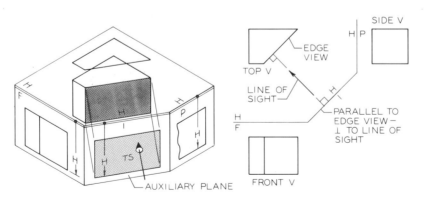
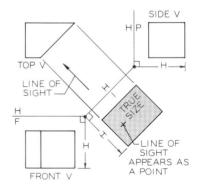

FIG. 7.3 Auxiliary from the top—folding-line method

Given An object with an inclined surface.
Required Find the inclined surface true size by an auxiliary view.

Step 1 Construct a line of sight perpendicular to the edge view of the inclined surface. Draw the H-1 fold line parallel to the edge, and draw the H-F fold line between the top and front views.

Step 2 Project the four corners of the edge parallel to the line of sight. Locate the corners by measuring perpendicularly from the horizontal plane with height (*H*) dimensions.

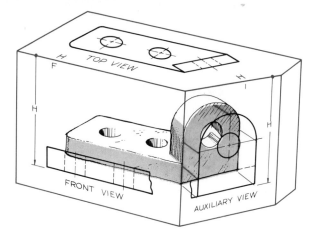

FIG. 7.4 A pictorial showing the relationship of the projection planes used to find the true-size view of the inclined surface.

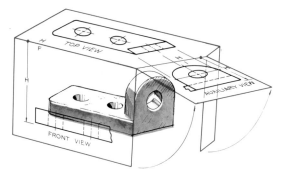

FIG. 7.5 The auxiliary plane is opened into the plane of the top view by revolving it about the H-1 fold line. This is how it would be positioned on your drawing surface.

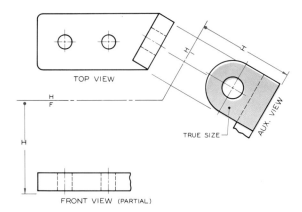

FIG. 7.6 When the object is drawn on a sheet of paper, it is laid out in this manner. The front view is drawn as a partial view since the omitted part is shown true size in the auxiliary view.

from the front view to the auxiliary view, since both are measured from the same horizontal plane.

The auxiliary plane is rotated about the H-1 fold line into the plane of the top view in Fig. 7.5.

When drawn on a sheet of paper, the drawing of this part would appear as shown in Fig. 7.6. The front view is shown as a partial view since the omitted portion would have been hard to draw and would not have been true size. The auxiliary view shows the true-size view of the inclined surface.

7.4
Constructing an auxiliary from the top view— folding-line method

The steps of constructing an auxiliary view that is projected from the top view are shown in Fig. 7.7.

The inclined surface can be found true size in a primary auxiliary view since it projects as an edge in the top view. The line of sight is drawn perpendicular to the edge view, and the **folding-line** is drawn parallel to the edge. The dimension that must be transferred from the front view is height (H). Height is perpendicular to the horizontal plane; consequently, it cannot be found in the top view. It must be taken from an adjacent view such as the front or side view.

The fold line H-1 is drawn as a thin, but black, line. It is also helpful to number or letter the points on the views. The projectors are construction lines, and they should be drawn with a hard pencil (3H–4H) just dark enough to be seen.

7.5
Auxiliaries from the top view—reference-plane method

A similar method of locating an auxiliary view is a method that uses a **reference plane** instead of the folding-line technique. An example of this type of auxiliary view is shown in Fig. 7.8, where an auxiliary view is projected from the top view.

Instead of placing a fold line between the top and front views, a reference plane is passed through the bottom of the front view, shown pictorially in A. The height dimensions (H) are measured upward from the

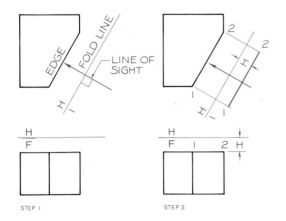

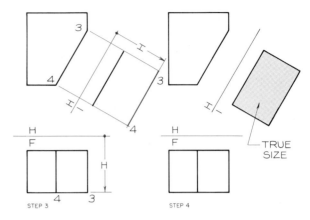

FIG. 7.7 Construction of an auxiliary view

Step 1 The line of sight is drawn perpendicular to the edge views of the inclined surface. The fold line is drawn parallel to the edge view of the inclined surface. An H-F fold line is drawn between the given views.

Step 2 Points 1 and 2 are found by transferring the height (H) dimensions from the front view to the auxiliary view. This locates Line 1–2.

Step 3 Points 3 and 4 are found in the same manner, using the dimensions of height (H) to locate Line 3–4.

Step 4 The corner points are connected to complete the true-size view of the inclined plane.

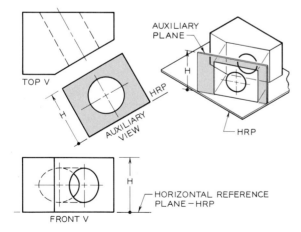

FIG. 7.8 A horizontal reference plane can be used instead of the folding-line technique to construct an auxiliary view. Instead of placing the reference plane between the top and auxiliary views, it is placed outside the auxiliary view (b).

reference plane instead of downward from a fold line.

In B, the reference plane is shown as a horizontal edge in the front view where it is labeled HRP (horizontal reference plane). This plane will appear as an edge that is parallel to the edge view of the inclined surface from which the auxiliary view is projected.

7.6
Auxiliaries from the front view—folding-line method

A plane that appears as an edge in the front view (Fig. 7.9) can be found true size in a primary auxiliary view projected from the front view. Fold line F-1 is drawn parallel to the edge view of the inclined plane in the front view at a convenient location.

The line of sight is drawn perpendicular to the edge view of the inclined plane in the front view. The frontal plane appears as an edge in the auxiliary view; consequently, the depth (D) dimensions that are perpendicular to the frontal plane will be seen true length. Depth dimensions are transferred from the top view to the auxiliary view with dividers.

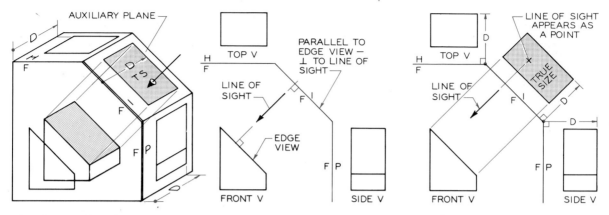

FIG. 7.9 Auxiliary from the front—folding-line method

Given An object with an inclined surface.
Required Find the inclined surface true size by an auxiliary view.

Step 1 Draw a line of sight perpendicular to the edge view of the inclined surface. Draw the F-1 fold line parallel to the edge, and draw the H-F and/or F-P fold lines between the given views.

Step 2 Project the corners of the edge view parallel to the line of sight. Locate the corners of the true-size view by measuring perpendicularly from the frontal plane with depth (D) dimensions.

The inclined surface will be true size in Step 2. Note that the **line of sight** will appear as a **point** in this view.

A practical application of this type of auxiliary view is the part shown in Fig. 7.10 where a true-size view of the inclined surface is needed. It is enclosed in a glass box where an auxiliary plane is constructed parallel to the inclined plane of the part.

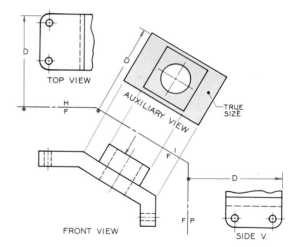

FIG. 7.11 The layout and construction of an auxiliary view of the object in Fig. 7.10 as it would appear on your drawing paper.

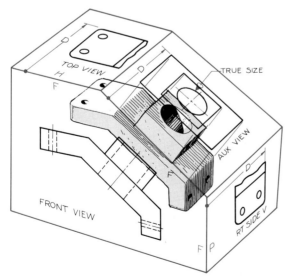

FIG. 7.10 A pictorial showing the relationship of the projection planes used to find the true-size view of the inclined surface of an object.

When this arrangement is drawn on a sheet of paper it will appear as shown in Fig. 7.11. The top and side views are drawn as partial views since the auxiliary view eliminates a need for more than is shown of them. The auxiliary view is located by using the depth dimension measured from the edge view of the frontal projection plane.

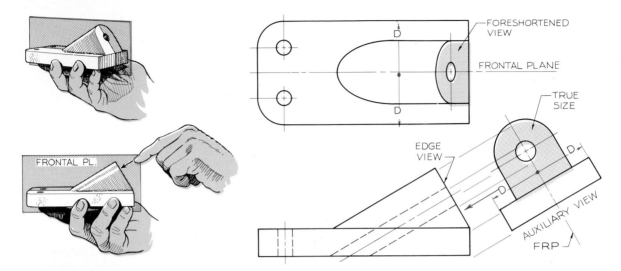

FIG. 7.12 Since the inclined surface of this part is symmetrical, it would be advantageous to use a frontal reference plane (FRP) that is passed through the object. The auxiliary view is projected perpendicularly from the edge view of the inclined plane in the front view. The FRP appears as an edge in the auxiliary view, and depth (D) dimensions are made on each side of it to locate points on the true-size view of the inclined surface.

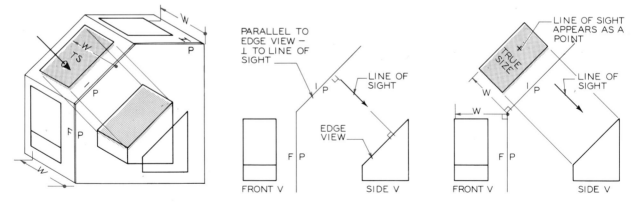

FIG. 7.13 Auxiliary from the side view—folding-line method

Given An object with an inclined surface.
Required Find the inclined surface true size by an auxiliary view.

Step 1 Draw a line of sight perpendicular to the edge view of the inclined surface. Draw the P-1 fold line parallel to the edge view, and draw the F-P fold line between the given views.

Step 2 Project the corners of the edge parallel to the line of sight. Locate the corners by measuring perpendicularly from the P-1 fold line with width (W) dimensions.

7.7
Auxiliaries from the front view—reference-plane method

The object in Fig. 7.12 has an inclined surface that appears as an edge in the front view. Consequently, this plane can be found true size in a primary auxiliary view.

Since it is symmetrical, it is advantageous to use a reference plane that passes through the center of the top view. The reference plane is a frontal plane, so it will be called a **frontal reference plane** (FRP) in the auxiliary view.

The FRP is located parallel to the edge view of the inclined plane. The reference plane will pass through the center of the view (Fig. 7.13) and parallel to the edge view of the inclined surface.

Whereas the folding line would be placed between the top and auxiliary view, the FRP is located through the center of the auxiliary view.

7.8
Auxiliaries from the profile view—folding-line method

Since the inclined surface in Fig. 7.13 appears as an edge in the profile plane, it can be found true size in a primary auxiliary view projected from the profile view. The auxiliary fold line, P-1, is drawn parallel to the edge view of the inclined surface.

A line of sight that is perpendicular to the auxiliary plane will see the profile plane as an edge. Consequently, dimensions of width (W) will appear true length in the auxiliary view. Width dimensions are transferred from the front view to the auxiliary view to draw the auxiliary view.

7.9
Auxiliaries from the profile—reference-plane method

The object in Fig. 7.14 has two inclined surfaces that appear as edges in the right-side view, the profile view. These are found true size by using a **profile reference plane** (PRP) that is a vertical edge in the front view.

The two auxiliary views are drawn to find the true-sizes of the inclined surfaces. The profile reference planes are positioned at the far edges of the auxiliary views instead of between the profile and auxiliary views as in the folding-line method. The views are found by transferring the width (W) dimensions from the edge view of the PRP in the front view to the auxiliary view.

7.10
Auxiliaries of curved shapes

When an auxiliary view is drawn to show a curve that is not a true arc, a series of points must be plotted and connected. The cylinder in Fig. 7.15 has a beveled surface that appears as an edge in the front view. When found true size, this surface will be elliptical in shape.

Since the cylinder is symmetrical, it is beneficial to use an FRP through the object so that dimensions can be measured on both sides of it. Points are located about the circular right-side view and projected to the edge view of the surface in the front view.

The FRP is located parallel to the edge view of the plane, and the points are projected perpendicularly from the edge view of the plane. Dimensions A and B are shown as examples for plotting points in the auxiliary view. More points are needed than are shown to construct a smooth curve.

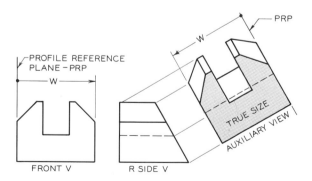

FIG. 7.14 Two auxiliary views are projected from the right-side view using a profile reference plane (PRP). The auxiliary views show the true-size views of the inclined surfaces.

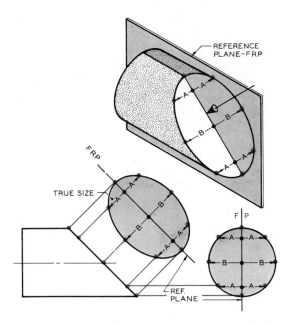

FIG. 7.15 This auxiliary view requires that a series of points be plotted. Since the object is symmetrical, the reference plane is positioned through its center.

An irregular curve appears as an edge in the front view of the object in Fig. 7.16. Points are located on the curve in the top view and are projected to the front view. These points are found in the auxiliary view by locating each point using the depth dimension (*D*) transferred from the top view.

7.11
Partial views

Since an auxiliary view is a supplementary view, some views of an orthographic arrangement can be drawn as partial views. It is understood that the omitted features in one view will be shown in another view, perhaps in the auxiliary view.

The object in Fig. 7.17 is shown by a complete front view and partial auxiliary and side views. These views are easier to draw, and they are more functional without sacrificing clarity.

A similar example is given in Fig. 7.18, where the front view is a complete view and the other two views are partial views. A frontal reference plane is passed through the center of the side view. Note that even

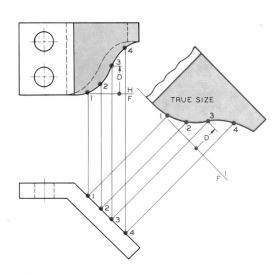

FIG. 7.16 The auxiliary view of this surface required that a series of points be located in the given view and then projected to the auxiliary view. The curve is drawn using an irregular curve through the plotted points.

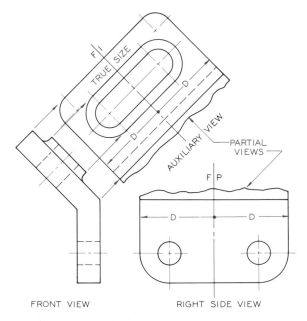

FRONT VIEW   RIGHT SIDE VIEW

FIG. 7.17 This auxiliary view was drawn as a partial view since it is supplemented by the true-size partial view shown in the right-side view. Note that the reference planes are labeled the same as a folding line, which is permissible.

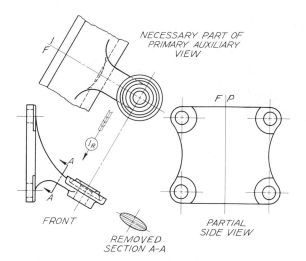

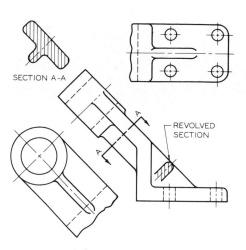

FIG. 7.18 This object is represented by a series of partial views. The hub, which would appear elliptical, is not shown in the side view at all.

FIG. 7.20 Auxiliary section A-A is projected from the cutting plane that is labeled A-A, to show the cross section of the object.

FIG. 7.19 A photograph of the guide bracket that was drawn orthographically in the previous figure.

though it is a reference plane, it is labeled F-1, which is a permissible practice. The guide bracket that was drawn in Fig. 7.18 is shown in Fig. 7.19.

7.12
Auxiliary sections

To better describe a part, auxiliary sections can be drawn to show the cross sections through it. A section through a part that is projected as an auxiliary view is shown in Fig. 7.20.

The section is labeled as Section A-A, and a cutting plane was passed through the object and labeled A-A. This cutting plane was used to show where the sectional view was projected from and where the object was cut to show the section.

The auxiliary section gives a good description of the part that cannot be as readily understood from the given principal views.

7.13
Secondary auxiliary views

A **secondary auxiliary view** is an auxiliary view that has been projected from a primary auxiliary view. (Fig. 7.21).

The inclined plane is found as an edge view in the primary auxiliary view, and then a line of sight is established that is perpendicular to the edge. This second auxiliary view shows the inclined surface true size.

Note that it has been necessary in all cases to project from an edge view of a plane before it can be found true size.

The problem in Fig. 7.22 is a secondary auxiliary projection in which the point view of a diagonal of a

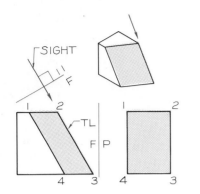

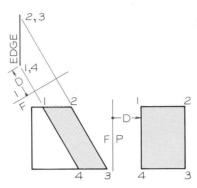

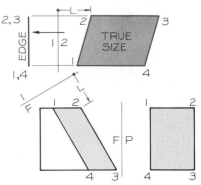

FIG. 7.21 Secondary auxiliary views

Step 1 A line of sight is drawn parallel to the true-length view of a line on the oblique surface. The folding line F-1 is drawn perpendicular to the line of sight.

Step 2 The primary auxiliary view of the oblique surface is an edge view. Dimension depth (D) is used to locate points in the primary auxiliary view.

Step 3 A line of sight is drawn perpendicular to the edge view, and a secondary auxiliary view is projected in this direction. The dimension L is used to locate a point in the true-size view.

cube is found. The three surfaces of the cube are equally foreshortened. This view is an **isometric projection,** and the basis for isometric pictorial drawing.

Since auxiliary views are supplementary views, they can be partial views if all features are sufficiently shown. The object in Fig. 7.23 is shown as a series of orthographic views, some of which are partial views.

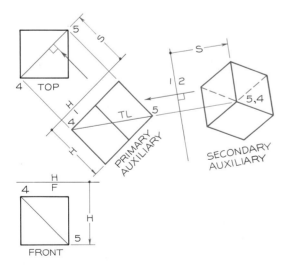

FIG. 7.22 A point view of a diagonal (4–5) of a cube is found in the secondary auxiliary view. This line is found true length in the primary auxiliary view and then as a point in the secondary auxiliary view. The secondary auxiliary view is an isometric projection of the cube.

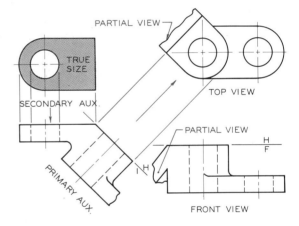

FIG. 7.23 An example of a secondary auxiliary view projected from a partial auxiliary view.

7.14
Elliptical features

There are occasions when circular shapes will project as ellipses. Ellipses can be drawn using any of the techniques introduced in Chapter 4 once the necessary points have been plotted.

The most convenient method of drawing ellipses is with the use of an ellipse guide (template). The angle of the ellipse guide is the angle the line of sight

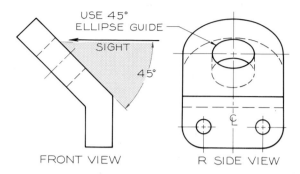

FIG. 7.24 The ellipse guide angle is the angle that the line of sight makes with the edge view of the circular feature. The ellipse guide for the right-side view is 45°.

makes with the edge view of the circular feature. In Fig. 7.24 the angle is found to be 45°. The right-side view is drawn as a 45° ellipse.

A more complex problem that involves a secondary auxiliary view is shown in Fig. 7.25. The inclined surface is found true size in the secondary auxiliary view by projecting from the edge view of the surface found in the primary auxiliary view.

Since the line of sight makes a 30° angle with the edge, this is the angle of the ellipse template to be used for drawing the top view.

The elliptical features in the front view can be found by locating a box around the ellipses in the top view and projecting them to the front view. The front view of these points is located by transferring dimensions of height (*H*) from the primary auxiliary view to the front view. This gives the conjugate diameters that

can be used for constructing an ellipse as covered in Chapter 4.

The auxiliary views in this problem could not have been drawn unless you knew what the dimensions and specifications of the object were before you started drawing the views as if you were its designer.

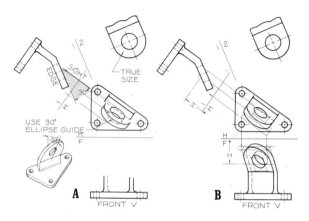

FIG. 7.25 Ellipses in secondary auxiliary views

A. The inclined surface is found true size in a secondary auxiliary view. The edge view of the plane in the primary auxiliary view makes a 30° angle with the line of sight from the top view. This is the angle of the ellipse guide to use in drawing the circular features in the top view.

B. The elliptical features in the front view can be completed by constructing a box about the circular features as shown in the top view. When transferred to the front view, these boxes are used to find the conjugate diameters of the ellipses, and then the ellipses are constructed.

Problems

The following problems are to be solved on Size A or Size B sheets as assigned by your instructor (refer to Section 2.19).

1–13. (Fig. 7.26) Using the example layout, change the top and front views by substituting the top views given at the right in place of the one given

in the example. The angle of inclination in the front view is 45° for all problems and the height is 38 mm in the front view. Construct auxiliary views that show the inclined surface true size. Draw four problems per Size A sheet when drawn full size, or one per sheet if drawn double size.

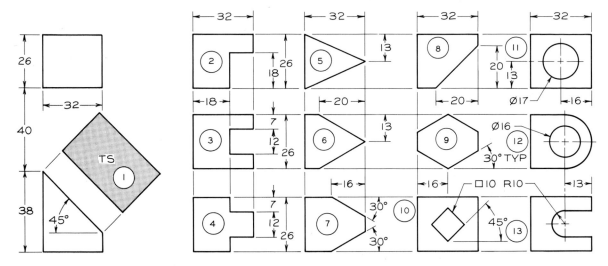

FIG. 7.26 Problems 1–13: Primary auxiliary views.

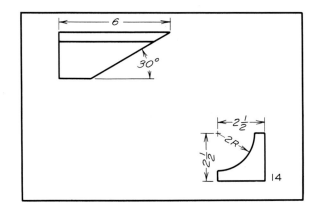

FIG. 7.27 Problems 14–20: Primary auxiliary views.

14–20. (Fig. 7.27) Using the example layout, change the top and right-side views using the side views given at the right of the example problem. Complete the front view and the auxiliary view. Draw two problems per Size AV sheet, showing each half size.

21–42. (Fig. 7.28–Fig. 7.49) Draw the necessary primary views and auxiliary views to describe the parts shown. Drawn one per Size A or Size B sheet as assigned. Adjust the scale of each to fit the space on the sheet.

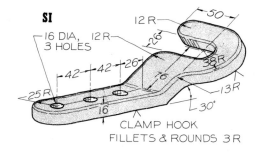

FIG. 7.28 Problem 21: Clamp hook.

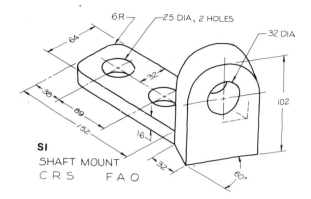

FIG. 7.29 Problem 22: Shaft mount.

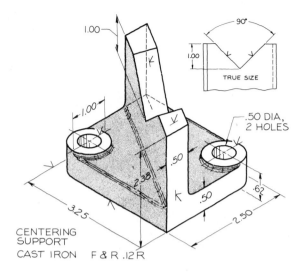

FIG. 7.32 Problem 25: Centering support.

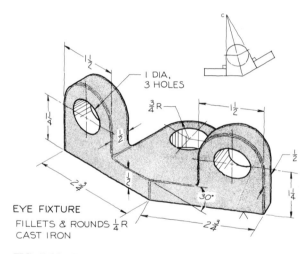

FIG. 7.30 Problem 23: Eye fixture.

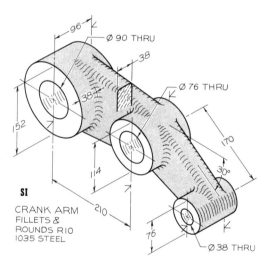

FIG. 7.33 Problem 26: Crank arm.

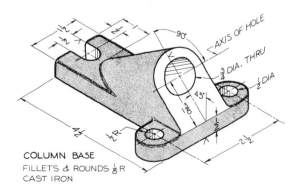

FIG. 7.31 Problem 24: Column base.

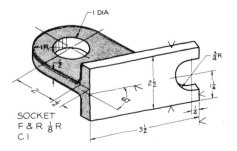

FIG. 7.34 Problem 27: Socket.

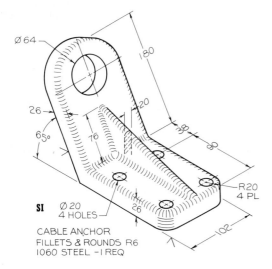

Ø 64

180

26

65°

76

20

38

90

SI Ø 20
4 HOLES

26

102

R20
4 PL

CABLE ANCHOR
FILLETS & ROUNDS R6
1060 STEEL –1 REQ

FIG. 7.35   Problem 28: Cable anchor.

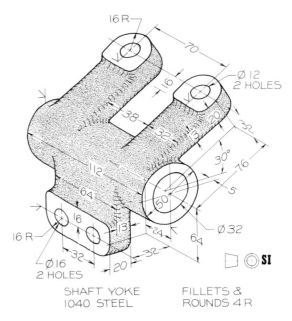

16 R

70

16

Ø 12
2 HOLES

38

32

20

32

30°

76

112

5

64

50

13

24

64

Ø 32

16 R

16

13

Ø 16
2 HOLES

32

20

32

SI

SHAFT YOKE
1040 STEEL

FILLETS &
ROUNDS 4 R

FIG. 7.38   Problem 31: Shaft yoke.

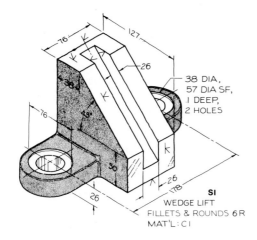

76

127

26

38 DIA,
57 DIA SF,
1 DEEP,
2 HOLES

38

76

45°

30

26

178

26

SI
WEDGE LIFT
FILLETS & ROUNDS 6R
MAT'L : C I

FIG. 7.36   Problem 29: Wedge lift.

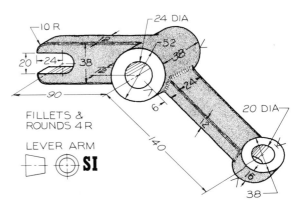

10 R

24 DIA

52

20

24

38

38

24

90

6

12

140

20 DIA

16

38

FILLETS &
ROUNDS 4 R

LEVER ARM

SI

FIG. 7.39   Problem 32: Lever arm.

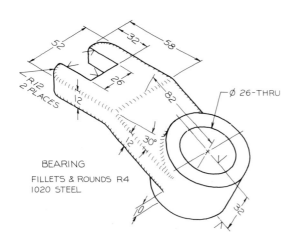

52

32

58

26

R12
2 PLACES

12

82

Ø 26-THRU

12

30°

10

32

BEARING
FILLETS & ROUNDS R4
1020 STEEL

FIG. 7.37   Problem 30: Bearing.

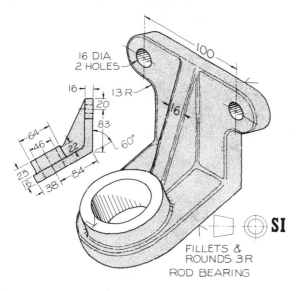

16 DIA
2 HOLES

100

16

13 R

16

16

20

83

64

46

22

60°

25

6

38

84

FILLETS &
ROUNDS 3 R
ROD BEARING

SI

FIG. 7.40   Problem 33: Rod bearing.

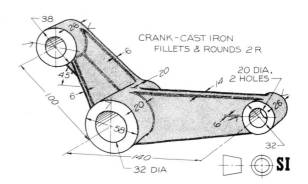

38

26

CRANK-CAST IRON
FILLETS & ROUNDS 2 R

6

20

20 DIA,
2 HOLES

45°

100

6

14

20

26

58

6

32

140

32 DIA

SI

FIG. 7.43   Problem 36: Crank.

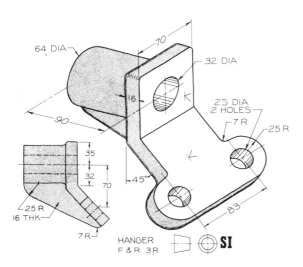

64 DIA

70

32 DIA

16

90

25 DIA
2 HOLES

7 R

25 R

35

32

70

45°

25 R

16 THK

7 R

83

HANGER
F & R 3 R

SI

FIG. 7.41   Problem 34: Hanger.

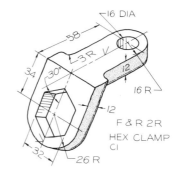

16 DIA

58

3 R

34

30°

12

16 R

12

32

26 R

F & R 2R
HEX CLAMP
CI

FIG. 7.44   Problem 37: Hexagon angle.

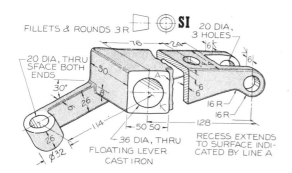

FILLETS & ROUNDS 3 R

SI

20 DIA,
3 HOLES

76

24

6

6

20 DIA, THRU
SFACE BOTH
ENDS

50

A

6

30°

6

26

6

16 R

114

16 R

50 SQ

128

36 DIA, THRU
FLOATING LEVER
CAST IRON

32

26

RECESS EXTENDS
TO SURFACE INDI-
CATED BY LINE A

FIG. 7.42   Problem 35: Floating lever.

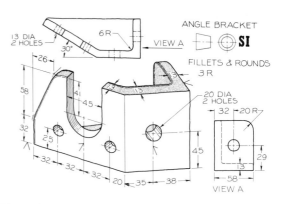

ANGLE BRACKET

13 DIA
2 HOLES

6R

VIEW A

SI

30°

FILLETS & ROUNDS
3 R

26

58

41

13

13

45

20 DIA
2 HOLES

32

32

25

20

35

38

45

32

32

32

32 20 R

29

13

58

VIEW A

FIG. 7.45   Problem 38: Angle bracket.

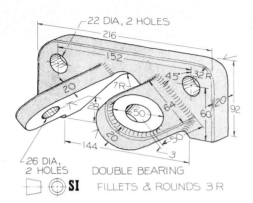

FIG. 7.46 Problem 39: Double bearing.

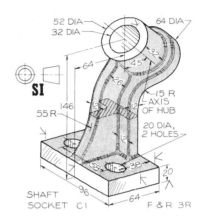

FIG. 7.48 Problem 41: Shaft socket.

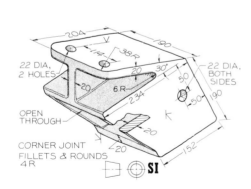

FIG. 7.47 Problem 40: Corner joint.

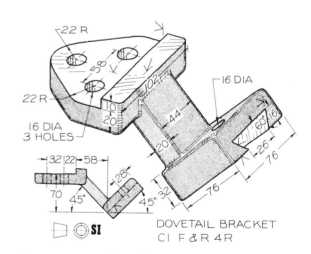

FIG. 7.49 Problem 42: Dovetail bracket.

43–44. (Fig. 7.50 and Fig. 7.51) Lay out these orthographic views on Size B sheets and complete the auxiliary and primary views.

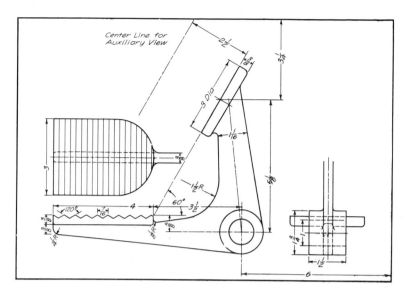

FIG. 7.50 Problem 43: Clutch pedal.

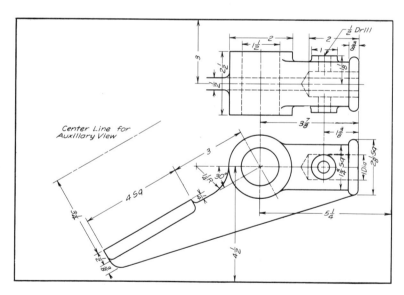

FIG. 7.51 Problem 44: Adjustment pedal.

45–47. (Fig. 7.52–Fig. 7.54) Construct orthographic views of the given objects and draw auxiliary views that give the true-size views of the inclined surfaces using secondary auxiliary views. Draw one per Size B. Sheet.

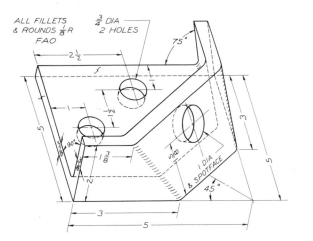

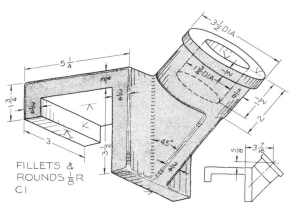

FIG. 7.52 Problem 45: Corner connector.

FIG. 7.53 Problem 46: Shaft bearing.

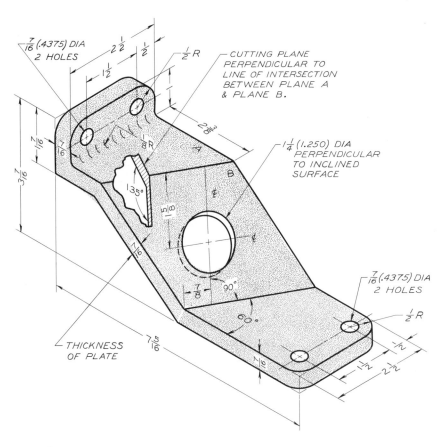

FIG. 7.54 Problem 47: Oblique support.

CHAPTER 8

Sections

8.1

Sections

The standard orthographic views that show all hidden lines may not effectively reveal the true details of an object. This shortcoming can often be improved by using a technique of cutting away part of the object and looking at the cross sectional view. Such a cutaway view is called a **section.**

A section is shown pictorially in Fig. 8.1A, where an imaginary cutting plane is passed through the object in order to show its internal features. The front view has been converted to a **full section,** at C, and the cut portion is cross-hatched or section-lined. Hidden lines have been omitted since they are not needed. The cutting plane is drawn in the top view as a heavy line with short dashes at intervals; this can be thought of as a knife-edge cutting through the object.

By referring to the top view and the front sectional view, you have no hidden lines to interpret, and

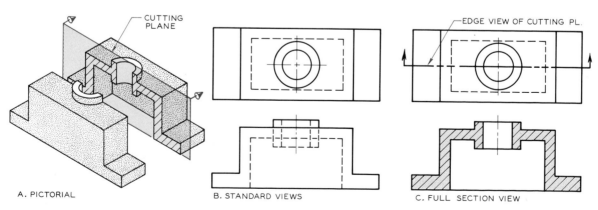

A. PICTORIAL   B. STANDARD VIEWS   C. FULL SECTION VIEW

FIG. 8.1 A comparison of a regular orthographic view with a full-section view of the same object, showing the internal features as well as the external features.

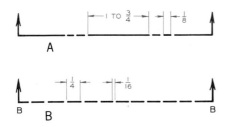

FIG. 8.2 Typical cutting plane lines used to represent sections. The cutting planes marked B-B will produce a section that will be labeled Section B-B.

you can understand the cross-sectional view of the object more easily.

Two types of cutting planes are shown in Fig. 8.2. Either is acceptable, although the upper example is more commonly used. The spacing of the dashes depends upon the size of the drawing. The weight of the cutting plane is the same as that of a visible object line. Letters can be placed at each end of the cutting plane to label the sectional view, such as Section B-B, wherever it is drawn.

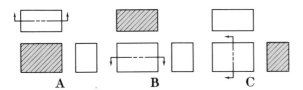

FIG. 8.3 The three standard positions of cutting planes that pass through given views to result in sectional views in the front, top, and side views. The arrows point in the direction of the line of sight for each section.

The three basic views that may appear as sections are shown in Fig. 8.3 with their respective cutting planes. Each cutting plane has perpendicular arrows with the ends pointing in the direction of the line of sight for the section. For example, the cutting plane at A passes through the top view, the front of the top view is removed, and the line of sight is toward the remaining portion of the top view. The top view will appear as a section when the cutting plane passes through the front view and the line of sight is downward (B). When the cutting plane passes through the front view (C), the right-side view will be a section.

8.2
Sectioning Symbols

The symbols that are used to distinguish between different materials in a section are shown in Fig. 8.4. Although the symbols alone can be used to indicate the materials within a section, it is good practice to provide supplementary notes specifying the materials.

> The **cast-iron** symbol of evenly spaced section lines can be used to represent **any material** and is the sectioning symbol used most frequently.

Cast-iron symbols are usually drawn with a 2H pencil with lines that are slanted at a 45° angle, or any other standard angle such as 30° or 60°, and spaced about $\frac{1}{16}$ in. apart by eye.

The proper spacing of section lines for cast iron is shown at the top left of Fig. 8.5, where the lines are evenly spaced. Common errors of section-lining are shown in the remainder of the figure.

Extremely thin parts such as sheet metal, washers, or gaskets (Fig. 8.6) are sectioned by blacking in the areas completely rather than using section lines. Large parts are sectioned with an **outline section** to save time and effort. Section lines are drawn closer together in small parts than in larger parts.

> Sectioned areas should be section-lined with line symbols that are neither parallel nor perpendicular to the outlines of the parts (Fig. 8.7).

Parallel and perpendicular section lines may be confused as serrations or other machining treatments of the surface.

8.3
Sectioning assemblies

When an assembly of several parts is sectioned to give the relationship of the parts, section lines should be drawn at varying angles to distinguish the parts (Fig.

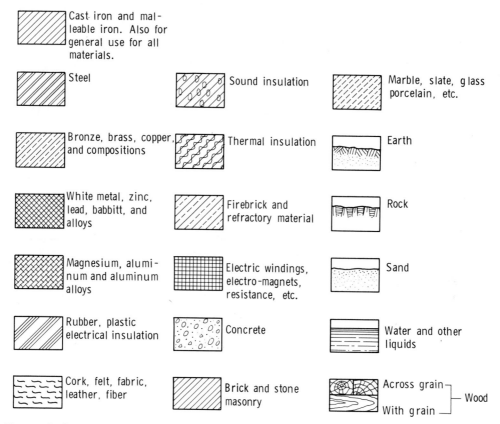

Cast iron and malleable iron. Also for general use for all materials.

Steel

Sound insulation

Marble, slate, glass porcelain, etc.

Bronze, brass, copper, and compositions

Thermal insulation

Earth

White metal, zinc, lead, babbitt, and alloys

Firebrick and refractory material

Rock

Magnesium, aluminum and aluminum alloys

Electric windings, electro-magnets, resistance, etc.

Sand

Rubber, plastic electrical insulation

Concrete

Water and other liquids

Cork, felt, fabric, leather, fiber

Brick and stone masonry

Across grain — Wood
With grain

FIG. 8.4 The symbols used for section-lining parts in a section. Note that the cast-iron symbol can be used for all materials.

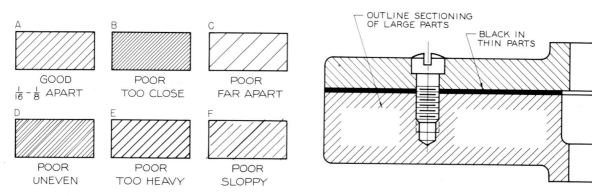

A GOOD $\frac{1}{16} - \frac{1}{8}$ APART

B POOR TOO CLOSE

C POOR FAR APART

D POOR UNEVEN

E POOR TOO HEAVY

F POOR SLOPPY

FIG. 8.5 Good section lines are drawn as thin lines and are spaced about 1/16 in. to 1/8 in. apart. Some typical section-lining errors are shown here.

OUTLINE SECTIONING OF LARGE PARTS

BLACK IN THIN PARTS

FIG. 8.6 To save time and effort, thin parts are blacked in and large areas are section-lined around their outlines.

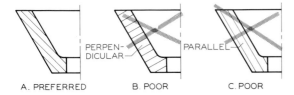

FIG. 8.7 Section lines should be drawn so that they are not parallel or perpendicular to the outlines of a part.

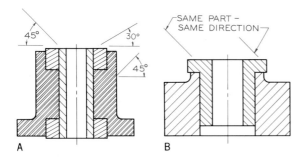

FIG. 8.8 The section lines of the same part should be drawn in the same direction. Section lines of different parts should be drawn at varying angles to separate the parts.

8.8). The use of different material symbols, when the assembly is composed of parts made from different materials, is helpful in distinguishing the parts from one another. The same part should be cross-hatched at the same angle and with the same symbol even though the part may be separated into different areas, as shown at Fig. 8.8B.

An assembly is illustrated in Fig. 8.9 with section lines effectively used to identify the parts of the assembly.

8.4
Full sections

A **full section** is a sectional view formed by passing a cutting plane completely through an object and removing half of it to give a view of its internal features.

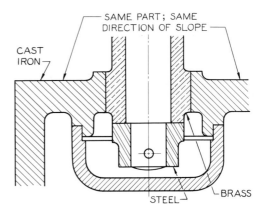

FIG. 8.9 A typical assembly in section with well-defined parts and correctly drawn section lines.

The object shown at Fig. 8.10 is drawn as two orthographic views in which hidden lines are shown. The front view can be drawn as a full section by passing a cutting plane fully through the top view and removing the front portion. The cutting plane is drawn

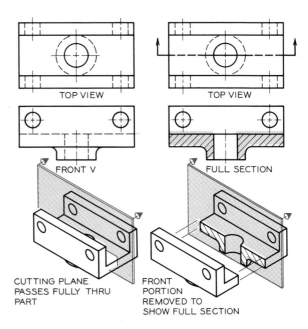

FIG. 8.10 A full section is a section formed by a cutting plane that passes completely through it. The cutting plane is shown passing through the top view, and the direction of sight is indicated by the arrows at each end. The front view is converted to a sectional view to give a clear understanding of the internal features.

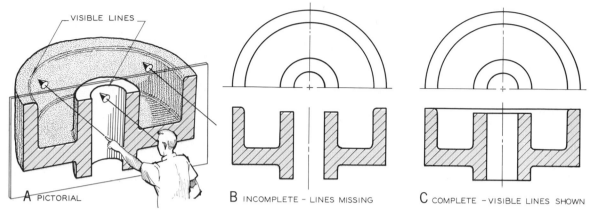

FIG. 8.11 Full section—cylindrical part

A. When a full section is passed through an object, you will see lines behind the sectioned area.

B. If only the sectioned area were shown, the view would be incomplete.

C. Visible lines behind the sectioned area must be shown also.

through the top view with arrows indicating the direction of sight. The front view is then section-lined to give the full section.

A full section through a cylindrical part is shown in Fig. 8.11, where half of the object is removed. A common mistake in constructing sectional views is the omission of the visible lines behind the cutting plane (B); C shows the correctly drawn sectional view. Hidden lines are omitted in all sectional views unless they are necessary to provide a clear understanding of the

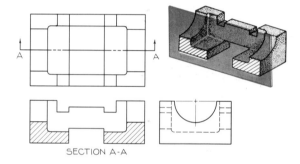

SECTION A-A

FIG. 8.13 A full section, Section A-A, is used to supplement the given views of the object.

view. Visible lines behind the sectional view also are omitted if they confuse the view.

Figure 8.12 is an example of a part whose right-side view is shown as a full section. Likewise, the part in Fig. 8.13 illustrates a front view that appears as a full section. Note that the lines behind the cutting plane are shown as visible lines.

8.5
Parts not section-lined

Many standard parts are not section-lined even though the cutting plane passes through them. Examples of such parts in Fig. 8.14 are nuts and bolts, riv-

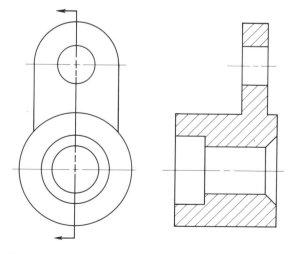

FIG. 8.12 A full section with the cutting plane shown

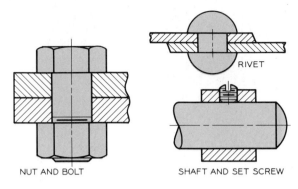

NUT AND BOLT

RIVET

SHAFT AND SET SCREW

FIG. 8.14 These parts are not cross-sectioned even though the cutting plane passes through them.

ets, shafts, and set screws. These parts have no internal features, and sections through them would be of no value.

Other parts not section-lined are roller bearings, ball bearings, gear teeth, shafts, dowels, pins, set screws, and washers (Fig. 8.15).

8.6
Ribs in section

Ribs are not section-lined when the cutting plane passes flatwise through them as shown in Fig. 8.16A, since this would give a misleading impression of the

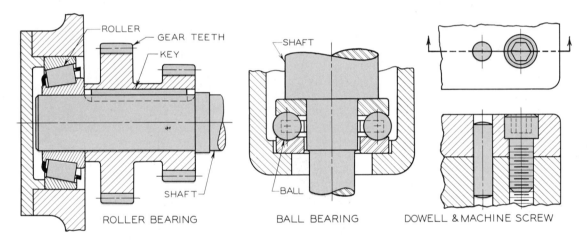

ROLLER
GEAR TEETH
KEY

SHAFT

ROLLER BEARING

SHAFT

BALL

BALL BEARING

DOWELL & MACHINE SCREW

FIG. 8.15 These parts are not section-lined even though cutting planes pass through them.

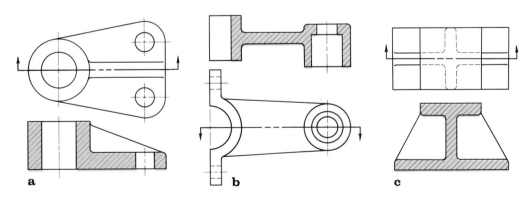

a

b

c

FIG. 8.16 A rib that is cut in a flatwise direction by a cutting plane is not section-lined. Ribs are section-lined when cutting planes pass perpendicularly through them, as shown at (b) and (c).

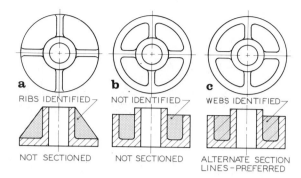

FIG. 8.17 Outside ribs in a section are not section-lined. Webs that are poorly identified, as at (b), should be identified by cross-hatching them with alternate section lines, as shown at (c). This calls attention to the fact that webs are present.

rib. A similar example is the section shown in Fig. 8.16C. However, a rib is section-lined when the cutting plane passes through it showing its true thickness (Fig. 8.16B).

An alternate method of section-lining webs and ribs is shown in Fig. 8.17. At (a) the ribs are not section-lined since the cutting plane passes through them in a flatwise direction. The webs at (b) are symmetrically spaced about the hub. Webs are not cross-hatched, but this would leave the webs unidentified. Therefore it is better to use the **alternate sectioning** technique and extend every other section line through the webs to call attention to them, as shown at (c).

The ribs in Fig. 8.18 are not section-lined, and thus afford a more descriptive view of the part. If the

ribs had been section-lined, the section would have given the impression that the part was solid and conical in shape (b). Note that the top views are partial views and the portion nearest the sectional view is the portion that has been omitted.

8.7
Half-sections

A **half-section** is a view that results from passing a cutting plane halfway through an object, removing a quarter of it.

Half-sections are most often used with symmetrical parts, cylinders in particular. A cylindrical part at Fig. 8.19A is shown as a pictorial half-section at B. The method of drawing the orthographic half-section is shown at C. In this view, both the internal and external features can be seen; the hidden lines are omitted in the sectional view.

A similar half-section is given in Fig. 8.20. Hidden lines are omitted unless they are essential to understand or dimension the object. The half-section in Fig. 8.21 has been drawn without showing the cutting plane. This is permissible if it is obvious where the cutting plane was passed through the view to give the section.

8.8
Partial views

A conventional method of representing symmetrical views is shown in Fig. 8.22. A half-view is sufficient when it is drawn adjacent to the sectional view, as at A. In full sections, the removed half is the portion nearest the section (Part A). When drawing half-views that are associated with views (nonsectional views), the removed half of the partial view will be the half that is away from the adjacent view (Fig. 8.22B).

When the partial views are drawn in conjunction with half-sections, either the near or far halves of the partial views can be omitted, as shown in Fig. 8.22C and D. Partial views are faster to draw and require less space.

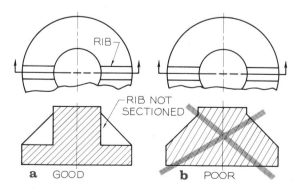

FIG. 8.18 It can be seen that, when ribs are not section-lined, the view is more descriptive of the part. The front part of the partial top view is removed when the front view is a section.

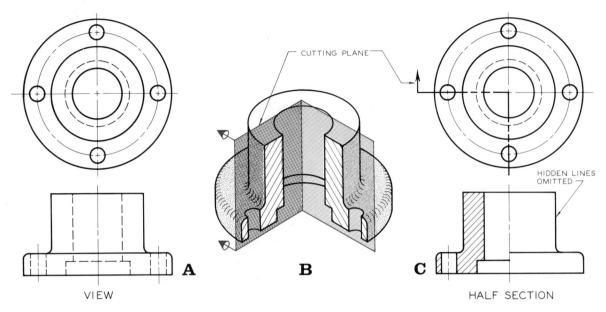

FIG. 8.19 The cutting plane of a half-section passes halfway through the object, which results in a sectional view that shows half of the outside and half of the inside of the object. Hidden lines are omitted unless they are necessary to clarify the view.

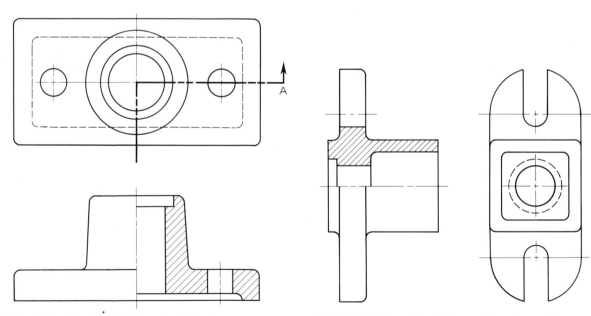

FIG. 8.20 A typical half-section drawing

FIG. 8.21 When it is obvious where the cutting plane is located, it is unnecessary to show it, as in this example.

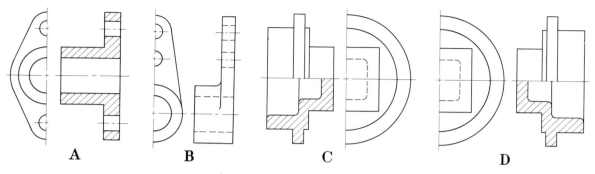

FIG. 8.22 Half-views can be used for symmetrical parts to conserve space and time. A. The omitted portion of the view is toward the full section. B. The omitted portion of the view is away from the adjacent view when it is not sectioned. C and D. The omitted half can be toward or away from the section in the case of a half-section.

8.9
Offset sections

An **offset section** is a type of full section in which the cutting plane is offset to pass through important features that would be missed by the usual full section formed by a flat plane.

An offset section is shown pictorially in Fig. 8.23, with the plane offset to pass through the large hole and one of the small holes. The cut formed by the offset is not shown in the section since this is an imaginary cut.

An offset section is used to describe a part in Fig. 8.24. As previously discussed, the ribs are not section-lined.

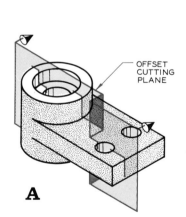

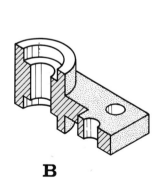

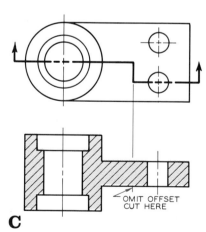

FIG. 8.23 Offset section

Step 1   An offset section may be necessary to show all the typical features of a sectioned part.

Step 2   When the front portion is removed, the internal features can be seen. Note that the cutting plane has been offset.

Step 3   In an offset section, the offset cut is not shown. The section is shown as if it were a typical full section.

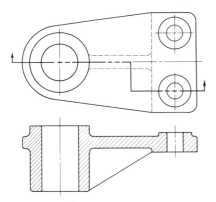

FIG. 8.24 An offset section

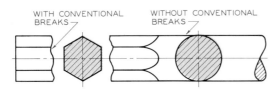

WITH CONVENTIONAL BREAKS

WITHOUT CONVENTIONAL BREAKS

FIG. 8.25 Examples of revolved sections with and without conventional breaks

8.10
Revolved sections

> A **revolved section** is used to describe a cross-section of a part, to eliminate the need for drawing an entirely separate view.

An example is given in Fig. 8.25, where revolved sections are used to indicate cross-sections of the parts. One of the revolved sections is positioned within the view, and conventional breaks are drawn on each side of the hexagonal section. The circular cross-section is drawn superimposed on the cylindrical portion of the part without using conventional breaks. Either of these methods can be used when drawing revolved sections.

A more advanced type of revolved section is illustrated in Fig. 8.26, where a cutting plane is passed through the object (Step 1). The plane is imagined to be revolved in the top view to give a true-size revolved section in the front view (Step 2). Note that the object lines do not pass through the revolved section in the front view. It would have been permissible to use con-

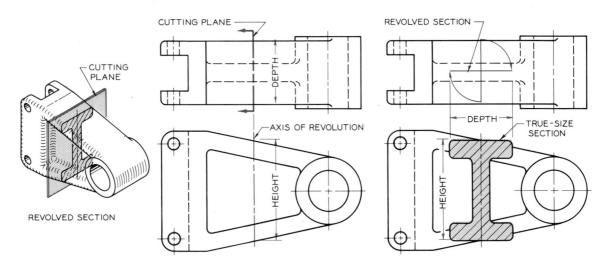

FIG. 8.26 Revolved section

Step 1   An axis of revolution is shown in the front view. The cutting plane would appear as an edge in the top view if it were shown.

Step 2   The vertical section in the top view is revolved so that the section can be seen true size in the front view. Object lines are not drawn through the revolved section.

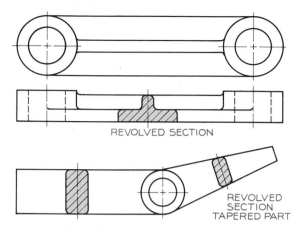

REVOLVED SECTION

REVOLVED SECTION TAPERED PART

FIG. 8.27 The revolved sections given here are helpful in describing the cross-sectional characteristics of the two parts. This is much more effective than using additional orthographic views.

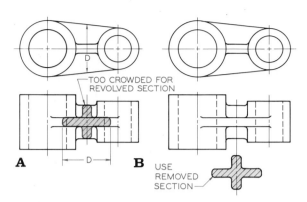

TOO CROWDED FOR REVOLVED SECTION

A D **B** USE REMOVED SECTION

FIG. 8.29 Removed sections can be used to good advantage where space does not permit the use of a revolved section.

ventional breaks on each side of the revolved section, as were used in Fig. 8.25.

Typical revolved sections are shown in Fig. 8.27 at A and B. These sections provide a direct method of giving a part's cross-section without relying upon another complete orthographic view.

8.11
Removed sections

A **removed section** is a revolved section that has been removed from the view where it was revolved, as shown in Fig. 8.28.

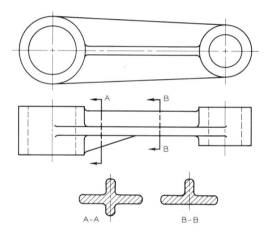

A-A B-B

FIG. 8.30 Sections can be lettered at each end of a cutting plane such as A-A. This removed section can then be shown elsewhere on the drawing, where it is designated as Section A-A.

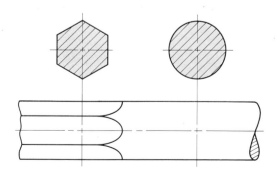

FIG. 8.28 Removed sections are similar to revolved sections, but they have been removed outside the object along an axis of revolution.

Centerlines are used as axes of rotation to show from where the sections were taken. Removed sections may be necessary where room does not permit revolution on the given view (Fig. 29A); instead, the cross-section must be removed from the view, as shown at B.

Removed sections do not have to be positioned directly along an axis of revolution adjacent to the view from where the sections were taken. Instead, cutting planes can be labeled at each end, as shown in Fig. 8.30, to specify the sections. For example, the plane labeled with an A at each end is used to label Section A-A. Section B-B is found in a similar manner.

CUTTING PLANE

FIG. 8.31 It may be necessary to remove a section to another page in a set of drawings. In this case, each end of the cutting plane can be labeled with a letter and a number. The letters refer to Section A-A and the numbers mean that this section is located on page 7.

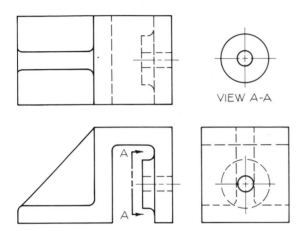

VIEW A-A

FIG. 8.32 A removed view (not a section) can also be used to view a part from an unconventional direction that would be more effective than an entire orthographic view. The plane of sight is located in much the same manner as a cutting plane in a section.

Removed sections may be put on different sheets when a set of drawings is composed of many pages. If this is necessary, a cutting plane may be labeled as shown in Fig. 8.31. The A at each end identifies the section as Section A-A, and the numerals indicate on which page of the set of drawings the section is located.

Removed views can also be used to provide inaccessible orthographic views (nonsectional views) when these are needed, as shown in Fig. 8.32.

8.12
Broken-out sections

A **broken-out section** is a convenient method used to show interior features without drawing a separate view as a section.

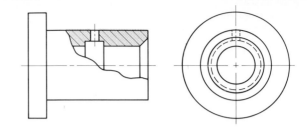

FIG. 8.33 A broken-out section is used to show an internal feature by using a conventional break.

A portion of the part in Fig. 8.33 is broken out to reveal details of the wall thickness to better explain the drawing. This reduces the need for hidden lines, which may then be omitted if desired.

Figure 8.34 is a similar example of a broken-out section where hidden lines are given. The irregular lines that are used to represent the break are called **conventional breaks;** these are discussed in Section 8.14.

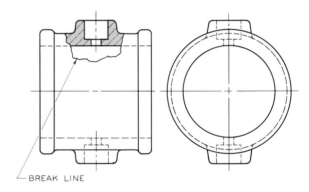

BREAK LINE

FIG. 8.34 A broken-out section that shows a section through the boss at the top of a collar.

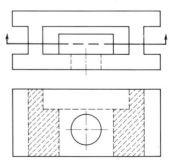

FIG. 8.35 Phantom sections give an "x-ray" view of an object. The section lines are shown as dashed lines, making it possible to show the section without removing the hole in the front of the part.

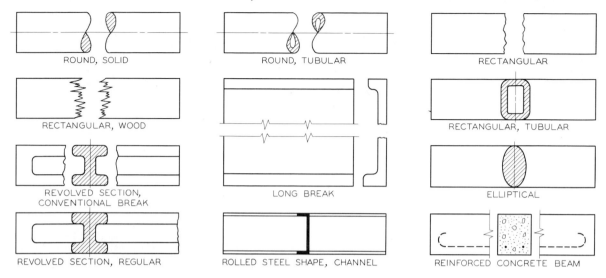

FIG. 8.36 These are often-used conventional breaks that are used to remove a portion of an object so it can be drawn at a larger scale.

8.13
Phantom (ghost) sections

> A **phantom section** or **ghost section** is used occasionally to depict parts as if they were viewed by an x-ray.

An example of a phantom section is shown in Fig. 8.35. The cutting plane is drawn in the usual manner, but the section lines are drawn as dashed lines. You can see that if the object had been shown as regular full section, the circular hole through the front surface could not have been shown in the same view.

8.14
Conventional breaks

Some of the previous examples have used conventional breaks to indicate that parts of an object have been removed. Examples of conventional breaks are illustrated in Fig. 8.36. The ''figure 8'' breaks that are used for cylindrical and tubular parts can be drawn freehand (Fig. 8.37), or when they are drawn to a

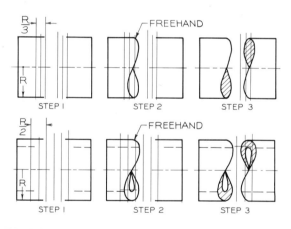

FIG. 8.37 Conventional breaks in cylindrical and tubular sections can be drawn freehand with the aid of the guidelines shown. The radius, R, is used to establish the width of both ''figure 8's.''

large scale they can be drawn by using a compass, as shown in Fig. 8.38.

One use of conventional breaks is to shorten a long piece that has a uniform cross-section. The long part in Fig. 8.39 has been shortened and drawn at a larger scale for more clarity by using the conventional breaks shown at (b). The dimension specifies the true length of the part, and the breaks indicate that a portion of the length has been removed.

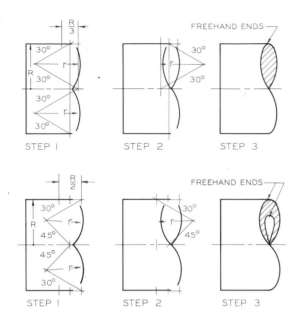

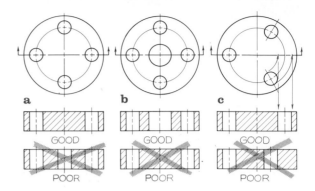

STEP 1   STEP 2   STEP 3

FIG. 8.38 The steps of constructing conventional breaks of solid and tubular shapes with instruments.

FIG. 8.40 Symmetrically spaced holes in a circular plate should be revolved in their sectional views to show them at their true radial distance from the center. At (a), no hole is shown at the center, but a hole is shown at the center in (b), since the hole is through the center of the plate. At (c), one of the holes is rotated to the cutting plane so that the sectional view will be symmetrical and more descriptive.

However, the hole at (b) does pass through the plate's center and is sectioned accordingly. At (c), the cutting plane does not pass through one of the symmetrically spaced holes in the top view, but the hole is revolved to the cutting plane in order to show the recommended full section.

When ribs are symmetrically spaced about a hub (Fig. 8.41) it is conventional practice to revolve them

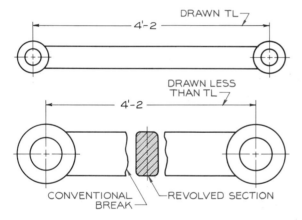

FIG. 8.39 By using conventional breaks and a revolved section, this part can be drawn on a larger scale that is easier to read.

8.15
Conventional revolutions

Three conventional sections are shown in Fig. 8.40. The center hole is omitted at (a) since it does not really pass through the center of the circular plate.

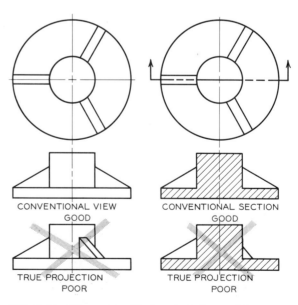

FIG. 8.41 Symmetrically located ribs are shown revolved in both orthographic and sectional views as a conventional practice.

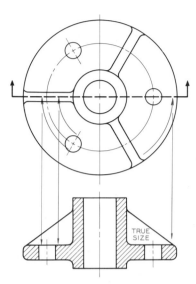

FIG. 8.42 A part with symmetrically located ribs and holes is shown in section with both ribs and holes rotated to the cutting plane.

plane can be used to show the path of the cutting plane in more complex parts.

Symmetrically spaced spokes are rotated and not section-lined, in the same manner as ribs in section. Figure 8.44 illustrates the preferred and poor-practice methods of representing spokes. Only the revolved, true-size spokes are drawn; the intermediate spokes are omitted.

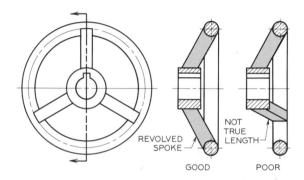

FIG. 8.44 Symmetrically positioned spokes are revolved to show the spokes true size in section. Spokes are not section-lined in section.

to where they will appear true size in either a view (B) or a section (D). A full section that shows both ribs and holes revolved to their true-size locations can be seen in Fig. 8.42.

The cutting plane can be positioned as shown in Fig. 8.43. Even though the cutting plane does not pass through the ribs and holes in (a), the sectional view should be drawn as at (b), where the cutting plane is shown revolved through the hole. The cutting plane can be drawn in either position. The revolved cutting

The reason for not section-lining spokes can be seen in Fig. 8.45. If the spokes at B had been section-lined, the cross section of the part would be confused with the part at A where there are no spokes, but a continuous web.

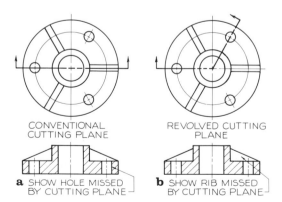

FIG. 8.43 Symmetrically located ribs are shown in section in revolved positions to show the ribs true size. Foreshortened ribs are omitted.

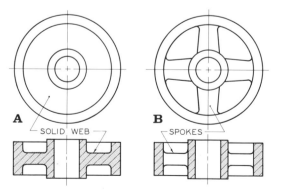

FIG. 8.45 Solid webs in sections of the type at A are section-lined. Spokes are not section-lined in B when the cutting plane passes through them in this manner.

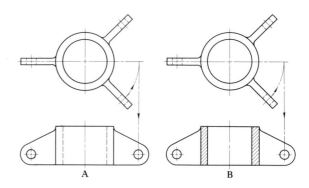

FIG. 8.46 Symmetrically spaced lugs (flanges) are revolved to show the front view and the sectional view as symmetrical.

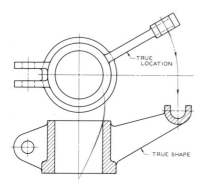

FIG. 8.47 A part with an oblique feature attached to the circular hub is revolved so it will appear true shape in the front view, which is a sectional view.

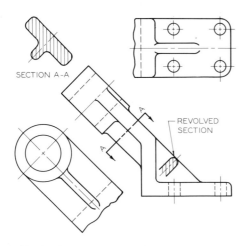

FIG. 8.48 Sectional views can be shown as auxiliary views for added clarity.

Lugs that are symmetrically positioned about the central hub of the object in Fig. 8.46 are revolved to show them true size in both views and in sections. A more complex object that involves the same principle of rotation can be seen in Fig. 8.47. The oblique arm is drawn in the section as if it had been revolved to the center line in the top view and then projected to the sectional view.

8.16
Auxiliary sections

Auxiliary sections can be used to supplement the principal views used in orthographic projections, as shown in Fig. 8.48. Auxiliary cutting plane A-A is passed through the front view, and the auxiliary view is projected from the cutting plane as indicated by the sight arrows. The sectional view, A-A, gives the cross-sectional description of the part.

In Fig. 8.49, Section A-A is drawn to clarify the front view where the cutting plane is positioned. The right-side view can be drawn as a partial view.

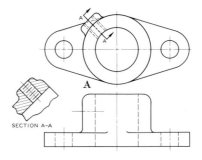

FIG. 8.49 An auxiliary section projected from the front view. The right-side view is a partial view since it is supplemented by the auxiliary section.

Problems

These problems can be solved on Size A or Size B sheets. Refer to Section 2.19 for typical layouts.

1–7. (Fig. 8.50) Full sections: Draw two of these problems per Size AV sheet. Each grid is equal to ¼ in. or 5 mm. Complete the front views as full sections.

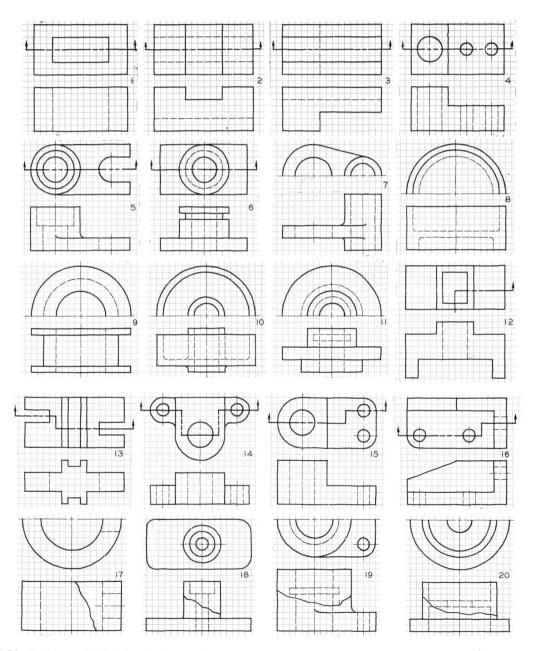

FIG. 8.50 Problems 1–20: Introductory sections.

8–12. (Fig. 8.50) Half sections: Draw two of these problems per Size AV sheet. Each grid is equal to ¼ in. or 5 mm. Complete the front views as half-sections.

13–16. (Fig. 8.50) Offset sections: Draw two of these problems per Size AV sheet. Each grid is equal to ¼ in. or 5 mm. Complete the front views as offset sections.

17–20. (Fig. 8.50) Broken-out sections: Draw two of these problems per Size AV sheet. Each grid is equal to ¼ in. or 5 mm. Complete the broken-out sections.

21–30. (Fig. 8.51, Fig. 8.52, Fig. 8.53) Half-sections: The views given are rectangular views of cylindri-

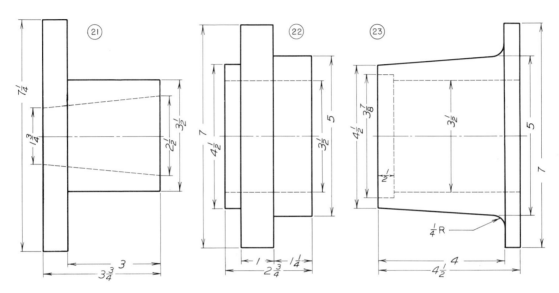

FIG. 8.51 Problems 21–23: Half-sections.

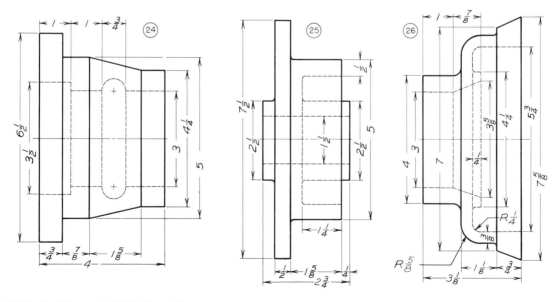

FIG. 8.52 Problems 24–26: Half-sections.

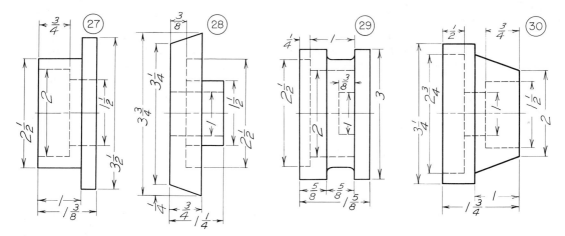

FIG. 8.53   Problems 27–30: Half-sections.

cal parts. Draw one problem per Size AV sheet. Show the circular view and draw the rectangular view as a half-section. Omit dimensions and show the cutting plane. Select the appropriate scale.

31–32. (Fig. 8.54) Full and offset sections: Draw the given views on Size AV sheets, two per sheet.

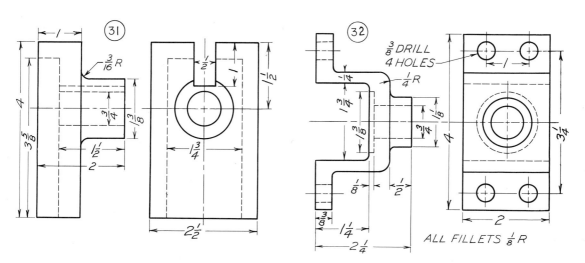

FIG. 8.54   Problems 31–32: Offset sections.

33–38. (Fig. 8.55–Fig. 8.58) Sections and conventions: Draw each problem on a Size B sheet. Complete the sectional views as indicated by the cutting planes or as good practice demands. Omit the dimensions.

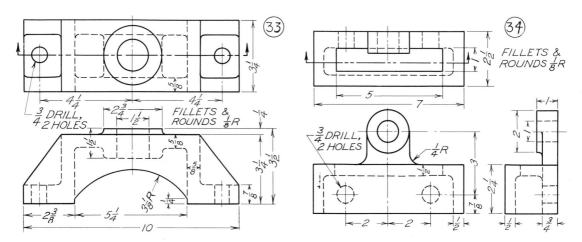

FIG. 8.55 Problems 33–34: Full sections.

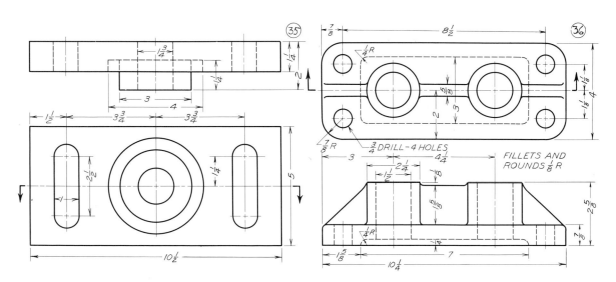

FIG. 8.56 Problems 35–36: Full sections.

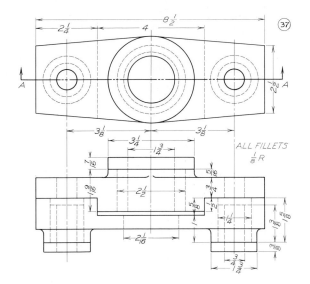

FIG. 8.57 Problem 37: Full section.

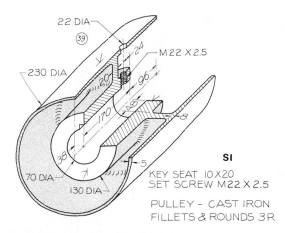

SI

KEY SEAT 10 X 20
SET SCREW M22 X 2.5

PULLEY – CAST IRON
FILLETS & ROUNDS 3 R

FIG. 8.59 Problem 39: Sections.

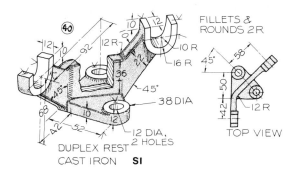

FILLETS &
ROUNDS 2 R

TOP VIEW

DUPLEX REST
CAST IRON   SI

FIG. 8.60 Problem 40: Sections.

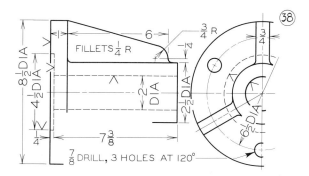

FIG. 8.58 Problem 38: Section and conventions.
(Show a full section through this part and show the
entire view.)

39–41. (Fig. 8.59–Fig. 8.61) Sections: Draw the necessary views to describe the parts using sections
and conventional practices. Draw one problem
per Size B sheet. Omit dimensions.

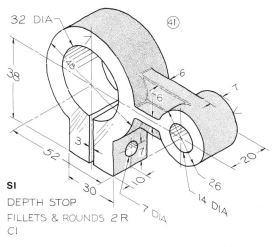

SI

DEPTH STOP
FILLETS & ROUNDS 2 R
CI

FIG 8.61 Problem 41: Sections.

CHAPTER

9

Screws, Fasteners, and Springs

9.1
Threaded fasteners

Screw threads provide fast and easy method of fastening two parts together and of exerting a force that can be used to adjust movable parts. For a screw thread to function, there must be two parts—an internal thread and an external thread. The internal threads may be tapped inside a part such as a motor block or, more commonly, they may be tapped inside a nut. The nuts and bolts used in industrial projects should be stock parts that can be obtained from many sources to manufacture expenses and improve the interchangeability of parts.

Threaded fasteners made in different countries or by different manufacturers may have threads of different specifications that will not match. Progress has been made toward establishing standards that will unify threads both in this country and abroad by the introduction of metric standards. Other efforts have led to the adoption of the **Unified Screw Thread** by the United States, Britain, and Canada (ABC Standards), which is a modification of both the American Standard thread and the Whitworth thread.

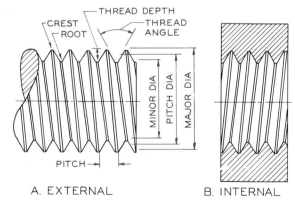

FIG. 9.1 Thread terminology

9.2
Definitions of thread terminology

Succeeding sections will discuss the uses and methods of representing screw threads. The terms defined below are illustrated in Fig. 9.1 and Fig. 9.2.

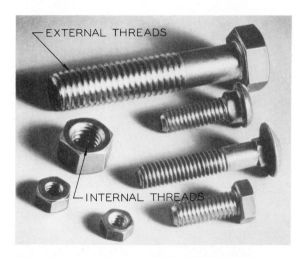

← EXTERNAL THREADS

← INTERNAL THREADS

FIG. 9.2 Examples of external threads (bolts) and internal threads (nuts). (Courtesy of Russell, Burdsall & Ward Bolt and Nut Co.)

External Thread A thread on the outside of a cylinder such as a bolt (Fig. 9.2).

Internal Thread A thread cut on the inside of a part such as a nut (Fig. 9.2).

Major Diameter The largest diameter on an internal or external thread.

Minor Diameter The smallest diameter that can be measured on a screw thread.

Pitch Diameter The diameter of an imaginary cylinder passing through the threads at the points at which the thread width is equal to the space between the threads.

Lead The distance that a screw will advance when turned 360°.

Pitch The distance between crests of threads. Pitch is found mathematically by dividing one inch by the number of threads per inch of a particular thread.

Crest The peak edge of a screw thread.

Thread Angle The angle between threads cut by the cutting tool.

Root The bottom of the thread cut into a cylinder.

Thread Form The shape of the thread cut into a threaded part.

Thread Series The number of threads per inch for a particular diameter, which results in three series: coarse, fine, and extra fine. Coarse series

provides for rapid assembly, and extra-fine series provides for fine adjustment.

Thread Class The closeness of fit between two mating threaded parts. Class 1 represents a loose fit, and Class 3 a tight fit.

Right-Hand Thread A thread that will assemble when turned clockwise. A right-hand thread will slope downward to the right on an external thread when the axis is horizontal, and in the opposite direction on an internal thread.

Left-Hand Thread A thread that will assemble when turned counterclockwise. A left-hand thread slopes downward to the left on an external thread when the axis is horizontal, and in the opposite direction on an internal thread.

9.3
Thread specifications (English system)

Form

A thread form is the shape of the thread cut into a part, as illustrated in Fig. 9.3. The Unified form, a combination of the American National and the British Whitworth, is the most widely used, since it is a standard in several countries. It is referred to as UN in abbreviations and thread notes. The American National is signified by the letter N.

A new thread form, the UNR, was introduced in the 1974 ANSI standards. This designation is specified only for external threads—there is no UNR designation for internal threads. Figure 9.4 gives a comparison of the profiles of the external UN and UNR threads. The UN form in Part A has a flat root (rounded root is optional), whereas the UNR thread *must* have a rounded root, formed by rolling as shown in Part B. The rounded root of the UNR thread is designed to reduce the wear of the threading tool and to improve the fatigue strength of the thread.

The transmission of power is achieved by the use of the **Acme, square, buttress,** and **worm** threads used in gearing and other pieces of machinery. The **sharp V** is used for set screws and in applications where friction in assembly is desired. The **knuckle** form is a fast-assembling thread used for light assemblies such as light bulbs and bottle caps.

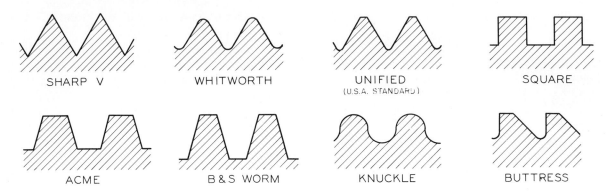

FIG. 9.3 Standard thread forms

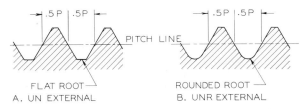

FIG. 9.4 The UN external thread (A) has a flat root (rounded root is optional); the UNR has a rounded root formed by rolling (B). The UNR form does not apply to internal threads.

Series

There are 11 standard series of threads listed under the American National form and the Unified National (UN/UNR) form. There are three series with graded pitches, with abbreviations coarse (C), fine (F), extra-fine (EF), and eight with constant pitches (4, 6, 8, 12, 16, 20, 28, and 32 threads per inch).

A Unified National form for a coarse-series thread is specified as UNC or UNRC, which is a combination of form and series in a single note. Similarly, an American National form for a coarse thread is written NC. The coarse-thread series (UNC/UNRC or NC) is suitable for bolts, screws, nuts, and general use with cast iron, soft metals, or plastics when rapid assembly is desired. The fine thread series (UNF/UNRF or NF) is suitable for bolts, nuts, or screws when a high degree of tightening is required. The extra fine series (UNEF/UNREF or NEF) is used for applications that will have to withstand high stresses, such as sheet metal, thin nuts, ferrules, or couplings when length of engagement is limited.

The 8 thread series (8 N or 8 UN), 12 thread series, (12 N or 12 UN/UNR), and 16 thread series (16 N or 16 UN/UNR) are threads with a uniform pitch for large diameters. The 8 N is used as a substitute for the coarse thread series on diameters larger than 1 in. when a medium pitch thread is required. The 12 UN is used on diameters larger than 1½ in., with a thread of a medium fine pitch as a continuation of the fine thread series. The 16 N series is used on diameters larger than 1¹¹⁄₁₆ in., with threads of a fine pitch as a continuation of the extra-fine series.

Class of fit

Thread classes are used to indicate the tightness of fit between a nut and a bolt, or between any two mating threaded parts. This fit is determined by the tolerances and allowances applied to threads. Classes of fit are indicated by the numbers 1, 2, or 3 followed by the letters A or B. For UN forms, the letter A represents an external thread, while the letter B represents an internal thread. These letters are omitted when the American National form is used.

Classes 1A and 1B are threads used on parts that require assembly with a minimum of binding.

Classes 2A and 2B are general-purpose threads for bolts, nuts, screws, and nominal applications in the mechanical field and are widely used in the mass production industries.

Classes 3A and 3B threads are used in precision assemblies where a close fit is needed to withstand stresses and vibration.

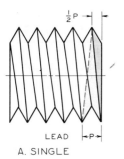
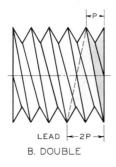
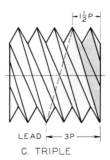

LEAD ├─P─┤   LEAD ├──2P──┤   LEAD ├── 3P ──┤

A. SINGLE   B. DOUBLE   C. TRIPLE

FIG. 9.5 Single and multiple threads

Single and multiple threads

A **single thread** (Fig. 9.5A) is a thread that will advance the distance of its pitch in one full revolution of 360°. In other words, its pitch is equal to its lead. In a drawing of a single thread, the crest line of the thread will slope ½P, since only 180° of the revolution is visible in a single view. A double thread is composed of two threads, resulting in a lead equal to 2P, meaning that the threaded part will advance a distance of 2P in a single revolution of 360° (Fig. 9.5B). The crest line of a double thread will slope a distance equal to P in the view in which 180° can be seen. Similarly, a triple thread will advance 3P in 360° with a crest line slope of 1½P in the view in which 180° of the cylinder is visible (Fig. 9.5C). The lead of a double thread is 2P; that of a triple thread, 3P. Although power on multiple threads is somewhat limited, they are used wherever quick motion is required.

Thread notes

Drawings of threads are only symbolic representations that are inadequate unless thread notes are applied to give the thread specifications (Fig. 9.6). The major diameter is given first, followed by the number of threads per inch, the form and series, the class of fit, and a letter denoting whether the thread is external or internal. This completes the note for a single, right-hand thread. In addition, the word DOUBLE or TRIPLE is included in the note for a multiple thread, and the letters LH are used for a left-hand thread.

The UNR thread note is shown for the external thread in Fig. 9.7A (UNR does not apply to internal threads). When inches are used as the unit of measurement, fractions can be written as decimals or as common fractions. Decimal fractions are preferred, as shown in Part B.

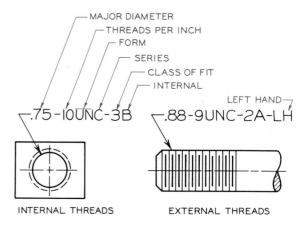

FIG. 9.6 Parts of a thread note for an external thread

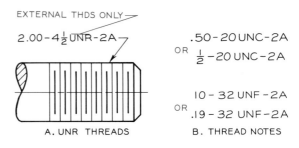

A. UNR THREADS   B. THREAD NOTES

FIG. 9.7 The UNR thread notes apply to external threads only. Notes can be given as decimal fractions or as common fractions, as shown in B.

9.4
Using thread tables

The UN/UNR thread table is given in Appendix 15. A portion of this table is shown in Table 9.1. If an external thread (bolt) with a 1½-in. diameter is to have a

TABLE 9.1
AMERICAN NATIONAL STANDARD UNIFIED INCH SCREW THREADS (UN AND UNR THREAD FORM)*

| Sizes | | Basic Major Diameter | Series with Graded Pitches | | | Threads Per Inch Series with Constant Pitches | | | | | | | | Sizes |
Primary	Second-ary		Coarse UNC	Fine UNF	Extra fine UNEF	4 UN	6 UN	8 UN	12 UN	16 UN	20 UN	28 UN	32 UN	
1		1.0000	8	12	20	—	—	UNC	UNF	16	UNEF	28	32	1
	1 1/16	1.0625	—	—	18	—	—	8	12	16	20	28	—	1 1/16
1 1/8		1.1250	7	12	18	—	—	8	UNF	16	20	28	—	1 1/8
	1 3/16	1.1875	—	—	18	—	—	8	12	16	20	28	—	1 3/16
1 1/4		1.2500	7	12	18	—	—	8	UNF	16	20	28	—	1 1/4
	1 5/16	1.3125	—	—	18	—	—	8	12	16	20	28	—	1 5/16
1 3/8		1.3750	6	12	18	—	UNC	8	UNF	16	20	28	—	1 3/8
	1 7/16	1.4375	—	—	18	—	6	8	12	16	20	28	—	1 7/16
1 1/2		1.5000	6	12	18	—	UNC	8	UNF	16	20	28	—	1 1/2
	1 9/16	1.5625	—	—	18	—	6	8	12	16	20	—	—	1 9/16
1 5/8		1.6250	—	—	18	—	6	8	12	16	20	—	—	1 5/8
	1 11/16	1.6875	—	—	18	—	6	8	12	16	20	—	—	1 11/16
1 3/4		1.7500	5	—	—	—	6	8	12	16	20	—	—	1 3/4
	1 13/16	1.8125	—	—	—	—	6	8	12	16	20	—	—	1 13/16
1 7/8		1.8750	—	—	—	—	6	8	12	16	20	—	—	1 7/8
	1 15/16	1.9375	—	—	—	—	6	8	12	16	20	—	—	1 15/16

*By using this table, a diameter of 1 1/2 inches that is to be threaded with a fine thread would have the following thread note: 1 1/2–12 UNF-2A. (Courtesy of ANSI; B1.1–1974.)

"fine" thread, it will have 12 threads per inch. Therefore, the thread note can be written

1 1/2-12 UNF-2A   or   **1.500-12 UNF-2A.**

If the thread had been an internal one, the thread note would have been the same, but the letter B would have been used instead of the letter A.

A constant-pitch thread series can be selected for large diameters. The constant-pitch thread notes are written with the C, F, or EF omitted. For example, a 1 3/4-in. diameter bolt with a fine thread could be noted in constant-pitch series as

1 3/4-12 UN-2A   or   **1.750-12 UN-2A.**

This table can also be used for the UNR thread form (for external threads only) by substituting UNR in place of UN; for example, UNEF can be written as UNREF.

9.5
Metric thread specifications (ISO)

Metric thread specifications are recommended by the ISO (**International Organization** for **Standardization**). Thread specifications can be given using a **basic designation** that is suitable for general applications, or the **complete designation** can be used where detailed specifications are needed.

Basic designation

Examples of metric screw thread notes are shown in Fig. 9.8. Each note begins with the letter M, which designates the note as a metric note, followed by the diameter in millimeters and the pitch in millimeters separated by the "×" sign. The pitch can be omitted

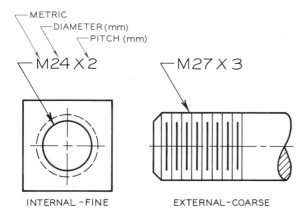

FIG. 9.8 Basic designations for metric threads

but in addition it has a tolerance class designation separated by a dash. The 5g represents the pitch diameter tolerance and 6g represents the crest diameter tolerance.

The numbers 5 and 6 are **tolerance grades.** Grade 6 is commonly used for a medium, general-purpose thread that is nearly equal to the 2A and 2B classes of fit specified under the Unified system. Grades with numbers smaller than 6 are used for "fine" quality fits and short lengths of engagement. Grades with numbers greater than 6 are recommended for "coarse" quality fits and long lengths of engage-

in notes for coarse threads, but it is preferred by U.S. standards that it be shown. The commercially available ISO threads recommended for general use are given in Table 9.2. Additional ISO specifications are given in Appendix 18.

Complete designation

For some applications it is necessary to show a complete thread designation, as shown in Fig. 9.9. The first part of this note is the same as the basic designation,

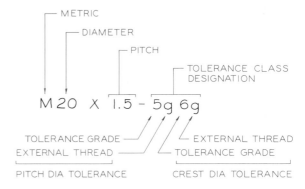

FIG. 9.9 A complete designation note for metric threads

TABLE 9.2
BASIC THREAD DESIGNATIONS FOR COMMERCIAL SERIES OF ISO METRIC THREADS

Nominal Size (mm)	Pitch, P (mm)	Basic Thread Designation*	Nominal Size (mm)	Pitch, P (mm)	Basic Thread Designation*	Nominal Size (mm)	Pitch, P (mm)	Basic Thread Designation*
1.6	0.35	M1.6	8 <	1.25	M8	22 <	2.5	M22
1.8	0.35	M1.8		1	M8 × 1		1.5	M22 × 1.5
2	0.4	M2	10 <	1.5	M10	24 <	3	M24
2.2	0.45	M2.2		1.25	M10 × 1.25		2	M24 × 2
2.5	0.45	M2.5	12 <	1.75	M12	27 <	3	M27
3	0.5	M3		1.25	M12 × 1.25		2	M27 × 2
3.5	0.6	M3.5	14 <	2	M14	30 <	3.5	M30
4	0.7	M4		1.5	M14 × 1.5		2	M30 × 2
4.5	0.75	M4.5	16 <	2	M16	33 <	3.5	M33
5	0.8	M5		1.5	M16 × 1.5		2	M33 × 2
6	1	M6	18 <	2.5	M18	36 <	4	M36
7	1	M7		1.5	M18 × 1.5		3	M36 × 3
			20 <	2.5	M20	39 <	4	M39
				1.5	M20 × 1.5		3	M39 × 3

*U.S. practice is to include the pitch symbol even for the coarse pitch series. Basic descriptions shown are as specified in ISO Recommendations.
Source: Courtesy of Greenfield Tap and Die Corporation.

TABLE 9.3
TOLERANCE GRADES, ISO THREADS

External thread		Internal thread	
Major diameter (d_1)	Pitch diameter (d_2)	Minor diameter (D_1)	Pitch diameter (D_2)
—	3	—	—
4	4	4	4
—	5	5	5
6	6	6	6
—	7	7	7
8	8	8	8
—	9	—	—

Grade 6 is medium; smaller numbers are finer, and larger numbers are coarser. (Courtesy of ANSI B1-1972.)

TOLERANCE POSITIONS

EXTERNAL THREADS LOWER-CASE LETTERS	INTERNAL THREADS UPPER-CASE LETTERS
e = LARGE ALLOWANCE	G = SMALL ALLOWANCE
g = SMALL ALLOWANCE	H = NO ALLOWANCE
h = NO ALLOWANCE	

LENGTH OF ENGAGEMENT

S = SHORT   N = NORMAL   L = LONG

FIG. 9.10 Symbols used to represent tolerence grade, position, and class.

ment. The tolerance grades for internal and external threads for the pitch diameter and the major and minor diameters are given in Table 9.3.

The letters following the grade numbers designate **tolerance position.** Lower-case letters represent external threads (bolts), as shown in Fig. 9.10. The lower-case letters e, g, and h represent large allowance, small allowance, and no allowance, respectively. Upper-case letters are used to designate internal threads (nuts); G designates small allowance and H designates no allowance. The letters are placed after the tolerance grade number. For example, 5g designates a medium tolerance with small allowance for the pitch diameter of an external thread, and 6H designates a medium tolerance with no allowance for the minor diameter of an internal thread.

Tolerance classes are fine, medium, and coarse, as listed in Table 9.4. These classes of fit are combinations of tolerance grades, tolerance positions, and lengths of engagement—short (S), normal (N), and long (L). The length of engagement can be determined by referring to Appendix 17. Once it has been decided to use either a fine, medium, or coarse class of fit for a particular application, the specific designation should be selected first from the classes shown in large print in Table 9.4, second from the classes shown in medium-size print, and third from the classes shown in small print. Classes shown in boxes are for commercial threads.

Variations in the complete designation thread notes are shown in Fig. 9.11. The tolerance class sym-

TABLE 9.4
PREFERRED TOLERANCE CLASSES, ISO THREADS*

Quality	External threads (bolts)									Internal threads (nuts)					
	Tolerance position e (large allowance)			Tolerance position g (small allowance)			Tolerance position h (no allowance)			Tolerance position G (small allowance)			Tolerance position H (no allowance)		
	Length of engagement			Length of engagement			Length of engagement			Length of engagement			Length of engagement		
	Group S	Group N	Group L	Group S	Group N	Group L	Group S	Group N	Group L	Group S	Group N	Group L	Group S	Group N	Group L
Fine							3h4h	4h	5h4h				4H	5H	6H
Medium		6e	7e6e	5g6g	6g	7g6g	5h6h	6h	7h6h	5G	6G	7G	5H	6H	7H
Coarse					8g	9g8g					7G	8G		7H	8H

*In selecting tolerance class, select first from the large bold print, second from the medium-size print, and third from the small-size print. Classes shown in boxes are for commercial threads.

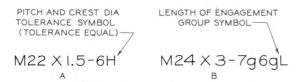

PITCH AND CREST DIA
TOLERANCE SYMBOL
(TOLERANCE EQUAL)

LENGTH OF ENGAGEMENT
GROUP SYMBOL

M22 X 1.5 - 6H

A

M24 X 3 - 7g 6gL

B

FIG. 9.11 When both pitch and crest diameter tolerance grades are the same, the tolerance class symbol is shown only once (A). Letters S, N, and L are used to indicate the length of the thread engagement (B).

bol is written as 6H if the crest and pitch diameters have identical grades (Part A). Since an upper-case H is used, this is an internal thread. Where considered necessary, the length-of-engagement symbol may be added to the tolerance class designation, as shown in Part B.

Designations for the desired fit between mating threads can be specified as shown in Fig. 9.12. A slash is used to separate the tolerance class designations of the internal and external threads.

Additional information pertaining to ISO threads may be obtained from *ISO Metric Screw Threads,* a booklet of standards published by ANSI in 1972. These standards were used as the basis for most of this section.

EXTERNAL
THREAD

INTERNAL
THREAD

M6 X 1 - 6H/6g

A

EXTERNAL
THREAD

INTERNAL
THREAD

M20 X 2 - 6H/5g6g

B

FIG. 9.12 A slash mark is used to separate the tolerance class designations of mating internal and external threads.

9.6
Thread representation

The three major types of thread representations are (A) simplified, (B) schematic, and (C) detailed (Fig. 9.13). The detailed presentation is the most realistic approximation of the true appearance of a thread, while the simplified representation is the most symbolic.

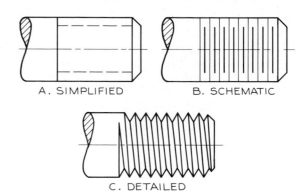

A. SIMPLIFIED   B. SCHEMATIC

C. DETAILED

FIG. 9.13 Three major types of thread representations

9.7
Detailed UN/UNR threads

Examples of detailed representations of internal and external threads are shown in Fig. 9.14. Instead of helical curves, straight lines are used to indicate crest and root lines. In this form of representation, internal threads in section can be indicated in two ways. Thread notes are applied in all cases, regardless of the representation used.

The construction of a detailed representation is shown in Fig. 9.15. The pitch is found by dividing 1 in. by the number of threads per inch. This can be done graphically, as shown in Step 1. However, in most cases, this construction is unnecessary, since the pitch can be approximated by using a calibration close to the true pitch taken directly from an existing scale or by using dividers for spacing. Note in Step 4 that a

> Where threads are close, they should purposely be drawn at a larger spacing to facilitate the drawing process.

45° chamfer is used to indicate a bevel of the threaded end to improve ease of assembly of the thread parts.

Metric threads would be drawn in the same manner; however, the pitch would not need to be computed. It is given in the metric thread table in millimeters.

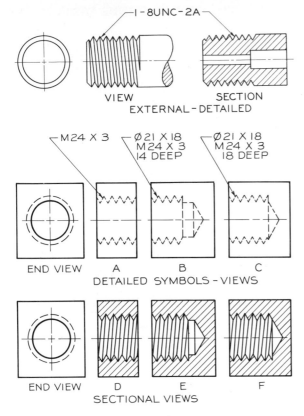

FIG. 9.14 Examples of detailed thread representations

9.8
Detailed square threads

The method of drawing a detailed representation of a square thread is shown in four steps in Fig. 9.16. This method gives an approximation of the true projection of a square thread.

In Step 1, the major diameter is laid off. The number of threads per inch is taken from the table in Appendix 19 for this size of thread. The pitch (*P*) is found by dividing 1 in. by the number of threads per inch. Distances of *P*/2 are marked off with dividers.

Steps 2, 3, and 4 are completed and a thread note is added to complete the thread representation.

Square internal threads are drawn in the same manner, as shown in Fig. 9.17. Note that the threads in the section view are drawn in a slightly different way. The thread note for an internal thread is placed in the circular view whenever possible, with the leader pointing toward the center.

When a square thread is rather long, it need not be drawn continuously, but can be represented using the symbol shown in Fig. 9.18.

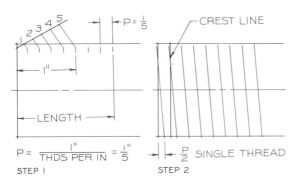
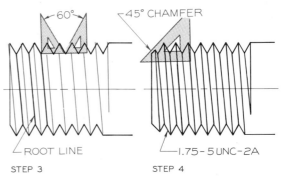

FIG. 9.15 Detailed thread representation

Step 1   To draw a detailed representation of a 1.75–5 UNC–2A thread, the pitch is determined by dividing 1 in. by the number of threads per inch, 5 in. this case. The pitch is laid off the length of the thread.

Step 2   Since the thread is a right-hand thread, the crest lines slope downward to the right equal to ½P (for single threads). The crest lines will be final lines drawn with an H or F pencil.

Step 3   The root lines are found by constructing 60° vees between the crest lines. The root lines are drawn from the bottom of the vees. Notice that root lines are parallel to each other, but not to crest lines.

Step 4   A 45° chamfer is constructed at the end of the thread from the minor diameter. Strengthen all lines and add a thread note.

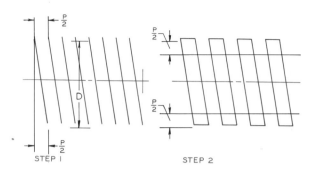 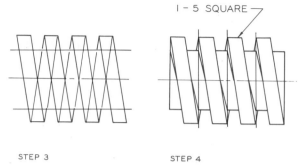

FIG. 9.16 Drawing the square thread

Step 1 Lay out the major diameter. Space the crest lines ½P apart. Slope them downward to the right for right-hand threads.

Step 2 Connect every other pair of crest lines. Find the minor diameter by measuring ½P inward from the major diameter.

Step 3 Connect the opposite crest lines with light construction lines. This will establish the profile of the thread form.

Step 4 Connect the inside crest lines with light construction lines to locate the points on the minor diameter where the thread wraps around the minor diameter.

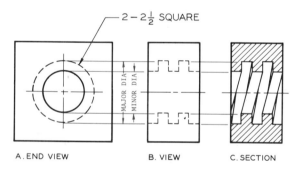

FIG. 9.17 Internal square threads

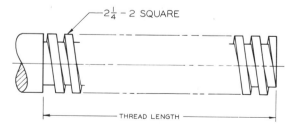

FIG. 9.18 Conventional method showing square threads without drawing each thread.

9.9
Detailed Acme threads

The method used in preparing detailed drawings of Acme threads is shown in Fig. 9.19 in four steps.

The length and the major diameter are laid off with light construction lines. From the table in Appendix 19, the pitch is found by dividing the number of threads per inch into 1 in. to begin Step 1.

Steps 2, 3, and 4 complete the thread representation. The thread note is added in the last step.

Internal Acme threads are shown in Fig. 9.20. Note that in the sectioned view, left-hand internal threads are sloped so that they look the same as right-hand external threads.

Figure 9.21 shows a shaft that is being threaded on a lathe. These Acme threads are being cut as the tool travels the length of the shaft.

9.10
Schematic representation

Examples of schematic representations of internal and external threads are shown in Fig. 9.22. Note that the threads are indicated by parallel, nonsloping lines that

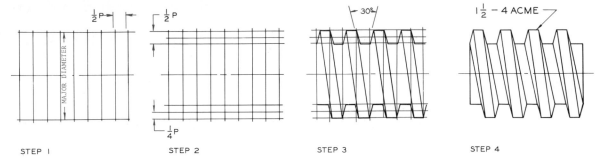

STEP 1   STEP 2   STEP 3   STEP 4

FIG. 9.19   Drawing the Acme thread

Step 1   Lay out the major diameter, the thread length, and divide the shaft into equal divisions ½P apart.

Step 2   Locate the minor diameter a distance ½P inside the major diameter. Locate the pitch diameter between the major and minor diameters.

Step 3   Draw lines at 15° with the vertical using the construction lines as shown to make a total angle of 30°. Draw the crest lines and the thread profile.

Step 4   Darken the lines, draw the root lines, and add the thread note to complete the drawing.

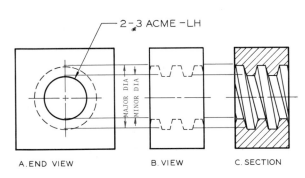

A. END VIEW   B. VIEW   C. SECTION

FIG. 9.20   Internal Acme threads

FIG. 9.21   Cutting An Acme thread on a lathe. (Courtesy of Clausing Corp.)

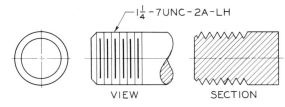

A External—Schematic

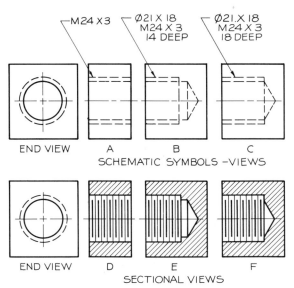

B Internal—Schematic

FIG. 9.22   Schematic representations of external threads are shown in Part A. Schematic representations of internal threads are shown in Part B.

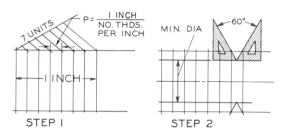

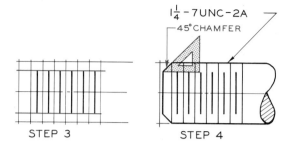

FIG. 9.23 Drawing schematic threads

Step 1 Lay out the major diameter and divide the shaft into equal divisions a distance *P* apart. These crest lines should be drawn as thin lines.

Step 2 Find the minor diameter by drawing a 60° angle between two crest lines.

Step 3 Draw heavy root lines between the crest lines.

Step 4 Chamfer the end of the thread and give a thread note.

do not show whether the threads are right-hand or left-hand. This information is given in the thread note. Since this representation is easy to construct and gives a good symbolic representation of threads, it is the most generally used thread symbol.

The method of constructing schematic threads is illustrated in Fig. 9.23 in four steps.

The method of drawing a schematic representation of metric threads is shown in Fig. 9.24. The pitch (in millimeters) can be taken directly from the metric thread tables as the distance used to separate the crest lines.

9.11
Simplified threads

Figure 9.25 illustrates the use of simplified representations with notes to specify thread details. Of the three types of thread representations covered, this is the easiest to draw. Hidden lines are positioned by eye to approximate the minor diameter.

The steps involved in constructing a simplified thread drawing are shown in Fig. 9.26.

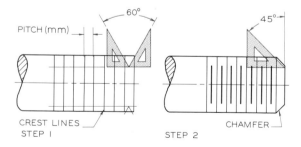

FIG. 9.24 Schematic metric threads

Step 1 The pitch of metric threads can be taken directly from the metric tables, which can be used to find the minor diameter.

Step 2 Draw heavy root lines between the crest lines. The end of the thread is chamfered.

9.12
Drawing small threads

Instead of using exact measurements to draw small threads, minor diameters can be drawn smaller by eye in order to separate the root and crest lines, as illustrated in Fig. 9.27. This procedure makes the drawing easier to draw and more readable. Exactness is unnecessary, since the drawing is only a symbolic representation of a thread.

For both internal and external threads, a thread note is added to the symbolic drawing to give the necessary specifications and to complete the description of the threaded part.

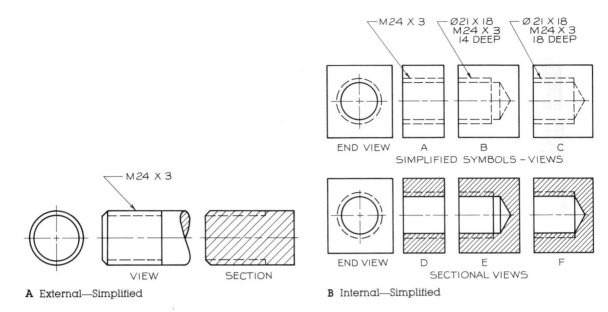

A External—Simplified

B Internal—Simplified

FIG. 9.25 Simplified thread representations of external threads (A). Simplified thread representations of internal threads (B).

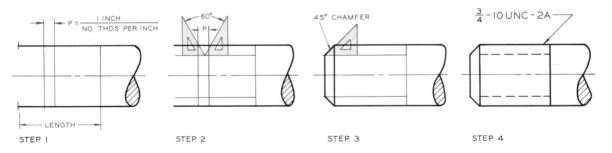

FIG. 9.26 Drawing simplified threads

Step 1 Lay out the major diameter. Find the pitch (*P*) and lay out two lines a distance of *P* apart.

Step 2 Find the minor diameter by constructing a 60° angle between the two lines.

Step 3 Draw a 45° chamfer from the minor diameter to the major diameter.

Step 4 Show the minor diameter as a dashed line. Add a thread note.

FIG. 9.27 Simplified and schematic threads should be drawn using approximate dimensions if the actual dimensions would result in lines drawn too close together.

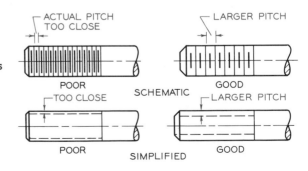

FIG. 9.28 Examples of nuts and bolts (Courtesy of Russell, Burdsall & Ward Bolt and Nut Co.)

9.13
Nuts and bolts

Nuts and bolts come in many forms and sizes for different applications (Fig. 9.28). Drawings of the more common types of threaded fasteners are shown in Fig. 9.29. A **bolt** is a threaded cylinder with a head and a nut for holding two parts together (Fig. 9.29A). A **stud** does not have a head, but is screwed into one part with a nut attached to the other end (Fig. 9.29B). A **cap screw** is similar to a bolt, but it does not have a nut; instead, it is screwed into a member with internal threads for greater strength (Fig. 9.29C). A **machine screw** is similar to a cap screw, but it is smaller. A **set screw** is used to adjust one member

with respect to another, usually to prevent a rotational movement.

The types of heads used on standard bolts and nuts are illustrated in Fig. 9.30. These heads are used on both types of bolts: **regular** and **heavy.** Heavy-series bolts have the thicker heads and are used at points where bearing loads are heaviest. Bolts and nuts are classified as **finished** and **unfinished.** Figure 9.30 shows an unfinished head; that is, none of the surfaces of the head are machined. The finished head has a washer face that is 1/64 in. thick to provide a circular boss on the bearing surface of the bolt head or the nut.

Other standard forms of bolt and screw heads are shown in Fig. 9.31. These heads are used primarily on cap screws and machine screws. Notice that it is standard practice to draw the slots at a 45° angle in the top views and to show the slots in the front views. Standard types of nuts are illustrated in Fig. 9.30. A hexagon jam nut does not have a washer face, but it is chamfered on both sides.

Although ANSI tables are provided in Appendixes 21–25 to indicate the standard bolt lengths and their corresponding thread lengths, the following can be used as a general guide for square and hexagon head bolts.

Hexagon bolt lengths are available in increments of 1/4 in. up to 8 in. in length, in 1/2 in. increments from 8 in. to 20 in. of length, and in 1 in. increments from 20 in. to 30 in. long.

Square head bolt lengths are available in increments of 1/8 in. from lengths of 1/2 in. to 3/4 in. long, 1/4-in. increments from 3/4 in. to 5 in. long, 1/2 in. increments from 5 in. to 12 in. long, and 1 in. increments from 12 in. to 30 in. long.

The lengths of the threads on both hexagon and square head bolts up to 6 in. in length can be found by the formula:

$$\text{thread length} = 2D + \tfrac{1}{4}\text{ in.,}$$

FIG. 9.29 Types of threaded bolts and screws.

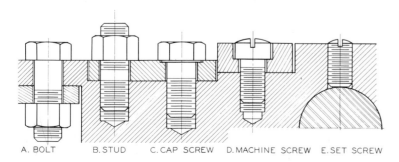

A. BOLT   B. STUD   C. CAP SCREW   D. MACHINE SCREW   E. SET SCREW

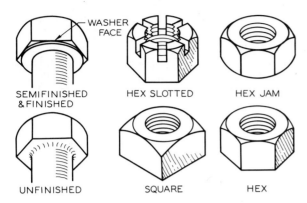

FIG. 9.30 Types of finishes for bolt heads and types of nuts

material. It is understood that these will have a Class 2 fit.

Nuts are designated by notes in the following form:

½ - 13 SQUARE NUT–STEEL, or

¾ - 16 HEAVY HEX NUT, SAE GRADE 5–STEEL, or

1.00 - 8 HEX THICK SLOTTED NUT–CORROSION RESISTANT STEEL

When not noted as HEAVY, nuts are assumed to be REGULAR nuts. The class of fit is assumed to be 2B for nuts when not noted.

where *D* is the diameter of the bolt. The threaded length for bolts over 6 in. in length can be found by the formula:

$$\text{thread length} = 2D + \tfrac{1}{2} \text{ in.}$$

The threads for bolts can be coarse, fine, or 8 pitch threads. It is understood that the class of fit for bolts and nuts will be 2A and 2B if unspecified.

Square and hexagon head bolts are designated by notes in the following form:

⅜ - 16 × 1½ SQUARE BOLT–STEEL, or

½ - 13 × 3 HEX CAP SCREW, SAE GRADE 8 STEEL, or

0.75 × 5.00 HEX LAG SCREW–STEEL

The numbers represent bolt diameter, threads per inch (omit for lag screws), length, name of screw, and

9.14
Drawing the square head bolt

Detailed tables are available in Appendix 20 and in published standards for various types of threaded parts. In most cases it is sufficient to draw nuts and bolts using only general proportions.

The first step in drawing a bolt head or a nut is to determine whether it is to be shown **across corners** or **across flats.** In other words, are the outlines of either side of the view going to represent corners, or are they going to be edge views of flat surfaces of the part? The head in Fig. 9.32 is drawn across corners. Nuts and bolts should be drawn across corners whenever possible; this type of drawing gives a better representation than drawing across flats.

9.15
Drawing the hexagon bolt head

It is preferable to draw nuts and bolts across corners, since this gives a better impression of the parts. An example of constructing the head of a bolt is shown in Fig. 9.33.

Note that the diameter of the bolt is *D.* The thickness of the head is drawn equal to ⅔*D.* The top view of the head is drawn as a circle with a radius of ¾*D.* For most applications, this proportionality based on *D* is sufficient for drawing bolt heads.

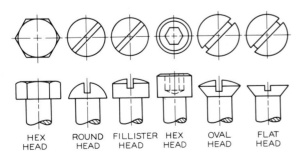

FIG. 9.31 Common types of bolt and screw heads

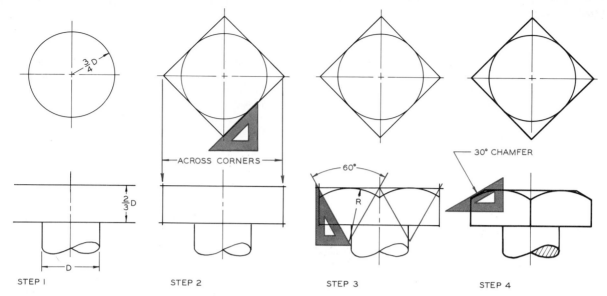

FIG. 9.32 Drawing the square head

Step 1 Draw the diameter of the bolt. Use this to establish the head diameter and thickness.

Step 2 Draw the top view of the square head with a 45° triangle to give an across-corners view.

Step 3 Show the chamfer in the front view by using a 30°–60° triangle to find the centers for the radii.

Step 4 Show a 30° chamfer tangent to the arcs in the front view. Strengthen the lines.

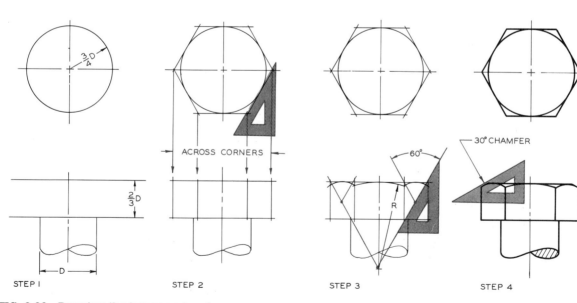

FIG. 9.33 Drawing the hexagon head

Step 1 Draw the diameter of the bolt. Use this to establish the head diameter and thickness.

Step 2 Construct a hexagon with a 30–60° triangle to give an across-corners view.

Step 3 Find arcs in the front view to show the chamfer of the head.

Step 4 Draw a 30° chamfer that is tangent to the arcs in the front view. Strengthen the lines.

177

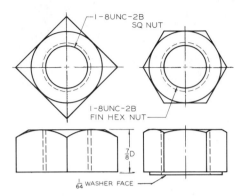

Fig. 9.34 Drawings of hexagon and square nuts are constructed in the same manner as drawings of bolt heads. Standard notes are added to give nut specifications.

9.16
Drawing nuts

The construction of a drawing of a square nut or a hexagon nut across corners is exactly the same as the construction of a drawing of a bolt head across corners. The only variation is the thickness of the nut. The regular nut thickness is $7/8D$, and for the heavy nut, the thickness is equal to the diameter (D).

Examples of square and hexagon nuts drawn across corners are shown in Fig. 9.34. Hidden lines are shown in the front view to indicate threads. Since it is understood that nuts are threaded, these hidden lines may be omitted in general applications.

Note that a $1/64$ in. washer face is shown on the hexagon nut. This is usually drawn thicker than $1/64$ in. so that the face will be more noticeable in the drawing. Thread notes are placed in the circular views rather than the front views where possible. In the case of the square nut, the note tells us that the major diameter of the thread is 1 in., that the nut has 8 threads per inch, that the thread is of the Unified National form and coarse series, with a fit of 2, and that the nut is a regular square nut since it is not labeled as HEAVY. The hexagon nut is similar except that it is a finished hexagon nut.

The leader from the note is directed toward the center of the circular view, but the arrow stops at the first visible circle it makes contact with.

Nuts can be drawn across flats in situations where doing so improves the drawing. Examples of nuts drawn across flats are shown in Fig. 9.35.

For regular nuts, the distance across flats is $1\frac{1}{2} \times D$ (D is the major diameter of the thread). For heavy nuts this distance is increased by $1/8$ in. The top views are drawn in the same manner as in across-corners drawings except that they are positioned to give different front views.

The hexagon nut drawn across flats looks more like a nut in the front view than does the square nut. Still, the hexagon nut drawn across corners is a better representation (Fig. 9.34). Notes are added with leaders to complete the representation of the nuts. A washer face should be added to a nut if it is finished, except in the case of square nuts which are always unfinished.

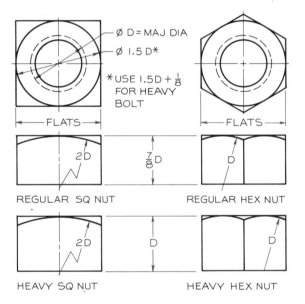

FIG. 9.35 Examples of hexagon and square nuts drawn across flats. Notes are added to give nut specifications.

9.17
Drawing nuts and bolts in combination

The same rules followed in drawing nuts and bolts separately apply when drawing nuts and bolts in assembly. Examples are shown in Fig. 9.36.

The diameter of the bolt is used as the basis for other dimensions. The note is added to give the spec-

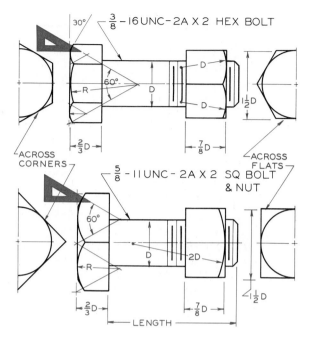

FIG. 9.36 Construction of nuts and bolts in assembly

ifications of the nut and bolt. In the figure, the bolt heads are drawn across corners and the nuts across flats. The end views have been included to show how the front views were found by projection.

9.18
Cap screws

Cap screws are used to hold two parts together without the use of a nut. One of these two parts has a threaded cylindrical hole and thus serves the same function as the nut. The other part is drilled with an oversize hole so that the cap screw will pass through it freely. When the cap screw is tightened, the two parts are held securely together.

The standard types of cap screws are illustrated in Fig. 9.37. Tables are available in Appendixes 22 and 23 to give the dimensions of several of these types of cap screws.

The cap screws in Fig. 9.37 are drawn on a grid to show the proportions of each type. The proportions shown here can be used for drawing cap screws of all sizes. These types of cap screws range in diameter from ¼ in. to 1½ in.

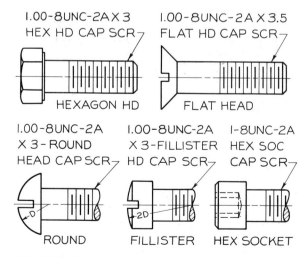

FIG. 9.37 The proportions of the standard types of cap screws are shown here for drawing cap screws of all sizes. Typical notes are given to provide specifications.

9.19
Machine screws

Machine screws are smaller than most cap screws, usually less than 1 in. in diameter. The machine screw is used to attach parts together; it is screwed either into another part or into a nut. Machine screws are threaded their full length when their length is 2 in. or shorter.

Drawings of common machine screws and their notes are given in Fig. 9.38. Many other styles are available in addition to these types. The dimensions of round-head machine screws are given in Appendix 24.

The four types of machine screws pictured in Fig. 9.38 are drawn on a grid to give the proportions of the head in relation to the major diameter of the screw. Machine screws range in size from No. 0 (0.060 in. in diameter) to a diameter of ¾ in.

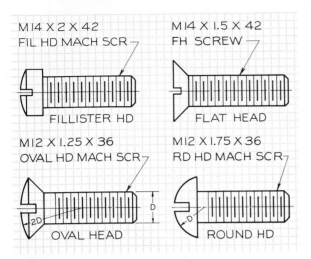

FIG. 9.38 Standard types of machine screws. The proportions shown here can be used for drawing machine screws of all sizes.

9.20
Set screws

Parts such as wheels or pulleys are commonly attached to shafts, using set screws or keys. Examples of various types of set screws are shown in Fig. 9.39.

The dimensions of the various features of the set screws shown in Fig. 9.39 are given in Table 9.5. Drawings of set screws need not employ these dimensions precisely; like the other fasteners discussed in this chapter, set screw threads can be drawn as approximations.

Set screws are available in any desired combination of point and head. The shaft against which the set screw is tightened may have a flat surface machined to give a good bearing surface for the set screw point. In this case, a dog point or a flat point would be most effective to press against the flat surface. The cup point gives good friction when applied to a round shaft.

Specifications for set screws are given in Appendixes 27, 28, and 29.

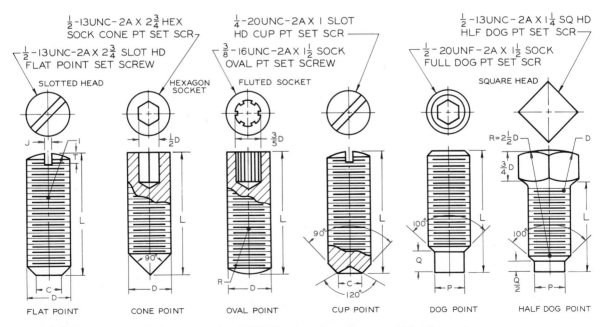

FIG. 9.39 Types of set screws. Set screws are available with various combinations of heads and points. Notes give their specifications. Dimensions are given in Table 9.5.

TABLE 9.5
DIMENSIONS FOR THE SET SCREWS SHOWN IN FIG. 9.40 (ALL DIMENSIONS GIVEN IN INCHES)

D		I	J	T	R	C		P		Q	q
Nominal size		Radius of headless crown	Width of slot	Depth of slot	Oval point radius	Diameter of cup and flat points		Diameter of dog point		Length of dog point	
						Max	Min	Max	Min	Full	Half
5	0.125	0.125	0.023	0.031	0.094	0.067	0.057	0.083	0.078	0.060	0.030
6	0.138	0.138	0.025	0.035	0.109	0.074	0.064	0.092	0.087	0.070	0.035
8	0.164	0.164	0.029	0.041	0.125	0.087	0.076	0.109	0.103	0.080	0.040
10	0.190	0.190	0.032	0.048	0.141	0.102	0.088	0.127	0.120	0.090	0.045
12	0.216	0.216	0.036	0.054	0.156	0.115	0.101	0.144	0.137	0.110	0.055
1/4	0.250	0.250	0.045	0.063	0.188	0.132	0.118	0.156	0.149	0.125	0.063
5/16	0.3125	0.313	0.051	0.078	0.234	0.172	0.156	0.203	0.195	0.156	0.078
3/8	0.375	0.375	0.064	0.094	0.281	0.212	0.194	0.250	0.241	0.188	0.094
7/16	0.4375	0.438	0.072	0.109	0.328	0.252	0.232	0.297	0.287	0.219	0.109
1/2	0.500	0.500	0.081	0.125	0.375	0.291	0.270	0.344	0.344	0.250	0.125
9/16	0.5625	0.563	0.091	0.141	0.422	0.332	0.309	0.391	0.379	0.281	0.140
5/8	0.625	0.625	0.102	0.156	0.469	0.371	0.347	0.469	0.456	0.313	0.156
3/4	0.750	0.750	0.129	0.188	0.563	0.450	0.425	0.563	0.549	0.375	0.188

Source: Courtesy of ANSI; B18.6.2–1956.

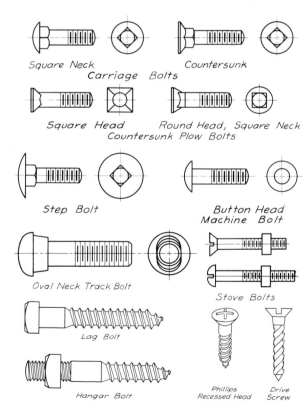

FIG. 9.40 Miscellaneous types of bolts and screws

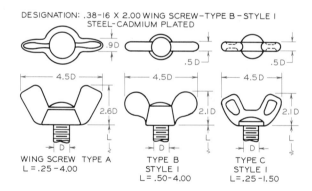

FIG. 9.41 Wing screw proportions are shown for screw diameters of about 5/16″. These proportions can be used to draw wing screws of any size diameter. Type A is available in screw diameters of 4, 6, 8, 10, 12, 0.25, 0.313, 0.375, 0.438, 0.50, and 0.625. Type B is available in diameters of 10 to 0.625; and Type C in diameters of 6 to 0.375.

9.21
Miscellaneous screws

A few of the many types of specialty screws are shown in Fig. 9.40, each having its own special application.

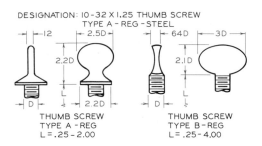

DESIGNATION: 10-32 X 1.25 THUMB SCREW
TYPE A - REG - STEEL

THUMB SCREW
TYPE A - REG
L = .25 - 2.00

THUMB SCREW
TYPE B - REG
L = .25 - 4.00

FIG. 9.42 Thumb screw proportions are given for screw diameters of about ¼″. These proportions can be used to draw thumb screws of any screw diameter. Type A is available in diameters of 6, 8, 10, 12, 0.25 , 0.313, and 0.375. Type B thumb screws are available in diameters of 6 to 0.50.

Wing screws of three types are shown in Fig. 9.41. They are available in incremental lengths of ⅛ in. Wing screws are used to join parts that are assembled and disassembled by hand.

Thumb screws of two types are shown in Fig. 9.42. These serve the same purpose as wing screws.

Wing nuts of three types are shown in Fig. 9.43. They can be used with screws of types other than wing screws or thumb screws.

The manufacture and design of specialized thread fasteners is a career field within itself, with an ever-growing need for additional fasteners of various types.

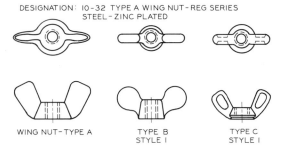

DESIGNATION: 10-32 TYPE A WING NUT - REG SERIES
STEEL - ZINC PLATED

WING NUT - TYPE A

TYPE B
STYLE I

TYPE C
STYLE I

FIG. 9.43 Wing nut proportions are given for screw diameters of ⅜″. These proportions can be used to draw thumb screws of any size. Type A wing nuts are available in screw diameters (in inches) of 3, 4, 5, 6, 8, 10, 12, 0.25, 0.313, 0.375, 0.438, 0.50, 0.583, 0.625, and 0.75. Type B nuts are available in sizes from 5 to 0.75, and Type C nuts in sizes from 4 to 0.50.

9.22
Wood screws

A wood screw is a pointed screw having a sharp thread of coarse pitch for insertion in wood. The three most common types of wood screws are shown in Fig. 9.44. These are drawn on a grid to show the proportions of the various heads in relation to the major diameter of the screw. Detailed dimensions for wood screws are given in tables published by the American National Standards Institute.

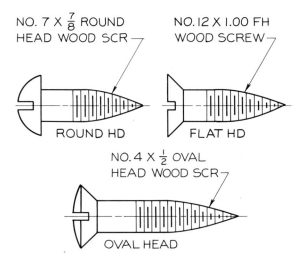

NO. 7 X $\frac{7}{8}$ ROUND
HEAD WOOD SCR

ROUND HD

NO.12 X 1.00 FH
WOOD SCREW

FLAT HD

NO. 4 X $\frac{1}{2}$ OVAL
HEAD WOOD SCR

OVAL HEAD

FIG. 9.44 Standard types of wood screws. The proportions shown here can be used for drawing wood screws of all sizes.

Sizes of wood screws are specified by single numbers such as 0, 6, or 16. From 0 to 10 each digit represents a different size. Beginning at 10, only even-numbered sizes are standard, i.e., 10, 12, 14, 16, 18, 20, 22, and 24. The following formula can be used to relate these numbered sizes to the actual diameter of the screws:

Actual DIA = 0.060 + screw number × 0.013.

For example, the actual diameter of a No. 5 screw is

$$0.060 + 5(0.013) = 0.125.$$

9.23
Use of templates

Templates are available for drawing threads, nuts, and threaded fasteners. They are available for a range of sizes that is satisfactory for most applications, since thread representations are approximations at best.

Two typical templates are shown in Fig. 9.45. The template is laid on the drawing and the threaded features are drawn using the template as a guide. Templates are also available for drawing nuts and bolts in pictorial.

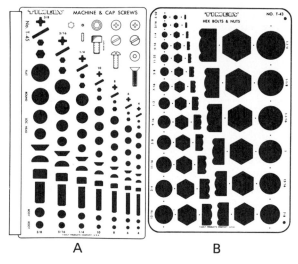

FIG. 9.45 Examples of templates that can be used for drawing threaded fasteners. (Courtesy of Timely Products Co.)

9.24
Tapping a hole

A threaded hole is called a **tapped hole,** since the tool used to cut the threads is called a tap. The types of taps available for threading small holes by hand are shown in Fig. 9.46.

The **taper, plug,** and **bottoming hand taps** are identical in size, length, and measurements, their only difference being the chamfered portion of their ends. The taper tap has a long chamfer (8 to 10 threads),

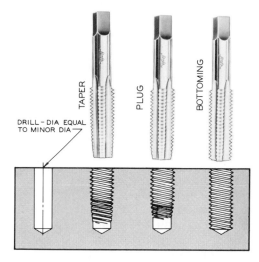

FIG. 9.46 Three types of taps for cutting internal threads. The drill point has an angle of 120°.

the plug tap has a chamfer of 3 to 5 threads, and the bottoming tap has a short chamfer of only 1 to 1½ threads.

When tapping by hand in open or "through" holes, the taper should be used for coarse threads, since in ensures straighter starting. The taper tap is also recommended for the harder metals. The plug tap can be used in soft metals or for fine-pitch threads. When it is desirable to tap a hole to the very bottom, all three taps—taper, plug, and bottoming—should be used in this order.

Notes are added to specify the depth of the drilled hole and the depth of the threads. For example, a note reading ⅞ DRILL-3 DEEP-1-8 UNC-2A-2 DEEP means that the hole will be drilled deeper than it is threaded and the last usable thread will be 2 in. deep in the hole. Note that the drill point has an angle of 120°.

9.25
Washers, lock washers, and pins

Washers, called **plain washers,** are used with nuts and bolts to improve the assembly and the strength of the fastening. Plain washers are noted on a working drawing in the following manner:

0.938 × 1.750 × 0.134 TYPE A PLAIN WASHER.

These numbers represent the inside diameter, outside diameter, and the thickness, in that order. These dimensions can be found in Appendix 36.

A lock washer is a washer that prevents a nut or cap screw from loosening as a result of vibration or movement. The more common types of lock washers are shown in Fig. 9.47.

Tables for spring lock washers are given in Appendix 37 for regular and extra heavy-duty helical-spring lock washers. They are designed on drawings with notes in the following form:

HELICAL-SPRING LOCK WASHER—¼ REGULAR-PHOSPHOR BRONZE

(the ¼ is the washer's inside diameter).

Tooth lock washers are designated with notes in the following form:

INTERNAL-TOOTH LOCK WASHER—¼-TYPE A-STEEL,

or

EXTERNAL-TOOTH LOCK WASHER—.562-TYPE B-STEEL.

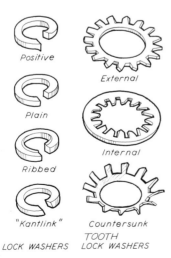

FIG. 9.47 Types of lock washers and locking devices

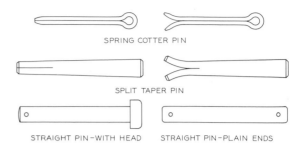

FIG. 9.48 Types of pins used to fix parts together

Straight pins and *taper pins* are used to fix parts together in a specified alignment. Dimensions for these are given in Appendix 34.

Other locking devices are *cotter pins* and *split taper pins* (Fig. 9.48). Tables of specifications for cotter pins are given in Appendix 38.

9.26
Pipe threads

Pipe threads are used in connecting pipes, tubing, lubrication fittings, and other applications. The most commonly used pipe thread is tapered at a ratio of 1 to 16, but straight pipe threads (without a taper) are available. Since pipe threads are usually tapered, the threads will only engage for an effective length determined by the formula

$$L = (0.80D + 6.8)P,$$

where D is the outside diameter of the pipe and P is the pitch.

Methods of representing tapered threads are shown in Fig. 9.49. Notice that a taper of 1 to 16 is shown to call attention to the fact that the threads are tapered.

The abbreviations associated with pipe threads are:

N = National
P = Pipe
T = Taper
C = Coupling
S = Straight
F = Fuel and oil
G = Grease

I = Internal
M = Mechanical
L = Locknut
H = Hose coupling
R = Railing fittings

These abbreviations are used in combination for the following ANSI symbols.

NPT = National pipe taper
NPTF = National pipe thread (dryseal—for pressure-tight joints)
NPS = Straight pipe thread
NPSC = Straight pipe thread in couplings
NPSI = National pipe straight internal thread
NPSF = Straight pipe thread (dryseal)
NPSM = Straight pipe thread for mechanical joints
NPSL = Straight pipe thread for locknuts and locknut pipe threads

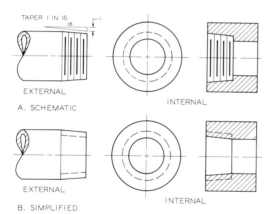

FIG. 9.49 Schematic and simplified techniques of representing external and internal pipe threads

NPSH = Straight pipe thread for hose couplings and nipples
NPTR = Taper pipe thread for railing fittings

To specify a pipe thread in note form, the nominal pipe diameter (the internal diameter), the number of threads per inch, and the symbol which denotes the type of thread are given. For example,

1¼–11½ NPT   or   **3–8 NPTR.**

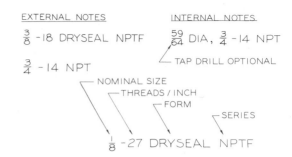

FIG. 9.50 Typical pipe-thread notes

These specifications can be taken from Appendix 10. Examples of external and internal thread notes are shown in Fig. 9.50. The dryseal thread is used in applications where a pressure-tight joint is required without the use of a lubricant or sealer. Dryseal threads may be straight or tapered. No clearance between the mating parts of the joint is permitted, giving the highest quality fit.

The tap drill is sometimes given in the internal pipe thread note (Fig. 9.50), but this is optional.

9.27
Keys

Keys are used to attach parts to shafts in order to transmit power to pulleys, gears, or cranks. Several types of keys are shown pictorially and orthographically in Fig. 9.51. The four types illustrated here are the most commonly used keys. To specify a key, notes must be given for the keyway, the key, and the keyseat, as shown in Fig. 9.51A, C, E, and G. The notes given are typical of the notes used to give key specifications. These dimensions may be found in Appendixes 31 and 32 for various types of keys.

9.28
Rivets

Rivets are fasteners used to join thin materials in a permanent joint. The rivet is inserted in the hole and the headless end is formed into the specified shape by applying extreme pressure to the projecting end. This forming operation is done when the rivets are either

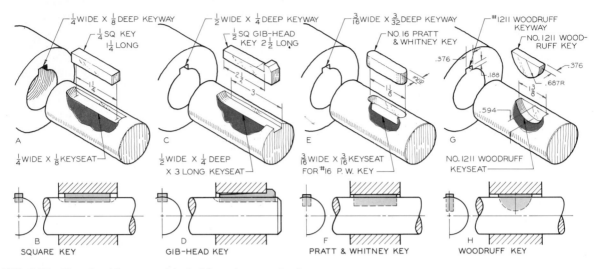

FIG. 9.51 Standard keys used to hold parts on a shaft

hot or cold, depending on the application.

Typical shapes and proportions of small rivets are shown in Fig. 9.52. These rivets vary in diameter from $\frac{1}{16}$ in. to $1\frac{3}{4}$ in. Rivets are used extensively in pressure-vessel fabrication, in heavy structures such as bridges and buildings, and in construction with sheet metal.

The proportions for large rivets are shown in Fig. 9.53. These proportions are based upon the diameters of the rivet bodies. Many ANSI tables of standard dimensions are available for sizing rivets.

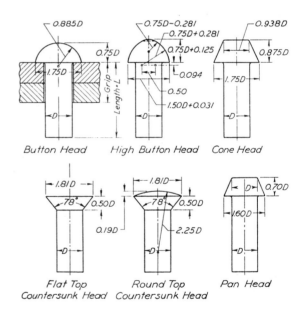

FIG. 9.53 Types and proportions of large rivets. Large rivets have shank diameters of $\frac{1}{2}$ in. DIA to $1\frac{3}{4}$ in. DIA.

Three types of lap joints are shown in Fig. 9.54, where the joints are held secure by one, two, or three rivets as shown in the sectional view. Note that the bodies of the rivets are drawn as hidden circles in the top views.

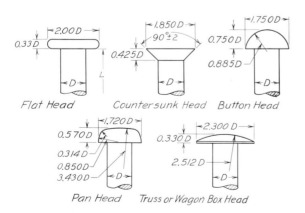

FIG. 9.52 Types and proportions of small rivets. Small rivets have shank diameters of up to $\frac{1}{2}$ in.

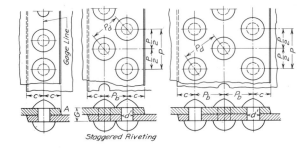

FIG. 9.54 Examples of lap joints using single rivets, double rivets, and triple rivets. The fourth example shows three plates that are fastened by double rivets.

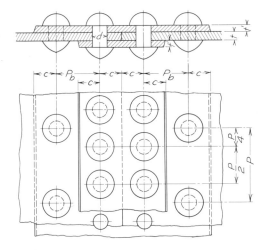

FIG. 9.55 Single and double riveted butt joints

Single and double riveted butt joints are illustrated in Fig. 9.55.

The standard symbols recommended by ANSI for representing rivets are shown in Fig. 9.56. Rivets that are driven in the shop are called shop rivets, and those assembled at the job site are called field rivets.

9.29
Springs

Of the many types of springs that are available, some of the more commonly used types are (1) **compression**, (2) **torsion**, (3) **extension**, (4) **flat**, and (5) **constant force**. Figure 9.57 shows the single-line, conventional representation of the first three types. Also shown are the types of ends that can be used on compression springs and the simplified, single-line representation of coil springs.

A typical working drawing of a compression spring is shown in Fig. 9.58. The ends of the spring are drawn using the double-line representation, and conventional lines are used to indicate the undrawn portion of the spring. Only the diameter and the free length of the spring are given on the drawing itself. The remaining specifications are given in tabular form near the drawing.

A working drawing of an extension spring (Fig. 9.59) is very similar to that of a compression spring. In a drawing of a helical torsion spring (Fig. 9.60), angular dimensions must be shown to specify the initial and ·final positions of the spring as torsion is applied to it. All types of springs require a table of specifications to describe their details.

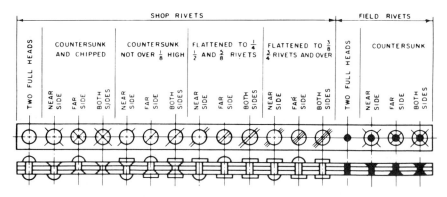

FIG. 9.56 The symbols used to represent rivets in a drawing. (Courtesy of ANSI 14.14.)

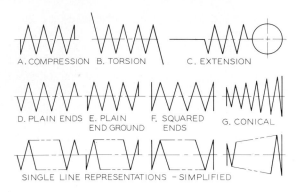

FIG. 9.57   Single-line representations of various types of springs

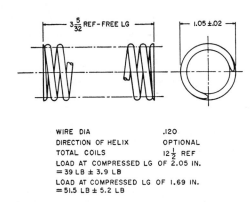

WIRE DIA   .120
DIRECTION OF HELIX   OPTIONAL
TOTAL COILS   12½ REF
LOAD AT COMPRESSED LG OF 2.05 IN.
= 39 LB ± 3.9 LB
LOAD AT COMPRESSED LG OF 1.69 IN.
= 51.5 LB ± 5.2 LB

FIG. 9.58   A conventional double-line drawing of a compression spring and its specifications. (Courtesy of the U.S. Department of Defense.)

9.30
Drawing springs

Springs may be drawn as schematic representations using single lines to represent the springs (Fig. 9.61). Each is drawn by laying out the diameter of the coils and the lengths of the springs, and then the number of active coils are drawn by using the diagonal-line method.

In Part B, the two end coils are "dead" coils and only four are active. An extension spring is drawn at Part C.

For applications where more realism is desired, a double-line drawing of a spring can be made as shown in Fig. 9.62. The end result is a good approximation of the spring.

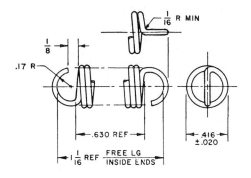

WIRE DIA   .042
DIRECTION OF HELIX   OPTIONAL
TOTAL COILS   14 REF
RELATIVE POSITION OF ENDS   180° ± 20°
EXTENDED LG INSIDE ENDS
WITHOUT PERMANENT SET   2.45 IN. (MAX)
INITIAL TENSION   1.0 LB ± .10 LB
LOAD   4 LB ± .4 LB AT 1.56 IN.
EXTENDED LG INSIDE ENDS
LOAD   6.3 LB ± .63 LB AT 1.95 IN.
EXTENDED LG INSIDE ENDS

FIG. 9.59   A conventional double-line drawing of an extension spring and its specifications. (Courtesy of the U.S. Department of Defense.)

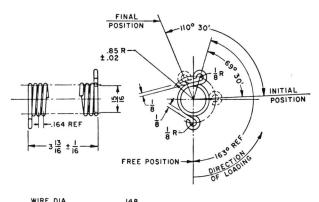

WIRE DIA   .148
DIRECTION OF HELIX   LEFT HAND
TOTAL COILS   20.55 REF
TORQUE   15 LB IN. ± 1.5 LB IN. AT INITIAL POSITION
TORQUE   33 LB IN. ± 3.3 LB IN. AT FINAL POSITION
MAXIMUM DEFLECTION WITHOUT SET BEYOND FINAL POSITION   56°
SPRING RATE   .16 LB IN. / DEG REF

FIG. 9.60   A conventional double-line drawing of a helical torsion spring and its specifications. (Courtesy of the U.S. Department of Defense.)

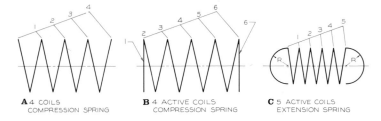

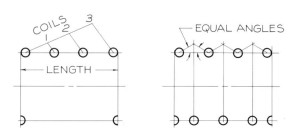

FIG. 9.61 In Part A, a schematic drawing of a spring with four active coils is shown. Once the length is laid out, the diagonal line method is used to divide it into four equally spaced coils. The spring at B has six coils, but only four of them are active coils. An extension spring with five active coils is shown at C.

FIG. 9.62 Detailed drawing of a spring

Step 1 Lay out the diameter and the length of the spring, and locate the coils by the diagonal-line technique.

Step 2 Locate the coils on the lower side along the bisectors of the spaces between the coils on the upper side.

Step 3 Connect the coils on each side. This is a right-hand coil; a left-hand spring would slope in the opposite direction.

Step 4 Construct the back side of the spring and the end coils, to complete the detailed drawing of a compression spring.

Problems

These problems are to be completed on Size A sheets in accordance with the specifications given in Section 2.33. Problems 1 through 16 are laid out one problem per sheet. Problems 17 through 30 are to be laid out two problems per sheet. The boxes drawn around the figures representing the problems are approximately 6 in. wide × 5 in. high, which equals about half of a Size A sheet.

Fasteners

1. The layout in Fig. 9.63 is to be used for constructing a detailed representation of an Acme thread with a major diameter of 3 in. The thread note specifications are 3–1½ ACME. Show both external and internal thread representations. Show the thread note. Use inches or millimeters as instructed.

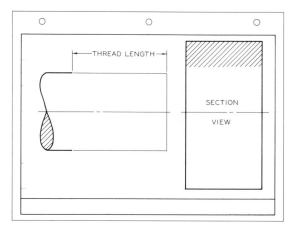

FIG. 9.63 Construction of thread symbols (Problems 1, 2, and 3).

2. Repeat Problem 1, but draw internal and external detailed representations of a square thread that is 3 in. in diameter. The note specifications are 3–1½ SQUARE. Apply notes to both parts.

3. Repeat Problem 1, but draw internal and external detailed representations of an American National thread form. The major diameter of each part is 3 in. The note specifications are 3–4 NC–2. Apply notes to both parts.

4. Notes are given in Fig. 9.64 to specify the depth of the holes that are to be drilled and the threads that are to be tapped in the holes. Following these notes, draw detailed representations of the threads as views according to specifications.

5. Repeat Problem 4, but use schematic representations.

6. Repeat Problem 4, but use simplified thread representations.

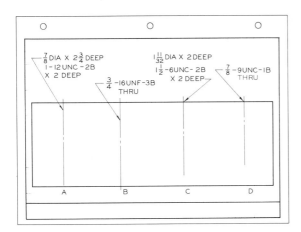

FIG. 9.64 Internal threads (Problems 4, 5, and 6).

7. Figure 9.65 shows a layout of two external threaded parts and their end views. Also shown is a piece into which the external threads will be screwed. Complete all three views of each of the parts. Use detailed threads and apply notes to the internal and external threads. Use the table in Appendix 15 for thread specifications. Use UNC threads with a 2A fit.

8. Repeat Problem 7, but use schematic thread representations.

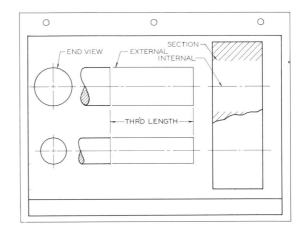

FIG. 9.65 Internal and external threads (Problems 7, 8, and 9).

9. Repeat Problem 7, but use simplified thread representations.

10. Referring to Fig. 9.66, complete the drawing with instruments as a finished hexagon bolt and nut. The bolt head is to be drawn across corners. The nut is a heavy nut drawn across corners. Use detailed thread representations. Show notes to specify the parts of the assembly. Thread specifications are 1½–6 UNC–3 or M36 × 4.

11. Referring to Fig. 9.66, complete the drawing with instruments as an unfinished square-head bolt

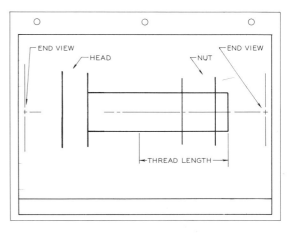

FIG. 9.66 Nuts and bolts in assembly (Problems 10, 11, and 12).

and nut. The bolt head is drawn across corners. The regular nut is to be drawn across corners. Use schematic thread representations. Show notes to specify the parts. Use the table in Appendix 15 or 18 for thread specifications (English or SI).

12. Referring to Fig. 9.66, complete the drawing with instruments as a finished hexagon nut and bolt. The regular bolt and nut are to be drawn across flats. Use simplified thread representations. Show notes to specify the parts. Use the table in Appendix 15 or 18 for specifications.

13. The notes in Fig. 9.67 apply to machine and cap screws that are to be drawn in the section view of the two parts. The holes in which the screws are to be drawn are through holes. Complete the drawings and show the notes as given. Show the remaining section lines. Use detailed thread symbols.

14. Repeat Problem 13, but use schematic thread symbols.

15. Repeat Problem 13, but use simplified thread symbols.

FIG. 9.68 Design involving threaded parts (Problem 16).

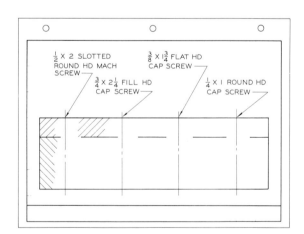

FIG. 9.67 Cap screws and machine screws (Problems 13, 14, and 15).

16. *Design.* The pencil pointer shown in Fig. 9.68 has a shaft of ¼ in. that fits into a bracket designed to clamp onto a desk top. A set screw holds the shaft in position. Make a drawing of the bracket, estimating its dimensions. Show the details and the method of using the set screw to hold the shaft. Give the specifications for the set screw.

17. (Fig. 9.69). On axes A and B, construct ⅝ in. (16 mm) hexagonal head bolts across flats with a coarse thread, UNC. Convert the view to a half-section to show how the parts are assembled together.

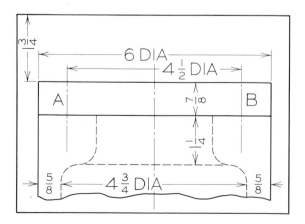

FIG. 9.69 Problem 17.

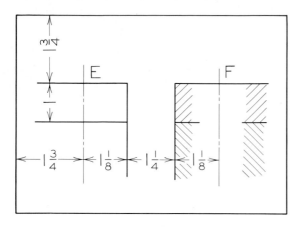

FIG. 9.70 Problem 18.

18. (Fig. 9.70). On axes E and F, draw a stud with a hexagon head nut shown across corners. The stud is to have a diameter of ⅞ in. (22 mm) and should be a UNC form and series.

19. (Fig. 9.71). Draw a ⅞ in. (22 mm) DIA special stud with a collar that is 1¼ in. DIA. Show a plain washer and a regular nut at end B. On axis CD, draw a machine screw of your selection to hold the two parts together.

20. (Fig. 9.72). Draw a cap screw on axis **AB** to fasten the parts together. The following dimensions are bolt DIA 1¼ in. or 32 mm, C—1.25 in., E—1.12 in., H—0.25 in., hexagon head across corners, G—4.50 in., and D—2.00 in.

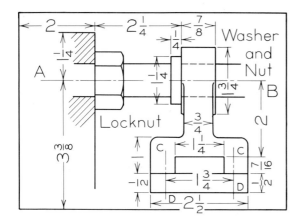

FIG. 9.71 Problem 19.

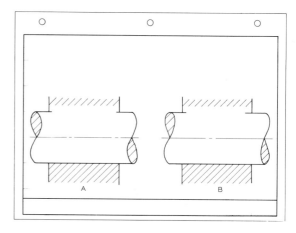

FIG. 9.73 Problems 21 and 22.

Keys

21. Figure 9.73 shows two parts assembled on a cylindrical shaft. These parts are to be held in position by a square key in Part A and a gib-head key in Part B. Show the necessary notes to specify the key, keyway, and keyseat.

22. Repeat Problem 21, but use a No. 16 Pratt & Whitney key in Part A and a No. 1211 Woodruff key in Part B. Show the necessary notes to specify the key, the keyway, and the keyseat.

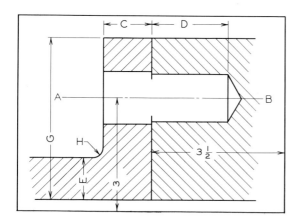

FIG. 9.72 Problem 20.

23–26. (Fig. 9.74 and Table 9.6). Using the double-line method, draw helical springs, two per Size A sheet. Use the specifications in the table for each.

27–30. (Fig. 9.74). Same as Problems 23–26, except use single-line representations for the springs.

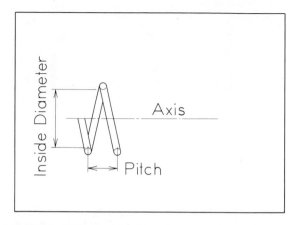

FIG. 9.74 Problems 23–30.

TABLE 9.6
PROBLEMS 23–26

Problem	No. of Turns	Pitch	Size Wire	Inside Dia	Outside Dia	
23	4	1	No. 4 = 0.2253	3		RH
24	5	¾	No. 6 = 0.1920		2	LH
25	6	⅝	No. 10 = 0.1350	2		RH
26	7	¾	No. 7 = 0.1770		1¾	LH

10

CHAPTER

Gears and Cams

10.1
Introduction to gears

Gears are toothed wheels that mesh together to transmit force and motion from one gear to the next. They are linked together by teeth cut into their circumferences that contact each other.

The more common types of gears (Fig. 10.1) are (A) spur gears, (B) bevel gears, and (C) worm gears.

A   B   C

FIG. 10.1 The three most basic types of gears are (A) spur gears, (B) bevel gears, and (C) worm gears. (Courtesy of the Process Gear Co.)

Each of these types of gears will be covered in this chapter. The many other more specialized types of gears deserve more coverage than is possible in this text.

10.2
Spur gear terminology

The spur gear is a circular gear with teeth cut around its circumference. Two mating spur gears can transmit power from one shaft to another parallel shaft.

When the two meshing gears are unequal in diameter, the smaller gear is called the **pinion** and the larger one is called the **gear.**

The following terms are used to describe the parts of a spur gear. Many of these features are illustrated and labeled in Figs. 10.2 and 10.3. The corresponding formulas for each feature are given.

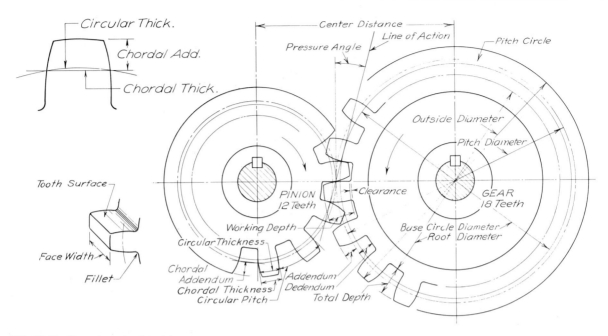

FIG. 10.2 Gear terminology for spur gears

PITCH CIRCLE (PC) The imaginary circle of a gear if it were a friction wheel without teeth that contacted the pitch circle of another friction wheel.

PITCH DIAMETER (PD) The diameter of the pitch circle; $PD = N/DP$ (where N = number of teeth and DP = diametrical pitch).

DIAMETRICAL PITCH (DP) The ratio between the number of teeth on a gear and its pitch diameter.

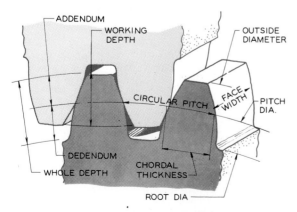

FIG. 10.3 Gear terminology for spur gears

For example, a gear with 20 teeth and a 4 in. pitch diameter will have a diametrical pitch of 5, which means that there are 5 teeth per inch of diameter; $DP = N/PD$ (where N = number of teeth).

CIRCULAR PITCH (CP) The circular measurement from one point on a tooth to the corresponding point on the next tooth measured along the pitch circle; $CP = 3.14/DP$.

CENTER DISTANCE (CD) The distance from the center of a gear to its mating gear's center; $CD = (N_p + N_s)/(2DP)$ (where N_p and N_s are the number of teeth in the pinion and spur, respectively).

ADDENDUM (A) The height of a gear above its pitch circle; $A = 1/DP$.

DEDENDUM (D) The depth of a gear below the pitch circle; $D = 1.157/DP$.

WHOLE DEPTH (WD) The total depth of a gear tooth; $WD = A + D$.

WORKING DEPTH (WKD) The depth to which a tooth fits into a meshing gear; $WKD = 2/DP$; or $WKD = 2A$.

CIRCULAR THICKNESS (CRT) The circular distance across a tooth measured along the pitch circle; $CRT = 1.57/DP$.

CHORDAL THICKNESS (CT) The straight-line distance across a tooth at the pitch circle; $CT = PD (\sin 90°/N)$ (where N = number of teeth).

FACE WIDTH (FW) The width across a gear tooth parallel to its axis. This is a variable dimension, but it is usually 3 to 4 times the circular pitch; $FW = 3$ to $4(CP)$.

OUTSIDE DIAMETER (OD) The maximum diameter of a gear across its teeth; $OD = PD + 2A$.

ROOT DIAMETER (RD) The diameter of a gear measured from the bottom of its gear teeth. $RD = PD - (2D)$.

PRESSURE ANGLE (PA) The angle between the line of action and a line perpendicular to the center line of two meshing gears. Angles of 14.5° and 20° are standard angles for involute gears.

BASE CIRCLE (BC) The circle from which an involute tooth curve is generated or developed; $BC = PD (\cos PA)$.

10.3
Tooth forms

The most common gear tooth is an **involute tooth** with a 14.5° pressure angle. The 14.5° angle is the angle of contact between two gears when the tangents of both gears pass through the point of contact. Gears with pressure angles of 20° and 25° are also used. Gear teeth with larger pressure angles are wider at the base and thus are stronger than the standard 14.5° teeth.

The standard gear face is an involute that keeps the meshing gears in contact as the gear teeth are revolved past one another. The principle of constructing an involute is illustrated in Fig. 10.4.

An involute curve can be thought of as the path of a string that is kept taut as it is unwound from the base arc. It is unnecessary to use this procedure in drawing gear teeth since most detail drawings employ only approximations of gear teeth, if teeth are shown at all.

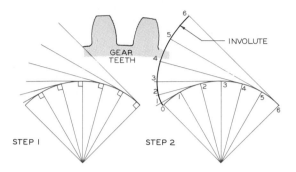

FIG. 10.4 Construction of an involute

Step 1 The base arc is divided into equal divisions with radial lines from the center. Tangents are drawn perpendicular to the radial lines on the arc.

Step 2 The chordal distance from 1 to 0 is used as the radius and 1 as the center to find Point 1 on the involute curve. The distance from 2 to newly found 1 is revolved to the tangent line through 2 to locate a second point, and the process is continued.

10.4
Gear ratios

The diameters of two meshing spur gears establish ratios that are important to the function of the gears. Examples of these ratios are given in Fig. 10.5.

If the radius of a gear is twice that of its pinion (the small gear), then the diameter is twice that of the pinion, and the gear has twice as many teeth as the pinion. In this case, the pinion must make twice as many turns as the larger gear. In other words, the revolutions per minute (RPM) of the pinion is twice that of the larger gear.

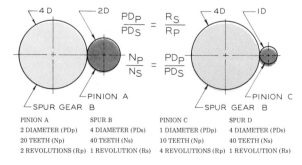

PINION A	SPUR B	PINION C	SPUR D
2 DIAMETER (PDp)	4 DIAMETER (PDs)	1 DIAMETER (PDp)	4 DIAMETER (PDs)
20 TEETH (Np)	40 TEETH (Ns)	10 TEETH (Np)	40 TEETH (Ns)
2 REVOLUTIONS (Rp)	1 REVOLUTION (Rs)	4 REVOLUTIONS (Rp)	1 REVOLUTION (Rs)

FIG. 10.5 Ratios between meshing spur gears

When the diameter of the gear is four times the diameter of the pinion, there must be four times as many teeth on the gear as on the pinion, and the number of revolutions of the pinion will be four times that of the larger gear.

The relationship between two meshing spur gears can be developed in formula form by finding the velocity of a point on the small gear that is equal to $\pi PD \times$ RPM of the pinion. The velocity of a point on the large gear is equal to: $\pi PD \times$ RPM of the spur. Since the velocity of points on each gear must be equal, the equation may be written as

$$\pi PD_p \,(\text{RPM}) = \pi PD_s \,(\text{RPM});$$

therefore

$$\frac{PD_p}{PD_s} = \frac{\text{RPM}_s}{\text{RPM}_p}.$$

If the radius of the pinion is 1 in., the radius of the spur is 4 in., and the RPM of the pinion is 20 RPM, then the RPM of the spur can be found as follows:

$$\frac{2(1)}{2(4)} = \frac{\text{RPM}_s}{20\ \text{RPM}_p}; \quad \text{RPM}_s = \frac{2(20)}{2(4)} = 5\ \text{RPM}$$

(one-fourth of the revolutions per minute of the pinion).

The number of teeth on each gear is also proportional to the radii and diameters of a pair of meshing gears. This relationship can be written:

$$\frac{N_p}{N_s} = \frac{PD_p}{PD_s}$$

(where N_p and N_s are the numbers of teeth on the spur and pinion, respectively, and PD_p and PD_s are their pitch diameters).

10.5
Gear calculations

Before a working drawing of a gear can be started, the drafter must perform a series of calculations to determine the dimensions of the gear using the definitions

and formulas previously introduced. The following problem is given as an example.

PROBLEM 1 Calculate the dimensions for a spur gear that has a pitch diameter of 5 in., a diametrical pitch of 4, and a pressure angle of 14.5°. (The diametrical pitch is the same for meshing gears.)

SOLUTION

No. of teeth $= PD \times DP = 5 \times 4 = 20$

Addendum $= 1/4 = 0.25$

Dedendum $= 1.157/4 = 0.2893$

Circular thickness $= 1.5708/4 = 0.3927$

Outside diameter $= (20 + 2)/4 = 5.50$

Root diameter $= 5 - 2(0.2893) = 4.421$

Chordal thickness $= 5(\text{sine } 90°/20) = 5(0.079)$
$\qquad\qquad\qquad = 0.392$

Chordal addendum $= 0.25 + (0.3927^2/(4 \times 5))$
$\qquad\qquad\qquad\qquad = 0.2577$

Face width $= 3.5(0.79) = 2.75$

Circular pitch $= 3.14/4 = 0.785$

Working depth $= 0.6366 \times (3.14/4) = 0.4997$

Whole depth $= 0.250 + 0.289 = 0.539$

These dimensions can be used to draw the spur gear and to provide specifications necessary for its manufacture.

A second problem is given as an example of the method of determining the design information for two meshing gears when their working ratios are known.

PROBLEM 2 Find the number of teeth and other specifications for a pair of meshing gears with a driving gear that turns at 100 RPM and a driven gear that turns at 60 RPM. The diametrical pitch for each is 10. The center to center distance between the gears is 6 in.

SOLUTION

Step 1 Find the sum of the teeth on both gears:

Total teeth $= 2 \times (\text{C to C dist.}) \times DP$
$\qquad\qquad = 2 \times 6 \times 10$
$\qquad\qquad = 120 \text{ teeth}$

Step 2 Find the number of teeth for the driving gear by two steps:

A. $\dfrac{\text{Driver RPM}}{\text{Driven RPM}} + 1 = \dfrac{100}{60} + 1 = 2.667$

B. $\dfrac{\text{Total teeth}}{\dfrac{100}{60} + 1} = \dfrac{120}{2.667} = 45 \text{ teeth*}$

Step 3 Find the number of teeth for the driven gear.

Total teeth minus teeth on driver = teeth on driven
$$120 - 45 = 75 \text{ teeth}$$

Step 4 The other specifications for the gears can be computed as shown in Problem 1 using the formulas in Section 10.2.

10.6
Drawing spur gears

A conventional drawing of a spur gear is shown in Fig. 10.6. The teeth need not be drawn since this is time-consuming and unnecessary. It is possible to omit the circular view and show only a sectional view of the gear with a table of dimensions that have been calculated. Dimensions of this type are called "cutting data."

When the circular view is drawn, circular center lines are drawn to represent the root circle, pitch circle, and outside circle of the gear.

The table of dimensions, as shown in Fig. 10.7, is a necessary part of a gear drawing. This data can be calculated by formula or taken from tables of standards in gear handbooks such as *Machinery's Handbook*.

10.7
Bevel gear terminology

Bevel gears are gears whose axes intersect at angles. Although the angle of intersection is usually 90°, other

---

*The number of teeth must be a whole number since there cannot be fractional teeth on a gear. It may be necessary to adjust the center distance to yield a whole number of teeth.

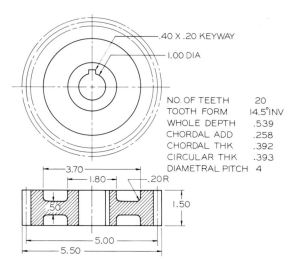

NO. OF TEETH	20
TOOTH FORM	14.5°INV
WHOLE DEPTH	.539
CHORDAL ADD	.258
CHORDAL THK	.392
CIRCULAR THK	.393
DIAMETRAL PITCH	4

.40 X .20 KEYWAY
1.00 DIA
3.70
1.80
.20R
1.50
.50
5.00
5.50

FIG. 10.6 A detail drawing of a spur gear with a table of values to supplement the dimensions shown on the drawing.

angles are used. The smaller of the two bevel gears is called the **pinion,** as in spur gearing.

The terminology of bevel gearing is illustrated in Fig. 10.8.

Pitch Angle of Pinion (small gear) (PA_p) is found by the following formula:

$$\text{Tan } PA_p = \frac{N_p}{N_g};$$

(N_g and N_p are the number of teeth on the gear and pinion, respectively).

Pitch Angle of Gear (PA_g) is found by the following formula:

$$\text{Tan } PA_g = \frac{N_g}{N_p}.$$

Pitch Diameter (PD) is the number of teeth (N) divided by the diametrical pitch (DP).

$$PD = N/P$$

Addendum (A) is measured at the large end of the tooth: $A = 1/DP$.

Dedendum (D) is measured at the large end of the tooth: $D = 1.157/DP$.

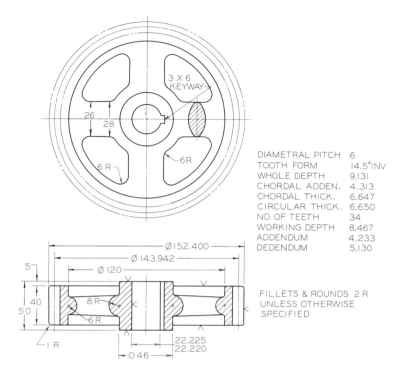

DIAMETRAL PITCH	6
TOOTH FORM	14.5°INV
WHOLE DEPTH	9.131
CHORDAL ADDEN.	4.313
CHORDAL THICK.	6.647
CIRCULAR THICK.	6.650
NO. OF TEETH	34
WORKING DEPTH	8.467
ADDENDUM	4.233
DEDENDUM	5.130

FILLETS & ROUNDS 2 R
UNLESS OTHERWISE
SPECIFIED

FIG. 10.7 A detail drawing of a spur gear that has been converted to metric units by multiplying dimensions in inches by 25.4 to yield millimeters.

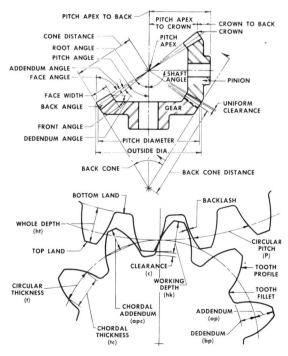

FIG. 10.8 The terminology and definitions of bevel gears. (Courtesy of Philadelphia Gear Corp.)

Whole Tooth Depth (WD): $WD = 2.157/DP$.

Thickness of Tooth (TT) at pitch circle: $TT = 1.571/DP$.

Dimetrical Pitch: $DP = N/PD$ (N = number of teeth).

Addendum Angle (AA) is the angle formed by the addendum and the pitch cone distance.

$$\text{Tan } AA = \frac{A}{PCD}$$

Pitch Cone Distance: $PCD =$

$$PD/(2 \times \sin PA).$$

Dedendum Angle (DA) is the angle formed by the dedendum and the pitch cone distance.

$$\text{Tan } DA = \frac{D}{PCD}$$

Face Angle (FA) is the angle between the gear's centerline and the top of its teeth: $FA = 90° - (PCD + AA)$.

Cutting Angle (or Root Angle) (CA) is the angle between the gear's axis and the roots of the teeth: $CA = PCD - D$.

Outside Diameter (OD) is the greatest diameter of a gear across its teeth: $OD = PD + 2A$.

Apex to Crown Distance (AC) is the distance from the crown of the gear to the apex of the cone measured parallel to the axis of the gear:

$$AC = OD/(2 \tan FA).$$

Chordal Addendum (CA) is

$$A + \frac{TT^2 \cos PA}{4(PD)}.$$

Chordal Thickness (CT) at the large end of the tooth is found by the following formula:

$$CT = PD \times \sin\frac{90°}{N}.$$

Face Width (FW) can vary, but it is recommended that it be approximately equal to the pitch cone distance divided by 3: $FW = PCD/3$.

Gear handbooks can be used for finding many of these dimensions from tables rather than using the formulas given above.

10.8
Bevel gear calculations

The following example of calculating the specifications for two bevel gears is given to demonstrate how the formulas in the previous section are used. You will note that some of the formulas result in the same specifications that apply to both the gear and pinion.

PROBLEM 3 Two bevel gears intersect at right angles. They have a diametrical pitch of 3, 60 teeth on the gear, 45 teeth on the pinion, and a face width of 4 inches. Find the dimensions of the gear.

SOLUTION

Pitch cone angle of gear: Tan $PCA = 60/45 = 1.33$; $PCA = 53° 7'$.

Pitch cone angle of pinion: Tan $PCA = 45/60$; $PCA = 36° 52'$.

Pitch diameter of gear: 60/3 = 20.00″.

Pitch diameter of pinion: 45/3 = 15.00″.

The following formulas are the same for both the gear and pinion.

Addendum: 1/3 = 0.333″.

Dedendum: 1.157/3 = 0.3857″.

Whole depth: 2.157/3 = 0.719″.

Tooth thickness on pitch circle: 1.571/3 = 0.5237.

Pitch cone distance: 20/(2 sin 53° 7′) = 12.5015.

Addendum angle: tan AA = 0.333/12.5015 = 1° 32′.

Dedendum angle: DA = 0.3857/12.5015 = 0.0308 = 1° 46′.

Face width: $PCD/3$ = 4.00″.

The remainder of the formulas must be applied separately to the gear and pinion.

Chordal addendum of gear:

$$0.333 + \frac{0.5237^2 \times \cos 53° 7'}{4 \times 20} = 0.336''.$$

Chordal addendum of pinion:

$$0.333 + \frac{0.5237^2 \times \cos 36° 52'}{4 \times 15} = 0.338''.$$

Chordal thickness of gear:

$$\sin \frac{90°}{60} \times 20 = 0.524''.$$

Chordal thickness of pinion:

$$\sin \frac{90°}{45} \times 15 = 0.523''.$$

Face angle of gear: 90° − (53° 7′ + 1° 32′) = 35° 21′.

Face angle of pinion:

$$90° - (36° 52' + 1° 32') = 51° 36'.$$

Cutting angle of gear: 53° 7′ − 1° 46′ = 51° 21′.

Cutting angle of pinion: 36° 52′ − 1° 46′ = 35° 6′.

Angular addendum of gear:

$$0.333 \times \cos 53° 7' = 0.1999".$$

Angular addendum of pinion:

$$0.333 \times \cos 36° 52' = 0.2667".$$

Outside diameter of gear: 20 + 2 (0.1999) = 20.4000".

Outside diameter of pinion: 15 + 2 (0.2667) = 15.533".

Apex to crown distance of gear: $\dfrac{20.400}{2} \times$ tan 35° 7' = 7.173".

Apex to crown distance of pinion: $\dfrac{15.533}{2} \times$ tan 51° 36' = 9.800".

10.9
Drawing bevel gears

The dimensions calculated above are used to lay out the bevel gears in a detail drawing. Many of the calculated dimensions would be difficult to measure on a drawing with a high degree of accuracy; therefore, it is important to provide a table of "cutting data" for each gear.

The steps involved in drawing the bevel gears are shown in Fig. 10.9. The finished drawings are shown with a combination of dimensions and a table of dimensions, which is not shown.

10.10
Worm gears

A worm gear is composed of a thread shaft called a **worm** and a circular gear called a **spider.** (Fig. 10.10). The worm is revolved in a continuous motion, causing the spider to revolve about its axis.

The following terminology is illustrated in Figs. 10.10 and 10.11. The following formulas can be used to calculate the dimensions associated with these terms.

Worm specifications and formulas

Linear Pitch (P) The distance from one thread to the next measured parallel to the worm's axis: $P = L/N$ (where N is number of threads: 1 if a single thread, 2 if a double thread, etc.)

Lead (L) The distance that a thread advances in a turn of 360°.

Addendum of Tooth: $AW = 0.3183\ P$.

Pitch Diameter: $PDW = OD - 2AW$ (OD is the outside diameter).

Whole Depth of Tooth: $WDT = 0.6866 \times P$.

Bottom Diameter of Worm: $BD = OD - 2\ WDT$.

Width of Thread at Root: $WT = 0.31P$.

Minimum Length of Worm: $MLW = \sqrt{8PDS \times AW}$ (PDS is the pitch diameter of the spider).

Helix Angle of worm:

$$\text{Cot } \beta = \frac{3.14PDW}{L}.$$

Outside Diameter: $OD = PD + 2A$.

Spider specifications and formulas

Pitch Diameter of Spider: $PDS = \dfrac{N(P)}{3.14}$ (N is the number of teeth on the spider).

Throat Diameter of Spider: $TD = PDS + 2A$.

Radius of Spider Throat:

$$RST = \frac{OD \text{ of worm}}{2} - 2A.$$

Face Angle (FA) may be selected to be between 60° and 80° for the average application.

Center to Center Distance (between the worm and spider):

$$CD = \frac{PDW + PDS}{2}.$$

Outside Diameter of Spider:

$$ODS = TD + 0.4775\ P.$$

Face Width of gear: $FW = 2.38\ (P) + 0.25$.

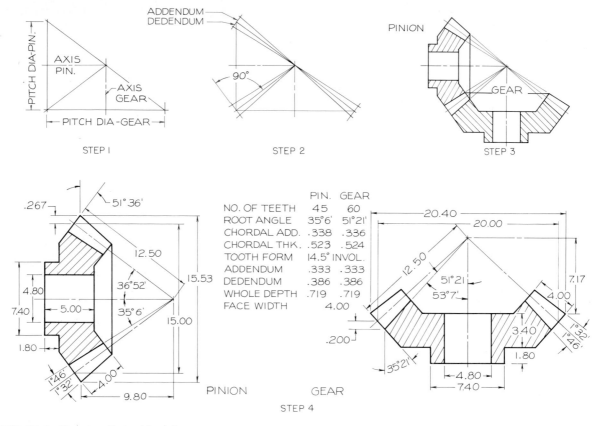

	PIN.	GEAR
NO. OF TEETH	45	60
ROOT ANGLE	35°6'	51°21'
CHORDAL ADD.	.338	.336
CHORDAL THK.	.523	.524
TOOTH FORM	14.5° INVOL.	
ADDENDUM	.333	.333
DEDENDUM	.386	.386
WHOLE DEPTH	.719	.719
FACE WIDTH	4.00	

FIG. 10.9 Construction of bevel gears

Step 1 Lay out the pitch diameters and axes of the two bevel gears

Step 2 Draw construction lines to establish the limits of the teeth by using the addendum and dedendum dimensions.

Step 3 Draw the pinion and the gear using the specified dimensions or those that were calculated by formula.

Step 4 Complete the detail drawings of both gears and provide a table of cutting data.

FIG. 10.10 The terminology and definitions of worm gears

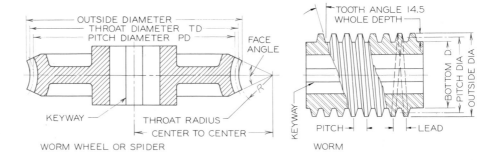

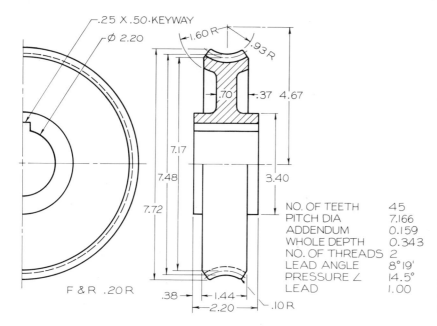

NO. OF TEETH	45
PITCH DIA	7.166
ADDENDUM	0.159
WHOLE DEPTH	0.343
NO. OF THREADS	2
LEAD ANGLE	8° 19'
PRESSURE ∠	14.5°
LEAD	1.00

FIG. 10.11 A detail drawing of worm gear (spider) and the table of cutting data

10.11
Worm gear calculations

The following is an example problem that has been solved for a worm gear using the formulas given above.

PROBLEM 4 Calculate the specifications for a worm and worm gear (spider). The gear has 45 teeth and the worm has an outside diameter of 2.50″. The worm has a double thread and a pitch of 0.5″.

SOLUTION

Lead: $L = 0.5 \times 2 = 1''$.

Worm addendum: $AW = 0.3183P = 0.1592''$.

Pitch diameter of worm: $PDW = 2.50 - 2(0.1592) = 2.1818''$.

Pitch diameter of gear: $PDS = (45 \times 0.5)/3.14 = 7.166''$.

Center distance between worm and gear:

$$CD = \frac{(2.182 + 7.166)}{2} = 4.674''.$$

Whole depth of worm tooth: $WDT = 0.687 \times 0.5 = 0.3433''$.

Bottom diameter of worm: $BD = 2.50 - 2(0.3433) = 1.813''$.

Helix angle of worm:

$$\text{Cot } \beta = \frac{3.14(2.1816)}{1} = 8° \ 19'.$$

Width of thread at root: $WT = 0.31 \ (1) = 0.155''$.

Minimum length of worm: $MLW = 8(0.1592) \ (7.1656) = 3.02''$.

Throat diameter of gear: $TD = 7.1656 + 2 \ (0.1592) = 7.484''$.

Radius of gear throat: $RST = (2.5/2) - (2 \times 0.1592) = 0.9318''$.

Face width: $FW = 2.38 \ (0.5) + 0.25 = 1.44''$.

Outside diameter of gear:

$$ODS = 7.484 + 0.4775 \ (0.5) = 7.723''.$$

10.12
Drawing worm gears

The worm and worm wheel (spider) are drawn and dimensioned as shown in Figs. 10.11 and 10.12. Each gear must be dimensioned and supplemented with cutting data.

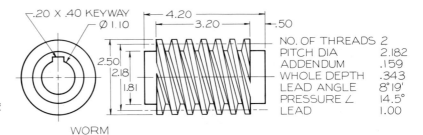

FIG. 10.12 A detail drawing of a worm using the dimensions that were calculated.

NO. OF THREADS	2
PITCH DIA	2.182
ADDENDUM	.159
WHOLE DEPTH	.343
LEAD ANGLE	8°19'
PRESSURE ∠	14.5°
LEAD	1.00

WORM

The specifications derived by the formulas in the previous section must be used for scaling and laying out the drawings.

10.13
Cams

Cams are irregularly shaped machine elements that produce motion in a single plane, usually up and down (Fig. 10.13). As the cam revolves about its center, the variation in the cam's shape produces a rise or fall in the follower that is in contact with it. The shape of the cam is determined graphically prior to the preparation of manufacturing specifications.

Only plate cams are covered in the brief review of this type of mechanism. Cams utilize the principle of the inclined wedge, with the surface of the cam causing a change in the slope of the plane, thereby producing the desired motion.

FIG. 10.13 Examples of machined cams. (Courtesy of Ferguson Machine Co.)

10.14
Cam motion

Cams are designed to produce (1) uniform or linear motion, (2) harmonic motion, (3) gravity motion, or (4) combinations of these. Some cams are designed to serve special needs that do not fit these patterns, but are instead based on particular design requirements.

Uniform motion

Uniform motion is shown in the displacement diagram in Fig. 10.14A. Displacement diagrams represent the motion of the cam follower as the cam rotates through 360°. The uniform-motion curve has sharp corners, indicating abrupt changes of velocity at two points, causing the follower to bounce. Hence this motion is usually modified with arcs that smooth this change of velocity. The radius of the modifying arc is varied up to a radius of one-half the total displacement of the follower, depending on the speed of operation.

Harmonic motion

Harmonic motion, plotted in Fig. 10.14B, is a smooth continuous motion based on the change of position of the points on the circumference of a circle. At moderate speeds, this displacement gives a smooth operation.

Gravity motion

Gravity motion (uniform acceleration), plotted in Fig. 10.14C, is used for high-speed operation. The variation of displacement is analogous to the force of gravity exerted on a falling body, with the difference in displacement being 1, 3, 5, 5, 3, 1, based on the square of the number. For instance, $1^2 = 1$; $2^2 = 4$; $3^2 = 9$. This same motion is repeated in reverse order for the remaining half of the motion of the follower. Intermediate points can be found by squaring fractional increments, such as $(2.5)^2$.

FIG. 10.14 Displacement diagrams

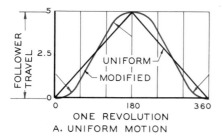

A. UNIFORM MOTION

Uniform-motion diagrams are modified with arcs of one-fourth to one-third of total displacement to smooth out the velocity of the follower at these points of abrupt change.

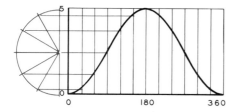

B. HARMONIC MOTION

Harmonic motion is plotted by projecting from a semicircle whose diameter is equal to the rise of the follower. The semicircle must be divided into the same number of sectors as the divisions on the x-axis of the graph to the point of maximum rise.

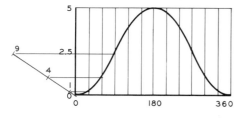

C. GRAVITY MOTION

Gravity-motion diagrams are constructed so that the rise of the follower is relative to the square of the units on the x-axis: 1^2, 2^2, 3^2, etc.

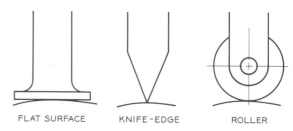

FLAT SURFACE   KNIFE-EDGE   ROLLER

FIG. 10.15 Three basic types of cam followers—the flat surface, the roller, and the knife edge

Cam followers

Three basic types of cam followers are (A) the flat surface, (B) the roller, and (C) the knife edge, as shown in Fig. 10.15. The flat-surface and knife-edge followers are limited to use with slow-moving cams, where a minimum of force will be exerted by the friction that is caused during rotation. The roller can withstand higher speeds and transmit greater forces.

10.15
Construction of a cam

The steps of constructing a plate cam with harmonic motion are shown in Fig. 10.16. The drafter must know the following before designing a cam: the motion of the follower, the rise of the follower, the size of the follower, the diameter of the base circle, and the direction of rotation.

The specifications for the cam pictured in Fig. 10.16 are given graphically. The four steps of construction are followed as illustrated to construct the cam profile in Step 4.

Plate cam—uniform acceleration

The steps of constructing a cam with uniform acceleration are performed in the same manner as the previous example except for a different displacement diagram and a knife-edge follower.

The graphic layout of the problem is shown in Fig. 10.17. The profile of the cam is found by following the four steps of construction.

FIG. 10.16 Plate cam with harmonic motion

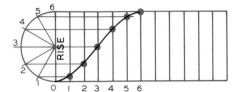

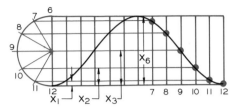

Step 1 Construct a semicircle whose diameter is equal to the rise of the follower. Divide the semicircle into the same number of divisions as there are between 0° and 180° on the horizontal axis of the displacement diagram. Plot half of the displacement curve in the displacement diagram.

Step 2 Continue the process of plotting points by projecting from the semicircle, starting from the top of the semicircle and proceeding to the bottom. Complete the curve symmetrical to the left half.

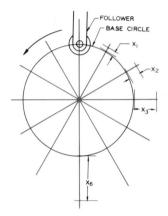

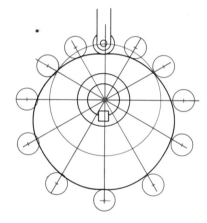

Step 3 Construct the base circle and draw the follower. Divide the base circle into the same number of sectors as there are divisions on the displacement diagram. Transfer distances from the displacement diagram to the respective radial lines of the base circle, measuring outward from the base circle.

Step 4 Draw circles to represent the positions of the roller as the cam revolves in a counterclockwise direction. Draw the cam profile tangent to all the rollers to complete the drawing.

Plate cam—combination

In Fig. 10.18 a knife-edge follower is used with a plate cam to produce a 4 in. rise with harmonic motion from 0° to 180°, a 4 in. fall with a uniform acceleration from 180° to 300°, and dwell (no follower motion) from 300° to 360°. You are to draw the cam that will give this motion from the base circle.

The displacement diagram is drawn with a harmonic curve with a full rise of 4 in. The curve is then drawn with a uniform acceleration drop of 4 in. The values of rise must be known before beginning the problem. The dwell is a horizontal line to complete the diagram of the 360° rotation of the cam.

The profile of the plate cam is found by completing Steps 2 and 3 in the same manner as was done in the previous two examples.

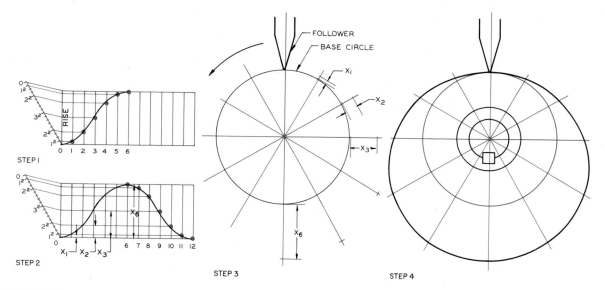

FIG. 10.17. Plate cam with uniform acceleration

Step 1 Construct a displacement diagram to represent the rise of the follower. Divide the horizontal axis into angular increments of 30°. Draw a construction line through point 0; locate the 1^2, 2^2, and 3^2 divisions and project them to the vertical axis to represent half of the rise. The other half of the rise is found by laying off distances along the construction line with descending values.

Step 2 Use the same construction to find the right half of the symmetrical curve.

Step 3 Construct the base circle and draw the knife-edge follower. Divide the circle into the same number of sectors as there are divisions in the displacement diagram. Transfer distances from the displacement diagram to the respective radial lines of the base circle, measuring outward from the base circle.

Step 4 Connect the points found in Step 3 with a smooth curve to complete the cam profile. Show also the cam hub and keyway.

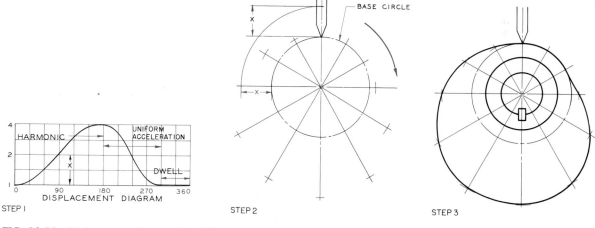

FIG. 10.18 Plate cam with combination motions

Step 1 The cam is to rise 4 in. in 180° with harmonic motion, fall 4 in. in 120° with uniform acceleration, and dwell for 60°. These motions are plotted on the displacement diagram.

Step 2 Construct the base circle and draw the knife-edge follower. Transfer distances from the displacement diagram to the respective radial lines of the base circle, measuring outward from the base circle.

Step 3 Draw a smooth curve through the points found in Step 2 to complete the profile of the cam. Show also the cam hub and keyway.

207

10.16

Construction of a cam with an offset follower

The cam in Fig. 10.19 is required to produce harmonic motion through 360°. This motion is plotted directly from the follower rather than using a displacement diagram, since there are no combinations of motion involved.

A semicircle is drawn with its diameter equal to the total motion of the follower. The base circle is drawn to pass through the center or roller of the follower. The centerline of the follower is extended downward, and a circle is drawn tangent to the extension with its center at the center of the base circle. The small circle is divided into 30° intervals to establish points through which construction lines will be drawn tangent to the circle.

The distances from the tangent points to the position points along the path of the follower are laid out along the tangent lines that were drawn at 30° intervals. These points can also be located by measuring from the base circle, as shown in the example, where Point 3 was located distance X from the base circle.

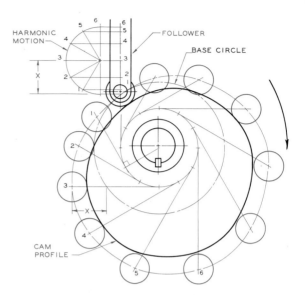

FIG. 10.19 Construction of a plate cam with an offset roller follower.

Draw the circular roller in all views and construct the profile of the cam tangent to the roller positions.

Problems

Gears

Use Size A (8½″ × 11″) sheets for the following gear problems. Select the most appropriate scale in order for the drawings to utilize the available space.

1–5. Calculate the following dimensions for the following spur gears and make a detail drawing of each. Give the dimensions and cutting data for each gear.

Problem	Gear teeth	Diametrical pitch	14.5° Involute
1	20	5	″
2	30	3	″
3	40	4	″
4	60	6	″
5	80	4	″

Provide any other dimensions that are needed, using your judgment.

6–10. Calculate the gear sizes and number of teeth (similar to Problem 2 in Section 10.5), using the ratios and data below.

Problem	RPM pinion	RPM gear	Center to center	Diametrical pitch
6	100 (driver)	60	6.0″	10
7	100 (driver)	50	8.0″	9
8	100 (driver)	40	10.0″	8
9	100 (driver)	35	12.0″	7
10	100 (driver)	25	14.0″	6

11–20. Make detail drawings of each of the gears for which calculations were made in Problems 6–10. Provide a table of cutting data and other dimensions that are needed to complete the specifications.

21–25. Calculate the specifications for the bevel gears that intersect at 90°, and make detail drawings of each with the necessary dimensions and cutting data.

Problem	Diametrical pitch	No. of teeth on pinion	No. of teeth on gear
21	3	60	15
22	4	100	40
23	5	100	60
24	6	100	50
25	7	100	30

26–30. Calculate the specifications for worm gears, and make detail drawings of each, providing the necessary dimensions and cutting data.

Problem	No. of teeth in spider gear	Outside DIA of worm	Pitch of worm	Thread of worm
26	45	2.50	0.5	double
27	30	2.00	0.80	single
28	60	3.00	0.80	double
29	30	2.00	0.25	double
30	80	4.00	1.00	single

Cams

Draw the following on Size B sheets (11″ × 17″) with the following standard dimensions: base circle—3.50 in.; roller follower—0.60 in. diameter; shaft—0.75 in. diameter; hub—1.25 in. diameter; direction of rotation—clockwise. The follower is positioned vertically over the center of the base circle except in Problems 40 and 41. Lay out the problems and displacement diagrams as shown in Fig. 10.20.

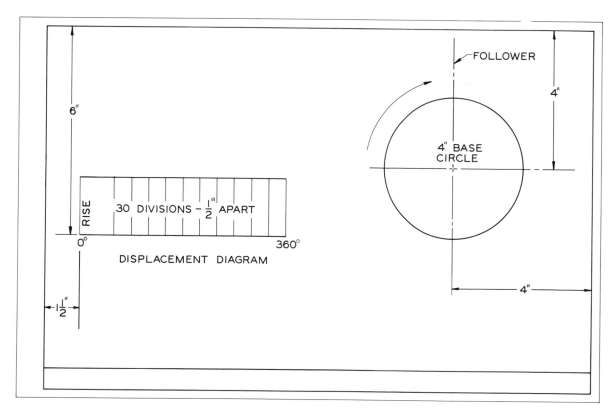

FIG. 10.20 Problem layout for the cam problems on Size B sheets.

31. Draw a plate cam with a knife-edge follower for uniform motion and a rise of 1.00 in.

32. Draw a displacement diagram and a cam that will give a modified uniform motion to a knife-edge follower with a rise of 1.7 in. Modify the uniform motion with an arc of one-quarter of the rise in the displacement diagram.

33. Draw a displacement diagram and a cam that will give a harmonic motion to a roller follower with a rise of 1.60 in.

34. Draw a displacement diagram and a cam that will give a harmonic motion to a knife-edge follower with a rise of 1.00 in.

35. Draw a displacement diagram and a cam that will give uniform acceleration to a knife-edge follower with a rise of 1.70 in.

36. Draw a displacement diagram and a cam that will give a uniform acceleration to a roller follower with a rise of 1.40 in.

37. Draw a displacement diagram and a cam that will give the following motion to a knife-edge follower: Rise 1.25 in. with harmonic motion in 120°, dwell for 120°, and fall 1.25 in. with uniform acceleration.

38. Draw a displacement diagram and a cam that will give the following motion to a knife-edge follower: Dwell for 70°; rise 1 in. with a modified uniform motion in 100°; fall 1 in. with a harmonic motion in 100°; and dwell for 90°.

39. Draw a displacement diagram and a cam that will give the following motion to a roller follower: Rise 1.25 in. with a harmonic motion in 120°; dwell for 120°; and fall 1.25 in. with a uniform acceleration in 120°.

40. Repeat Problem 32, but offset the follower 0.60 in. to the right of the vertical centerline.

41. Repeat Problem 33, but offset the follower 0.60 in. to the left of the vertical centerline.

11

Materials and Processes

11.1
Introduction

This chapter presents an overview of the materials and processes used in manufacturing (Fig. 11.1).

The study of metals, called **metallurgy**, is a highly complex area that is constantly changing as new processes and alloys are being developed. The guidelines for designations of various types of metals have been standardized by three associations: the American Iron and Steel Institute (**AISI**), the Society of Automotive Engineers (**SAE**), and the American Society for Testing Materials (**ASTM**).

11.2
Iron

Tom Pollock credit

Metals that contain iron, even in small quantities, are called **ferrous metals.** The three types of iron are **gray iron, white iron,** and **ductile iron.** Iron is used in the production of machine parts by the casting

process; consequently, iron is often called **cast iron.** Iron is cheaper than steel and it is easy to machine, but it does not have the ability to withstand the shock and forces that steel can withstand.

FIG. 11.1 This furnace operator is pouring an aluminum alloy of manganese into ingots (shown at the right) that will be remelted and cast. (Courtesy of the Aluminum Company of America.)

TABLE 11.1

ATSM Grade (1000 psi)	SAE Grade	Typical Uses
ASTM 25 CI	G 2500 CI	Small engine blocks, pump bodies, transmission cases, clutch plates
ASTM 30 CI	G 3000 CI	Auto engine blocks, flywheels, heavy casting
ASTM 35 CI	G 3500 CI	Diesel engine blocks, tractor transmission cases, heavy and high-strength parts
ASTM 40 CI	G 4000 CI	Diesel cylinders, pistons, cam shafts

*This section on iron was developed by Dr. Tom Pollock, a metallurgist at Texas A&M University.

GRAY IRON   contains flakes of graphite, which results in low strength and ductility but makes the material easy to machine. Gray iron will resist vibrations better than will other types if iron. Types of gray iron with two designations and their typical applications are given in Table 11.1.

WHITE IRON   contains carbon in the form of hard carbide particles that are very hard and brittle, which enables it to withstand wear and abrasion. There are no designated grades of white iron, but there are differences in composition from one supplier to another. White iron is used for parts on grinding and crushing machines, digging teeth on earthmovers and mining equipment, and wear plates on reciprocating machinery used in textile mills.

DUCTILE IRON   also called **nodular** or **spheroidized** iron, contains carbon in the form of tiny graphite spheres. It is usually stronger and tougher than **gray iron** of a similar composition, but it is more expensive to produce. The numbering system for ductile iron is given by three sets of numbers, as shown below:

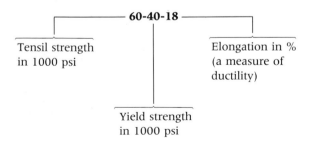

Commonly used alloys of ductile iron and their applications are shown in Table 11.2.

TABLE 11.2

Grade	Typical Uses
60–40–18 CI	Valves, steam fittings, chemical plant equipment, pump bodies
65–45–12 CI	Machine components that are shock loaded, disc brake calipers
80–55–6 CI	Auto crankshafts, gears, rollers
100–70–3 CI	High-strength gears and machine parts
120–90–2 CI	Very high strength gears, rollers, and slides

MALLEABLE IRON   is made from **white iron** by a heat-treatment process that converts carbides into carbon nodules (similar to ductile iron). The numbering system for designating the grades of malleable iron is shown below:

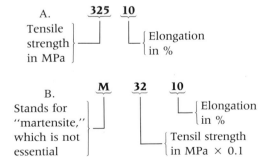

Some of the commonly used grades of malleable iron and their applications are given in Table 11.3.

TABLE 11.3

ASTM Grade	Typical Uses
35018 CI	Marine and railroad valves and fittings, "black-iron" pipe fittings (similar to 60–40–18 ductile CI)
45006 CI	Machine parts (similar to 80–55–6 ductile CI)
M3210 CI	Low-stress components, brackets
M4504 CI	Crankshafts, hubs
M7002 CI	High-strength parts, connecting rods, universal joints
M8501 CI	Wear-resistant gears and sliding parts

11.3
Steel

Steel is an alloy of iron with the addition of other materials, but carbon is the ingredient that has the greatest effect on the grade of the steel. The three broad types of steel are **plain carbon steels, free-cutting carbon steels,** and **alloy steels.** The types of steels

TABLE 11.4
NUMBERING AND APPLICATIONS OF TYPES OF STEEL

Type of Steel	Number	Application
Carbon steels		
Plain carbon	10XX	Tubing, wire, nails
Resulphurized	11XX	Nuts, bolts, screws
Manganese steel	13XX	Gears, shafts
Nickel steel	23XX	Keys, levers, bolts
	25XX	Carburized parts
Nickel-chromium	31XX	Axles, gears, pins
	32XX	Forgings
	33XX	Axles, gears
Molybdenum steel	40XX	Gears, springs
Chromium-molybdenum	41XX	Shafts, tubing
Nickel-chromium	43XX	Gears, pinions
Nickel-molybdenum	46XX	Cams, shafts
	48XX	Roller bearings, pins
Chromium steel	51XX	Springs, gears
	521XX	Ball bearings
Chromium vanadium	61XX	Springs, forgings
Silicon manganese	92XX	Leaf springs

Source: Courtesy of the Society of Automotive Engineers

and their designations by four-digit numbers are shown in Table 11.4.

The number designations begin with a digit that indicates the type of steel: 1 is carbon steel, 2 is nickel steel, and so on. The second digit gives the percentage content of the material represented by the first digit. The last two digits give the percentage of carbon in the alloy, where 100 is equal to 1% and 50 is equal to 0.50%.

Some of the more often used SAE (Society of Automotive Engineers) steels are: 1010, 1015, 1020, 1030, 1040, 1070, 1080, 1111, 1118, 1145, 1320, 2330, 2345, 2515, 3135, 3130, 3240, 3310, 4023, 4042, 4063, 4140, and 4320.

11.3
Copper

Copper was one of the first metals discovered. It is a soft metal that can be easily formed and bent without breakage. Since it has a high resistance to corrosion and because of its high level of conductance, it is used in the manufacture of pipes, tubing, and electrical wiring. It is an excellent roofing and screening material since it withstands the weather well.

Copper has a number of alloys, including brasses, tin bronzes, nickel silvers, and copper-nickel alloys. A few of the numbered designations of wrought copper are: C11000, C11100, C11300, C11400, C11500, C11600, C10200, C12000, and C12200. Brass is an alloy of copper and zinc, and bronze is an alloy of copper and tin.

Copper and copper alloys can be easily finished by buffing or plating. All of these can be joined by soldering, brazing, and welding, and can be easily machined and used for casting.

11.5
Aluminum

Aluminum is a corrosion-resistant, lightweight metal that has applications for many industrial products. Most of the materials that are called aluminum are actually alloys of aluminum which possess greater strength than the pure metal.

The types of wrought aluminum alloys are designated by four digits, as shown in Table 11.5. (Wrought

TABLE 11.5
NUMBERING DESIGNATIONS FOR WROUGHT
ALUMINUM AND ALUMINUM ALLOYS

Composition	Alloy Number	Applications
Aluminum (99% pure)	1XXX	Tubing, tank, cars
Aluminum alloys		
Copper	2XXX	Aircraft parts, screws, rivets
Manganese	3XXX	Tanks, siding, gutters
Silicon	4XXX	Forging, wire
Magnesium	5XXX	Tubes, welded vessels
Magnesium and silicon	6XXX	Auto body, pipe
Zinc	7XXX	Aircraft structures
Other elements	8XXX	

aluminum has properties that permit it to be formed by hammering.) The first digit, from 2 through 9, indicates the alloying element that is combined with aluminum. The last two digits identify the other alloying materials or indicate the aluminum purity. The second digit indicates modifications of the original alloy or impurity limits.

A four-digit numbering system, with the last digit to the right of the decimal point, is used to designate

TABLE 11.6
ALUMINUM CASTING AND
INGOT DESIGNATIONS

Composition	Alloy Number
Aluminum, (99 % pure)	1XX.X
Aluminum alloys	
Copper	2XX.X
Silicon with copper and/or magnesium	3XX.X
Silicon	4XX.X
Magnesium	5XX.X
Zinc	7XX.X
Tin	8XX.X
Other elements	9XX.X

Source: Society of Automotive Engineers.

types of cast aluminum and aluminum alloys. (When used for castings, the aluminum is melted and poured into a mold to form it.) The first digit indicates the alloy group as shown in Table 11.6. The next two digits identify the aluminum alloy or the aluminum purity. The numeral 1 to the right of the decimal point represents ingot aluminum, and 0 represents aluminum for casting. **Ingots** are blocks of cast metal that are to be remelted. **Billets** are castings of aluminum that are to be formed by forging.

11.6
Magnesium

Magnesium is a light metal that is available in an inexhaustible supply since it is extracted from seawater and natural brines. It is approximately half the weight of aluminum, and therefore it is an excellent material for aircraft parts, clutch housing, crankcases for air-cooled engines, and other applications where lightness of weight is desirable.

Magnesium and its alloys can be joined by bolting, riveting, and welding. Some numbered designations of magnesium alloys are: M10100, M11630, M11810, M11910, M11912, M12390, M13320, M16410, and M16620.

Magnesium is used for die casting, sand castings, extruded tubing, sheet metal, and forging.

11.7
Properties of materials

All materials have different properties that the designer must use to his or her best advantage. Consequently, it is important to be familiar with the terms used to describe these properties.

DUCTILITY is a softness present in some materials, such as copper and aluminum, that permits them to be formed by stretching (drawing) or hammering without breaking. Wire is made of ductile materials that can be drawn through a die.

BRITTLENESS is a characteristic of metals that will not stretch without breaking, such as cast irons and hardened steels.

MALLEABILITY is the ability of a metal to be rolled or hammered without breaking.

HARDNESS is a metal's ability to resist being dented when it receives a blow.

TOUGHNESS is the property of being shock-resistant while remaining malleable and resisting cracking or breaking.

ELASTICITY is the characteristic of a metal to return to its original shape after being bent or stretched.

11.8
Heat treatment of metals

The properties of different metals can be changed by various forms of heat treating. Steels are affected to a greater extent by heat treating than other materials.

HARDENING of steel is performed by heating the material to a prescribed temperature depending upon its content and then quenching the hot steel in oil or water.

QUENCHING is the process of rapidly cooling heated metal by immersing it in liquids, gases, or solids (such as sand, limestone, or asbestos).

TEMPERING is the process of reheating previously hardened steel and then cooling it, usually by air. This increases the steel's toughness.

ANNEALING is the process of heating and cooling metals to soften them, to release their internal stresses, and to make them easier to machine and work with.

NORMALIZING is achieved by heating metals and letting them cool in air to room temperature, relieving their internal stresses.

CASE HARDENING is the process of hardening a thin outside layer of a metal. In this process the outer layer is placed in contact with carbon or nitrogen compounds that become absorbed by the metal as it is heated. Afterwards, it is quenched to complete the case hardening.

FLAME HARDENING is the method of heating a metal to a prescribed range with a flame and then quenching the metal.

11.9
Castings

Two major methods of forming shapes are **casting** and **pressure forming.** Casting involves the preparation of a mold inside which is poured molten metal that cools and forms the part. The types of casting are **sand casting, permanent-mold casting, die casting,** and **investment casting.** These types of casting vary by the method in which the molds are made before pouring the metal.

Sand casting

Sand casting is the most commonly used method. In the first step, a form or **pattern** is made that is representative of the final part that is to be cast. It is made of wood or metal. The pattern is placed in a metal box called a **flask**, and molding sand is packed around the pattern. When the pattern is withdrawn from the sand it leaves a void that forms the mold. The molten metal is poured into the mold through **sprues**, or gates. After cooling, the casting is removed and cleaned (Fig. 11.2).

Cores are parts that are formed in sand and placed within a mold to leave holes or hollow portions within the finished casting (Fig. 11.3). Once the casting has been formed and set, the sand cores can be broken apart and removed, leaving behind the desired void within the casting, thereby reducing its weight. Cores add considerably to the cost of a casting, and they should not be used unless they are adequately offset by savings in materials.

Since the patterns must be placed in sand and then withdrawn before the metal is poured, it is necessary that the sides of the patterns be tapered for ease of withdrawal from the sand (Fig. 11.4). This taper is called **draft.** The amount of draft depends upon the depth of the pattern in the sand, and it varies from 2° to 8° for most applications. Also, patterns are made oversize to compensate for shrinkage that will occur when the metal cools.

The sand casting has a rough surface that is not desirable for contacting other moving parts or surfaces.

FIG. 11.2 A large casting of a landing-gear piston for an aircraft is being removed from its mold. (Courtesy of Cameron Iron Works.)

Consequently, it is common practice to machine portions of a casting by drilling, grinding, shaping, or other machining operations (Fig. 11.5). The pattern should be made larger in the machining areas to provide for removal of metal.

Fillets and rounds at all intersections will increase the strength of a casting. Also, fillets and rounds are used since it is difficult to form square corners by the sand casting process.

Permanent-mold castings

Permanent molds are often made for the mass production of parts. The molds are generally made of cast iron and are coated to prevent fusing with the molten metal that is poured into them. An example of a permanent mold is shown in Fig. 11.6.

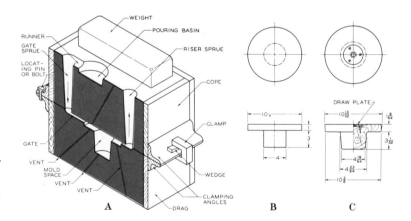

FIG. 11.3 A typical sand mold. The opening that will be filled with molten metal is formed by a wood or metal pattern that is pressed into the sand.

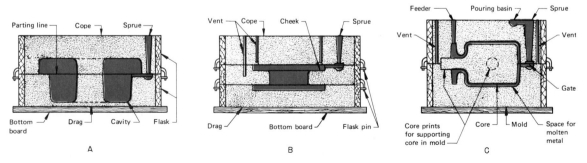

FIG. 11.4 Three types of sand molds: (A) two-section mold, (B) three-section mold, (C) two-section mold with a core (Courtesy of General Motors Corp.)

FIG. 11.5 This casting of the outer cylinder of an aircraft's landing gear is being bored on a horizontal boring mill. (Courtesy of Cameron Iron Works.)

Die castings

Die castings are generally used for the mass production of parts made of aluminum, magnesium, zinc alloys, and copper; however, other materials are used

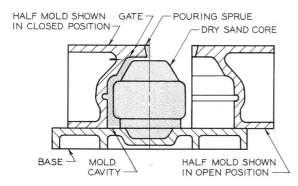

FIG. 11.6 Permanent molds are made of metal for repetitive usage when parts are mass produced. In this example a sand core is made from another mold and is placed in the permanent mold to give a hollow void within the casting.

also. Die castings are made by forcing molten metal into dies (or molds) under pressure. Dies are permanent molds that are used over and over again. They can be produced at a low cost, a high rate of production, at close tolerances, and with good surface qualities.

The same general principles recommended for sand castings—using fillets and rounds, allowing for shrinkage, and specifying draft angles—apply to die castings also Fig. 11.7.

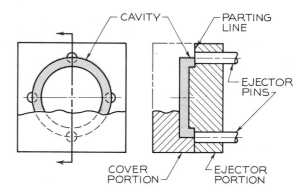

FIG. 11.7 A die for casting a simple part. Unlike the sand casting, the metal is forced into the die to form the die casting.

Investment casting

Investment casting is a process used to produce complicated parts that would be difficult to form with uniformity by any other method. This technique is even used to form the complicated forms of artistic sculptures.

A mold or die is made to cast a master pattern that is made of wax since a new pattern must be used for each investment casting. The wax pattern will be identical to the finished casting. The wax pattern is placed inside a container and a plaster mixture or sand is poured (or invested) around the wax pattern.

Once the investment has cured, the wax pattern is melted to remove the wax and leave a hollow cavity that will serve as the mold for the molten metal. When filled and set, the plaster or sand is broken away from the finished investment casting.

A working drawing of a casting is shown in Fig. 11.8.

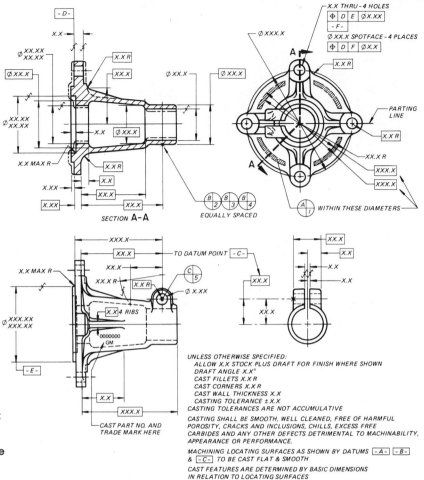

FIG. 11.8 A detail drawing of a die casting. Note that the part has been cast oversize to allow for finished surfaces that must be machined. (Courtesy of General Motors Corp.)

UNLESS OTHERWISE SPECIFIED:
ALLOW X.X STOCK PLUS DRAFT FOR FINISH WHERE SHOWN
DRAFT ANGLE X.X°
CAST FILLETS X.X R
CAST CORNERS X.X R
CAST WALL THICKNESS X.X
CASTING TOLERANCE ± X.X
CASTING TOLERANCES ARE NOT ACCUMULATIVE

CASTING SHALL BE SMOOTH, WELL CLEANED, FREE OF HARMFUL
POROSITY, CRACKS AND INCLUSIONS, CHILLS, EXCESS FREE
CARBIDES AND ANY OTHER DEFECTS DETRIMENTAL TO MACHINABILITY,
APPEARANCE OR PERFORMANCE.

MACHINING LOCATING SURFACES AS SHOWN BY DATUMS -A- -B-
& -C- TO BE CAST FLAT & SMOOTH

CAST FEATURES ARE DETERMINED BY BASIC DIMENSIONS
IN RELATION TO LOCATING SURFACES

FIG. 11.9 Three stages of manufacturing a turbine fan are shown here. The blank is first formed by forging; it is machined; and the fan blades are attached in their machined slots. (Courtesy of Avco Lycoming.)

11.10
Forgings

Forging is the process of shaping or forming metal by hammering or squeezing the heated metal into a die. The resulting forging possesses high strength and a resistance to loads and impacts.

When preparing forging drawings, the following must be considered: (1) draft angles and parting lines, (2) fillets and rounds, (3) forging tolerances, (4) allowance of extra material for machining, and (5) heat treatment of the finished forging (Fig. 11.9). Some of the standard steels used for forging are designated by the following SAE numbers: 1015, 1020, 1025, 1045, 1137, 1151, 1335, 1340, 4620, 5120, and 5140. Other materials that can be forged are iron, copper, and aluminum.

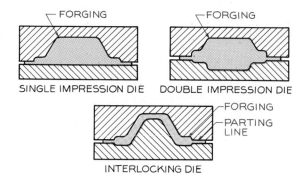

SINGLE IMPRESSION DIE   DOUBLE IMPRESSION DIE

INTERLOCKING DIE

FIG. 11.10 Three types of forging dies: (A) single-impression die, (B) double-impression die, and (C) interlocking die.

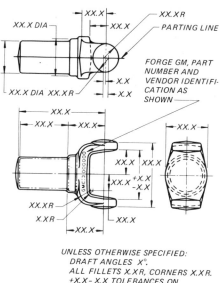

UNLESS OTHERWISE SPECIFIED:
DRAFT ANGLES X°.
ALL FILLETS X.XR, CORNERS X.XR.
+X.X - X.X TOLERANCES ON
FORGING DIM.

SNAG AND REMOVE SCALE.

SAMPLE FORGINGS ARE TO BE
APPROVED BY METALLURGICAL
AND ENGRG DEPTS FOR GRAIN
FLOW STRUCTURE.

FORGING DRAWING

FIG. 11.12 A typical detail drawing of a forging. The blank is forged oversize to allow for machining operations that will remove metal from it. (Courtesy of General Motors Corp.)

Drop forges and press forges are used to hammer the metal (called billets) into the forging dies by multiple blows or forces.

Examples of dies are shown in Fig. 11.10. A single-impression die gives an impression on one side of the parting line between the mating dies; the double-impression die gives an impression on both sides of the parting line. The interlocking dies result in a forging whose impression may cross the parting line on either side.

An example of an object that is forged with auxiliary rams to hollow the forging is shown in Fig. 11.11. A working drawing of a forged part is illustrated in Fig. 11.12.

Rolling

Rolling is a type of forging in which the stock is rolled between two rolls to give it a desired shape. Rolling can be done at right angles to the axis of the part or parallel to its axis (Fig. 11.13). The stock is usually heated before rolling if a high degree of shaping is required. If the forming requires only a slight change in configuration, the rolling can be performed when the metal is cold. This is called cold rolling.

11.11
Stamping

Stamping is an economical method of forming flat metal stock into three-dimensional shapes. The first step of stamping is to cut out the shapes, called **blanks,** that are to be bent.

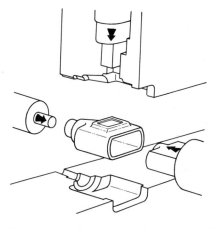

FIG. 11.11 Auxiliary rams can be used to form internal features on a part. (Courtesy of General Motors Corp.)

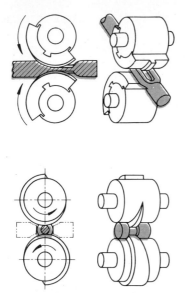

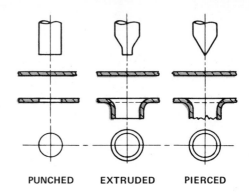

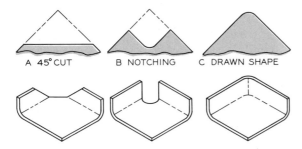

FIG. 11.13 In these examples, parts are being rolled parallel and perpendicular to their axes. (Courtesy of General Motors Corp.)

PUNCHED   EXTRUDED   PIERCED

FIG. 11.16 Three methods of forming holes in sheet metal (Courtesy of General Motors Corp.)

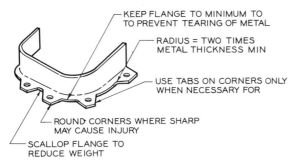

A 45° CUT   B NOTCHING   C DRAWN SHAPE

FIG. 11.14 Box-shaped parts formed by stamping. (A) A corner cut of 45° permits folding flanges and may require no further trim. (B) Notching has the same effect as the 45° cut, and is often more attractive. (C) A continuous corner flange requires that the blank be developed so that it can be drawn into shape.

Blanks are formed into shape by bending and pressing them against forms. Examples of box-shaped parts are shown in Fig. 11.14. An example of a flange stamping is illustrated in Fig. 11.15. Holes in stampings are made by punching, extruding, or piercing, as shown in Fig. 11.16.

11.12
Machining operations

After the metal has been formed into the shape of the final product, machining operations usually have to be performed to complete the part. The machining oper-

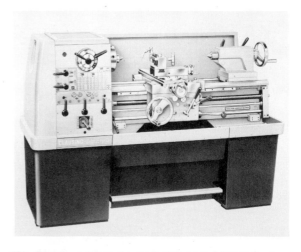

FIG. 11.17 A typical small-size metal lathe that holds the work piece between centers as the piece is rotated about its axis. (Courtesy of the Clausing Corp.)

KEEP FLANGE TO MINIMUM TO TO PREVENT TEARING OF METAL

RADIUS = TWO TIMES METAL THICKNESS MIN

USE TABS ON CORNERS ONLY WHEN NECESSARY FOR

ROUND CORNERS WHERE SHARP MAY CAUSE INJURY

SCALLOP FLANGE TO REDUCE WEIGHT

FIG. 11.15 A sheet metal flange design, with notes calling attention to design details.

ations involve several basic types of equipment: the **lathe, drill press, milling machine, shaper,** and **planer.**

The lathe

The lathe is a machine that shapes cylindrical parts by rotating the work piece between the centers of the lathe (Fig. 11.17). The more fundamental operations performed on the lathe are **turning, facing, drilling, boring, reaming, threading,** and **undercutting** (Fig. 11.18).

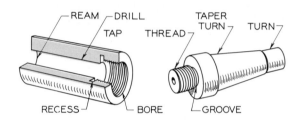

FIG. 11.18 The fundamental operations that are performed on a lathe are illustrated on the two parts shown above.

TURNING is the process of forming a cylinder by a tool that advances against the cylinder being turned and moves parallel to its axis (Fig. 11.19).

FACING is the process of forming flat surfaces that are perpendicular to the axis of rotation of the part being rotated by the lathe, such as the end of a cylinder.

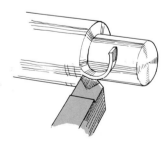

FIG. 11.19 Turning is the most basic of all operations performed on the lathe. A continuous chip is removed by a cutting tool as the part is rotated.

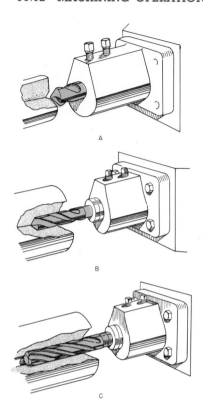

FIG. 11.20 Three steps of drilling a hole in the end of a cylinder involve (A) start drilling, (B) twist drilling, and (C) core drilling. These steps should be performed in sequence. Core drilling enlarges the previously drilled hole to the required size.

DRILLING is performed by mounting a drill in the tail stock of the lathe and rotating the work piece while the bit is advanced into the part (Fig. 11.20).

BORING is the process of making large holes that are too big to be drilled. Large holes are bored by enlarging smaller holes or cored in a casting (Fig. 11.21).

REAMING is the removal of only a few thousandths of an inch of material inside a drilled hole to bring it to its required level of tolerance. Conical as well as cylindrical reaming can be performed on the lathe (Fig. 11.22).

THREADING of external and internal holes can be done on the lathe. The die used for cutting internal holes is called a **tap** (Fig. 11.23).

The turret lathe is a programmable lathe that can perform sequential operations on the same part, such

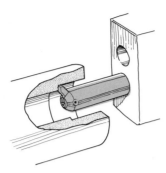

FIG. 11.21 Boring is the method of enlarging holes that are usually larger than available drill bits. The cutting tool is attached to the boring bar on the lathe.

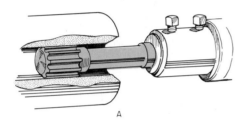

A

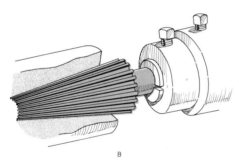

B

FIG. 11.22 Fluted reamers can be used to finish inside cylindrical and conical holes within a few thousandths of an inch.

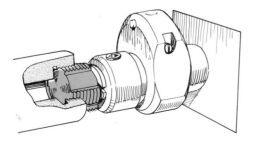

FIG. 11.23 External and internal threads (shown here) can be cut on a lathe with a tap. Note that a recess has been formed at the end of the threaded hole prior to threading.

FIG. 11.24 A turret lathe that performs a series of operations in succession as the turret head rotates.

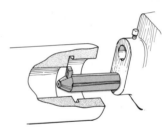

FIG. 11.25 A recess can be formed by using the boring bar with the cutting tools attached as shown. As the boring bar is moved off center of the axis, the tool will form the recess.

as drilling a series of holes, boring them, and then reaming them in succession (Fig. 11.24).

UNDERCUTTING is a method of cutting a recess inside of a cylindrical hole with a tool mounted on a boring bar. The groove is cut as the tool advances from the center of the axis of revolution into the part (Fig. 11.25).

The drill press

The drill press is used to drill small and medium size holes into stock that is held on the bed of the press by a fixture or a clamp (Fig. 11.26).

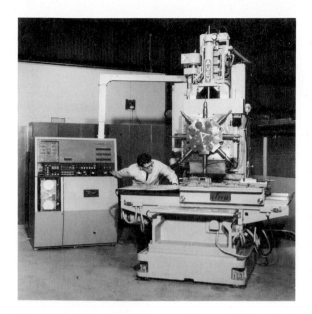

FIG. 11.26 A multiple-head drill press that can be programmed to perform a series of drill press operations in a desired sequence.

In addition to drilling holes, the drill press can be used for **counterdrilling, countersinking, counterboring, spotfacing,** and **threading** Fig. 11.27.

BROACHING Cylindrical holes can be converted into square holes or hexagonal holes by using a tool called a *broach.*

The broach has a series of teeth along its axis, beginning with teeth that are near the size of the hole to be broached, and tapering to the size of the finished hole that is to be broached. The broach is forced through the hole with each tooth cutting more from the hole as it passes through.

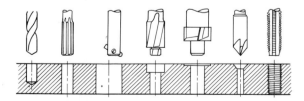

FIG. 11.27 The basic operations that can be performed on the drill press are (left to right): drilling, reaming, boring, counterboring, spotfacing, countersinking, and tapping (threading).

Milling machine

The milling machine uses a series of cutting tools that are rotated about a shaft (Fig. 11.28). The work piece is passed under the cutters and in contact with them to remove the metal.

The milling machine has a wide variety of cutters to form different grooved slots, threads, and gear teeth. Irregular grooves in cams can be cut by the milling machine. It can be used to finish a surface on a part within a high degree of tolerance.

FIG. 11.28 This small milling machine is being used to cut a slot in the work piece. (Courtesy of the Clausing Corp.)

The shaper

The shaper is a machine that holds the work stationary while the cutter on the machine passes back and forth across the work to finish the surface or to cut a groove, one stroke at a time (Fig. 11.29). With each stroke of the cutting tool, the material is shifted slightly so as to align the part for the next overlapping stroke.

The planer

The planer is similar to the shaper except that the work is passed under the cutters of the planer rather than the work remaining stationary as in the case of the shaper (Fig. 11.30). The planer can cut grooves or slots and finish surfaces that must meet tolerance specifications.

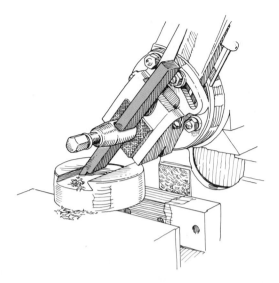

FIG. 11.29 The shaper moves back and forth across the part, removing metal as it advances. It can be used to finish surfaces, cut slots, and for many other operations.

FIG. 11.30 The planer has stationary cutters, and the work is fed past them to finish larger surfaces. This planer has a 30 foot bed. (Courtesy of Simmons Machine Tool Corp.)

11.13
Surface finishing

The process of finishing a surface to the desired uniformity is called surface finishing. It may be accomplished by several methods including **grinding, polishing, lapping, buffing,** and **honing.**

GRINDING is the finishing of a flat surface by holding it against a rotating abrasive wheel (Fig. 11.31). Grinding is used to smooth surfaces and to sharpen edges that are used for cutting, such as drill bits.

POLISHING is performed in the same manner as grinding except the polishing wheel is flexible since it is made of felt, leather, canvas, or fabric.

LAPPING is used to produce very smooth surfaces by holding the part against another large flat surface, called a **lap,** which has been coated with a fine abrasive powder. As the lap rotates, the surface is finished to a high degree of smoothness. Lapping is done only after the surface has been previously finished by a less accurate technique such as grinding or polishing. Cylindrical parts can be lapped by using a lathe in conjunction with the lapping surface.

BUFFING is a method of removing scratches from a surface with a rotating buffer wheel that is made of wool, cotton, or fabric. Sometimes the buffer is a cloth or felt belt that is applied to the surface being buffed. An abrasive mixture is applied to the buffed surface from time to time to enhance the buffing.

HONING is the process of finishing the outside or the inside of cylinders within a high degree of tolerance. The honing tool is rotated as it is passed through the holes to give the sort of finishes found in gun barrels, engine cylinders, and similar products where a high degree of smoothness is required.

FIG. 11.31 The upper surface of this part is being ground to a smooth finish by a grinding wheel. (Courtesy of the Clausing Corp.)

CHAPTER

12

Dimensioning

12.1
Introduction

Working drawings are dimensioned drawings used to describe the details of a part or a project so that construction can be performed in accordance with desired specifications.

When properly applied, dimensions and notes will supplement the drawings so they can be used as legal contracts for construction.

The techniques of dimensioning presented in this chapter are based primarily on the standards of the American National Standards Institute (ANSI) and especially their standards. Y14.5M, *Dimensioning and Tolerancing for Engineering Drawings.* Various industrial standards from major corporations, such as the General Motors Corp. have also been used.

12.2
Dimensioning terminology

The Guide Slide in Fig. 12.1 is used as an example to identify some of the terms of dimensioning.

DIMENSION LINES are thin lines (2H-4H pencil) with arrows at each end. Numbers placed near their midpoints specify a part's size.

EXTENSION LINES are thin lines (2H-4H pencil) that extend from a view of an object for dimensioning the part. The arrowheads of dimension lines end at these lines.

CENTERLINES are thin lines (2H-4H pencil) that are used to locate the centers of cylindrical parts, such as cylindrical holes.

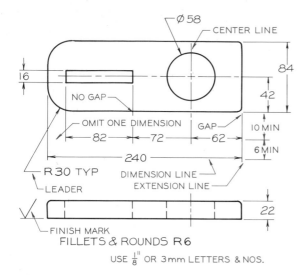

FIG. 12.1 This typical working drawing is dimensioned in millimeters.

LEADERS are thin lines (2H-4H pencil) drawn from a note to the feature to which the note applies.

ARROWHEADS are placed at the ends of dimension lines and leaders to indicate the endpoints of these lines. Arrowheads are drawn the same length as the height of the letters or numerals, ⅛ in. in most cases (Fig. 12.2).

DIMENSION NUMBERS are placed near the middle of the dimension line and are usually ⅛ in. in height; units (″, IN, or mm) are omitted.

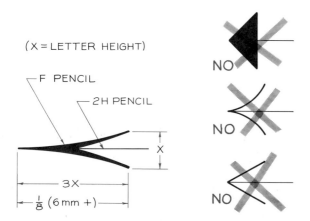

FIG.12.2 Arrowheads are drawn as long as the height of the letters used on a drawing. They are one-third as wide as they are long.

12.3
Units of measurement

The two most commonly used units of measurement are the **decimal inch**, in the English (imperial) system, and the **millimeter**, in the metric (SI) system.

The inch in its common fraction form can be used, but it is preferable not to. Common fractions make additional, division, and general arithmetic very hard to perform. A comparison of dimensions in millimeters with those in inches is shown in Fig. 12.3.

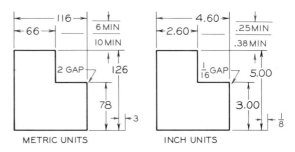

FIG. 12.3 Dimension lines should be placed at least ⅜ in. (10 mm) from an object. Other rows of dimensions should be located at least ¼ in. (6 mm) apart.

Examples are shown in Fig. 12.4 where the units are given in millimeters, decimal inches, and fractional inches. Dimensions in millimeters are usually rounded off to whole numbers **without decimal fractions**. When a metric dimension is less than a millimeter, a zero precedes the decimal point.

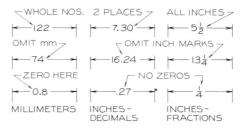

FIG. 12.4 When using SI units, dimensions are usually given to the nearest whole millimeter. When decimal inches are used, the fractions are carried to two decimal places. If fractions are given as common fractions, the fractions are twice as tall as whole numbers.

Units of measurement are omitted from dimension numbers, and the units are normally understood to be in millimeters or inches. For example: 112, not 112 mm; and 67, not 67″ or 5′–7″.

Architects use a combination of feet and inches, but the inch units can be omitted, where 7′–2 to represent seven feet and two inches. Feet and decimal fractions of feet are used by engineers to dimension large-scale projects such as road designs. These dimensions will be expressed as 252.7′, for example.

When using decimal inches, show all dimensions with two-place decimal fractions even though the last numbers are zeros. For dimensions of less than an inch, no zero precedes the decimal point.

12.4
English/metric conversions

Dimensions in inches can be converted to millimeters by multiplying by 25.4. Similarly, dimensions in millimeters can be converted to inches by dividing by 25.4.

For most applications, the millimeter does not need more than a one-place decimal when it is found by conversion from inches. The last digit retained in a conversion of either mm or inches is unchanged if it is followed by a number less than 5. For example, 34.43 is rounded off to 34.4 millimeters.

The last digit to be retained is increased by one if it is followed by a number greater than 5. The number 34.46 is rounded off to become 34.5.

The last digit to be retained is unchanged if it is even, and is followed by exactly 5. The number 34.45 is rounded off to become 34.4.

The last digit to be retained is increased by one if it is odd and is followed by exactly 5. The number 34.75 is rounded off to become 34.8.

12.5
Dual dimensioning

Some drawings require that both metric and English units be shown on each dimension. This method of dimensioning is called **dual dimensioning.**

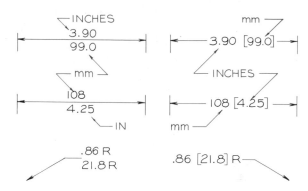

FIG. 12.5 Some dimensions are dual-dimensioned, with both inches and millimeters given on a single dimension line. If the drawing was originally made in inches, then the equivalent measurement in millimeters is placed under the inches or to the right in brackets. If the drawing was originally made in millimeters, then the inch equivalents would be placed under or to the right. When inches are converted to millimeters, the millimeters may need to be written as decimal fractions.

Dual dimensioning can be accomplished by placing the inch equivalent of millimeters either under or over the other units, as shown in Fig. 12.5. If the drawing was originally dimensioned in inches, then the inch dimensions are placed on top and the equivalent millimeters are given under them.

If the drawing was originally dimensioned in millimeters and then converted to inches, the millimeters would be placed over the equivalent in inches.

The other method of dual dimensioning involves the use of brackets placed around the converted dimensions (Fig. 12.5). It is important that you be consistent with whichever method you decide to use.

12.6
Metric designation

A comparison of the metric system with the English system is shown in Fig. 12.6. The metric system is the **Système International d'Unités** and is denoted by the letters SI. This system uses the European system of the first-angle of projection, which locates the front view over the top view and the right side view to the left of the front view.

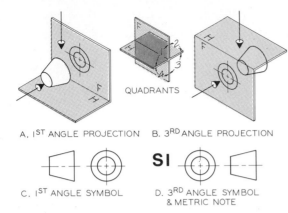

A. 1ST ANGLE PROJECTION   B. 3RD ANGLE PROJECTION

C. 1ST ANGLE SYMBOL   D. 3RD ANGLE SYMBOL & METRIC NOTE

FIG. 12.6 The European system of orthographic projection places the top view under the front view, the opposite of the American system. The system used on a drawing should be indicated by one of the symbols shown in Parts C and D. If metric units are used, the large letters SI should be placed near the title block, or else the word METRIC should be used on the drawing.

When drawings are made for international circulation, it is customary to use one of the symbols at (C) or (D) in Fig. 12.6 to designate the angle of projection used. The large letters **SI** are used to incidate that the measurements are metric, or the word **METRIC** can be written prominently on the drawing or in the title block.

12.7
Aligned and unidirectional numbers

The two methods of positioning dimension numbers on a dimension line are the **aligned** and **unidirectional** methods.

The unidirectional system is more widely accepted since it is easier to read numerals as shown in Fig. 12.7B.

The aligned system places the numerals in alignment with the dimension lines (Fig. 12.7A). The numbers must be readable from the bottom or the right side of the page.

Avoid placing aligned dimensions in the "trouble zone" since these numerals would read from the left instead of the right side of the sheet (Fig. 12.8).

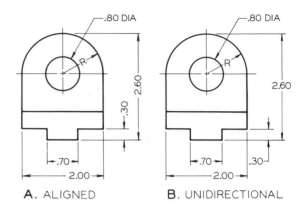

A. ALIGNED   B. UNIDIRECTIONAL

FIG. 12.7 (A) Aligned dimensions are positioned to read from the bottom and right side of the sheet. (B) When the dimensions are positioned so all of them read from left to right, they are unidirectional.

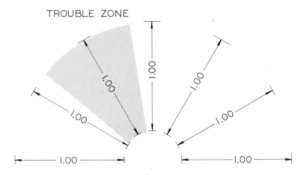

TROUBLE ZONE

FIG. 12.8 Numbers on angular dimension lines should not be placed in the trouble zone. This causes them to be read from the left side of the sheet rather than the right and bottom.

12.8
Placement of dimensions

Dimension the views that are most descriptive.

You can see that the front view of Fig. 12.9 is more descriptive than the top view, and therefore the front view is the view that should be dimensioned.

The dimensions should be applied to the views in an organized manner, such as those in Fig. 12.10. Locate the dimension lines by beginning with the smaller ones to avoid crossing the dimension and extension lines.

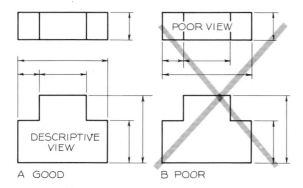

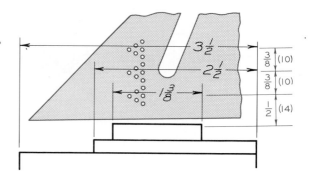

FIG. 12.9 Place dimensions on the most descriptive views, where the true contour of the object can be seen.

FIG. 12.12 When common fractions are used, the center holes of the triangular arrangements on the Braddock-Rowe triangle are aligned with the dimension lines to automatically space the lines as well as to draw the guide lines.

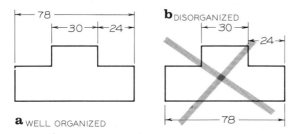

FIG. 12.10 Dimensions should be placed on the views in a well organized manner to make them as readable as possible.

The Braddock-Rowe triangle can be used to space the dimension lines and as a lettering guide (Fig. 12.12).

When a row of dimensions is placed on a drawing, one of the dimensions is omitted, as in Fig. 12.13A, since the overall dimension supplements the omitted dimension. If it is felt that the dimension needs to be given as a reference dimension, it is placed in parentheses to indicate that it is a reference dimension. In some cases the abbreviation REF is placed after a dimension to indicate a reference dimension.

The recommended techniques of dimensioning features are shown in Fig. 12.14.

Examples of the placement of extension lines are shown in Fig. 12.15. Extension lines may cross other extension lines or object lines with no gaps. Extension lines are also used to locate theoretical points outside of curved surfaces (Fig. 12.16).

Always leave at least 0.4 in. (10 mm) between the object and the first row of dimensions (Fig. 12.11). The successive rows of dimensions should be at least 0.25 in. (6 mm) apart. If greater spaces are used, these same general proportions should be applied.

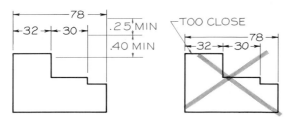

FIG. 12.11 The first row of dimensions should be placed at least 0.40 in. (10 mm) from the view, and successive rows should be at least 0.25 in. (6 mm) from the first row. If greater spaces are used, these general proportions should be used.

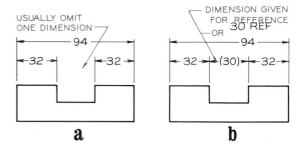

FIG. 12.13 One intermediate dimension is customarily omitted since the overall dimension provides this measurement, (A). If all the intermediate dimensions are given, one should be placed in parentheses to indicate that it has been given as a reference dimension.

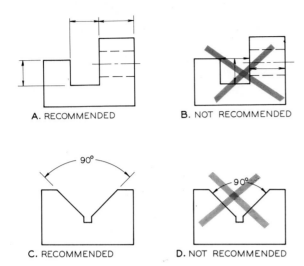

FIG. 12.14 Dimensioning rules

It is preferred practice to place the dimensions outside of a part. Dimension lines should not be used as extension lines, as at B.

(C and D) Angles are dimensioned with arcs and extension lines rather than showing the angle inside the angular cut, as shown at D.

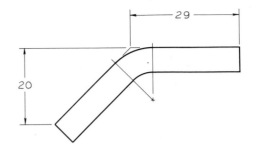

FIG. 12.16 A curved surface is dimensioned by locating the theoretical point of intersection with extension lines.

12.9
Dimensioning in limited spaces

Several examples of dimensioning in limited spaces are shown in Fig. 12.17. Regardless of space limitations, the numerals should not be drawn smaller than they appear elsewhere on the drawing.

When numbers are crowded where dimension lines are closely grouped, they should be staggered to make them more readable (Fig. 12.18).

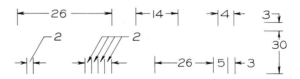

FIG. 12.17 Where room permits, the numerals and arrows should be placed inside the extension lines.

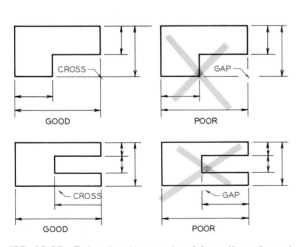

FIG. 12.15 Extension lines extend from the edge of an object leaving a small gap. They do not have gaps where they cross object lines or other extension lines.

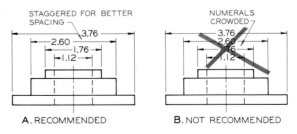

FIG. 12.18 When close spacing tends to crowd dimensioning numerals, they should be staggered for better spacing.

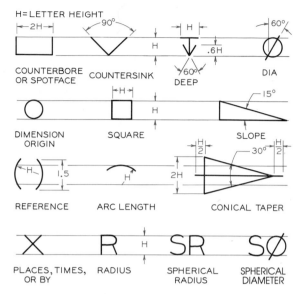

COUNTERBORE OR SPOTFACE · COUNTERSINK · DEEP · DIA

DIMENSION ORIGIN · SQUARE · SLOPE

REFERENCE · ARC LENGTH · CONICAL TAPER

PLACES, TIMES, OR BY · RADIUS · SPHERICAL RADIUS · SPHERICAL DIAMETER

FIG. 12.19 These symbols can be used to dimension parts. The proportions of the symbols are based on the letter height, H, which is usually ⅛ in.

12.10
Dimensioning symbology

A number of symbols used in dimensioning are shown in Fig. 12.19. The sizes of the symbols are based on the height of the lettering that is used in dimensioning an object, usually one-eighth inch. Symbols are used to reduce time in preparing notes while adding to the clarity of a dimension.

12.11
Dimensioning prisms

The following rules of dimensioning simple parts are illustrated in Fig. 12.20.

1/ Dimensions should extend from the most descriptive view (Fig. 12.20A).

2/ Dimensions that apply to two views should be placed between these two views (Fig. 12.20A).

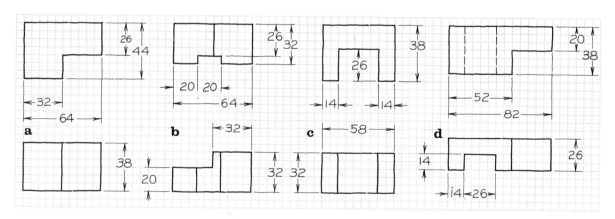

FIG. 12.20 Dimensioning prisms (metric)

A. Dimensions should extend from the most descriptive view and be placed between the views to which they apply.

B. One intermediate dimension is not given. Extension lines may cross object lines.

C. It is permissible to dimension a notch inside the object if this improves clarity.

D. Whenever possible, dimensions should be placed on visible lines and not hidden lines.

3/ The first row of dimension lines should be placed a minimum of 0.40 in. (10 mm) from the view. Successive rows are placed at least 0.25 in. (5 mm) apart.

4/ Extension lines may cross, but dimension lines should not cross another line unless absolutely necessary.

5/ In order to dimension each measurement in its most descriptive view, you may have to place dimensions in more than one view (Fig. 12.20B).

6/ Notches may be dimensional by placing the dimension line inside the notch (Fig. 12.20C).

7/ Whenever possible, dimensions should be applied to visible lines rather than hidden lines (Fig. 12.20D).

8/ Dimensions should not be repeated nor should unnecessary information be given.

12.12
Dimensioning angles

Angles can be dimensioned by using either coordinates to locate the ends of angular lines or planes, or angular measurements in degrees (Fig. 12.21). The two methods should not be mixed when dimensioning the same angle since they may not agree.

Units for angular measurements are degrees, minutes, and seconds, as shown in Fig. 12.21C. There are 60 minutes in a degree, and 60 seconds in a minute. Seldom will angular measurements need to be measured to the nearest second.

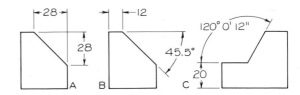

FIG. 12.21 Angular planes can be dimensioned by using coordinates (A), or an angle measured in degrees from the located vertex (B). The angles can be measured in decimal fractions (B) or in degrees, minutes, and seconds (C).

12.13
Dimensioning cylinders

Cylinders that are dimensioned may be either solid cylinders or cylindrical holes. All diametric dimensions should be preceded by the symbol ∅ (Fig. 12.22).

> Solid cylinders are dimensioned in their rectangular views using diameters (not radii) since diameters are much easier to measure.

Parts composed of several concentric cylinders are dimensioned with diameters, beginning with the smallest cylinder (Fig. 12.22B). Parts of this type are assumed to be concentric unless otherwise noted.

A cylindrical part may be represented with only one view since ∅ or DIA is used with the diametrical dimension (Fig. 12.22C).

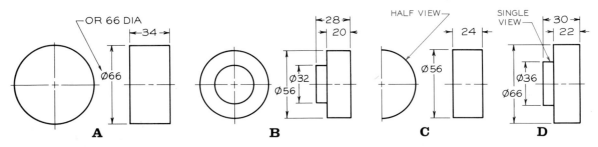

FIG. 12.22 (A) It is preferred that cylinders be dimensioned in their rectangular views using a diameter rather than a radius. (B) Dimensions should be placed between the views when possible. (C) The circular view can be omitted since ∅ is written before the dimensions of the diameters.

12.14

Measuring cylindrical parts

Cylindrical parts are dimensioned with diameters rather than radii because diameters are easier to measure. An internal cylindrical hole is measured with an internal micrometer caliper that has a built-in gauge permitting greater accuracy of measurement (Fig. 12.23).

Likewise, an external micrometer caliper can be used for measuring the outside diameters of a part (Fig. 12.24). The choice of the diameter rather than a radius makes it possible to measure diameters during machining when the part is held between centers on a lathe.

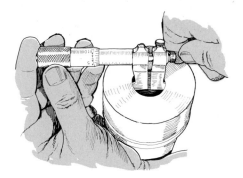

FIG. 12.23 A micrometer caliper is used to measure internal cylindrical diameters. This is why holes are dimensioned with diameters instead of radii.

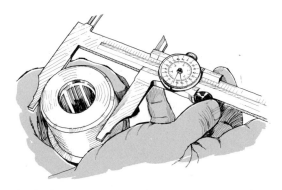

FIG. 12.24 An outside micrometer caliper with a built-in gauge for measuring the diameter of a cylinder.

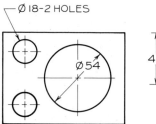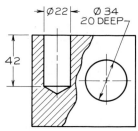

FIG. 12.25 Several acceptable methods of dimensioning cylindrical holes and shapes. The symbol $\varnothing$ is always placed in front of the diametrical dimension.

12.15

Cylindrical holes

Cylindrical holes may be dimensioned by one of the methods shown in Fig. 12.25.

The preferred method of dimensioning cylindrical holes is to draw a leader from the circular view, and then add the dimension, preceded by $\varnothing$, to indicate that the dimension is a diameter (Fig. 12.26). The English system often uses the DIA after the diametrical dimension.

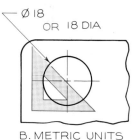

A. ENGLISH UNITS   B. METRIC UNITS

FIG. 12.26 The preferable method of dimensioning cylindrical holes is with a leader, the symbol $\varnothing$ to indicate that the dimension is a diameter, and a dimension. Previous standards recommended the abbreviation DIA after the dimension.

Sometimes the note DRILL or BORE is added to specify the shop operation, but current standards prefer the use of $\varnothing$ instead.

A part containing cylindrical features is dimensioned in Fig. 12.27 in both the circular and rectangular views to illustrate various methods of dimen-

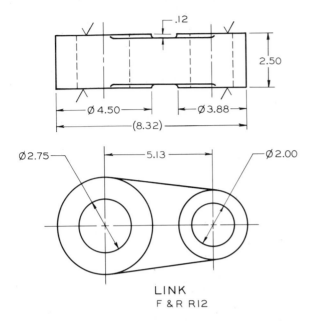

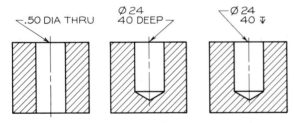

FIG. 12.27 This part, composed of cylindrical features, has been dimensioned using several approved methods. (F&R R.12 means that fillets and rounds have a 0.12 radius.)

FIG. 12.28 Three methods of dimensioning holes, with combinations of notes and symbols.

sioning. Three methods of dimensioning holes are shown in Fig. 12.28. The depth of a hole is its usable depth, not the depth to the point left by the drill bit.

12.16
Pyramids, cones, and spheres

Three methods of dimensioning pyramids are shown in Fig. 12.29A, B, and C. The pyramids at B and C are truncated (portions removed). Note that in all three examples the apex of the pyramid is located in the rectangular view.

Two acceptable methods of dimensioning cones are shown in Fig. 12.29D and E.

A sphere is dimensioned using its diameter if it is a complete sphere (Fig. 12.29F); a radius is used if it is less than a full sphere (Fig. 12.29G). Only one view is necessary to describe a sphere.

12.17
Leaders

Leaders are used to apply notes and dimensions to a feature that they describe. Leaders are drawn at a standard angle of a triangle, as illustrated in Fig. 12.26.

Examples of notes using leaders are shown in Fig. 12.30. The leader should be drawn from either the first word of the note or the last word of the note, and should begin with a short horizontal line near the note.

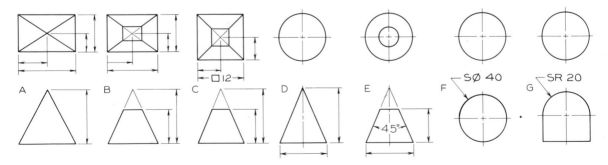

FIG. 12.29 Methods of dimensioning pyramids, cones, and spheres are shown here.

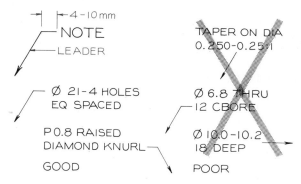

FIG. 12.30 Leaders from notes should begin with a horizontal bar from the first or last word of the note, not from the middle of the note.

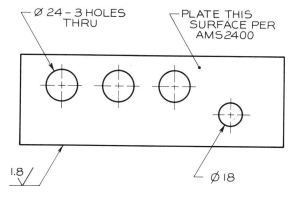

FIG. 12.31 Examples of notes applied to a part with leaders

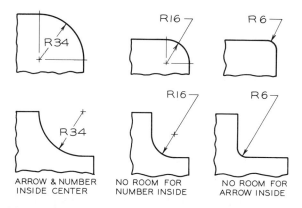

FIG. 12.32 When space permits, the dimension and arrow should be placed between the center and the arc. If room is not available for the number, the arrow is placed between the center and the arc with the number on the outside. If there is not room for the arrow, then both the dimension and the arrow are placed outside the arc with a leader.

Applications of leaders on a part are shown in Fig. 12.31. Notice that a dot is used instead of an arrowhead when the note is applied to a surface that does not appear as an edge.

12.18
Dimensioning arcs

Cylindrical parts that are less than a full circle are dimensioned with radii Fig. 12.32. Current standards recommend that radii be dimensioned with an *R* preceding the dimension, such as *R* 10. Previous standards recommended that the *R* follow the dimension, such as 10 *R*. Consequently, both methods will be seen in practice.

When the arc being dimensioned is a long arc, it may be dimensioned with a false radius, as shown in Fig. 12.33A. A "zigzag" is placed in the radius to indicate that it is not a true radius.

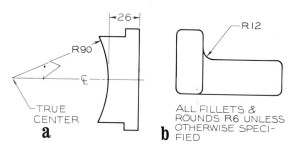

FIG. 12.33 When a radius is very long, it may be shown with a false radius with a "zigzag" to indicate that it is not true length. It should end on the centerline that passes through the true center. Fillets and rounds may be noted, to reduce repetitive dimensions of small arcs.

12.19
Fillets, rounds, and TYP

Fillets and **rounds** are rounded corners used on castings. A fillet is an internal rounding and a round is an external rounding.

When fillets and rounds are equal in radii, a note may be placed on the drawing to eliminate the need

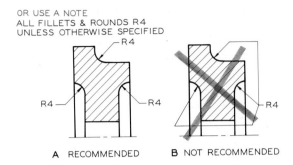

FIG. 12.34 When several arcs are dimensioned, it is preferable that separate leaders be used rather than extending the leaders as shown at B.

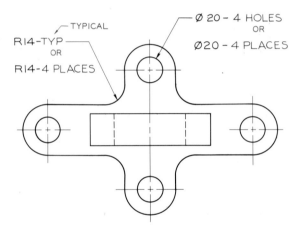

FIG. 12.35 Notes can be used to indicate that similar features and dimensions are repeated on drawings without having to dimension them individually.

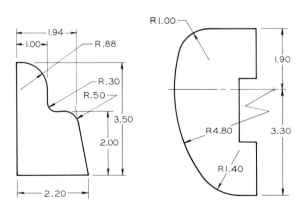

FIG. 12.36 Examples of dimensioned parts that are composed of tangent arcs.

for repetitive dimensioning: ALL FILLETS AND ROUNDS R6. If most, but not all, of the fillets and rounds have equal radii, the following note may be used: ALL FILLETS AND ROUNDS R6 UNLESS OTHERWISE SPECIFIED. In this case, only the fillets and rounds of different radii are dimensioned (Fig. 12.33B). The notes may be abbreviated such as F & R R10.

Fillets and rounds that are dimensioned as shown in Fig. 12.34A, rather than using long, confusing leaders as shown at B.

Repetitive features on a drawing may be noted as shown in Fig. 12.35. The note **TYPICAL** or **TYP** means that although only one of these features is dimensioned, these dimensions are typical of those that are undimensioned. The term **PLACES** is sometimes used to specify the number of places that a similar feature appears. Similarly, the number of holes sharing the same dimension may be indicated by including the number of holes in the note.

12.20
Curved surfaces

An irregular shape composed of a number of tangent arcs of varying sizes (Fig. 12.36) can be dimensioned by using a series of radii.

When the curve is irregular rather than composed of arcs (Fig. 12.37), the coordinate method can be used to locate a series of points along the curve from

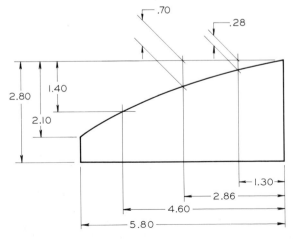

FIG. 12.37 This object with an irregular curve is dimensioned using coordinates.

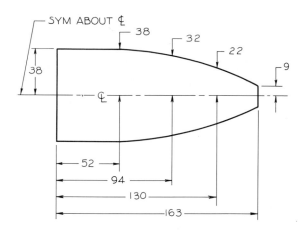

FIG. 12.38 This symmetrical part is dimensioned about its centerline.

two datum lines. The drafter must use judgment to determine the proper spacing for the points. Extension lines may be placed at an angle to provide additional space for showing dimensions.

A special case of an irregular curve is the symmetrical curve shown in Fig. 12.38. Note that dimension lines are used as extension lines, in violation of a previously established rule.

12.21
Symmetrical objects

Symmetrical objects may be dimensioned as shown in Fig. 12.39A, where a centerline is used with the initials **CL** to identify it. In this case, it is assumed that

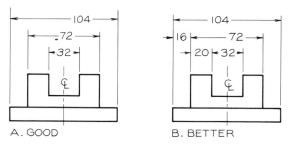

FIG. 12.39 Symmetrical parts may be dimensioned about their centerlines as shown in A. It is better to dimension symmetrical parts of this type as shown at B.

the dimensions are each centered about the centerline.

The better method is the one shown in Fig. 12.39B, where the need for an assumption is eliminated. All dimensions are located with respect to each other.

12.22
Finished surfaces

Many parts are formed as castings in a mold that gives their exterior surfaces a rough finish. If the part is designed to come in contact with another surface, the rough finish must be machined by grinding, shaping, lapping, or a similar process.

To indicate that a surface is to be finished, finish marks are drawn on the surface where it appears as an edge (Fig. 12.40).

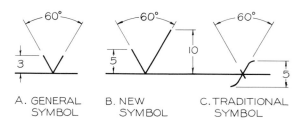

FIG. 12.40 Finish marks of these types can be used to indicate that a surface has been machined to a smooth surface. The traditional V can be used for general applications, but the new finish mark at B is suggested as the new symbol. The f shown at C is also acceptable to indicate finished surfaces.

> **Finish marks** should be repeated in every view where the finished surface appears as an edge, even if it is a hidden line.

Three methods of drawing finish marks are shown in Fig. 12.40. The simple V mark is preferred in general cases. The uneven V (Fig. 12.40B) is a newly recommended symbol that is related to surface texture and will be discussed in the next chapter. The steps of constructing the traditional f-mark are shown in Fig. 12.41.

When an object is finished on all surfaces, the note FINISHED ALL OVER or FAO is placed on the drawing.

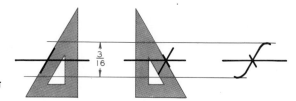

FIG. 12.41 The steps of drawing the f-mark

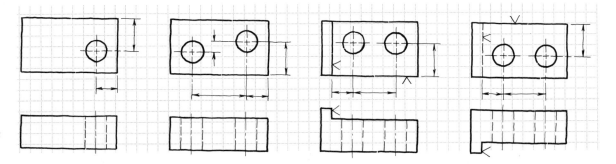

FIG. 12.42 Location dimensions

A. Cylindrical holes should be located in their circular views from two surfaces of the object.

B. When more than one hole is to be located, the other holes should be located in relation to the first from center to center.

C. Holes should be located from finished surfaces.

D. Holes should be located in the circular view and from finished surfaces even if the finished surfaces are hidden.

12.23
Location dimensions

Location dimensions are used to locate the positions, not the sizes, of geometric elements such as cylindrical holes (Figure 12.42). Since location dimensions do not involve the sizes of various features, the dimensions of the holes are omitted for clarity in Fig. 12.42. Centers of the holes are located with coordinates in the circular view when possible.

When finished surfaces appear on the object, it is desirable that holes be located from these surfaces since they can be located more accurately from a smooth, machined surface than from an unfinished surface. This rule is followed even if the finished surface is a hidden line, as in Fig. 12.43.

Only a single corner of a prism need be located with respect to another when the sides of the prisms are parallel (Fig. 12.44).

Location dimensions should be placed on views where both dimensions can be shown, the top view in

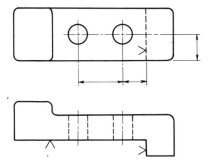

FIG. 12.43 Cylindrical holes are located from center to center in their circular view and from a finished surface, if one is available.

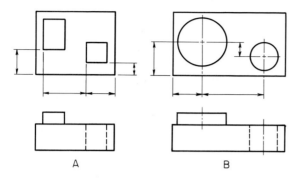

FIG. 12.44 Prisms and cylinders are located with coordinates in the view where both coordinates can be seen.

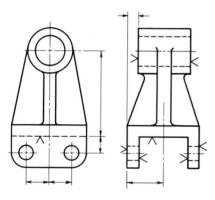

FIG. 12.45 Location dimensions applied to a part to locate its geometric features.

Fig. 12.44. Cylinders are located in their circular views (Fig. 12.44B and Fig. 12.45).

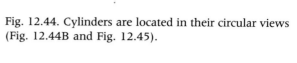

12.24
Location of holes

When holes must be located accurately, the dimensions should originate from a common reference plane on the part to reduce the accumulation of errors in measurement (Figs. 12.46A and B).

When several holes in a series are to be equally spaced, as in Fig. 12.46C, a note specifying that they are equally spaced can be used to locate the holes. The first and last holes of the series are determined by the usual location dimensions.

Holes through circular plates (Fig. 12.47) may be located by coordinates or by a note. When a note is used, the diameter of the circle passing through the centers of the hole must be given. This circle is referred to as the "bolt circle," or circle of centers.

A similar method of locating holes is the polar system illustrated in Fig. 12.48. The radial distances from the point of concurrency and their angular measurements (in degrees) between the holes are used to locate the centers.

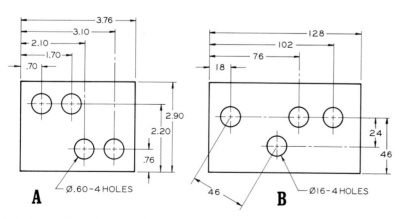

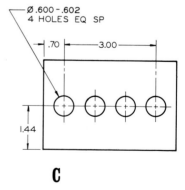

FIG. 12.46 Location of holes

A. Holes can be more accurately located if common datum planes are used, from which all measurements are made.

B. A diagonal dimension can be used to locate a hole of this type from another hole's center.

C. A note can be used to specify the spacing between the centers of holes in an equally spaced series.

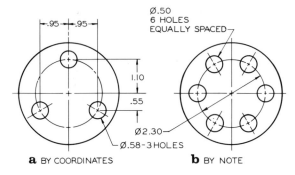

a BY COORDINATES   **b** BY NOTE

FIG. 12.47 Holes may be located in circular plates by coordinates (A) or by a note (B).

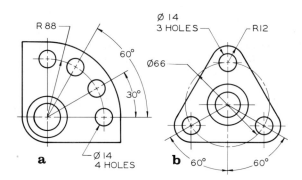

FIG. 12.48 Methods of locating cylindrical holes on concentric arcs

12.25
Objects with rounded ends

Objects with rounded ends should be dimensioned from end to end, as shown in Fig. 12.49. The radius is shown with the letter R without a dimension to specify that the end is formed by an arc. Since the height

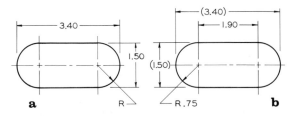

FIG. 12.49 The preferred method of dimensioning objects with rounded ends is shown at (a). A less desirable method is shown at (b).

is given, the radius is understood to be half of the height (1.50 in this case).

If the object is dimensioned from center to center of the rounded ends, the overall dimension should be given as a reference dimension (3.40) to eliminate the calculations required to determine the overall dimension. In this case the dimension of the radius must be given.

A part with partially rounded ends is dimensioned as shown in Fig. 12.50A. The overall dimension and the radii are given in order that their centers may be located.

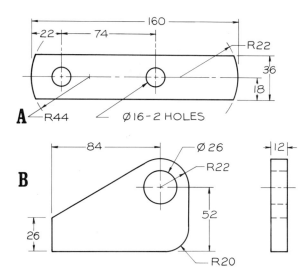

FIG. 12.50 Examples of dimensioned parts consisting of rounded ends and cylindrical features.

When an object has a rounded end that is less than a full semicircle (Fig. 12.50B), location dimensions must be used to locate the center of the arc.

Slots with rounded ends are dimensioned in Fig. 12.51. Only one slot is dimensioned in Part A, with a note indicating that there are two slots. The slot at B is dimensioned giving the overall dimension and the two arcs, which are understood to apply to both ends. The distance between the centers is given as a reference dimension.

The dimensioned views of the tool holder table in Fig. 12.52 show examples of arcs and slots. To prevent crossing of dimension lines, it is often necessary to place dimensions in a view that is not as descriptive as might be desired.

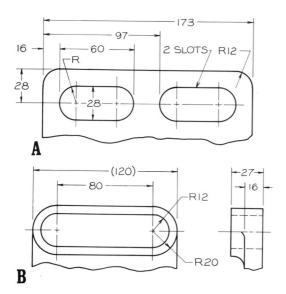

FIG. 12.51 Methods of dimensioning slots

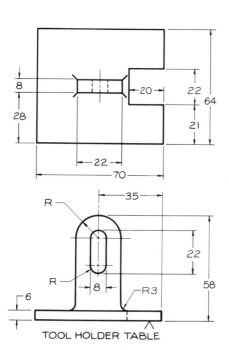

FIG. 12.52 This part is dimensioned using the previously presented principles.

12.26
Machined holes

Machined holes are holes that are made or refined by a machine operation, such as drilling or boring are specified by notes and leaders (Fig. 12.53).

It is preferred to give the diameter of the hole with the symbol ∅ in front of the dimension, with no reference to the machining method used. However, the note 32 DRILL may be used in some cases instead of ∅ 32. In practice, you will see diameters dimensioned as 32 DIA, since this was recommended by the previous standards.

DRILLED HOLES The depth of drilled holes can be specified in the note or it may be dimensioned in the rectangular view (Fig. 12.53). Note that the depth of a drilled hole is dimensioned as the usable part of the hole; the conical point is disregarded, as shown in Part B.

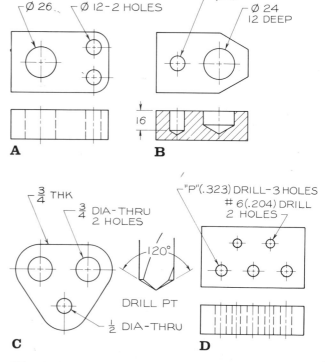

FIG. 12.53 Cylindrical holes may be dimensioned by either of the methods shown at A and B. Note that when one view is given at C, it is necessary to indicate in the note that the holes are "THRU" holes, since this cannot be seen otherwise.

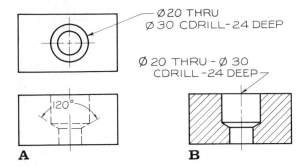

FIG. 12.54 Counterdrilling notes give the specifications for a larger hole drilled inside a smaller hole. The 120° angle is not required as a dimension since this is the standard angle of a drill point.

COUNTERDRILLED HOLES A large hole drilled inside a smaller drilled hole to enlarge it is a counterdrilled hole, as illustrated in Fig. 12.54. The 120° dimension is given to indicate the angle of the drill point and it need not be shown on the drawing.

COUNTERSUNK HOLES Countersinking is the process of forming a hole for a conical head (Fig. 12.55). The diameter of the countersink hole (the maximum diameter on the surface) and the angle of the countersink are given in the notes.

Countersinking is also used to provide center holes in shafts, spindles, and other cylindrical parts that must be held between the centers of a lathe (Fig. 12.56).

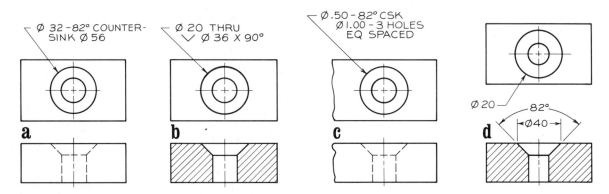

FIG. 12.55 Examples of notes specifying countersunk holes are shown here in various acceptable forms.

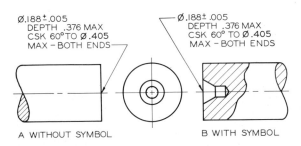

FIG. 12.56 Methods of specifying countersinks in the ends of cylinders for mounting them on a lathe between centers.

SPOTFACED HOLES This is a machining process used to finish the surface around the top of a hole in order to provide a seat for a washer or a fastener head (Fig. 12.57a and b).

A spotfacing tool is shown in Fig. 12.58, where it has spotfaced a **boss** (a raised cylindrical element).

BORED HOLES Boring is a machine operation for making large holes that is usually performed on a lathe with a boring bar (Fig. 12.59).

COUNTERBORED HOLES Counterbored holes are made by boring holes within smaller holes (Fig. 12.57). Note that the bottoms of the counterbored holes are flat, with no taper as in counterdrilled and countersunk holes.

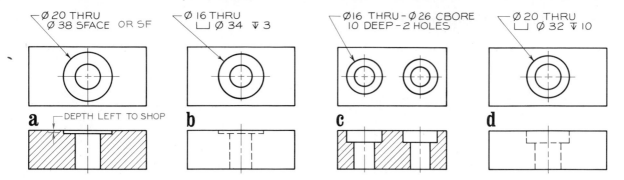

FIG. 12.57 Spotfaces can be specified as shown at (a) and (b). The depth of the spotface can be specified as shown if needed. Counterbores are dimensioned by giving the diameters of both holes and the depth of the larger hole (c and d).

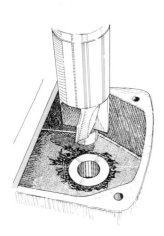

FIG. 12.58 This spotfacing tool is being used to spotface the cylindrical boss to provide a smooth seat for a bolt head.

FIG. 12.59 Boring a large hole on a lathe with a boring bar. (Courtesy of Clausing Corp.)

REAMED HOLES These are drilled or bored holes that have been finished or slightly enlarged using a **ream**, which is similar to a drill bit.

12.27
Chamfers

Chamfers are beveled edges formed on cylindrical parts such as shafts and threaded fasteners to eliminate rough edges and facilitate assembly.

When a chamfer angle is 45°, a note can be used in either of the forms shown in Fig. 12.60A. For other

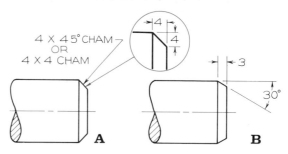

FIG. 12.60 Chamfers can be dimensioned by using either type of note shown at A when the angle is 45°. If the angle is other than 45°, it should be dimensioned as shown at B.

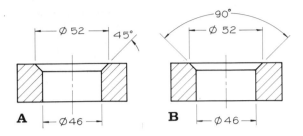

FIG. 12.61 Inside chamfers are dimensioned by using one of these methods.

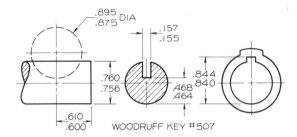

FIG. 12.62 Methods of dimensioning slots in a shaft and a slot for a Woodruff key that will hold the part on the shaft are shown here. The dimensions for these features are given in Appendix 32.

angles, the angle and the length of the chamfer are given, as shown in Part B.

Chamfers can also be specified at the openings of holes, as shown in Fig. 12.61.

12.28
Keyseats

A keyseat is a slot cut into a shaft for the purpose of aligning the shaft with a part mounted on it such as may be a pulley or a collar. The method of dimensioning a keyway and keyseat is shown in Fig. 12.62. The double dimensions are tolerances that will be discussed in the next chapter.

12.29
Knurling

Knurling is the operation of cutting diamond-shaped or parallel patterns on cylindrical surfaces for gripping, for decoration, or for a press fit between two parts that will be permanently assembled.

A diamond knurl and a straight knurl are drawn and dimensioned in Fig. 12.63. Knurls should be dimensioned with specifications that give type, pitch, and diameter.

The "96 DP" means diametrical pitch, which is the ratio of the number of grooves on the circumference (N) to the diameter (D), and is found by the equation $DP = N/D$. The preferred diametrical pitches for knurling are 64DP, 96DP, 128DP, and 160DP. For diameters of one inch, knurling of 64DP, 96DP,

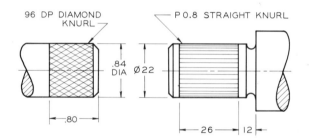

FIG. 12.63 A diamond knurl with a diametrical pitch of 96 is shown at A. The diametrical pitch (DP) is the number of teeth about the circumference divided by the diameter. A straight knurl is shown at B where the linear pitch (P) is 0.8 mm. Pitch is the distance between the grooves on the circumference.

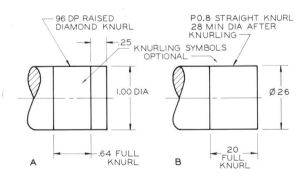

FIG. 12.64 Knurls need not be drawn if they are dimensioned as shown here.

128DP, and 160DP will have 64, 96, 128, and 160 teeth, respectively, on the circumference. The note P 0.8 means that the knurling grooves are 0.8 mm apart. (Note: Calculations must be made using inches. Conversion to millimeters can be made afterward.)

Knurls for press fits are specified with diameters before knurling and the minimum diameter after knurling. A simplified method of representing knurls is shown in Fig. 12.64, where notes are used and the knurls are not drawn.

12.30
Necks and undercuts

A **neck** is a recess cut into a cylindrical part used where cylinders of different diameter join (Fig. 12.65). The neck ensures that a part assembled on the smaller shaft will fit flush against the shoulder of the larger cylinder.

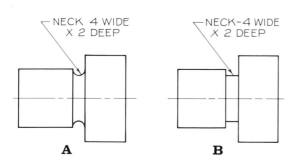

FIG. 12.65 Necks are recesses in cylinders that are used where cylinders of different sizes join together.

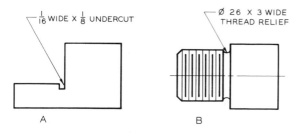

FIG. 12.66 An undercut can be dimensioned as shown at A. A thread relief, which is a type of neck, is dimensioned at B.

Undercuts are similar to necks and serve the same purpose (Fig. 12.66A). An undercut ensures that a part fitting in the corner of the surface will fit flush against both surfaces, and it permits space for trash to drop out of the way. An undercut could also be a recessed neck on the inside of a cylindrical hole. A thread relief (Fig. 12.66B) is used to square the threads where they intersect a larger cylinder.

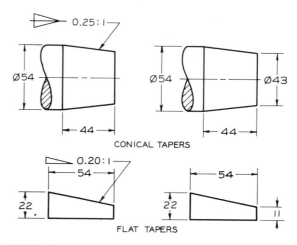

FIG. 12.67 Examples of flat and conical tapers. Taper is the ratio of diameters (or heights) at each end of a sloping surface to the length of the taper.

12.31
Tapers

Tapers can be either conical surfaces or flat planes. (Fig. 12.67). A taper can be dimensioned by (1) the diameter or width at each end of the taper, (2) the length of the tapered feature, or (3) the rate of taper.

Taper is the ratio of the difference in the diameter of a cone to the distance between two diameters (Fig. 12.67). Taper can be expressed as inches per inch (.25 per inch), inches per foot (3.00 per foot), or millimeters per millimeter (0.25:1).

Flat taper is the ratio of the difference in the heights at each end of a feature to the distance be-

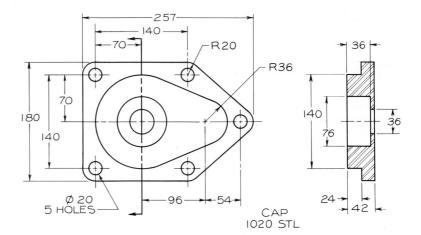

FIG. 12.68 An example of a part dimensioned with a variety of dimensioning principles.

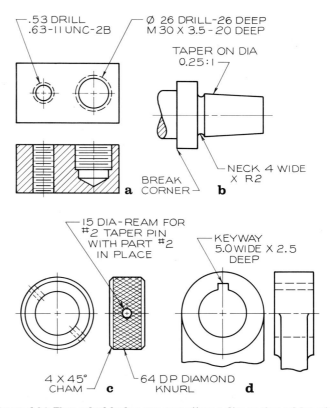

FIG. 12.69 (a and b) Threaded holes are sometimes dimensioned by giving the tap drill size in addition to the thread specifications. The tap drill size is not required, but it is permissible. The part at (b) is dimensioned to indicate a neck, a taper, and a break corner, which is a slight round to remove the sharpness from a corner. The collar at (c) has a knurl note, a chamfer note, and a note indicating the insertion of a #2 taper pin. The part at (d) gives a note for dimensioning a keyway.

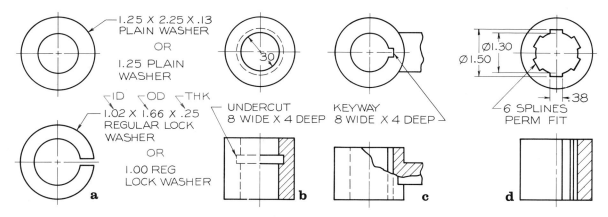

FIG. 12.70 (a and b) Methods of dimensioning washers and lock washers are shown. These dimensions can be found in the tables in Appendixes 36 and 37. An undercut is dimensioned at (b).

(c and d) A keyway is dimensioned at (c) and a spline inside a hole is dimensioned at (d).

tween the heights. Flat tapers can be expressed as inches per inch (.20 per inch), inches per foot (2.40 per foot), or millimeters per millimeter (0.20:1), as shown in Fig. 12.67.

12.32
Dimensioning sections

Sections are dimensioned in the same manner as regular views (Fig. 12.68). Most of the principles of dimensioning covered in this chapter have been applied to.

12.33
Miscellaneous notes

A variety of notes are used on detail drawings to provide information and specifications that would be dif-

ficult to represent by other means (Figs. 12.69 and 12.70).

Notes are placed horizontally on the sheet, with short dashes between the lines in the notes if they lie on the same horizontal line. The common abbreviations that are used in these notes can be used to save space. Refer to Appendix 1 for the approved abbreviations.

Leaders for the notes originate at the first or the last word of the note.

12.34
Tabular drawings

When parts have features that are similar except for variations in size, a table of values can be given to provide the dimensions, (Fig. 12.71). The part numbers are given in the tables with their respective dimensions A through Z. Tabular dimensions shorten drafting time while providing the essential information.

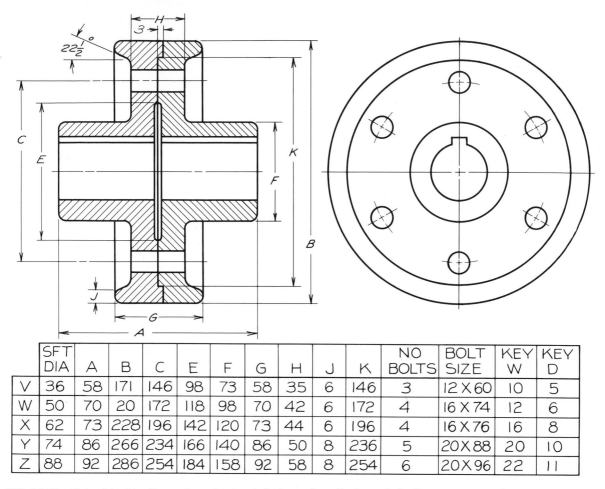

	SFT DIA	A	B	C	E	F	G	H	J	K	NO BOLTS	BOLT SIZE	KEY W	KEY D
V	36	58	171	146	98	73	58	35	6	146	3	12 X 60	10	5
W	50	70	20	172	118	98	70	42	6	172	4	16 X 74	12	6
X	62	73	228	196	142	120	73	44	6	196	4	16 X 76	16	8
Y	74	86	266	234	166	140	86	50	8	236	5	20 X 88	20	10
Z	88	92	286	254	184	158	92	58	8	254	6	20 X 96	22	11

FIG. 12.71 This object is dimensioned using tabular values that apply to the same drawing. The letters x on each dimension line are given in the table below the drawing.

Problems

1–50. The problems shown in Figs. 12.72 and 12.73 are to be solved on Size A paper if they are drawn half-size. If drawn full size, use Size B paper. Note that the dimensions of the views are found by using your dividers to transfer drawing sizes to the scales beneath the problems.

You will need to vary the spacing between the views to provide adequate room for the dimensions. It would be desirable to sketch the views and their dimensions in order to determine the spacing required before laying out the problems with instruments.

Refer to Section 2.19 for typical layouts.

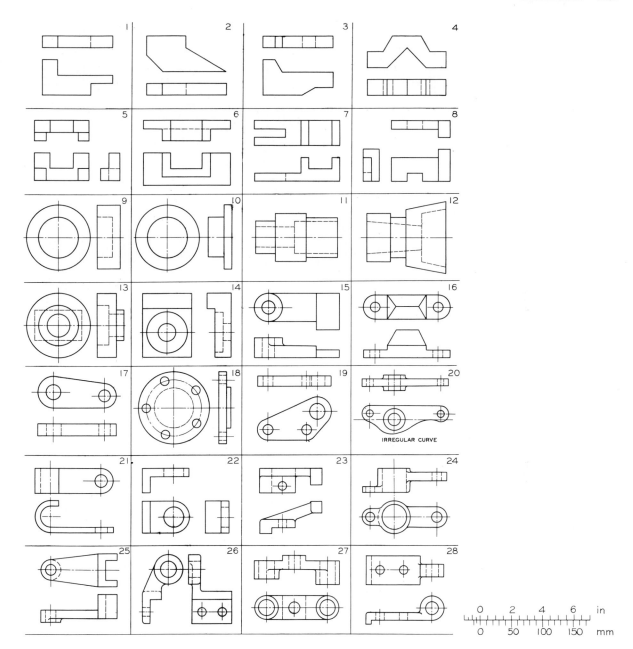

FIG. 12.72 Problems 1–28.

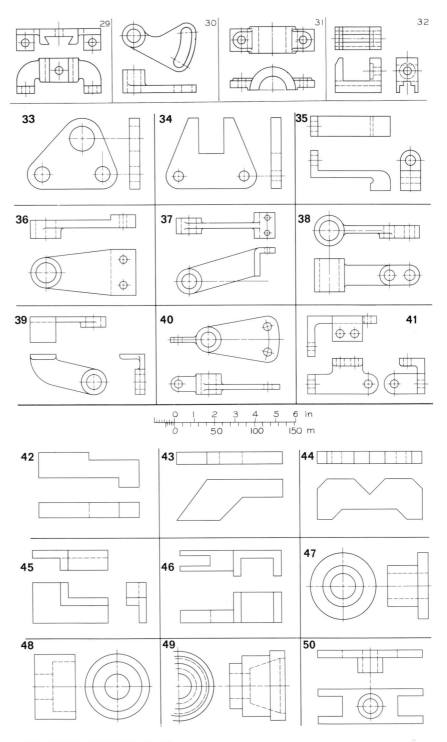

FIG. 12.73 Problems 29–50.

13

Tolerances

13.1
Introduction

Parts produced today require more accurate dimensions than did those produced in the past because many of today's parts are made by different companies in different locations. Therefore these parts must be specified so they will be interchangeable.

The techniques of dimensioning parts to ensure interchangeability is called **tolerancing.** Each dimension is allowed a certain degree of variation within a specified zone, called a tolerance. For example, a part's dimension might be expressed as 100 ± 0.50, which yields a tolerance of 1.00 mm.

All dimensions be given as large a tolerance as possible without interfering with the function of a part. This reduces production costs, since manufacturing to close tolerances is expensive.

13.2
Tolerance dimensions

Several acceptable methods of specifying tolerances are shown in Fig. 13.1. When ''plus-and-minus'' tolerancing is used, tolerances are applied to a basic di-

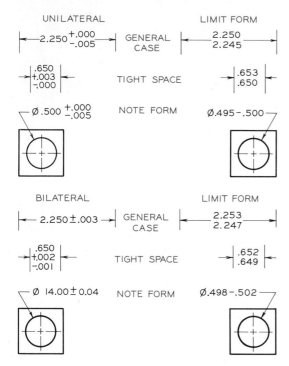

FIG. 13.1 Methods of positioning and indicating tolerances in unilateral, bilateral, and limit forms.

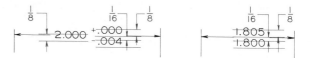

FIG. 13.3 Positioning and spacing of numerals used to specify tolerances.

and spacing of numerals of toleranced dimensions are shown in Fig. 13.3.

Tolerances may be given in the form of limits; that is, two dimensions are given that represent the largest and smallest sizes permitted for a feature of the part. When limits for dimensions are compared with their counterparts in plus-and-minus form, we see that both methods result in the same tolerance.

mension. When dimensions allow variation in only one direction, the tolerancing is **unilateral.** Tolerancing that permits variation in either direction from the basic dimension is **bilateral.**

The customary methods of indicating toleranced dimensions are shown in Fig. 13.2. The positioning

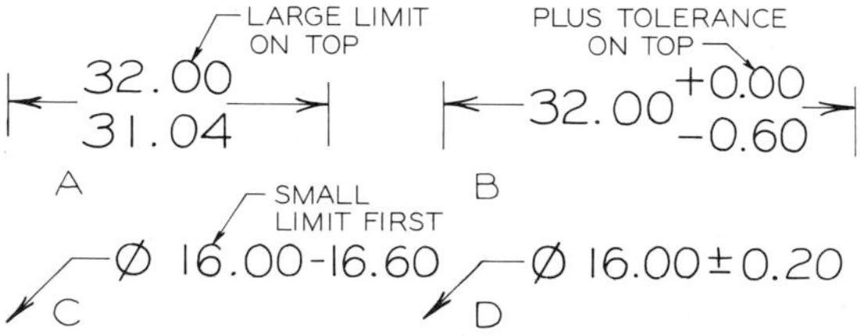

13.3
Mating parts

Mating parts are parts that fit together within a prescribed degree of accuracy (Fig. 13.4). The upper piece

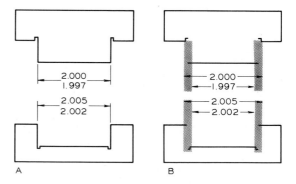

FIG. 13.2 When limit dimensions are given, the large limits are placed either above the small limits or to their right. The plus limits are placed above the minus limits in plus-and-minus tolerancing.

FIG. 13.4 Each of these mating parts has a tolerance of 0.003 in. (variation in size). The allowance between the assembled parts (tightest fit) is 0.002 in.

is dimensioned with two measurements that indicate the upper and lower limits of the size. The notch is slightly larger, allowing the parts to be assembled with a clearance fit.

An example of mating cylindrical parts is shown in Fig. 13.5A. Part B of the figure illustrates the meaning of the tolerance dimensions. The size of the shaft can vary in diameter from 1.500 in. (its maximum size) to 1.498 in. (its minimum size). The difference between these limits on a single part is a tolerance, 0.002 in. in this case. The dimensions of the hole in Part A are given with limits of 1.503 and 1.505, for a tolerance of 0.002 (the difference between the limits as illustrated in Part B).

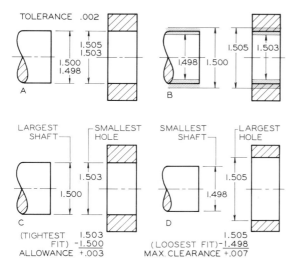

FIG. 13.5 The allowance (tightest fit) between these assembled parts is +0.003 in. The maximum clearance is 0.007 in.

13.4
Terminology of tolerancing

The meaning of most of the terms used in tolerancing can be seen by referring to Fig. 13.5.

TOLERANCE The difference between the limits prescribed for a single part. The tolerance of the shaft in Fig. 13.5 is 0.002 in.

LIMITS OF TOLERANCE The extreme measurements permitted by the maximum and minimum sizes

of a part. The limits of tolerance of the shaft in Fig. 13.5 are 1.500 and 1.498.

ALLOWANCE The tightest fit between two mating parts. The allowance between the largest shaft and the smallest hole in Fig. 13.5 is 0.003 (negative for an interference fit).

NOMINAL SIZE An approximate size that is usually expressed with common fractions. The nominal sizes of the shaft and hole in Fig. 13.5 are 1.50 in. or 1½ in.

BASIC SIZE The exact theoretical size from which limits are derived by the application of plus-and-minus tolerances. There is no basic diameter if this is expressed with limits.

ACTUAL SIZE The measured size of the finished part.

FIT This signifies the type of fit between two mating parts when assembled. The types of fit are clearance, interference, transition, and line fits.

CLEARANCE FIT A fit that gives a clearance between two assembled mating parts. The fit between the shaft and the hole in Fig. 13.5 is a clearance fit that permits a minimum clearance of 0.003 in. and a maximum clearance of 0.007 in.

INTERFERENCE FIT A fit that results in an interference between the two assembled parts. The shaft in Fig. 13.6A is larger than the hole, which requires a force or press fit that has an effect similar to that of a weld between the two parts.

TRANSITION FIT Can result in either an interference or a clearance. The shaft in Fig. 13.6B can be either smaller or larger than the hole and still be within the prescribed tolerances.

LINE FIT Can result in a contact of surfaces or a clearance between them. The shaft in Fig. 13.6C can have contact or clearance when the limits are approached.

SELECTIVE ASSEMBLY A method of selecting and assembling parts by trial and error. Using this method, parts can be assembled that have greater tolerances and consequently can be produced at a reduced cost. This process of hand assembly represents a compro-

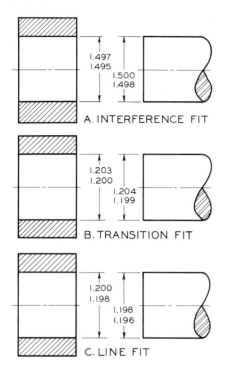

A. INTERFERENCE FIT

B. TRANSITION FIT

C. LINE FIT

FIG. 13.6 Types of fits between mating parts. The clearance fit is not shown.

mise between a high degree of manufacturing accuracy and ease of assembly of interchangeable parts.

SINGLE LIMITS Dimensions that are designated by MIN or MAX (minimum or maximum) instead of being labeled by both (Fig. 13.7). Depths of holes, lengths, threads, corner radii, chamfers, and so on, are dimensioned in this manner in some cases. Caution should be exercised to prevent substantial deviaitons from the single limit.

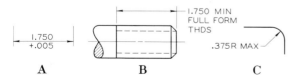

FIG. 13.7 Single tolerances can be given in some applications in MAX or MIN form.

13.5
Basic hole system

The basic hole system is a widely used system of dimensioning holes and shafts to give the required allowance between the two assembled parts. The smallest hole is taken as the basic diameter from which the limits of tolerance and allowance are applied.

The hole is used because many of the standard drills, reamers, and machine tools are designed to give standard hole sizes. Therefore it is advantageous to use this diameter as the basic dimension.

If the smallest diameter of a hole is 1.500 in., the allowance (0.003 in this example) can be subtracted from this diameter to find the diameter of the largest shaft (1.497 in.). The smallest limit for the shaft can then be found by subtracting the tolerance from 1.497 in.

13.6
Basic shaft system

Some industries use the basic shaft system of applying tolerances to dimensions of shafts, since many shafts come in standard sizes. In this system, the largest diameter of the shaft is used as the basic diameter from which the tolerances and allowances are applied.

For example, if the largest permissible shaft is 1.500 in., the allowance can be added to this dimension to yield the smallest possible diameter of the hole into which the shaft must fit. Therefore, if the parts are to have an allowance of 0.004 in., the smallest hole would have a diameter of 1.504 in.

13.7
Metric limits and fits

This section will cover the metric system as recommended by the International Standards Organization (ISO), which has been presented in ANSI B4.2-1978. These fits usually apply to cylinders—holes and shafts. However, these tables can also be used to determine the fits between any parallel surfaces, such as a key in a slot.

Metric definitions of limits and fits

Most of the definitions given below are illustrated in Fig. 13.8.

BASIC SIZE   The size from which the limits or deviations are assigned. Basic sizes, usually diameters, should be selected from Table 13.1 under the heading of "first choice."

TABLE 13.1
PREFERRED SIZES

Basic Size, mm		Basic Size, mm		Basic Size, mm	
First Choice	Second Choice	First Choice	Second Choice	First Choice	Second Choice
1		10		100	
	1.1		11		110
1.2		12		120	
	1.4		14		140
1.6		16		160	
	1.8		18		180
2		20		200	
	2.2		22		220
2.5		25		250	
	2.8		28		280
3		30		300	
	3.5		35		350
4		40		400	
	4.5		45		450
5		50		500	
	5.5		55		550
6		60		600	
	7		70		700
8		80		800	
	9		90		900
				1000	

DEVIATION   The difference between the hole or shaft size and the basic size.

UPPER DEVIATION   The difference between the maximum permissible size of a part and its basic size.

LOWER DEVIATION   The difference between the minimum permissible size of a part and its basic size.

FUNDAMENTAL DEVIATION   The deviation that is closest to the basic size. In the note 40H7, the letter *H* (an uppercase letter) represents the fundamental de-

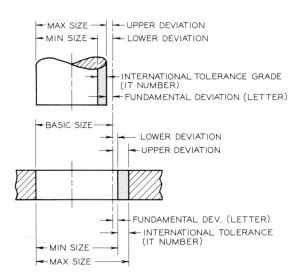

FIG. 13.8   Definition of terms related to metric fits

viation for a hole. In the note 40g6, the letter *g* (a lowercase letter) represents the fundamental deviation for a shaft.

TOLERANCE   The difference between the maximum and minimum allowable sizes of a single part.

INTERNATIONAL TOLERANCE (IT) GRADE   A group of tolerances that vary in accordance with the basic size and provide a uniform level of accuracy within a given grade. In the note 40H7, the number 7 represents the IT grade. There are 18 IT grades: IT01, IT0, IT1, . . . , IT16.

TOLERANCE ZONE   The zone that represents the tolerance grade and its position in relation to the basic size. This is a combination of the fundamental deviation (represented by a letter) and the International Tolerance grade (IT number). For example, in note 40H8, the H8 portion indicates the tolerance zone.

HOLE BASIS   A system of fits based on the minimum hole size as the basic diameter. The fundamental deviation for a hole basis system is an uppercase letter, *H*, for example (Fig. 13.9).

SHAFT BASIS   A system of fits based on the maximum shaft size as the basic diameter. The fundamental deviation for a shaft basis system is a lowercase letter, *f*, for example (Fig. 13.9).

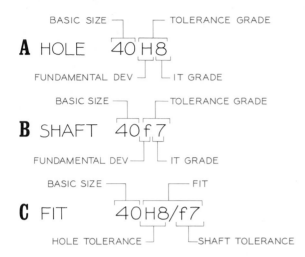

FIG. 13.9 Symbols and their definitions as applied to holes and shafts

CLEARANCE FIT A fit that results in a clearance between the two assembled parts under all tolerance conditions.

INTERFERENCE FIT A fit between two parts that requires that they be forced together when assembled because of their interference with each other.

TRANSITION FIT A fit that results in either a clearance or an interference fit between two assembled parts.

TOLERANCE SYMBOLS Notes that are used to communicate the specifications of tolerance and fit (Fig. 13.9). The basic size is the primary dimension from which the tolerances are determined; therefore it is the first part of the symbol. It is followed by the fundamental position letter and the IT number to give the tolerance zone. Uppercase letters are used to indicate the fundamental deviation for holes, and lowercase letters for shafts.

Three methods of specifying the tolerance information are shown in Fig. 13.10. The information in

TABLE 13.2
DESCRIPTION OF PREFERRED FITS

	ISO Symbol		
	Hole Basis	Shaft Basis	Description
Clearance fits ↑	H11/c11	C11/h11	*Loose running* fit for wide commercial tolerances or allowances on external members.
	H9/d9	D9/h9	*Free running* fit—not for use where accuracy is essential, but good for large temperature variations, high running speeds, or heavy journal pressures.
	H8/f7	F8/h7	*Close running* fit for running on accurate machines and for accurate location at moderate speeds and journal pressures.
	H7/g6	G7/h6	*Sliding* fit—not intended to run freely, but to move and turn freely and locate accurately.
Transition fits ↕	H7/h6	H7/h6	*Locational clearance* fit provides snug fit for locating stationary parts, but can be freely assembled and disassembled.
	H7/k6	K7/h6	*Locational transition* fit for accurate location, a compromise between clearance and interference.
	H7/n6	N7/h6	*Locational transition* fit for more accurate location where greater interference is permissible.
Interference fits ↓	H7/p6[1]	P7/h6	*Locational interference* fit for parts requiring rigidity and alignment with prime accuracy of location but without special bore pressure requirements.
	H7/s6	S7/h6	*Medium drive* fit for ordinary steel parts or shrink fits on light sections, the tightest fit usable with cast iron.
	H7/u6	U7/h6	*Force* fit suitable for parts that can be highly stressed or for shrink fits where the heavy pressing forces required are impractical.

[1]Transition fit for basic sizes in range from 0 through 3 mm.

$$40H8 \qquad 40H8\binom{40.039}{40.000} \qquad \frac{40.039}{40.000}(40H8)$$

A   **B**   **C**

FIG. 13.10 Three methods of giving tolerance symbols. The numbers in parentheses are for reference.

parentheses indicates that it is for reference only. The upper and lower limits are found in the tables in Appendix 44.

13.8
Preferred sizes and fits

The preferred basic sizes for computing tolerances are shown in Table 13.1. Under the "First Choice" heading, the numbers increase by about 25% of the preceding numbers. Those in the "Second Choice" column increase at approximately 12% increments.

Where possible, you should select basic diameters from the first column since these correspond to standard sizes for round, square, and hexagonal metal products. This makes it possible to use stock sizes and reduce expenses.

Preferred fits for clearance, transition, and interference fits are shown in Table 13.2 for hole basis and shaft basis fits. Where possible, fits should be taken from this table for mating parts. The tables in Appendixes 45–48 correspond to these fits.

PREFERRED FITS—HOLE BASIS SYSTEM Figure 13.11 illustrates the symbols used to show the combinations of fits that are possible when using the hole basis system. There is a clearance between the two parts at A, a transition fit at B, and an interference fit at C. This technique of representing fits is used in Fig. 13.12 to show a series of fits for a hole basis system. Note that the lower deviation of the hole is zero. In other words, the smallest size of the hole is the basic size.

The sizes of the shafts are varied to give a variety of fits from c11 to u6, where there is a maximum of interference. These fits correspond to those given in Table 13.2.

PREFERRED FITS—SHAFT BASIS SYSTEM Figure 13.13 illustrates the preferred fits based on the shaft basis system, where the largest shaft size is the basic diameter. The variation in the fit between the parts is caused by varying the size of the holes. This results in a range of fits from a clearance fit of C11/h11 to an interference fit of U7/h6.

13.9
Example problems—metric system

The following problems are given and solved as examples of determining the sizes and limits and the application of the proper symbols to mating parts. The

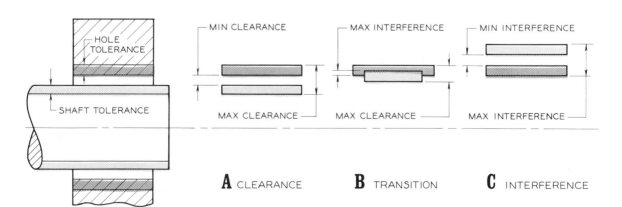

FIG. 13.11 Types of fits (A) A clearance fit; (B) a transition fit where there can be an interference or a clearance; and (C) an interference fit, where the parts must be forced together.

FIG. 13.12 The preferred fits for a hole basis system. These fits correspond to those given in Table 13.2.

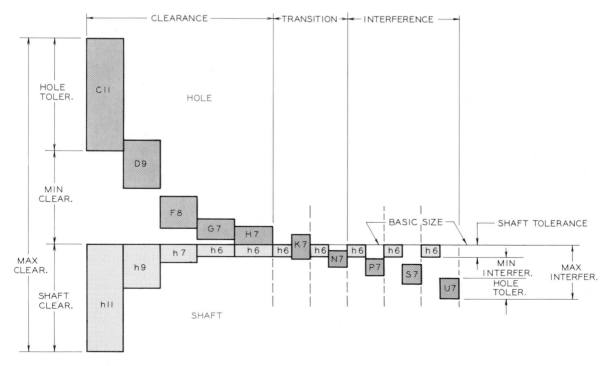

FIG. 13.13 The preferred fits for a shaft basis system. These fits correspond to those given in Table 13.2.

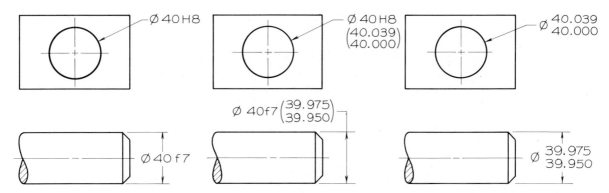

FIG. 13.14 Any of these three methods can be used to apply symbols to a detail drawing of two mating parts.

solution of these problems requires the use of the tables in Appendix 45, the table of preferred sizes (Table 13.1), and the table of preferred fits (Table 13.2).

EXAMPLE 1 (Fig. 13.14) Given: Hole basis system, close running fit, basic diameter = 39 mm.

Solution: Use a basic diameter of 40 mm (Table 13.1) and fit of H8/f7 (Table 13.2).

Hole: Find the upper and lower limits of the hole in Appendix 45 under H8 and across from 40 mm. These limits are 40.000 and 40.039 mm.

Shaft: The upper and lower limits of the shaft are found under f7 and across from 40 mm in Appendix 45. These limits are 39.950 and 39.975 mm.

Symbols: The methods of noting the drawings are shown in Fig. 13.14. Any of these methods is appropriate.

EXAMPLE 2 (Fig. 13.15) Given: Hole basis system, locational transition fit, basic diameter = 57 mm.

Solution: Use a basic diameter of 60 mm (Table 13.1) and a fit of H7/k6 (Table 13.2).

Hole: Find the upper and lower limits of the hole in Appendix 46 under H7 and across from 60 mm. These limits are 60.000 and 60.030 mm.

Shaft: The upper and lower limits of the shaft are found under k6 and across from 60 mm in Appendix 46. These limits are 60.021 and 60.002 mm.

Symbols: Two methods of applying the tolerance symbols to a drawing are shown in Fig. 13.15.

EXAMPLE 3 (Fig. 13.16) Given: Hole basis system, medium drive fit, basic diameter = 96 mm.

Solution: Use a basic diameter of 100 mm (Table 13.1) and a fit of H7/s6 (Table 13.2).

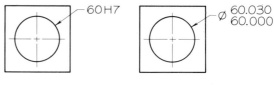

EXAMPLE 2: LOCATIONAL TRANSITION FIT – H7/k6

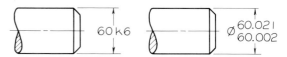

FIG. 13.15 Examples of tolerance symbols applied to a transition fit. Both methods are acceptable.

Hole: Find the upper and lower limits of the hole in Appendix 46 under H7 and across from 100 mm. these limits are 100.035 and 100.000 mm.

Shaft: The upper and lower limits of the shaft are found under s6 and across from 100 mm in Appendix 46. These limits are found to be 100.093 and 100.071

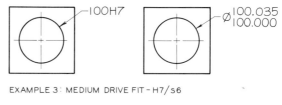

EXAMPLE 3: MEDIUM DRIVE FIT – H7/s6

FIG. 13.16 Examples of tolerance symbols applied to an interference fit. Both methods are acceptable.

mm. From the table, you can see that the tightest fit is an interference of 0.093 mm and the loosest fit is an interference of 0.036 mm. An interference is indicated by a minus sign in front of the numbers.

EXAMPLE 4 (Fig. 13.17) Given: Shaft basis system, loose running fit, basic diameter = 116 mm.

Solution: Use a basic diameter of 120 mm (Table 13.1) and a fit of C11/h11 (Table 13.2).

Hole: Find the upper and lower limits of the hole in Appendix 47 under C11 and across from 120 mm. These limits are 120.400 and 120.180.

Shaft: The upper and lower limits of the shaft are found under h11 and across from 120 mm in Appendix 47. These limits are 119.780 and 120.000 mm.

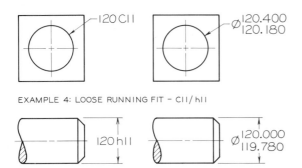

FIG. 13.17 Examples of tolerance symbols applied to a clearance fit. Both methods are acceptable.

13.10
Preferred metric fits—nonpreferred sizes

Limits of tolerances for preferred fits, shown in Table 13.2, can be computed for nonstandard sizes. Limits of tolerances appear in Appendix 49 for nonstandard hole sizes and in Appendix 50 for nonstandard shaft sizes.

The hole and shaft limits for an H8/f7 fit and a 45-mm DIA are computed in Fig. 13.18. The tolerance limits of 0.000 and 0.046 for an H8 hole are taken from Appendix 49 across from the size range of 40–50 mm. The tolerance limits 0.000 and 0.050 mm are taken from Appendix 50. The limits of sizes for the hole and shaft are computed by applying these limits of tolerance to the 45-mm basic diameter.

FIT: H8/f7	Ø 45 BASIC	FROM APPENDIX	
		H8	f7
HOLE LIMITS	45.046 / 45.000	0.046 / 0.000	−0.025 / −0.050
SHAFT LIMITS	44.975 / 44.950	HOLE	SHAFT

FIG. 13.18 The limits of a nonstandard diameter, 45 mm and an H8/f7 fit are calculated by using values from Appendixes 49 and 50.

13.11
Standard fits (English units)

The ANSI B4.1-1955 standard specifies a series of fits between cylindrical parts that are based on the basic hole system in inches. The types of fit covered in this standard are:

RC Running and sliding fits

LC Clearance locational fits

LT Transition locational fits

LN Interference locational fits

FN Force and shrink fits

Each of these types of fit is listed in tables in Appendixes 39 to 43. There are several classes of fit under each of the types given above.

RUNNING AND SLIDING FITS (RC) Fits for which limits of clearance are specified to provide a similar running performance, with suitable lubrication allowance throughout the range of sizes. The clearance for the first two classes (RC 1 and RC 2), used chiefly as slide fits, increases more slowly with diameter size than that of the other classes, so that accurate location is maintained even at the expense of free relative motion.

LOCATIONAL FITS (LC) Fits intended to determine only the location of the mating parts; they may provide rigid or accurate location (interference fits) or some freedom of location (clearance fits). They are divided into three groups: clearance fits (*LC*), transition fits (*LT*), and interference fits (*LN*).

FORCE FITS (FN) Special types of interference fits, typically characterized by maintenance of constant

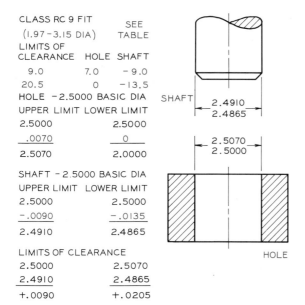

CLASS RC 9 FIT
(1.97 – 3.15 DIA)   SEE TABLE

LIMITS OF CLEARANCE	HOLE	SHAFT
9.0	7.0	– 9.0
20.5	0	–13.5

HOLE – 2.5000 BASIC DIA

UPPER LIMIT	LOWER LIMIT
2.5000	2.5000
.0070	0
2.5070	2.0000

SHAFT – 2.5000 BASIC DIA

UPPER LIMIT	LOWER LIMIT
2.5000	2.5000
–.0090	–.0135
2.4910	2.4865

LIMITS OF CLEARANCE

2.5000	2.5070
2.4910	2.4865
+.0090	+.0205

FIG. 13.19 The method of calculating limits and allowances for an RC9 fit between a shaft and a hole. The basic diameter is 2.5000 in.

bore pressures throughout the range of sizes. The interference therefore varies almost directly with diameter, and the difference between its minimum and maximum values is small, to maintain the resulting pressures within reasonable limits.

Figure 13.19 illustrates how one uses the values from the tables in Appendix 39 for an RC 9 fit. The basic diameter for the hole and shaft is 2.5000 in., which is between the range of 1.97 and 3.15 given in the last column of the table. Since all limits are given in thousandths, the values can be converted by moving the decimal point three places to the left. For example, + 0.7 is + 0.0007 in.

The upper and lower limits of the shaft (2.4910 in. and 2.4865 in.) are found by subtracting the two limits (– 0.0090 in. and – 0.0135 in.) from the basic diameter. The upper and lower limits of the hole (2.5007 in. and 2.5000 in.) are found by adding the two limits (+ 0.007 in. and 0.000 in.) to the basic diameter.

When the two parts are assembled, the tightest fit (+ 0.0205 in.) and the loosest fit (+ 0.0090 in.) are found by subtracting the maximum and minimum sizes of the holes and shafts. These values are provided

in the second column of the table as a check on the limits.

The same method (but different tables) is used for calculating the limits for all types of fit. Plus values of clearance indicate that there is clearance, and minus values that there will be interference between the assembled parts.

These values should be converted to millimeters when using the metric system by multiplying inches by 25.4 or by using corresponding metric tables.

13.12
Chain dimensions

When parts are dimensioned to locate surfaces or geometric features by a chain of dimensions (Fig. 13.20A), variations may occur that exceed the tolerances specified. As successive measurements are made, with each based on the preceding one, the tolerances may accumulate, as shown in Fig. 13.20A. For exam-

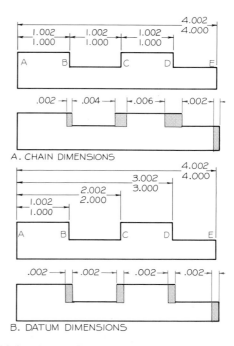

FIG. 13.20 When dimensions are given as chain dimensions, the tolerances can accumulate to give a variation of 0.006 in. at D instead of 0.002 in. When dimensioned from a single datum plane, the variations of X and Y cannot deviate more than the specified 0.002 in. from the datum.

ple, the tolerance between Surfaces A and B is 0.002; between A and C, 0.004; and between A and D, 0.006.

This accumulation of tolerances can be eliminated by measuring from a single plane called a datum plane. A datum plane is usually a plane on the object, but it could be a plane on the machine used to make the part.

Note that the tolerances between the intermediate planes in Fig. 13.20B are uniform since each of the planes was located with respect to a single datum. In our example, this is 0.002, which represents the maximum tolerance.

13.13
Origin selection

In some cases, there will be a need to specify a surface as the origin for locating another surface. An example is illustrated in Fig. 13.21, where the origin surface is the shorter one at the base of an object. A smaller angular variation for the longer surface than if the longer surface had been used as the origin surface.

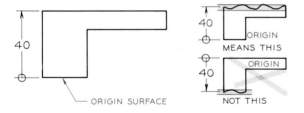

FIG. 13.21 This method is used to indicate the origin surface for locating one feature of a part with respect to another.

13.14
Conical tapers

A method of specifying a conical taper is shown in Fig. 13.22 by giving a basic diameter and a basic taper. The basic diameter of 20 mm is located midway in the length of the cone with a toleranced dimension. Taper is a ratio of the difference in the diameters of two circular sections of a cone to the distance between the

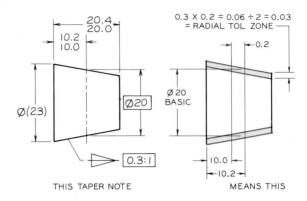

FIG. 13.22 Taper is indicated with a combination of tolerances and taper symbols. The variation in diameter at any point is 0.06 mm or 0.03 mm in radius.

sections. The radial tolerance zone can be found as shown in Fig. 13.22.

13.15
Tolerance notes

All dimensions on a drawing are toleranced either by the previously covered rules or by a general note on a drawing placed in or near the title block. For example, the note **TOLERANCE ± 1/64** (or its decimal equivalent, 0.40 mm) might be given on a drawing for the less critical dimensions.

Some industries may give dimensions in inches where decimals are carried out to two, three, and four places. A note might be given on the drawing in this manner: TOLERANCES XX.XX ± 0.10; XX.XXX ± 0.005. Tolerances of four places would be given directly on the dimension lines, but those dimensions with two and three decimal places would have the tolerances indicated in the note.

The most common method of noting tolerances is to use as large a tolerance as feasible expressed in a note such as TOLERANCES ± 0.05 (± 1 mm when using metrics), and give the tolerances on the dimension lines for the mating dimensions that require smaller tolerances.

Angular tolerances should be given in a general note in or near the title block, such as ANGULAR TOLERANCES ± 0.5° or ± 30′. When angular tolerances less than this are specified, they should be given on the drawing where these angles are dimensioned.

FIG. 13.23 Angles can be toleranced by any of these methods using limits or the plus-and-minus method.

Techniques of tolerancing angles are shown in Fig. 13.23.

13.16
General tolerances—metric

All dimensions on a drawing are understood to have tolerances, and the amount of tolerance must be noted. This section is based on the metric system, with the millimeter as the unit of measurement, as outlined in ANSI B4.3-1978.

Tolerances may be specified by (1) applying tolerances directly to the dimensions, (2) giving toler-

ances in specification documents that are associated with the drawings, or (3) applying a general note on the drawing.

LINEAR DIMENSIONS may be toleranced by indicating ± one-half of an International Tolerance (IT) grade as given in Appendix 44. When the nature of the application is known, the appropriate IT grade can be selected from the graph in Fig. 13.24.

You can see that IT grades for mass produced items range form IT12 through IT16. When the machining process is known, the IT grades can be selected from Table 13.3.

FIG. 13.24 The International Tolerance grades and their applications

TABLE 13.3
IT GRADES AND THEIR RELATIONSHIP TO MACHINING PROCESSES

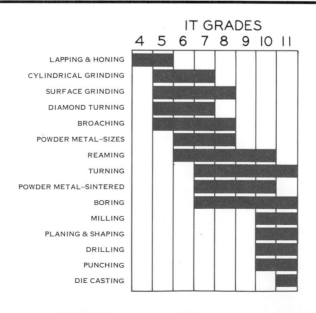

General tolerances using IT grades may be expressed in a note as follows:

UNLESS OTHERWISE SPECIFIED ALL

UNTOLERANCED DIMENSIONS ARE $\pm \dfrac{\text{IT14}}{2}$.

This means that tolerance of $\pm$ 0.700 mm is allowed for a dimension between 315 and 400 mm. The value of the tolerance is listed in Appendix 44 as 1.400.

Table 13.4 shows recommended tolerances for fine, medium, and coarse series for dimensions of graduated sizes. A medium tolerance, for example, can be specified by the following note:

**GENERAL TOLERANCE SPECIFIED IN
ANSI B4.3 MEDIUM SERIES APPLY.**

This same information can be given on the drawing in table form by selecting the grade—medium in this example—from Table 13.4 and presenting it on the drawing as shown in Fig. 13.25.

DIMENSIONS IN mm							
GENERAL TOLERANCE UNLESS OTHERWISE SPECIFIED THE FOLLOWING TOLERANCES ARE APPLICABLE							
LINEAR	OVER TO	0.5 6	6 30	30 120	120 315	315 1000	1000 2000
TOL	$\pm$	0.1	0.2	0.3	0.5	0.8	1.2

FIG. 13.25 This table is a medium series of values taken from Table 14.4. It is placed on the working drawing to provide the tolerances for various ranges of sizes.

DIMENSIONS IN mm					
GENERAL TOLERANCE UNLESS OTHERWISE SPECIFIED THE FOLLOWING TOLERANCES ARE APPLICABLE					
LINEAR	OVER TO	$-$ 120	120 315	315 1000	1000 $-$
TOL	ONE DECIMAL $\pm$	0.3	0.5	0.8	1.2
	NO DECIMALS $\pm$	0.8	1.2	2	3

FIG. 13.26 This table of tolerances can be placed on a drawing to indicate the tolerances for dimensions with one decimal and those with no decimals.

General tolerances can be expressed in a table of values that gives the tolerances for dimensions with one or no decimal places. An example of this table is shown in Fig. 13.26. Another method of giving general tolerances is a note in this form:

**UNLESS OTHERWISE SPECIFIED ALL
UNTOLERANCED DIMENSIONS ARE** $\pm$ 0.8 mm.

ANGULAR TOLERANCES are expressed (1) as an angle in decimal degrees or in degrees and minutes, (2) as a taper expressed in percentage (number of millimeters per 100 mm), or (3) as milliradians. A milliradian is found by multiplying the angle in degrees by 17.45. The suggested tolerances for each of these units are shown in Table 13.5. The angular tolerances are based on the length of the shorter leg of the angle.

TABLE 13.4
FINE, MEDIUM, AND COARSE SERIES: GENERAL TOLERANCE—LINEAR DIMENSIONS

Variations in mm								
Basic Dimensions, mm		0.5 to 3	Over 3 to 6	Over 6 to 30	Over 30 to 120	Over 120 to 315	Over 315 to 1,000	Over 1,000 to 2,000
Permissible variations	Fine series	± 0.05	± 0.05	± 0.1	± 0.15	± 0.2	± 0.3	± 0.5
	Medium series	± 0.1	± 0.1	± 0.2	± 0.3	± 0.5	± 0.8	± 1.2
	Coarse series		± 0.2	± 0.5	± 0.8	± 1.2	± 2	± 3

TABLE 13.5
GENERAL TOLERANCE—ANGLES AND TAPERS

Length of the Shorter Leg mm		Up to 10	Over 10 to 50	Over 50 to 120	Over 120 to 400
Permissible variations	In degrees and minutes	±1°	±0°30'	±0°20'	±0°10'
	In millimeters per 100 mm	±1.8	±0.9	±0.6	±0.3
	In milliradians	±18	±9	±6	±3

ANGULAR TOLERANCE				
LENGTH OF SHORTER LEG – mm	UP TO 10	OVER 10 TO 50	OVER 50 TO 120	OVER 120 TO 400
TOL	±1°	±0° 30'	±0° 20'	±0° 10'

FIG. 13.27 This table can be placed on a drawing to indicate the general tolerances for angles. These values were extracted from Table 13.5.

General angular tolerances may be given on the drawing with a note in the following form:

UNLESS OTHERWISE SPECIFIED THE GENERAL TOLERANCES IN ANSI B4.3 APPLY.

A second method shows the portion of Table 13.5 on the drawing using the units desired, as shown in Fig. 13.27.

A third method is a note with a single tolerance such as:

UNLESS OTHERWISE SPECIFIED THE GENERAL ANGULAR TOLERANCES ARE ±0° 30' (or ±0.5°).

13.17
Geometric tolerances

Geometric tolerancing is a general term used to describe tolerances that specify and control form, profile, orientation, location, and runout on a dimensioned part. The basic principles of this area of tolerancing are standardized by the *ANSI Y14.5M-1982 Standards and the Military Standards (Mil-Std)* of the U.S. Department of Defense.

These standards are based on the metric system and thus use the millimeter as the unit of measurement. However, inch units with decimal fractions can be used instead of millimeters if needed.

13.18
Symbology of geometric tolerances

The various symbols that are used to specify geometric characteristics of dimensioned drawings are shown in Fig. 13.28. Additional features and their proportions are shown in Fig. 13.29, where the letter height (H) is used as the basis of the proportions. On most drawings the letter height is recommended to be ⅛ in. or 3 mm high.

Other examples of feature control symbols are given in Fig. 13.30, where the dimensions are generally the same as those specified in Fig. 13.29.

13.19
Limits of size

Three terms that are used to specify the limits of size of a part when applying geometric tolerances are **maximum material condition** (MMC), **least ma-**

	TOLERANCE	CHARACTERISTIC	SYMBOL
INDIVIDUAL FEATURES	FORM	STRAIGHTNESS	—
		FLATNESS	▱
		CIRCULARITY	○
		CYLINDRICITY	⌭
INDIVIDUAL OR RELATED FEATURES	PROFILE	PROFILE OF A LINE	⌒
		PROFILE OF A SURFACE	⌓
RELATED FEATURES	ORIENTATION	ANGULARITY	∠
		PERPENDICULARITY	⊥
		PARALLELISM	//
	LOCATION	POSITION	⌖
		CONCENTRICITY	◎
	RUNOUT	CIRCULAR RUNOUT	↗
		TOTAL RUNOUT	↗↗

FIG. 13.28 These symbols are used to specify the geometric characteristics of a dimensioned part.

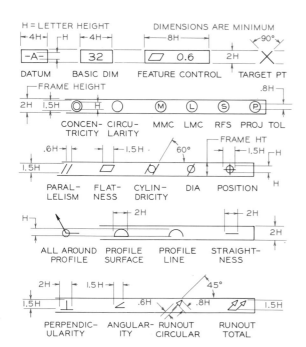

FIG. 13.29 The general proportions of notes and symbols that are used in feature control symbols. H is height of lettering that is used.

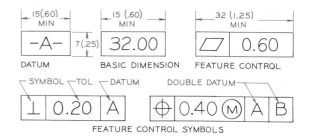

FIG. 13.30 Examples of symbols used to indicate datum planes, basic dimensions, and feature control symbols.

terial condition (LMC), and **regardless-of-feature size** (RFS).

MMC is the condition in which a part is made with the maximum amount of material. For example, the shaft in Fig. 13.31 is at MMC when it has the largest permitted diameter of 24.6 mm. On the other hand, a hole is at MMC when it has the most material, or the smallest diameter of 25.0 mm.

LMC is the condition in which a part has the least amount of material. The shaft in Fig. 13.31 is at LMC when it has the smallest diameter of 24.0 mm. The hole is at LMC when it has the largest diameter of 25.6 mm.

RFS is a term used to indicate that tolerances apply to a geometric feature **regardless of the size** it may be, from MMC to LMC.

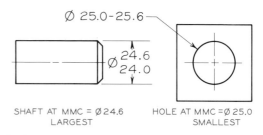

FIG 13.31 A shaft is at MMC when it is at its largest size that is permitted by its tolerance. A hole is at MMC when it is at its smallest size.

13.20
Three rules of tolerances

There are three general rules of tolerancing geometric features that should be followed in this type of dimensioning.

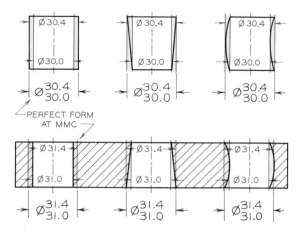

FIG. 13.32 When only a tolerance of size is specified on a part, the limits prescribe the form of the part, as shown in these examples of shafts and holes with the same limits of tolerance.

RULE 1 (INDIVIDUAL FEATURE SIZE) When only a tolerance of size is specified on a part, the limits of size prescribe the amount of variation permitted in its geometric form. In Fig. 13.32, you can see how the forms of the shaft and hole are permitted to vary within the tolerance of size indicated by the dimensions.

RULE 2 (TOLERANCES OF POSITION) When a tolerance of position is specified on a drawing. RFS, MMC, or LMC must be specified with respect to the tolerance, the datum, or both. You can see that the specification of symmetry of the part in Fig. 13.33 is based on a tolerance at RFS from a datum at RFS.

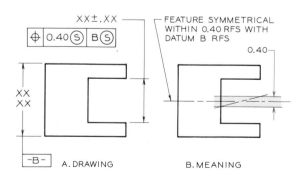

FIG. 13.33 Tolerances of position should include a note of M, S, or L to indicate maximum material condition, regardless of feature size, or least material condition.

RULE 3 (ALL OTHER GEOMETRIC TOLER-ANCES) RFS applies for all other geometric tolerances for individual tolerances and datum references, if no modifying symbol is given in the feature control symbol. If a feature is to be at maximum material condition, MMC must be specified.

13.21
The three-datum plane concept

A datum plane is used as the origin of a part's dimensioned features that have been toleranced. Datum planes are usually associated with the manufacturing equipment, such as machine tables, or with the locating pins.

Three mutually perpendicular datum planes are used to dimension a part accurately. For example, the part in Fig. 13.34 is placed in contact with the primary datum plane at its base where three points must make contact with the datum. The part is further related to the secondary plane with two contacting points. The third (tertiary) datum is contacted by a single point on the object.

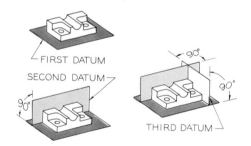

FIG. 13.34 When a part is referenced to a primary datum plane, the object contacts the plane with its three highest points. The vertical surface contacts the secondary vertical datum plane with two points. The third datum plane is contacted by one point on the object. The datum planes are listed in this order in the feature control symbol.

The priority of these datum planes is noted on the drawing of the part by feature control symbols, as shown in Fig. 13.35. The primary datum is surface P, the secondary is surface S, and the tertiary is surface T. Examples of feature control symbols are given in

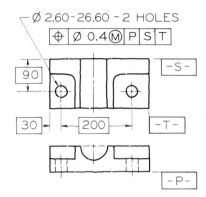

FIG. 13.35 The sequence of the three-plane reference system (shown in Fig. 13.34) is labeled where the planes appear as edges. Note that the primary datum plane, P, is listed first in the feature control symbol; the secondary plane, S, next; and the third plane, T, last.

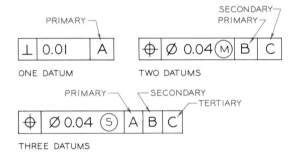

FIG. 13.36 Feature control symbols may have from one to three datum planes indicated. These are listed in order of priority.

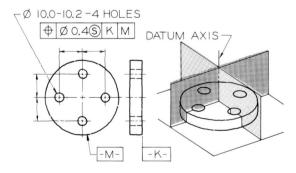

FIG. 13.37 These true-position holes are located with respect to primary Datum K and Datum M. Since datum M is a circle, this implies that the holes are located about two intersecting datum planes formed by the crossing center lines in the circular view. This satisfies the three-plane concept.

Fig. 13.36, where the primary, secondary, and tertiary datum planes are listed in order of priority in the symbol.

13.22
Cylindrical datum features

Figure 13.37 illustrates a part with a cylindrical datum feature that is the axis of a true cylinder. Primary datum *K* establishes the first datum. Datum *M* is associated with two theoretical planes—the second and third in a three-plane relationship.

The two theoretical planes are represented on the drawing by perpendicular center lines. The intersection of the centerline planes coincides with the datum axis. All dimensions originate from the datum axis, which is perpendicular to datum *K*; the two intersecting datum planes are used to indicate the direction of measurements in an *x* and *y* direction.

The sequence of the datum reference in the feature control symbol is significant to the manufacturing and inspection processes. The part in Fig. 13.38 is dimensioned with an incomplete feature control symbol; it does not specify the primary and secondary datum planes. The schematic drawing at B illustrates the effect of specifying Diameter *A* as the primary and Surface *B* as the secondary datum plane. This means that the part is centered about Cylinder *A* by mounting the part in a chuck, mandrel, or centering device on the processing equipment, which centers the part at *RFS* (regardless of feature size). Surface *B* is assembled to contact at least one point of the third datum plane.

If Surface *B* were specified as the primary datum feature, it would be assembled to contact datum plane *B* in at least three points. The axis of datum feature *A* will be gauged by the smallest true cylinder that is perpendicular to the first datum that will contact Surface *A* at *RFS*.

In Fig. 13.28D, Plane *B* is specified as the primary datum feature, and Cylinder *A* is specified as the secondary datum feature at *MMC* (maximum material condition). The part is mounted on the processing equipment where at least three points of Feature *B* are in contact with Datum *B*. The second and third planes intersect at the datum axis to complete the three-plane relationship. This utilization of the modifier to specify MMC gives a more liberal tolerance zone than would be otherwise acceptable when RFS was specified.

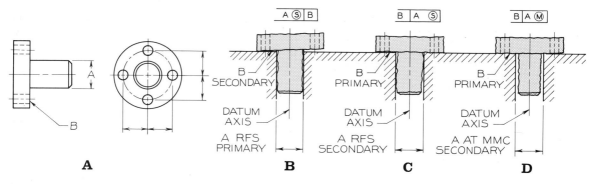

FIG. 13.38 Three examples are shown to illustrate the effects of selection of the datum planes in order of priority and the effect of RFS and MMC.

13.23
Datum features at RFS

When dimensions of size are applied to a part at RFS, the datum is established by contact between surfaces on the processing equipment with surfaces of the part. Variable machine elements, such as chucks or center devices, are adjusted to fit the external or internal features of a part and thereby establish datums.

PRIMARY DIAMETER DATUMS For an external cylinder (shaft), the datum axis is the axis of the smallest circumscribed cylinder that contacts the cylindrical feature of the part. In other words, the largest

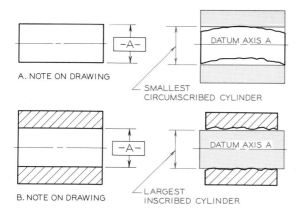

FIG. 13.39 The datum axis of a shaft is the smallest circumscribed cylinder that contacts the shaft. The datum axis of a hole is the centerline of the largest inscribed cylinder that contacts the hole.

diameter of the part will make contact with the smallest contacting cylinder of the machine element that holds the part (Fig. 13.39).

For an internal cylinder (hole), the datum axis is the axis of the largest inscribed cylinder that contacts the inside of the hole. In other words, the smallest diameter of the hole will make contact with the largest cylinder of the machine element inserted in the hole (Fig. 13.39).

PRIMARY EXTERNAL PARALLEL DATUMS The datum for external features is the center plane between two parallel planes at their minimum separation that contact the planes of the object (Fig. 13.40A). These planes are planes of a viselike device that holds the part; therefore the planes of the part are at maximum separation, while the planes of the device are at minimum separation.

PRIMARY INTERNAL PARALLEL DATUMS The datum for internal features is the center plane between two parallel planes at their maximum separation (Fig. 13.40B) that contact the planes of the object. This is the condition in which the slot is at its smallest opening size.

SECONDARY DATUMS The secondary datum (axis or center plane) for both external and internal diameters or distances between parallel planes is found as covered in the previous two paragraphs, but with an additional requirement. This requirement is that the contacting cylinder or the contacting parallel planes must be oriented perpendicular to the primary datum. Figure 13.41 illustrates how datum *B* is the axis of a cylinder. This principle also can be applied for parallel planes in addition to cylinders.

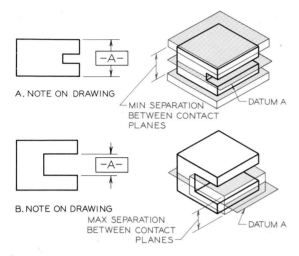

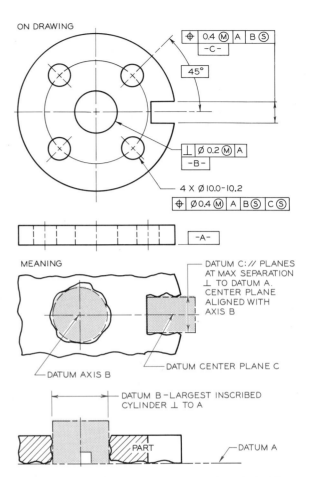

ON DRAWING

MEANING

FIG. 13.40 The datum plane for external parallel surfaces is the center plane between two contacting parallel planes at their minimum separation. The datum plane for internal parallel surfaces is the center plane between two contacting parallel surfaces at maximum separation.

FIG. 13.41 The illustration of a part located with respect to primary, secondary, and tertiary datum planes.

TERTIARY DATUMS The third datum (axis or center plane) for both external and internal features is found in the same manner as covered in the previous three paragraphs, but with an additional requirement. This requirement is that the contacting cylinder or parallel planes must be angularly oriented with respect to the secondary datum. Datum *C* in Fig. 13.41 is the tertiary datum plane.

13.24
Datum targets

Instead of using a plane surface as a datum, specified datum targets are indicated on the surface of a part where the part is supported by spherical or pointed locating pins. The symbol *X* is used to indicate target points that are supported by locating pins at specified points (Fig. 13.42). Datum target symbols are placed outside the outline of the part with a leader that is directed toward the target. When the target is on the near (visible) surface, the leaders are solid lines; when the target is on the far (invisible) surface, as shown in Fig. 13.44, the leaders are hidden.

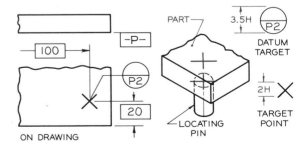

FIG. 13.42 Target points from which a datum point is established, are located with an "X" and a target symbol.

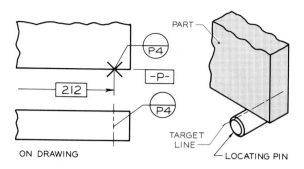

FIG. 13.43 An "X" and a phantom line are used to locate target lines on a drawing.

Three target points are required to establish the primary datum plane, two for the secondary, and one for the tertiary. Notice that the target symbol in Fig. 13.42 is labeled *P2* to match the designation of the primary datum, *P*. The other two points (not shown) would be labeled *P1* and *P2* to establish the primary datum.

A datum target line is specified in Fig. 13.43 for a part that is supported on a datum line instead of a datum point. An *X* and a phantom line are used to locate the line of support.

Target areas are specified for cases where spherical or pointed locating pins are inadequate to support a part. The diameters of the targets are specified with crosshatched circles surrounded by phantom lines, as shown in Fig. 13.44. The target symbols give both the diameter of the targets and their number designations.

The *X* symbol could be used as an alternative method of indicating targets with areas.

The part is located on is datum plane by placing it on the three locating pins with 30-mm diameters, as shown in Fig. 13.44. The leaders from the target areas to the target symbols are hidden to indicate that the targets are on the hidden side of the object.

13.25
Tolerances of location

Tolerances of location deal with **position, concentricity,** and **symmetry.**

True-position tolerancing

Whereas toleranced location dimensions give a square tolerance zone for locating the center of a hole, **true-position dimensions** locate the exact position of a hole's center about which a circular tolerance zone is given (Fig. 13.45). **Basic dimensions** are used to locate true positions. Basic dimensions are exact untoleranced dimensions which are indicated by boxes drawn around them (Part B).

In both methods, the diameters of the holes are toleranced by notes. The true-position method (Part B) uses a feature control symbol to specify the diameter of the circular tolerance zone inside of which the center of the hole must lie. A circular zone gives a

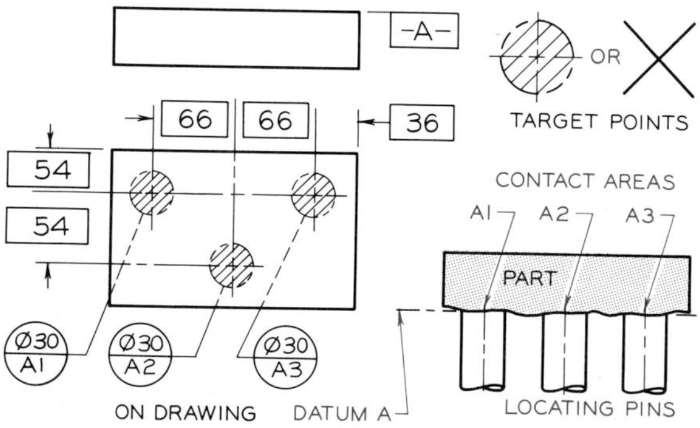

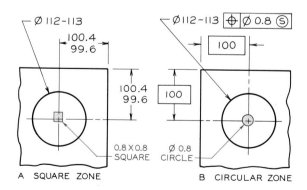

FIG. 13.44 Target points with areas are located with basic dimensions and target symbols that give the diameters of the targets. Hidden leaders indicate that the targets are on the hidden side of the plane.

FIG. 13.45 The dimensions at A give a square tolerance zone for the axis of the hole. At B, basic dimensions locate the true center of the circle about which a circular tolerance zone of 0.8 mm is specified.

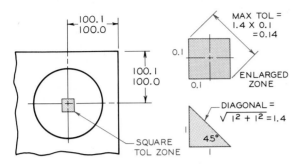

FIG. 13.46 The coordinate method of tolerancing gives a square tolerance zone. The diagonal of the square exceeds the specified tolerance by a factor of 1.4.

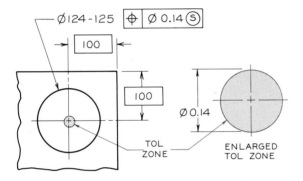

FIG. 13.47 The true-position method of tolerancing gives a circular tolerance zone with equal variations in all directions from the true axis of the hole.

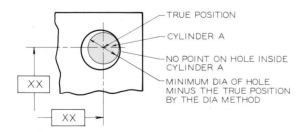

FIG. 13.48 When a hole is located at true position at MMC, no element of the hole shall be inside the imaginary cylinder, Cylinder A.

more uniform tolerance of the hole's true position than a square zone.

In Fig. 13.46, you can see an enlargement of the square tolerance zone, the diagonal across the square zone is greater than the specified tolerance by a factor of 1.4. The true-position method shown in Fig. 13.47 can have a larger circular tolerance zone by a factor of 1.4 and still have the same degree of accuracy of position as the specified 0.1 square zone.

If the coordinate method could accept a variation of 0.014 across the diagonal of the square tolerance zone, then the true-position tolerance should be acceptable with a circular zone of 0.014, which is a greater tolerance than the square zone permitted (Fig. 13.46). True-position tolerances can be applied by symbol, as in Fig. 13.47.

The circular tolerance zone specified with the circular view of a hole is assumed to extend the full depth of the hole. Therefore the tolerance zone for the centerline of the hole is a cylindrical zone inside which the axis must lie. The size of the hole and its position are both toleranced; consequently, these two tolerances are used to establish the diameter of a gauge cylinder that can be used to check that the dimensions conform to the specifications (Fig. 13.48).

The circle is found by subtracting the true position tolerance form the hole at maximum material condition (the smallest permissible hole). This zone represents the least favorable condition when the part is gauged or assembled with a mating part. When the hole is not at MMC, it is larger and permits a greater tolerance and easier assembly.

Gauging a two-hole pattern

Gauging is a technique of checking dimensions to determine whether they have met the specifications of tolerance (Fig. 13.49).

The two holes are positioned 26.00 mm apart with a basic dimension. The holes have limits of 12.70 and 12.84 for a tolerance of 0.14. They are located at true position within a diameter of 0.18. The gauge pin diameter is calculated to be 12.52 mm (the smallest hole's size minus the true-position tolerance), as illustrated in Part B. This means that two pins with diameters of 12.52 mm that are spaced exactly 26.00 mm apart could be used to check the diameters and positions of the holes at MMC, the most critical size. If the

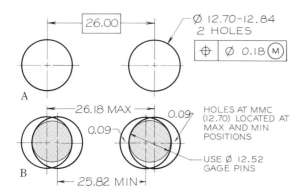

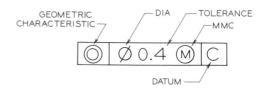

FEATURE CONTROL FRAME

FIG. 13.51 A typical feature control symbol. This one indicates that a surface is concentric to datum C within a diameter of 0.4 mm at maximum material condition.

FIG. 13.49 When two holes are located at true position at MMC, they may be gauged with pins 12.52 mm in diameter that are located 26.00 mm apart.

Concentricity

Feature control symbols will be used to specify concentricity and other geometric characteristics throughout the remainder of this chapter (Fig. 13.51).

Concentricity is a feature of **location** because it specifies the relationship of one cylinder with another; that is, both share the same axis. In Fig. 13.52, the large cylinder is "flagged" as datum *A*. This means that the large diameter is used as the datum for measuring the variation of the smaller cylinder's axis.

pins can be inserted in the holes, then the holes are properly sized and located.

When the holes are not at MMC (when they are larger than their minimum size), these gauge pins will permit a greater range of variation (Fig. 13.50). When the holes are at their maximum size of 12.84 mm, they can be located as close as 25.68 from center to center or as far apart as 26.32 center to center. When not specified, true-position tolerances are assumed to apply at MMC.

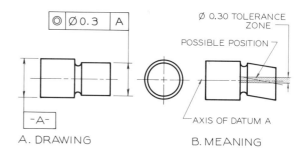

A. DRAWING   B. MEANING

FIG. 13.52 Concentricity is a tolerance of location. The feature control symbol specifies that the smaller cylinder should be concentric to Cylinder *A* within 0.3 mm about axis A.

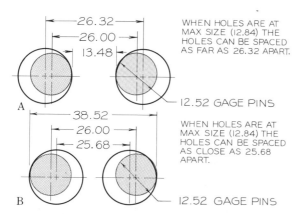

FIG. 13.50 When two holes are at their maximum size, the centers of the holes can be spaced as far as 26.32 mm apart and still be acceptable (A). The holes can be placed as close as 25.68 apart when the holes are at maximum size (B).

Symmetry

Symmetry is a feature of **location.** A part or a feature is symmetrical when it has the same contour and size on opposite sides of a central plane. A symmetry tolerance locates features with respect to a datum plane.

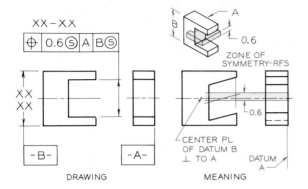

FIG. 13.53 Symmetry is a tolerance of position that specifies that a part's feature be symmetrical about the center plane between parallel surfaces on the part.

The method of noting symmetry is shown in Fig. 13.53. The feature control symbol notes that the notch is symmetrical about Datum *B* within a zone of 0.6 mm.

13.26
Tolerances of form

Flatness

A surface is flat when all its elements are in one plane. In Fig. 13.54, a feature control symbol is used to specify flatness within a 0.4 mm zone RFS. No point on the surface may vary more than 0.40 from the highest to the lowest point on the surface.

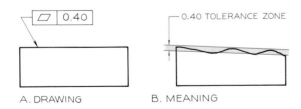

FIG. 13.54 Flatness is a tolerance of form that specifies two parallel planes inside of which the object's surface must lie.

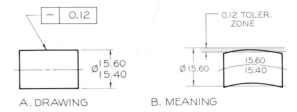

FIG. 13.55 Straightness is a tolerance of form that indicates that the elements of a surface are straight lines. The symbol must be applied where the elements appear as straight lines.

Straightness

A surface is straight if all its elements are straight lines. In Fig. 13.55, a feature control symbol is used to specify straightness of a cylinder. A total of 0.12 mm RFS is permitted as the elements are gauged in a vertical plane parallel to the axis of the cylinder.

Roundness

A surface of revolution (a cylinder, cone, or sphere) is round when all points on the surface intersected by a plane are equidistant from the axis. In Fig. 13.56, a feature control symbol is used to specify roundness of a cone and cylinder. This symbol permits a tolerance of 0.34 mm RFS on the radius of each part. The roundness of a sphere is specified in Fig. 13.57.

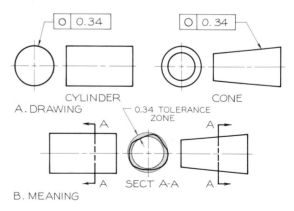

FIG. 13.56 Roundness is a tolerance of form that indicates that a cross section through a surface of revolution is round and lies within two concentric circles.

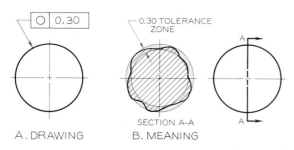

A. DRAWING

SECTION A-A

B. MEANING

FIG. 13.57 Roundness of a sphere is indicated in this manner, which means that any cross section through it is round within the specified tolerance in the symbol.

Cylindricity

A surface of revolution is cylindrical when all its elements form a cylinder. A cylindricity tolerance zone is specified in Fig. 13.58, where a tolerance of 0.54 mm RFS is permitted on the radius of the cylinder. Cylindricity is a combination of tolerances of roundness and straightness applied to a cylindrical object.

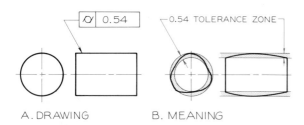

A. DRAWING **B. MEANING**

FIG. 13.58 Cylindricity is a tolerance of form that indicates that the surface of a cylinder lies within an envelope formed by two concentric cylinders.

13.27
Tolerances of profile

Profile tolerancing is used to specify tolerances about a contoured shape formed by arcs or by irregular curves. Profile can apply to a single line or to a surface.

The surface in Fig. 13.59 is given a profile tolerance that is unilateral. (It can only be smaller than the points located.) Examples of specifying bilateral and unilateral tolerance zones are shown at B.

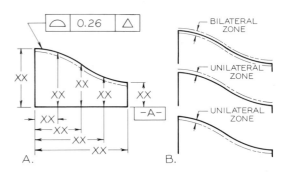

FIG. 13.59 Profile is a tolerance of form that is used to tolerance irregular curves of planes. The curving plane is located by coordinates, and the tolerance is located by any of the methods shown at B.

A profile tolerance for a single line can be specified as shown in Fig. 13.60. In this example, the curve is formed by tangent arcs whose radii are given as basic dimensions. The radii are permitted to vary by plus or minus 0.10 mm about the basic radii.

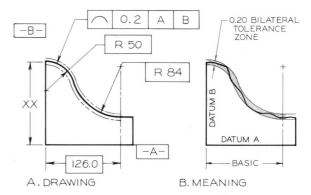

A. DRAWING **B. MEANING**

FIG. 13.60 Profile of a line is a tolerance of form that specifies the variation allowed from the path of a line. In this case, the line is formed by tangent arcs. The tolerance zone may be either bilateral or unilateral, as shown in Fig. 13.59B.

13.28
Tolerances of orientation

Tolerances of orientation include **parallelism, perpendicularity,** and **angularity.**

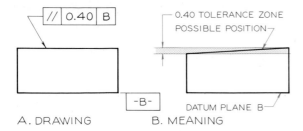

A. DRAWING   B. MEANING

FIG. 13.61 Parallelism is a tolerance of form that specifies that a plane is parallel to another within specified limits. Plane *B* is the datum plane in this case.

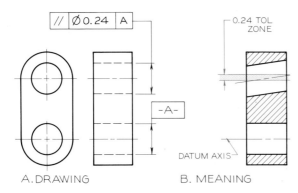

A. DRAWING   B. MEANING

FIG. 13.62 Parallelism of one centerline to another can be specified by using the diameter of one of the holes as the datum.

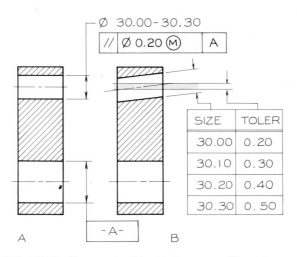

SIZE	TOLER
30.00	0.20
30.10	0.30
30.20	0.40
30.30	0.50

FIG. 13.63 The most critical tolerance will exist when features are at MMC. In this example, the upper hole must be parallel to the hole used as datum *A* within 0.20 DIA. As the hole approaches its maximum size of 30.30 mm, the tolerance zone approaches 0.50 mm.

Parallelism

A surface or a line is parallel when all its points are equidistant from a datum plane or axis. Two types of parallelism are:

1. A tolerance zone between planes parallel to a datum plane within which the axis or surface of the feature must lie (Fig. 13.61). This tolerance also controls flatness.

2. A cylindrical tolerance zone parallel to a datum feature within which the axis of a feature must lie (Fig. 13.62).

The effect of specifying parallelism at MMC can be seen in Fig. 13.63, where the modifier *M* is given in the feature control symbol. Tolerances of form apply regardless of feature size when not specified. Specifying parallelism at MMC means that the axis of the cylindrical hole must vary no more than 0.20 mm when the holes are at there smallest permissible size.

As the hole approaches its upper limit of 30.30, the tolerance zone increases until it reaches 0.50 DIA. Therefore a greater variation is given at MMC than at RFS.

Perpendicularity

The perpendicularity of two planes is specified in Fig. 13.64. Note that datum plane *C* is "flagged," and the feature control symbol is applied to the perpendicular surface.

A hole is specified as perpendicular to a surface in Fig. 13.65, where Surface *A* is indicated as the datum plane.

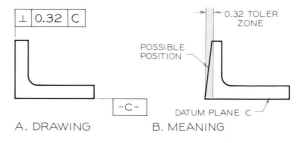

A. DRAWING   B. MEANING

FIG. 13.64 Perpendicularity is a tolerance of form that gives a tolerance zone for a plane that is perpendicular to a specified datum plane.

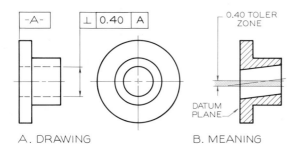

FIG. 13.65 Perpendicularity can apply to the axis of a feature, such as the centerline of a cylinder.

Angularity

A surface or line is angular when it is at a specified angle (other than 90°) from a datum or an axis. The angularity of a surface is specified in Fig. 13.66, where the angle is given a basic dimension of 30°. The angle is permitted to vary within a tolerance zone of 0.25 mm about the angle.

13.29
Tolerances of runout

Runout tolerance is a means of controlling the functional relationship between one or more parts to a common datum axis. The features controlled by runout are surfaces of revolution about an axis and surfaces that are perpendicular to the axis.

The datum axis is established by using a functional cylindrical feature that rotates about the axis, such as Diameter *B* in Fig. 13.67. When the part is rotated about this axis, the features of rotation must fall within the prescribed tolerance at **full indicator movement** (FIM).

The two types of runout are circular runout and total runout. One arrow in the feature control symbol indicates circular runout, and two arrows indicate total runout.

CIRCULAR RUNOUT (one arrow) of an object is measured by rotating it about its axis for 360° to determine whether a circular cross section at any point exceeds the permissible runout tolerance. This same technique is used to measure the amount of wobble that exists in surfaces that are perpendicular to the axis of rotation.

TOTAL RUNOUT (two arrows) is used to specify cumulative variations of circularity, straightness, coaxiality, angularity, taper, and profile of a surface (Fig. 13.68). Total runout is applied to all circular and profile positions as the part is rotated 360°. When applied to surfaces that are perpendicular to the axis, total runout controls variations in perpendicularity and flatness of the surface.

The part shown in Fig. 13.69 is dimensioned by using a composite of several techniques of geometric tolerancing.

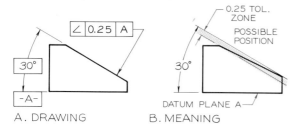

FIG. 13.66 Angularity is a tolerance of form that specifies the tolerance for an angular surface with respect to a datum plane. The 30° angle is a true angle, a basic angle. The tolerance of 0.25 mm is from this basic angle.

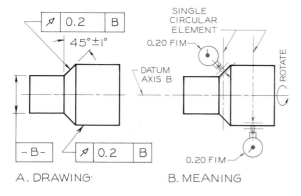

FIG. 13.67 Runout tolerance of a surface is a tolerance of form that is a composite of several form characteristics. It is used to specify concentric cylindrical parts. The part is mounted on one of the axes, the datum axis, and the part is gauged as it is rotated about the datum axis.

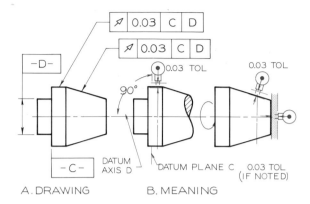

FIG. 13.68 The runout tolerance in this example is measured by mounting the object on the primary datum plane, Surface *C*, and the secondary datum plane, Cylinder *D*. The cylinder and conical surface is gauged to determine if it conforms to a tolerance zone of 0.03 mm. The end of the cone could have been noted to specify its runout (perpendicularity to the axis).

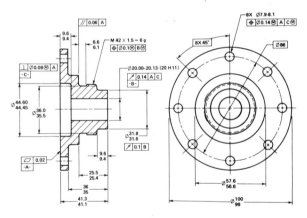

FIG. 13.69 A part dimensioned with a combination of notes and symbols to describe its geometric features. (Courtesy of ANSI Y14.5M-1982.)

13.30
Surface texture

The surface texture of a part will affect its function; consequently, this must be more precisely specified than by the general V that does not elaborate on the finish desired. Most of the terms of surface texture defined below are shown in Fig. 13.70.

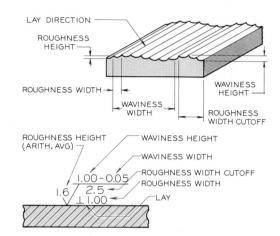

FIG. 13.70 Characteristics of surface texture

SURFACE TEXTURE   The variation in the surface, including roughness, waviness, lay, and flaws.

ROUGHNESS   This describes the finest of the irregularities in the surface, which are usually caused by the manufacturing process used to smooth the surface.

ROUGHNESS HEIGHT   The average deviation from the mean plane of the surface. It is measured in microinches (μin.), or micrometers (μm), which are millionths of an inch or micrometer.

ROUGHNESS WIDTH   The width between successive peaks and valleys that form the roughness measured in microinches or micrometers.

ROUGHNESS WIDTH CUTOFF   The largest spacing of repetitive irregularities that includes average roughness height (measured in inches or millimeters). When not specified, a value of 0.8 mm (0.030 in.) is assumed.

WAVINESS   A widely spaced variation that exceeds the roughness width cutoff. Roughness may be considered as superimposed on a wavy surface. Waviness is measured in inches or millimeters.

WAVINESS HEIGHT   The peak-to-valley distance between waves. It is measured in inches or millimeters.

WAVINESS WIDTH   The spacing between peaks or wave valleys measured in inches or millimeters.

LAY The direction of the surface pattern, which is determined by the production method used.

FLAWS Irregularities or defects that occur infrequently or at widely varying intervals on a surface. These include cracks, blow holes, checks, ridges, scratches, and the like. Unless otherwise specified, the effect of flaws are not included in roughness height measurements.

A. BASIC SURFACE TEXTURE SYMBOL: Surface may be produced by any method except when the bar or circle (B or D) is specified.

B. MATERIAL REMOVAL BY MACHINING REQUIRED: The horizontal bar indicates that material removal by machining is required to produce the surface and that material must be provided for that purpose.

C. MATERIAL REMOVAL ALLOWANCE: The number indicates the amount of stock to be removed by machining in millimeters (or inches). Tolerances may be added to the basic value shown in a general note.

D. MATERIAL REMOVAL PROHIBITED: The circle in the vee indicates that the surface must be produced by processes such as casting, forging, hot finishing, cold finishing, die casting, powder metallurgy, or injection molding without subsequent removal of material.

E. SURFACE TEXTURE SYMBOL: To be used when any surface characteristics are specified above the horizontal line or to the right of the symbol. Surface may be produced by any method except when the bar or circle (B or D) is specified.

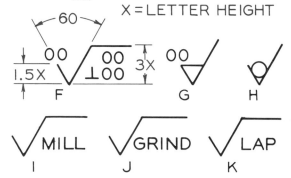

FIG. 13.71 Surface control symbols for specifying surface finish. General notes (I, J, and K) can be used to specify the recommended machining operation for finishing a surface.

Roughness average rating is placed at the left of the long leg. One rating indicates the maximum value in micrometers or microinches.

The maximum and minimum roughness average values are specified in this manner. (Microinches or micrometers.)

Maximum waviness height rating is the first rating above the horizontal extension in millimeters or inches.

Maximum waviness spacing is the second rating above the horizontal extension and to the right of the waviness height rating. This is specified in millimeters or inches.

The amount of stock provided for material removal is specified at the left of the short leg of the symbol in millimeters or inches.

Removal of material is prohibited

Lay is designated by the symbol at the right of the long leg. This means that the lay is perpendicular to this edge of the surface.

Roughness sampling length or cutoff rating is placed below the horizontal extension. When no value is shown, 0.80 mm (0.030 inches) applies. Specify in mm or inches.

Maximum roughness spacing shall be placed at the right of the lay symbol. Specify in mm or inches.

FIG. 13.72 Values can be added to surface control symbols for more precise specifications. These may be in combinations other than those shown in these examples.

CONTACT AREA The surface that will make contact with its mating surface.

The symbols used to specify surface texture are given in Fig. 13.71. The point of the V must contact the edge view of the surface being specified, an extension line from the surface, or a leader that points to the surface.

In Fig. 13.72, values of surface texture that can be applied to surface texture symbols, individually or in combination, are given. The roughness height values are related to manufacturing processes used to finish the surface (Fig. 13.73).

Lay symbols that indicate the direction of texture of a surface are given in Fig. 13.74. These symbols can be incorporated into surface texture symbols, as shown in Fig. 13.75. An example of a part with a variety of surface texture symbols applied to it is shown in Fig. 13.76.

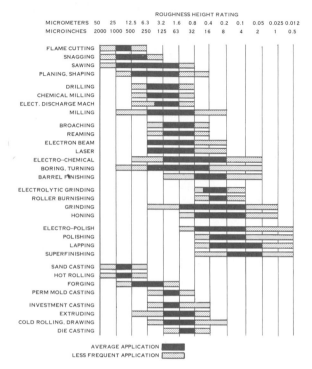

ROUGHNESS HEIGHT RATING

MICROMETERS	50	25	12.5	6.3	3.2	1.6	0.8	0.4	0.2	0.1	0.05	0.025	0.012	
MICROINCHES	2000	1000	500	250	125	63	32	16	8	4	2	1	0.5	

FLAME CUTTING
SNAGGING
SAWING
PLANING, SHAPING

DRILLING
CHEMICAL MILLING
ELECT. DISCHARGE MACH
MILLING

BROACHING
REAMING
ELECTRON BEAM
LASER
ELECTRO–CHEMICAL
BORING, TURNING
BARREL FINISHING

ELECTROLYTIC GRINDING
ROLLER BURNISHING
GRINDING
HONING

ELECTRO–POLISH
POLISHING
LAPPING
SUPERFINISHING

SAND CASTING
HOT ROLLING
FORGING
PERM MOLD CASTING

INVESTMENT CASTING
EXTRUDING
COLD ROLLING, DRAWING
DIE CASTING

AVERAGE APPLICATION
LESS FREQUENT APPLICATION

FIG. 13.73 The surface roughness heights produced by various types of production methods are shown here in micrometers (microinches). (Courtesy of the General Motors Corp.)

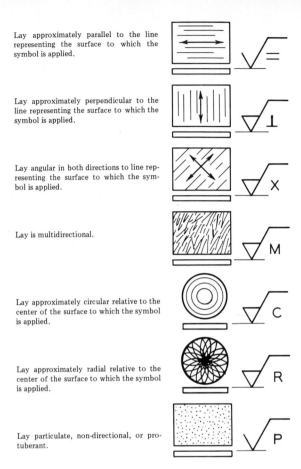

Lay approximately parallel to the line representing the surface to which the symbol is applied.

Lay approximately perpendicular to the line representing the surface to which the symbol is applied.

Lay angular in both directions to line representing the surface to which the symbol is applied.

Lay is multidirectional.

Lay approximately circular relative to the center of the surface to which the symbol is applied.

Lay approximately radial relative to the center of the surface to which the symbol is applied.

Lay particulate, non-directional, or protuberant.

FIG. 13.74 These symbols are used to indicate the direction of lay with respect to the surface where the control symbol is placed. (Courtesy of ANSI B46.1.)

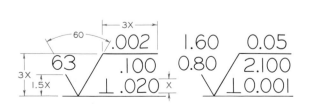

FIG. 13.75 Examples of fully specified surface control symbols

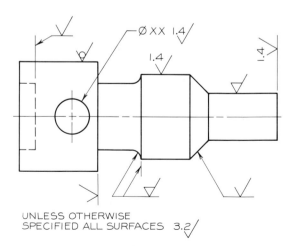

UNLESS OTHERWISE SPECIFIED ALL SURFACES 3.2/

FIG. 13.76 This drawing illustrates the techniques of applying surface texture symbols to a part.

280

Problems

These problems can be solved on Size A sheets. The problems are laid out on a grid of 0.20 in. (5 mm).

Cylindrical fits

1. Construct the drawing of a shaft and hole as shown in Fig. 13.77 (it need not be drawn to scale), give the limits for each diameter, and complete the table of values. Use a basic diameter of 1.00 in. (25 mm) and a class RC 1 fit, or a corresponding metric fit.

2. Same as Problem 1, but use a basic diameter of 1.75 in. (45 mm) and a class RC 9 fit, or a corresponding metric fit.

3. Same as Problem 1, but use a basic diameter of 2.00 in. (51 mm) and a class RC 5 fit, or a corresponding metric fit.

4. Same as Problem 1, but use a basic diameter of 12.00 in. (305 mm) and a class LC 11 fit, or a corresponding metric fit.

5. Same as Problem 1, but use a basic diameter of 3.00 in. (76 mm) and a class LC 1 fit, or a corresponding metric fit.

6. Same as Problem 1, but use a basic diameter of 8.00 in. (203 mm) and a class LC 1 fit, or a corresponding metric fit.

7. Same as Problem 1, but use a basic diameter of 102 in. (2591 mm) and a class LN 3 fit, or a corresponding metric fit.

8. Same as Problem 1, but use a basic diameter of 11.00 in. (279 mm) and a class LN 2 fit, or a corresponding metric fit.

9. Same as Problem 1, but use a basic diameter of 6.00 in. (152 mm) and a class FN 5 fit, or a corresponding metric fit.

10. Same as Problem 1, but use a basic diameter of 2.60 in. (66 mm) and a class Fn 1 fit, or a corresponding metric fit.

Tolerances of position

11. Make an instrument drawing on Size A paper of the part shown in Fig. 13.78. Locate the two

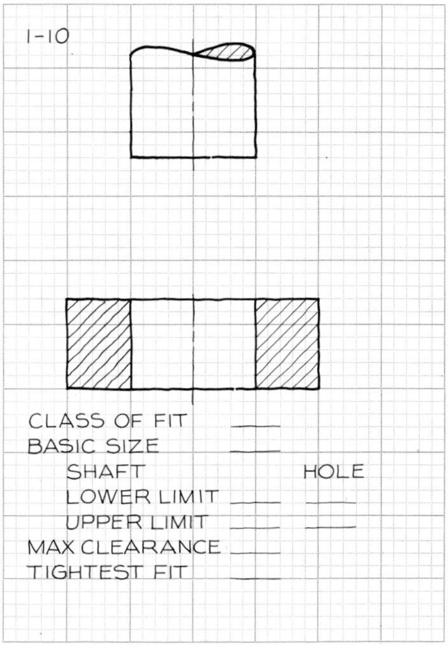

I-10

CLASS OF FIT ____
BASIC SIZE ____
   SHAFT   HOLE
   LOWER LIMIT ____ ____
   UPPER LIMIT ____ ____
MAX CLEARANCE ____
TIGHTEST FIT ____

FIG. 13.77 Problems 1–10.

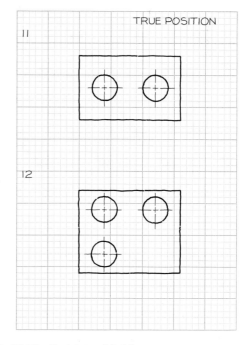

TRUE POSITION

11

12

FIG. 13.78 Problems 11–13.

holes with a size tolerance of 1.00 mm and a true-position tolerance of 0.50 DIA. Show the proper symbols and dimensions for this arrangement.

12. Same as Problem 11, except locate three holes using the same tolerances for size and position.

13. Give the specifications for a two-pin gauge that can be used to gauge the correctness of the two holes specified in Problem 11. Make a sketch of the gauge and show the proper dimensions on it.

14. Using true positioning, locate the holes and properly note them to provide a size tolerance of 1.50 mm and a locational tolerance of 0.60 DIA (Fig. 13.79).

15. Same as Problem 14, except locate six equally spaced holes of the same size using the same tolerances of position.

16. Using a feature control symbol and the necessary dimensions, indicate that the notch is symmetrical within 0.60 mm to the left-hand end of the part in Fig. 13.80.

17. Using a feature control symbol and the necessary dimensions, indicate that the small cylinder is concentric with the large one (the datum cylinder) within a tolerance of 0.80 (Fig. 13.80).

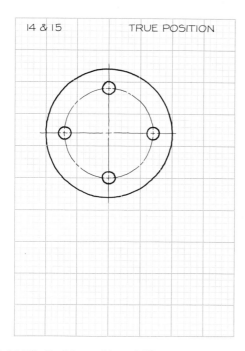

FIG. 13.79 Problems 14 and 15.

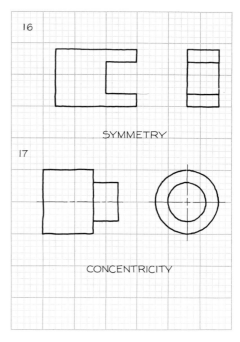

FIG. 13.80 Problems 16 and 17.

18. Using a feature control symbol and the necessary dimensions, indicate that the elements of the cylinder are straight within a tolerance of 0.20 mm (Fig. 13.81).

19. Using a feature control symbol and the necessary dimensions, indicate that the upper surface of the object is flat within a tolerance of 0.08 mm (Fig. 13.81).

20–22. Using feature control symbols and the necessary dimensions, indicate that the cross sections of the cylinder, cone, and sphere are round within a tolerance of 0.40 mm (Fig. 13.82).

23. Using a feature control symbol and the necessary dimensions, indicate that the profile of the irregular surface of the object lies within a tolerance zone of 0.40 mm, either bilateral or unilateral (Fig. 13.83).

24. Using a feature control symbol and the necessary dimensions, indicate that the profile of the line formed by tangent arcs lies within a tolerance zone of 0.40 mm, either bilateral or unilateral (Fig. 13.83).

25. Using a feature control symbol and the necessary dimensions, indicate that the cylindricity of the cylinder is 0.90 mm (Fig. 13.84).

26. Using a feature control symbol and the necessary

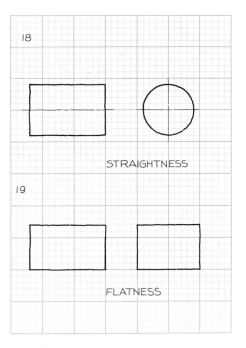

FIG. 13.81 Problems 18 and 19

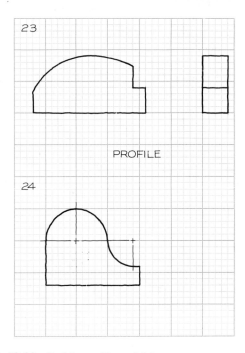

FIG. 13.83 Problems 23 and 24.

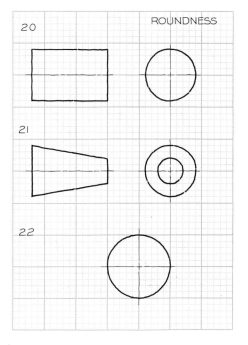

FIG. 13.82 Problems 20–22.

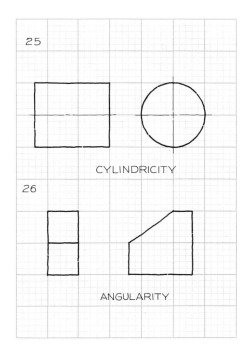

FIG. 13.84 Problems 25 and 26.

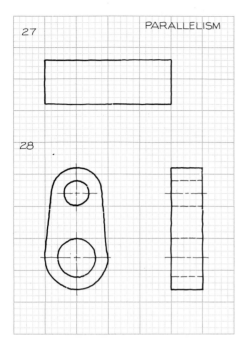

FIG. 13.85 Problems 27 and 28.

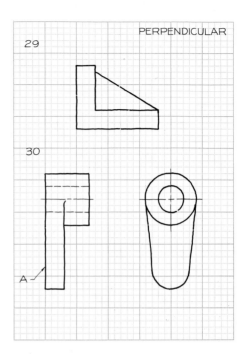

FIG. 13.86 Problems 29 and 30.

dimensions, indicate that the angularity tolerance of the inclined plane is 0.7 mm from the bottom of the object, the datum plane (Fig. 13.84).

27. Using a feature control symbol and the necessary dimensions, indicate that the upper surface of the object is parallel to the lower surface, the datum, within 0.30 mm (Fig. 13.85).

28. Using a feature control symbol and the necessary dimensions, indicate that the small hole is parallel to the large hole, the datum, within a tolerance of 0.80 mm (Fig. 13.85).

29. Using a feature control symbol and the necessary dimensions, indicate that the vertical surface is perpendicular to the bottom of the object, the datum, within a tolerance of 0.20 mm (Fig. 13.86).

30. Using a feature control symbol and the necessary dimensions, indicate that the hole is perpendicular to datum *A* within a tolerance of 0.08 mm (Fig. 13.86).

31. Using a feature control symbol and Cylinder *A* as the datum, indicate that the conical feature has a runout of 0.80 mm (Fig. 13.87).

32. Using a feature control symbol with Cylinder *B* as the primary datum and Surface *C* as the secondary datum, indicate that Surfaces *D*, *E*, and *F* have a runout of 0.60 mm (Fig. 13.87).

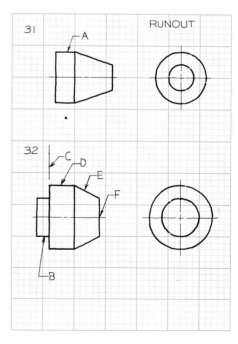

FIG. 13.87 Problems 31 and 32.

CHAPTER 14

Welding

14.1

Introduction

Welding is the process of joining metal by heating a joint to a suitable temperature with or without the application of pressure, and with or without the use of filler material. Welding is used to permanently join assemblies when it will be unnecessary to disassemble them for maintenance or other purposes.

The welding practices presented in this chapter are in compliance with the standards developed by the American Welding Society and the American National Standards Institute (ANSI). Reference is also made to the drafting standards used by General Motors Corp.

Some of the advantages of welding over other methods of fastening are: (1) simplified fabrication, (2) economy, (3) increased strength and rigidity, (4) ease of repair, (5) creation of gas- and liquid-tight joints, and (6) reduction in weight and/or size.

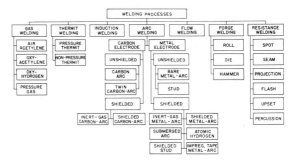

FIG. 14.1 The types of welding processes available to the designer. The three major types are gas welding, arc welding, and resistance welding. (Courtesy of General Motors Corp.)

The various welding processes are given in Fig. 14.1. The three main types of welding arc **gas welding, arc welding,** and **resistance welding.**

GAS WELDING A process in which gas flames are used to melt and fuse metal joints. Gases such as acetylene or hydrogen are mixed in a torch and are

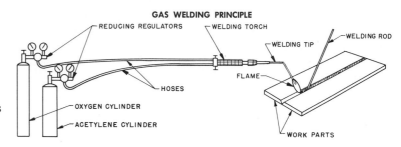

FIG. 14.2 The gas welding process burns gases, such as oxygen and acetylene, in a torch to apply heat to a joint while the welding rod supplies the filler material. (Courtesy of General Motors Corp.)

burned with air or oxygen (Fig. 14.2). The oxyacetylene method is the best-known process of gas welding, and it is widely used for repair work and field construction.

Most oxyacetylene welding is done manually with a minimum of equipment. Filler material in the form of welding rods is used to deposit metal at the joint as it is heated. Most metals, except for low- and medium-carbon steels, will require fluxes that aid in the process of melting and fusing the metals together.

ARC WELDING A process that uses an electric arc to heat and fuse the joints (Fig. 14.3). In some cases pressure may be applied in addition to the heat, and in others, pressure will not be required. The filler material is supplied by a consumable electrode through which the electric arc is transmitted or through a nonconsumable electrode to fuse the metals. Metals that are well-suited to arc welding are wrought iron, low-

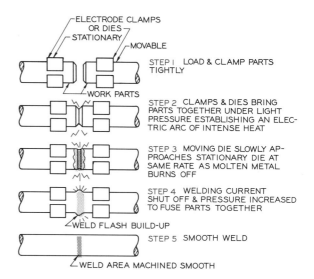

FIG. 14.4 Flash welding is a type of arc welding that uses a combination of electric current and pressure to fuse two parts together.

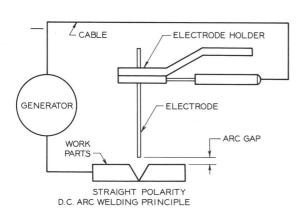

FIG. 14.3 The arc welding process can use either DC or AC current that is passed through an electrode to heat the joint to be welded.

and medium-carbon steels, stainless steel, copper, brass, bronze, aluminum, and some nickel alloys.

FLASH WELDING A form of arc welding that is similar to resistance welding in that both pressure and an electric current are used to join two pieces (Fig. 14.4). The two parts are brought together and an electric current causes a heat build-up between them. As the metal burns, the current is turned off and the pressure between the parts is increased to fuse the parts together.

RESISTANCE WELDING A group of processes where metals are fused together by heat produced from the resistance of the parts to the flow of electric current, and by the application of pressure. Fluxes and filler materials are normally not used. All resistance welds are either lap- or butt-type welds.

Figure 14.5 illustrates how resistance spot welding is performed on a lap joint. The two parts are lapped, pressed together, and an electric current fuses the part together where they join. A series of spots spaced at intervals, called **spot welds,** are used to secure the parts. Table 14.1 suggests the processes of welding that can be used for various types of materials.

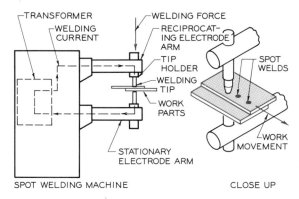

FIG. 14.5 Resistance welding can be used to join lap and butt joints to fuse metals by heat generated by passing an electrical current through them.

TABLE 14.1
RECOMMENDED RESISTANCE WELDING PROCESSES

Material	Spot Welding	Flash Welding
Low Carbon Mild Steel:		
SAE 1010	R	R
SAE 1020	R	R
Medium Carbon Steel:		
SAE 1030	R	R
SAE 1050	R	R
Wrought Alloy Steel:		
SAE 4130	R	R
SAE 4340	R	R
High Alloy Austenitic Stainless Steel:		
SAE 30301–30302	R	R
SAE 30309–30316	R	R
Ferritic and Martensitic Stainless Steel:		
SAE 51410–51430	S	S
Wrought Heat Resisting Alloys:*		
19–9–DL	S	S
16–25–6	S	S
Cast Iron	NA	NR
Gray Iron	NA	NR
Aluminum & Aluminum Alloys	R	S
Nickel & Nickel Alloys	R	S

S–Satisfactory NA–Not applicable
R–Recommended NR–Not recommended
*–For composition see American Society of Metals Handbook.
Source: Courtesy of General Motors Corporation

14.2
Weld joints

The five standard weld joints are illustrated in Fig. 14.6. The **butt joint** can be joined with the following types of welds: square groove, V-groove, bevel groove, U-groove, and J-groove. The **corner joint** can be joined with the same welds as the butt joint, but with the addition of the fillet weld as well.

The **tee joint** can be joined with the following welds: bevel groove, J-groove, and fillet welds. Welds used to join **lap joints** are fillet, bevel groove, J-groove, slot, plug, spot, projection, and seam welds. The **edge joint** uses the same welds that are used for lap joints with the addition of the square groove, V-groove, U-groove, and seam welds.

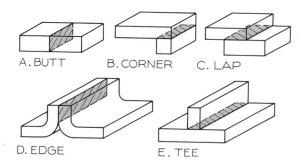

FIG. 14.6 The standard types of joints encountered in the welding process.

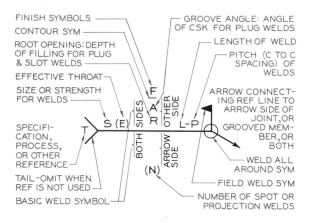

FIG. 14.7 The welding symbol It is not necessary to show the entire symbol in all applications. It may be used in modified form.

14.3
Welding symbols

The specification of welds on a working drawing is done by the application of symbols. If a drawing has a general note such as ALL JOINTS WELDED or WELDED THROUGHOUT, the designer has transferred the design responsibility to the welder. Welding is too important to be left to chance; it must be completely specified.

A welding symbol, as shown in Fig. 14.7, is used to provide specifications on a drawing. This example gives the symbol in its entirety, which is seldom needed. Instead, the symbol is usually modified to a

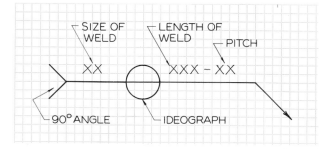

FIG. 14.8 When the grid is drawn as a full-size ⅛ in. (3 mm) grid, the size of the welding symbol can be determined. It can be drawn smaller at these same proportions when necessary.

simpler form when not all of the specifications are necessary for a particular application.

The scale of the welding symbol is shown in Fig. 14.8, where it is drawn on a 3 mm (⅛ in.) grid. Its size can be scaled down using these same proportions where space is limited. The lettering used is the standard height of ⅛ in. or 3 mm.

> The **ideograph** is the symbol used to denote the type of weld desired. In general, the ideograph depicts the cross section of the type of weld used.

The more often used ideographs are drawn to scale on the ⅛ in. (3 mm) grid to represent their full size when added to the welding symbol (Fig. 14.9).

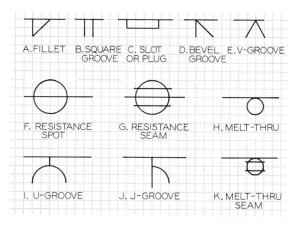

FIG. 14.9 The sizes of the ideographs are shown on the ⅛ in. (3 mm) grid. These sizes should be used in conjunction with the symbol shown in the previous figure.

14.4
Types of welds

The more commonly used welds are shown in Fig. 14.10 along with their corresponding ideographs. The **fillet weld** is a built-up weld at the angular intersection between two surfaces. The **square, bevel, V-groove, J-groove,** and **U-groove** welds involve grooves inside of which the weld is placed. **Slot** and **plug** welds have intermittent holes or openings where

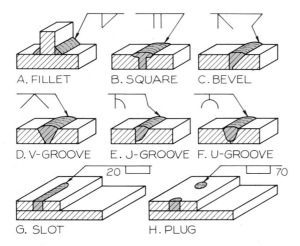

FIG. 14.10 The standard types of welds and their corresponding ideographs.

the parts are welded. Holes are unnecessary when resistance welding is used. Note that the symbols (ideographs) are symbolic of the cross sections of the welds and grooves.

14.5

Application of symbols

Fillet welds are applied to two parts in Fig. 14.11 to indicate three methods of welding. At A, the fillet ideograph is placed on the lower side of the horizontal

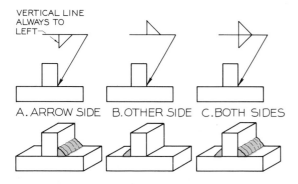

FIG. 14.11 Fillet welds are indicated by abbreviated symbols. When the ideograph is on the lower side it refers to the arrow side, and to the other side when it is placed above the horizontal line.

line of the symbol, which indicates that the weld is to be at the joint on the **arrow side,** the side of the arrow.

> The vertical leg of the ideograph is **always** on the left side.

By placing the ideograph on the upper side of the horizontal, the weld is specified to be on the **other side,** the joint on the other side of the part away from the arrow. Again, the vertical leg is on the left side.

When the part is to be welded on both sides of the vertical part, the ideograph at C is used. Note that the tail has been omitted from the symbol along with other written specifications. This is permitted when detailed specifications are given in another form to specify the details of the welds used.

A single arrow can be used to specify a weld that is to be all around two joining parts (Fig. 14.12). A circle of 6 mm diameter placed at the bend in the leader of the symbol gives this specification. If this process is to be done **in the field,** a black flag can be used to denote this joint. This means that the parts will be assembled on the site rather than in a shop as part of the manufacturing process.

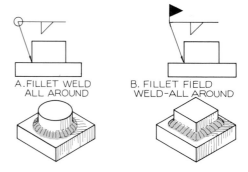

FIG. 14.12 Symbols for indicating fillet welds all around two types of parts.

When a fillet weld is to be the full length of the two parts, it may be specified as shown in Fig. 14.13. Since the ideograph is on the lower side, the weld will be on the arrow side. If the weld is to be less than full length, it can be specified as shown at B where the number, 40, represents its length in millimeters, and it is centered about the approximate location of the arrow.

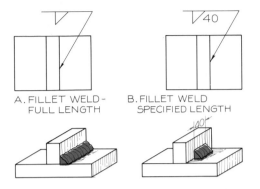

FIG. 14.13 Symbols for indicating full-length fillet welds and fillet welds less than full length.

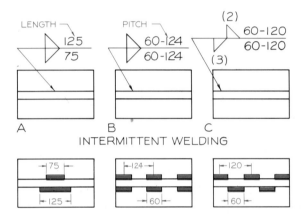

FIG. 14.14 Symbols for specifying varying and intermittent welds

When fillet welds are of different lengths and are positioned on both sides, they may be specified as shown in Fig. 14.14A. The dimensions on the lower side of the horizontal give the length of the weld on the arrow side and the number on the upper side gives the length on the other side.

Intermittent welds are welds of a given length that are spaced uniformly apart from center to center by a distance called the **pitch.** Since the welds are on both sides, 60 mm long, and have pitches of 124 mm, this can be indicated by a symbol as shown on Fig. 14.14B. If the intermittent welds can be specified by the symbols shown in Part C.

14.6
Groove welds

The more standard groove welds are illustrated in Fig. 14.15. When the depth of the grooves, the angle of the chamfer, and the root openings are not given on a symbol, this information needs to be specified elsewhere on the drawings or in supporting documents. At B and E, the depth of the chamfer of the prepared joint is given in parentheses above the size dimension of the weld, which takes into account the penetration of the weld beyond this chamfer. The size of the joint is equal to the depth of the prepared joint when only one number is given.

When the chamfer is different on each side of the

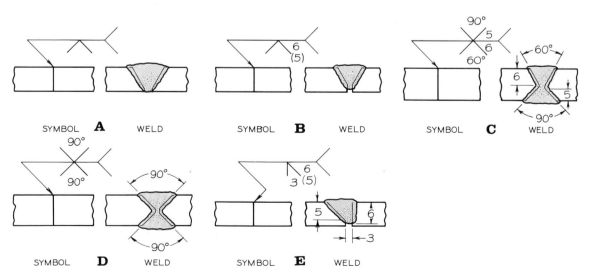

FIG. 14.15 Types of groove welds and their general specifications

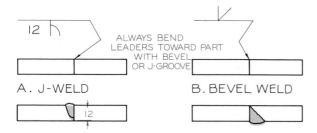

A. J-WELD   B. BEVEL WELD

FIG. 14.16 J-welds and bevel welds are specified by bent arrows pointing to the side of the joint that is to be grooved.

joint, it can be noted with a symbol as shown at Fig. 14.15C. If the spacing between the two parts, the **root opening,** is to be specified, this is done by placing its dimensions in millimeters between the groove angle number and the weld ideograph, Part E.

A bevel weld is groove beveled from one of the parts being joined; consequently the symbol must indicate which is to be beveled, as shown in Fig. 14.15E. To call attention to this operation, the leader from the symbol is bent and aimed toward the piece to be beveled. This practice also applies to J-welds where one side is grooved and the other is not (Fig. 14.16).

14.7
Example welds

A series of joints and welds that incorporates the previously covered principles is shown in Fig. 14.17. Letters are used instead of numerals to represent specified dimensions.

14.8
Surface contoured welds

Contour symbols are used to indicate which of the three types of contours is desired on the surface of the weld: **flush, concave,** or **convex.** Flush welds are those that are smooth with the surface or flat across the hypotenuse of a fillet weld. A concave contour is a weld that bulges inward with a curve, and a convex contour is one that bulges outward with a curve (Fig. 14.18).

In many cases, it is necessary to finish the weld by a supplementary process to bring it to the desired

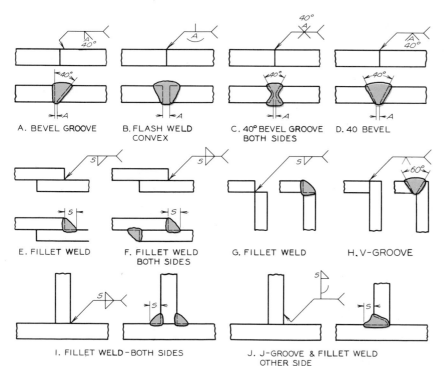

A. BEVEL GROOVE

B. FLASH WELD CONVEX

C. 40° BEVEL GROOVE BOTH SIDES

D. 40 BEVEL

E. FILLET WELD

F. FILLET WELD BOTH SIDES

G. FILLET WELD

H. V-GROOVE

I. FILLET WELD-BOTH SIDES

J. J-GROOVE & FILLET WELD OTHER SIDE

FIG. 14.17 An assortment of welds and grooves with their accompanying symbols. Letters are used instead of numbers to give the weld specification.

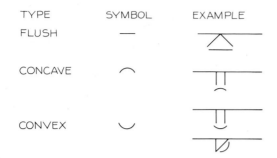

TYPE	SYMBOL	EXAMPLE
FLUSH	—	
CONCAVE	⌒	
CONVEX	⌣	

FIG. 14.18 The contour symbols that are used to specify the surface finish of a weld.

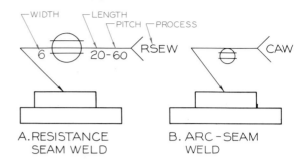

A. RESISTANCE SEAM WELD

B. ARC-SEAM WELD

FIG. 14.20 Resistance and arc seam welds with the processes are indicated in the tail of the symbol for each. The arc weld must specify arrow side or other side in the symbol.

contour. These processes may be added to the contour symbols to specify their operations in finishing the welds. These processes and their letter specifications are: **chipping** (C), **grinding** (G), **hammering** (H), **machining** (M), **rolling** (R), and **peening** (P). Examples of these contour symbols are given in Fig. 14.19.

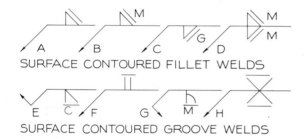

SURFACE CONTOURED FILLET WELDS

SURFACE CONTOURED GROOVE WELDS

FIG. 14.19 Examples of contoured surface symbols and letters of finishing applied to them.

14.9
Seam welds

A **seam weld** is a weld that joins two lapping parts with a continuous weld or a series of closely spaced spot welds. The process used for seam welds must be given by abbreviations placed in the tail of the weld symbol. The ideograph for a resistance weld is about 12 mm in diameter and it is placed with the horizontal line of the symbol through its center (Fig. 14.20). The weld's width, length, and pitch are indicated by numbers.

When the seam weld is made by arc welding, the diameter of the ideograph is about 6 mm and it is placed on the upper or lower side of the horizontal bar of the symbol to indicate whether the seam will be on the arrow or the other side when applied (Part B). When the numeral that represents the length of the weld is omitted from the symbol, it is understood that the seam weld extends between abrupt changes in the seam or as it is dimensioned.

Spot welds are specified in a similar manner with ideographs, and specifications given by diameter, number of welds, and pitch between the welds (Fig. 14.21). The process, **resistance spot welding** (RSW), is noted in the tail of the symbol. For arc welding, the arrow side or other side must be indicated by a symbol as shown in Part B.

(The abbreviations of various welding processes can be seen in Table 14.2).

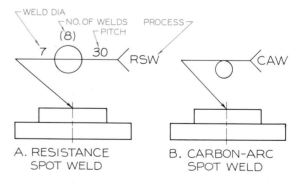

A. RESISTANCE SPOT WELD

B. CARBON-ARC SPOT WELD

FIG. 14.21 Resistance and arc spot welds with the process indicated in tail of the symbol. The arc weld must specify arrow or other side location in the symbol.

TABLE 14.2
WELDING PROCESS SYMBOLS

CAW	Carbon arc welding	FRW	Friction welding	PGW	Pressure gas welding
CW	Cold welding	FW	Flash welding	RB	Resistance brazing
DB	Dip brazing	GMAW	Gas metal arc welding	RPW	Projection welding
DFW	Diffusion welding	GTAW	Gas tungsten welding	RSEW	Resistance seam welding
EBW	Electron beam welding	IB	Induction brazing	RSW	Resistance spot welding
ESW	Electroslag welding	IRB	Infrared brazing	RW	Resistance welding
EXW	Explosion welding	OAW	Oxyacetylene welding	TB	Torch brazing
FB	Furnace brazing	OHW	Oxyhydrogen welding	UW	Upset welding
FOW	Forge welding				

14.10
Built-up welds

When the surface of a part is to be enlarged by welding, called **building up,** this can be indicated by the symbol shown in Fig. 14.22. The width of the built-up weld is dimensioned in the view, and the height of the weld above the surface is specified in the symbol to the left of the ideograph. The radius of the circular segment is 6 mm.

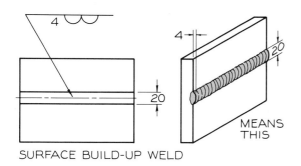

SURFACE BUILD-UP WELD

FIG. 14.22 The method of applying a symbol to a built-up weld on a surface.

14.11
Welding standards

Figure 14.23 gives an overview of the welding symbols and specifications that have been covered in the previous paragraphs. The chart was prepared by the American Welding Society of Miami, Florida. It can be used as a single reference for most general types of welding and their associated symbols.

14.12
Brazing

Brazing is method of joining pieces of metal that is similar to welding. It is a process in which the joints are heated above 800° F and a nonferrous filler material with a melting point below the base materials is distributed by capillary action between the closely fit parts.

Prior to brazing, the parts to be brazed must be cleansed, and flux is added to the joints. The brazing filler is also added prior to or just as the joints are heated beyond the melting point of the filler. There are two basic joints for brazing: lap and butt joints, as shown in Fig. 14.24. The filler material is allowed to flow between the parts to form the joint after it has melted.

Brazing is used to hold parts together, to provide gas- and liquid-tight joints, to assure electrical conductivity, and to aid in repair and salvage. Brazed joints will withstand more stress, higher temperature, and more vibration than will soft-soldered joints.

14.13
Soft soldering

Soldering is the process of joining two metal parts with another metal that melts below the temperature of the metals being joined. Solders are alloys of non-

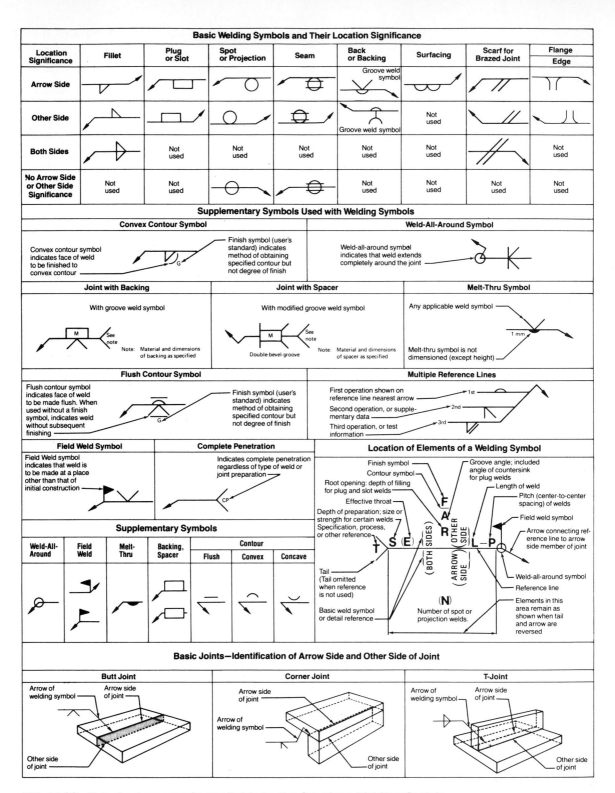

FIG. 14.23 This chart was made available by the American Welding Society.

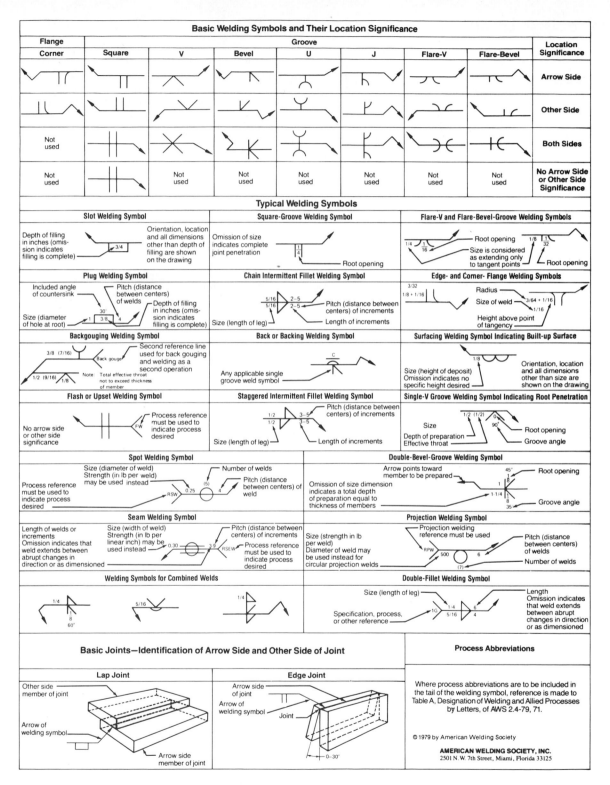

It will serve as a review of the principles covered in this chapter.

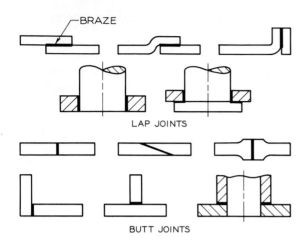

FIG. 14.24 Examples of the two basic types of brazing joints: lap joints and butt joints.

the type shown in Fig. 14.25. The iron is placed on the joint to heat it and to melt the solder that fuses the joint. The method of indicating a soldered joint is shown in Fig. 14.25, where a heavy dark line is used with a note.

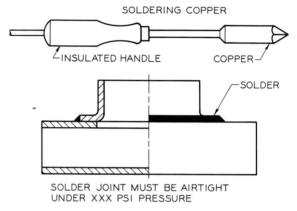

FIG.14.25 A typical hand-held soldering iron used to soft-solder two parts together, and the method of indicating a soldered joint on a drawing.

ferrous metals that melt below 800° F. This method of joining is widely used in the automotive and electrical industries.

Soldering is one of the basic techniques of welding and is often done by hand with a soldering iron of

---

Problems

Make working drawings of the following problems given in other chapters. Wherever feasible, change the joints of the features of the parts to be welded instead of joined by one-piece casting. These drawings should be made on Size B sheets.

1. Use Fig. 15.36.
2. Use Fig. 15.40.
3. Use Fig. 15.45.
4. Use Fig. 15.47.
5. Use Fig. 15.51.
6. Use Fig. 15.53.
7. Use Fig. 15.67.
8. Use Fig. 15.78, Part 1.
9. Use Fig. 15.84, Part 2.
10. Use Fig. 15.84, Part 1.

15

Working Drawings

15.1
Introduction

Working drawings are drawings from which a design is constructed. The set may contain any number of sheets, from one to over one hundred, depending on the complexity of the project. The written instructions that accompany working drawings are called **specifications.** When the project can be represented on several drawing sheets, the written specifications are often written on the drawings to consolidate the information into a single format. Although much of the work in preparing working drawings is done by the drafter, the designer, who is most often an engineer, is responsible for their correctness.

A working drawing is often called a **detail drawing** because it describes and gives the dimensions of the details of the parts being presented.

All the principles of orthographic projection, sections, conventions, dimensioning, tolerancing, pictorials, and all other types of graphical techniques are used to communicate the details in a working drawing.

15.2
Working drawings—inch system

An example of a working drawing is shown in Fig. 15.1, where the base-plate mount is detailed in three orthographic views. Dimensions and notes are used to give the information necessary to construct the piece without misinterpretation. This particular drawing is dimensioned with decimal inches.

Decimal inches are preferable to common fractions, although both systems are still in widespread use. The English system, which is based on the inch,

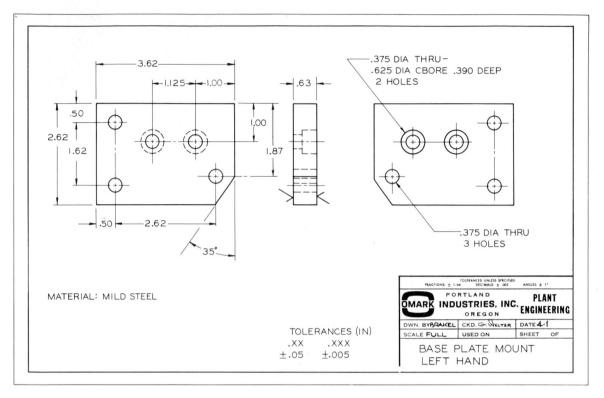

FIG. 15.1 A working drawing of a single part dimensioned in inches. (Courtesy of Omark Industries, Inc.)

FIG. 15.2 A revolving clamp assembly manufactured to hold parts stationary while they are being machined. (Courtesy of Jergens, Inc.)

is giving way to the metric system, which is based on the millimeter. Decimal inches make it possible to handle arithmetic with much greater ease than is possible with fractions. Inch marks such as X:XX'' are omitted from dimensions on a working drawing since it is understood that these dimensions are in inches.

An example of a working drawing dimensioned with common fractions using the inch as the unit of measurement is shown in Fig. 15.2. The detail drawings of the parts of the assembly are shown in Figs. 15.3–15.5. Several dimensioned parts are shown on each sheet as orthographic views. The arrangement of these parts on the sheet has no relationship to how they fit together. They are simply positioned to take advantage of the available space. Each part is given a number and a part name for identification purposes. The material that each part is made of is indicated along with notes to explain fully any necessary manufacturing procedures.

Each sheet is numbered in the title block, and the other title block information is completed.

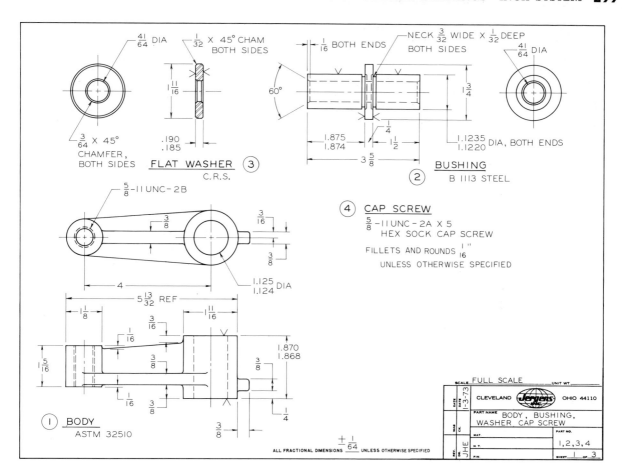

FIG. 15.3 Detail drawings showing parts of the clamp assembly. (Courtesy of Jergens, Inc.)

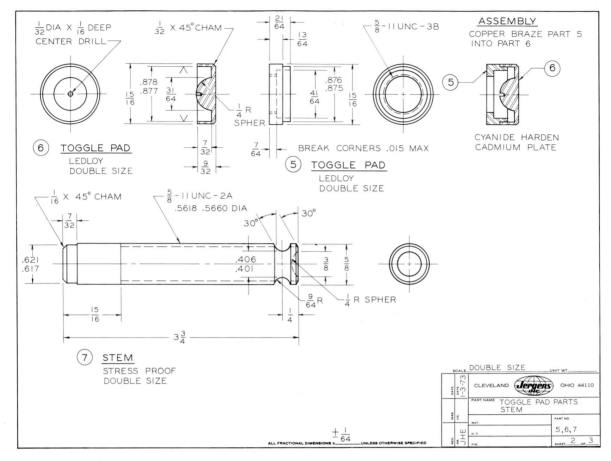

FIG. 15.4 A continuation of Fig. 15.3. (Courtesy of Jergens, Inc.)

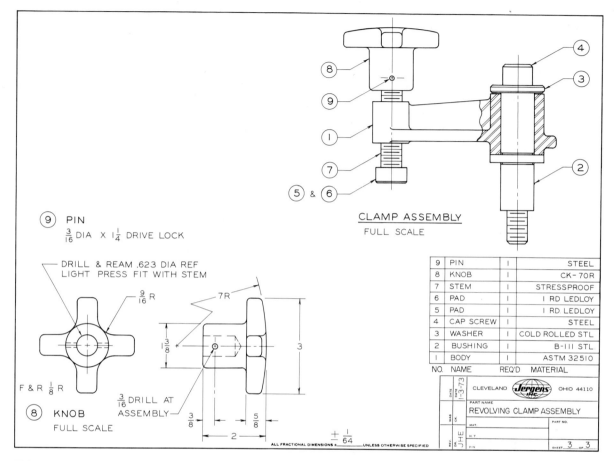

FIG. 15.5 A detail drawing and an orthographic assembly of the clamp assembly. (Courtesy of Jergens, Inc.)

An orthographic assembly is given on Sheet 3 (Fig. 15.5), which explains how the parts fit together. The parts are numbered to correspond to the part numbers in the parts list, which serves as a bill of materials.

15.3
Working drawings—metric system

The **millimeter** is the basic unit of the metric system, which is the recommended system. A single part, a back tool post, is detailed in Fig. 15.6. All dimensions

are measured to the nearest whole millimeter, with no decimal fractions except where tolerances are shown. Metric units, such as XX, mm are omitted from dimensions on a working drawing since it is understood from the title block that the units are in millimeters. Remember that the width of the fingernail of your index finger is about 10 millimeters wide. A single part is dimensioned in Fig. 15.7 with the metric system.

The lifting device that is used to level heavy equipment (Fig. 15.8) is detailed in working drawings shown in Fig. 15.9 and 15.10. All dimensions are given in millimeters, indicated by the SI symbol near the title block along with the symbol for the third angle of projection.

In Fig. 15.10, the parts are shown in an assembly that is a full section. This explains how the parts are

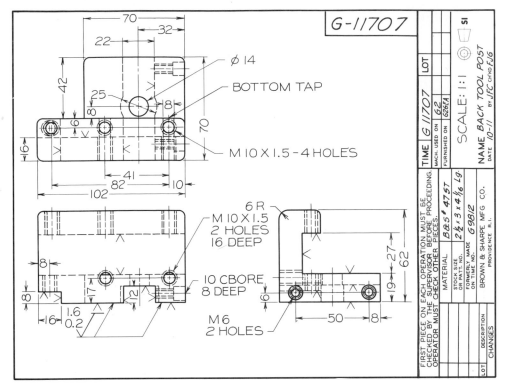

FIG. 15.6 A detail drawing of a back tool post. All dimensions are in millimeters.

FIG. 15.7 A single part dimensioned with the metric system (SI units).

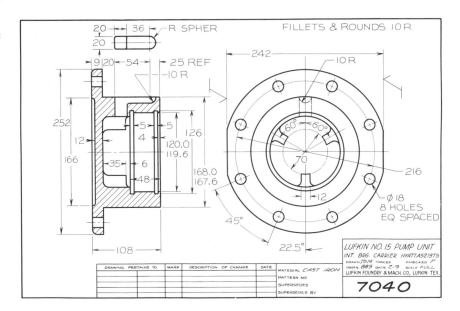

FIG. 15.8 A Lev-L-ine lifting device used to level heavy machinery. This product is the basis of the working drawings in Fig. 15.9 and 15.10. (Courtesy of Unisorb Machinery Installation Systems.)

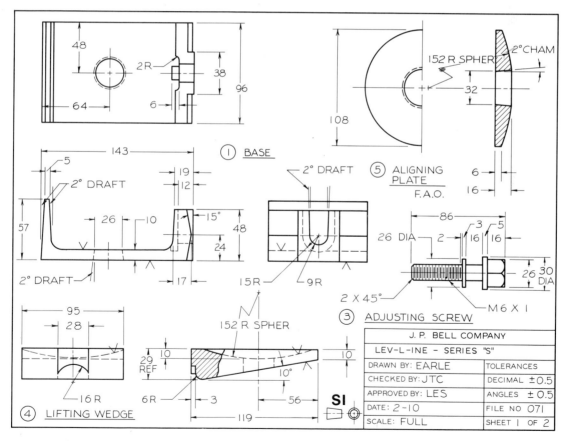

FIG. 15.9 A working drawing of the lifting device dimensioned in SI units. (Courtesy of Unisorb Machinery Installation Systems.)

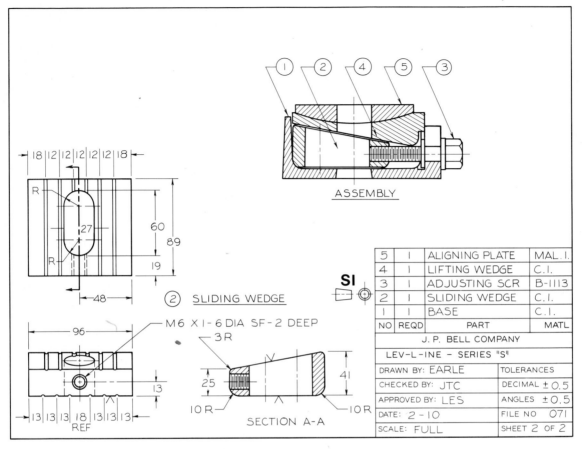

FIG. 15.10 A working drawing and assembly of the lifting device dimensioned in SI units. (Courtesy of Unisorb Machinery Installation Systems.)

FIG. 15.11 A dual-dimensioned drawing. The dimensionse are given in inches, with their equivalents in millimeters given in parentheses. (Courtesy of General Motors Corp.)

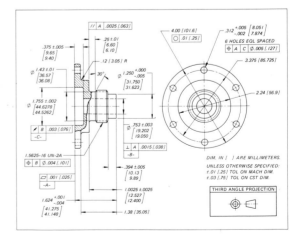

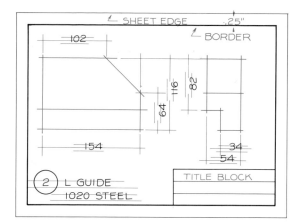

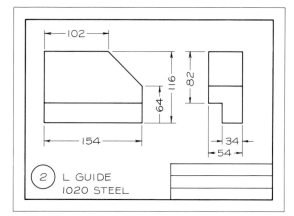

FIG. 15.12 Laying out a detail drawing

Step 1 The border and title strip are drawn on a preliminary sheet that will be traced. The views are positioned to allow adequate room for their dimensions. Guidelines are constructed for all dimensions and notes.

Step 2 The layout is overlaid with vellum or film. Then the lines are drawn to their proper weights, the dimensions and notes are lettered, and the title block is completed to finish the detail drawing.

assembled after they have been constructed. Assemblies are not dimensioned since they have been dimensioned and noted elsewhere in the drawings. A parts list is given on the same page with the assembly, above the title block for easy reference to the assembly.

15.4
Working drawings— dual dimensions

Some working drawings are dimensioned in both inches and millimeters for those who may work in both systems. A typical example is shown in Fig. 15.11, where the dimensions in parentheses are millimeters and the dimensions above the parentheses are inches. Converting from one unit to the other will result in fractional units that must be rounded off.

The units may also be given in millimeters first and then converted to inches. Usually, the converted units are shown in parentheses. An explanation of the system used should be noted in the title block.

15.5
Laying out a working drawing

The working drawing is laid out by beginning with the border, if a printed border is not provided. At least 0.25 in. (7 mm) should be allowed for the border at the edge of the sheet (Fig. 15.12, Step 1). The title block is drawn to size at the lower right corner of the sheet, and views and dimensions are drawn lightly to take advantage of the available space. When using tracing paper or film, it is more efficient to lay out the views and dimensions on a different sheet of paper and then overlay this drawing with vellum or film for tracing the final working drawing. Guidelines for lettering are drawn for each dimension line.

The lines are darkened to their proper weight, and the drawing is completed as shown in Step 2. Properly positioning the views to provide adequate space is one of the major concerns of the drafter in laying out a drawing.

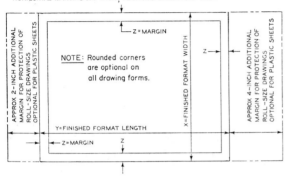

FLAT SIZES				ROLL SIZES				
SIZE DES LTR	X WIDTH	Y LENGTH	Z MARGIN	SIZE DES LTR	X WIDTH	Y MIN LENGTH	Y MAX LENGTH	Z MARGIN
A(HORIZ)	8.50	11	.25 & .38*	G	11	42	144	.38
A(VERT)	11	8.50	.25 & .38*	H	28	48	144	.50
B	11	17	.38	J	34	48	144	.50
C	17	22	.50	K	40	48	144	.50
D	22	34	.50					
E	34	44	.50					
F	28	40	.50					

*HORIZONTAL MARGINS .38-INCH, VERTICAL MARGIN .25-INCH

FIG. 15.13 The standard sheet sizes for working drawings dimensioned in inches. (Courtesy of the U.S. Department of Defense.)

The standard sheet sizes of working drawings are shown in Fig. 15.13. Papers, films, cloths, and reproduction materials are available in these modular sizes.

15.6
Title blocks and parts lists

A parts list and a title block suitable for student assignments are shown in Fig. 15.14. These blocks are usually located in the lower right-hand corner of the drawing sheet, against the borders. You will notice from observation that practically all title blocks contain the following information: title or part name, drafter, date, scale, company, and sheet number. Other information such as tolerances, checkers, and materials may be shown in more elaborate title blocks.

The standard parts list is shown in Fig. 15.14 with the usual elements.

> The **parts list** should be placed directly over and in contact with the title block in the lower right corner of the drawing.

A title block used by General Motors Corp. is shown in Fig. 15.15. A note to the left of the title block lists John F. Brown as an inventor. One or two associates are asked to date and sign the drawings as witnesses of the work of the inventor. This procedure establishes the ownership of the ideas and dates the time of their developement, in case this becomes an issue in obtaining a patent at a later date.

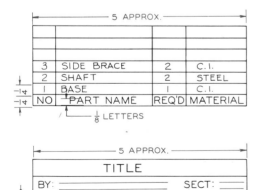

FIG. 15.14 A typical parts list and title strip suitable for most student assignments.

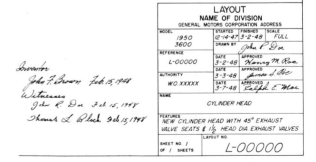

FIG. 15.15 A title block typical of those used in industry. (Courtesy of General Motors Corp.)

REVISIONS	FISHER BODY DIVISION

FISHER BODY DIVISION
GENERAL MOTORS CORPORATION
LANSING PLANT
PROCESS ENGINEERING DEPARTMENT

TITLE

SCALE   DR. BY   FILE NO.

DATE   CK'D. BY

FIG. 15.16 An example of a title block with a revision block used in industry. (Courtesy of General Motors Corp.)

Another example of a title block is shown in Fig. 15.16 which is typical of those that are printed on sheets by various industries. Revision blocks are used to indicate modifications of parts at later dates to improve the design.

It is important to give the number of sheets in the set on each sheet. For example: sheet 2 of 6, sheet 3 of 6, and so on.

15.7
Scale specification

The scale of a working drawing should be indicated in or near the title block when all drawings are the same scale. If several drawings are made at different scales, the scales used should be indicated in the drawings.

Several methods of indicating scales are shown in Fig. 15.17. When the colon is used, such as 1:2, the metric system is implied, whereas the English system is implied when the equal sign is used: 1 = 2. The symbol **SI** or the word **METRIC** on a drawing clearly specifies that the units of measurement are millimeters.

In some cases, a graphical scale is given with calibrations that permit the interpretation of linear units

SCALE: 1 = 2 (IMPLIES INCHES)
SCALE: 1:2 (IMPLIES mm)
SI OR
METRIC } (SPECIFIES SI UNITS)
(GRAPHICAL)

FIG. 15.17 Methods of specifying scales and metric units on working drawings

by transfering dimensions from the drawing with your dividers to the scale. This is commonly used for scaling distances on maps.

15.8
Tolerances

General notes can be given on working drawings to specify the tolerances of dimensions. A table of values is shown in Fig. 15.18, where a check can be made to indicate whether the units will be in inches or in millimeters.

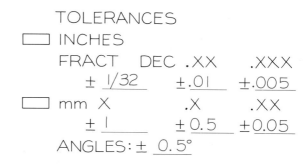

TOLERANCES
☐ INCHES
FRACT   DEC .XX   .XXX
± 1/32   ±.01   ±.005
☐ mm X   .X   .XX
± 1   ±0.5   ±0.05
ANGLES: ± 0.5°

FIG. 15.18 General tolerance notes given on working drawings to specify the tolerances permitted on dimensions.

Plus-or-minus tolerances are given in the blanks under the number of digits under each decimal fraction. For example, this table specifies that each dimension with two-place decimals will have a tolerance of ±0.01 in.

Angular tolerances can be given in general notes also. Refer to Chapter 13 for more detailed examples of using general tolerance notes.

15.9
Part names

Each part should be given a name and a number, as shown in Fig. 15.19. The letters should be ⅛ in. (3 mm) high. The part numbers are placed inside circles, called **balloons,** which are drawn approximately four times the height of the numbers.

The part numbers should be placed close to the parts on the working drawing so that it is clear which

FIG. 15.19 Each part of a working drawing should be named and numbered for listing in the parts list.

part they are associated with. Balloons are especially important on assembly drawings since the numbers of the parts refer to the same numbers in the parts list.

15.10
Checking a drawing

All drawings must be checked before they are released for production, since a slight mistake could prove very expensive when many parts are made. The people who check drawings have special qualifications that enable them to suggest revisions and modifications that will result in a better product at less cost. The checker may be a chief drafter who is experienced in this type of work, or the engineer or designer who

originated the project. In larger companies, the drawings are reviewed by the various shops involved to determine whether the most efficient methods of production are specified for each particular part.

The checker never checks an original drawing but instead checks a **diazo print** (a blue-line print). The checker marks the print with a colored pencil, making notes and corrections that he or she feels are desirable. The print is returned to the drafter for revision of the original drawing, and another print is made for approval.

In Fig. 15.20, the various modifications made by checkers are labeled with letters that are circled and placed near the revisions. The changes are listed and dated in the revision record by the drafter.

Note that several drafters and checkers were involved in the approval and preparation of the drawing. Tolerances and general information are printed in the title block to ensure uniformity in the production of similar parts.

Checkers are responsible for checking the soundness of the design and its functional characteristics. They are also responsible for the completeness of the drawing, the quality of the drawing, its readability, lettering, drafting techniques, and clarity. A poorly drawn view must be redrawn so that it will reproduce well and be understood by those using it. Quality of lettering is very important, since the shop person must rely on lettered notes and dimensions for the information.

The best method for students to use in checking their drawings is to make a scale drawing of the part from the working drawings. It is easier to find another's mistakes rather than one's own.

A grading scale for checking working drawings prepared by students is given in Fig. 15.21. This general list can be used as an outline for reviewing working drawings to ensure that the major requirements have been met.

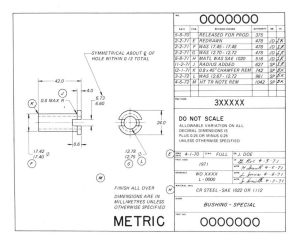

FIG. 15.20 The revisions of this working drawing are noted near the revisions with letters in balloons that are cross-referenced in a table of revisions. (Courtesy of General Motors Corp.)

15.11
Drafter's log

Drafters should keep a record called a **log** to show all changes that were made during the project. Changes, dates, and the people involved should be recorded for references as the project progresses, and as a review of the finished project.

	Max Value	Points Earned
TITLE BLOCK		
Student's name	1	
Checker	1	
Date	1	
Scale	1	
Sheet number	1	
REPRESENTATION OF DETAILS		
Selection of views	5	
Assembly drawings	10	
Positioning of views	5	
DRAFTING PRACTICES		
Line quality	8	
Lettering	8	
Proper dimensioning	10	
Proper use of sections	5	
Proper use of auxiliary views	5	
DESIGN INFORMATION		
Indication of tolerances	5	
General tolerance notes	5	
Fillets and rounds notes	5	
Finish marks	5	
Parts list	8	
Thread notes and symbols	5	
PRESENTATION		
Properly trimmed	2	
Properly folded	2	
Properly stapled	2	
	100	

grade

FIG. 15.21 A checklist for evaluating a working-drawing assignment

DRAFTSMAN'S DESIGN LOG   Sheet No. / of / Sheet

Detailed Description: *Layout and details of new transmission low speed gear and mainshaft combination. Low speed gear to have 10 of splines accurately ground with respect to gear teeth, and mainshaft to have three ground lands for mounting low speed gear on these surfaces.*

Job no. *9344-97*
Job name *Trans Mainshaft First and Reverse Gear*
Models *1950*
Engineer *Poe*
Job started *3-25-48*
Job finished *4-7-48*
Layout numbers *L-36042*

Job Objective: *To eliminate selective fit on mating parts.*

References: *L-33827*

Progress, Decisions and Authority:
3-30-48 - Messrs. Poe and Poe decided to change from a 22 tooth basic spline with 3 unevenly spaced lands to both a 24 tooth basic spline with 6 evenly spaced lands and a 24 tooth basic spline with 8 evenly spaced lands.
Engineers also requested study of a longer hub for 1st reverse gear to reduce runout.
4-6-48 After preliminary investigation Messrs. Poe and Poe decided to cancel the 8 lands construction and the elongated hub.
4-7-48 Mr. Poe decided to have an additional layout made of a 22 tooth basic spline with alternating teeth and lands.

Calculations and sketches to be dated and attached.   *R. Doe*   Signature

FIG. 15.22 A drafter's log should be kept as a record of the project to explain actions taken that might otherwise be forgotten. (Courtesy of General Motors Corp.)

An example of a drafter's log is shown in Fig. 15.22. The description of the project and its objectives are given first. Each change and the reason for it are tabulated under "Progress, Decisions, and Authority." The people responsible for the changes are mentioned by name.

Calculations are often made during the process of preparing a drawing. If they are lost or if they are poorly done, it may be necessary to make them again. Consequently, they should be made a permanent part of the log and attached to the log.

15.12
Assembly drawings

When the parts have been made according to the specifications of the working drawings, they will be assembled (Fig. 15.23). This requires a drawing called an **assembly drawing.**

FIG. 15.23 An assembly drawing is used to explain how the parts of a product such as this Ford tractor are assembled. (Courtesy of Ford Motor Co.)

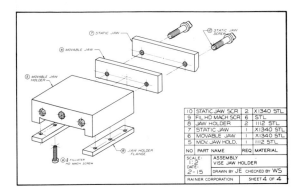

FIG. 15.24   An exploded pictorial assembly. Each part is listed by number in the parts list.

Two general types of assembly drawings are drawn by using (1) pictorial techniques or (2) orthographic projection.

An assembly drawing in pictorial is shown in Fig. 15.24. The parts are numbered and cross-referenced to the parts list, where more information about each part is given. This assembly shows the parts separated, or **exploded** apart, along their centerlines of assembly. The pictorial assembly offers the most understandable view of the relationship of the parts. This is more important when the assemblies are complex and difficult to visualize in orthographic views.

Another exploded assembly drawing is shown in Fig. 15.25, where a solenoid control valve is illustrated and its parts list given. Drawings of this type are used in catalogs and maintenance manuals for ease of interpretation.

FIG. 15.25   An exploded pictorial assembly drawing of the parts of a solenoid control valve. (Courtesy of Jenkins Bros.)

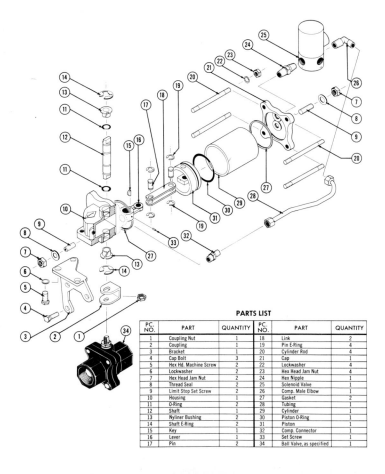

PARTS LIST

PC. NO.	PART	QUANTITY	PC. NO.	PART	QUANTITY
1	Coupling Nut	1	18	Link	2
2	Coupling	1	19	Pin E-Ring	4
3	Bracket	1	20	Cylinder Rod	4
4	Cap Bolt	3	21	Cap	1
5	Hex Hd. Machine Screw	2	22	Lockwasher	4
6	Lockwasher	2	23	Hex Head Jam Nut	4
7	Hex Head Jam Nut	2	24	Hex Nipple	1
8	Thread Seal	2	25	Solenoid Valve	1
9	Limit Stop Set Screw	1	26	Comp. Male Elbow	1
10	Housing	1	27	Gasket	2
11	O-Ring	2	28	Tubing	1
12	Shaft	1	29	Cylinder	1
13	Nyliner Bushing	2	30	Piston O-Ring	1
14	Shaft E-Ring	2	31	Piston	1
15	Key	1	32	Comp. Connector	1
16	Lever	1	33	Set Screw	1
17	Pin	2	34	Ball Valve, as specified	1

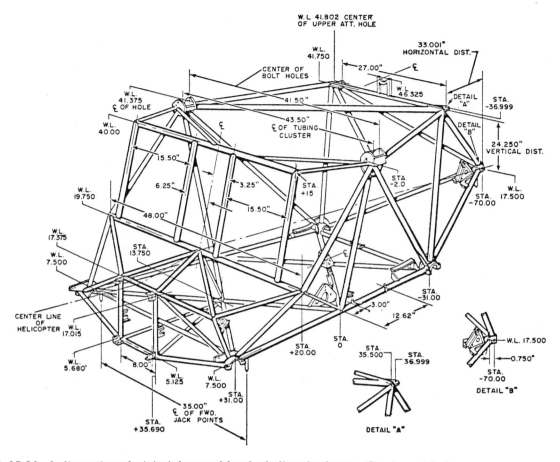

FIG. 15.26 A dimensioned pictorial assembly of a helicopter frame. (Courtesy of Bell Helicopter Co.)

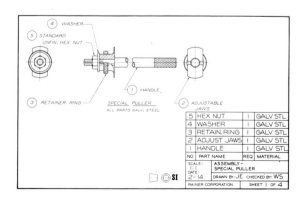

FIG. 15.27 An assembled orthographic assembly of a special puller. Note that sections can be used to clarify the assembly.

It is unnecessary to dimension assemblies in most cases, since the parts have been dimensioned individually in the working drawings prior to drawing the assembly.

An example of a dimensioned assembly is shown in Fig. 15.26, where a helicopter frame is drawn as a pictorial. More clarity is provided by this assembly than would be possible in an orthographic assembly.

An orthographic assembly of a special puller is shown in Fig. 15.27. These parts are completely assembled with two end views and a half-section to explain their relationship. A similar orthographic assembly is the subassembly of the gear-cutting fixture in Fig. 15.28. This is a partially exploded assembly.

The outline assembly drawing illustrated in Fig. 15.29 is used to show how various components are

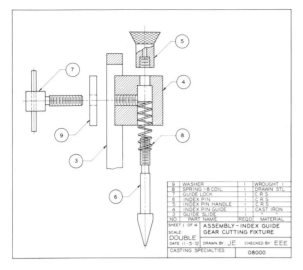

FIG. 15.28 A partially exploded orthographic assembly drawing of an index guide of a cutting fixture.

connected. Each part is composed of subassemblies that are not shown in detail. A computer-drawn assembly is shown as a half-section in Fig. 15.30.

15.13
Freehand working drawings

A freehand sketch can serve the same purpose as an instrument drawing, provided the part is simple and the essential dimensions are given (Fig, 15.31). The same principles of working-drawing construction should be followed as when instruments are used.

15.14
Castings and forged parts

Two parts are shown in Fig. 15.32 to illustrate the difference between a **forged part** and a machined part. You can see that a forging is a rough form that is made

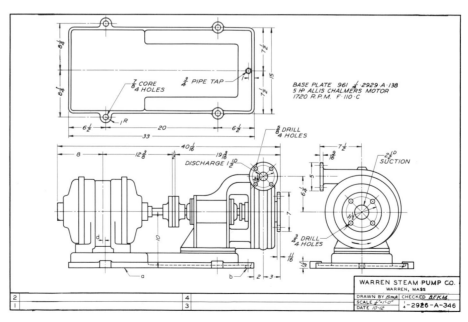

FIG. 15.29 An outline assembly that shows the general relationship of the parts of the assembly and their overall dimensions.

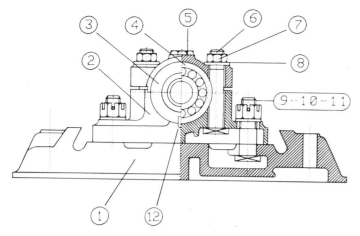

FIG. 15.30 A computer-drawn half-section of an assembly of parts

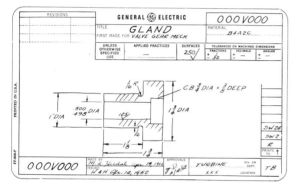

FIG. 15.31 A freehand working drawing with the essential dimensions can be adequate as a detail drawing. (Courtesy of General Electric Co.)

FIG. 15.32 The upper part is a "blank" that has been forged. When the forging has been machined, it will appear as shown in the lower photograph.

oversize by hammering the metal into shape or by pressing it between two forms (called **dies**) until the general shape is obtained. The forging is then machined to its finished dimensions and tolerances.

A **casting** is a general shape, much like that of a forging, that must be machined so that it will fit with other parts in an assembly. The casting is formed by pouring molten metal into the void in a mold formed by a pattern that is made slightly larger than the finished part (Fig. 15.33). In order for the part's pattern

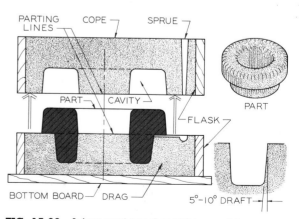

FIG. 15.33 A two-part sand mold is used to produce a casting. The draft of 5° or 10° is necessary to permit withdrawl of the pattern from the sand. The casting must be machined to size it within specified tolerances.

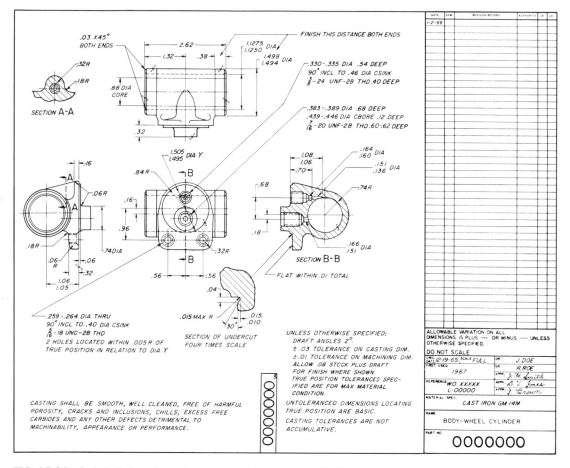

FIG. 15.34 A detail drawing of a casting that shows both the machining operations and the specifications. (Courtesy of General Motors Corp.)

FIG. 15.35 A detail drawing of a sheet metal part. It is shown as a flat pattern and as orthographic views when bent into shape.

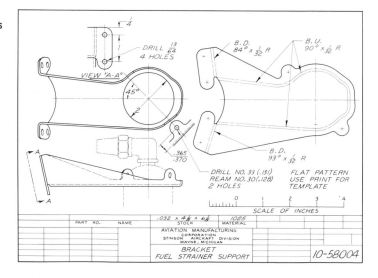

to be removable from the sand that forms the mold, its sides must be tapered. This taper of 5° to 10° is called the **draft** (Fig. 15.33).

In some industries, casting and forging drawings are made separately from machine drawings. These are more often combined into one drawing, with the understanding that the forgings and castings must be made with additional material to allow for the removal of excess by machining to meet the final design specifications. The body-wheel cylinder drawing in Fig. 15.34 dimensions both the casting and the machining operations in one drawing.

15.15
Sheet metal drawings

Parts that are made of thin metal (sheet metal) are formed by bending or stamping. The flat metal patterns for these parts must be developed graphically.

An example of a sheet metal part is shown in Fig. 15.35. The bend lines are shown on the flat pattern with the angles of bend and the radii of the bends given. The note B.D. means to bend downward, and B.U. means to bend upward.

Problems

The following problems (Figs. 15.36–15.85) are to be drawn on the sheet sizes assigned or those suggested. Each problem should be drawn with the appropriate dimensions and notes to fully describe the parts and assemblies being drawn.

Working drawings may be made on film or tracing vellum in ink or in pencil. Select a suitable title block and complete it using good lettering practices. Some problems will require more than one sheet to show all the parts properly.

Assemblies should be prepared with a parts list where there are several parts. These may be ortho-

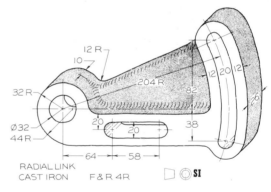

RADIAL LINK
CAST IRON   F & R 4R   □ ◎ **SI**

FIG. 15.37 Make a detail drawing on a Size B sheet.

graphic or pictorial assemblies, either exploded, assembled, or partially exploded.

The dimensions given in the problems do not always represent good dimensioning practices because of space limitations; but the dimensions given are usually adequate for you to complete the detail drawings. In some cases, there may be omitted dimensions that you must approximate using your own judgment. When making the detail drawings, strive to provide all the necesssary information, notes, and dimensions to describe the views completely. Utilize any of the previously covered principles, conventions, and techniques to present the views with the maximum of clarity and simplicity.

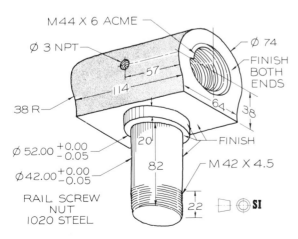

FIG. 15.36 Make a detail drawing on a Size B sheet.

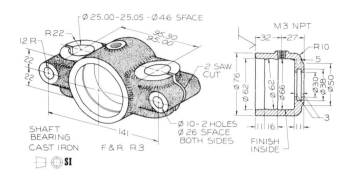

FIG. 15.38 Make a detail drawing on a Size C sheet.

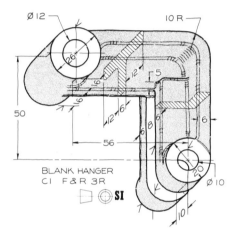

FIG. 15.39 Make a detail drawing on a Size B sheet.

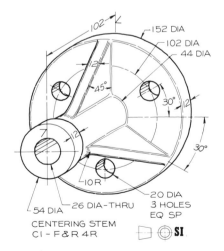

FIG. 15.40 Make a detail drawing on a Size B sheet.

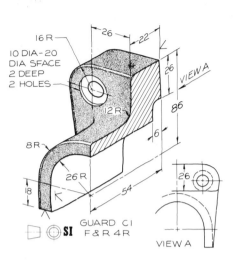

FIG. 15.41 Make a detail drawing on a Size B sheet.

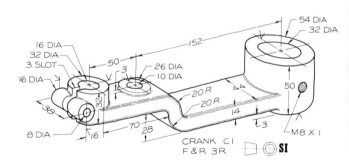

FIG. 15.42 Make a detail drawing on a Size B sheet.

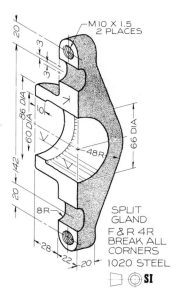

M10 X 1.5
2 PLACES

86 DIA

60 DIA

66 DIA

48R

8R

SPLIT
GLAND
F & R 4R
BREAK ALL
CORNERS
1020 STEEL

SI

FIG. 15.43 Make a detail drawing on a Size B sheet.

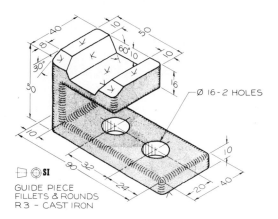

Ø 16 - 2 HOLES

SI
GUIDE PIECE
FILLETS & ROUNDS
R 3 - CAST IRON

FIG. 15.45 Make a detail drawing on a Size B sheet.

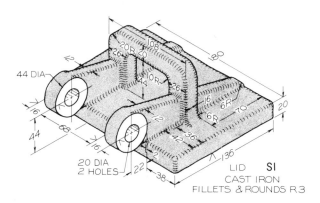

44 DIA

20 DIA
2 HOLES

LID   SI
CAST IRON
FILLETS & ROUNDS R 3

FIG. 15.46 Make a detail drawing on a Size B sheet.

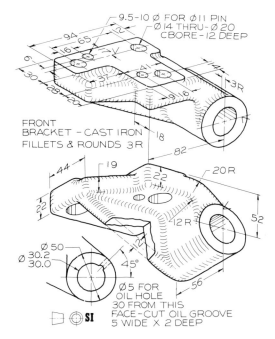

9.5-10 Ø FOR Ø11 PIN
Ø 14 THRU- Ø 20
CBORE - 12 DEEP

3R

FRONT
BRACKET - CAST IRON
FILLETS & ROUNDS 3R

18

82

44

19

22

20 R

22

12 R

52

Ø 50
Ø 30.2
30.0

45°

56

Ø 5 FOR
OIL HOLE
30 FROM THIS
FACE-CUT OIL GROOVE
5 WIDE X 2 DEEP

SI

FIG. 15.44 Make a detail drawing on a Size B sheet.

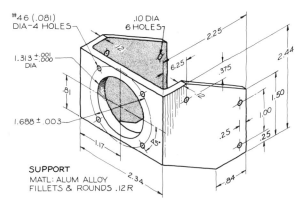

#46 (.081)
DIA-4 HOLES

.10 DIA
6 HOLES

2.25

2.44

1.313 ±.001
DIA

6.25

.375

1.50

1.00

.81

1.688 ± .003

.25

.25

45°

1.17

SUPPORT
MATL: ALUM ALLOY
FILLETS & ROUNDS .12 R

2.34

.84

FIG. 15.47 Make a detail drawing on a Size B sheet using (A) decimal inches, or (B) inches converted to millimeters.

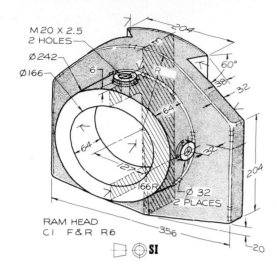

M 20 X 2.5
2 HOLES
Ø 242
Ø 166
6
R
204
60°
38
32
64
64
127
32
166R
Ø 32
2 PLACES
204
RAM HEAD
C1 F&R R6
356
20

FIG. 15.48 Make a detail drawing on a Size B sheet.

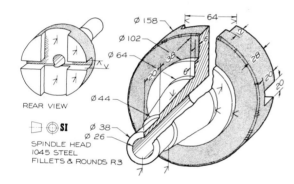

Ø 158
64
Ø 102
Ø 64
38
28
Ø 44
REAR VIEW
Ø 38
Ø 26
SPINDLE HEAD
1045 STEEL
FILLETS & ROUNDS R3

FIG. 15.49 Make a detail drawing on a Size B sheet.

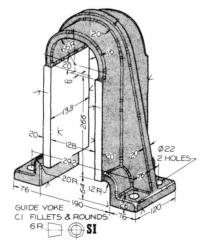

20
66R
16
20
6
133
266
20
20
128
20
20R 12R
129
Ø22
2 HOLES
76
76
64
190
76
120
GUIDE YOKE
C1 FILLETS & ROUNDS
6R

FIG. 15.50 Make a detail drawing on a Size B sheet.

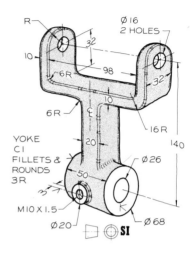

R
Ø 16
2 HOLES
32
10
6R
98
32
6R
16R
140
YOKE
C1
FILLETS &
ROUNDS
3R
20
Ø 26
50
M10 X 1.5
Ø 20
Ø 68

FIG. 15.51 Make a detail drawing on a Size B sheet.

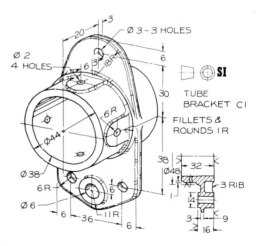

20
3
Ø 3 - 3 HOLES
Ø 2
4 HOLES
16
6R
6
SI
30
TUBE
BRACKET C1
FILLETS &
ROUNDS 1R
6R
Ø44
38
Ø48
32
Ø 38
6R
6
3 RIB
Ø 6
6
36
11R
6
3
9
16

FIG. 15.52 Make a detail drawing on a Size B sheet.

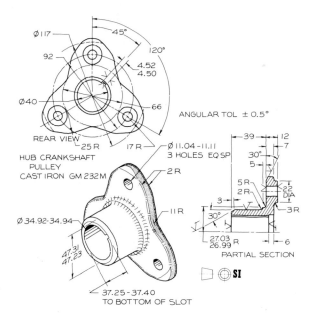

Ø 117
45°
92
120°
4.52
4.50
Ø40
66
ANGULAR TOL ± 0.5°
REAR VIEW
25 R
17 R
Ø 11.04 - 11.11
3 HOLES EQ SP
HUB CRANKSHAFT
PULLEY
CAST IRON GM 232 M
2 R

39
12
7
30°
5
5 R
2 R
22
DIA
3
3 R
30°
27.03
26.99 R
6
PARTIAL SECTION

Ø 34.92 - 34.94
11 R

47.31
47.23

SI

37.25 - 37.40
TO BOTTOM OF SLOT

FIG. 15.53 Make a detail
drawing on a Size B sheet.

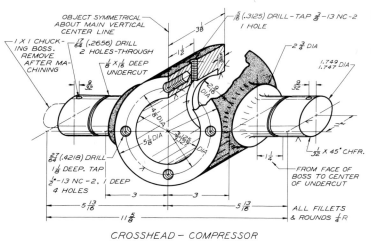

OBJECT SYMMETRICAL
ABOUT MAIN VERTICAL
CENTER LINE

$\frac{5}{16}$ (.3125) DRILL - TAP $\frac{3}{8}$ -13 NC - 2
1 HOLE

FIG. 15.54 Make a detail
drawing on a Size B sheet using
(A) decimal inches, or (B) inches
converted to millimeters.

1 X 1 CHUCK-
ING BOSS.
REMOVE
AFTER MA-
CHINING

$\frac{17}{64}$ (.2656) DRILL
2 HOLES-THROUGH

$\frac{1}{8}$ X $\frac{1}{16}$ DEEP
UNDERCUT

$3\frac{1}{8}$

$1\frac{1}{2}$

2 $\frac{3}{4}$ DIA

1.749
1.747 DIA

$\frac{9}{32}$

$\frac{9}{32}$

$\frac{1}{32}$ X 45° CHFR.

$1\frac{1}{4}$

FROM FACE OF
BOSS TO CENTER
OF UNDERCUT

$\frac{27}{64}$ (.4218) DRILL
1 $\frac{1}{8}$ DEEP. TAP
$\frac{1}{2}$ -13 NC - 2, 1 DEEP
4 HOLES

3

3

5 $\frac{13}{16}$

5 $\frac{13}{16}$

11 $\frac{5}{8}$

ALL FILLETS
& ROUNDS $\frac{1}{4}$ R

CROSSHEAD — COMPRESSOR

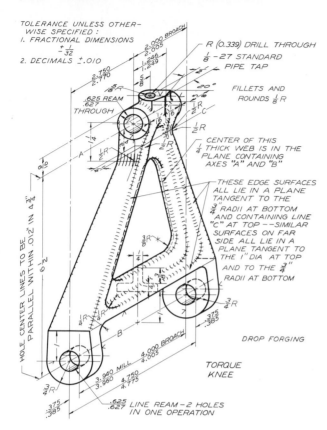

TOLERANCE UNLESS OTHER-
WISE SPECIFIED :
1. FRACTIONAL DIMENSIONS $\pm\frac{1}{32}$
2. DECIMALS $\pm$.010

R (0.339) DRILL THROUGH
$\frac{1}{8}$ - 27 STANDARD
PIPE TAP

FILLETS AND
ROUNDS $\frac{1}{8}$ R

.625 REAM
.627 THROUGH

CENTER OF THIS
$\frac{1}{4}$ THICK WEB IS IN THE
PLANE CONTAINING
AXES "A" AND "B"

THESE EDGE SURFACES
ALL LIE IN A PLANE
TANGENT TO THE
$\frac{3}{4}$" RADII AT BOTTOM
AND CONTAINING LINE
"C" AT TOP --SIMILAR
SURFACES ON FAR
SIDE ALL LIE IN A
PLANE TANGENT TO
THE 1" DIA AT TOP
AND TO THE $\frac{3}{4}$"
RADII AT BOTTOM

HOLE CENTER LINES TO BE
PARALLEL WITHIN .012" IN 4 $\frac{3}{4}$"

2.000 BROACH
2.005
1.236
1.249
2.750
2.770

.375
.385

DROP FORGING

TORQUE
KNEE

3.940 MILL
3.960
4.750
4.775

4.000 BROACH
4.005

.625 LINE REAM -2 HOLES
.627 IN ONE OPERATION

.375
.385

FIG. 15.55 Make a detail
drawing on a Size B sheet using
(A) decimal inches, or (B) inches
converted to millimeters.

FIG. 15.56 Make a detail
drawing on a Size B sheet using
(A) decimal inches, or (B) inches
converted to millimeters.

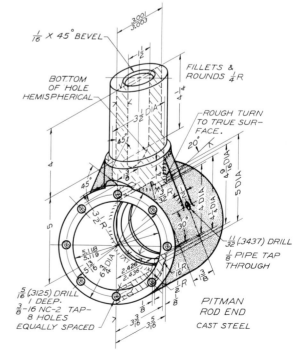

3.001
3.003

$\frac{1}{16}$ X 45° BEVEL

FILLETS &
ROUNDS $\frac{1}{4}$ R

BOTTOM
OF HOLE
HEMISPHERICAL

ROUGH TURN
TO TRUE SUR-
FACE.

20°

$\frac{11}{32}$ (.3437) DRILL
$\frac{1}{8}$ PIPE TAP
THROUGH

5.118
5.119

2.426
2.436

$\frac{5}{16}$ (.3125) DRILL
1 DEEP.
$\frac{3}{8}$ -16 NC-2 TAP-
8 HOLES
EQUALLY SPACED

PITMAN
ROD END
CAST STEEL

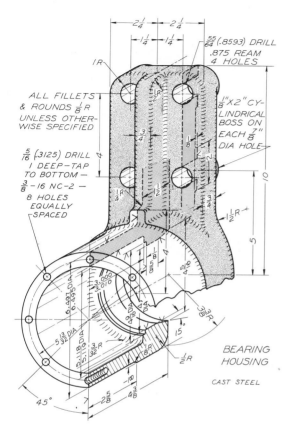

FIG. 15.57 Make a detail drawing on a Size B sheet using (A) decimal inches, or (B) inches converted to millimeters.

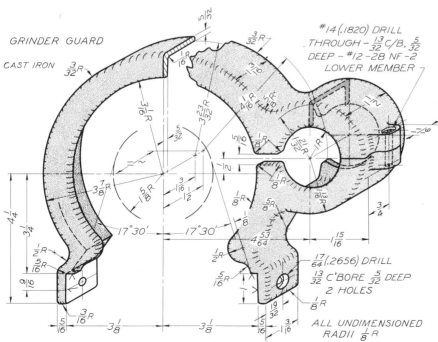

FIG. 15.58 Make a detail drawing on a Size B sheet using (A) decimal inches, or (B) inches converted to millimeters.

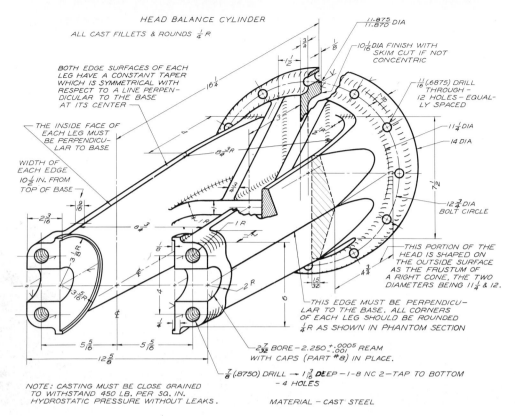

HEAD BALANCE CYLINDER

ALL CAST FILLETS & ROUNDS $\frac{1}{4}$ R

BOTH EDGE SURFACES OF EACH
LEG HAVE A CONSTANT TAPER
WHICH IS SYMMETRICAL WITH
RESPECT TO A LINE PERPEN-
DICULAR TO THE BASE
AT ITS CENTER

THE INSIDE FACE OF
EACH LEG MUST BE
PERPENDICU-
LAR TO BASE

WIDTH OF
EACH EDGE
$10\frac{1}{2}$ IN. FROM
TOP OF BASE

11.875 / 11.870 DIA

$10\frac{1}{16}$ DIA FINISH WITH
SKIM CUT IF NOT
CONCENTRIC

$\frac{11}{16}$ (.6875) DRILL
THROUGH –
12 HOLES – EQUAL-
LY SPACED

$11\frac{1}{4}$ DIA

14 DIA

$12\frac{3}{4}$ DIA
BOLT CIRCLE

THIS PORTION OF THE
HEAD IS SHAPED ON
THE OUTSIDE SURFACE
AS THE FRUSTUM OF
A RIGHT CONE, THE TWO
DIAMETERS BEING $11\frac{1}{4}$ & 12.

THIS EDGE MUST BE PERPENDICU-
LAR TO THE BASE. ALL CORNERS
OF EACH LEG SHOULD BE ROUNDED
$\frac{1}{4}$ R AS SHOWN IN PHANTOM SECTION

$2\frac{7}{32}$ BORE – 2.250 $^{+.0005}_{-.001}$ REAM
WITH CAPS (PART #8) IN PLACE.

$\frac{7}{8}$ (.8750) DRILL → $1\frac{3}{16}$ DEEP – 1-8 NC 2–TAP TO BOTTOM
– 4 HOLES

NOTE: CASTING MUST BE CLOSE GRAINED
TO WITHSTAND 450 LB. PER SQ. IN.
HYDROSTATIC PRESSURE WITHOUT LEAKS.

MATERIAL – CAST STEEL

FIG. 15.59 Make a detail drawing on a Size B sheet using (A) decimal inches, or
(B) inches converted to millimeters.

FIG. 15.60 Make a detail
drawing on a Size C sheet using
(A) decimal inches, or (B) inches
converted to millimeters.

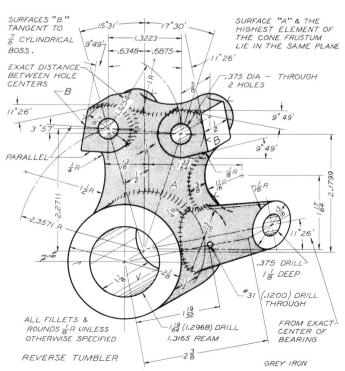

SURFACES "B"
TANGENT TO
$\frac{7}{8}$ CYLINDRICAL
BOSS.

EXACT DISTANCE
BETWEEN HOLE
CENTERS

PARALLEL

SURFACE "A" & THE
HIGHEST ELEMENT OF
THE CONE FRUSTUM
LIE IN THE SAME PLANE

.375 DIA – THROUGH
2 HOLES

.375 DRILL
$1\frac{1}{8}$ DEEP

#31 (.1200) DRILL
THROUGH

FROM EXACT
CENTER OF
BEARING

ALL FILLETS &
ROUNDS $\frac{1}{8}$ R UNLESS
OTHERWISE SPECIFIED

$1\frac{19}{64}$ (1.2968) DRILL
1.3165 REAM

REVERSE TUMBLER

GREY IRON

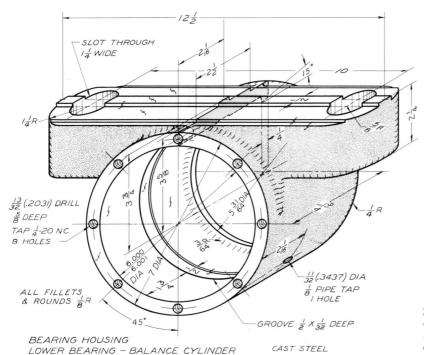

SLOT THROUGH
1¼ WIDE

12½

2¹⁄₁₆

2½

15°

10

1¼ R

13/32 (.2031) DRILL
⅝ DEEP
TAP ¼-20 NC
8 HOLES

6.000
6.001
DIA

7 DIA

3¾

3⅜

5 31/64 DIA

ALL FILLETS
& ROUNDS ⅛ R

45°

GROOVE ½ X 1/32 DEEP

11/32 (.3437) DIA
⅛ PIPE TAP
1 HOLE

¼ R

2¹⁄₁₆

BEARING HOUSING
LOWER BEARING − BALANCE CYLINDER

CAST STEEL

FIG. 15.61 Make a detail
drawing on a Size C sheet using
(A) decimal inches, or (B) inches
converted to millimeters.

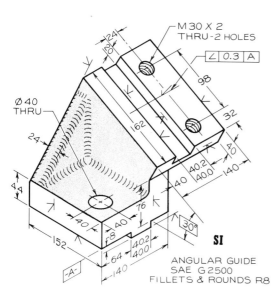

M 30 X 2
THRU-2 HOLES

∠ 0.3 A

Ø 40
THRU

24

162

98

32

40.2
40.0

140

76

40

40.2
400

64

140

152

30°

SI

-A-

ANGULAR GUIDE
SAE G 2500
FILLETS & ROUNDS R8

FIG. 15.62 Make a detail drawing on a Size B
sheet using (A) decimal inches, or (B) millimeters.

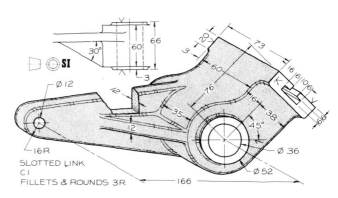

30°

60

66

3

SI

Ø 12

12

12

35

16R

SLOTTED LINK
C1
FILLETS & ROUNDS 3R

166

20

73

60

76

6

38

45°

Ø 36

Ø 52

FIG. 15.63 Make a detail drawing on a Size B
sheet using (A) decimal inches, or (B) millimeters.

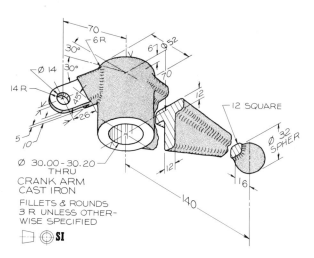

FIG. 15.64 Make a detail drawing on a Size B sheet.

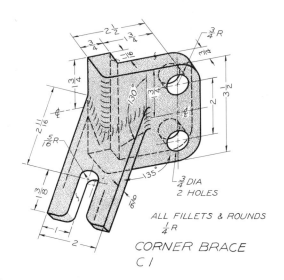

FIG. 15.66 Make a detail drawing on a Size B sheet. Convert the fractional inches to (A) decimal inches, or (B) millimeters.

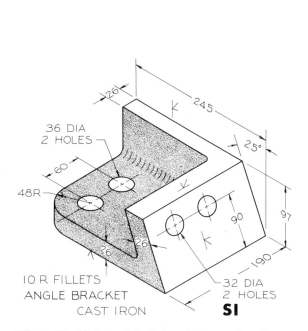

FIG. 15.65 Make a detail drawing on a Size A sheet.

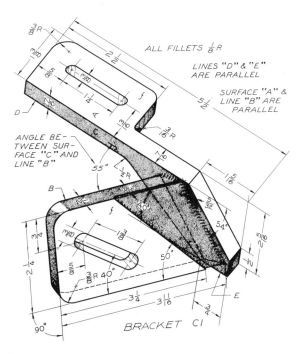

FIG. 15.67 Make a detail drawing on a Size C sheet. Convert the fractional inches to (A) decimal inches, or (B) millimeters.

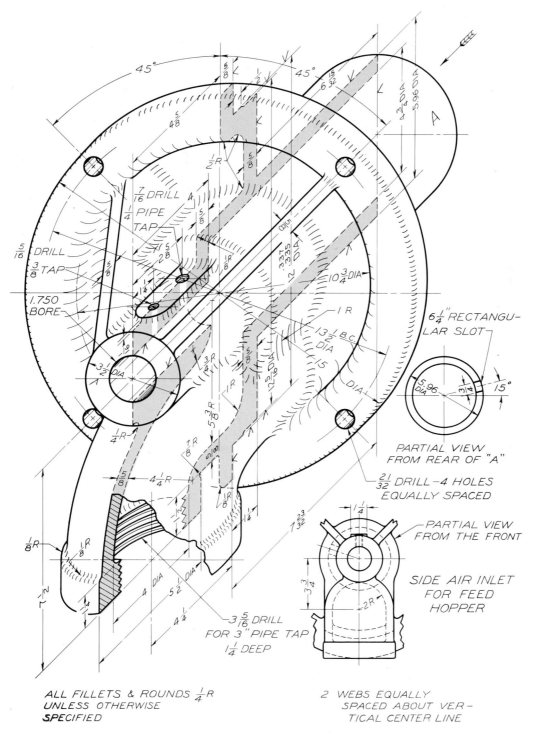

FIG. 15.68 Make a detail drawing on a Size C sheet. Convert the fractional inches to (A) decimal inches or (B) millimeters.

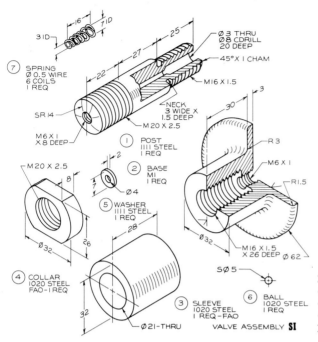

FIG. 15.69 Make a detail drawing of the parts of the valve assembly on Size B sheets. Draw an assembly and provide a parts list.

FIG. 15.70 Make a detail drawing of the parts of the cut-off crank on Size B sheets. Convert the fractional dimensions to (A) decimal inches or (B) millimeters. Draw an assembly and provide a parts list.

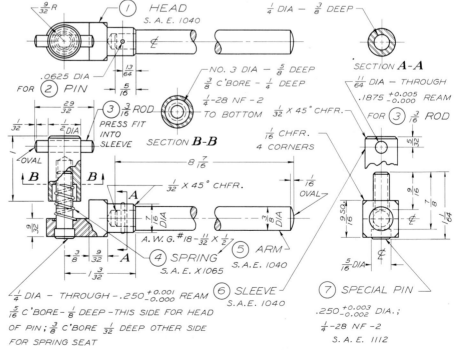

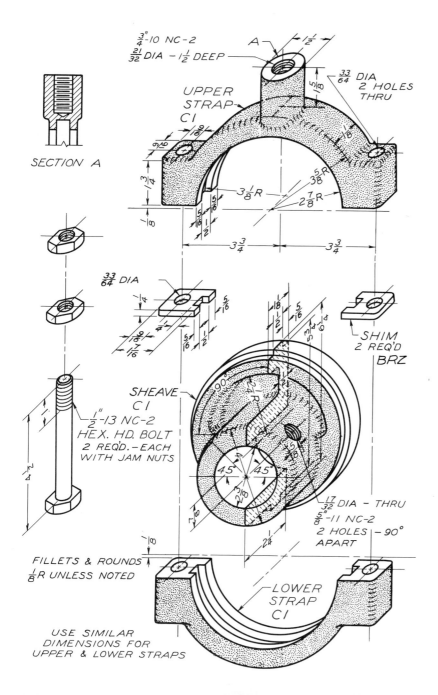

$\frac{3''}{4}$-10 NC-2
$\frac{21}{32}$ DIA - $1\frac{1}{2}$ DEEP

A

$1\frac{1}{2}$

$\frac{33}{64}$ DIA
2 HOLES
THRU

UPPER
STRAP
C1

$\frac{5}{8}$

$\frac{9}{16}$

$\frac{1}{16}$

SECTION A

$\frac{3}{4}$

$3\frac{5}{8}$ R

$3\frac{1}{8}$ R

$2\frac{7}{8}$ R

$\frac{1}{8}$

$\frac{5}{16}$

$\frac{5}{16}$

$1\frac{1}{2}$

$3\frac{3}{4}$

$3\frac{3}{4}$

$\frac{33}{64}$ DIA

$\frac{1}{4}$

$\frac{5}{16}$

$\frac{9}{16}$

$\frac{7}{16}$

$\frac{5}{16}$

$1\frac{1}{2}$

SHIM
2 REQ'D
BRZ

SHEAVE
C1

$\frac{1''}{2}$-13 NC-2
HEX. HD. BOLT
2 REQ'D.—EACH
WITH JAM NUTS

45°   45°

$\frac{17}{32}$ DIA - THRU
$\frac{5}{8}$"-11 NC-2
2 HOLES - 90°
APART

$\frac{1}{8}$

FILLETS & ROUNDS
$\frac{1}{8}$ R UNLESS NOTED

$2\frac{1}{4}$

LOWER
STRAP
C1

USE SIMILAR
DIMENSIONS FOR
UPPER & LOWER STRAPS

FIG. 15.71 Make a detail
drawing of the parts of the
journal assembly on Size B
sheets. Convert the fractional
dimensions to (A) decimal
inches or (B) millimeters. Draw
an assembly and provide a
parts list.

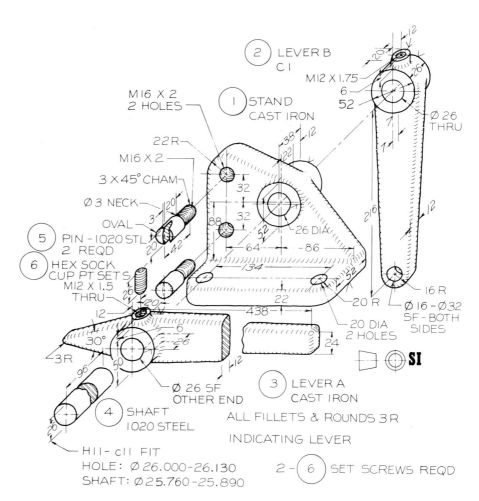

FIG. 15.72 Make a detail drawing of the parts of the indicating lever on Size B sheets, and draw an assembly with a parts list.

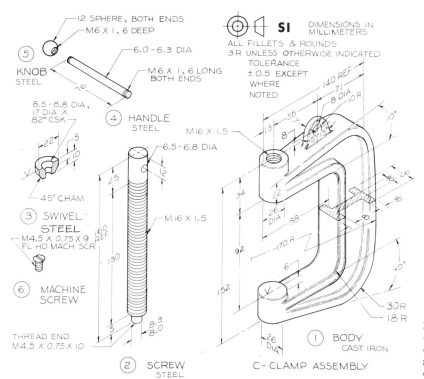

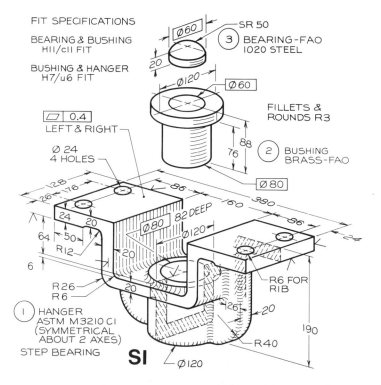

FIG. 15.73 Make a detail drawing of the parts of the C-clamp assembly on Size B sheets; show an assembly of the parts and provide a parts list.

FIG. 15.74 Make a detail drawing of the parts of the step bearing on Size B sheets, and draw an assembly with a parts list.

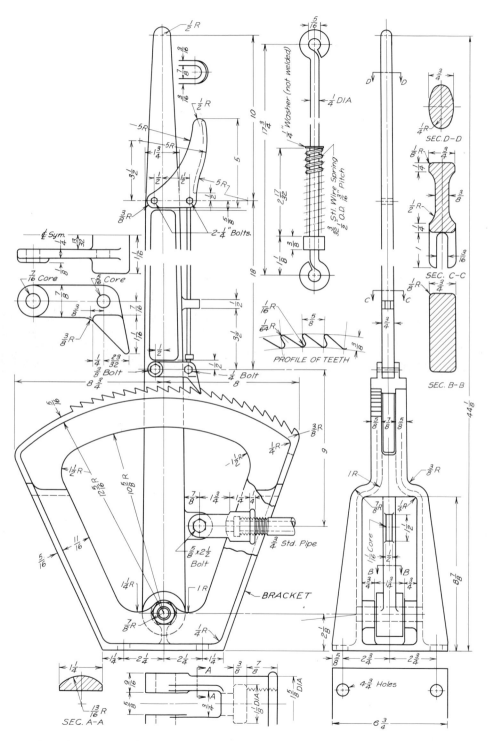

FIG. 15.75 Make detail drawings of the parts of the brake lever on Size B sheets. Convert the fractional inches to (A) decimal inches, or (B) millimeters. Draw an assembly of the parts and provide a parts list.

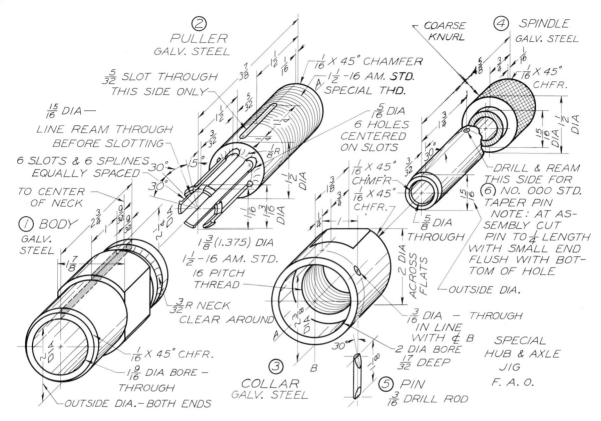

② PULLER
GALV. STEEL

$\frac{5}{32}$ SLOT THROUGH
THIS SIDE ONLY

$\frac{15}{16}$ DIA—
LINE REAM THROUGH
BEFORE SLOTTING

6 SLOTS & 6 SPLINES
EQUALLY SPACED
TO CENTER
OF NECK

① BODY
GALV.
STEEL

$\frac{1}{16}$ X 45° CHAMFER
$1\frac{1}{2}$-16 AM. STD.
SPECIAL THD.

$\frac{5}{16}$ DIA
6 HOLES
CENTERED
ON SLOTS

$\frac{1}{16}$ X 45°
CHMFR.
$\frac{1}{16}$ X 45°
CHFR.

$1\frac{3}{8}$(1.375) DIA
$1\frac{1}{2}$-16 AM. STD.
16 PITCH
THREAD
$\frac{3}{32}$R NECK
CLEAR AROUND

$\frac{1}{16}$ X 45° CHFR.
$1\frac{9}{16}$ DIA BORE—
THROUGH
OUTSIDE DIA.—BOTH ENDS

③ COLLAR
GALV. STEEL

COARSE
KNURL

④ SPINDLE
GALV. STEEL
$\frac{1}{16}$ X 45°
CHFR.

DRILL & REAM
THIS SIDE FOR
⑥ NO. 000 STD.
TAPER PIN
NOTE: AT AS-
SEMBLY CUT
PIN TO $\frac{1}{4}$ LENGTH
WITH SMALL END
FLUSH WITH BOT-
TOM OF HOLE
OUTSIDE DIA.

$\frac{5}{8}$ DIA
THROUGH

2 DIA
ACROSS
FLATS

$\frac{3}{16}$ DIA — THROUGH
IN LINE
WITH ₵ B
2 DIA BORE
$\frac{17}{32}$ DEEP

⑤ PIN
$\frac{3}{16}$ DRILL ROD

SPECIAL
HUB & AXLE
JIG
F. A. O.

FIG. 15.76 Make detail drawings of the parts of the hub and axle jig on Size B sheets. Convert the fractional inches to (A) decimal inches, or (B) millimeters. Draw an assembly of the parts and provide a parts list.

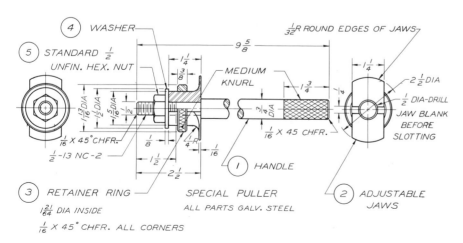

④ WASHER
⑤ STANDARD $\frac{1}{2}$
UNFIN. HEX. NUT

$\frac{1}{32}$R ROUND EDGES OF JAWS
$9\frac{5}{8}$

MEDIUM
KNURL

$\frac{1}{16}$ X 45° CHFR.

$2\frac{1}{2}$ DIA
$\frac{1}{2}$ DIA-DRILL
JAW BLANK
BEFORE
SLOTTING

$\frac{1}{16}$ X 45°CHFR.
$\frac{1}{2}$ -13 NC-2

① HANDLE

③ RETAINER RING
$1\frac{21}{64}$ DIA INSIDE
$\frac{1}{16}$ X 45° CHFR. ALL CORNERS

SPECIAL PULLER
ALL PARTS GALV. STEEL

② ADJUSTABLE
JAWS

FIG. 15.77 Make a detail drawing of the parts of the special puller on Size B sheets. Convert the fractional inches to (A) decimal inches, or (B) millimeters. Draw a pictorial assembly of the parts and give a parts list.

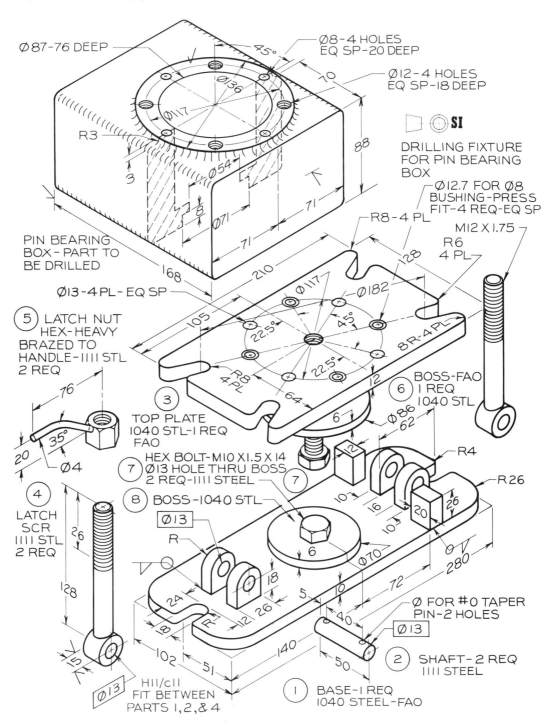

FIG. 15.78 Make detail drawings of the parts of the drilling jig and crank pin bearing box. Draw an assembly of the parts and provide a parts list.

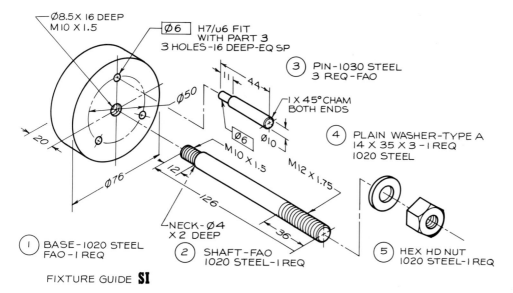

Ø8.5 X 16 DEEP
M10 X 1.5

Ø6   H7/u6 FIT
WITH PART 3
3 HOLES–16 DEEP–EQ SP

③ PIN–1030 STEEL
3 REQ–FAO

Ø50

Ø6

Ø10

1 X 45° CHAM
BOTH ENDS

④ PLAIN WASHER–TYPE A
14 X 35 X 3 –1 REQ
1020 STEEL

20

Ø76

11

44

M10 X 1.5

M12 X 1.75

12

126

36

NECK–Ø4
X 2 DEEP

① BASE–1020 STEEL
FAO–1 REQ

② SHAFT–FAO
1020 STEEL–1 REQ

⑤ HEX HD NUT
1020 STEEL–1 REQ

FIXTURE GUIDE **SI**

FIG. 15.79 Make detail drawings of the parts of the fixture guide. Draw an assembly and give a parts list.

FIG. 15.80 Make a detail drawing of the parts of the centering point. Draw an assembly and give a parts list.

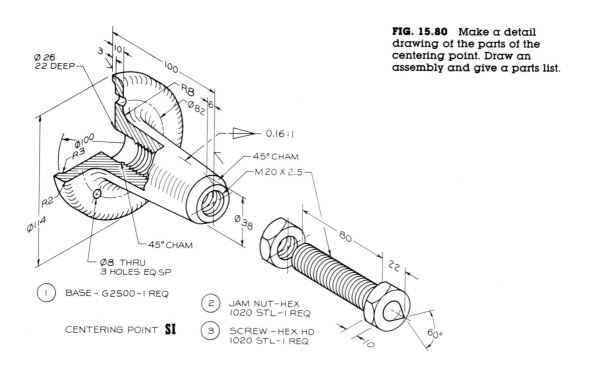

Ø 26
22 DEEP

3

10

100

R8

Ø82

6

0.16:1

45° CHAM

M20 X 2.5

Ø100

R3

R2

Ø114

45° CHAM

Ø38

Ø8 THRU
3 HOLES EQ SP

80

22

① BASE – G2500 –1 REQ

② JAM NUT–HEX
1020 STL–1 REQ

③ SCREW –HEX HD
1020 STL–1 REQ

10

60°

CENTERING POINT **SI**

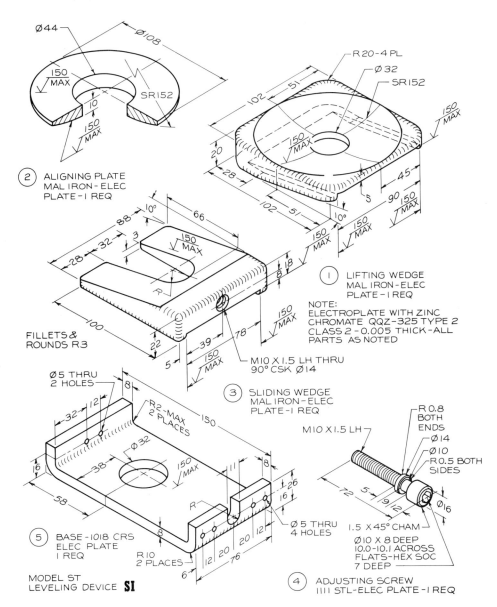

Ø44

Ø108

$\frac{150}{MAX}$

SR152

10

$\frac{150}{MAX}$

(2) ALIGNING PLATE
MAL IRON – ELEC
PLATE – I REQ

R 20 – 4 PL

Ø 32

SR152

$\frac{150}{MAX}$

102

51

20

28

102

51

10°

$\frac{150}{MAX}$

$\frac{150}{MAX}$

$\frac{150}{MAX}$

5

90

45

$\frac{150}{MAX}$

10°

88

66

32

3

28

$\frac{150}{MAX}$

R

100

22

5

8
8

39

78

$\frac{150}{MAX}$

$\frac{150}{MAX}$

FILLETS &
ROUNDS R3

(1) LIFTING WEDGE
MAL IRON – ELEC
PLATE – I REQ

NOTE:
ELECTROPLATE WITH ZINC
CHROMATE QQZ–325 TYPE 2
CLASS 2 – 0.005 THICK – ALL
PARTS AS NOTED

M10 X 1.5 LH THRU
90° CSK Ø14

(3) SLIDING WEDGE
MAL IRON – ELEC
PLATE – I REQ

Ø5 THRU
2 HOLES

8

32

12

R2 – MAX
2 PLACES

150

Ø 32

$\frac{150}{MAX}$

11

8

16

38

R

26

16

58

8

Ø 5 THRU
4 HOLES

12

20

20

12

76

6

R10
2 PLACES

(5) BASE – 1018 CRS
ELEC PLATE
I REQ

MODEL ST
LEVELING DEVICE **SI**

M10 X 1.5 LH

R 0.8
BOTH
ENDS

Ø14

Ø10

R0.5 BOTH
SIDES

72

5

9
12

Ø16

1.5 X 45° CHAM

Ø10 X 8 DEEP
10.0 – 10.1 ACROSS
FLATS – HEX SOC
7 DEEP

(4) ADJUSTING SCREW
IIII STL – ELEC PLATE – I REQ

FIG. 15.81 Make a detail drawing of the parts of the leveling device. Draw an
assembly and give a parts list.

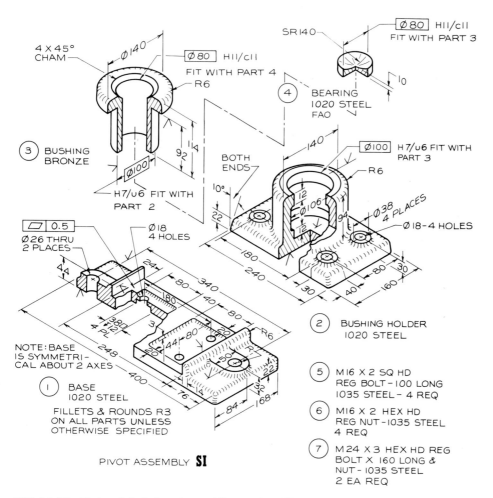

4 X 45° CHAM

Ø140

Ø80 H11/c11
FIT WITH PART 4

R6

SR140

Ø80 H11/c11
FIT WITH PART 3

10

4 BEARING
1020 STEEL
FAO

3 BUSHING
BRONZE

Ø100

114

92

H7/u6 FIT WITH
PART 2

BOTH
ENDS

10°

22

140

Ø100 H7/u6 FIT WITH
PART 3

R6

12

Ø106

12

94

Ø38
4 PLACES

Ø18-4 HOLES

180

240

30

40

80

30

160

0.5

Ø26 THRU
2 PLACES

Ø18
4 HOLES

44

24

340

80

40

80

R6

180

3

38
4 PL
12

44

80

80

R

22

32

2 BUSHING HOLDER
1020 STEEL

NOTE: BASE
IS SYMMETRI-
CAL ABOUT 2 AXES

248

400

76

14

84

168

1 BASE
1020 STEEL

FILLETS & ROUNDS R3
ON ALL PARTS UNLESS
OTHERWISE SPECIFIED

PIVOT ASSEMBLY **SI**

5 M16 X 2 SQ HD
REG BOLT – 100 LONG
1035 STEEL – 4 REQ

6 M16 X 2 HEX HD
REG NUT – 1035 STEEL
4 REQ

7 M24 X 3 HEX HD REG
BOLT X 160 LONG &
NUT – 1035 STEEL
2 EA REQ

FIG. 15.82 Make detail drawings of the parts of the pivot assembly. Draw an assembly and give a parts list.

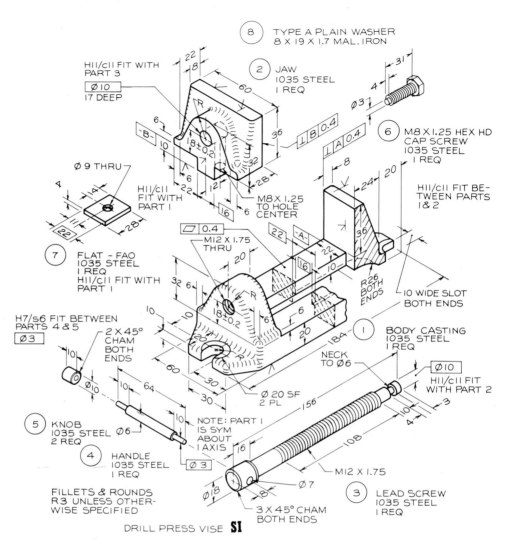

FIG. 15.83 Make detail drawings of the parts of the drill press vise. Draw an assembly and give a parts list.

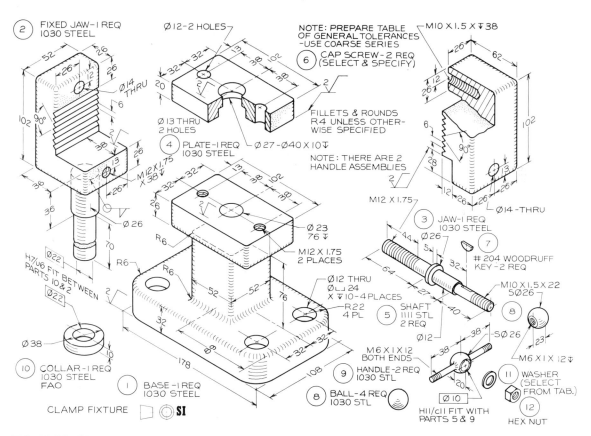

FIG. 15.84 Make detail drawings of the parts of the clamp fixture. Draw an assembly and give a parts list.

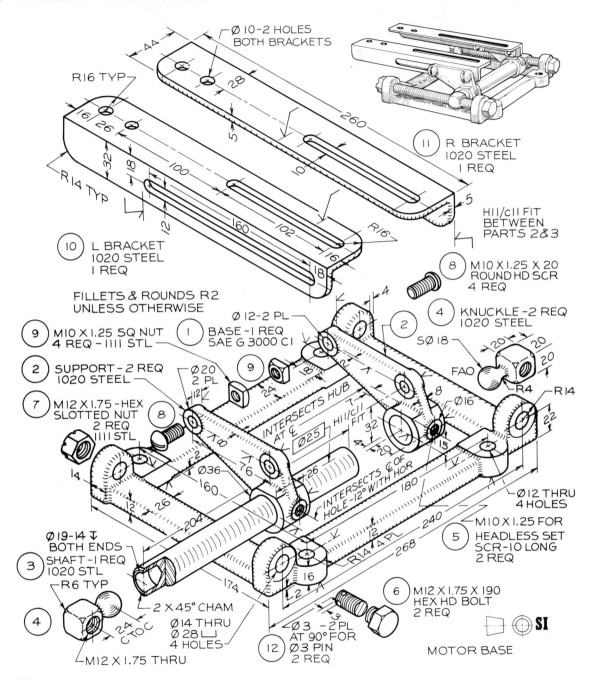

Ø 10-2 HOLES
BOTH BRACKETS

R16 TYP

44

28

260

16
26

32

18

R14 TYP

R16

100

5

10

160 102 18

R16

5

(11) R BRACKET
1020 STEEL
1 REQ

H11/C11 FIT
BETWEEN
PARTS 2&3

(10) L BRACKET
1020 STEEL
1 REQ

FILLETS & ROUNDS R2
UNLESS OTHERWISE

(8) M10 X1.25 X 20
ROUND HD SCR
4 REQ

(9) M10 X1.25 SQ NUT
4 REQ – 1111 STL

(1) BASE – 1 REQ
SAE G 3000 C1

Ø 12-2 PL

(4) KNUCKLE – 2 REQ
1020 STEEL

SØ 18

20 20

20

FAO

R4 R14

(2) SUPPORT – 2 REQ
1020 STEEL

Ø20
2 PL

12

(9)

18

24

2

INTERSECTS HUB
AT ℄

Ø25

H11/C11
FIT

8

Ø16

(7) M12 X1.75 – HEX
SLOTTED NUT
2 REQ
1111 STL

(8)

8

2

32

4

15

22

14

Ø36

160

6

26

INTERSECTS ℄ OF
HOLE-12° WITH HOR

180

240

M10 X1.25 FOR

12

26

204

(5) HEADLESS SET
SCR-10 LONG
2 REQ

Ø 19-14
BOTH ENDS

(3) SHAFT-1 REQ
1020 STL

R6 TYP

(4)

12

174

16

2

R14 4PL

268

Ø 12 THRU
4 HOLES

24
C TO C

Ø14 THRU
Ø 28
4 HOLES

M12 X1.75 THRU

2 X 45° CHAM

3

(6) M12 X1.75 X 190
HEX HD BOLT
2 REQ

Ø 3 – 2PL
AT 90° FOR

(12) Ø3 PIN
2 REQ

SI

MOTOR BASE

FIG. 15.85 Make a detail drawing of the parts of the motor base. Draw an assembly
and give a parts list.

CHAPTER 16

Reproduction Methods and Drawing Shortcuts

16.1
Introduction

So far this text has dealt with the processes of preparing drawings and specifications to communicate ideas in a technical manner. This has progressed through the working-drawing stage where a detailed drawing is completed on tracing film or paper. Now the drawing must be reproduced, folded, and prepared for filing or transmittal to the users of the drawings. These steps will be discussed in this chapter.

16.2
Reproduction of working drawings

A drawing made by a drafter is of little use in its original form. It would be impractical for the original to be handled by checkers and, even more so, by workers in the field or in the shop. The drawing would quickly be damaged or soiled and no copy would be available as a permanent record of the job. Consequently, reproduction of drawings is necessary so that copies can be available for use by the various people concerned. A checker can mark corrections on a work copy without damaging the original drawing. The drafter in turn can make the corrections on the original from the work copy.

Several methods of reproduction are used for making the copies that have traditionally been called "blueprints." This term comes from the original reproduction process that gave a blue background with white lines. The term blueprint is still used, although incorrectly, to describe almost all reproduced working drawings regardless of the process.

The most often used processes of reproducing engineering drawings are: (1) **diazo printing,** (2) **blueprinting,** (3) **microfilming,** (4) **xerography,** and (5) **photostatting.**

Diazo printing

The diazo print is more correctly called a "whiteprint" or a "blue-line print" than a blueprint, since it has a white background and blue lines. Other colors of lines are available depending on the type of paper used. The white background makes notes and corrections drawn on the drawing more clearly visible than does the blue background of the blueprint.

> Both blueprinting and diazo printing require that the original drawing be made on semitransparent tracing paper, cloth, or film that will allow light to pass through the drawing.

The paper on which the copy is made, the diazo paper, is chemically treated so that it has a yellow tint on one side. This paper must be stored away from heat and light to prevent spoilage.

The tracing paper or film drawing is placed face up on the yellow side of the diazo paper and is run through the diazo-process machine, which exposes the drawing to a built-in light. The light passes through the tracing paper and burns out the yellow chemical on the diazo paper except where the drawing lines have shielded the paper from the light. After exposure to light, the diazo paper is a duplicate of the original drawing except that the lines are light yellow and are not permanent. The diazo paper is then passed through the developing unit of the diazo machine where the yellow lines are developed into permanent blue lines by exposure to ammonia fumes. Diazo printing is a completely dry process.

A typical diazo printer-developer, sometimes called a whiteprinter, is shown in Fig. 16.1. This machine will take sheets up to 42 in. wide.

The speed at which the drawing passes under the light determines the darkness of the copy. A slow speed burns out more of the yellow and produces a clear white background; however, some of the lighter lines of the drawing may be lost. Most diazo copies are made at a somewhat faster speed to give a light tint of blue in the background and stronger lines in the copy. Ink drawings give the best reproductions since the lines are uniform in quality.

A print will not be clear and readable unless drawn lines of the drawing are dark and dense. Light will pass through gray lines and the result will be a fuzzy print that will not be satisfactory.

Blueprinting

Blueprints are made with paper that is chemically treated on one side. As in the diazo process, the tracing-paper drawing is placed in contact with the chemically treated side of the paper and exposed to light. The exposed blueprint paper is washed in clear water for a few seconds and is coated with a solution of potassium dichromate. The print is washed again and dried. The wet sheets can be hung on a line to dry or can be dried by special equipment made for this purpose.

This process is still used but to a lesser degree than in the past. Since it is a wet process, it requires more time than does the diazo process.

Microfilming

Microfilming is a photographic process that converts large drawings into film copies—either aperture cards or roll film. Drawings must be photographed on either 16 mm or 35 mm film. A camera and copy table are shown in Fig. 16.2.

The roll film or aperture cards can be placed in a microfilm enlarger-printer (Fig. 16.3), where the individual drawings can be viewed on a built-in screen. The selected drawings can then be printed from the film to give standard-size drawings. The range of enlargement varies with the equipment used. Microfilm copies are usually smaller than the original drawings; this saves paper and makes the drawings more manageable and easier to use.

Microfilming makes it possible to eliminate large, bulky files of drawings, since hundreds of drawings can be stored in miniature size on a small amount of film. The aperture cards shown in Fig. 16.3 are data processing cards that can be cataloged and recalled by a computer to make them accessible with a minimum of effort.

Fig. 16.1 A typical whiteprinter that operates on the diazo process. (Courtesy of Blu-Ray, Inc.)

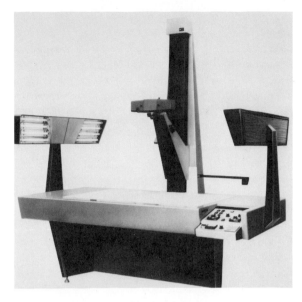

Fig. 16.2 The Micro-Master 35 mm camera and copy table are used for microfilming engineering drawings. (Courtesy of Keuffel & Esser Co., Morristown, N.J.)

Xerographic reproduction

Xerography is an electrostatic process of duplicating drawings on ordinary, unsensitized paper. This process was developed originally for office duplication uses, but has recently been used for the reproduction of engineering drawings.

Fig. 16.3 The Bruning 1200 microfilm enlarger-printer makes drawings up to 18″ × 24″ from aperture cards and roll film. (Courtesy of Bruning Co.)

An advantage of the xerographic process is the possibility of making copies of drawings at a reduced size (Fig. 16.4). The Xerox 2080 reduces drawings as large as 24″ × 36″ directly from the original to paper sizes ranging from 8″ × 10″ to 14″ × 18″.

Fig. 16.4 This Xerox 7080 Engineering Print System accepts original drawings sized A to E, makes prints sized A to C as fast as 58 per minute, and will stamp, fold, and sort prints automatically,

Photostatic reproduction

A method of enlarging or reducing drawings using a camera is called the **photostatic process.** The combination camera and processor shown in Fig. 16.5 is used for photographing drawings and producing high-contrast photographic copies. This process is sometimes called the PMT process (photo-mechanical process).

The drawing or artwork is placed under the glass of the exposure table (Fig. 16.6), which is lit by built-in lamps. The image can be seen on the glass inside the darkroom where it is exposed on photographically sensitive paper. The negative paper that has been exposed to the image is placed in contact with receiver paper and the two are fed through the developing solution to obtain a photostatic copy.

These high-contrast reproductions are often used to prepare artwork that is to be printed by offset printing presses. This process can also be used to make reproductions on transparent films, and for the reproduction of photographs with tones of gray, called halftones.

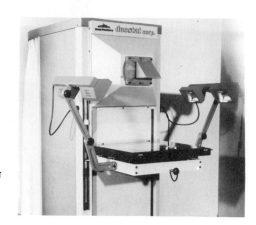

FIG. 16.5 A combination camera-processor for enlarging and reducing drawings to be reproduced as photostats. (Courtesy of the Duostat Corp.)

FIG. 16.6 The steps in making a photostat with the Duostat camera-processor,.

Step 1   The drawing or artwork is placed under the glass on the copyboard.

Step 2   The image is projected inside the darkroom compartment onto photographically sensitive negative paper.

Step 3   The exposed negative paper is placed in contact with the receiver paper and the two are fed through the developing chemicals in the processor.

16.3
Folding the drawing

Once the prints have been finished, the original drawings should be stored in a flat file for future use and updating. The original drawings should not be folded, and handlling of any kind should be kept to a minimum.

The printed drawings, on the other hand, are usually folded for transmittal from office to office, often through the mail. The methods of folding Sizes B, C, D, and E sheets are shown in Fig. 16.7. In each case, the final size after folding is 8½″ × 11″ (or 9″ × 12″) to fit most mailing envelopes and file cabinets.

Note that the drawings are folded with the title blocks positioned to be visible at the top and the lower right of the drawing. This is essential in order for a drawing to be easily retrieved from a file cabinet.

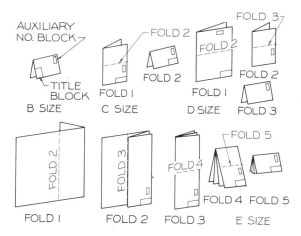

FIG. 16.7 Standard folds for engineering drawing sheets. The final size in each case is 8½″ × 11″.

16.4
Overlay drafting techniques

Valuable drafting time can be saved by taking advantage of current processes and materials that utilize a series of overlays to separate parts of a single drawing. For example, engineers and architects often work from a single site plan or floor plan on which a variety of drawings will be made that all utilize the same base plan. The floor plan of a building will be used for the

FIG.16.8 In the pin system, separate overlays are aligned by seven pins mounted on metal strips. (Courtesy of Keuffel & Esser Co., Morristown, N.J.)

electrical plan, furniture arrangement plan, air-conditioning plan, floor materials plan, etc. You can see that it would be expensive to retrace the plan for each application.

A series of overlays can be used in a system referred to as **pin drafting,** where accurately spaced holes are punched in the polyester drafting film at the top edge of the sheets. These holes are aligned on pins attached to a metal strip that match the holes that were punched in the film (Fig. 16.8). This method ensures accurate alignment or registration of a series of sheets, and the polyester film ensures stability of the material since it does not stretch or sag with changes in humidity.

The steps in using the pin drafting system are shown in Fig. 16.9. The title block can be printed on all drawing film sheets along with the border lines. The overlay sheets need not have borders or a title block. The base plan is the sheet that will be common to several drawings.

The composite of the various overlays is shown in the third part of Fig. 16.9. Note that the base plan is printed in gray, which means that it has been screened to the desired percentage of black by a photographic process. This makes the additional information provided by the subcontractor or consultant on the plan more noticeable and easier to read from the base plan.

The set of overlays could be attached by the alignment pins or taped together and run through a diazo machine for full-size prints. Another option of reproduction is the use of a flat-bed process camera (Fig. 16.10) to photograph and reduce the drawings to a standard 8½″ × 11″ size.

When a large number of prints are needed, reproductions may be printed by offset lithography. This permits them to be printed in multicolors to highlight certain features on drawings.

16.5
Paste-on photos

The engineer who is laying out a manufacturing plant will need to represent many identical machines on a drawing along with other features that are repeated. The architect encounters the same situation when locating furniture or common details such as offices, bathrooms, or window details.

When a number of repetitive drawings are necessary, it is often more economical to use the photo-

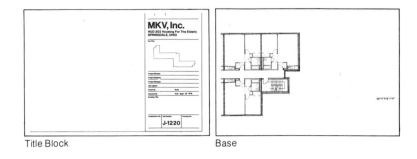

Title Block   Base

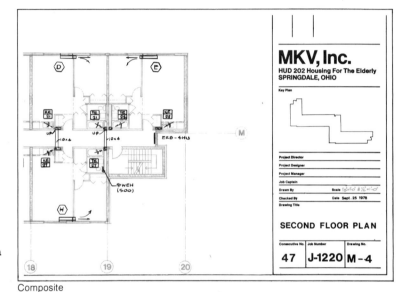

FIG. 16.9 Examples of overlays that are overlaid and reproduced to give a combination of several sheets in the final reproduced drawing. (Courtesy of Keuffel & Esser Co., Morristown, N.J.)

Composite

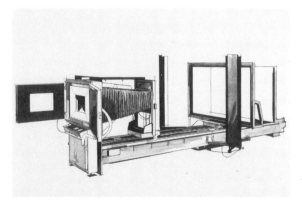

FIG. 16.10 The process camera that is used to reduce and enlarge engineering drawings is the heart of the pin system. (Courtesy of Keuffel & Esser Co., Morristown, N.J.)

FIG. 16.11 When drawings of parts or arrangements are to be used repetitively on a set of drawings, it may be more economical to photographically reproduce them than draw them. (Courtesy of Eastman Kodak Co.)

FIG. 16.12 The photographically reproduced drawing features are taped in position to complete the overall drawing. (Courtesy of Eastman Kodak Co.)

graphic process to make a number of reproductions of the drawings on transparent film. These features can be "pasted" into position on the master drawing. (The term "pasted" is commonly used to describe this attachment, but in reality, tape is used when the drawings are reproduced on transparent film. If the reproductions are made as opaque photostats, then rubber cement may be used.)

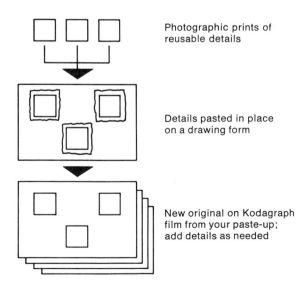

Photographic prints of reusable details

Details pasted in place on a drawing form

New original on Kodagraph film from your paste-up; add details as needed

FIG. 16.13 The steps in photographically composing an engineering drawing. (Courtesy of Eastman Kodak Co.)

The office arrangements shown in Fig. 16.11 are transparencies that have been photographically duplicated from a single drawing. The architect in Fig. 16.12 is composing an entire drawing sheet with paste-on images of repetitive features that were previously drawn.

The steps in preparing final drawings from paste-on photos are schematically shown in Fig. 16.13. Photographic prints of the repetitive details are made, and then are pasted into position on the new drawing. In the third step, the drawings are reproduced on film or paper.

16.6
Photo revisions

When a previously made drawing is in need of revision, drawing time can be saved by photographically modifying the drawing. If you wish to change the old drawing in Fig. 16.14 to look like the example shown, this can be done by making a clear film reproduction of the parts, cutting them out, and taping them into position on a new form. The new drawing is then photographed onto a new film on which additional notes and lines can be provided to complete the drawing. This new drawing can be used as a master for making diazo prints.

Opaquing is another method of revising a drawing. A photographic negative is made from the original drawing. The area to be removed is opaqued out by using a brush and an ink-like opaquing solution (Fig. 16.15). You can see that it is advantageous to reduce the negative to a smaller size to lessen the area and effort of opaquing.

The negative is then used to make a positive on polyester drafting film, on which the revision can be drawn and noted in the conventional manner. You now have a new master from which photographic or diazo prints can be made.

16.7
Stick-on materials

A number of companies market stick-on symbols, screens, and lettering that can be applied to drawings to economize on time and improve the appearance of

346

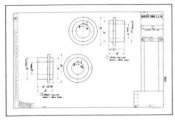

Say you have an existing drawing:

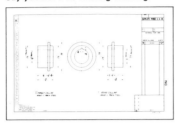

and you want to revise it like this.

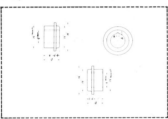

Step 1: First you make a *clear* film reproduction of the original and cut out the elements.

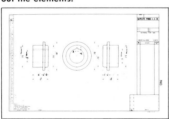

Step 2: Then tape the elements in their new positions on a new form.

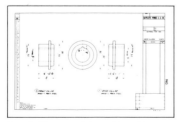

Step 3: Photograph it on film with a matte finish (the tapes and film edges will disappear). Draw in whatever extra detail you want—and you have a new original drawing.

FIG. 16.14 The steps of photographically revising an engineering drawing. (Courtesy of Eastman Kodak Co.)

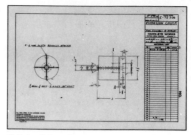

Original drawing

Step 1: Make a reduced-size negative and opaque the area to be revised. (The small size makes it quicker.)

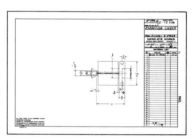

Step 2: Then make a positive on matte film, enlarged to original size.

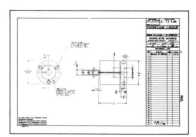

Step 3: Draw in the new detail, and you have a new second original.

FIG. 16.15 The method of modifying a drawing by opaquing the negative. (Courtesy of Eastman Kodak Co.)

drawings. The three standard types of materials are **stick-ons, burnish-ons,** and **tape-ons.**

The stick-on symbols or letters are printed on thin plastic sheets that are cut out with a razor sharp blade and are transferred to the drawing, where the cutout is burnished permanently to it (Fig. 16.16). This material is available in glossy and matte finishes.

FIG. 16.16 Stick-on lettering can be applied to a drawing by cutting the letters from the plastic sheet, applying them to the drawing surface in alignment with a guideline, and burnishing them to the sheet. (Courtesy of Graphic Products Corp.)

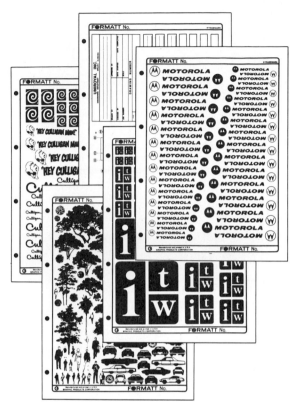

FIG. 16.17 Stick-on symbols are available in a wide range of designs, and they can be custom-printed to suit the needs of the client. (Courtesy of Graphic Products Corp.)

The burnish-on symbols are applied by placing the entire sheet over the drawing and burnishing the desired symbol into place with a rounded-end object such as the end of a pencil cap. The symbol thus is transferred from the plastic sheet to the drawing surface.

Sheets of symbols can be custom-printed for users who have repetitive needs for trademarks and other often-used symbols (Fig. 16.17). Title blocks are sometimes printed in this manner for application to drawings, to reduce drawing time. A number of symbols that are used on architectural plans are shown in Fig. 16.18.

Colored adhesive tapes in varying widths are available for the preparation of charts and graphs (Fig. 16.19). Tapes are also used to represent wide lines on large drawings. Although tapes can be burnished on tightly, they can be removed for modification when this is desired.

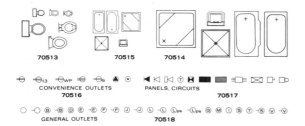

FIG. 16.18 Examples of architectural symbols that can be used on drawings by sticking them to the drawing surface. (Courtesy of Zip-a-Tone Inc.)

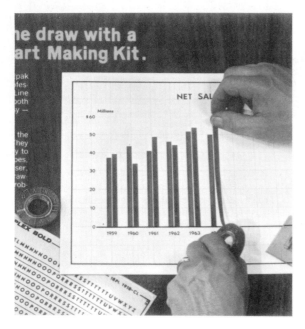

FIG. 16.19 Colored tapes in various widths can be used to speed the process of preparing charts and graphs. (Courtesy of Chartpak Rotex.)

FIG. 16.20 This especially-made typewriter can be used to type notes on a drawing, thereby relieving the drafter of this chore. (Courtesy of Vari-Typer.)

A matte finished sheet is available from several sources that can be typed on with a standard typewriter and then transferred to the drawing where it is attached with its adhesive backing to become a permanent part of the drawing.

A specially designed typewriter can be used to type notes directly on the surface of the drawing, shortening the time that would be required to hand-letter the notes (Fig. 16.20).

16.8
Photo drafting

An example of a photodrawing is shown in Fig. 16.21, where an assembly of parts has been noted. In some cases it would be worthwhile to even build a model for photographing in order to clarify assembly details. This is especially true in the piping industry, where complex refineries are built as models, then photographed, noted, and reproduced as photodrawings.

The steps of preparing a photodrawing are shown in Fig. 16.22, where it is desired to specify certain parts of a sprocket-and-chain assembly. In Step 1, a halftone print of the photograph is made. (This is the process of screening the photograph, or representing it by a series of dots that give varying tones of gray.) In Step 2 the halftone is taped to a white drawing sheet, which is then photographed to give a negative. The negative is used to produce a positive on polyester drafting film. The notes can be lettered on this drawing film to complete the master drawing (Step 4). The master drawing can then be used to make diazo prints, or it can be microfilmed or reproduced photographically to the desired size.

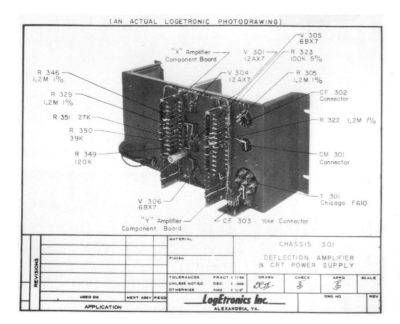

(AN ACTUAL LOGETRONIC PHOTODRAWING)

FIG. 16.21 This assembly is an example of a photodrawing that has saved a considerable amount of drafting time. (Courtesy of LogEtronics, Inc., and Eastman Kodak Co.)

Step 1: Make a halftone print of the photograph.

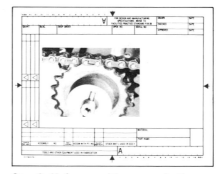

Step 3: Make a positive reproduction on matte film.

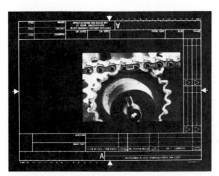

Step 2: Tape the halftone print to a drawing form, and photograph it to produce a negative.

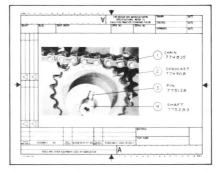

Step 4: Now draw in your callouts—and the job is done.

FIG. 16.22 The steps in making a photodrawing. (Courtesy of Eastman Kodak Company.)

17 CHAPTER

The Design Process

17.1
The design process

Design is the act of devising an original solution to a problem using a combination of principles, resources, and products. Design is the most distinctive responsibility that separates the engineer from the scientist and the technician.

The six steps of the design process as shown in Fig. 17.1 are:

1. problem identification,
2. preliminary ideas,
3. problem refinement,
4. analysis,
5. decision, and
6. implementation.

Although the designer will sequentially work from step to step, he or she may recycle to previous steps while progressing.

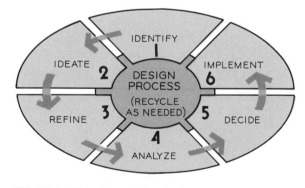

FIG. 17.1 The steps of the design process

17.2
Problem identification

Most engineering problems are not clearly defined; consequently they must be identified before an attempt is made to solve them (Fig. 17.2). For example, a concern today is air pollution. Before this problem can be solved, you must identify what air pollution is and what causes it. Is pollution caused by automobiles, factories, atmospheric conditions that harbor impurities, or geographic features that contain impure atmospheres?

Problem identification requires considerable study beyond a simple problem statement like "solve air pollution." You will need to gather data of several types: field data, opinion surveys, historical records, personal observations, experimental data, and physical measurements and characteristics (Fig. 17.2).

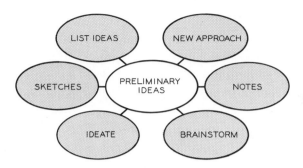

FIG. 17.3 Preliminary ideas are developed after the identification step has been completed. All possibilities should be listed and sketched to give the designer a broad selection of ideas from which to work.

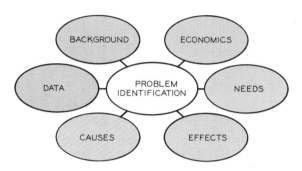

FIG. 17.2 Problem identification requires the accumulation of as much information concerning the problem as possible before a solution is attempted by the designer.

17.3
Preliminary ideas

Once the problem has been identified, the next step is to accumulate as many ideas for solutions as possible (Fig. 17.3) Many rough sketches of preliminary ideas should be made and retained (Fig. 17.4).

Preliminary ideas can be gathered from several methods, including brainstorming, market analysis, and research of existing designs.

17.4
Refinement

Rough sketches are converted to scale drawings that will permit spatial analysis, measurement, and calculation of areas and volumes affecting the design (Fig.

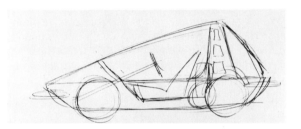

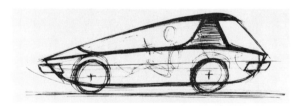

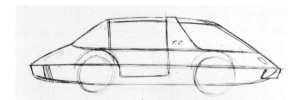

FIG. 17.4 A sequence of sketches is shown that was used to arrive at the final body configuration of the Pacer. (Courtesy of American Motors Corp.)

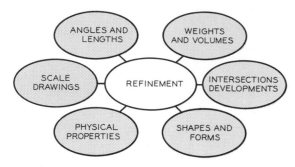

FIG. 17.5 Refinement begins with the construction of scale drawings of the better preliminary ideas. Descriptive geometry and graphical methods are used to find necessary geometric characteristics.

17.5). Consideration is given to geometric relationships, angles between planes, lengths of structural members, and intersections of surfaces and planes. Descriptive geometry is a very valuable tool for determining information of this type.

To refine the design of the lunar vehicle shown in Fig. 17.6, the designer made scale drawings of its features. It was necessary to determine fundamental lengths, angles, and specifications that are related to the fabrication of the landing gear. The length of each leg of the landing apparatus and the angles between the members at the point of junction had to be found in order to design connectors.

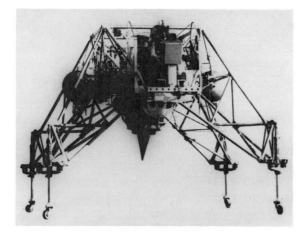

FIG. 17.6 The refinement of the lunar vehicle required the use of descriptive geometry and other graphical methods. (Courtesy of Ryan Aeronautical Co.)

17.5
Analysis

Analysis is the step of the design process where engineering and scientific principles are used most (Fig. 17.7). Analysis involves the study of the best designs to determine the merits of each with respect to cost, strength, function, and market appeal. Graphical solutions to analytical problems offer a readily available means of determining forces by using vectors. Graphical methods can also be applied in converting functions of mechanisms to a form that will permit ease of analysis. Data can be gathered and graphically analyzed that would otherwise be difficult to interpret by mathematical means.

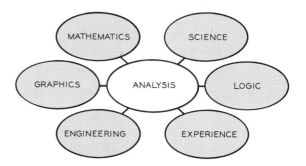

FIG. 17.7 The analysis phase of the design process is the application of all available technological methods from science to graphics in evaluating the refined designs.

17.6
Decision

A decision must be made at this stage to select a single design as the solution to the design problem (Fig. 17.8). The several designs that have been refined and analyzed will offer unique features, and it will probably be impossible to include all of these in a single final solution. In many cases, the final design is a compromise that offers as many of the best features as possible.

The outstanding aspects of each design usually lend themselves to presentation in the form of graphs that compare manufacturing costs, weights, opera-

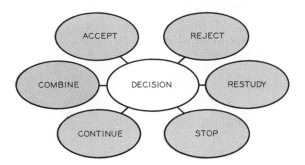

FIG. 17.8 Decision is the selection of the best design or design features to be implemented.

tional characteristics, and other data to be considered in arriving at the final decision.

17.7
Implementation

The final design concept must be presented in the form of working drawings and specifications (Fig. 17.9). Engineering graphics fundamentals must be used to convert all preliminary designs and data into the language of the manufacturer who will be responsible for converting the ideas into a reality.

Designers and engineers must be sufficiently knowledgeable to be able to supervise the preparation of working drawings even though they may not be involved in the mechanics of producing them. They must approve all plans and specifications prior to their release for production.

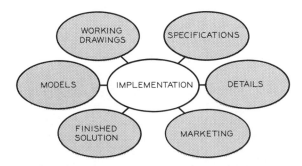

FIG. 17.9 Implementation is the final step of the design process, where drawings and specifications are prepared from which the final product can be constructed.

17.8
Problem identification: hunting seat

The following example is used to illustrate the problem identification step of designing a hunting seat, which is typical of a problem that might be assigned as a class project.

Hunting seat

Many hunters, especially deer hunters, hunt from trees to obtain a better vantage point. Design a seat that would provide the hunter with comfort and safety while hunting from a tree and that would meet the general requirements of economy and hunting limitations.

WORK SHEET COMPLETION: The work sheet in Fig. 17.10 is typical of the information that is needed by the designer to understand the background of the problem.

Project Title and Problem Statement The title of the project is recorded along with a brief problem statement that could be easily understood by anyone.

Requirements and Limitations The requirements and limitations are listed along with any sketches that would aid in a better understanding of the problem. In some cases, it might be necessary to list the requirements as questions for which you have no answer at the time. Later, after further investigation, you should list the limits, such as "must cost between $60 and $100." It is better to give a price or weight range rather than attempting to be exact in your estimates.

A source for information such as sales prices, weights, and sizes are catalogs dealing with similar products. If catalogs must be written for, the correspondence should be mailed as early as possible to give the necessary time for a response, so as not to delay your progress.

Needed Information You will wish to have data that would tell you more about your problem, as listed in Fig. 17.10. How many hunters are there? How many hunt from trees or elevated blinds? You could get this and similar information from your state game office.

PROBLEM IDENTIFICATION

1. Project title

SEAT FOR HUNTING FROM A TREE

2. Problem statement

MANY HUNTERS SIT ON TREE LIMBS WHILE HUNTING, WHICH IS UNSAFE AND UNCOMFORTABLE. A TREE'S ELEVATION PROVIDES AN EXCELLENT VANTAGE POINT FOR HUNTING. A SEAT IS TO BE DESIGNED TO FILL THIS NEED.

3. Requirements and limitations

A. MUST BE CARRIED TO SITE BY HUNTER
B. MUST PROVIDE SAFETY & COMFORT
C. 7'-12' ABOVE GROUND
D. PROTECTION FROM WEATHER
E. METHOD OF ASCENDING TO DESIRED HEIGHT
F. METHOD OF CARRYING – COULD DOUBLE AS A BACKPACK
G. PRICE: $20- $100
H. WEIGHT: 20 LBS MAX
I. FIT IN TRUNK OF CAR

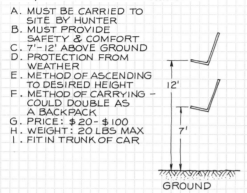

12'

7'

GROUND

4. Needed information

A. NUMBER OF HUNTERS WHO HUNT FROM TREES? IN STATE? NATION? – CONTACT STATE GAME COMMISSION.
B. WHAT HAPPENS ON A TYPICAL HUNTING TRIP? SURVEY HUNTERS
C. HOW DO HUNTERS HUNT FROM TREES WITHOUT SEATS? INTERVIEW HUNTERS
D. LAWS CONCERNING HUNTING FROM TREES – WRITE STATE GAME COMMISSION
E. HOW MANY BOW & ARROW HUNTERS ARE THERE? CHECK LIBRARY FOR SOURCES
F. HOW LONG & WHEN ARE THE VARIOUS HUNTING SEASONS? CONTACT GAME COMMISSION
G. EQUIPMENT CARRIED BY HUNTER?

5. Market Considerations

A. WOULD SPORTING GOODS RETAILERS LIKE A TREE SEAT? INTERVIEW DEALERS
B. WHAT IS COMPETITION? REVIEW ADS IN HUNTING MAGAZINES
C. WHAT IS BEST PRICE RANGE? – INTERVIEW DEALERS
D. HOW MUCH DO HUNTERS SPEND PER SEASON? INTERVIEW HUNTERS
E. FEATURES DESIRED BY HUNTERS? – INTERVIEW HUNTERS
F. POSSIBLE MARKET OUTLETS? VISIT STORES

FIG. 17.10 A work sheet for the problem identification step of the design process for the hunting seat design.

FIG. 17.11 An example model that shows the breakdown of expenses and costs involved in arriving at the retail price of a product.

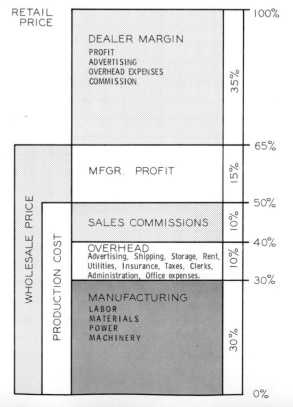

RETAIL PRICE

DEALER MARGIN
PROFIT
ADVERTISING
OVERHEAD EXPENSES
COMMISSION

100%

35%

65%

MFGR. PROFIT

15%

50%

WHOLESALE PRICE

SALES COMMISSIONS

10%

40%

PRODUCTION COST

OVERHEAD
Advertising, Shipping, Storage, Rent, Utilities, Insurance, Taxes, Clerks, Administration, Office expenses.

10%

30%

MANUFACTURING
LABOR
MATERIALS
POWER
MACHINERY

30%

0%

What is the average income of the hunter? How much do they spend on their hobby per year? Sporting-goods dealers could help you here by sharing their experiences, and perhaps they could direct you to other sources for answers to these questions.

Market Considerations The designer must think about cost control at all stages of the project.

By referring to Fig. 17.11, you can see how an item is priced from wholesale to retail. The percentages will vary by product: There is less profit in retail food sales than in furniture sales. It is customary for the designer to think in terms of what the product should retail for, and thus work backward to the necessary wholesale price.

Another method of collecting information would be to conduct a survey among hunters you know to gather firsthand opinions of the merits of introducing a hunting seat on the market. The results of a survey of this type are listed on the work sheet in Fig. 17.12. These data show that 40% of fifty people surveyed

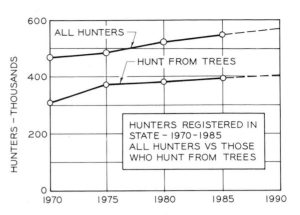

FIG. 17.13 Survey data is plotted in this graph to describe the population trends of hunters who are potential customers for the hunting seat.

gave the market potential of the seat a high ranking. A dealer survey shows that about $390 is spent by each hunter per season.

Graphs Tabular data is easier to interpret and to understand if presented in the form of a graph. For example, the number of hunters is compared with those who hunt from trees in Fig. 17.13. A thorough problem identification will include graphs, sketches, and schematics.

The problem identification of this problem is not yet complete; it is difficult to decide when enough information has been gathered in any project. However, you should be able to follow the method in this example and incorporate your own personal innovations to arrive at a satisfactory problem identification.

17.9
Preliminary ideas: hunting seat

Ideas are gathered from a brainstorming session with classmates. All of the ideas are listed on a work sheet (Fig. 17.14). Remember, wild ideas are encouraged to stimulate the act of forming ideas.

The better ideas are listed on a second work sheet (Fig. 17.15) that summarizes the features that would be desirable in the final design of your project. You may list more features than would be possible to include in a

PROBLEM IDENTIFICATION

NUMBER OF HUNTING LICENSES SOLD STATEWIDE AND AN ESTIMATE OF NUMBER OF THOSE WHO HUNT FROM TREES:

YR	TOTAL	FROM TREES
1970	467,000	305,000
1975	481,000	370,000
1980	520,000	380,000
1985	542,000	392,000

CONSUMER SURVEY
QUESTIONNAIRE GIVEN TO 50 HUNTERS
RANKING: 1-HIGH, 4-LOW

OPINION OF SEAT IDEA:

RANK	NO	%
1	20	40
2	15	30
3	8	16
4	7	14
	50	100

RETAILER SURVEY
   4 DEALERS WERE POSITIVE ABOUT THE PROSPECTS OF A HUNTING SEAT. SUGGESTED $95 PRICE.

ANNUAL EXPENDITURES BY HUNTERS:

AMMUNITION	$50
CLOTHING	70
LEASES	100
GUN	100
TRAVEL	70
	$390

FIG. 17.12 A work sheet for collecting data

PRELIMINARY IDEAS

1. Brainstorming ideas

    A. USE LAWN CHAIR
    B. CHAIR ON STILTS
    C. INFLATABLE CHAIR
    D. PROVIDE FOOTREST
    E. ROOF FOR RAIN
    F. PADDED SEAT
    G. HEADREST
    H. RIFLE REST
    I. AMMUNITION COMPARTMENT
    J. REFRESHMENT COMPARTMENT
    K. ENTERTAINMENT COMPARTMENT
    L. TV ACCESSORY
    M. RADIO
    N. CB RADIO
    O. HOIST SYSTEM
        1. PULLEYS & CABLES
        2. TREE CLIMBER
        3. LADDER
        4. STEPS
    P. PLATFORM FOR STANDING
    Q. PLATFORM FOR SLEEPING
    R. LIGHTS FOR NIGHT
    S. SAFETY BELT
    T. SAFETY BELT FOR ARCHER
    U. SAFETY BELT FOR RIFLEMAN
    V. DOUBLES AS BACKPACK
    W. DOUBLES AS CAMP CHAIR
    X. DOUBLES AS TENT
    Y. CARRYING CASE FOR SEAT
    Z. SEAT ON WHEELS
    A1. MOTORIZED SEAT
    A2. TELESCOPE MOUNT
    A3. HEATING SYSTEM

FIG. 17.14 A worksheet that lists the brainstorming ideas that were recorded by a design team.

FIG. 17.15 A work sheet that lists the better ideas that were selected from the original brainstorming ideas.

2. Description of best ideas

    A. SEAT WITH FOOTREST

    B. NEED METHOD OF ASCENDING - CABLE AND PULLEY SYSTEM

    C. ACCESSORIES

        1. STANDING PLATFORM

        2. RIFLE RACK

        3. BOW RACK

        4. CLIMBING ACCESSORY

    D. SAFETY BELT

    E. FOLDING SEAT

    F. WEIGH UNDER 10 LBS

3. Attach sketches

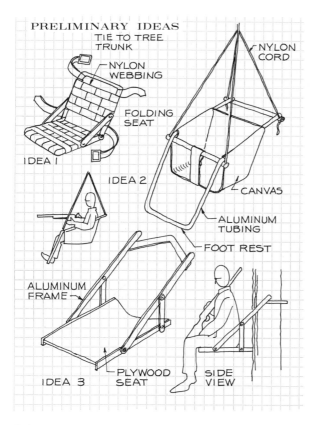

PRELIMINARY IDEAS

TIE TO TREE TRUNK

NYLON WEBBING

FOLDING SEAT

IDEA 1

IDEA 2

NYLON CORD

CANVAS

ALUMINUM TUBING

FOOT REST

ALUMINUM FRAME

PLYWOOD SEAT

SIDE VIEW

IDEA 3

FIG. 17.16 A work sheet for the presentation of preliminary ideas for the development of a hunting seat. Notes and sketches are used to supplement the sketches.

FIG. 17.17 Additional preliminary ideas are recorded to illustrate design concepts for the hunting seat problem.

single design, but be sure that no ideas are forgotten or lost at this stage.

Sketches of preliminary ideas are drawn on additional work sheets, using rapid freehand techniques. Orthographic views and pictorial methods are used in combination. Lettering and sketching techniques need not be highly detailed or precisely executed, but readable and understandable.

In Fig. 17.16, ideas have been adapted from various types of known chairs, lawn chairs in particular. Each idea is numbered for identification purposes.

Another work sheet (Fig. 17.17) gives other ideas. The upper part of the sheet shows sketches that identify the need for tilting the seat designs for comfort.

Many other ideas of this type need to be developed and sketched before leaving this step of the design process. No ideas are discarded; all are retained as part of the work-sheet file.

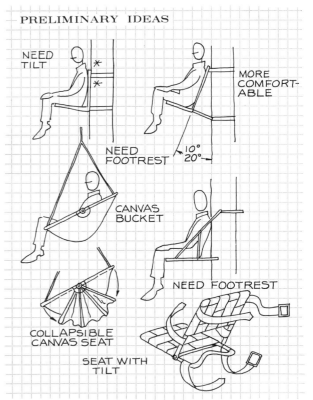

PRELIMINARY IDEAS

NEED TILT

MORE COMFORT-ABLE

NEED FOOTREST

10° 20°

CANVAS BUCKET

NEED FOOTREST

COLLAPSIBLE CANVAS SEAT

SEAT WITH TILT

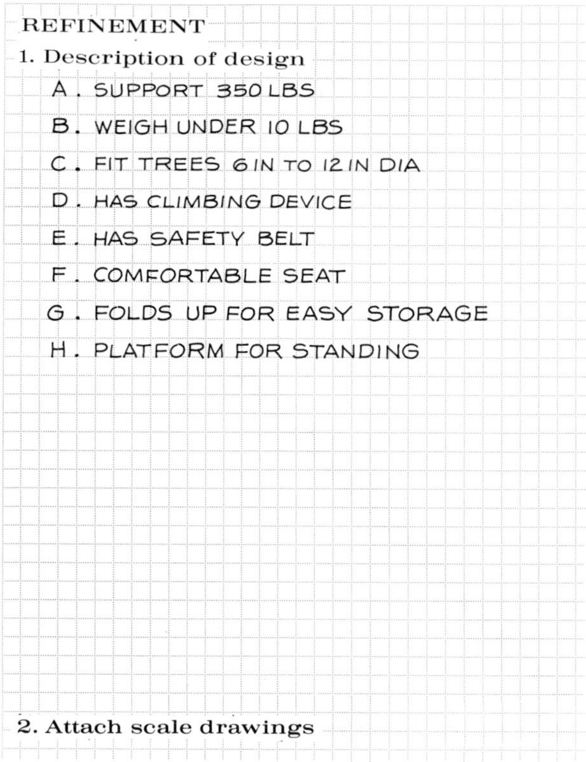

REFINEMENT

1. Description of design

A. SUPPORT 350 LBS

B. WEIGH UNDER 10 LBS

C. FIT TREES 6 IN TO 12 IN DIA

D. HAS CLIMBING DEVICE

E. HAS SAFETY BELT

F. COMFORTABLE SEAT

G. FOLDS UP FOR EASY STORAGE

H. PLATFORM FOR STANDING

FIG. 17.18 A list of a design's specifications and desirable features are listed on a work sheet of this type. In this example, the hunting seat is refined.

2. Attach scale drawings

FIG. 17.19 A refinement drawing of Idea #2 for a hunting seat. Only general dimensions are given on the scale drawing.

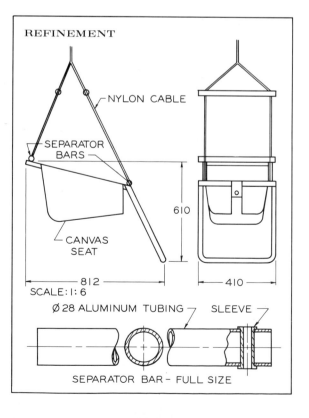

REFINEMENT

NYLON CABLE

SEPARATOR BARS

CANVAS SEAT

610

812

410

SCALE: 1:6

Ø 28 ALUMINUM TUBING   SLEEVE

SEPARATOR BAR - FULL SIZE

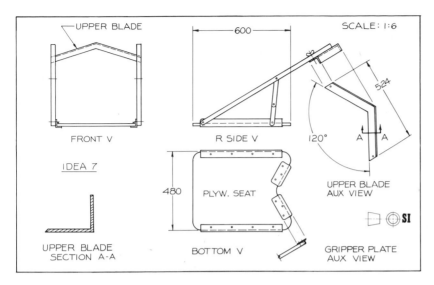

FIG. 17.20 Another design concept for a hunting seat is shown as a refinement drawing.

17.10
Refinement: hunting seat

A list of the design features that are to be incorporated into the design are listed on a work sheet (Fig. 17.18).

Idea 2 is refined in Fig. 17.19, where a scale drawing of the seat is given in orthographic views. Tubular parts, such as the separator bars, are blocked in to expedite the drawing process. Also, some hidden lines are omitted.

Refinement drawings should be made to scale to give an accurate proportion of the design and serve as a basis for finding angles, lengths, shapes, and other geometric specifications. Specific details are shown for the separator bar to explain an idea for a sleeve to protect the nylon cord from being cut.

Another design concept is shown developed as a refinement drawing in Fig. 17.20. Only the major dimensions are given on the scaled instrument drawing.

These two example sheets do not represent a complete refinement of the design; they are merely examples of the type of drawings required in this step of the design process.

17.11
Analysis: hunting seat

The major areas of analysis are listed on the three work sheets in Fig. 17.21. Additional sheets should be used to elaborate on each of these as required to cover each category of design thoroughly.

To determine the strength of the hunting seat's support system in one design, the loads are determined by using graphical vectors (Fig. 17.22). The loads in each support cable are found, from which the proper size cable will be selected.

For further analysis, models are constructed at a reduced scale and full size (Fig. 17.23). The adaptability of the seat as a backpack can be tested to evaluate comfort and other human engineering factors (Fig. 17.24).

A commercial version of the seat is illustrated in Fig. 17.25, where it is tested to measure its functional features, including the method of using it to climb a tree.

An analysis drawing of the hunting seat, shown in Fig. 17.26, illustrates the operation of the linkage

ANALYSIS

1. Function

    A. PROVIDES A METHOD CLIMBING TREES

    B. PROVIDES COMFORTABLE SEATING

    C. PROVIDES DECK FOR STANDING

2. Human engineering

    A. SAFETY BELT
    B. 360° VISION
    C. FOOTREST FOR COMFORT
    D. EASE OF CLIMBING TREE
    E. PORTABLE: 10-15 LBS
    F. SHOULDER STRAPS FOR CARRYING

3. Market & consumer acceptance

    A. POTENTIAL MARKET
      1. STATE: 40,000
      2. NATION: 1,800,000

    B. CHEAPER THAN DEER STAND BY 100-500%

    C. AFFORDABLE AT $100

    D. ADVERTISE IN SPORTS MAGAZINES

    E. RETAIL THROUGH SPORTING GOODS STORES.

FIG. 17.21 Typical work sheets used to analyze the various features of a design. They are used here to analyze the hunting seat.

4. Physical description

    A. PLATFORM 19"X 24"-STAINED

    B. FITS TREES 5"DIA-18"DIA

    C. STRAPS FOR CARRYING ON BACK

    D. HAND CLIMBER DEVICE INCLUDED

    E. SEAT FOLDS FLAT TO 20"X 36"

    F. WEIGHT - 10 LBS

    G. SAFETY BELT

5. Strength

    A. SUPPORTS 300 LBS

    B. SAFETY BELT SUPPORTS 300 LBS

    C. SHOULDER STRAPS SUPPORT 300 LBS

    D. HAND CLIMBER SUPPORTS 400 LBS

6. Production procedures

    A. STRUCTURAL MEMBERS CUT FROM STANDARD ALUMINUM CHANNELS

    B. JOINTS CONNECTED WITH NUTS & BOLTS

    C. METAL EDGES SMOOTHED BY GRINDING

    D. SEAT- 3/4" PLYWOOD - STAINED GREEN

    E. SEAT COVERED WITH BLACK VINYL - BUCKLED ON

7. Economic analysis

MATERIALS	$10
LABOR	21
SHIPPING	5
WAREHOUSING	4
TOTAL	$40
SALES COMMISSION	5
PROFIT	20
WHOLESALE PRICE	65
RETAIL PRICE	$90

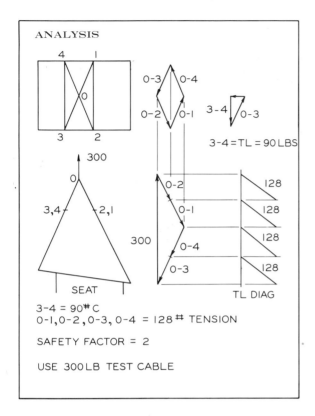

ANALYSIS

3-4 = TL = 90 LBS

3-4 = 90# C
0-1, 0-2, 0-3, 0-4 = 128# TENSION

SAFETY FACTOR = 2

USE 300 LB TEST CABLE

FIG. 17.22 A work sheet on which the support system of a proposed hunting seat is analyzed with graphical vectors.

FIG. 17.23 Full-size and half-size scale models were built by Keith Sherman and Larry Oakes to aid them in analyzing their design.

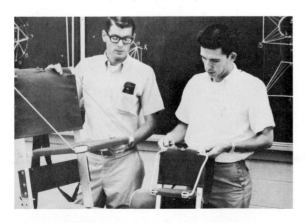

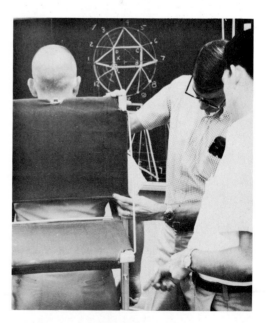

FIG. 17.24 The full-size model of the hunting seat was tested for its adaptation as a backpack.

(a) (b) (c)

(d) (e) (f)

FIG. 17.25 The hunting seat of Baker Manufacturing Co.

(a) The hunter hugs the tree and lifts the hunting seat with his feet. (b) The hunting seat used as a platform for standing. (c) The hunting seat used for sitting. (d) An accessory can be used to assist the hunter in climbing the tree. (e) The hunter pulls upward and lifts the seat, repeating this until the proper height is attained. (f) The hunting seat can be used for sitting or standing.

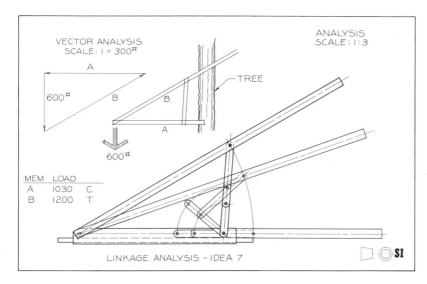

VECTOR ANALYSIS
SCALE: 1 = 300#

ANALYSIS
SCALE: 1:3

A

600#

B   B

A

TREE

600#

MEM	LOAD	
A	1030	C
B	1200	T

LINKAGE ANALYSIS - IDEA 7

FIG. 17.26 The drawing is used to analyze the linkage system of the hunting seat, and a vector diagram is used to determine the forces in the members when the seat is loaded to maximum.

This is the Original Patented Baker Tree Stand

BAKER TREE STAND FEATURES:
1. Platform 19'' x 24'' (456 sq. in.) stained.
2. Back pack wt. 10 lbs.
3. Tested to hold 560 lbs.
4. Fits trees 5'' to 18'' diameter.
5. Riveted assembly folds flat for carrying.
6. Hand Climber fits inside frame.
7. Back packs with Strap Assembly.
8. Safety Belt with Extension and a Tie Down (for safety strap) included.

FIG. 17.27 A summary of the features and physical properties of the Baker Tree Stand. (Courtesy of Baker Manufacturing Co.)

system that permits the seat to collapse into a single plane for carrying ease. The forces in the members are also found graphically by using vector analysis.

The overall features and the physical properties of the Baker Tree Stand are shown in Fig. 17.27. These features are helpful to a consumer in making a purchase.

17.12
Decision: hunting seat

The chart shown in Fig. 17.28 can be used to compare the various designs that are available to choose from for implementation. Each idea is listed and given a number for identification, such as Design 1, Design 2, and so forth.

Next, maximum values for the various factors of analysis are assigned in order for the total of all factors to be 10 points. You must use your judgment to determine these values since they will vary from product to product. Using the maximum values as guide, you can now evaluate each factor of the competing designs.

The vertical columns of numbers are summed to determine the design with the lowest total, the best design, perhaps. However, your instincts may disagree with your numerical analysis. If this is the case, you should have enough faith in your judgment not to be restricted by your numerical decision.

DECISION

1. Decision table for evaluation

DESIGN 1: FOLDING SEAT

DESIGN 2: CANVAS SEAT

DESIGN 3: PLATFOM SEAT

DESIGN 4:

DESIGN 5:

DESIGN 6:

Maximum value	Factors for analysis	1	2	3	4	5	6
3.0	Function	2.0	2.3	2.5			
2.0	Human factors	1.6	1.4	1.7			
.5	Market analysis	.4	.4	.4			
1.0	Strength	1.0	1.0	1.0			
.5	Production proced.	.3	.2	.4			
1.0	Cost	.7	.6	.8			
1.5	Profitability	1.1	1.0	1.3			
.5	Appearance	.3	.4	.4			
10	TOTALS	7.4	7.3	8.5			

FIG. 17.28 A work sheet with a decision table that is used to evaluate the developed design alternatives.

In short, the availability of facts, an outstanding presentation, and a well-analyzed design will not ensure a profitable product. Decision will always remain the most subjective part of the design process.

Conclusion Once a decision has been made, it should be clearly stated, along with the reasons of its acceptance (Fig. 17.29). Additional information may be given, such as number to be produced initially; selling price per unit; expected profit per unit; expected sales during the first year, second, etc.; number that must be sold to break even; and most marketable features.

It is possible that you would recommend that a design not be implemented. If this is the case, the design process should not be considered a failure. A negative decision could save an investor from large losses by recognizing a poor venture at the outset.

17.13

Implementation: hunting seat

Four working drawing sheets (Figs. 17.30–17.33) have been prepared to present the details of the hunting seat design that was selected for implementation. The fifth sheet, Fig. 17.34, is an assembly drawing with a parts list that illustrates how the parts are assembled once they have been made as individual pieces. (This particular design was developed, patented, and is marketed by Baker Manufacturing Co., Valdosta, Georgia. It is the Baker Favorite Seat, Patent No. 3460649.)

CONCLUSIONS

IMPLEMENT AND PRODUCE THE FLAT FOLDING SEAT. THIS DESIGN HAS THE BEST MARKET POTENTIAL. IT WILL BE MARKETED THROUGH SPORTING GOODS DEALERS

SALES PRICE (RETAIL)	$98
SHIPPING COSTS	5
NO. TO BREAK EVEN	1500
ESTIMATED PROFIT	$20
MANUFACTURING COST	
PER CENT RETURN	29%

FIG. 17.29 The decision is summarized on this work sheet to give the designer's conclusion and recommendation concerning the next step, implementation.

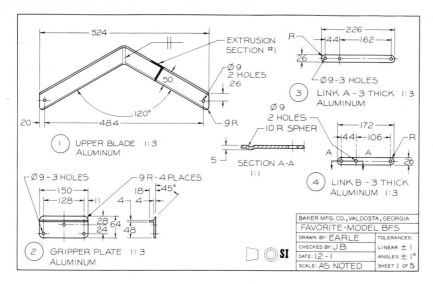

FIG. 17.30 A working drawing sheet of parts of a hunting seat design, Sheet 1 of 5.

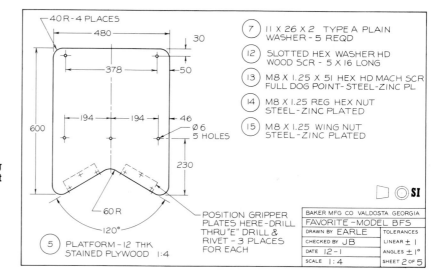

FIG. 17.31 A working drawing sheet of parts of a hunting seat design, Sheet 2 of 5.

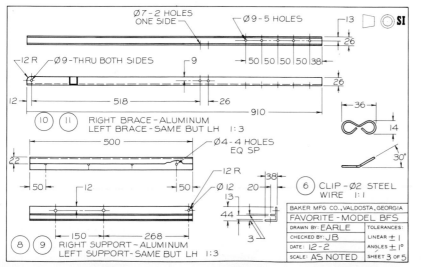

FIG. 17.32 A working drawing sheet of parts of a hunting seat design, Sheet 3 of 5.

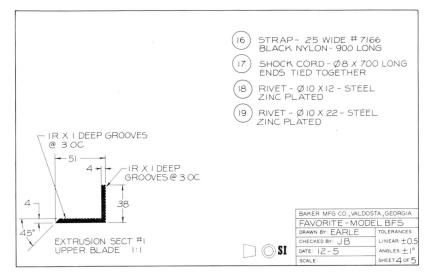

(16) STRAP - 25 WIDE #7166
BLACK NYLON - 900 LONG

(17) SHOCK CORD - Ø8 X 700 LONG
ENDS TIED TOGETHER

(18) RIVET - Ø10 X 12 - STEEL
ZINC PLATED

(19) RIVET - Ø10 X 22 - STEEL
ZINC PLATED

1R X 1 DEEP GROOVES
@ 3 OC

1R X 1 DEEP
GROOVES @ 3 OC

EXTRUSION SECT #1
UPPER BLADE   1:1

BAKER MFG CO., VALDOSTA, GEORGIA	
FAVORITE - MODEL BFS	
DRAWN BY: EARLE	TOLERANCES:
CHECKED BY: JB	LINEAR: ±0.5
DATE: 12-5	ANGLES: ±1°
SCALE:	SHEET 4 OF 5

FIG. 17.33 A working drawing sheet of parts of a hunting seat design, Sheet 4 of 5.

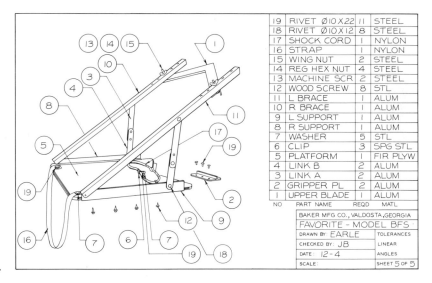

19	RIVET Ø10 X 22	11	STEEL
18	RIVET Ø10 X 12	8	STEEL
17	SHOCK CORD	1	NYLON
16	STRAP	1	NYLON
15	WING NUT	2	STEEL
14	REG HEX NUT	4	STEEL
13	MACHINE SCR	2	STEEL
12	WOOD SCREW	8	STL
11	L BRACE	1	ALUM
10	R BRACE	1	ALUM
9	L SUPPORT	1	ALUM
8	R SUPPORT	1	ALUM
7	WASHER	5	STL
6	CLIP	3	SPG STL
5	PLATFORM	1	FIR PLYW
4	LINK B	2	ALUM
3	LINK A	2	ALUM
2	GRIPPER PL	2	ALUM
1	UPPER BLADE	1	ALUM
NO	PART NAME	REQD	MATL
BAKER MFG CO., VALDOSTA, GEORGIA			
FAVORITE - MODEL BFS			
DRAWN BY: EARLE		TOLERANCES	
CHECKED BY: JB		LINEAR	
DATE: 12-4		ANGLES	
SCALE:		SHEET 5 OF 5	

FIG. 17.34 An assembly drawing that demonstrates how the parts of the hunting seat design are assembled, Sheet 5 of 5.

The assembly drawing (Fig. 17.34) is a pictorial with the different parts identified by balloons attached to leaders. Each part is listed in the parts list by number with general information to describe it.

The drawings given in this example design give only the details for the basic Baker Favorite Seat. Additional accessories are available as shown in Fig. 17.35.

Packaging   The Baker Favorite Seat is packaged in a corrugated cardboard box that is 20″ × 38″ × 2.5″ in size, and the package weighs approximately 10 lbs when it contains the hunting seat (Fig. 17.36). When shipped by motor freight, the shipping costs are computed by hundred-pound multiples (Cwt). Consequently, shipping costs are reduced when shipping in volume.

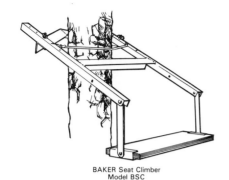

BAKER Seat Climber
Model BSC

A Conversion Kit, Model CK, will
convert Hand Climbers to Seat
Climbers.

An accessory to the Seat Climber
is the Padded Pouch — Model
PP.

Secure Seat Climber and Tree
Stand with tie down for added
safty while hunting.

FIG. 17.35 Examples of accessories that have been designed to accompany the basic hunting seat. (Courtesy of Baker Manufacturing Co., Valdosta, Georgia.)

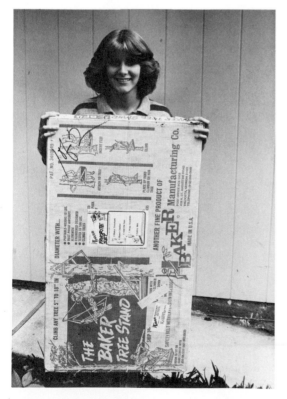

FIG. 17.36 The hunting seat is shipped in a cardboard box to the retail outlet or to the consumer.

Storage An inventory of seats must be maintained to meet the regular seasonal orders that are anticipated throughout the year. The periods prior to hunting seasons will require more inventory than during the rest of the year.

An inventory of seats that are waiting to be sold adds to the cost of overhead in the form of interest payments, warehouse rent, warehouse personnel, and loading equipment. Costs of this nature are just as real as those expended in the product's manufacture.

Shipping The shipping cost for a Baker Favorite Seat with its accessories is $5.75 when shipped one at a time by United Parcel Service. The cost per unit is reduced by about 50% when they are shipped in bundles of ten to the same destination by truck.

The retail price for the Baker Seat is about $90 when purchased by the consumer. The retail price is about five or six times higher than the material and labor cost necessary to manufacture the seat. Retailers are given approximately a 40% margin, and the area distributors who take the orders from the retailers earn about 10%. The remainder of the overhead is advertising expenses and the other areas of overhead mentioned in this article. All expenses, such as 15% interest rates, must be absorbed by the consumer who ultimately purchases the product.

Problems

Design problem specifications

The following specifications are those an individual or design team must consider when preparing a design proposal or outlining their assignments. The specifications, or part of them, may be assigned by the instructor or may be agreed to by the team.

1. Completed work sheets illustrating the development of the design process.
2. A freehand sketch of the design for implementation.
3. An instrument drawing of the proposed design.
4. A dimensioned-instrument working drawing of the proposed design.
5. A pictorial sketch (or one made with instruments) illustrating the design.

The following are short problems that can be completed in less than two periods. These are usually assigned as individual assignments.

1. **Teaching aid.** Design an apparatus that can be used by a teacher to illustrate the basic principles of orthographic projection. Investigate the market potential of such a product.
2. **Cup dispenser.** Design a papercup dispenser that can be attached to a wall. This dispenser should hold a series of cups 2 in. in diameter that measure 6 in. tall when stacked together.
3. **Drawer handle.** Design a handle that would be satisfactory for a standard file cabinet drawer.
4. **Paper dispenser.** Design a dispenser that will hold a 6 in. diameter by 24 in. wide roll of wrapping paper. The paper will be used on a table top for wrapping packages.

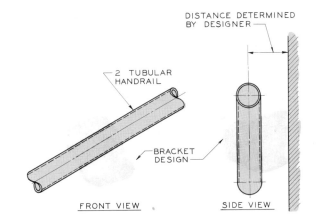

FIG. 17.37 Problem 5. A handrail bracket.

5. **Handrail bracket.** Design a bracket that will support a tubular handrail that will be used on a staircase (Fig. 17.37). Consider the weight that the handrail must support.
6. **Latch-pole hanger.** Design a hanger that can be used to support a latch-pole from a vertical wall. It should be easy to install and use (Fig. 17.38).
7. **Pipe clamp.** A pipe with a 4 in. diameter must be supported by angles (Fig. 17.39) that are spaced 8 ft apart. Design a clamp that will support the pipe without drilling holes in the angles.
8. **TV yoke.** Design a yoke that can support a TV set from the ceiling of a classroom and permit it to be adjusted at the best position for viewing.

FIG. 17.38 Problem 6. A latch-pole hanger.

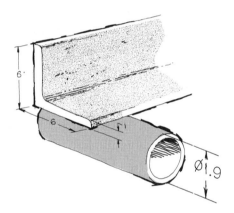

FIG. 17.39 Problem 7. A pipe clamp.

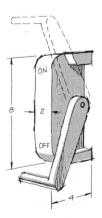

FIG. 17.40 Problem 13. A safety lock.

9. **Flag pole socket.** Design a socket for flags that is to be attached to a vertical wall. Determine the best angle of inclination for the flag pole.

10. **Crutches.** Design a portable crutch that could be used by a person with a temporary leg injury.

11. **Cup holder.** Design a holder that will support a soft drink can or bottle on the interior of an automobile.

12. **Gate hinge.** Design a hinge that could be attached to a 3 in. diameter tubular post to support a 3 ft wide gate.

13. **Safety lock.** Design a safety lock that will hold a high voltage power switch in either the "off" or "on" positions to prevent an accident (Fig. 17.40).

14. **Tubular hinge.** Design a hinge that can be used to hinge 2.5 in. OD high strength aluminum pipe in the manner shown in Fig. 17.41. A hinge of this type is needed for portable scaffolding.

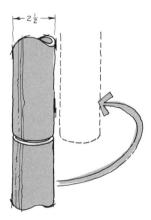

FIG. 17.41 Problem 14. A tubular hinge.

15. **Miter jig.** Design a jig that can be used for assembling wooden frames at 90° angles. The stock for the frames is to be rectangular in cross-sections that vary from 0.75″ × 1.5″ to 1.60″ × 3.60″. Outside dimensions vary from 10 in. to 24 in. (Fig. 17.42).

16. **Base hardware.** Design the hardware needed at the points indicated for a standard volleyball net. The 7 ft pipes are supported by crossing two-by-fours. Design the hardware needed at points *a, b,* and *c* (Fig. 17.43).

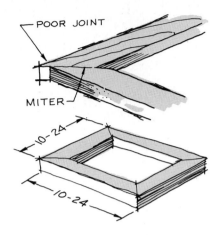

FIG. 17.42 Problem 15. A miter jig.

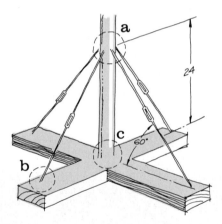

FIG. 17.43 Problem 16. Base hardware for a volleyball net.

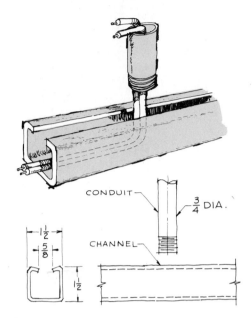

FIG. 17.44 Problem 17. A conduit connector/hanger.

17. **Conduit connector/hanger.** Design a support that will attach to a 0.75 in. conduit that will support a channel that is used as an adjustable raceway for electrical wiring. Your design should permit ease of adjustment. (Fig. 17.44).

18. **Fixture design.** Design a fixture that will permit a small-scale manufacturer to saw the corner of the block as shown in Fig. 17.45.

19. **Drum truck.** Design a truck that can be used for moving a 55-gallon drum of turpentine (7.28 pounds per gallon). The drum will be kept in a horizontal position, but it would be advantageous to incorporate a feature into the truck that would permit it to be set in an upright position (Fig. 17.46).

Product design

A product design involves the development of a device that will perform a specific function and will be mass-produced and sold to a broad market.

20. **Hunting blind.** Hunters of geese and ducks must remain concealed while hunting. Design a portable hunting blind to house two hunters. This blind should be completely portable so that it can be carried in separate sections by each of the hunters. Specify its details and how it is to be assembled and used.

21. **Convertible drafting table.** Many laboratories are equipped with drawing tables that have drafting machines attached to them. This arrangement clutters the table top surface when the tables are used for classes that do not require drafting machines. Design a drafting table that can enclose a drafting machine, thereby concealing it and protecting it when it is not needed.

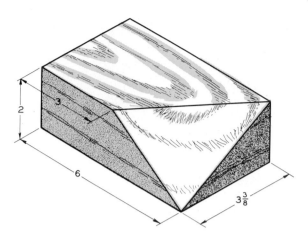

FIG. 17.45 Problem 18. A fixture design.

FIG. 17.47 Problem 22. Folding chair with a writing tablet attached.

22. Writing tablet for a folding chair. Design a writing tablet arm for a folding chair that could be used in an emergency or when a class needs more seating. To allow easy storage, the tablet arm must fold with the chair (Fig. 17.47). Use a folding chair available to you for dimensions and specifications.

23. Portable truck ramp. Delivery trucks need ramps to load and unload supplies and materials at their destinations (Fig. 17.48). Assume that the bed of the truck is 20 in. from ground level. Design a portable ramp that would permit the unloading of goods with the use of a hand truck and reduce manual lifting.

24. Sensor retaining device. The Instrumentation Department of the Naval Oceanographic Office uses underwater sensors to learn more about the ocean. These sensors are submerged on cables from a boat on the surface. The winch used to retrieve the sensor frequently over-runs (continues pulling when it has been retrieved), causing the cable to break and the sensor to be lost. Design a safety device that will retain the 75 lb sen-

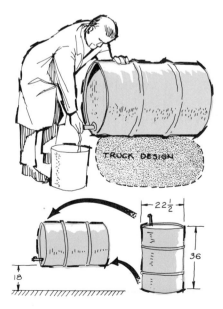

FIG. 17.46 Problem 19. A drum truck.

FIG. 17.48 Problem 23. Delivery ramp for unloading goods.

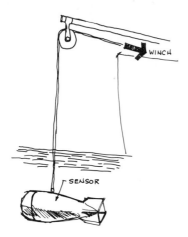

FIG. 17.49 Problem 24. Underwater sensor.

sor if the cable is broken when the sensor reaches a pulley (Fig. 17.49).

25. **Flexible trailer hitch.** In combat zones, vehicles must tow trailers where terrain may be very uneven and hazardous. Design a trailer hitch that will provide for the most extreme conditions possible. Study the problem requirements and limitations to identify the parameters within which your design must function.

26. **Worker's stilts—human engineering.** Workers who apply gypsum board and other types of wallboards to the interiors of buildings and homes must work on scaffolds or wear some type of stilts to be able to reach the ceiling to nail the 4′ × 8′ boards into position.

    Design stilts that will provide the worker with access to a ceiling 8 ft high while permitting him or her to perform the job of nailing ceiling panels with comfort. The stilts should be adjustable to accommodate workers of various weights, sizes, and heights.

27. **Pole-vault standards.** Many pole vaulters are exceeding the 18 ft height in track meets, which introduces a problem for the officials of this event. The pole vault uprights must be adjusted for each vaulter by moving them forward or backward plus or minus 18 in. Also, the crossbar must be replaced with great difficulty at these heights by using forked sticks and ladders, a process that is crude and inefficient. Develop a more efficient set of uprights that can be easily repositioned and will allow the crossbar to be replaced with greater ease.

28. **Sportsman's chair.** Analyze the need for a sportman's chair that could be used for camping, fishing from a bank or boat, at sporting events, and for as many other purposes as you can think of. The need is not for a special-purpose chair, but for a chair that is suitable for a wide variety of uses to fully justify it as a marketable item.

29. **Portable toilet.** Design a portable toilet unit for the camper and outdoorsperson. This unit should be highly portable, with consideration given to the method of waste disposal. Evaluate the market potential for this product.

30. **Child carrier for a bicycle.** Design a seat that can be used to carry a small child as a passenger on a bicycle. Determine the age of the child who would probably be carried as a passenger. Design the seat for safety and comfort.

31. **Lawn sprinkler control.** Design a sprinkler that can be used to water irregularly shaped yards while giving a uniform coverage. This sprinkler should be adjustable so that it can be adapted, within its range, to yards of any shape. Also consider a method of cutting the water off at certain sprinkler positions to prevent the watering of patios or other areas that are to remain dry.

32. **Power lawn fertilizer attachment.** The rotary-power lawn mower emits a force through its outlet caused by the air pressure from the rotating blades. This force might be used to distribute fertilizer while the lawn is being mowed. Design an attachment for a power mower that could spread fertilizer while the mower is performing its usual cutting operation.

33. **Car and window washer.** The force of water coming from a hose provides a source of power that could operate a mechanism that could be used to wash windows or cars. Design an attachment for the typical garden hose that would apply water and agitation (for optimum action) to the surface being cleaned. Consider other applications of the force exerted by water pressure in the performance of yard and household chores that involve water and require agitation that could be provided by the water force.

34. **Projector cabinet.** Many homes have slide projectors, but each showing of the family slides

must be preceded by the time-consuming effort of setting up the equipment. Design a cabinet that could serve as an end table or some other function while also housing a slide projector ready for use at any time. The cabinet might also serve as storage for slide trays. It should have electrical power for the projector. Evaluate the market for a multipurpose cabinet of this type.

35. Heavy appliance mover. Design a device that can be used for moving large appliances, such as stoves, refrigerators, and washers, about the house. This product would not be used often—only for rearrangement, cleaning, and for servicing of the equipment.

36. Car jack. The conventional car jack is a somewhat dangerous means of changing tires on any surface other than a horizontal one. The average jack does not attach itself adequately to the automobile's frame or bumper, introducing a severe safety problem. Design a jack that would be an improvement over existing jacks and and possibly employ a different method of applying a lifting force to a car. Consider the various types of terrain on which the device must serve.

37. Map holder. The driver of an automobile who is traveling alone in an unfamiliar part of the country must frequently refer to a map. Design a system that will give the driver a ready view of the map in a convenient location in the car. Provide a means of lighting the map that will not distract the driver during night driving. Consider all possibilities of making the map's information more available to the driver.

38. Bicycle-for-two adapter. Design the parts and assembly required to convert the typical bicycle into a bicycle built for two (tandem) when mated with another bicycle of the same make and size. Work from an existing bicycle, and consider, among other things, how each rider can equally share in the pedaling. Determine the cost of your assembly and its method of attachment to the average bicycle. Use existing stock parts when possible, to reduce special machining.

39. Automobile unsticker. All drivers have had the experience of getting their car stuck in sand, mud, or snow and are familiar with the lack of traction and the sound of spinning wheels. Design a kit to be carried in the car trunk in a minimum of space that will contain the items required by the driver who must get his or her car "unstuck" when no other help is available. This kit can be composed of one or several items.

40. Stump remover. Assume that a number of tree stumps must be removed from the ground to clear land for construction. The stumps are dead with partially deteriorated root systems and require a force of approximately 2000 lbs to remove them. Design an apparatus that could be attached to the bumper of a car that could be used to remove the stumps by either pushing or pulling. The device should be easy to attach to the stump and to the car.

41. Gate opener. An aggravation to farmers and ranchers is the necessity of opening and closing gates when driving from one fenced area to the next. Design a manually operated gate that could be opened and closed without getting out of the vehicle.

42. Paint mixer. Paint purchased from the shelf of a paint store must be mixed by stirring or some other form of agitation that will bring it to a consistent mixture. Design a product that could be used by the paint store or the paint contractor to quickly mix paint in the store or on the job. Determine the standard size paint cans for which your mixer will be designed.

43. Mounting for an outboard motor on a canoe. Unlike a square-end boat, the pointed-end canoe does not provide a suitable surface for attaching an outboard motor. Design an attachment that will adapt an outboard motor to a canoe. Indicate how the motor will be controlled by the operator in the canoe.

44. Automobile coffee maker. Adequate heat is available in the automobile's power system to prepare coffee quickly. Design an attachment as an integral system of an automobile that will serve coffee from the dashboard area. Consider the type of coffee to be used, instant or regular, method of changing or adding water, the spigot system, and similar details.

45. Baby seat (cantilever). Design a child's chair that can be attached to a standard table top and will support the child at the required height. The chair should be designed to ensure that the child cannot crawl out of or detach it from the table top. A possible solution could be a design that would cantilever from the table top, using the child's body as a means of applying the force necessary to grip the table top. The design would be further improved if the chair were collapsible or suitable for other purposes. Determine the age group that would be most in need of the chair and base your design on the dimensions of a child of this age.

46. **Miniature-TV support.** Miniature television sets for close viewing are available with a screen size of 5″ × 5″. An attachment is needed that would support sets ranging in size from 6″ × 6″ to 7″ × 7″ for viewing from a bed. Determine the placement of the set with respect to a viewer for best results. Provide adjustments on the support that will be used to position the set properly. Analyze the method of concealing electrical wires within the apparatus.

47. **Panel applicator.** A worker who applies 4′ × 8′ gypsum board or paneling must be assisted by a helper who holds the panel in position while it is being nailed to the ceiling. This helper is only partially efficient and adds to the cost of labor.

   Design a device that could be used in this capacity; it should be collapsible for easy transportation, economical, and versatile. Consider the most efficient way in which the operator could control the mechanism and list the features that would be advantageous to the marketing of the device. The average ceiling height is 8 ft, but provide adjustments that would adapt the device to lower or higher ceilings.

48. **Backpack.** Design a backpack that can be used by the outdoorsperson who must carry supplies while hiking. Your design should be based on the analysis of the supplies that would be carried by the average outdoorsperson. A major portion of your design effort should be devoted to adapting the backpack to the human body for maximum comfort and the best leverage for carrying a load over an extended period of time. Can other uses be made of your design?

49. **Automobile controls.** Design driving controls that can be attached to the standard automobile that will permit an injured person to drive a car without the use of his or her legs. This device should be easy to attach and to operate with the maximum of safety.

50. **Bathing apparatus.** Design an apparatus that would assist a wheelchair-bound person, who

does not have use of his or her legs, to get in and out of a bathtub without assistance from others.

51. **Adjustable TV base.** Design a TV base to support full-size TV sets that would allow the maximum of adjustment: up and down, rotation about a vertical axis and about a horizontal axis. Design the base to be as versatile as possible.

52. **Trailer jack.** Design a trailer jack that can be used to repair flat tires that may occur on a boat trailer. It may be possible to build in jack devices as permanent features of the trailer.

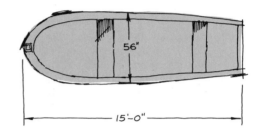

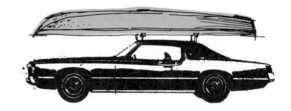

FIG. 17.50 Problem 54. Boat specifications.

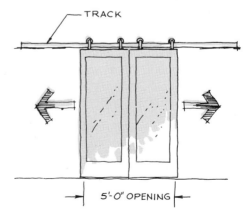

FIG. 17.51 Problem 55. Door dimensions.

53. **Projector cabinet.** Design a portable cabinet that could be left permanently in a classroom that would house a slide projetor and a movie projector. The cabinet should provide both convenience and security from vandalism and theft.

54. **Boat loader.** Design a rack and a system whereby one person could load a boat on top of a car for transporting from site to site. Use the boat specifications in Fig. 17.50; the boat weighs 110 lbs.

55. **Door opener.** Design a method whereby a trucker at a loading dock could open the warehouse door without having to get out of the truck. The doors are dimensioned in Fig. 17.51, and the dock extends 8 ft from the doors.

18 CHAPTER

Pictorials

18.1
Introduction

A pictorial is an effective means of communicating an idea. This is especially true if a design is unique or if the person to whom it is being explained has difficulty interpreting multiview drawings.

Pictorials are also helpful in developing a design since they enable the designer or drafter to work in three dimensions.

Pictorials, sometimes called technical illustrations, are widely used to describe various products in catalogs, parts manuals, and maintenance publications. Pictorials are used in industry for the purpose of putting parts together properly.

This chapter will cover the basic types of pictorial methods: (1) oblique drawing, (2) isometric drawing, (3) axonometric projection, and (4) perspective projection.

18.2
Types of pictorials

The three commonly used forms of pictorials are (1) obliques, (2) axonometrics/isometrics, and (3) perspectives. Examples of these are shown in Fig. 18.1.

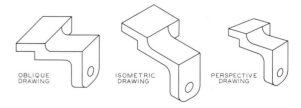

FIG. 18.1 The three standard pictorial systems: oblique, isometric, and perspective.

OBLIQUE pictorials are three-dimensional pictorials made on a plane of paper by projecting from the object with parallel projectors that are **oblique** to the picture plane (Fig. 18.2B).

AXONOMETRIC (ISOMETRIC) projection is a three-dimensional pictorial on a plane of paper drawn by projecting from the object to the picture plane as illustrated in Fig. 18.2A. The parallel projectors are perpendicular to the picture plane.

PERSPECTIVE PICTORIALS are drawn with projectors that converge at the viewer's eye and make varying angles with the picture plane (Fig. 18.2C).

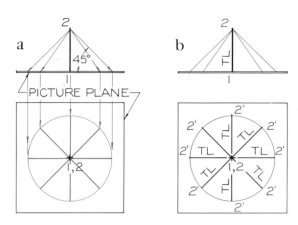

FIG. 18.3 The underlying principle of the cavalier oblique can be seen here, where a series of projectors form a cone. Each element makes a 45° angle with the picture plane. Consequently, the projected lengths of 1–2′ are true length and are equal in length to Line 1–2 that is perpendicular to the picture plane.

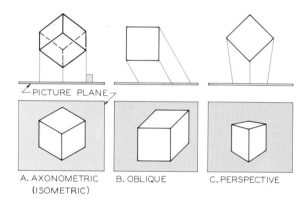

FIG. 18.2 Types of projection systems for pictorials. (A) Axonometric pictorials are formed by parallel projectors that are perpendicular to the picture plane. (B) Obliques are formed by parallel projectors that are oblique to the picture plane. (C) Perspectives are formed by converging projectors that make varying angles with the picture plane.

plane can be seen in the front view (b). Each of these projections of 1–2′ is equal in length to the true length of 1–2.

This is called a **cavalier oblique** projection because the projectors make 45° with the picture plane, and measurements along the receding axis can be made true length and in any direction.

Examples of cavalier obliques are shown in Fig. 18.4. The front surface is usually positioned parallel to

18.3
Oblique projections

The system of oblique **projection** is used as the basis of oblique **drawings;** however, oblique projections are seldom used. In Fig. 18.3A, a number of lines of sight are drawn through point 2 of Line 1–2. Each line of sight makes a 45° angle with the picture plane, which creates a cone with its apex at 2, and each element on the cone makes a 45° angle with the plane. A variety of projections of Line 1–2′ on the picture

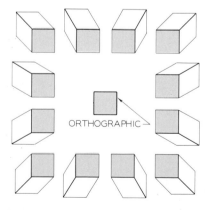

FIG. 18.4 A cavalier oblique is drawn with one surface as a true-size orthographic view. The dimensions along the receding axis are true length and the axes are drawn at any angle.

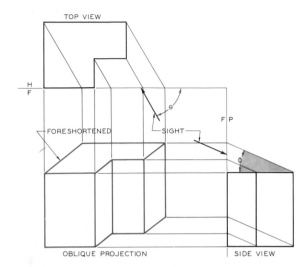

FIG. 18.5 An oblique projection can be drawn at any angle of sight to obtain an oblique pictorial. However, the line of sight should not make an angle less than 45° with the picture plane. This would result in a receding axis longer than true length, thereby distorting the pictorial.

the picture plane; therefore it will appear true size as an orthographic view.

The top and side views are given as orthographic views, and the front view is drawn as a projection by using the two views of a selected line of sight (Fig. 18.5). Projectors are drawn from the object parallel to the lines of sight to locate their respective points in the oblique view.

The dimensions along the receding axes are less than true length and greater than half-length; therefore this is called a **general oblique.**

If the angle between the line of sight and the picture plane is less than 45°, the measurements along the receding axes will be greater than true length. This is objectionable since the oblique will be distorted.

18.4
Oblique drawings

Oblique projections are seldom used in the manner just illustrated.

Instead, three basic types of **oblique drawings** are used that are based on these principles. The three types are: (1) **cavalier,** (2) **cabinet,** and (3) **general** (Fig. 18.6).

In each case, the angle of the receding axis can be at any angle between 0° and 90° (Fig. 18.6). Measurements along the receding axes of the **cavalier oblique** are true length (full scale). The **cabinet oblique** has measurements along the receding axes reduced to half-length. The **general oblique** has measurements along the receding axes reduced to between half and full length.

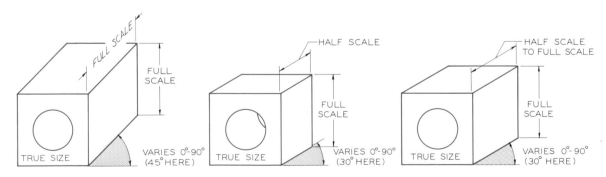

FIG. 18.6 Types of obliques

A. The *cavalier oblique* can be drawn with a receding axis at any angle, but measurements along this axis are true length.

B. The *cabinet oblique* can be drawn with a receding axis at any angle, but the measurements along this axis are half scale.

C. The *general oblique* can be drawn with a receding axis at any angle, but the measurements along this axis can vary from half to full scale.

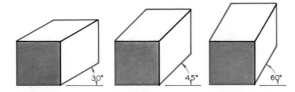

FIG. 18.7 The cavalier oblique is usually drawn with the receding axis at the standard angles of the drafting triangles. Each gives a different view of a cube.

Three examples of cavalier obliques of a cube are shown in Fig. 18.7. Each has a different angle for the receding axes, but the measurements along the receding axes are true length.

A comparison of cavalier and cabinet obliques is given in Fig. 18.8. The cabinet oblique reduces the distortion of an object with a long depth, thereby giving a more pleasing appearance.

18.5
Construction of an oblique

An oblique should be drawn by constructing a box using the overall dimensions of height, width, and depth with light construction lines. In Fig. 18.9, the front view is drawn true size in Step 1. In Step 2, the receding axis is drawn at 30° and the depth dimensions are measured true length. This will be a cavalier oblique. True measurements can be made parallel to the three axes.

The notches are removed from the blocked-in construction box to complete the oblique. These measurements are transferred from the given orthographic views with your dividers.

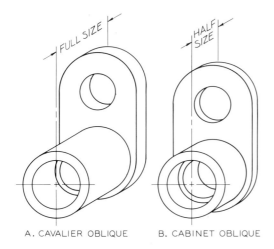

FIG. 18.8 Measurements along the receding axis of a cavalier oblique are full size, and those in a cabinet oblique are half-size.

18.6
Angles in oblique

Angular measurements can be made on the true-size plane of an oblique that is parallel to the picture plane.

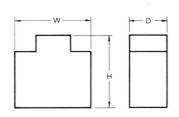

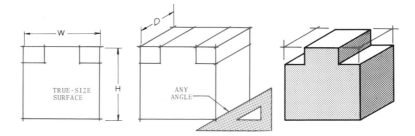

FIG. 18.9 Oblique Construction

Step 1 The front surface of the oblique is drawn as a true-size plane. The corners are removed.

Step 2 The receding axis is selected and the true dimensions are measured along this axis.

Step 3 The finished cavalier oblique is strengthened to complete the drawing.

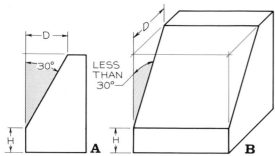

FIG. 18.10 Angles in oblique must be located by using coordinates. They cannot be measured true size except on a true-size plane.

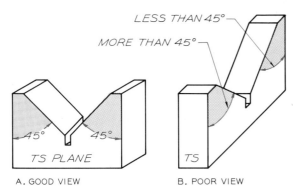

A. GOOD VIEW   B. POOR VIEW

FIG. 18.11 The best view is the view that takes advantage of the ease of construction offered by oblique drawings. The view in (B) is less descriptive and more difficult to construct than the view in (A).

However, angular measurements will not be true-size on the other two planes of the oblique.

To construct an angle in an oblique, coordinates must be used, as shown in Fig. 18.10. The sloping surface of 30° must be found by locating the vertex of the angle H distance from the bottom. The inclination is found by measuring the distance of D along the receding axis to establish the slope. This angle is not equal to the 30° angle that was given in the orthographic view.

You can see in Fig. 18.11 that a true angle can be measured on a true-size surface. At B, angles along the receding planes are either smaller or larger than their true angles.

It requires less effort and gives a better appearance when obliques are drawn where angles will appear true size, as shown in Fig. 18.11A, rather than as shown at B.

18.7
Cylinders in oblique

The major advantage of obliques is that circular features can be drawn as true circles when they are parallel to the picture plane. This is illustrated in Fig. 18.12, where an oblique of a cylinder is drawn.

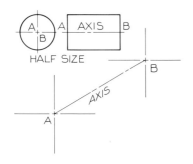
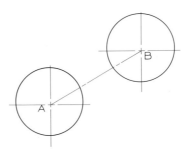
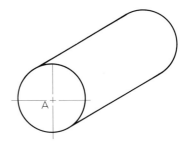

FIG. 18.12 A cylinder in oblique

Step 1 The centerlines of the circular ends are located at *A* and *B*. The axis is drawn at any angle and since it is true length, this is a cavalier oblique.

Step 2 Circles are drawn using Centers *A* and *B*. These are true circles, since they lie in the orthographic plane of the oblique.

Step 3 Lines are drawn parallel to the axis and tangent to the end circles. Hidden lines are omitted. Centerlines may also be omitted.

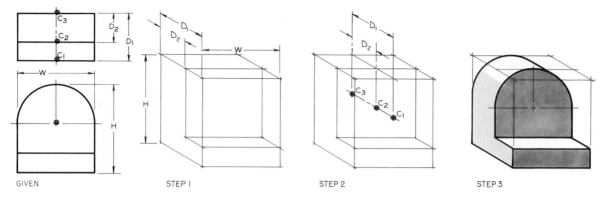

FIG. 18.13 Construction of an oblique

Step 1 The overall dimensions are used to block in the oblique pictorial. The notch is removed.

Step 2 The three centers, C_1, C_2, and C_3, are located on each of the planes.

Step 3 The three centers found in Step 2 are used to draw the semicircular features of the oblique. Lines are strengthened.

The centerlines of the circular end at A are drawn and the receding axis is drawn at any desired angle. The end at B is located by measuring along the axis. Circles are drawn at each end using centers A and B, are connected with tangent lines parallel to the axis.

These same principles are used to construct an object with semicircular features (Fig. 18.13). The oblique is positioned to take advantage of the option of drawing circular features as true circles. Centers C_2 and C_3 are located for drawing two semicircles (Step 3).

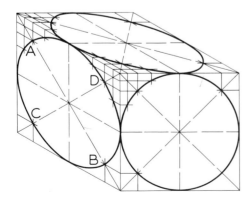

FIG. 18.14 Circular features on the faces of a cavalier oblique of a cube appear as two ellipses and one true circle.

18.8
Circles in oblique

Although circular features will be true size on a true-size plane of a cavalier oblique pictorial, circular features on the other two planes will appear as ellipses (Fig. 18.14).

A more frequently used technique of drawing elliptical views in oblique is the four-center ellipse technique, which is shown in Fig. 18.15. A rhombus is drawn that would be tangent to the circle at four points. Perpendicular construction lines are drawn from where the centerlines cross the sides of the rhombus. This construction is performed as shown in Steps 2 and 3 to locate four centers that are used to draw four arcs that join together to form the ellipse in oblique.

The four-center ellipse method will not work for the cabinet or the general oblique. In these cases, coordinates must be used to locate a series of points on the curve. Coordinates in Fig. 18.16 are shown in orthographic (Part A) and then in the cabinet oblique (B). The coordinates that are parallel to the receding axis are reduced to half-size in the oblique. Those in a horizontal direction are drawn full size. The ellipse can be drawn with an irregular curve, or with an ellipse template that approximates the plotted points.

Whenever possible, oblique drawings of objects with circular features should be positioned with circles

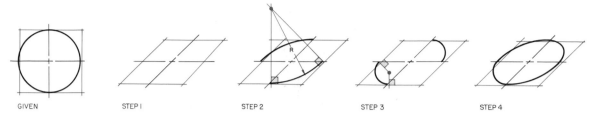

FIG. 18.15 Four-center ellipse in oblique

Step 1 The circle that is to be drawn in oblique is blocked in with a square which is tangent to the circle at four points. This square will appear as a rhombus on the oblique plane.

Step 2 Construction lines are drawn perpendicularly from the points of tangency to locate the centers for drawing two segments of the ellipse.

Step 3 The centers for the two remaining arcs are located with perpendiculars drawn from adjacent tangent points.

Step 4 When the four arcs have been drawn, the final result is an approximate ellipse.

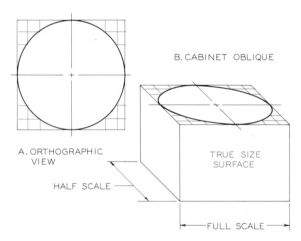

FIG. 18.16 The four-center ellipse technique cannot be used to locate circular shapes on the foreshortened surface of a cabinet oblique. These ellipses must be plotted with coordinates.

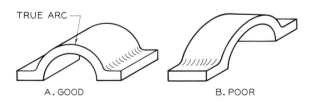

FIG. 18.17 An oblique should be positioned to enable circular features to be drawn with the greatest of ease.

on true-size planes, so they can be drawn as true circles. In Fig. 18.17, the view at A is better than the one at B since it gives a more descriptive view of the part and it was easier to draw.

18.9
Curves in oblique

Irregular curves in oblique must be plotted point by point using coordinates (Fig. 18.18). The coordinates are transferred from the orthographic view to the

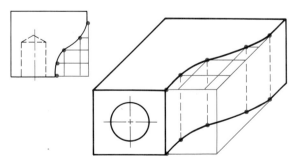

FIG. 18.18 Coordinates are used to establish irregular curves in oblique. The lower curve is found by projecting the points downward a distance equal to the height of the oblique.

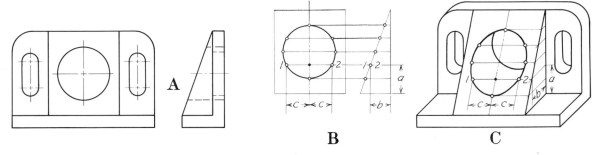

FIG. 18.19 The construction of a circular feature on an inclined surface must be found by plotting points using three coordinates, A, B, and C, to locate Points 1 and 2 in this example. The plotted points are connected to complete the elliptical feature.

oblique view and the curve is drawn through these points with an irregular curve.

If the object has a uniform thickness, the lower curve can be found by projecting vertically downward from the upper points a distance equal to the height of the object.

In Fig. 18.19 the elliptical feature on the inclined surface was found by using a series of coordinates to locate the points along its curve. These points are then connected with an irregular curve or by using an ellipse template of approximately the same size.

18.10
Oblique sketching

An understanding of the mechanical principles of oblique construction is essential for sketching obliques.

As shown in Fig. 18.20, the use of light guidelines is helpful in developing a sketch. These guidelines should be drawn lightly so they will not need to be erased when the finished lines of the sketch are darkened.

When sketching on tracing vellum, it is helpful if a printed grid is placed under the vellum to provide guidelines.

18.11
Dimensioned obliques

A dimensioned oblique full section is given in Fig. 18.21 where the interior features and the dimensions of the part are shown.

When dimensions and notes are given in oblique, the numerals and lettering should be applied using

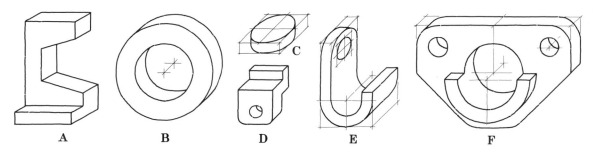

FIG. 18.20 Obliques may be drawn as freehand sketches by utilizing the same principles that were used for instrument pictorials. It is beneficial to use light construction lines to locate the more complex features.

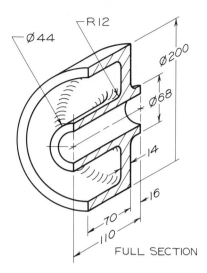

FULL SECTION

FIG. 18.21 Oblique pictorials can be drawn as sections and dimensioned to serve as working drawings.

one of the methods shown in Fig. 18.22. In the aligned method, the numerals are aligned with the dimensioned lines. In the unidirectional method, the numerals are all positioned in a single direction regardless of the direction of the dimension lines. Notes that are connected with leaders are positioned horizontally in both methods.

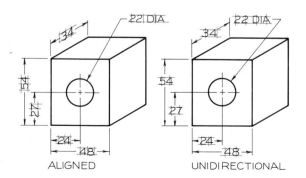

ALIGNED   UNIDIRECTIONAL

FIG. 18.22 Oblique pictorials can be dimensioned by using either of these methods of applying numerals to the dimension lines.

18.12
Isometric pictorials

An **isometric projection** is a type of axonometric projection in which parallel projectors are perpendicular to the picture plane and the diagonal of a cube is seen as a point (Fig. 18.23). The three axes are spaced 120° apart and the sides are foreshortened to 82% of their true length.

> The term isometric, which means "equal measurement," is used to describe this type of pictorial since the planes are equally foreshortened.

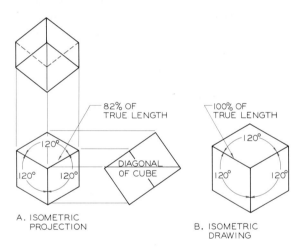

A. ISOMETRIC PROJECTION

B. ISOMETRIC DRAWING

FIG. 18.23 A true isometric projection is found by constructing a view that shows the diagonal of a cube as a point. An isometric drawing is not a true projection since the dimensions are drawn true size rather than reduced in size as in projection.

An **isometric drawing** is similar to an **isometric projection** except that it is not a true axonometric projection, but an approximate method of drawing a pictorial. Instead of reducing the measurements along the axes to 82%, they are drawn true length (Fig. 18.23B).

A comparison between an isometric projection and an isometric drawing is shown in Fig. 18.24. By using the isometric drawing instead of the isometric

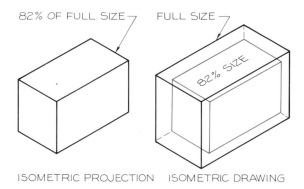

82% OF FULL SIZE   FULL SIZE

82% SIZE

ISOMETRIC PROJECTION   ISOMETRIC DRAWING

FIG. 18.24 The isometric projection is foreshortened to 82% of full size. The isometric drawing is drawn full size for convenience.

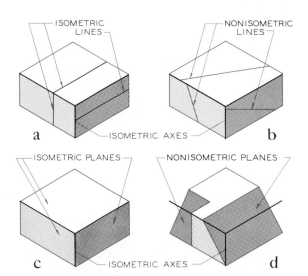

ISOMETRIC LINES   NONISOMETRIC LINES

a   ISOMETRIC AXES   b

ISOMETRIC PLANES   NONISOMETRIC PLANES

c   ISOMETRIC AXES   d

FIG. 18.26 (A) True measurements can be made along isometric lines (parallel to the three axes). (B) Nonisometric lines cannot be measured true length. (C) The three isometric planes are indicated; there are no nonisometric planes in this drawing. (D) Nonisometric planes are planes that are inclined to any of the three isometric planes of a cube.

projection, pictorials can be measured using standard scales, the only difference being the 18% increase in size. Isometric drawings are used more often than isometric projections.

The axes of isometric drawings are separated by 120° (Fig. 18.25). Although one of the axes is usually drawn vertically, this is not necessary.

Construction of an isometric drawing

An isometric drawing is begun by drawing three axes 120° apart. Lines that are parallel to these axes are called **isometric lines** (Fig. 18.26A). True measure-

ments can be made along isometric lines but along nonisometric lines (Part B).

The three surfaces of a cube in isometric are called **isometric planes** (Fig. 18.26C). Planes that are parallel to these planes are isometric planes and planes that are not parallel to them are called nonisometric planes.

To draw an isometric, you will need a scale and a 30°–60° triangle (Fig. 18.27). Begin by constructing a plane of the isometric using the dimensions of height (H) and depth (D). In Step 3, the third dimension, width (W), is used to complete the isometric drawing.

It is recommended that all isometric drawings be blocked in using light guidelines, as shown in Fig. 18.28, and overall dimensions of W, D, and H. Other dimensions can be taken from the given views and measured along the isometric axes to locate notches and portions that are removed from the "blocked-in" drawing.

A more complex isometric is shown in Fig. 18.29. Again, the object is blocked in using H, W, and D, and portions of the block are removed to complete the isometric drawing.

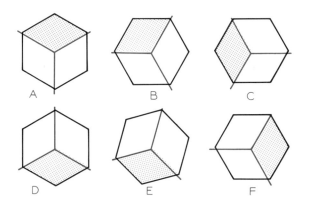

A   B   C

D   E   F

FIG. 18.25 Isometric axes are spaced 120° apart, but they can be revolved into any position.

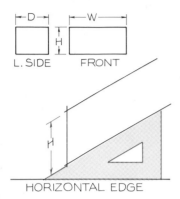
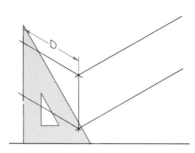
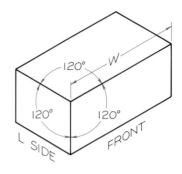

FIG. 18.27 Construction of an isometric drawing

Step 1 A 30°–60° triangle is used in combination with a horizontal straightedge to begin the isometric. The dimensions of *W*, *D*, and *H* from the orthographic views were doubled in this case.

Step 2 The depth dimension, *D*, in the left side view is laid off in isometric. The parallel lines appear parallel in isometric drawings, as well as in orthographic views.

Step 3 The final sides of the isometric are drawn and the lines are strengthened. The 30°–60° triangle automatically separates the axes by 120°.

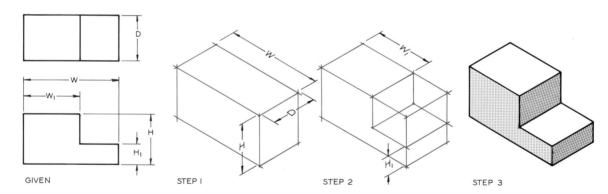

FIG. 18.28 Layout of an isometric drawing (simple)

Step 1 The object is blocked in using the overall dimensions. The notch is removed.

Step 2 The notch is located by establishing its end points.

Step 3 The lines are strengthened to complete the drawing.

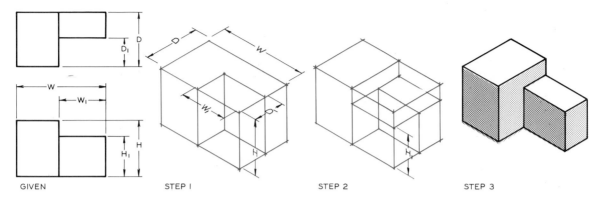

FIG. 18.29 Layout of an isometric drawing (complex)

Step 1 The overall dimensions of the object are used to lightly block in the object. One notch is removed.

Step 2 The second notch is removed using dimension H_1.

Step 3 The final lines of the isometric are strengthened.

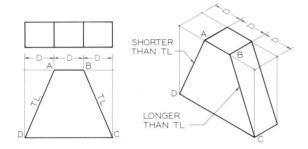

FIG. 18.30 Inclined surfaces must be found by using coordinates measured along the isometric axes. The lengths of angular lines will not be true length in isometric.

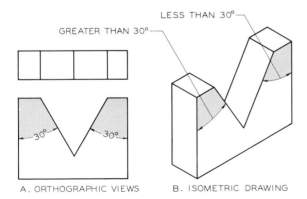

FIG. 18.31 Angles in isometric must be found by using coordinates. Angles will not appear true size in isometric.

18.13
Angles in isometric

Angles cannot be measured true size in an isometric drawing since the surfaces of an isometric are not true size. Angles must be located by using coordinates measured along isometric lines as shown in Fig. 18.30. Lines AD and BC are equal in length in the orthographic view, but they are shorter and longer than true length in the isometric drawing.

A similar example can be seen in Fig. 18.31, where two angles are drawn in isometric. The equal size angles in orthographic are less and greater than true size in the isometric drawing.

An isometric drawing of an object with an inclined surface is drawn in three steps in Fig. 18.32. The object is blocked in pictorially and portions are removed.

18.14
Circles in isometric

Three methods of constructing circles in isometric drawings are: (1) point plotting, (2) four-center ellipse construction, and (3) ellipse templates.

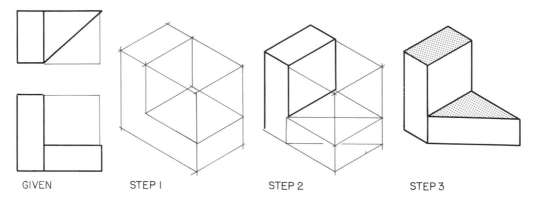

GIVEN   STEP 1   STEP 2   STEP 3

FIG. 18.32 Construction of an isometric drawing with an inclined plane

Step 1   The object is blocked in using the overall dimensions. The notch is removed.

Step 2   The inclined plane is located by establishing its end points.

Step 3   The lines are strengthened to complete the drawing.

Circles: point plotting

A series of points located on a circle can be located in an isometric drawing by using two dimensions that are parallel to the isometric axes. These two dimensions are called **coordinates** (Fig. 18.33).

The cylinder is blocked in and drawn pictorially, with the centerlines added in Step 1. Coordinates A, B, C, and D are used in Step 2 to locate points on the ellipse and are connected using an irregular curve.

The lower ellipse located on the bottom plane of the cylinder can be found by using a second set of coordinates. The most efficient method is by measuring the distance, E, vertically beneath each point that was located on the upper ellipse (Step 3). A plotted ellipse is a true ellipse and it is equivalent to a 35° ellipse on an isometric plane.

An example of a design that is composed of circular features that were drawn in isometric is the handwheel shown in Fig. 18.34.

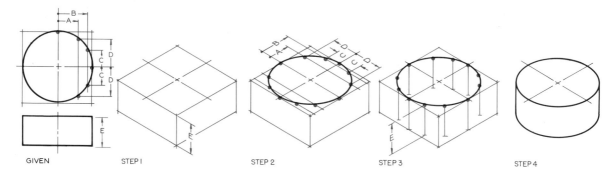

GIVEN   STEP 1   STEP 2   STEP 3   STEP 4

FIG. 18.33 Plotting circles

Step 1   The cylinder is blocked in using the overall dimensions. The centerlines locate the points of tangency of the ellipse.

Step 2   Coordinates are used to locate points on the circumference of the circle.

Step 3   The lower ellipse is found by dropping each point a distance equal to the height of the cylinder E.

Step 4   The two ellipses can be drawn with an irregular curve and connected with tangent lines to complete the cylinder.

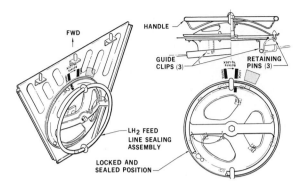

FIG. 18.34 An example of parts that have been drawn using ellipses in isometric to represent circles. This is a handwheel that was proposed for use in an orbital workshop. (Courtesy of NASA.)

Circles: four-center ellipse construction

The **four-center ellipse** method can be used to construct an approximate ellipse in isometric by using four arcs that are drawn with a compass (Fig. 18.35).

The four-center ellipse is drawn by blocking in the orthographic view of the circle with a square that is tangent to the circle at four points. The four centers are found by constructing perpendiculars to the sides of the rhombus at the midpoints of the sides (Step 2). The four arcs are drawn to give the completed four-center ellipse (Step 3). This method can be used to draw ellipses on any of the three isometric planes, since each is equally foreshortened, (Fig. 18.36).

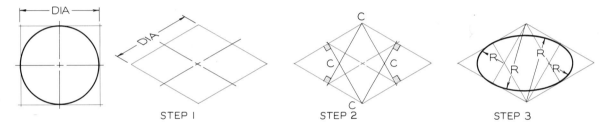

FIG. 18.35 The four-center ellipse

Step 1   The diameter of the circle is used to construct a rhombus that is tangent to the ellipse.

Step 2   Perpendicular lines are drawn from the midpoints of the sides of the rhombus to locate four centers.

Step 3   Using the four centers and two radii, the four-center ellipse is drawn tangent to the rhombus.

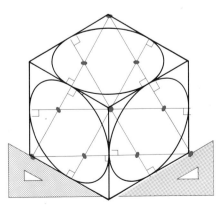

FIG. 18.36 Four-center ellipses can be drawn on all three surfaces of an isometric drawing.

You can see in Fig. 18.37 that the four-center ellipse is only an approximate ellipse when it is compared with a true ellipse.

Circles: ellipse templates

A specially designed **ellipse template** can be purchased for drawing ellipses in isometric. A typical example is shown in Fig. 18.38.

The diameters of the ellipses on the template are measured along the direction of the isometric lines, since this is how diameters are measured in an isometric drawing (Fig. 18.39). The maximum diameter that can be measured across the ellipse is the **major diameter,** which is a true diameter. Consequently,

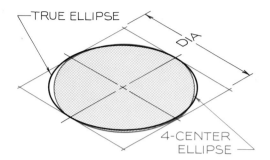

FIG. 18.37 The four-center ellipse is not a true ellipse, but an approximate ellipse.

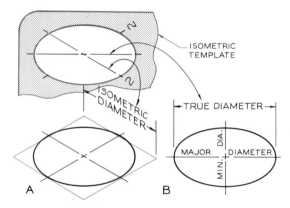

FIG. 18.38 The isometric ellipse template is a special template designed to reduce drafting time. Note that the true diameters of the circles are not the major diameters of the ellipses, but the diameters that are parallel to the isometric axes.

FIG. 18.39 The diameter of a circle in isometric is measured along the direction of the isometric axes. Therefore the major diameter of an isometric ellipse is greater than the labeled diameter. The minor diameter is perpendicular to the major diameter.

the size of the diameter marked on the template is less than the ellipses' major diameter, the true diameter.

The isometric ellipse template can be used to draw an ellipse by constructing the centerlines of the ellipse in isometric and aligning the ellipse template with these isometric lines (Fig. 18.39A).

18.15
Cylinders in isometric

A cylinder can be drawn in isometric by using the four-center ellipse method, as shown in Fig. 18.40. A rhombus is drawn at each end of the cylinder's axis with the centerlines drawn as isometric lines (Step 1). The ellipses are drawn using the four-center ellipse method at each end (Step 2). The ellipses are connected with tangent lines and the lines are darkened in Step 3.

A cylinder can also be drawn using the ellipse template, as illustrated in Fig. 18.41. The axis of the cylinder is drawn and perpendiculars are constructed at each end (Step 1). Since the axis of a right cylinder is perpendicular to the major diameter of its elliptical end, the ellipse template is positioned as shown in Step 2. The ellipses are drawn at each end and are connected with tangent lines (Step 3).

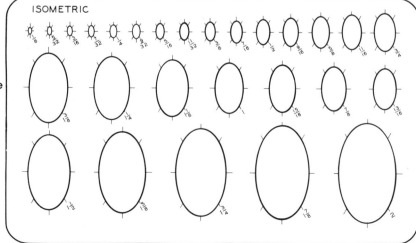

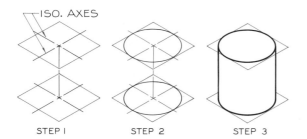

FIG. 18.40 Cylinder: four-center method

Step 1   A rhombus is drawn in isometric at each end of the cylinder's axis.

Step 2   A four-center ellipse is drawn within each rhombus.

Step 3   Lines are drawn tangent to each rhombus to complete the isometric drawing.

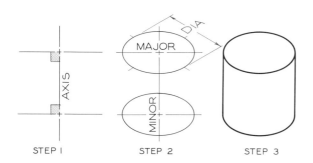

FIG. 18.41 Cylinder: ellipse template method

Step 1   The axis of the cylinder is drawn its proper length, and perpendiculars are drawn at each end.

Step 2   The elliptical ends are drawn by aligning the major diameter with the perpendiculars at the ends of the axis. The isometric diameter of the isometric ellipse template is given along the isometric axis.

Step 3   The ellipses are connected with tangent lines to complete the isometric drawing. Hidden lines are omitted.

To construct a cylindrical hole in the block (Fig. 18.42), begin by locating the center of the hole on the isometric plane. The axis of the cylinder is drawn parallel to the isometric axis that is perpendicular to this plane through its center (Step 2). The ellipse template is aligned with the major and minor diameters to complete the elliptical view of the cylindrical hole (Step 3).

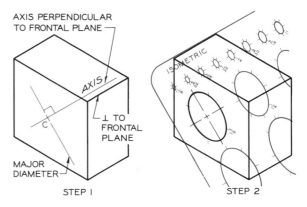

FIG. 18.42 Cylinders in isometric

Step 1   The center of the hole with a given diameter is located on a face of the isometric drawing. The axis of the cylinder from the center parallel to the isometric axis that is perpendicular to the plane of the circle. The major diameter is drawn perpendicular to the axis.

Step 2   The 1⅜″ ellipse template is used to draw the ellipse by aligning the major and minor diameters with the guidelines on the template.

18.16
Partial circular features

When an object has a semicircular end, as in Fig. 18.43, the four-center ellipse method can be used with only two centers to draw half of the circle (Step 2). To draw the lower ellipse at the bottom of the object, the centers are projected downward a distance of H, the height of the object. These centers are used with the same radii that were used on the upper surface to draw the arcs.

In Fig. 18.44, an object with rounded corners is blocked in and the centerlines are located at each rounded corner (Step 1.) An ellipse template is used to construct the rounded corners at Part B. The

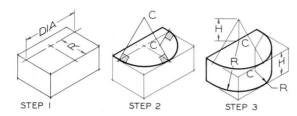

FIG. 18.43 Semicircular features

Step 1 Objects with semicircular features can be drawn by blocking in the objects as if they had square ends. The centerlines are drawn to locate the centers and tangent points.

Step 2 Perpendiculars are drawn from each point of tangency to locate two centers. These are used to draw half of a four-center ellipse.

Step 3 The lower surface can be drawn lowering the centers by the distance of H, the thickness of the part. The same radii are used with these new centers to complete the isometric.

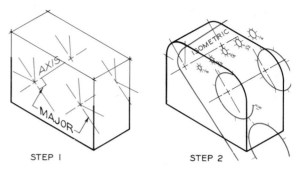

FIG. 18.44 To construct rounded corners on an object, the centerlines of the ellipse are drawn at Step 1. The elliptical corners are drawn with an ellipse template at Step 2.

rounded corners could have been constructed by the four-center ellipse method or by plotting points on the arcs.

A similar drawing involving the construction of ellipses is the conical shape in Fig. 18.45. The ellipses are blocked in at the top and bottom surfaces (Step 1), and the half ellipses are drawn in Step 2 by using a template or the four-center method.

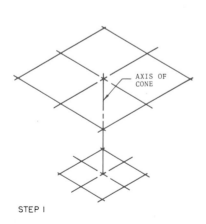

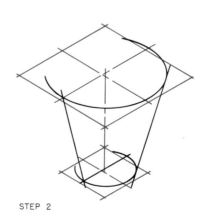

FIG. 18.45 A cone in isometric

Step 1 The axis of the cone is constructed. Each circular end of the cone is blocked in.

Step 2 An ellipse guide is used for constructing the circles in isometric at each end. These ends are connected to give the outline of the object.

Step 3 The remaining details of the screen storage provisions are added to complete the isometric. (Courtesy of NASA.)

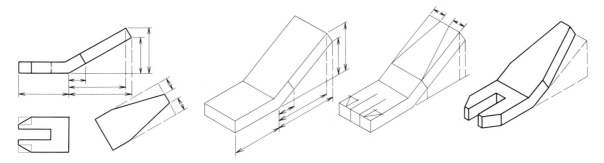

FIG. 18.46 Inclined surfaces in isometric must be located by using coordinates laid off parallel to the isometric axes. True angles cannot be measured in isometric drawings.

18.17
Measuring angles

Angles in isometric may be located by coordinates, as shown in Fig. 18.46, since angles will not appear true in an isometric.

A second method of measuring and locating angles is the ellipse template method shown in Fig. 18.47. Since the hinge line is perpendicular to the

path of revolution, an ellipse is drawn in Step 1 with the major diameter perpendicular to the hinge line. A true circle is drawn with a diameter that is equal to the major diameter of the ellipse.

In Step 2, Point *A* is located on the ellipse and is projected to the circle. This locates the direction of a horizontal line. From this line, the angle of revolution of the hinged part can be measured true size, 120° in this example (Step3). Point *B* is projected to the ellipse to locate a line at 120° in isometric.

The thickness of the revolved part is found by

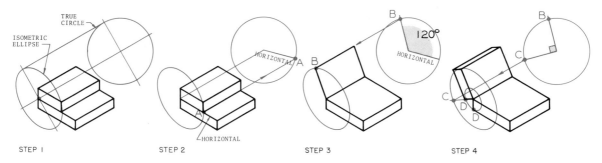

FIG. 18.47 Measuring angles with an ellipse template

Step 1   An ellipse is drawn with the major diameter perpendicular to the hinge line of the two parts. Any size of ellipse can be used. A true circle is drawn with its center on the projection of the hinge line and with a diameter equal to the major diameter of the ellipse.

Step 2   Point *A* is projected to the circle to locate the direction of the horizontal in the circular view.

Step 3   The position of rotation is measured 120° from the horizontal to locate Point *B*, which is then projected to the ellipse to locate the position of the revolved surface.

Step 4   To locate the perpendicular to the surface, a 90° angle is drawn in the circular view and Point *C* is projected to the ellipse; a line is drawn from Point *C* on the ellipse to the center of the ellipse. A smaller ellipse is drawn to pass through Point *D*. The point where this ellipse intersects the line from *C* to the center of the ellipse establishes the thickness of the part.

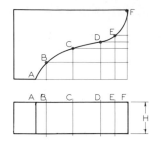
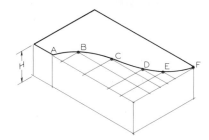
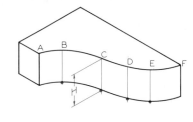

FIG. 18.48 Plotting irregular curves

Step 1   Draw two coordinates to locate a series of points on the irregular curve. These coordinates must be parallel to the standard *W*, *D*, and *H* dimensions.

Step 2   Block in the shape using overall dimensions. Locate points on the irregular curve using the coordinates from the orthographic views.

Step 3   The lower curve can be found by projecting downward the distance *H* from the upper points. Connect the points with an irregular curve.

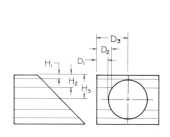
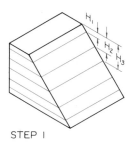
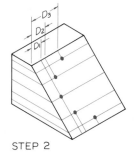
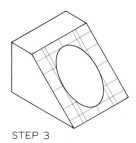

STEP 1   STEP 2   STEP 3

FIG. 18.49 Construction of ellipses on an inclined plane

Step 1   Coordinates are established in the orthographic views.

Step 2   One set of coordinates is transferred to the isometric view.

Step 3   The second set of coordinates is transferred to the isometric to establish points on the ellipse.

Step 4   The plotted points are connected with an elliptical curve. An ellipse template can serve as a guide for connecting the points.

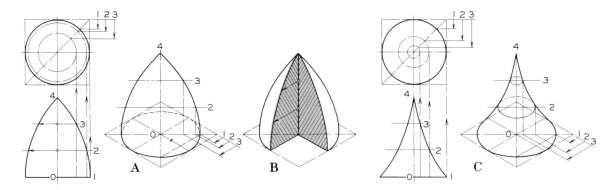

FIG. 18.50 Surfaces of revolution must be constructed by drawing a series of cross sections that are connected to give their overall shapes. The cross sections in these examples are circles that are drawn as ellipses in isometric.

394

drawing Line *C* perpendicular to Line *A* and projecting back to the ellipse. A smaller ellipse through Point *D* is drawn to locate the thickness of the revolved part in the isometric. The remaining lines are drawn parallel to these key lines.

18.18
Curves in isometric

Irregular curves must be plotted point by point, using coordinates to locate each point. Points *A* through *F* are located in the orthographic view with coordinates of width and depth (Fig. 18.48). These coordinates are transferred to the isometric view of the blocked-in part (Step 1) and are connected with an irregular curve.

Each point on the upper curve is projected downward for a distance of *H*, the height of the part, to locate points on the lower curve. The points are connected with an irregular curve to complete the isometric.

18.19
Ellipses on nonisometric planes

When an ellipse lies on a nonisometric plane such as the one shown in Fig. 18.49, points on the ellipse can be plotted to locate the ellipse.

Coordinates are located in the orthographic views and then transferred to the isometric as shown in Steps 1 and 2. The plotted points can be connected with an irregular curve, or an ellipse template can be selected that will approximate the plotted points.

18.20
Surfaces of revolution

A surface of revolution is a solid form made by revolving a plane about an axis of revolution. Two examples of surfaces of revolution are shown in Fig. 18.50. To draw these in isometric, the axes were drawn through Point 0 to Point 4. Circular cross sections were located along this axis using the correct elliptical diameters taken from the orthographic views for each.

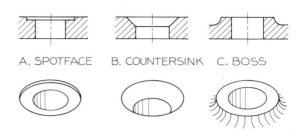

A. SPOTFACE   B. COUNTERSINK   C. BOSS

FIG. 18.51 Examples of circular features drawn in isometric. These can be drawn by using ellipse templates.

The elliptical cross sections are connected with tangent lines to find the outlines of the objects in isometrics. The more cross sections that are used, the more accurate will be the location of the tangent lines.

18.21
Machined parts in isometric

Orthographic and isometric views of a spotface, countersink, and boss are shown in Fig. 18.51. The isometric drawings of these features could be drawn by point-plotting the circular features, by the four-center method, or by using an ellipse template. The template method is by far the easiest method.

A threaded shaft can be drawn in isometric as shown in Fig. 18.52, by first drawing the cylinder that

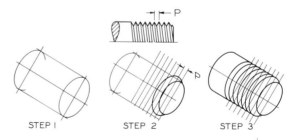

STEP I   STEP 2   STEP 3

FIG. 18.52 Threads in isometric

Step 1 Draw the cylinder that is to be threaded by using an ellipse template.

Step 2 Lay off a number of perpendiculars that are spaced by a distance equal to the pitch of the thread, *P*.

Step 3 Draw a series of ellipses to represent the threads. The chamfered end is drawn using an ellipse whose major diameter is equal to the root diameter of the threads.

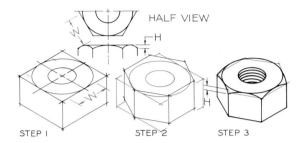

FIG. 18.53 Construction of nut

Step 1 The overall dimensions of the nut are used to block in the nut.

Step 2 The hexagonal sides are constructed at the top and bottom.

Step 3 The chamfer is drawn with an irregular curve. The internal threads are drawn to complete the isometric.

is to be threaded in isometric (Step 1). In Step 2, the major diameters of the crest lines of the thread are drawn separated at a distance of P, the pitch of the thread. In Step 3, ellipses are drawn by aligning the major diameter of the ellipse template with the perpendiculars to the cylinder's axis. Note that the 45° chamfered end is drawn using a smaller ellipse at the end.

A hexagon head nut (Fig. 18.53) is drawn in three steps using an ellipse template. The nut is blocked in and an ellipse drawn tangent to the rhombus. The hexagon is constructed by locating the distance across a flat, W, parallel to the isometric axes. The other sides of the hexagon are found in Step 2 by drawing lines tangent to the ellipse. The distance H is laid off at each corner to establish the amount of chamfer at each corner (Step 3).

A hexagon head bolt is drawn in two positions in Fig. 18.54. The washer face can be seen on the lower side of the head and the chamfer on the upper side of the bolt head.

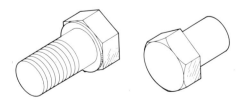

FIG. 18.54 Isometric drawings of the upper and lower sides of a hexagon-head bolt.

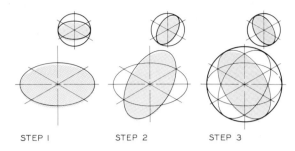

FIG. 18.55 An isometric sphere

Step 1 The three intersecting isometric axes are drawn. The ellipse template is used to draw the horizontal elliptical section.

Step 2 The isometric ellipse template is used to draw one of the vertical elliptical sections.

Step 3 The third vertical elliptical section is drawn and the center is used to draw a circle tangent to the three ellipses.

The sphere in isometric is constructed as shown in Fig. 18.55. Three ellipses are drawn as isometric planes with a common center. The center is used to construct a circle that will be tangent to each ellipse, as shown in Step 3. The sphere is larger in isometric than in a true axonometric projection.

A portion of a sphere is used to draw a round-head screw in Fig. 18.56. A hemisphere is constructed in Step 1. The centerline of the slot is located along one of the isometric planes. The thickness of the head is measured as distance of E from the highest point on the sphere.

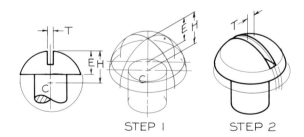

FIG. 18.56 Spherical features

Step 1 An isometric ellipse template is used to draw the elliptical features of a round-head screw.

Step 2 The slot in the head is drawn and the lines are darkened to complete the isometric of the head.

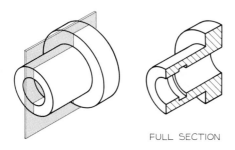

FIG. 18.57 Parts can be shown in isometric sections to clarify internal features, such as this full section.

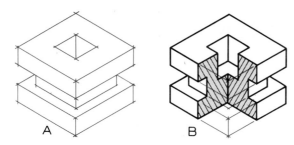

FIG. 18.58 An isometric drawing of a half-section can be constructed.

18.22
Isometric sections

A full section can be drawn in isometric to clarify internal details that might otherwise be overlooked (Fig. 18.57). Half-sections can also be used, as illustrated in Fig. 18.58. The same part is shown as a full section and a half-section in Fig. 18.59.

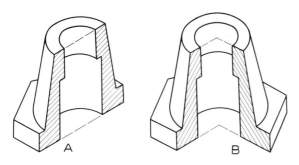

FIG. 18.59 A comparison of isometric full and half-sections of the same part

18.23
Dimensioned isometrics

When it is advantageous to dimension and note a part shown in isometric, either of the aligned or unidirectional methods illustrated in Fig. 18.60 can be used to apply the notes. In both cases, notes connected with leaders are positioned horizontally. Always use guidelines for your lettering and numerals.

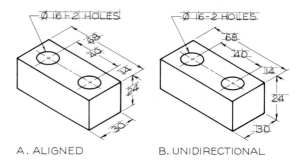

FIG. 18.60 Dimensions can be placed on isometric drawings using either of the techniques shown here. Guidelines should always be used for the lettering.

18.24
Fillets and rounds

Fillets and rounds in isometric can be represented by either of the methods shown in Fig. 18.61 to give added realism to a pictorial drawing. The examples at A and B show how intersecting guidelines are drawn equal in length to the radii of the fillets and rounds, and arcs are drawn tangent to these lines. These arcs can be drawn freehand or with an ellipse template. The method in Part C uses freehand lines drawn parallel or concentric with the directions of the fillets and rounds.

An example of these two methods of showing fillets and rounds is shown in Fig. 18.62. The stipple shading was applied by using an adhesive overlay film that can be purchased.

When fillets and rounds are illustrated as shown in Fig. 18.63 and dimensions are applied, it is much easier to understand the features of the part than

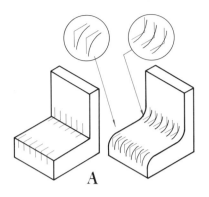
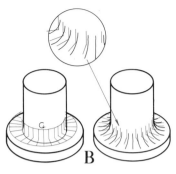

FIG. 18.61 Representation of fillets and rounds

A. Fillets and rounds can be represented by segments of an isometric ellipse if guidelines are constructed at intervals.

B. Fillets and rounds can be represented by elliptical arcs by constructing radial guidelines.

C. Fillets and rounds can be represented by lines that run parallel to the fillets and rounds.

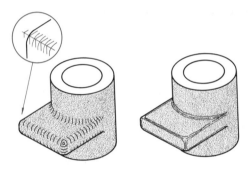

FIG. 18.62 Two methods of representing fillets and rounds on a part

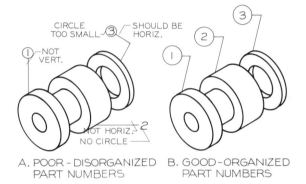

A. POOR – DISORGANIZED PART NUMBERS   B. GOOD – ORGANIZED PART NUMBERS

FIG. 18.64 Part numbers should be applied to assemblies as shown at B and by avoiding the errors made at A.

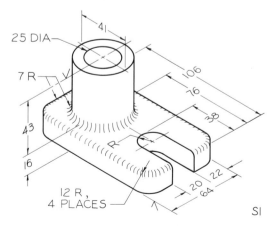

FIG. 18.63 An isometric drawing with complete dimensions and fillets and rounds represented.

when it is represented by orthographic views. The dimensions in this example were applied using the aligned method.

18.25
Isometric assemblies

Assemblies are used to explain how a series of parts is assembled. The more common mistakes in applying leaders to an assembly are shown in Fig. 18.64A. The more acceptable techniques are shown in Part B. The numbers in the circles ("balloons") refer to the number given to each part that appears in the parts list.

ELECTRIC OPERATORS for Jenkins Ball Valves
PARTS LIST

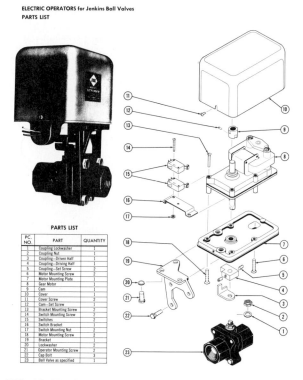

PARTS LIST

PC. NO.	PART	QUANTITY
1	Coupling Lockwasher	1
2	Coupling Nut	1
3	Coupling—Driven Half	1
4	Coupling—Driving Half	1
5	Coupling—Set Screw	2
6	Motor Mounting Screw	2
7	Motor Mounting Plate	1
8	Gear Motor	1
9	Cam	1
10	Cover	1
11	Cover Screw	2
12	Cam—Set Screw	1
13	Bracket Mounting Screw	2
14	Switch Mounting Screw	2
15	Switches	2
16	Switch Bracket	1
17	Switch Mounting Nut	2
18	Motor Mounting Screw	2
19	Bracket	1
20	Lockwasher	2
21	Operator Mounting Screw	2
22	Cap Bolt	3
23	Ball Valve as specified	1

FIG. 18.65 An industrial example of an assembly

An assembly used in a parts manual is shown in Fig. 18.65, along with its parts list. This assembly is "exploded" apart so it is more clear how they would be assembled. An exploded isometric assembly is given in Fig. 18.66.

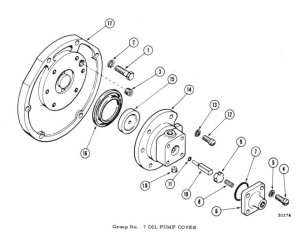

Group No. 7 OIL PUMP COVER

FIG. 18.66 An exploded isometric assembly

18.26
Piping systems in isometric

Piping layouts are often drawn as isometrics to clarify complex systems that are hard to interpret when shown in orthographic views. Some of the more often used piping connectors are shown in Fig. 18.67 as both orthographic and isometric views. Ellipse templates are used to represent circular arcs in isometric. These are single-line representations since the pipes are represented by single, heavy lines.

A portion of a pipe system is shown in Fig. 18.68 as an isometric drawing. Additional pipe symbols are given in Appendix 9 and Chapter 28.

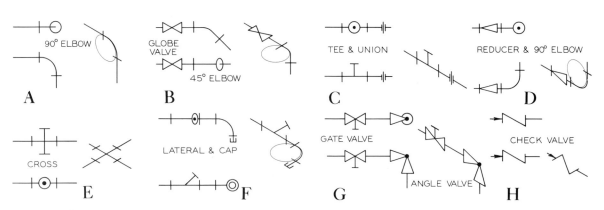

FIG. 18.67 A comparison of orthographic views and isometric pictorials of single-line representations of piping symbols. Note that the isometric template is used for constructing rounded corners in the isometric drawings.

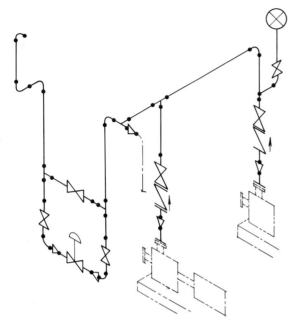

FIG. 18.68 A typical piping system drawn in isometric using the appropriate symbols.

18.27

Axonometric projection

An **axonometric projection** is a form of orthographic projection in which the pictorial view is projected perpendicularly onto the picture plane with parallel projectors. The object is positioned in an angular position with the picture plane so that its pictorial projection will be a three-dimensional view.

Three types of axonometric pictorials are possible: (1) isometric, (2) dimetric, or (3) trimetric. The iso-

metric **projection** is the view where the diagonal of a cube is viewed as point. The planes will be equally foreshortened and axes equally spaced 120° apart (Fig. 18.69A). The measurements along the three axes will be equal, but less than true length since this is true projection.

A **dimetric projection** is an axonometric projection in which two planes are equally foreshortened and two of the axes are separated by equal angles (Part B). The measurements along two of the axes are equal.

A **trimetric projection** is an axonometric projection where all three planes are unequally foreshortened and the angles between the three axes are different (Fig. 18.69C).

Figure 18.70 is a trimetric projection of a cold diffusion pump. Note that the angles between the three axes are unequal.

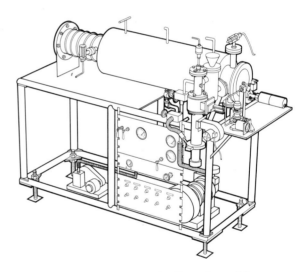

FIG. 18.70 A trimetric projection of a cold diffusion pump. (Courtesy of Aro, Inc.)

18.28

Axonometric construction

All axonometric constructions can be made in the same manner as the trimetric constructed in Fig. 18.71.

A cube should always be used instead of the object to be drawn as an axonometric. By using a cube, you can construct axonometric scales that can be used

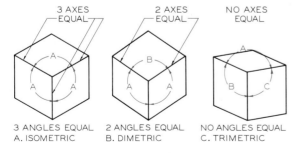

FIG. 18.69 The three types of axonometric projection

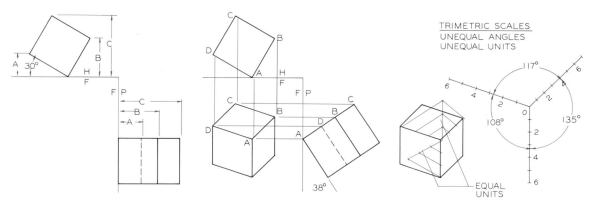

FIG. 18.71 Trimetric-scale construction

Step 1 Revolve the top view 30° clockwise. Find the side view by transferring dimensions A, B, and C from the top view. If the revolution had been 45° in the top view, the resulting projection would be either a dimetric or an isometric.

Step 2 Tilt the side view 38°. This will change the projection of the top view but not the width dimension; consequently it is unnecessary to change the top view. Determine the trimetric projection of the cube by projecting from the top and side views as in orthographic projection.

Step 3 All sides of a cube are equal; therefore divide each axis (or edge) into an equal number of units by proportional divisions as shown, even though the three axes in a trimetric have different lengths. The three axes can be extended and scaled into as many units as necessary.

to draw trimetrics of many objects, not just a single part. The trimetric scales found in Step 3 can be lengthened to accommodate any size part.

Trimetric scales are not complete unless the ellipse angles are found for each plane so that ellipse templates can be used. In a trimetric, a different ellipse angle will be used for each plane since each plane is foreshortened differently.

When a trimetric is given (Fig. 18.72) a true-size plane can be found by drawing Lines 1–2, 2–3, and 3–4 perpendicular to the three axes (Step 1). The side view of this true-size plane is found as a vertical edge in Step 2. Using these two views, the angles between the lines of sight and each plane can be found in Step 3. These angles are the ellipse angles for each surface.

The ellipses are positioned on each plane perpendicular to the axes intersecting each. This gives you a set of trimetric scales with calibrations, and the ellipse angles for each plane, Fig. 18.73A. These scales can be used to construct a trimetric by overlaying the scales with tracing vellum, as shown in Fig. 18.73B.

A second technique of constructing an axonometric is shown in Fig. 18.74. The three axes are located in a convenient position and Plane 1–2–3 is constructed with three true-length lines in Step 1. Semi-

circles are drawn with diameters equal to each of these true-length lines. Angles are inscribed inside of each to give a 90° angle at Point 0 for each semicircle (Step 2). The top, front, and side views of the object are located at Point 0 for each view, and each view is projected back where the three projectors converge at common points.

18.29
Perspectives

A perspective is a view that is normally seen by the eye or camera, and is the most realistic form of pictorial. All parallel lines converge at infinite vanishing points as they recede from the observer.

The three basic types of perspectives are: (1) one-point, (2) two-point, and (3) three-point, depending on the number of vanishing points used in their construction (Fig. 18.75).

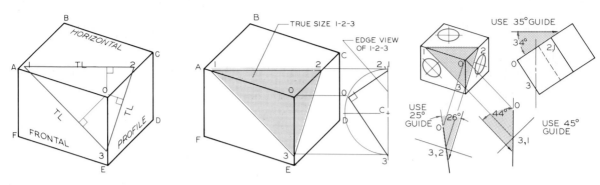

FIG. 18.72 Ellipse template angles for a trimetric scale

Step 1 Line *OA* is perpendicular to Plane *OCDE* and Line *OE* is perpendicular to Plane *ABCO*. On each plane, construct true-length lines, which are located perpendicular to the axis Lines *AO, CO,* and *EO.* The true-length lines are 1–2, 2–3, and 3–1. These lines will intersect at points on the axes forming Triangle 1–2–3.

Step 2 Since Plane 1–2–3 is composed of true-length lines, it is true size. Determine the side view of the plane, which is a frontal plane in the profile view. Find the 90° angle of the cube in the side view by constructing a semicircle, using the edge view of Plane 1–2–3 as the diameter. Project Point *O* to the semicircle where the 90° angle is inscribed.

Step 3 Determine the edge view of each principal plane by locating the point views of Lines 1–2, 2–3, and 3–1. The ellipse guide angle is the angle between the edge views of 1–0–2, 2–0–3, and 1–0–3 and the line of sight. Position the ellipse guides on each plane so that the major diameter is parallel to the true-length lines on that plane.

FIG. 18.73 The complete trimetric scales showing the units along the axes and the ellipse template angles. The object at B was drawn using the axonometric scales.

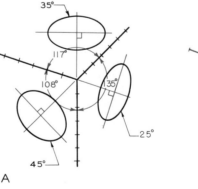

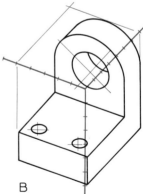

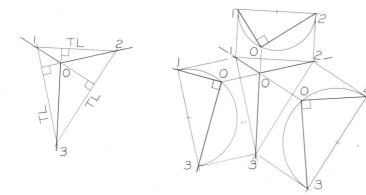

FIG. 18.74 Axonometric projection

Step 1 The three axes of an axonometric projection can be selected and drawn at the angle of your choice. True-length perpendiculars are drawn so they intersect at 1, 2, and 3.

Step 2 Semicircles are projected from each true-length line. Right angles are inscribed in the semicircle, 2–0–3 for example.

Step 3 The three orthographic views of the object are located with the same corner placed at the right angles found in Step 2. These views are projected back to intersect and form the trimetric.

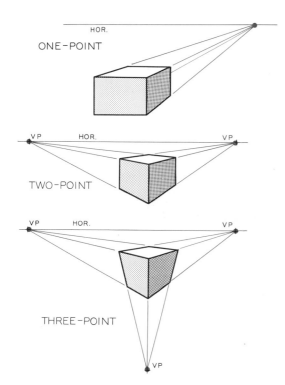

FIG. 18.75 A comparison of one-point, two point, and three-point perspectives

The **one-point perspective** has one surface of the object parallel to the picture plane; therefore it is true shape. The other sides vanish to a single point on the horizon, called a vanishing point (Part A).

A **two-point perspective** is a pictorial that is positioned with two sides at an angle to the picture plane; this requires two vanishing points (Fig. 18.75B). All horizontal lines converge at the vanishing points, but vertical lines remain vertical and have no vanishing point.

The **three-point perspective** utilizes three vanishing points since the object is positioned so that all sides of it are at an angle with the picture plane (Fig. 18.75C). The three-point perspective is used in drawing larger objects such as buildings.

18.30
One-point perspectives

The steps of drawing a one-point perspective are shown in Fig. 18.76. Here are given the top and side views of the object, the picture plane, station point (S.P.), the horizon, and the ground line.

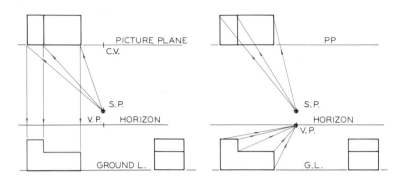

FIG. 18.76 Construction of a one-point perspective

Step 1 Since the object is parallel to the picture plane, there will be only one vanishing point, which will be located on the horizon below the station point. Projections from the top view and side view establish the front plane of the object. This surface is true size, since it lies in the picture plane.

Step 2 Draw projectors from the station point to the rear points of the object in the top view and from the front view to the vanishing point on the horizon. In a one-point perspective, the vanishing point is the front view of the station point.

Step 3 Construct vertical projectors to the front view from the points where the projectors from the station point cross the picture plane. These projectors intersect the lines leading to the vanishing point to establish the complete perspective. This is a one-point perspective.

THE PICTURE PLANE is the plane on which the perspective is projected. It appears as an edge in the top view.

THE STATION POINT is the location of the observer's eye in the plan view. The front view of the station point will always lie on the horizon.

THE HORIZON is a horizontal line in the front view that represents an infinite horizontal, such as the surface of the ocean.

THE GROUND LINE is an infinite horizontal line in the front view that passes through the base of the object being drawn.

THE CENTER OF VISION (C.V.) is a point that lies on the picture plane in the top view and on the horizon in the front view. In both cases, it is on the line from the station point that is perpendicular to the picture plane.

When drawing any perspective, the station point (SP) should be located far enough away from the object so that the perspective can be contained in a cone of vision that is not more than 30° (Fig. 18.77). If a larger cone of vision is required, the perspective will be distorted.

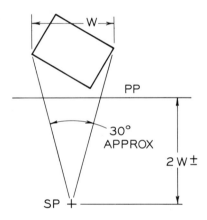

FIG. 18.77 The station point (SP) should be placed far enough away from the object to permit the cone of vision to be less than 30° to reduce distortion.

Measuring points for one-point perspectives

A measuring point is an additional vanishing point that is used to locate measurements along the receding lines that vanish to the horizon. In Fig. 18.78, the measuring point of a one-point perspective is found by revolving Line 0–2 into the picture plane to 0–2' and drawing a construction line from the station point to

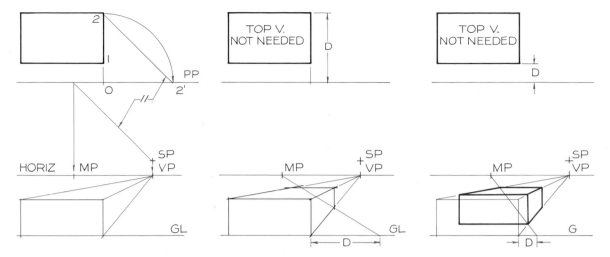

FIG. 18.78 One-point perspective—measuring points

Step 1   Line 0–2 is revolved into the picture plane to locate Point 2'. A line is drawn parallel to 2'–2 through SP to the PP. The measuring point is located on the horizon.

Step 2   Distance D is laid off along the gound line from the front corner of the perspective. This distance is projected to the measuring point to locate the rear corner of the perspective.

Step 3   The front of the object is located by laying off Distance D from the corner of the perspective and projecting to the measuring point. This locates the front surface of the object behind the picture plane.

the picture plane parallel to 2–2'. The measuring point is located on the horizon by projection from the picture plane (Step 1).

Since the distance 0–2 is equal to 0–2', depth dimensions can be laid off along the ground line and then projected to the measuring point. This locates the rear corner of the one-point perspective, Step 2. The depth from the picture plane to the front of the object is found in the same manner.

The use of the measuring-point method eliminates the need for placing the top view in the customary top-view position. Instead the dimensions can be transferred to the ground line in a more convenient manner.

18.31
Two-point perspectives

If two surfaces of an object are positioned at an angle to the picture plane, two vanishing points will be required to draw it as a perspective. Different views can be obtained by changing the relationship between the horizontal and the ground line, as illustrated in Fig. 18.79.

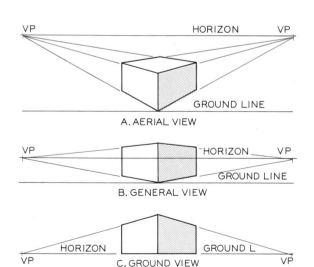

FIG. 18.79 Different perspectives can be obtained by locating the horizontal over, under, and through the object.

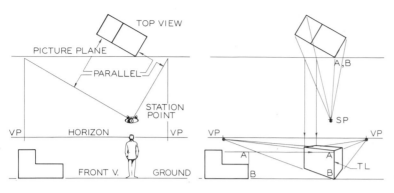

FIG. 18.80 Construction of a two-point perspective

Step 1 Construct projectors that extend from the station point to the picture plane parallel to the forward edges of the object. Project these points vertically to the horizon in the front view to locate vanishing points. Draw the ground line below the horizon and construct the known side view on the ground line.

Step 2 Line *AB* is true length. Consequently, Line *AB* is projected from the side view to determine its height. Project each end of *AB* to the vanishing points to determine two perspective planes. Draw projectors from the station point to the exterior edges of the top view. Project the intersections of these projectors with the picture plane to the front view.

Step 3 Find Point *C* in the front view by projecting from the side view to the true-length Line *AB*. Draw a projector from Point *C* to the left vanishing point. Point *D* will lie on this projector beneath the point where a projector from the station point to the top view of Point *D* crosses the picture plane. Complete the notch by projecting to the respective vanishing points.

An **aerial view** will be obtained when the horizon is placed above the ground line and the top of the object in the front view. When the ground line and the horizon coincide in the front view, a **ground-level view** will be obtained. A **general view** is one where the horizon is placed above the ground line and throught the object, usually equal to the height of a person (Part C).

The steps of constructing a two-point perspective are shown in Fig. 18.80. The horizon has been placed above the object to give a slight aerial view.

Since Line *AB* lies in the picture plane, it will be true length in the perspective. All height dimensions must originate at this vertical line because it is the only true-length line in the perspective. Points *C* and *D* are found by projecting to *AB,* and then projecting toward the vanishing points.

A typical two-point perspective is shown in Fig. 18.81. By referring to Fig. 18.80, you will be able to understand the development of the construction used.

The object in Fig. 18.82 does not contact the picture plane in the top view as in the previous examples. To draw a perspective of this object, the lines of the object must be measured where the extended plane intersects the picture plane. The height is measured,

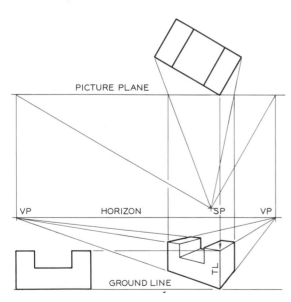

FIG. 18.81 A two-point perspective of an object

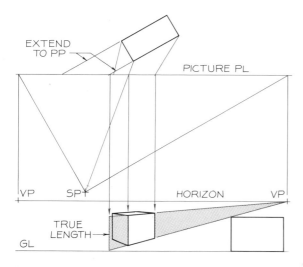

FIG. 18.82 A two-point perspective of an object that is not in contact with the picture plane.

and the infinite plane is drawn to the vanishing point.

The corner of the object can be located on this infinite plane by projecting the corner to the picture plane in the top view with a projector from the station point.

Two-point perspectives: measuring points

Measuring points are found in Fig. 18.83 to aid in the construction of two-point perspectives. The use of measuring points eliminates the need to have the top view in the top-view position, after the vanishing points have been found.

Two vanishing points are found for two-point perspectives. These are used to locate dimensions along the receding planes that vanish to the horizon, as shown in Step 2. Depth dimensions are laid off along the ground line from Point *B*, and secondary planes are passed from these points to their measuring points.

Measuring points are used in Fig. 18.84 to construct a two-point perspective. No top view is needed

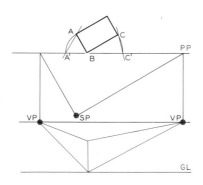
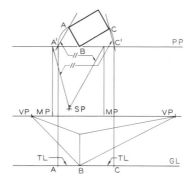
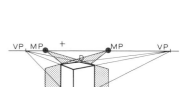

FIG. 18.83 Construction of measuring points

Step 1 In the top view, revolve Lines *AB* and *BC* into the picture plane using Point *B* as the center of revolution. Draw construction lines through Points *A–A'* and *C–C'*. These lines represent edge views of vertical planes passing through corner Points *A* and *C*.

Step 2 Draw lines from the station point parallel to Lines *A–A'* and *C–C'* to the picture plane. Project these points of intersection to the horizon to locate two measuring points. Distances *AB* and *BC* can be laid off true length on the ground line (GL), since they have been revolved into the picture plane in the top view.

Step 3 Extend planes from Points *A* and *C* on the ground line to their respective measuring points. These planes intersect the infinite planes, which are extended to the vanishing points, to locate the two corners of the block. Actual measurements can be laid off on the ground line and projected to measuring points to find corner points.

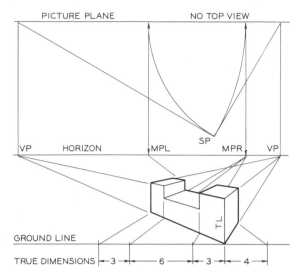

FIG. 18.84 A two-point perspective drawn with the use of measuring points

since the measuring point system is used to locate measurements along the receding lines.

Arcs in perspective

Arcs in two-point perspectives must be found by using coordinates to locate points along the curves in perspective (Fig. 18.85). Points 1 through 7 are located

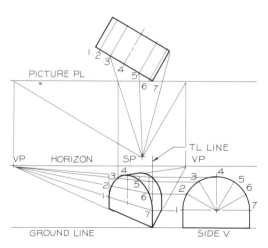

FIG. 18.85 A two-point perspective of an object with semicircular features

along the semicircular arc in the orthographic view. These same points are found in the perspective by projecting coordinates from the top and side views.

The points do not form a true ellipse, but an egg-shaped oval. An irregular curve is used to connect the points.

Sloping planes

A sloping plane that is not horizontal will not have its vanishing point on the horizon, but below or above the horizon. In Fig. 18.86, a roof slopes 25° with the horizontal in two directions. The vanishing points of these two planes are found in Step 2 by revolving AB into the picture plane to AB', and B' is projected to the horizon. From this point, two lines are drawn at 25° with the horizon to the vertical projector through the vanishing point found in Step 1. This locates the vanishing points of the two planes of the roof.

The planes of the roof are located in the perspective and are projected to their respective vanishing points.

18.32
Three-point perspectives

A three-point perspective can be constructed by positioning the object so that all of its sides make angles with the picture plane. In Fig. 18.87, the top and side views are orthographic views that have been tilted with respect to the picture plane. The station point is the same distance from the picture plane in the side view.

Three vanishing points are located by constructing lines from the two views of the station point that are parallel to the sides of object. The corner points in perspective are found by projecting to the picture plane and then to the perspective view where the lines vanish to their respective vanishing points.

In laying out a three-point perspective, you must begin by positioning the top and side views. The vanishing point for vertical lines will lie along a vertical construction line through the station point (SP).

A short-cut method of drawing a three-point perspective is shown in Fig. 18.88. An equilateral triangle with 60° angles is drawn and a convenient Point O is located within the triangle. The vertexes of the triangle are used as the three vanishing points.

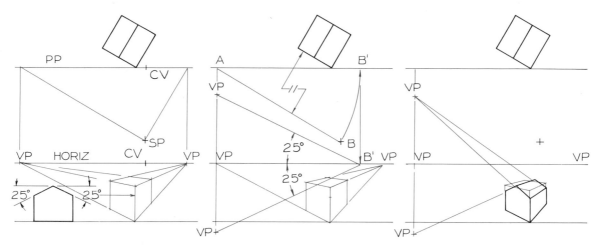

FIG. 18.86 Perspectives of sloping surfaces

Step 1 Vanishing points are located and the perspective of the object is drawn as if it had no sloping planes. The point where the sloping plane begins is found on the true-length vertical line by projecting from the side view.

Step 2 Line *AB* is rotated in the top view to locate *B'* on PP which is then projected to the horizon. Since both planes slope 25°, their vanishing points are drawn at 25° with the horizon from *B'* to a vertical line that passes through the left VP.

Step 3 Each sloping plane is found by using the two vanishing points found in Step 2 and projecting from points found on the vertical lines where the slopes begin.

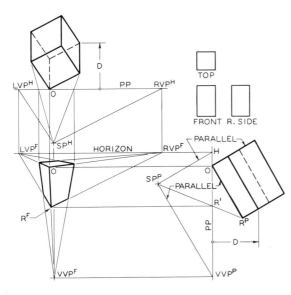

FIG. 18.87 The construction of a three-point perspective.

Width and depth are laid off on either side of Point *O*. Height is laid off along a construction line through *O* that is parallel to one on the other sides of the triangle. Height is projected back to the vertical line from *O* to the lower vanishing point.

18.33
Perspective charts

Perspective charts are time-saving grids that eliminate the need for constructing vanishing points and other tedious constructions (Fig. 18.89). The grid is drawn to scale so that it can be overlaid by tracing vellum and perspectives drawn by using the printed grid. Scales can be assigned to the grid for varying sizes of perspectives.

Perspective charts come in a variety of angles and views to suit most of your perspective needs. Perspective grids are excellent for preliminary sketches, since they give a general idea of the appearance of the finished perspective in the shortest time.

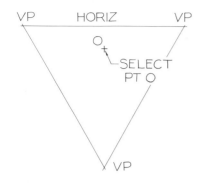

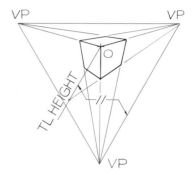

FIG. 18.88 Three-point perspective

Step 1 Draw an equilateral triangle. The vanishing points lie at each vertex. Locate Point *O* anywhere within this triangle and it will be the beginning point of the perspective.

Step 2 Draw a horizontal through Point *O*. The width and depth can be laid off from Point *O* and projectors drawn to the two upper vanishing points. This locates the top surface.

Step 3 The height is laid off along a line that is parallel to one of the triangle's sides. The height is projected back to where it intersects a line from Point *O* to the lower VP. The perspective is completed.

FIG. 18.89 Use of a perspective grid. (Courtesy of Graphicraft.)

18.34
Shades and shadows in orthographic

Shades and shadows are added to drawings to increase their realistic appearance. A surface is in shade when the light does not strike it. For example, if a block were facing the sunlight, its backside would not be in the direct rays of light and would be in **shade.**

When an object shields another surface from the sunlight, the dark area that is cast on this surface is called **shadow.**

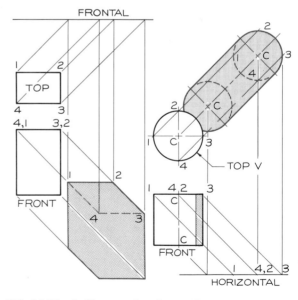

FIG. 18.90 A. The construction of the shadow of a triangular prism in orthographic on a vertical plane. B. The construction of the shadow of the cylinder in orthographic on a horizontal plane.

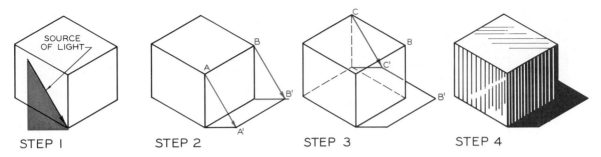

| STEP 1 | STEP 2 | STEP 3 | STEP 4 |

FIG. 18.91 Shades and shadows of a cube

Step 1 The light-source triangle is constructed for a ray of light that is parallel to the drawing paper.

Step 2 The shadows of the vertical corners passing through the Points A and B are found. The projectors from each end of the vertical corners are parallel to the sides of the light-source triangle.

Step 3 The shadow of the vertical line through Point C is found even though it is hidden. Line C'B' is drawn to complete the outline of the shadow.

Step 4 The object is shaded to indicate the surfaces in shade and the shadow.

The rectangular prism in Fig. 18.90A is drawn in orthographic projection. The rays of light are chosen to be 45° backward and downward in the two given views. The shadow of the object helps you instinctively visualize the shape of the object.

The shades and shadows of a cylinder are found in Fig. 18.90B. Points are located around the circular ends of the cylinder. These are projected to the horizontal plane, and upward to the projectors from the top view. The shadow will be an edge in the front view, but it can be seen true size in the top view. Since

the circular ends of the cylinder are parallel to the horizontal, their shadows will be true circles that are connected with tangent lines.

The area of shade is located where the ray of light is tangent to the circular view in the top view. The shadow gives the impression that the object is suspended above the the horizontal surface on which the shadow is cast.

18.35
Shades and shadows—
a block in perspective

A light ray is established in Fig. 18.91 that is parallel to the picture plane since the vertical and horizontal legs of the triangle are perpendicular and true length. The shadows of two vertical lines are found in Step 2 when the rays are passed through Points A and B. The shadows are found at A' and B' where horizontal lines from the vertical lines intersect the light rays. The process is continued to find the shadows A', B', and C', which are connected to complete the shadow and the shaded area is indicated.

These same principles are used to find the shades and shadows of a more complex object, as shown in Fig. 18.92.

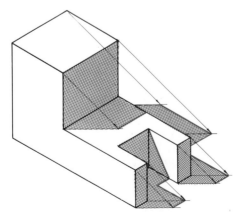

FIG. 18.92 Shades and shadows constructed for an object in isometric.

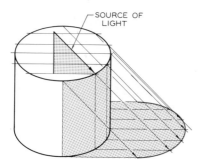

FIG. 18.93 Construction of the shades and shadows of a vertical cylinder. The shadows of each element are found one at a time. The light is parallel to the picture plane.

18.36
Shades and shadows—cylinders in isometric

The shadow of a cylinder is found by locating the shadows of a number of its elements as if they were individual lines (Fig. 18.93).

The light source is parallel to the picture plane in this example. The shadows of the points along the upper circle are connected to complete the shadow. The shaded area is located where the light source is tangent to the upper surface of the cylinder.

When a cylinder is in a horizontal position, the circular ends are blocked in and coordinates are used to project points to the horizontal surface (Fig. 18.94). This gives an elliptical shadow for each end. The area

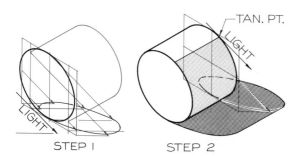

FIG. 18.94 The construction of shades and shadows of a horizontal cylinder. The light is parallel to the picture plane.

of shade is located where the ray of light is tangent to the circular end of the cylinder.

18.37
Shades and shadows—oblique light

It is unnecessary for the ray of light to be parallel to the picture plane. A triangle and source of light is established in Fig. 18.95, Step 1. The ray of light passes through the end of a vertical line at A and intersects with the projector that passes through the bottom of the line, parallel to the triangle (Step 2). This locates the shadow, A'.

Note that horizontal lines of the object and their shadows are parallel and equal in length.

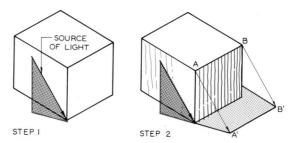

FIG. 18.95 The construction of the shadow of an isometric with a source of light that is oblique to the picture plane.

18.38
Shades and shadows—obliques

An oblique drawing can have shades and shadows applied in the same manner as has been used for isometrics. A triangle is formed to establish the ray of light in Step 1 of Fig. 18.96. The shadows of the vertical corner lines are found. Points C', B', and E' are connected to complete the outline of the shadow. Both the vertical surfaces of the block that are visible are in shade.

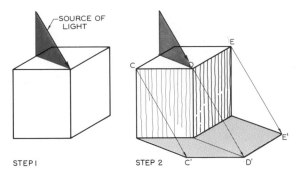

FIG. 18.96 The construction of the shadow of an oblique drawing of a cube with an angular light source.

18.39
Shades and shadows—perspectives

To find shades and shadows of perspectives, begin by establishing the ray of light. Since the coordinates of the ray of light in Fig. 18.97 intersect at 90°, the light is parallel to the picture plane.

The shadows of the vertical corners of the object are found by passing projectors through the upper corners that are parallel to the light source. They intersect on the horizontal plane where the projectors from the bottom of vertical lines are extended parallel to the horizontal leg of the light triangle.

The shadows of each upper point are found in this manner and are connected to form the outline of the shadow. Horizontal lines and their shadows converge to the same vanishing points, since they are parallel in reality.

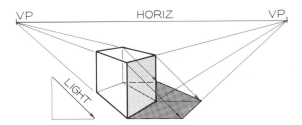

FIG. 18.97 The construction of shades and shadows of a perspective. The light source is parallel to the picture plane.

The shades and shadows of an object are shown in Fig. 18.98 where the source of light is at an angle with the picture plane. The light ray is drawn in the top and front views at selected angles. The vanishing points of horizontal and angular shadows are found on the horizontal and below the ground line.

The shadows of the vertical line project the VPL (vanishing point, light) where they intersect the shadows that converge at VPH (vanishing point, horizontal). This is repeated for each corner to complete the outline of the shadow.

18.40
Rendering techniques

Rendering or shading an object is a technique of adding realism to a pictorial. Examples of freehand and instrument rendering are shown on several basic shapes in Fig. 18.99. These illustrations were made using India ink, but a similar effect can be achieved by using black pencil lines.

18.41
Overlay film

Overlay film is an acetate film on which is printed a pattern that can be applied to a drawing to shade an illustration. It is applied as shown in Fig. 18.100. These films have adhesive backings and can be burnished permanetly to the surface with a firm rubbing pressure. It is trimmed to size using a pointed stylus or a razor blade.

Patterns, symbols, letters, numbers, and arrowheads, among other designs, are available on film. Arrowheads are applied to an assembly in Fig. 18.101. Overlay films are available with both dull and glossy surfaces.

Examples of various patterns and symbols are shown in ·Fig. 18.102. Film is also available in percentage screens that are labeled in percent and lines per inch (Fig. 18.103). The percentage indicates the percentage of solid black; the lines per inch represent the number of rows of dots per inch. A 27.5-line screen is an open screen, whereas a 55-line screen is composed of smaller dots that are twice as close together. Both have the same percentage regardless of how much they have been reduced or enlarged if they have the same percentage rating.

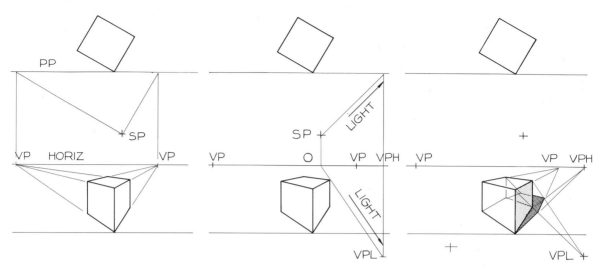

FIG. 18.98 Shades and shadows—oblique light source

Step 1 The perspective of the object is drawn in the conventional manner.

Step 2 The top and front views of the ray of light are established. The SP is projected to the horizon to locate Point *O*. The vanishing point of the light ray (VPL) is found by drawing a construction line parallel to the light ray in the front view from Point *O*.

Step 3 A shadow of a vertical line is found by locating the intersection between the projectors to VPL and VPH.

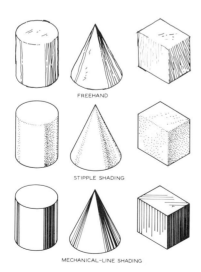

FIG. 18.99 Line and dot shading add realism to a pictorial.

FIG. 18.100 The steps of applying overlay film to shade an area. (Courtesy of Artype Inc.)

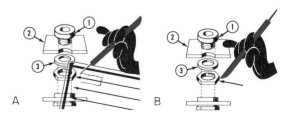

FIG. 18.101 The steps of applying leaders and numbers to an assembly drawing. (Courtesy of Artype Inc.)

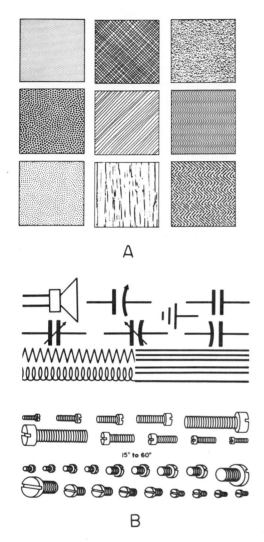

A

B

FIG. 18.102 Examples of a few of the (A) patterns and (B) symbols that are available in 9″ × 12″ sheets of overlay film. (Courtesy of Artype Inc.)

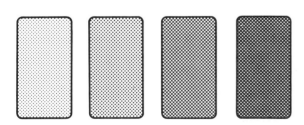

FIG. 18.103 Overlay film is available in screens at 10-percent increments. A 10-percent screen is 10 percent solid black, and a 70-percent screen is 70 percent solid black.

18.42
Photographic illustrations

Technical illustrations can be made from photographs as shown in Fig. 18.104. Drawings can emphasize certain aspects of the photograph that might not be otherwise clear. The photograph can be converted to a drawing by projecting it onto the drawing surface by

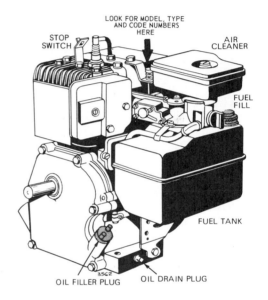

FIG. 18.104 An example of a line drawing made from a photograph to improve its reproduction when reduced to a small size. (Courtesy of Briggs & Stratton Corp.)

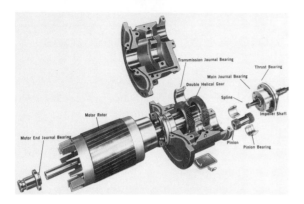

FIG. 18.105 A photographic assembly that has been retouched with an airbrush. (Courtesy of Carrier Air Conditioning Co.)

Photographs are usually retouched with an airbrush to improve their quality. An airbrush is a small spray gun that applies ink or paint to the drawing surface to give soft tones of gray.

An airbrush is shown in Fig. 18.106 as it is used to shade a drawing. The illustrator who uses an airbrush must use numerous masks to shield the portion of the drawing surface where shading is unwanted. The masks are called **friskets.** They are applied to the surface and windows are cut from them to expose the areas to be sprayed.

an opaque projector, where it can be outlined and then rendered.

Photographs of a large size can be overlaid with tracing vellum or polyester film and the drawing made by tracing from the photograph. A bold illustration is required if it is to be reduced to a much smaller size for publication.

Photographs can be used to illustrate parts such as the assembly in Fig. 18.105. These parts were positioned and photographed and the background was masked (whited out) to eliminate the props that were used to support the parts while it was being photographed.

FIG. 18.106 The airbrush is a device that allows the illustrator to apply a light spray of ink or paint on an illustration to obtain gradual tones.

Problems

The following problems are to be solved on Size A or B sheets as assigned by your teacher. Select the appropriate scale that will best take advantage of the space available on each sheet.

Obliques

1–31. Construct either cavalier, cabinet, or general obliques of the objects assigned.

Isometrics

1–31. Construct isometric drawings of the objects assigned.

Axonometrics

1–31. Construct axonometric scales and find the ellipse template angles for each surface by using a cube rotated into positions of your choice. Calibrate the scales to represent one-half in. intervals. Overlay the scales and draw axonometrics of the objects assigned.

Perspectives

32–39. Lay out these perspective problems on Size B sheets and complete the perspectives as assigned.
40–44. Construct three-point perspectives of Parts 1 through 5 as assigned. Select the most appropriate scale to take best advantage of the space available.

Shades and shadows

45–57. Construct pictorials (any type as assigned) of Parts 1–13 and find the shades and shadows of each. Establish the source of light that will best enhance the pictorial.

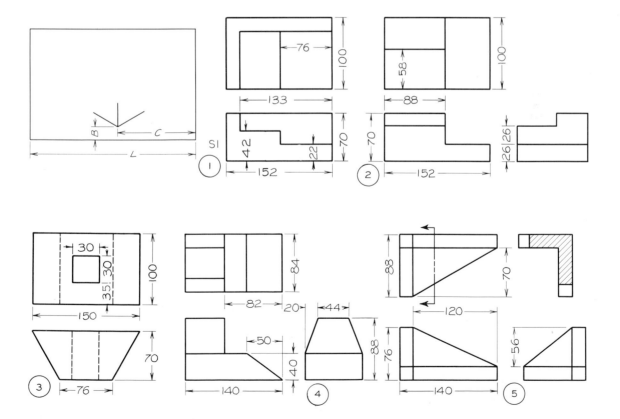

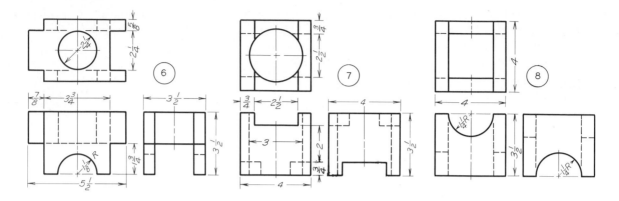

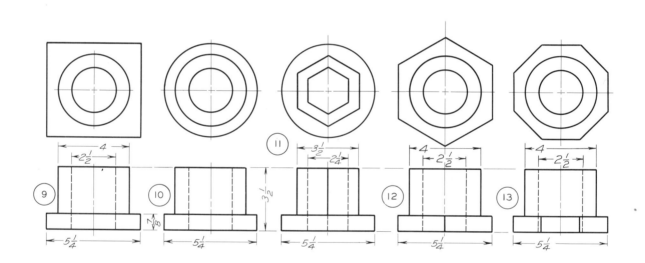

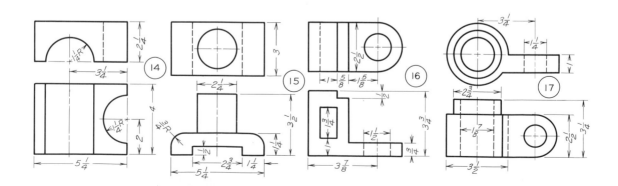

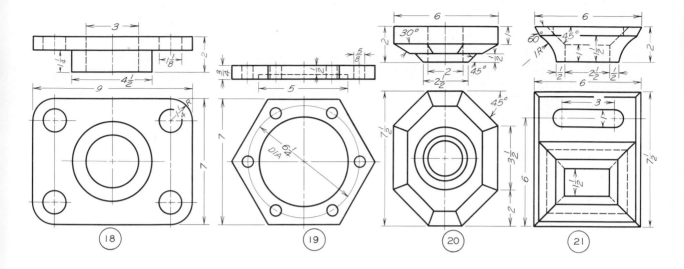

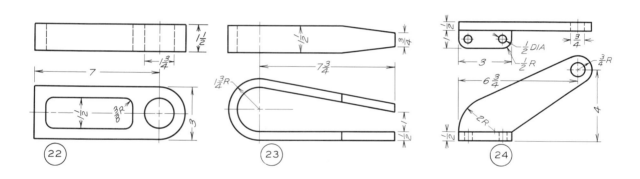

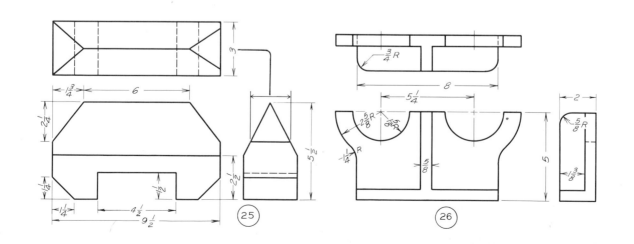

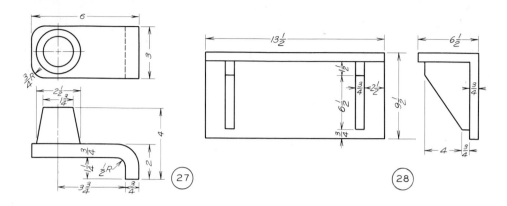

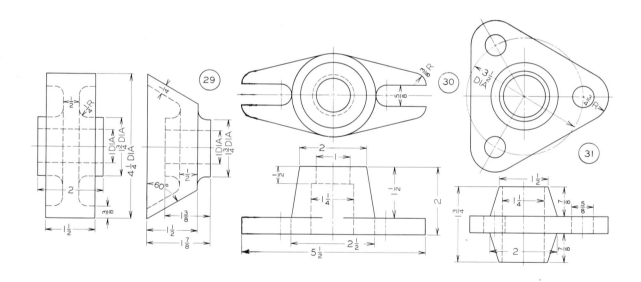

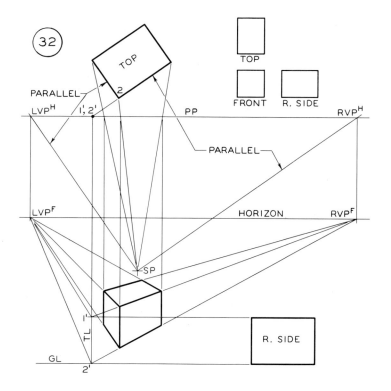

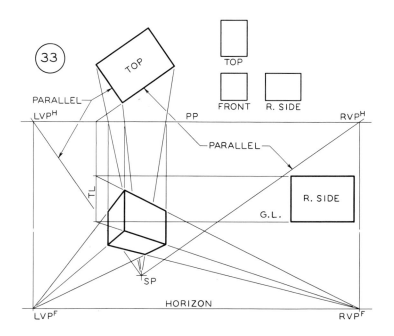

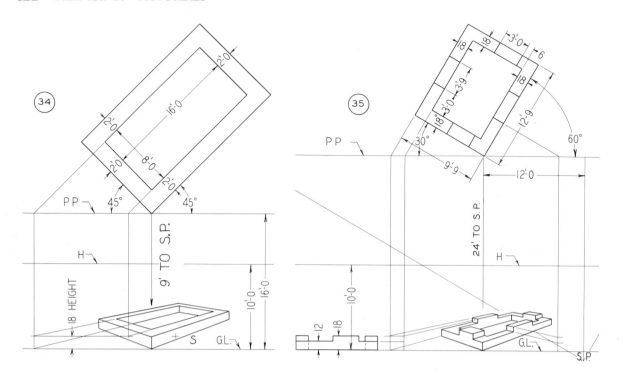

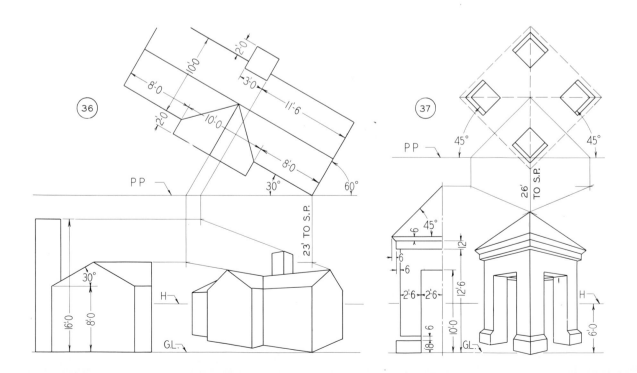

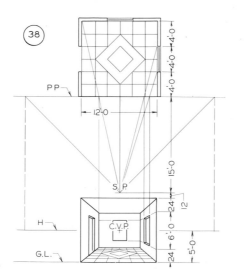

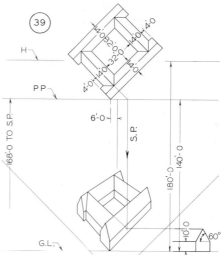

19

Points, Lines, and Planes

19.1

Introduction

Points, lines, and planes are the basic geometric elements that are used extensively in descriptive geometry, which is the discipline that deals with graphically describing three-dimensional geometry.

The following rules of solving and labeling descriptive problems are illustrated in Fig. 19.1.

LETTERING All points, lines, and planes should be labeled using ⅛ in. letters with guidelines. Lines should be labeled at each end and planes at each corner. Either letters or numbers can be used.

POINTS Points in space should be indicated by two short perpendicular dashes that form a cross, *not* a dot. Each dash should be approximately ⅛ in. long.

POINTS ON A LINE Points on a line should be indicated with a short perpendicular dash on the line, *not* a dot. Label the point with a letter or numeral.

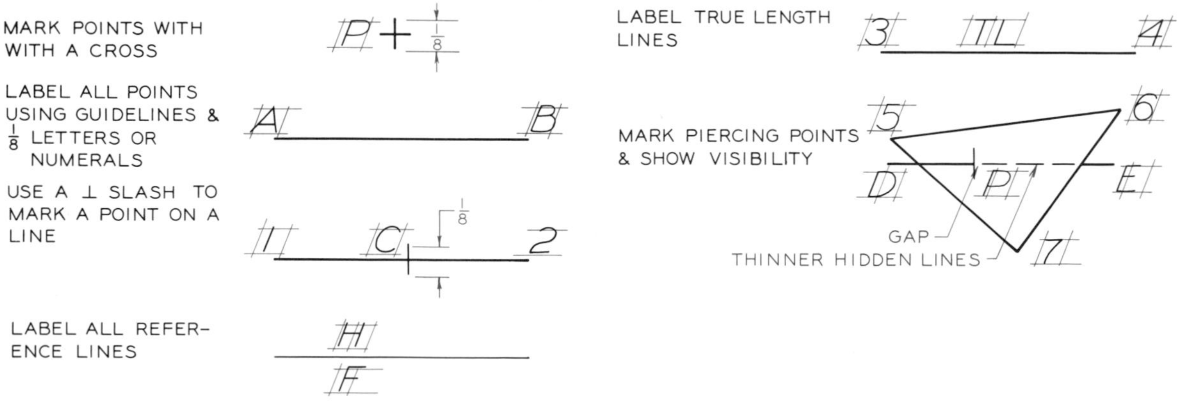

MARK POINTS WITH
WITH A CROSS

LABEL ALL POINTS
USING GUIDELINES &
⅛ LETTERS OR
NUMERALS

USE A ⊥ SLASH TO
MARK A POINT ON A
LINE

LABEL ALL REFER-
ENCE LINES

LABEL TRUE LENGTH
LINES

MARK PIERCING POINTS
& SHOW VISIBILITY

GAP
THINNER HIDDEN LINES

FIG. 19.1 Standard practices for labeling points, lines, and planes

424

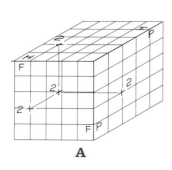
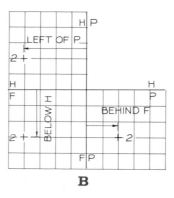
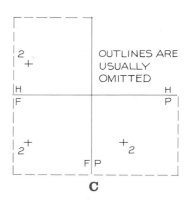

A   **B**   **C**

FIG. 19.2 The three projections of Point 2 are shown pictorially at A and orthographically at B where the projection planes are opened into the plane of the drawing paper. Point 2 is 4 units to the left of the profile, 3 below the horizontal, and 2 behind the frontal. The outlines of the projection planes are usually omitted in orthographic projection as shown at C.

REFERENCE LINES These are thin black lines that should be labeled in accordance with the text in Chapter 5.

OBJECT LINES Object lines are used to represent points, lines, and planes. They should be drawn heavier than reference lines, with an H or F pencil. Hidden lines are drawn thinner than visible lines.

TRUE-LENGTH LINES These should be labeled by the full note, TRUE LENGTH, or by the abbreviation TL.

TRUE-SIZE PLANES True-size planes should be labeled by a note, TRUE SIZE, or by the abbreviation TS.

PROJECTION LINES Projection lines that are used in constructing the solution to a problem should be precisely drawn with a 4H pencil. These should be thin gray lines, just dark enough to be visible. They need not be erased after the problem is completed.

A point must be projected perpendicularly onto at least two principal planes to establish its position (Fig. 19.2). Note that when the planes of the projection box at (a) are opened into the plane of the drawing surface in (b), the projectors from each view of Point 2 are perpendicular to the reference lines between the planes. The letters *H, F,* and *P* are used to represent the horizontal, frontal, and profile planes, respectively, which are the three principal projection planes.

A point can be located from verbal descriptions with respect to the principal planes. For example, Point 2 in Fig. 19.2 can be described as being (1) 4 units left of the profile plane, (2) 3 units below the horizontal plane, and (3) 2 units behind the frontal plane.

When looking at the front view, the horizontal and profile planes appear as edges. The frontal and profile planes appear as edges in the top view, and the frontal and horizontal planes appear as edges in the side view.

19.2
Orthographic projection of a point

A point is a theoretical location in space and it has no dimension. However, a series of points can establish areas, volumes, and lengths.

19.3
Lines

A line is a straight path between two points in space. A line can appear in three forms: (1) as a foreshortened line, (2) as a true-length line, or (3) as a point (Fig. 19.3).

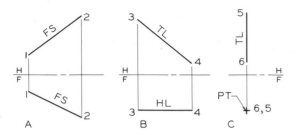

FIG. 19.3 A line in orthographic projection can appear as a point (PT), foreshortened (FS), or true length (TL).

OBLIQUE LINES Oblique lines are lines that are neither parallel nor perpendicular to a principal projection plane, as shown in Fig. 19.4. When Line 1–2 is projected onto the horizontal, frontal, and profile planes, it appears foreshortened in each view. This is the general case of a line.

PRINCIPAL LINES Principal lines are lines that are parallel to at least one of the principal projection planes.

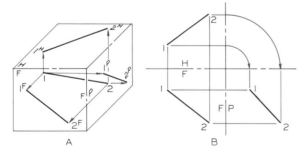

FIG. 19.4 A pictorial of the orthographic projection of a line is shown at A, and as a standard orthographic projection at B.

The three principal lines are (1) **horizontal,** (2) **frontal,** and (3) **profile** lines, since these are the three principal projection planes.

A principal line is true length in the view where the principal plane to which it is parallel appears true size.

A horizontal line is shown in Fig. 19.5A where it appears true size in the horizontal view, the top view. It may be shown in an infinite number of positions in

the top view and still appear true length provided it is parallel to the horizontal plane.

An observer cannot tell whether the line is horizontal or not when looking at the top view. This must be determined from looking at the front or side views. A horizontal line will be parallel to the edge view of the horizontal in the front and side views, which is the H-F fold line. A line that projects as a point in the front view is a combination horizontal and profile line.

A frontal line is parallel to the frontal projection plane and it appears true length in the front view since the observer's line of sight is perpendicular to it in this view. Line 3–4 in Fig. 19.5B is determined to be a frontal line by observing its top and side view where the line appears parallel to the edge view of the frontal plane.

Profile lines are parallel to the profile projection planes and they appear true length in the side views, the profile views. It is necessary to look at a view adjacent to the profile view to tell whether or not a line is a profile line. In Fig. 19.5C Line 5–6 is parallel to the edge view of the profile plane in the top and side views.

19.4
Location of a point on a line

The top and front views of line 1–2 are shown in Fig. 19.6. Point O is located on the line in the top view and it is required that the front view of the point be found.

Since the projectors between the views are perpendicular to the H-F fold line in orthographic projection, Point O is found by projecting in this same direction from the top view to the front view of the line.

If a point is to be located at the midpoint of a line, it will be at the line's midpoint whether the line appears true length or foreshortened.

19.5
Intersecting and nonintersecting lines

Lines that intersect have a point of intersection that lies on both lines and is common to both. Point O in

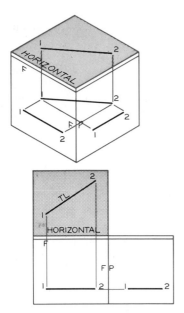

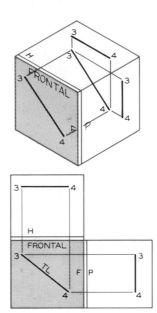

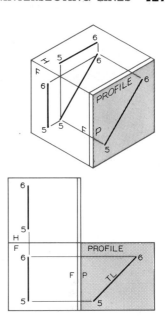

FIG. 19.5 Principal lines

A. The **horizontal line** is true length in the horizontal view (the top view). It will appear parallel to the edge view of the horizontal plane in the front and side views.

B. The **frontal line** is true length in the front view. It will appear parallel to the edge view of the frontal plane in the top and side views.

C. The **profile line** is true length in the profile view (the side view). It will appear parallel to the edge view of the profile plane in the top and front views.

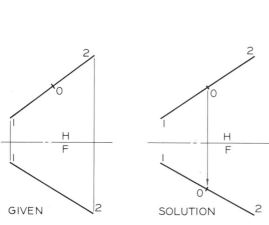

GIVEN

SOLUTION

FIG. 19.6 A point on a line that is shown orthographically can be found on the front view by projection. The direction of the projection is perpendicular to the reference line between the two views.

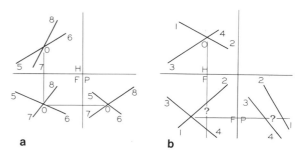

a

b

FIG 19.7 The lines at (b) cross in the top and front views, but they do not intersect. The common point of intersection does not project from view to view. The lines at (a) do intersect because the point of intersection O projects as a point of intersection in all views.

Fig. 19.7A is a point of intersection since it projects to a common crossing point in the three views given.

On the other hand, the crossing point of the lines in Fig. 19.7B in the front view is not a point of intersection. Point O does not project to a common crossing

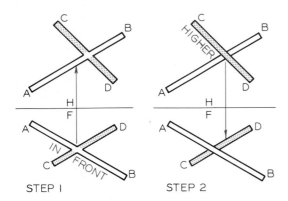

FIG 19.8 Visibility of lines

Given The top and front views of lines *AB* and *CD*.

Required Find the visibility of the lines in both views.

Step 1 Project the point of crossing from the front to the top view. This projector strikes *AB* before *CD*; therefore, line *AB* is in front and is visible in the front view.

Step 2 Project the point of crossing from the top view to the front view. This projector strikes *CD* before *AB*; therefore, line *CD* is above *AB* and is visible in the top view.

point in the top view; Point *O* is not aligned with the projector. Therefore, the lines do not intersect although they do cross.

19.6
Visibility of crossing lines

Lines *AB* and *CD* in Fig. 19.8 do not intersect, however; it is necessary to determine the visibility of the lines by analysis.

Select a crossing point in one of the views, the front view in step 1, and project it to the top view to determine which line is in front of the other.

This process of determining visibility is done by analysis rather than visualization. Two views must be utilized since it would be impossible if only one view were available.

19.7
Visiblility of a line and a plane

The principle of visibility of intersecting lines is used in determining the visibility for a line and a plane. In step 1 of Fig. 19.9, the intersections of *AB* and Lines 1–3 and 2–3 are projected to the top view to determine that the lines of the plane are in front of *AB* in the front view. Consequently, the line is shown as a dashed line in the front.

Similarly, the two intersections on *AB* in the top view are projected to the front view where line *AB* is found to be above the two lines of the plane, 2–3 and 1–3. Therefore, *AB* is drawn as visible in the top view since it is above the plane.

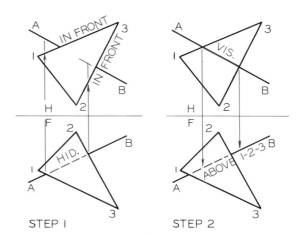

FIG. 19.9 Visibility of line and a plane

Given The front and top views of plane 1–2–3 and line *AB*.

Required Find the visibility of the plane and the line in both views.

Step 1 Project the points where *AB* crosses the plane in the front view to the top view. These projectors encounter lines 1–3 and 2–3 of the plane first; therefore, the plane is in front of the line, making the line invisible in the front view.

Step 2 Project the points where *AB* crosses the plane in the top view to the front view. These projectors encounter line *AB* first; therefore, the line is higher than the plane, and the line is visible in the top view.

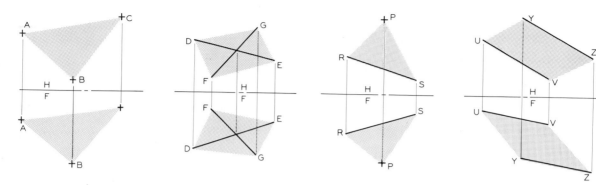

FIG. 19.10 Representations of a plane

A. Three points not in a straight line can be used to represent a plane.

B. Two intersecting lines can be used to represent a plane.

C. A line and a point not on the line or its extension can be used to represent a plane.

D. Two parallel lines can be used to represent a plane.

19.8
Planes

A plane can be represented by any of the four methods shown in Fig. 19.10. Planes in orthographic projection can appear in one of the forms shown in Fig. 19.11: (1) as an edge, (2) as true size, (C) as foreshortened.

OBLIQUE PLANES These are planes that are not parallel to principal projection planes in any view, as shown in Fig. 19.12.

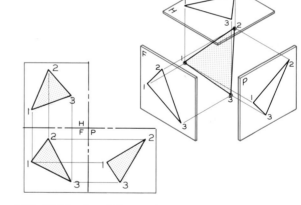

FIG. 19.12 An oblique plane is one that is not parallel or perpendicular to a projection plane; it can be called a general-case plane.

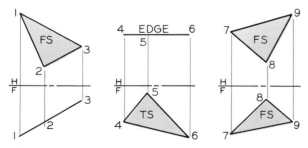

FIG. 19.11 A plane in orthographic projection can appear as an edge, true size (TS), or foreshortened (FS). If a plane is foreshortened in all principal views, it is an oblique plane.

PRINCIPAL PLANES Principal planes are parallel to the projection planes as shown in Fig. 19.13. The three types of principal planes are: **frontal, horizontal,** and **profile** planes.

A frontal plane is parallel to the frontal projection plane and it appears true size in the front view. To tell that the plane is frontal, you must observe the top or side views where its parallelism to the edge view of the frontal plane can be seen.

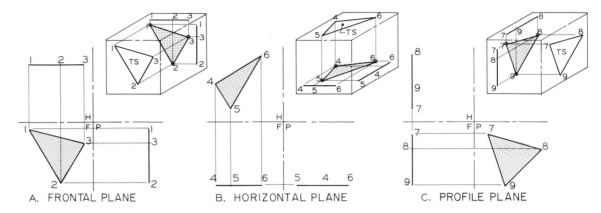

FIG. 19.13 Principal planes

A. Frontal planes are true size and shape (TS) in the front view. They will appear as edges parallel to the frontal plane in the top and side views.

B. Horizontal planes are true size in the horizontal views (top views). They appear as edges parallel to the horizontal plane in the front and side views.

C. Profile planes are true size in the profile views (side views). They appear as edges parallel to the profile plane in the top and front views.

A horizontal plane is parallel to the horizontal projection plane and it is true size in the top view. To tell that the plane is horizontal you must observe the front or side views where its parallelism to the edge view of the horizontal plane can be seen.

A profile plane is parallel to the profile projection plane and it is true size in the side view. To tell that the plane is a profile plane, you must observe the top or front views where its parallelism to the edge view of the profile plane can be seen.

19.9
A line on a plane

Line *AB* is given on the front view of the plane in Fig. 19.14. It is required to find the top view of the line. Points *A* and *B* that lie on Lines 1–4 and 2–3 of the plane can be projected to the top view to the same lines of the plane.

Points *A* and *B* are found in the top view and are connected to complete the top view of Line *AB*. This is an extension of the principle covered in Section 19.4.

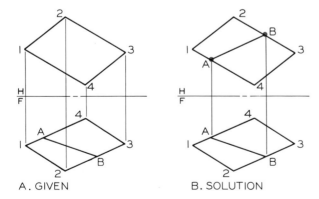

FIG. 19.14 If Line *AB* lying on the plane is given, the top view of the line can be found. Points *A* and *B* are projected to Lines 1–4 and 2–3, respectively, and are connected to form line *AB*.

19.10
A point on a plane

Point *O* is given on the front view of plane 4–5–6 in Fig. 19.15. It is required to locate the point on the plane in the top view.

In Step 1, a line in any direction other than vertical is drawn through the point to establish a line on the plane. The line is projected to the top view in Step 2, and the point is projected from the front view to the top view of the line.

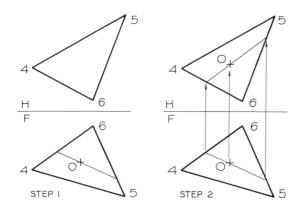

FIG. 19.15 Location of a point on a plane

Find the top view of Point O that lies on the plane.

Step 1 Draw a line through the given view of Point O in any convenient direction except vertical.

Step 2 Project the ends of the line to the top view and draw the line. Point O is projected to the line.

19.11
Principal lines on a plane

Principal lines—**horizontal, frontal,** and **profile**—may be found in any view of a plane when at least two views of the plane are given.

Horizontal lines are drawn in the front view of the plane in Fig. 19.16A that are parallel to the edge view of the horizontal projection plane. These horizontal lines are projected to the top view of the plane where they will be horizontal and true length.

Frontal lines are drawn parallel to the frontal projection plane in the top view in Fig. 19.16B. When projected to the front view, the lines are true length.

Profile lines are drawn parallel to the profile projection plane in the top and front views in Fig. 19.16C. When projected to the side view, the lines will appear true length.

An infinite number of principal lines can be drawn on a single plane.

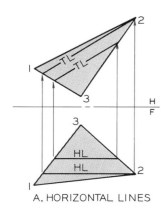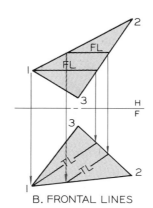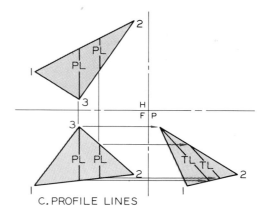

A. HORIZONTAL LINES   B. FRONTAL LINES   C. PROFILE LINES

FIG. 19.16 Principal lines on a plane

Construct the three principle lines on the views of the plane.

A. Horizontal lines are drawn first in the front view parallel to the edge view of the horizontal plane. These lines are found true length when projected to the top view.

B. Frontal lines are drawn first in the top view parallel to the edge view of the frontal plane. These lines are found true length when projected to the front view.

C. Profile lines are drawn first in the front view parallel to the edge view of the profile plane. These lines are found true length when projected to the profile view (side view).

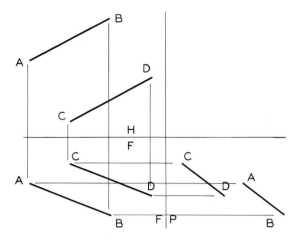

FIG. 19.17 When two lines are parallel, they will project as parallel in all orthographic views.

19.12
Parallelism of lines

If two lines are parallel, they will appear parallel in all views in which they are seen, except where both appear as points. Lines *AB* and *CD* appear parallel in three views in Fig. 19.17. Parallelism of lines in space cannot be determined if only one view is given; two or more views are required.

Using this principle, a line can be drawn parallel to a given line through a specified point as shown in Fig. 19.18.

19.13
Parallelism of a line and a plane

> A line is parallel to a plane if it is parallel to any line in the plane.

In Fig. 19.19 it is required that a line with its midpoint at Point *O* be drawn that is parallel to Plane 1–2–3. This is done by drawing Line *AB* parallel to a line in the plane, Line 1–3 in this case, in the top and front views.

The line could have been drawn parallel to *any* line in the plane; therefore, there are an infinite number of positions for lines that are parallel to a given plane.

A similar example is shown in Fig. 19.20 where it is required to draw a line with its midpoint at *O* that is parallel to the plane formed by two intersecting lines.

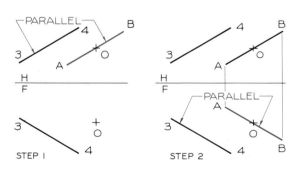

FIG. 19.18 A line parallel to a line

Draw a line through *O* that is parallel to the given line.

Step 1   Draw Line *AB* parallel to the top view of line 3–4 with its midpoint at *O*.

Step 2   Draw the front view of Line *AB* parallel to the front view of 3–4 through Point *O*.

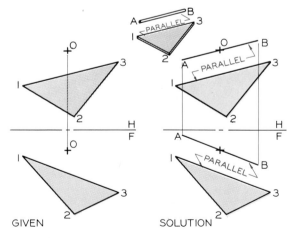

FIG. 19.19 A line can be drawn through Point *O* that is parallel to the given plane if the line is parallel to any line in the plane. Line *AB* is drawn parallel to Line 1–3 in the front and top views making it parallel to the plane.

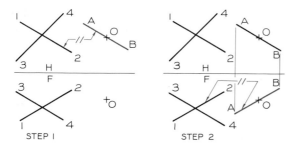

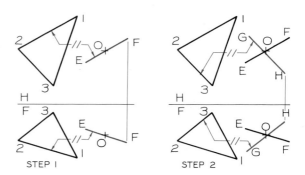

FIG. 19.20 A line parallel to plane

Draw a line parallel to the plane represented by two intersecting lines.

Step 1   Line *AB* is drawn parallel to Line 1–2 through Point *O*.

Step 2   Line *AB* is drawn parallel to the same line, line 1–2, in the front view which makes *AB* parallel to the plane.

19.14
Parallelism of planes

> Two planes are parallel when intersecting lines in one plane are parallel to intersecting lines in the other, as shown in Fig. 19.21.

It is easy to determine that planes are parallel when both appear as parallel edges in a view.

It is required that a line be drawn through Point *O* and parallel to Plane 1–2–3 in Fig. 19.22. In

FIG. 19.22   A plane through a point parallel to a plane.

Draw a plane through Point *O* that is parallel to the given plane.

Step 1   Draw Line *EF* parallel to any line in the plane, Line 1–2 in this case. Show the line in both views.

Step 2   Draw a second line parallel to Line 2–3 in the top and front views. These two intersecting lines represent a plane parallel to 1–2–3.

Step 1, *EF* is drawn through Point *O* and parallel to Line 1–2 in both the top and front views. In Step 2, a second line is drawn through Point *O* parallel to Line 2–3 of the plane in the front and top views. These two intersecting lines form a plane that is parallel to Plane 1–2–3.

19.15
Perpendicularity of lines

> When two lines are perpendicular, they will project at true 90° angles when one or both are true length (Fig. 19.23).

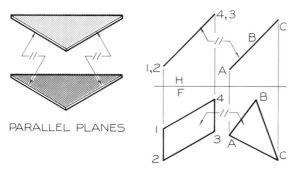

FIG. 19.21   Two planes are parallel when intersecting lines in one are parallel to intersecting lines in the other. When parallel planes appear as edges, the edges will be parallel.

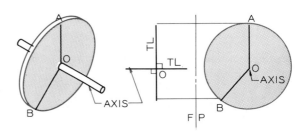

FIG. 19.23   Perpendicular lines will intersect at 90° angles in a view where one or both of the lines appear true length.

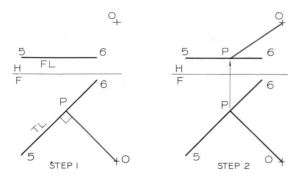

FIG. 19.24 A line perpendicular to a frontal line

Draw a line from Point *O* perpendicular to Line 5–6.

Step 1 Since Line 5–6 is a frontal line and is true length in the front view, a perpendicular from Point *O* will make a true 90° angle with it.

Step 2 Project Point *P* to the top view and connect it to Point *O*. Since neither of the lines is true length they will not intersect at 90° in the top view.

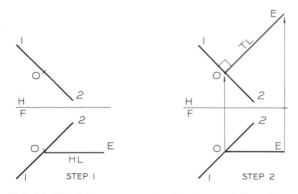

FIG. 19.25 A line perpendicular to an oblique line

Draw a line from Point *O* that is perpendicular to the given line, 1–2.

Step 1 Draw a horizontal line from Point *O* in the front view.

Step 2 Horizontal Line *OE* will be true length in the top view; therefore, it can be drawn perpendicular to Line 1–2 in this view.

It can be seen that the axis is true length in the front view; therefore, any spoke will be shown perpendicular to the axis in the front view. Spokes *OA* and *OB* are examples where one is true length and the other is foreshortened.

When two lines are perpendicular but neither is true length, they will not project with a true 90° angle.

19.16
A line perpendicular to a principal line

It is required in Fig. 19.24 to construct a line through Point *O* that is perpendicular to frontal Line 5–6 that is true length in the front view. In Step 1, *OP* is drawn perpendicular to 5–6 since it is true length. In Step 2, point *P* is projected to the top view of 5–6. Line *OP* in the top view cannot be drawn as a true 90° angle since neither of the lines is true length in this view.

19.17
A line perpendicular to an oblique line

It is required in Fig. 19.25 to construct a line from Point *O* that is perpendicular to oblique Line 1–2.

A line could have been found perpendicular to 1–2 in the front view by drawing a frontal line in the top view.

19.18
Perpendicularity involving planes

A line can be drawn perpendicular to a plane if it is drawn perpendicular to any two intersecting lines in the plane, as shown in Fig. 19.26A. Also, a plane is perpendicular to another plane if a line in one is perpendicular to the other. This is illustrated in Fig. 19.26B.

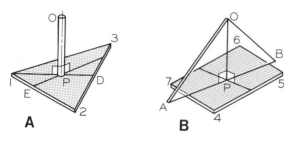

FIG. 19.26 (A) A line is perpendicular to a plane if it is perpendicular to two intersecting lines on the plane. (B) A plane is perpendicular to another plane if it contains a line that is perpendicular to the other plane.

19.19

A line perpendicular to a plane

It is required in Fig. 19.27 that a line be drawn from Point O on the plane perpendicular to the plane.

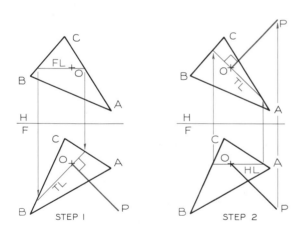

FIG. 19.27 A line perpendicular to a plane

Draw a line from Point O that is perpendicular to the plane.

Step 1 Construct a frontal line on the plane through O in the top view. This line is true length in the front view; therefore, Line OP can be drawn perpendicular to the true-length line.

Step 2 Construct a horizontal line through Point O in the front view. This line is true length in the top view; therefore, Line OP can be drawn perpendicular to it.

In Step 1, a frontal line is drawn on the plane in the top view and is projected to the front view where the line is true length. Line OP is drawn perpendicular to the true-length line.

In Step 2, a horizontal line is drawn in the front view and then in the top view of the plane through Point O perpendicular to the true-length line. This results in a line perpendicular to the plane since the line is perpendicular to two intersecting lines in the plane, a horizontal and a frontal line.

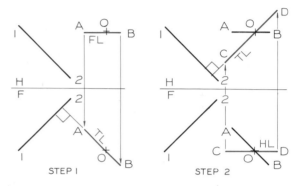

FIG. 19.28 A plane through a point perpendicular to a line

Draw a plane through O that is perpendicular to Line 1–2.

Step 1 Draw Line AB as a frontal line in the top view. Since it will be true length in the front view, it can be drawn perpendicular to Line 1–2.

Step 2 Draw Line CD as a horizontal line in the front view. Since it will be true length in the top view, it can be drawn perpendicular to Line 1–2. The intersecting lines form a plane that is perpendicular to the line.

19.20

A plane perpendicular to an oblique line

It is required in Fig. 19.28 to construct a plane through Point O that is perpendicular to Line 1–2.

In Step 1, AB is drawn as a frontal line in the top view. Since it will be true length in the front view, AB is drawn perpendicular to 1–2 in the front view.

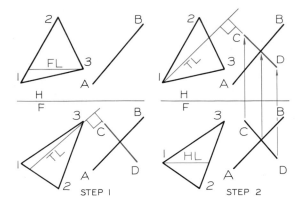

FIG. 19.29 A plane through a line perpendicular to a plane

Draw a plane through the line that is perpendicular to the plane.

Step 1 Draw a frontal line on the plane in the top view and find its true-length view in the front view. Line *CD* is drawn through Line *AB* and perpendicular to the true-length line.

Step 2 Draw a horizontal line on the plane in the front view and find its true-length view in the top view. Draw Line *CD* perpendicular to the true-length line and through Line *AB* at the point of intersection projected from the front view.

In Step 2, *CD* is drawn as a horizontal line in the front view. Line *CD* is drawn perpendicular to 1–2 in the top view since *CD* is true length in this view. These two intersecting lines, *AB* and *CD,* form a plane that is perpendicular to Line 1–2.

19.21
Perpendicularity of planes

In Fig. 19.29 it is required that a plane be constructed through Line *AB* that is perpendicular to Plane 1–2–3.

In Step 1, a true-length frontal line is found on the plane. Line *CD* is drawn through *AB* and perpendicular to the frontal line in the plane.

In Step 2, the intersection point between *AB* and *CD* is projected to the top view. A true-length horizontal line is found on the plane. The top view of *CD* is drawn through the point of intersection and perpendicular to the true-length line in the plane. These two intersecting lines, *AB* and *CD,* form a plane that is perpendicular to the plane since a single line in the plane, *CD,* is perpendicular to the plane.

Problems

Use Size A (8½″ × 11″) sheets for the following problems, and lay out the problems using instruments. Each square on the grid is equal to 0.20 in. or about 5 mm. The problems can be laid out on grid paper or plain paper. Label all reference planes and points in each problem with ⅛ in. letters or numbers, using guidelines.

1. (Fig. 19.30) Find the missing third views of the given points in Problems 1A through 1D. Find the missing third views of the lines between the given points in Problems 1E and 1F.

2. (Fig. 19.31) (2A) Find the front and top views of the profile line. (2B) Find the front and side views of the horizontal line. (2C) Find the side view of 5–6. (2D) Find the front and side views of 7–8.

(2E and 2F) Find the top and side views of the given lines.

3. (Fig. 19.32) (3A) Find the side view of the line. (3B) Find the front and side views of the horizontal line. (3C) Draw three views of a line from Point 2 to Point *O* on the line. (3D) Find the side view of the line and locate the midpoint of the line in each view. (3E and 3F) Find the piercing points between the lines and planes, show visibility, and complete the missing third view.

4. (Fig. 19.33) (4A) Find the front and side views of the horizontal plane. (4B) Draw the top and side views of the frontal plane. (4C) Draw the front and top views of the profile plane. (4D) Draw the top view of the plane. (4E) Find the side view of

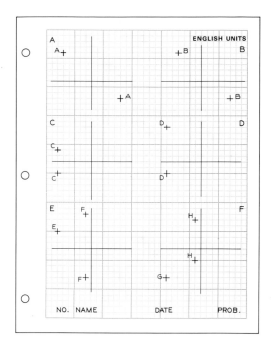

FIG. 19.30 Problems 1A through 1F.

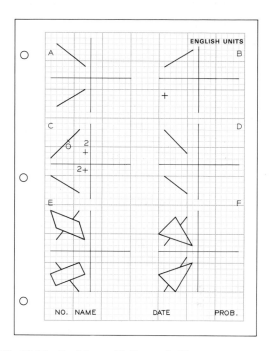

FIG. 19.32 Problems 3A through 3F.

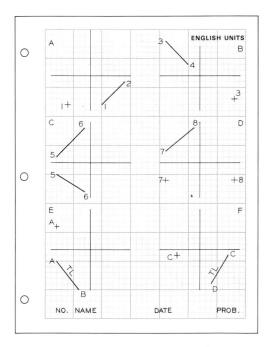

FIG. 19.31 Problems 2A through 2F.

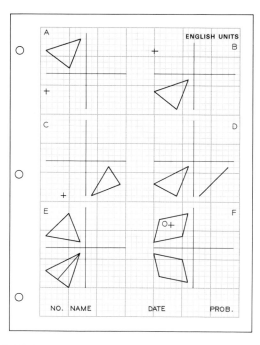

FIG. 19.33 Problems 4A through 4F.

the plane and locate the line upon it in each view. (4F) Find the missing view of the plane and locate Point *O* on each view of the plane.

5. (Fig. 19.34) (5A) Draw a line with its midpoint at *O* that is parallel to the line. (5B) Construct the front view of a second plane that is parallel to the given plane. (5C and 5D) In each problem, draw lines through Point *O* that are parallel to the given planes.

6. (Fig. 19.35) In each problem, draw lines through Point *O* that are perpendicular to the given lines.

7. (Fig. 19.36) (7A and 7B) Draw a line through each Point *O* that is perpendicular to the given planes. (7C) Draw a plane through Point *O* that is perpendicular to the given line. (7D) Draw a plane through the given line that is perpendicular to the plane.

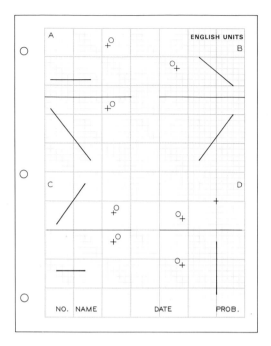

FIG. 19.35 Problems 6A through 6D.

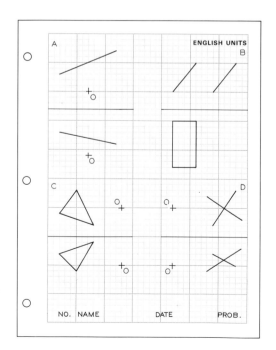

FIG. 19.34 Problems 5A through 5D.

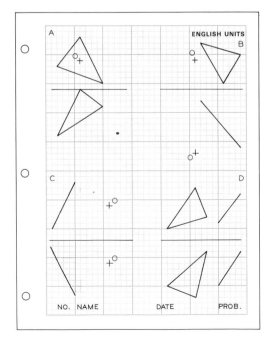

FIG. 19.36 Problems 7A through 7D.

20

Primary Auxiliary Views in Descriptive Geometry

20.1
Introduction

Descriptive geometry can be defined as the projection of three-dimensional figures onto a two-dimensional plane of paper in such a manner as to allow geometric manipulations to determine lengths, angles, shapes, and other geometric information by means of graphics. Orthographic projection is the system used for laying out descriptive geometry problems.

The primary auxiliary view is a powerful tool of descriptive geometry that permits the analysis of three-dimensional geometry that would be difficult by other means.

20.2
Primary auxiliary view of a line

The top and front views of Line 1–2 are shown pictorially and orthographically in Fig. 20.1. Since the line is not a principal line, it is not true length in the prin-

cipal views. To find its true-length view, a primary auxiliary view must be used.

At B the line of sight is drawn perpendicular to the front view of the line and Reference Line F-1 is drawn parallel to the frontal view. You can see in the pictorial that the auxiliary plane is parallel to the line

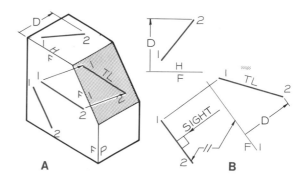

FIG. 20.1 A pictorial of Line 1–2 is shown inside a projection box where a primary auxiliary plane is drawn perpendicular to the frontal plane and parallel to the line. The orthographic arrangement of this auxiliary view is shown at B where the auxiliary view is projected from the front view to find 1–2 true length.

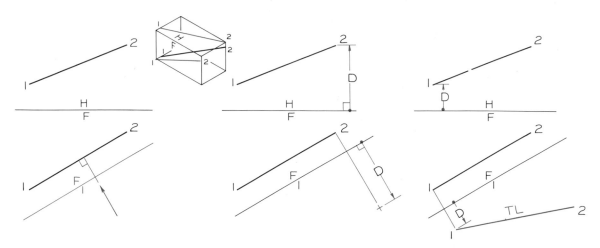

FIG. 20.2 True length of a line by a primary auxiliary view

Step 1 To find 1–2 true length, construct the line of sight perpendicular to the line. Draw reference line F-1 parallel to the front view of the line.

Step 2 Project Point 2 perpendicularly from the front view. Dimension D from the top view locates Point 2.

Step 3 Point 1 is located by transferring dimension D from the top view. Line 1–2 is true length in the auxiliary view.

and perpendicular to the frontal plane, which accounts for it being labeled as F-1.

The auxiliary view is found by projecting parallel to the line of sight and perpendicular to the F-1 reference line. Point 2 is found by transferring Distance D with your dividers to the auxiliary view, since the frontal plane appears as an edge in both the top and auxiliary views. Point 1 is located in the same

manner and the points are connected to find the true-length view of Line 1–2.

Figure 20.2 separates the sequential steps required to find the true length of an oblique line. It is beneficial to letter all reference planes using the notation suggested in the various steps with the exception of the dimensions (such as D) that are transferred from one view to another with your dividers.

A primary auxiliary view can result in a point view of the line if projected from a true-length view of the line in a principal view (Fig. 20.3). The auxiliary view that is projected from the front view of the line does not give a point view since the line is foreshortened in the front view.

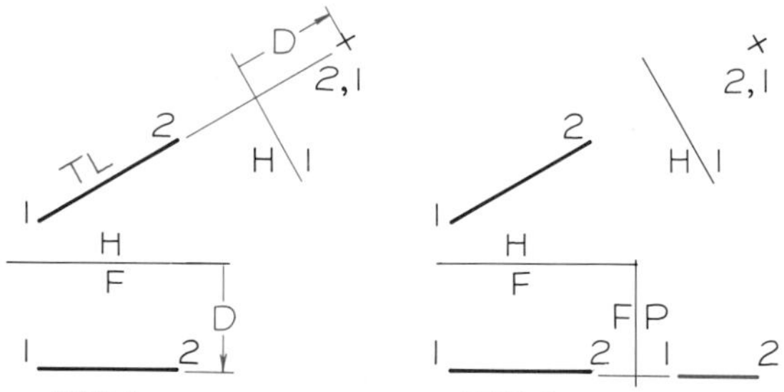

FIG. 20.3 Point view of a line

Step 1 The point view of a line can be found in a primary auxiliary view that is projected from the true-length view of the line.

Step 2 An auxiliary view projected from a foreshortened view of a line will result in a foreshortened view of the line; not a point view.

20.3
True length by analytical geometry

You can see in Fig. 20.4 that the length of a frontal line can be found in the front view by the application of analytical geometry (mathematics) and the Pythagorean theorem. The Pythagorean theorem states that the hypotenuse of a right triangle is equal to the square root of the sum of the squares of the other two sides.

The length of the line can be measured in the front view since it is true length in this view.

The line shown pictorially in Fig. 20.5 can be found true length by analytical geometry by determining the length of the front view where the x- and y-coordinates form a right triangle at A. A second right triangle at B, (1–0–2), is solved to find its hypotenuse, which is the true length of 1–2. The true length of an oblique line is the square root of the sum of the squares of the x-, y- and z-coordinates that correspond to width, height, and depth.

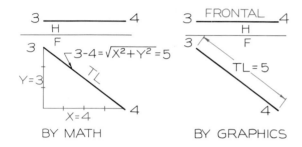

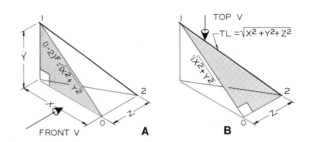

FIG. 20.5 A three-dimensional line that is not true length in a principal view can be found true length by the Pythagorean theorem in two steps. The frontal projection, 1–0, is found using the x- and y-coordinates. At B, the hypotenuse of right triangle 1–0–2 is found using x-, y-, and z-coordinates.

The steps in determining the true length of line 1–2 by analytical geometry are shown in Fig. 20.6.

20.4
The true-length diagram

FIG. 20.4 A line that appears true length in a view (the front view in this case) can have its length calculated by application of the Pythagorean theorem and mathematics. Since the line is a frontal line and is true length in the front view, its length can be measured graphically.

A true-length diagram is constructed with two perpendicular lines to find a line of true-length view of a line (Fig. 20.7). This method does not give a direction for the line, but merely its true length.

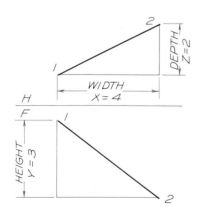

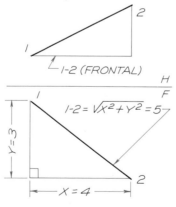

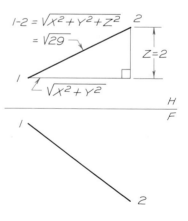

FIG. 20.6 True length of a line—analytical method

Step 1 Right triangles are drawn with Line 1–2 as the hypotenuse in the top and front views. The legs, of the right triangles are drawn parallel and perpendicular to the H-F reference line.

Step 2 The true length of the front view of 1–2 is found by the Pythagorean theorem. The resulting length is found to be 5 units.

Step 3 The true length of the line is found by combining the length of the front projection with the true length of the z-coordinate in the top view. The length is $\sqrt{29}$.

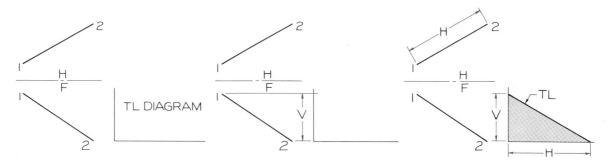

FIG. 20.7 True-length diagram

Required Find Line 1–2 true length in a TL diagram. Construct two perpendicular lines.

Step 1 Transfer the vertical distance between the ends of 1–2 to the vertical leg of the TL diagram.

Step 2 Transfer the horizontal length of the line in the top view to the horizontal leg of the TL diagram.

The two measurements that are laid out on the true-length diagram can be transferred from any two adjacent orthographic views. One measurement is the distance between the endpoints in one of the views. The other measurement, taken from the adjacent view, is measured between the endpoints in a direction perpendicular to the reference line between the two views.

20.5
Angles between lines and principal planes

To measure the angle between a line and a plane, the line must appear true length in the view where the

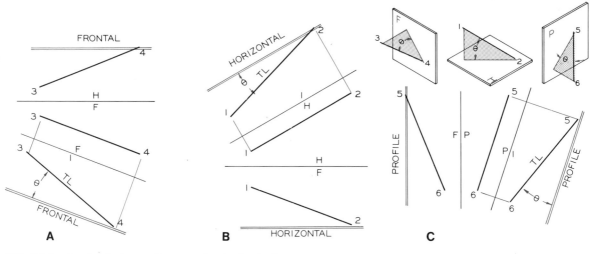

FIG. 20.8 Angles between lines and principal planes

A. When an auxiliary view is projected from the front view, the frontal plane appears as an edge and the line is true length. The angle with the frontal plane can be measured in this view.

B. When an auxiliary view is projected from the top view, the horizontal plane appears as an edge and the line is true length. The angle with the horizontal plane can be measured in this view.

C. When an auxiliary view is projected from the side view, the profile plane appears as an edge and the line is true length. The angle with the profile plane can be measured in this view.

plane appears as an edge. Since a principal plane will appear as an edge in a primary auxiliary view, the angle a line makes with this principal plane can be measured if the line is found true length in this view (Fig. 20.8).

20.6
Slope of a line

> **Slope** is defined as the angle a line makes with the horizontal plane.

It may be specified by either of the three methods in Fig. 20.9; it can be indicated as **slope angle, percent grade,** or **slope ratio.**

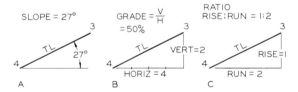

FIG. 20.9 The inclination of a line with the horizontal can be measured and expressed by (A) slope angle, (B) percent grade, or (C) slope ratio.

Slope angle

The slope of a line can be measured in a view where the line is true length and the horizontal plane appears as an edge. Consequently, the slope of *AB* in Fig. 20.10 can be measured in the front view where θ is found to be 31°. This angle can also be found by converting its tangent of 0.60 to 31° by using the trigonometric tables.

Percent grade

The percent grade of a line is found in the view where the line is true length and the horizontal plane appears as an edge. Grade is the ratio of the vertical (rise) divided by the horizontal (run) between the ends of a line expressed as a percentage.

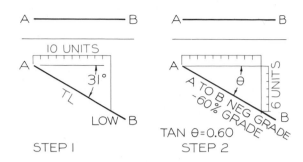

FIG. 20.10 Percent grade of a line

Step 1 The percent grade of a line can be measured in the view where the horizontal appears as an edge and the line is true length (the front view here). Ten units are laid off parallel to the horizontal from the end of the line.

Step 2 A vertical distance from *A* to the line is measured to be 6 units. The percent grade is 6 divided by 10 or 60%. This grade is negative when the direction is from *A* to *B*. The tangent of this slope angle is 6/10 or 0.60.

The percent grade of *AB* is found in Fig. 20.10 by using a combination of mathematics and graphics. Ten units are laid off parallel to the horizontal from *A* using a convenient scale with decimal units in Step 1. In Step 2 the vertical drop of the line after 10 units along the horizontal is measured as 6 units. Since 10 units were used along the horizontal, your arithmetic is simplified in finding the tangent of the angle to be 0.60, which is easily converted into −60% grade. The grade is negative from *A* to *B* since this is downhill. It would be positive from *B* to *A,* if it were uphill. Line *CD* has a positive fifty-percent grade (+50%) from *C* to *D* in Fig. 20.11A.

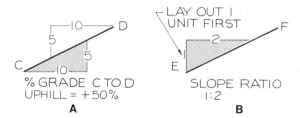

FIG. 20.11 The percent grade of a line is positive if uphill; negative if downhill (part A). The slope ratio is expressed as 1 : xx where 1 is the rise and xx is the horizontal distance (Part B).

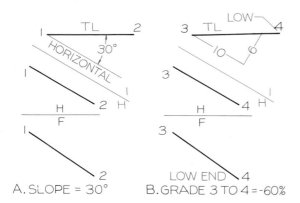

FIG. 20.12 Slope of an oblique line

A. The slope angle can be measured in a view where the horizontal appears as an edge and the line is true length. The slope of 30° is found in an auxiliary view projected from the top view.

B. The percent grade can be measured in a true-length view of the line projected from the top view. Line 3–4 has a −60% grade from 3 to the low end at 4.

Slope ratio

The first number of the slope ratio is always one, such as 1:10, 1:200, and so on. The first number is the rise (always one) and the second number is horizontal run (see Fig. 20.9).

The graphical method of finding the slope ratio is shown in Fig. 20.11B where the rise of one is laid off on the true-length view of *EF*. The corresponding horizontal is found to be 2, which results in a slope ratio of 1:2.

OBLIQUE LINES When a line is oblique and does not appear true length in the front view, it must be found true length by an auxiliary view projected from the top view. This auxiliary view shows the horizontal as an edge and the line true length making it possible to measure the slope angle (Fig. 20.12A).

Similarly, an auxiliary view projected from the top view must be used to find the percent grade of a line of an oblique line (Fig. 20.12B). Ten units are laid off horizontally, parallel to the H-1 reference line and the vertical distance is found to be 6, or a −60% grade downhill from 3 to 4.

20.7
Compass bearing of a line

Two types of bearings of a line's direction are (A) **compass bearings,** and (B) **azimuth bearings** (Fig. 20.13).

> Compass bearings always begin with the north or south directions and the angles with north and south are measured toward east or west.

The line in Part A that makes 30° with north has a bearing of N 30° W. A line making 60° with south toward the east has a compass bearing of south 60° east, or S 60° E. Since a compass can be read only when held horizontally, the compass bearings of a line can be determined only in the top view, the horizontal view.

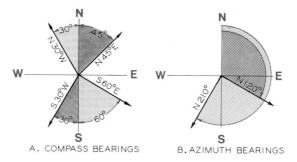

FIG. 20.13 A. Compass bearings are measured with respect to north and south directions on the compass. B. Azimuth bearings are measured with respect to north in a clockwise direction up to 360°.

An azimuth bearing is measured from north in clockwise direction to 360° (Fig. 20.13B). Azimuth bearings of a line are written as N 120°, N 210°, etc., with this notation indicating that the measurements are made from north.

> The compass bearing (direction) of a line is assumed to be toward the low end of the line unless otherwise specified.

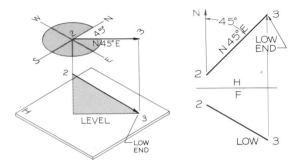

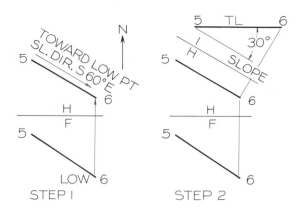

FIG. 20.14 The compass bearing of a line is measured in the top view toward its low end (unless specified toward the high end). Line 2–3 has a bearing of N 45° E toward the low end at 3.

FIG. 20.15 Slope and bearing of a line

Step 1 Slope bearing can be found in the top view toward its low end. Direction of slope is S 60° E.

Step 2 The slope angle of 30° is found in an auxiliary view projected from the top view where the line is found true length.

For example, Line 2–3 in Fig. 20.14 has a bearing of N 45° E since the line's low end is Point 3. It can be seen in the front view that point 3 is the lower end.

The compass bearing and slope of a line are found in Fig. 20.15. This information can be used to describe verbally the line as having a compass bearing of S 60° E and a slope of 30° from 5 to 6. When given verbal information and one point in the top and front views, a line can be drawn as shown in Fig. 20.16.

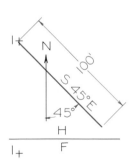

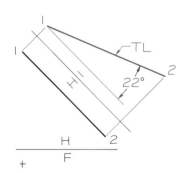

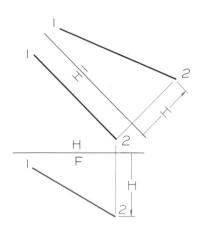

FIG. 20.16 A line from slope specifications

Step 1 It is required to draw a line through Point 1 that bears S 45° E for 100 ft horizontally and slopes 22°. The bearing and horizontal length are drawn in the top view.

Step 2 An auxiliary view is projected from the top view where the 22° slope angle is measured, where 1–2 is true length.

Step 3 The front view of 1–2 is found by locating Point 2 in the front view using dimension H.

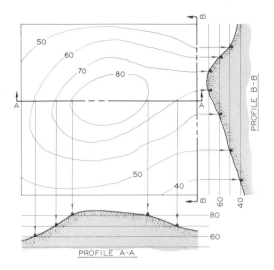

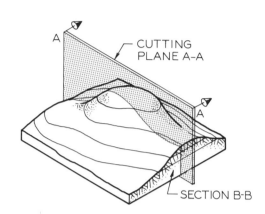

FIG. 20.17 A contour map uses contour lines to show variations in elevation on an irregular surface. Vertical sections taken through a contour map are called profiles.

20.8
Contour maps and profiles

The contour map is the method of representing irregular surfaces of the earth. A pictorial view of a sectional plane and a portion of the earth are shown in Fig. 20.17, along with the conventional orthographic views of the contour map and profiles.

CONTOUR LINES Contour lines are horizontal lines that represent constant elevations from a horizontal datum such as sea level. Contour lines can be thought of as the intersections of horizontal planes with the surface of the earth. The vertical interval of spacing between the contours in Fig. 20.17 is 10 feet.

CONTOUR MAPS Contour maps contain contour lines that are drawn to represent irregularities of the surface (Fig. 20.17). The closer the contour lines are to each other on the map, the steeper the terrain.

PROFILES Profiles are vertical sections through a contour map that are used to show the earth's surface at any desired location. Two profiles are shown in Fig. 20.17. Contour lines represent edge views of equally spaced horizontal planes in profiles. The true representation of a profile is drawn with the vertical scale equal to the scale of the contour map; however, the

vertical scale is often increased to emphasize changes in elevation.

CONTOURED SURFACES Airfoils, automobile bodies, ship hulls, and household appliances are contoured surfaces that must also be depicted on the drawing board by using contour lines. When applied to objects other than the earth's surface, this technique of representing contours is called **lofting.**

STATION NUMBERS Station numbers are used to locate distances on a contour map. Since the civil engineer uses a chain (metal tape) that is 100 ft long, stations are located 100 ft apart (Fig. 20.18). Station 7

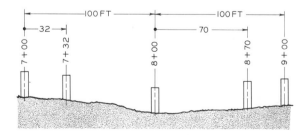

FIG. 20.18 Station points are located 100 ft apart. For example station 7 is 700 ft from the beginning point. A point that is 32 ft beyond station 7 is labeled as station 7 + 32. A point 870 ft from the origin is labeled as station 8 + 70.

is 700 ft from the beginning point, Station 0. A point that is 32 ft from station 7 toward station 8 is labeled and marked as station 7 + 32.

20.9
Vertical sections

In Fig. 20.19, a vertical section (called a profile) is passed through the top view of an underground pipe that is known to have an elevation of 90 ft at 1 and 60 ft at 2. An auxiliary view is projected perpendicularly from the top view, contour lines are located, and the top of the earth over the pipe is found in profile.

The same scale was used to draw the profile as was used to draw the contour map in order to measure the true lengths and angles of slope in the profile. The percent grade and compass bearing of the line are labeled on the contour map.

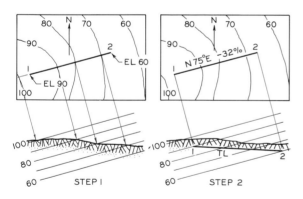

FIG. 20.19 Vertical sections (profiles)

Step 1 An underground pipe is known to have elevations of 90 ft and 60 ft at each end. An auxiliary view is projected perpendicularly from the top view and contours are drawn at 10 ft intervals to correspond to the plan view. The top of the ground is found by projecting from the contour lines in the plan view.

Step 2 Points 1 and 2 are located at elevations of 90 ft and 60 ft in the vertical section (profile). Since 1–2 is TL in the section, its slope or percent grade can be measured. The compass direction and percent grade are labeled in the top view on the line.

20.10
Plan-profiles

A plan-profile is a combination drawing that includes a plan with contours and a vertical section called a profile. A plan-profile is used to show an underground drainage system from manhole 1 to manhole 3 in Fig. 20.20 and Fig. 20.21.

The profile is drawn with an exaggerated vertical scale to emphasize the variations in the earth's surface and the grade of the pipe. Although the vertical scale is usually increased, it can be drawn at the same scale as is used in the plan if desired.

Manhole 1 is projected to the profile using orthographic projection, but the other points are not orthographic projections (Fig. 20.20). The points where the contour lines cross the top view of the pipe are transferred to their respective elevations in the profile with your dividers. These points are connected to show the surface of the earth over the pipe and the location of manhole centerlines.

The drop from manhole 1 to manhole 2 is found to be 5.20 ft by multiplying the horizontal distance of 260.00 ft by a −2.00% grade (Fig. 20.21).

Since the pipes intersect at manhole 2 at an angle, the flow of the drainage is disrupted at the turn; consequently, a drop of 0.20 ft is given from the inlet across the floor of the manhole to compensate for the loss of pressure (head) through the manhole.

The true lengths of the pipes cannot be accurately measured in the profile when the vertical scale is different from the horizontal scale. Trigonometry must be used for this computation.

20.11
Edge view of a plane

> The edge view of a plane can be found in any view where a line on the plane appears as a point.

A line can be found as a point by projecting from a view where the line is true length (Fig. 20.22).

A true-length line can be drawn on any plane by drawing the line parallel to one of the principal planes and projecting it to the adjacent view, as shown in Step 1 of Fig. 20.23. Since Line 1–0 is true length in

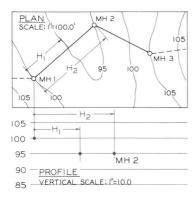

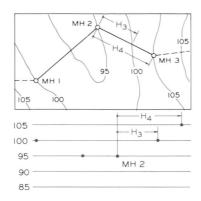

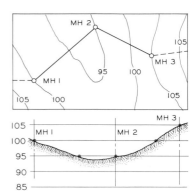

FIG. 20.20 Plan-profile

Required Find the profile of the earth over the underground drainage system.

Step 1 Distances H_1 and H_2 from manhole 1 are transferred to their respective elevations in the profile. This is not a true orthographic projection.

Step 2 Distances H_3 and H_4 are measured from manhole 2 in the plan and are transferred to their respective elevations on the earth above the pipe.

Step 3 The five points are connected with a freehand line and the drawing is crosshatched to represent the earth's surface. Center lines are drawn to show the locations of the three manholes.

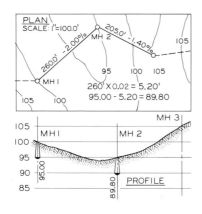

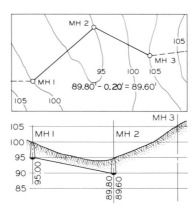

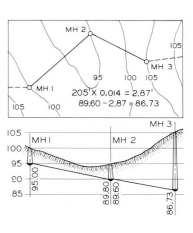

FIG. 20.21 Plan-profile, manhole location

Step 1 The horizontal distances from the manholes and the percent grade of flow lines of the pipes are given on the plan. The elevation of the bottom of manhole 2 is calculated by subtracting from the given elevation of manhole 1.

Step 2 The lower side of manhole 2 is 0.20 ft lower than the inlet side to compensate for loss of head (pressure) due to the turn in the pipeline. The lower side is found to be 89.60 ft and is labeled.

Step 3 The elevation of manhole 3 is calculated to be 86.73 ft. The flow line is drawn from manhole to manhole and the elevations are labeled. This profile shows the relationship of the pipe to the surface.

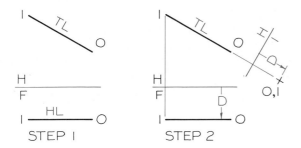

FIG. 20.22 Point view of a line

Step 1 Line 1–0 is horizontal in the front view and is therefore true length in the top view.

Step 2 The point view of 1–0 is found by projecting parallel from the top view to the auxiliary view.

the top view, its point view may be found and the plane will appear as an edge in this auxiliary view.

20.12
Dihedral angles

The angle between two planes is called a **dihedral angle.** This angle can be found in the view where the line of intersection between two planes appears as a point.

The line of intersection, 1–2, between the two planes in Fig. 20.24 is true length in the top view. This makes it possible to find the point view of Line 1–2 and the edge view of both planes in a primary auxiliary view.

20.13
Piercing points by projection

Figure 20.25 gives the sequential steps necessary to find the piercing point of Line 1–2 that passes through the plane. Cutting planes are passed through the line and plane in the top view. The trace of this cutting plane, Line *DE*, is then projected to the front view

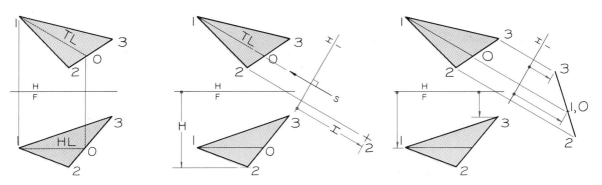

FIG. 20.23 Edge view of a plane

Step 1 To find Plane 1–2–3 as an edge, horizontal Line 1–0 is drawn on the plane in the front view. Line 1–0 is projected to the top view where it is true length.

Step 2 A line of sight is drawn parallel to the true-length Line, 1–0. H-1 is drawn perpendicular to the line of sight. Point 2 is found in the auxiliary view by transferring dimension *H* from the front view.

Step 3 Points 1 and 3 are found by transferring their height dimensions from the front view. These points lie in a straight line which is the edge of the plane. Line 1–0 will appear as a point in this view.

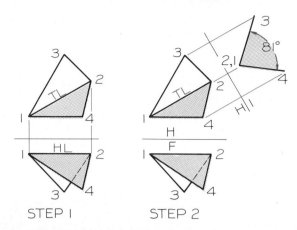

FIG. 20.24 Dihedral angle

Step 1 The line of intersection between the planes, 1–2, is true length in the top view.

Step 2 The angle between the planes (the dihedral angle) can be found in the auxiliary view where the line of intersection appears as a point.

where the piercing Point *P* is found in Step 1. The top view of *P* is located in Step 2, and the visibility of the line is found in Step 3.

20.14
Piercing points by auxiliary views

The piercing point of a line and a plane can be found by an auxiliary view as shown in Fig. 20.26. Piercing Point *P* can be seen in Step 2 where the plane is found as an edge. Point *P* is projected back to the line in the top and front views from the auxiliary view in Step 3. The location of *P* in the front view is checked by transferring dimension *H* from the auxiliary view with your dividers.

Visibility is easily determined in the top view since it can be seen that *AP* is higher than the plane in the auxiliary view, and is therefore visible in the top

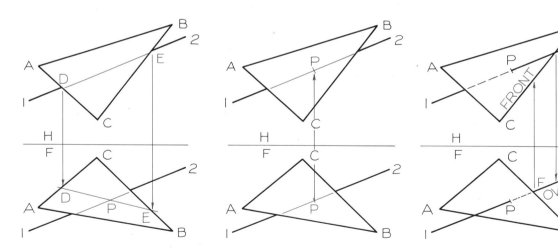

FIG. 20.25 Piercing points by projection

Step 1 A vertical cutting plane is passed through the top view of 1–2, the plane intersects *AC* and *BC* at *D* and *E*. The intersection of *DE* with 1–2 in the front view locates the piercing Point *P*.

Step 2 Point *P* is projected to the top view of Line 1–2 from the front view where it was first located.

Step 3 Lines *CB* and *P2* cross at Point *T* in the top view. By projecting downward from *T*, *P2* is found to be over *CB*; therefore, *PT* is visible in the top view. Intersection *F* can be used to find that Line *CB* is in front of *P2*; therefore, *PF* is hidden in the front view.

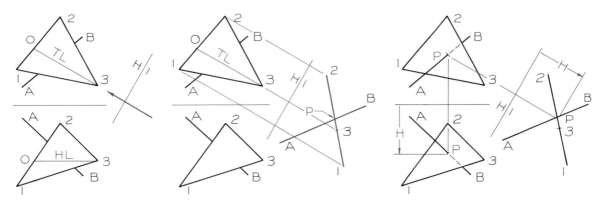

FIG. 20.26 Piercing points by auxiliary view

Step 1 Draw a horizontal line on the plane in the front view, project it to the top view to find *TL* Line 0–3 on the plane. The line of sight is drawn parallel to the *TL* line.

Step 2 Find the edge view of the plane and project the line *AB* to this view. Point *P* is the piercing point in the auxiliary view.

Step 3 Point *P* is projected to the top and front views. Point *A* of the auxiliary view is the highest point and *AP* will be visible in the top view. Point *B* in the top view is the farthest back and it is therefore hidden in the front view.

view. Analysis of the top view shows that endpoint *A* is the most forward point; and therefore *AP* is visible in the front view.

20.15
Perpendicular to a plane

In Fig. 20.27 it is required that a line be drawn from Point *O* that is perpendicular to the plane.

> A perpendicular line will appear true-length and perpendicular to a plane where the plane appears as an edge.

The true-length perpendicular is drawn in Step 2 to locate piercing point *P*. Point *P* is found in the top view by drawing Line *OP* parallel to the H-1 reference line. It must lie in this direction since *OP* is true length in the auxiliary view. It will also be perpendicular to a true-length line in the top view of the plane. The front view of Point *P* is found by projection in Step 3 along with its visibility.

20.16
Intersections by projection

The line of intersection between two planes can be found by locating the piercing points of two lines on one plane with the other. In Fig. 20.28 a cutting plane is used in Step 1 to find piercing Point *P* by projection. In Step 2, piercing Point *T* is found by the same method. Line *PT* is the line of intersection, which will always be visible.

The two lines that are selected to be analyzed for their piercing points should be lines of a plane that cross the other plane. Lines *AB* and 1–2 would be poor selections since they lie outside the other plane. Each line is then analyzed to find its piercing point and visibility as if it were a single line.

20.17
Intersections by auxiliary view

The intersection between planes can be found by finding the edge view of one of the planes as shown in Fig. 20.29, Step 1. Piercing Points *L* and *M* are pro-

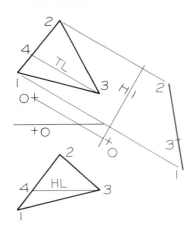
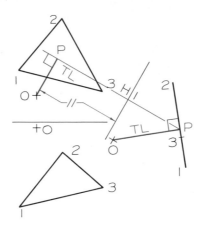
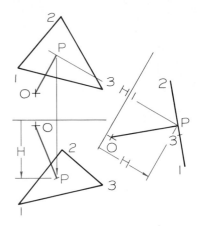

FIG. 20.27 Line perpendicular to a plane

Step 1 Find the edge view of the plane by finding the point view of a line on it. Project from either view. Project Point O.

Step 2 Line OP is drawn perpendicular to the edge view of the plane. Since OP is TL in the auxiliary view, it will be parallel to the H-1 reference line in the top view and perpendicular to a TL line on the plane.

Step 3 Piercing Point P is found in the front view by projecting from the top view. Point P is accurately located by transferring dimension H from the auxiliary view to the front view.

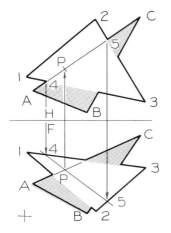
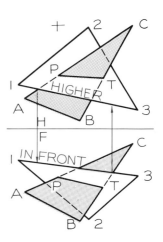

FIG. 20.28 Intersection by projection

Step 1 Pass a cutting plane through Line AC in the top view to locate Points 4 and 5 that are projected to the front view. Point P is found on 4–5 in the front view and Point P is projected to the top view.

Step 2 Pass a cutting plane through the top view of BC to locate Points 6 and 7 that are projected to the front view. Piercing Point T is located on 6–7 in the front view and then project Point T to the top view.

Step 3 Intersection PT is visible. The intersection between AC and 1–3 is projected to the front view to determine that 1–3 is higher in the top view. Line 1–3 is found to be in front of BC in the front view.

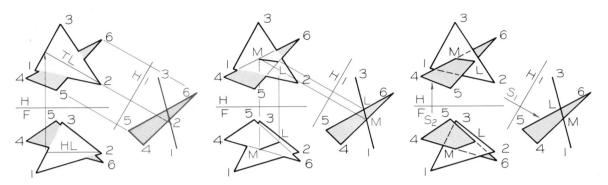

FIG. 20.29 Intersection by auxiliary view

Step 1 Find the edge view of one of the planes and project the other plane to this view also.

Step 2 Piercing Points *L* and *M* can be seen on the edge view of the plane. *LM* is projected to the top and front views.

Step 3 The line of sight strikes L–5 first in the auxiliary view, which makes this portion of the plane visible in the top view. Line 4–5 is farthest forward in the top view and is visible in the front view.

jected from the auxiliary view to their respective Lines, 5–6 and 4–6, in the top view in Step 2.

The visibility of Plane 4–5–6 in the top view is apparent by inspection of the auxiliary view, where sight Line S_1 has an unobstructed view of the 4–*L*–*M* portion of the plane. Plane 4–5–*L*–*M* is visible in the front view, since sight Line S_2 has an unobstructed view of the top view of this portion of the plane.

20.18
Slope of a plane

Planes can be established by using verbal specifications of **slope** and **direction of slope** of a plane as defined below.

SLOPE The slope of a plane is the angle its edge view makes with the edge of the horizontal plane.

DIRECTION OF SLOPE The direction of slope is the compass bearing of a line that is perpendicular to a true-length line in the top view of a plane toward its low side. This is the direction in which a ball would roll on the plane.

It can be seen in Fig. 20.30A that a ball would roll perpendicular to all horizontal lines of the roof toward the low side. The slope is seen when the roof and the horizontal are edges in a single view.

Figure 20.31 gives the steps of finding the direction of slope and the slope angle of a plane. An understanding of these terms enables you to verbally describe a sloping plane.

A three-dimensional plane can be established by working from slope and direction specifications (Fig. 20.32). The direction of slope is drawn in the given top view, which locates the direction of a true-length

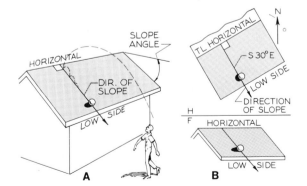

FIG. 20.30 The direction of slope of a plane is the compass bearing of a line on the plane that is perpendicular to a true-length line on the plane. Slope direction is measured in the top view toward the low side of the plane.

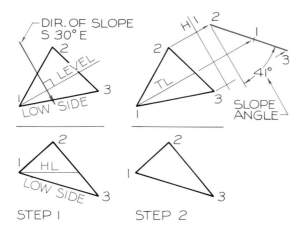

FIG. 20.31 Slope and direction of slope of a plane

Step 1 Slope direction can be found as perpendicular to a true-length level line in the top view toward the low side of the plane, S 30° E in this case.

Step 2 Slope is measured in an auxiliary view where the horizontal is an edge and the plane is an edge, 41° in this case.

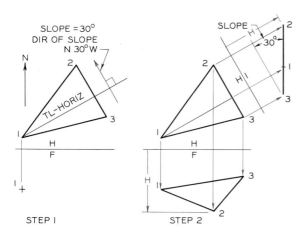

FIG. 20.32 Construction from slope specifications

Step 1 If the top view, front view of 1, and slope specifications are given, the front view can be completed. Draw the direction of slope in the top view. A true-length horizontal line on the plane is drawn perpendicular to the slope direction.

Step 2 A point view of the TL line is found in the auxiliary view to locate Point 1. The edge view of the plane is drawn through 1 at a slope of 30°, according to the specifications. The front view is completed by transferring height dimensions from the auxiliary view to the front view.

horizontal line on the plane in the top view (Step 1). In Step 2, the edge view of the plane is found by locating Point 1 and constructing a slope of 30°. Points 3 and 2 are then transferred to the front view to complete that view.

20.19
Cut and fill

A level roadway routed through irregular terrain or the embankment of a fill used to build a dam involve the principles of cut-and-fill (Fig. 20.33). Cut-and-fill is the process of cutting away equal amounts of the high ground to fill the lower areas to form a nearly level roadway where possible.

FIG. 20.33 The road across the top of this dam was built by applying the principles of cut and fill. (Courtesy of the Bureau of Reclamation, U.S. Department of the Interior.)

In Fig. 20.34, it is required that a level roadway of a given width and an elevation of 60 ft be constructed about the given center line in the contour map using the specified angles of cut and fill.

In Step 1 the roadway is drawn in the top view, and the contour lines in the profile view are drawn 10 ft apart since the contours in the top view are this far apart. The cut angles are measured and drawn on both sides of the roadway on the upper side, Step 2. The points on the various elevation lines crossed by the cut

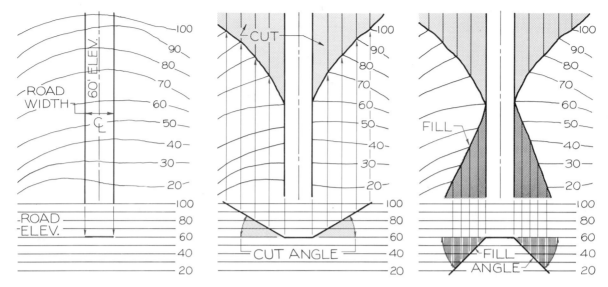

FIG. 20.34 Cut and fill of a level roadway

Step 1 Draw a series of elevation planes in the front view at the same scale as the map and label them to correspond to the contours on the map. Draw the width of the roadway in the top and front views at the given elevation, 60 ft in this case.

Step 2 Draw the cut angles on the upper sides of the road in the front view. The points of intersection between the cut angles and the contour planes in the front view are projected to their respective contour lines in the top view.

Step 3 Draw the fill angles on the lower sides of the road in the front view. The points in the front view where the fill angles cross the elevation planes are projected to their respective contour lines in the top view. Contour lines are changed in the cut-and-fill areas to show new contours.

embankments are projected to the top view to find the limits of cut in this view.

The fill angles are laid off in the profile in Step 3 from given specifications. The crossing points on the profile view of the elevation lines are projected to the top view to find the limits of fill. New contour lines are drawn in Step 3 inside the areas of cut and fill to indicate that they have been changed by the process of cut and fill.

20.20
Design of a dam

The basic definitions of terms associated with dams are shown in Fig. 20.35. These are (1) **crest,** the top of the dam, (2) **water level,** the level of water held by the dam, and (3) **freeboard,** the height of the crest above the water level.

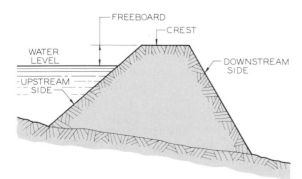

FIG. 20.35 The terms and symbols used in the construction drawing of a dam.

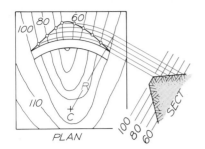
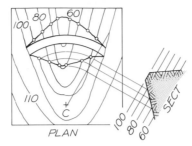
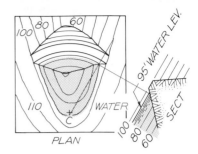

FIG. 20.36 Graphical design of a dam

Step 1 A dam in the shape of an arc with its center at *C* has an elevation of 100'. Draw radius *R* from *C* and project perpendicularly from this line and draw a section through the dam from specifications. The downstream side of the dam is projected to radial line, *R*. Using the radii from *C*, to locate points on their respective contour lines.

Step 2 The elevations of the dam on the upstream side of the section are projected to the radial line, *R*. Using the Center *C* and your compass, locate points on their respective contour lines in the plan view as they are projected from the section.

Step 3 The elevation of the water level is 95' in this case and is drawn in the section. The point where the water intersects the dam is projected to the radial line in the plan view, and is drawn as an arc using Center *C*. The limit of the water is drawn between the 90' and 100' contour lines in the top view.

FIG. 20.37 An aerial view of Hoover Dam and Lake Mead, which were built during the period 1931 to 1935. (Courtesy of the Bureau of Reclamation, U.S. Department of the Interior.)

An earthen dam is located on the contour map in Fig. 20.36. It makes an arc with its center at Point *C*. The top of the dam is specified to be level to provide a roadway. This method of drawing the top view of the dam and indicating the level of the water held by the dam are shown in the three steps of Fig. 20.36.

These same principles were used in the design and construction of the 726-ft high Hoover Dam that was built in the 1930s. Since this dam was made of concrete instead of earth, the dam was built in the shape of an arch that is bowed toward the water to take advantage of the compressive strength of concrete (Fig. 20.37).

20.21
Strike and dip

Strike and **dip** are terms used in geological and mining engineering to describe strata of ore under the surface of the earth.

STRIKE Strike is the compass bearing of a level line in the top view of a plane. It has two possible compass bearings since it is the direction of a level line.

DIP Dip is the angle the edge view of a plane makes with the horizontal plus its general compass direction, such as NW or SW. The dip angle is found in the pri-

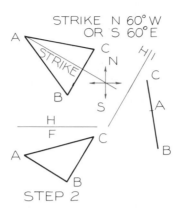 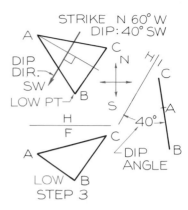 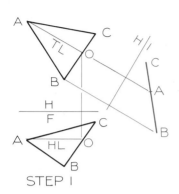

STEP I STEP 2 STEP 3

FIG. 20.38 Strike and dip of a plane

Step 1   Find the edge view of the plane by projecting from the top view.

Step 2   Strike is the compass direction of a level line on the plane and is measured in the top view. The strike of the plane is either N 60° W or S 60° E.

Step 3   Dip is the angle a plane makes with the horizontal (40°) plus the general compass direction toward the low side in the top view (SW). Dip direction is perpendicular to a TL line in the top view. Dip is written as 40° SW.

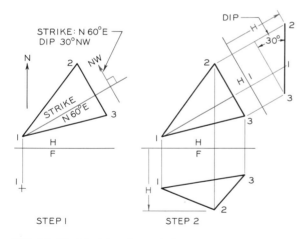

STEP I STEP 2

FIG. 20.39 Working from strike and dip specifications

Step 1   Strike is drawn on the top view of the plane, which is a TL horizontal line. The direction of dip is drawn perpendicular to the strike toward the NW, as specified.

Step 2   A point view of the strike line is found in the auxiliary view to locate Point 1. The edge view of the plane is drawn through 1 at a dip of 30°, according to the specifications. The front view is completed by transferring height dimensions from the auxiliary to the front view.

mary auxiliary view projected from the top view, and its dip direction is measured perpendicular to a level line in the top view toward the low side.

The steps of finding the strike and dip of a plane are given in Fig. 20.38. Strike can be measured in the top view by finding a true-length line on the plane in this view. The dip angle requires an auxiliary view that must be projected from the top view in order for the horizontal and the plane to appear as edges.

A plane can be constructed from strike and dip specifications as shown in Fig. 20.39. The strike is drawn to represent a true-length horizontal line on the plane. Dip direction is perpendicular to the strike (Step 1). In Step 2, the edge view of the plane is drawn through Point 1 at a dip of 30°. Points 2 and 3 are located in the front view.

20.22

Distances from a point to a plane

Descriptive geometry principles can be used to find various distances from a point to a plane. An example of this is shown in Fig. 20.40 where the distance from Point O on the ground to an ore vein under ground is found.

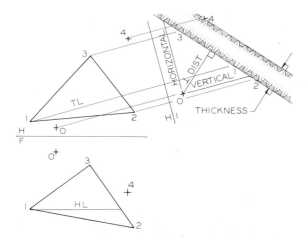

FIG. 20.40 The vertical, horizontal, and perpendicular distances from a point to an ore vein can be found in an auxiliary view projected from the top view. The thickness of an ore vein is perpendicular to the upper and lower planes of the ore vein.

Three points are located on the top plane of an ore vein. Point *O* is the point on the earth from which the tunnels are to be drilled to the vein for mining. Point 4 is a point on the lower plane of the vein.

The edge view of Plane 1–2–3 is found by projecting from the top view. The lower plane is drawn parallel to the upper plane through Point 4. The horizontal distance from Point *O* to the plane is drawn parallel to the H-1 reference line. The vertical distance

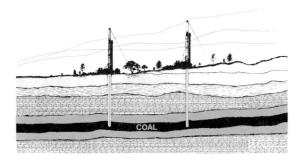

FIG. 20.41 Test wells are drilled into coal zones to determine which coal seams will contribute significantly to the production of gas. (Courtesy of Texas Eastern News.)

is perpendicular to the H-1 line. The shortest distance to the plane is perpendicular to the plane. Each of these lines is true length in this view where the ore vein appears as an edge.

The process of finding the distance from a point to a plane or a vein is a technique often used in solving mining and geological problems. Figure 20.41 illustrates test wells that are drilled into coal zones to learn more about them.

20.23
Outcrop

Strata of ore or rock formations usually approximate planes of a uniform thickness. This assumption is employed in analyzing data concerning the orientation of ore veins that are underground. A vein of ore may be inclined to the surface of the earth and may actually outcrop on its surface. Outcrops permit open-surface mining operations at a minimum of mining expense.

The steps of finding the outcrop of an ore vein are given in Fig. 20.42. The locations of sample drillings, *A*, *B*, and *C* are shown on the contour map and their elevations are located on the contours of the profile. These points are known to lie on the upper surface of the vein. Point *D* is known to lie on the lower plane of the vein.

The edge view of the ore vein can be found in an auxiliary view projected from the top view (Step 1). The points on the upper surface are projected back to their respective contour lines in the top view in Step 2. The points on the lower surface of the vein are projected to the top view in Step 3. If the ore vein does continue uniformly at its angle of inclination to the surface, the space between these two lines will be the edge of the vein on the surface of the earth.

20.24
Intersections between planes—cutting plane method

The top and front views of two planes are given in Fig. 20.43 where it is required to find the line of intersection between them if the planes are infinite in size.

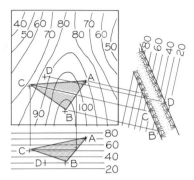
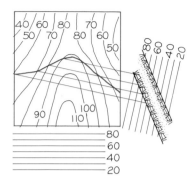
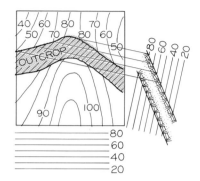

FIG. 20.42 Ore vein outcrop

Step 1 Using Points *A*, *B*, and *C* on the upper surface of the plane, find the edge view of the plane by projecting off the top view. The lower surface of the plane is found by drawing it parallel to the upper surface through Point *D*, a known point on the lower surface.

Step 2 Points of intersection between the upper surface and the contour lines in the auxiliary view are projected to their respective contour lines in the top view to find one line of the outcrop.

Step 3 Points from the lower surface in the auxiliary view are projected to their respective contour lines in the top view to find the second line of outcrop. Cross-hatch this area to indicate the outcrop area.

Cutting planes are passed through either view at any angle and projected to the adjacent view. The two Points *L* and *M* that are found in the top view are connected to form the top view of the line of intersection. The compass direction of the line can be used to describe its direction of slope toward its low end.

The front view of the line of intersection is found by projecting the points from the top view to their respective planes in the front view. This is the line on which all lines on the planes would intersect.

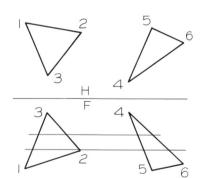
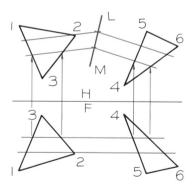
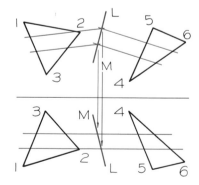

FIG. 20.43 Intersection of planes by cutting-plane method

Step 1 Pass cutting planes through the front view of the planes. These planes can be drawn in any direction.

Step 2 Project the intersections of the cutting planes in the front to the top view of the planes. Line of intersection, *LM*, is found in the top view.

Step 3 Points *L* and *M* are projected to their respective cutting planes in the front view to complete the solution.

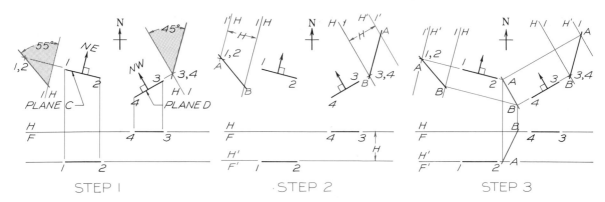

FIG. 20.44 Intersection between ore veins by auxiliary view

Given The strike and dip of two ore veins, Plane *C* and Plane *D*.

Required Find the line of intersection between the ore veins, assuming that each is continuous.

Step 1 Lines 1–2 and 3–4 are strike lines and are true length in the horizontal view. The point view of each strike line is found by an auxiliary view, using a common reference plane. The edge view of the ore veins can be found by constructing the dip angles with the H-1 line through the point views. The low side is the side of the dip arrow.

Step 2 A supplementary horizontal plane, H′–F′, is constructed at a convenient location in the front view. This plane is shown in both auxiliary views located *H* distance from the H-1 reference line. The H′–1′ plane cuts through each ore vein edge in the auxiliary views.

Step 3 Points *A*, which were established on each auxiliary view by the H′–1′ plane, are projected to the top view, and they intersect at Point *A*. Point *B* on the H-1 plane are projected to their intersection in the top view at Point *B*. Points *A* and *B* are projected to their respective planes in the front view. To establish the line of intersection, *AB*.

20.25
Intersection between planes—auxiliary method

In Fig. 20.44 two planes have been located and specified using strike and dip. Since the given strike lines are true-length level lines in the top view, the edge view of the planes can be found in the view where the strike appears as a point, Step 1. The edge views are drawn using the given dip angles and directions.

Horizontal datum planes H-F and H′-F′, are used to find lines on each plane that will intersect when projected from the auxiliary views to the top view. Points *A* and *B* are connected to determine the line of intersection between the two planes in the top view. These points are projected to the front view, where Line *AB* is found.

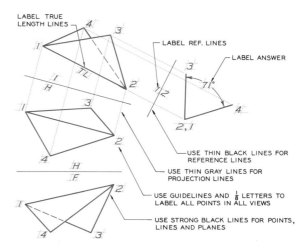

FIG. 20.45 Notational techniques that should be used in solving descriptive geometry problems.

20.26
Solution of descriptive geometry problems

Figure 20.45 illustrates the techniques of labeling and solving a descriptive geometry problem. Note that some of the lettering and numbering is aligned with inclined lines and reference lines to which the labeling applies, while other lettering is not aligned but is parallel to the edge of the paper. You may use either technique or a combination of the two.

Always use guidelines and ⅛ in. lettering for best results. Observe the difference in line qualities that are used in the problem solution. Guidelines and projection lines need be only dark enough to be seen and used as guides.

Problems

Use Size AV (8½″ × 11″) sheets for the following problems, and lay out the problems using instruments. Each square on the grid is equal to 0.20 in. or about 5 mm. The problems can be laid out on grid or plain paper. Label all reference planes and points in each problem with ⅛ in. letters and/or numbers, using guidelines.

1. Fig. 20.46. (1A–1D) Find the true length views of the lines as indicated by the given lines of sight by an auxiliary view.

2. Fig. 20.47. (2A–2D) Find the angles that these lines make with the respective principal planes indicated by the given auxiliary reference lines.

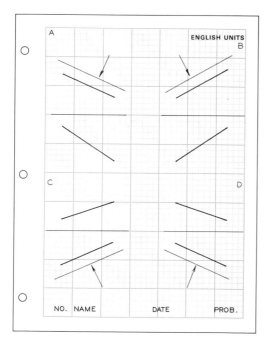

FIG. 20.46 Problems 1A through 1D.

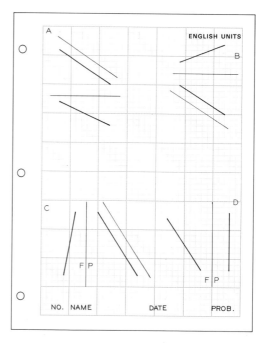

FIG. 20.47 Problems 2A through 2D.

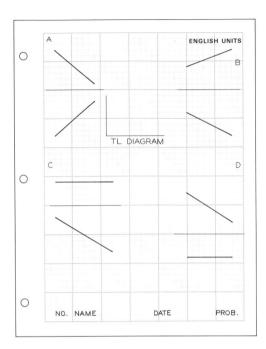

FIG. 20.48 Problems 3A through 3D.

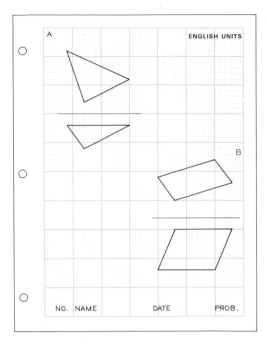

FIG. 20.50 Problems 5A and 5B.

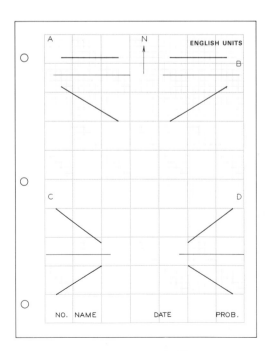

FIG. 20.49 Problems 4A through 4D.

3. Fig. 20.48. (3A and 3B) Find the lines' true length by the true-length diagram method, using the same diagram for both lines. (3C and 3D) Find the point views of the lines.

4. Fig. 20.49. (4A and 4B) Find the slope angle, tangent of the slope angle, and the percent grade of the four lines.

5. Fig. 20.50. (5A and 5B) Find the edge views of the two planes.

6. Fig. 20.51. (6A) Find the angle between the planes. (6B) Find the piercing point by projection. (6C) Find the piercing point by an auxiliary view.

7. Fig. 20.52. (7A) Construct a line perpendicular to the plane and through Point *O* on the plane by an auxiliary view. (7B) Construct a line perpendicular to the plane from Point *O* by an auxiliary view.

8. Fig. 20.53. Find the line of intersection between the planes by projection in Part A and by an auxiliary view in Part B. Show visibility.

9. Fig. 20.54. (9A and 9B) Find the direction of slope and slope of angle of the planes.

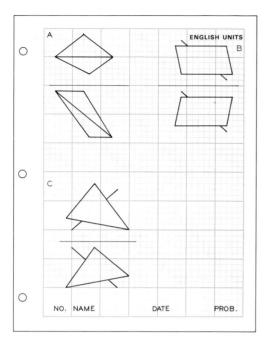

FIG. 20.51 Problems 6A through 6C.

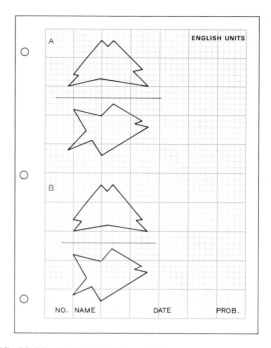

FIG. 20.53 Problems 8A and 8B.

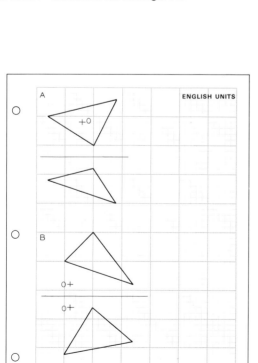

FIG. 20.52 Problems 7A and 7B.

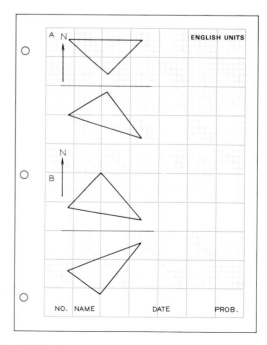

FIG. 20.54 Problems 9A, 9B, 10A, and 10B.

10. Fig. 20.54. (10A and 10B) Find the strike and dip of the planes.

11. Fig. 20.55. Find the shortest distance, the horizontal distance, and the vertical distance from Point *O* on the ground to the underground ore vein represented by the triangle. Point *B* is on the lower plane of the vein. Find the thickness of the vein.

12. Fig. 20.56. (12A) Find the line of intersection between the two planes by the cutting-plane method. (12B) Find the line of intersection between the two planes indicated by strike Lines 1–2 and 3–4. The plane with strike 1–2 has a dip of 30°, and the one with strike 3–4 has a dip of 60°.

13. Fig. 20.57. Find the limits of cut and fill in the plan view of the roadway. The roadway has a cut angle of 35° and a fill angle of 40°.

14. Fig. 20.58. Find the outcrop of the ore vein represented by the triangle (upper surface). Point *B* is on the lower surface.

15. Fig. 20.59. Complete the plan-profile drawing of the drainage system from manhole 1, through manhole 2, to manhole 3, using the grades indicated. Allow a drop of 0.20 ft. across each manhole to compensate for loss of pressure.

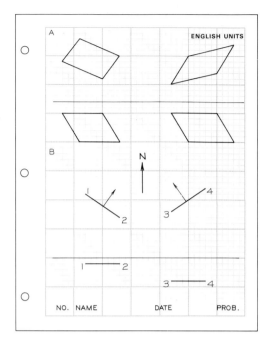

FIG. 20.56 Problems 12A and 12B.

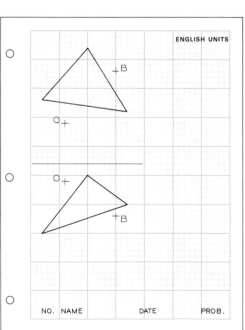

FIG. 20.55 Problem 11.

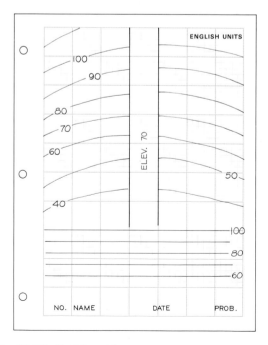

Fig. 20.57 Problem 13.

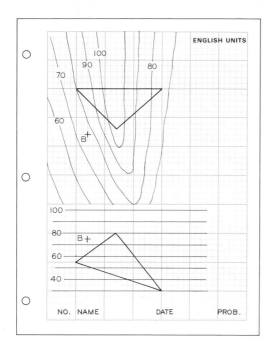

FIG. 20.58 Problem 14.

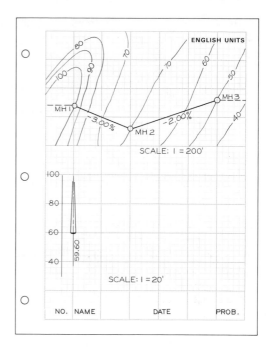

FIG. 20.59 Problem 15.

21 CHAPTER

Successive Auxiliary Views

21.1
Introduction

A design cannot be detailed with complete specifications necessary for construction unless its complete geometry has been determined. This usually requires the application of descriptive geometry principles. The Comsat satellite (Fig. 21.1) is an example of a design where various problems of geometry were solved by the use of succesive auxiliary views.

A **secondary auxiliary view** is an auxiliary view that is projected from a primary auxiliary view.
A **successive auxiliary view** is an auxiliary view of a secondary auxiliary view.

21.2
Point view of a line

When a line appears true length, its point view can be found by projecting an auxiliary view from it. In Fig. 21.2, Line 3–4 is true length in the top view since it is horizontal in the front view. It's point view is found in the primary auxiliary view by constructing reference line H-1 perpendicular to the true-length line. The height dimension, H, is transferred to the auxiliary view to locate the point view of 4–3.

The line in Fig. 21.3 is not true length in either view, which requires that the line be found true length by a primary auxiliary view. The line is found true length by projecting from the front view, but this view could have been projected from the top as well. The point view of the line is found by projecting from the

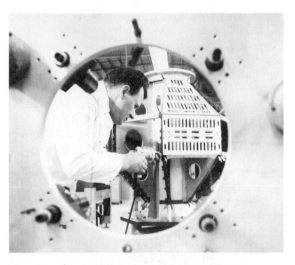

FIG. 21.1 Applications of descriptive geometry requiring successive auxiliary views are seen in this Comsat satellite. The angles between the planes and the true-size views of the surfaces were determined by applying the principles of descriptive geometry. (Courtesy of TRW Systems.)

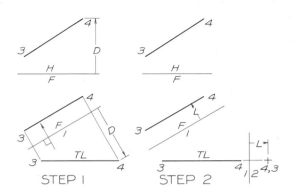

FIG. 21.3 Point of view of an oblique line

Step 1 A line of sight is drawn perpendicular to one of the views, the front view in this example. Line 3–4 is found true length by projecting perpendicularly from the front view.

Step 2 A secondary reference line 1–2 is drawn perpendicular to the true-length view of 3–4. The point view is found by transferring dimension L from the front view to the secondary auxiliary view.

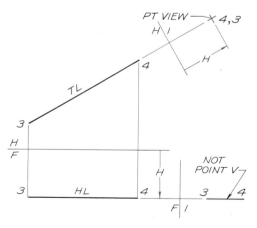

FIG. 21.2 The point view of a line can be found by projecting an auxiliary view from the true-length view of the line.

true-length line to a secondary auxiliary view. This point is labeled 2,1 since Point 2 is seen first.

21.3
Angle between planes

The angle between two planes is called a **dihedral angle.** This angle can be found in the view where the line of intersection appears as a point.

Since this view results in the point view of a line that lies on both planes, the planes will appear as edges.

The two planes in Fig. 21.4 represent a special case since the line of intersection, 1–2, is true length in the top view. This permits you to find its point view in a primary auxiliary view where the true angle can be measured.

A general case is given in Fig. 21.5 where the line of intersection between the two planes is not true length in either of the principal views.

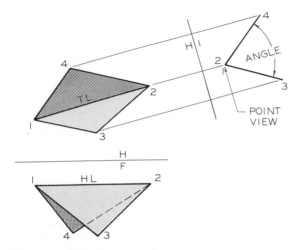

FIG. 21.4 The angle between two planes can be found in the view where the line of intersection between them projects as a point. Since the line of intersection, 1–2, is true length in the top view, it can be found as a point in a view that is projected from the top view.

The line of intersection, 1–2, is found true length in a primary auxiliary view, and the point view of the line is then found in the secondary auxiliary view. The dihedral angle is measured in the secondary auxiliary view.

It is apparent that this principle must be used to determine the angles between intersecting planes such as those shown in Fig. 21.6, where the side panels of a control tower join. This is necessary in order to design corner braces to hold the structure together.

21.4
True size of a plane

> A plane can be found true size in a view that is projected perpendicularly from an edge view of a plane.

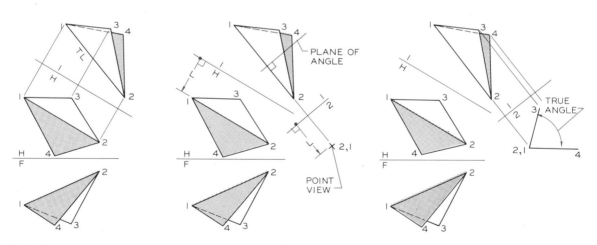

FIG. 21.5 Angle between two planes

Step 1 The angle between two planes can be seen in a view where the line of intersection appears as a point. The line of intersection is first found true length by projecting a primary auxiliary view perpendicularly from the top view, in this case.

Step 2 The point of view of the line of intersection is found in the secondary auxiliary view by projecting parallel to the true length of the line of intersection. The plane of the dihedral angle is an edge and perpendicular to the true-length line of intersection.

Step 3 The edge views of the planes are completed in the secondary auxiliary view by locating Points 3 and 4. The angle between the planes, the dihedral angle, can be measured in this view.

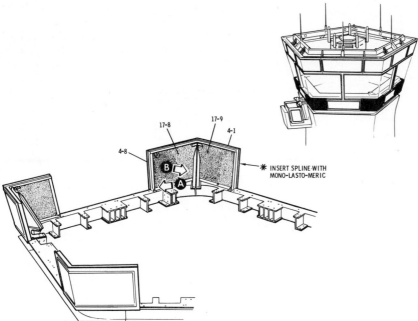

FIG. 21.6 The determination of the angle between the planes of the corner panels of the control tower utilized principles of descriptive geometry. (Courtesy of the Federal Aviation Agency.)

The front view of Plane 1–2–3 appears as an edge in the front view (Fig. 21.7). It can be found true size in a primary auxiliary view projected perpendicularly from the edge view.

In Fig. 21.8, the true size of Plane 1–2–3 is found by first finding the edge view of the plane in Step 1.

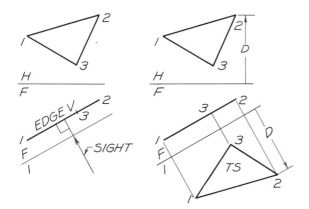

FIG. 21.7 True size of a plane

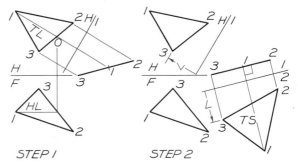

FIG. 21.8 True size of a plane

Step 1 Since Plane 1–2–3 appears as an edge in the front view, the line of sight is drawn perpendicular to the edge. The F-1 reference line is drawn parallel to the edge.

Step 2 The plane will appear true size in the primary auxiliary by locating the vertex points with depth (*D*) dimensions.

Step 1 The edge view of Plane 1–2–3 is found by finding the point view of TL line, 1–0, in a primary auxiliary view.

Step 2 A true-size view is found by projecting a secondary auxiliary view perpendicularly from the edge view of the plane. Dimension L is shown as a typical measurement used to complete the TS view.

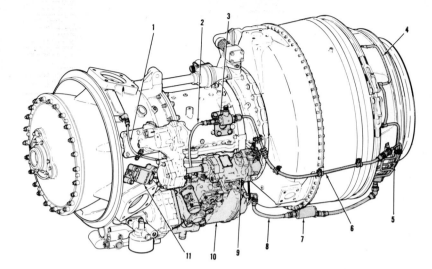

FIG. 21.9 The angles of bend in the fuel line were found by the application of the principle of finding the angle between two lines. (Courtesy of Avco Lycoming.)

In Step 2, the secondary auxiliary view is projected perpendicularly from the edge view. The result is a true-size view of the plane where each angle is true size.

This principle can be used to find the angle between lines such as bends in a fuel line of an aircraft engine (Fig. 21.9). A problem of this type is shown in Fig. 21.10 where the top and front views of intersecting lines are given. It is required that the angles be determined at each bend, and that a given radius of curvature be shown.

Angle 1–2–3 is found as an edge in the primary auxiliary view and as a true-size angle in the secondary view. The angle can be measured in this view and the radius of curvature drawn.

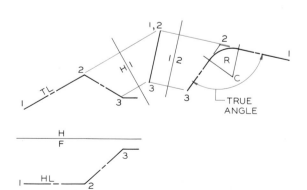

FIG. 21.10 The angle between two lines can be found by finding the plane of the lines true size.

21.5
Shortest distance from a point to a line

The shortest distance from a point to a line can be measured in the view where the line appears as a point.

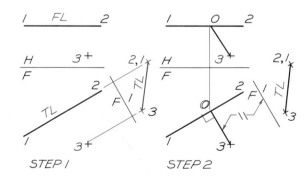

FIG. 21.11 Shortest distance from a point to a line

Step 1 The shortest distance from a point to a line will appear TL where the lines appears as a point. The TL line is found in the primary auxiliary view.

Step 2 For the connecting line to be TL in the auxiliary view, it must be parallel to the F–1 reference line in the preceding view, the front view. Line 3–0 is found and projected to the top view.

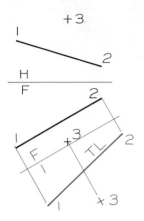
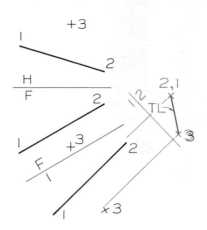
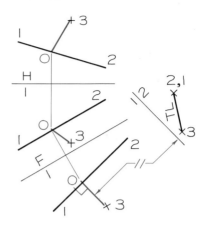

FIG. 21.12 Shortest distance from a point to a line

Step 1 The shortest distance from a point to a line can be found in the view where the line appears as a point. Line 1–2 is found true length by projecting from the front view.

Step 2 Line 1–2 is found as a point in a secondary auxiliary view projected from the true-length view of 1–2. The shortest distance appears true length in this view.

Step 3 Since 3–O is true length in the secondary auxiliary view, it must be parallel to the 1–2 reference line in the primary auxiliary view and perpendicular to the line. The front and top views of 3–O are found by projecting from view to view.

The shortest distance from Point 3 to Line 1-2 is found in a primary auxiliary view in Fig. 21.11 (Step 1). Since the line from 3 to the line is true length where the line appears as a point, it must be parallel to the reference line, F-1, in the front view.

This type of problem is solved in Fig. 21.12 by finding the Line 1-2 true length in a primary auxiliary view along with Point 3. The line is found as a point in the secondary auxiliary view, where the distance from Point 3 is true length. Since Line 0-3 is true length in this view, it must be parallel to the 1-2 reference line in the preceding view, the primary auxiliary view. It is also perpendicular to the true length view of Line 1-2 in the primary auxiliary view.

21.6
Shortest distance between skewed lines—line method

> The shortest distance between two skewed lines (randomly positioned lines) can be measured in the view where one of the lines appears as a point.

The shortest distance between two lines is perpendicular to both lines. The location of the shortest distance is both functional and economical, as demonstrated by the connector between two pipes in Fig. 21.13, since a standard connector is a 90° Tee.

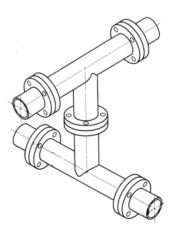

FIG. 21.13 The shortest distance between two lines, or pipes, is a line that is perpendicular to both. This is the most economical connection since perpendicular connectors are standard.

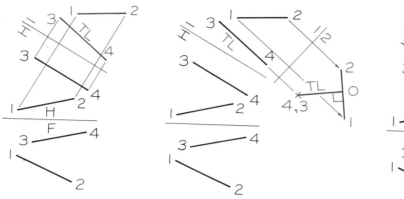
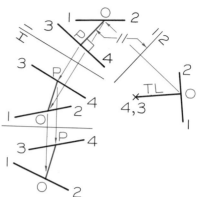

FIG. 21.14 Shortest distance between skewed lines—line method

Step 1 The shortest distance between two skewed lines can be found in the view where one line appears as a point. Line 3–4 is found true length by projecting from the top view.

Step 2 The point view of Line 3–4 is found in a secondary auxiliary view projected from the true-length view of 3–4. The shortest distance between the lines is drawn perpendicular to Line 1–2.

Step 3 Since the shortest distance is TL in the secondary auxiliary view, it must be parallel to the reference line in the preceding view. Points O and P are projected back to the given views.

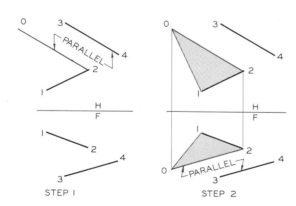

FIG. 21.15 A plane can be constructed through a line and parallel to another line by construction.

Step 1 Line 0–2 is drawn parallel to Line 3–4 to a convenient length.

Step 2 The front view of Line 0–2 is drawn parallel to the front view of Line 3–4. The length of the front view of 0–2 is found by projecting from the top view of 0. Plane 1–2–0 is parallel to Line 3–4.

A problem of this type is solved by the **line method** in Fig. 21.14. Line 3-4 is found as a point in the secondary auxiliary view where the shortest distance is drawn perpendicular to Line 1-2. Since the distance is true length in the secondary auxiliary view, it must be parallel to the 1-2 reference line in the primary auxiliary view. Point O is found by projection and OP is drawn perpendicular to Line 3-4. The line is projected back to the given principal views.

21.7
Shortest distance between skewed lines—plane method

The distance between skewed lines can be solved using the alternative **plane method.** This method involves the construction of a plane through one of the lines parallel to the other (Fig. 21.15). The top and front views of 0–2 are drawn parallel to their respective views of Line 3–4. Plane 1–2–0 is parallel to Line 3–4. Both lines will appear parallel in a view where 1–2–0 appears as an edge.

In Fig. 21.16, Plane 3–4–0 is constructed, its edge view is found, and both lines appear parallel (Step 1). A secondary auxiliary view is projected perpendicu-

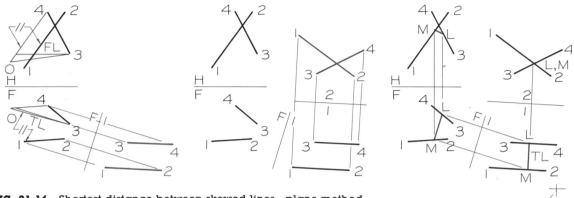

FIG. 21.16 Shortest distance between skewed lines—plane method

Step 1 Construct a plane through Line 3–4 that is parallel to Line 1–2. When this plane is found as an edge by projecting from the front view, the two lines will appear parallel in this view.

Step 2 The shortest distance will appear TL in the primary auxiliary view, where it will be perpendicular to both lines. Draw a secondary auxiliary view by projecting perpendicularly from the lines in the primary auxiliary view.

Step 3 The crossing point of the two lines is the point view of the perpendicular distance, *LM*, between the lines. This distance is projected to the primary auxiliary view, where it is TL, and back to the given views.

larly from these parallel lines to find the view where the lines cross (Step 2). This crossing point represents the point view of the shortest distance between the two lines. It will appear true length and perpendicular to the two lines when it is projected to the primary auxiliary view, where it is labeled as Line *LM*. It is projected back to the given views to complete the solution.

This principle was applied to the design of the separation of power lines shown in Fig. 21.17 where the clearance is critical.

FIG. 21.17 The determination of clearance between power lines is a critical problem for the electrical engineer. This is an application of the shortest distance between two lines. (Courtesy of the Tennessee Valley Authority.)

21.8
Shortest level distance between skewed lines

When it is required to find the shortest level (horizontal) distance between two skewed lines, the plane method must be used instead of the line method. Also, the primary auxiliary view must be projected from the top view in order to find a view where the horizontal plane appears as an edge.

In Fig. 21.18, Plane 3–4–*O* is drawn parallel to Line 1–2 and an edge view of the plane is found in the primary auxiliary view. The lines appear parallel in this view and the horizontal (H-1) appears as an edge. A line of sight is drawn parallel to the H-1, and the secondary reference line, 1–2, is drawn perpendicular to the H-1. The crossing point of the lines found in the secondary auxiliary view locates the point view of the shortest horizontal distance between the lines.

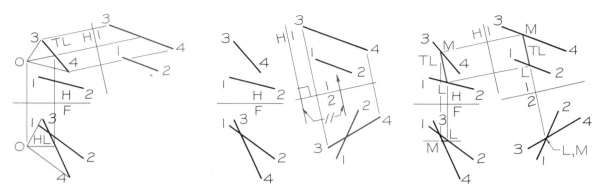

FIG. 21.18 Shortest level distance between skewed lines—plane method

Step 1 Construct Plane *O*–3–4 parallel to Line 1–2 by drawing Line *O*–4 parallel to 1–2. Find the edge view of *O*–3–4 by projecting off the top view. The lines will appear parallel in this view. The auxiliary view *must* be projected from the *top view* to find the horizontal plane as an edge.

Step 2 An infinite number of horizontal (level) lines can be drawn parallel to H-1 between the lines in the auxiliary view, but only the shortest level line will appear TL in the primary auxiliary view. Construct the secondary auxiliary view by projecting parallel to the horizontal (H-1).

Step 3 The crossing point of the two lines in the secondary auxiliary view establishes the point view of the level connector, *LM*. Project *LM* back to the given views. *LM* is parallel to the H-plane in the front view, which verifies that it is a level line.

This line, *LM*, is true length in the primary auxiliary view and parallel to the H-1 plane.

Line *LM* is projected back to the given views. As a check, this line must be parallel to the H-line in the front view since it is a level or horizontal line.

FIG. 21.19 These conveyors represent the application of skewed-line problems that must be solved using descriptive geometry principles.

21.9
Shortest grade distance between skewed lines

Many lines representing highways, power lines, or conveyors are connected by lines at a specified grade other than horizontal or perpendicular. Conveyors, such as the one shown in Fig. 21.19, are used to transport aggregates or grain for mixing.

If a 50% grade connector between two lines must be found between the two lines in Fig. 21.20, the plane method is used as in the two previous examples. A view where the lines appear parallel is constructed and a 50% grade line is constructed from the edge view of the horizontal (H-1). The auxiliary views must be projected from the top view in order to have an edge view of the horizontal from which the 50% grade is constructed.

The grade line can be constructed in two directions from the H-1 line, but the shortest distance will be the most nearly perpendicular to both lines (Step 2). The secondary auxiliary view is projected parallel with this 50% grade line to find the crossing point of the lines to locate the shortest connector, *LM*, at a 50% grade.

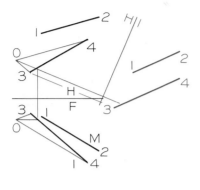 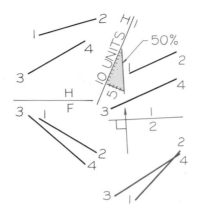 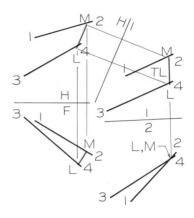

FIG. 21.20 Grade distance between skewed lines

Step 1   To find a line on a grade between two skewed lines, the primary auxiliary must be projected from the top view. Plane 3–4–O is constructed parallel to Line 1–2. The edge view of the plane is found where the lines appear parallel.

Step 2   Construct a 50-percent grade line with respect to the edge view of the H-1 line in the primary auxiliary view that is as nearly perpendicular to the lines as possible. Project the secondary auxiliary view parallel to the grade line. The shortest grade distance will appear true length in the primary auxiliary view.

Step 3   The point of crossing of the two lines in the secondary auxiliary view establishes the point view of the 50-percent grade line, LM. This line is projected back to the previous views in sequence.

Line LM is projected back to all views. This line appears true length in the primary auxiliary view where the lines appear parallel.

Any connector between skewed lines will appear true length in the view where the lines appear parallel. Perpendicular, horizontal, and grade lines are true length in this view.

21.10
Angular distance to a line

Standard connectors used to connect pipes and structural members are available in two standard angles—90° and 45°. Consequently, it is economical to incorporate these into a design rather than having to design specially made connectors.

In Fig. 21.21 it is required to locate the point of intersection on Line 1–2 of a line drawn from Point O at a 45° angle to the line. The plane of the line and point, 1–2–O, is found as an edge in the primary auxiliary, and as a true-size plane in the secondary auxil-

iary view. The angle can be measured in this view where the plane of the line and point is true size.

The 45° connector is drawn from O to the line toward Point 2 to slope downhill, or toward Point 1 if it is to slope uphill. This can be determined by referring to the front view where height can be easily seen. Line OP is projected back to the given views.

21.11
Angle between a line and a plane—plane method

The angle between a line and a plane can be measured in the view where the plane appears as an edge and the line true length.

In Fig. 21.22, the edge view of the plane is found in a primary auxiliary view projected from any primary view. The plane is then found true size in

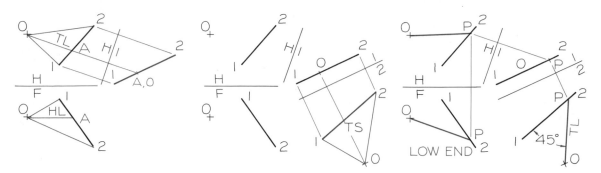

FIG. 21.21   Line through a point with a given angle to a line

Step 1   Connect Point O to each end of the line to form a Plane 1–2–O in both views. Draw a horizontal line in the front view of the plane and project it to the top view, where it is TL. Find the edge view of the plane by obtaining the point view of Line A–O.

Step 2   Find the TS of Plane 1–2–O by projecting perpendicularly from the edge view of the plane in the primary auxiliary view. The plane can be omitted in this view and only Line 1–2 and Point O shown.

Step 3   Line OP is constructed at the specified angle with Line 1–2. If the angle is toward Point 2 the line slopes downhill, and if toward Point 1 it slopes uphill. Point P is projected back to the other views in sequence to show the line, OP, in all views.

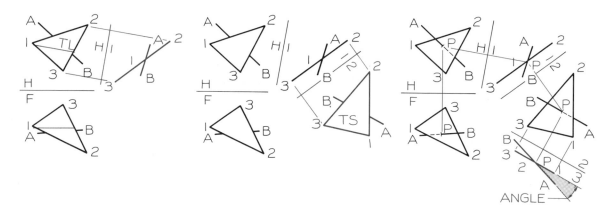

FIG. 21.22   Angle between a line and a plane—plane method

Step 1   The angle between a line and a plane can be measured in the view where the plane is an edge and line is TL. The plane is found as an edge by projecting off the top view. The line is not true length in this view.

Step 2   The plane is found true size by projecting perpendicularly from the edge view of the plane. A view projected in any direction from a TS plane will show the plane as an edge.

Step 3   A third successive auxiliary view is projected perpendicularly from Line AB. The line appears TL and the plane as an edge in this view where the angle is measured. The piercing points and visibility are shown in the views by projecting back in sequence.

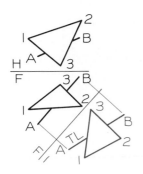
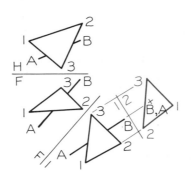
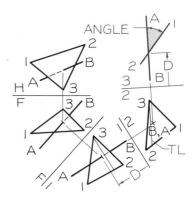

FIG. 21.23 Angle between a line and a plane—line method

Step 1 The angle between a line and a plane can be measured in the view where the plane appears as an edge and the line is true length. Find a view where Line *AB* is true length and project the plane to this view also.

Step 2 Construct a point view of Line *AB* in the secondary auxiliary view. Plane 1–2–3 does not appear true size in this view unless the line is perpendicular to the plane. The point view of the line in this view is the piercing point on the plane.

Step 3 An edge view of the plane is found by finding the point view of a line on the plane. *AB* will appear TL in this view since it was a point in the secondary auxiliary view. Measure the angle, project back to each view and determine the visibility.

Step 2 where the line is foreshortened. The line, *AB*, can be found true length in a third auxiliary view projected perpendicularly from the secondary auxiliary view of *AB*. The line appears true length and the plane as an edge in the third successive auxiliary view.

The piercing point is projected back to the views in sequence and the visibility is determined for each view.

21.12
Angle between a line and a plane—line method

To find the angle between a line and a plane by the line method, the line is first found as a point, and then true length is shown as in Fig. 21.23. The plane is foreshortened in Step 2 where the Line *AB* appears as a point.

The plane is found as an edge in a third auxiliary view by finding the point view of a line on the plane in this view. Since the view is projected from the point view of Line *AB*, the line will appear true length. This view satisfies the condition that the line be true length

and the plane an edge. The angle is measured in the third view. The piercing point is projected back to previous views and the visibility is determined to complete the solution.

21.13
Elliptical views of a circle

Circles appear as circles only when your line of sight is perpendicular to the plane of the circle. When the line of sight is oblique to the plane, the circle will appear as an ellipse. The following definitions are given to explain the terminology of ellipses (Fig. 21.24).

MAJOR DIAMETER Major diameter is the largest diameter measured across an ellipse. It is always true length.

MINOR DIAMETER The minor diameter is the shortest diameter across an ellipse and it is perpendicular to the major diameter at its midpoint.

ELLIPSE ANGLE The angle between the line of sight and the edge view of the plane of the circle is the ellipse angle.

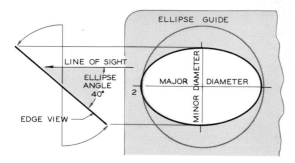

FIG. 21.24 The ellipse angle is the angle the line of sight makes with the edge view of a circle. An ellipse template can be used to construct the ellipse by aligning with the major and minor axes.

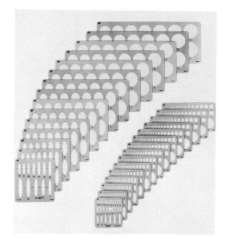

FIG. 21.25 Typical ellipse templates used for ellipse representation. (Courtesy of the A. Lietz Company.)

CYLINDRICAL AXIS The cylindrical axis is the center line of a cylinder that connects the centers of the circular ends of a right cylinder.

ELLIPSE TEMPLATE Ellipse templates are used to draw ellipses by aligning the major and minor diameters. The templates are available in 5° intervals (Fig. 21.25).

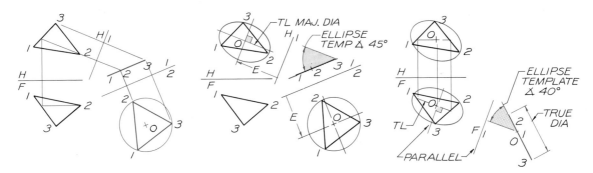

FIG. 21.26 Elliptical views of a circle

Required Construct a circle that will pass through each vertex of the plane. Show the circle in all views.

Step 1 Determine the true size of Plane 1–2–3. Draw a circle through the vertexes in the TS view. The center of the circle, *O*, is found at the intersection of the perpendicular bisectors of the triangle's sides.

Step 2 Draw diameters, *AB* and *CD*, parallel and perpendicular to the 1–2 line, respectively, in the secondary auxiliary view. Project these lines to the primary auxiliary and top views, where they will represent the major and minor diameters of an ellipse. Select the ellipse template for drawing the top view by measuring the angle between the line of sight and the edge view of the plane.

Step 3 Determine the ellipse template for drawing the ellipse in the front view by finding the edge view of the plane in an auxiliary view projected from the front view. The ellipse angle is measured in the auxiliary view. The major diameter is TL and parallel to a TL line on the plane in the front view. The minor diameter is perpendicular to the major diameter.

It is required in Fig. 21.26 to draw a circle through the three vertexes of the triangular plane, and to show the circle in all views. This is done by finding the true size of the triangle and then finding the center of the circle where the perpendicular bisectors of the sides intersect. The circle is drawn in this view.

The circle is projected to the primary auxiliary view where it appears as an edge. The major and minor diameters are projected back to the top view where they are parallel and perpendicular to the H-1 reference line. The ellipse guide angle is found to be 45° since this is the angle the line of sight makes with the edge view of the circle.

The elliptical view of the circle is found in the front view projecting an edge view of the plane from the front view. The major and minor diameters are drawn parallel and perpendicular to the F-1 reference line.

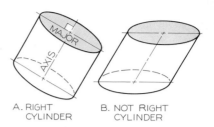

FIG. 21.27 The axis of the cylinder (the center line) is drawn perpendicular to the major diameter of its elliptical end, if it is a right cylinder. It should be apparent to you by inspection that the cylinder at B is not a right cylinder.

The elliptical ends of right cylinders will be perpendicular to the cylinder's axis (Fig. 21.27A). Consequently, the major diameters will be perpendicular to the axis. When the major diameter is not drawn perpendicular to the axis (Fig. 21.27B), it is apparent to even the untrained eye that the cylinder is not a right cylinder.

The angle of the ellipse template is the angle the line of sight makes with the edge view of the circle, 40° in this case.

Problems

Use Size A (8½″ × 11″) sheets for the following problems and lay out the problems using instruments. Each square on the grid is equal to 0.20 in. or about 5 mm. The problems can be laid out on grid or plain paper. Label all reference planes and points in each problem with ⅛″ letters and/or numbers, using guidelines.

The crosses marked "1" and "2" are to be used for placing the primary and secondary reference lines. The primary reference line should pass through "1" and the secondary through "2".

1—2. Find the point views of the line in Fig. 21.28.

3—4. Find the angles between the planes in Fig. 21.28.

5—6. Find the true-size views of the planes in Fig. 21.29. Project from the front view in Problem 5 and from the top view in Problem 6.

7—8. Find the angles between the lines in Fig. 21.30. Project from the top views of both problems.

9. Find the shortest distance from the point to the line in Fig. 21.31 and show the distance in all views. Use the plane method and project from the left side view. Scale: full size.

10. Find the shortest distance from point O to the line in Fig. 21.31 and show the distance in all views. Use the line method and project from the top view. Scale: full size.

11—12. Find the shortest distances between the two skewed lines in Fig. 21.32 using the line method, and show the distances in all views. Begin each problem by finding line 3–4 true length, using the cross marks given. Scale: full size.

13. Find the shortest horizontal distance between the

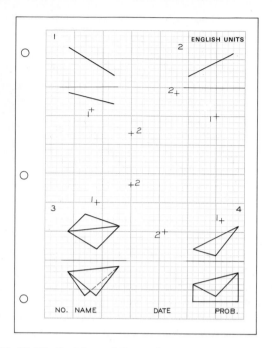

FIG. 21.28 Problems 1 through 4.

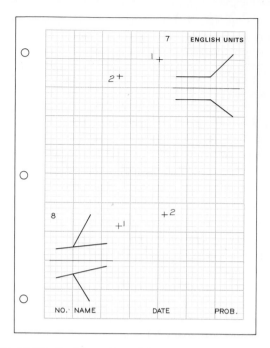

FIG. 21.30 Problems 7 and 8.

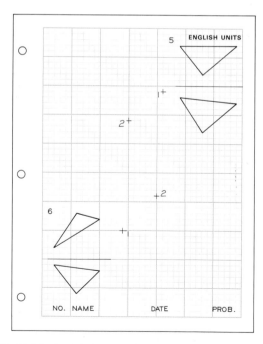

FIG. 21.29 Problems 5 and 6.

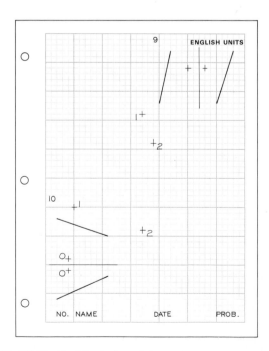

FIG. 21.31 Problems 9 and 10.

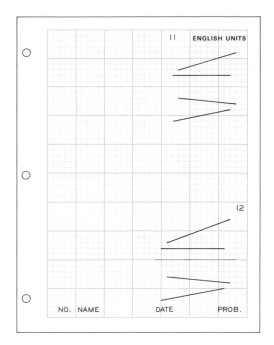

FIG. 21.32 Problems 11 and 12.

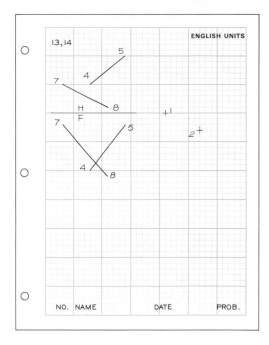

FIG. 21.33 Problems 13 and 14.

two lines in Fig. 21.33 by the plane method. Project from the top view. Scale: full size.

14. On a separate sheet of paper, redraw Problem 11 (Fig. 21.33) and find the shortest 20-percent grade between the two lines. Project the first view from the top view. Scale: full size.

15. Find the shortest 25-percent grade distance between the two lines in Fig. 21.34. Show the distance in all views. Scale: full size.

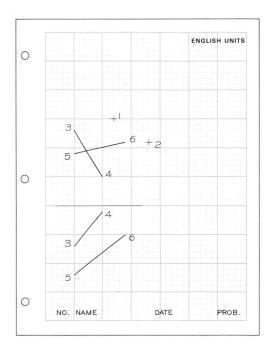

FIG. 21.34 Problem 15.

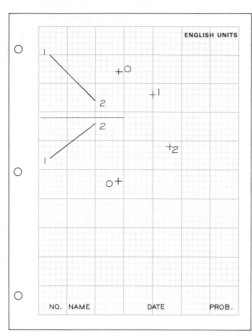

FIG. 21.35 Problem 16.

16. Find the connector from Point *O* that will intersect line 1–2 at a 60° angle (Fig. 21.35). Show this line in all views. Project from the top view. Scale: full size.

17. Find the angle between the line and the plane in Fig. 21.36 by using the plane method. Project from the front view and show the visibility in all views. Scale: full size.

18. Same as Problem 17, except use the line method.

19. Construct a circle that will pass through each vertex of the triangle in Fig. 21.37. Project from the top view and show the elliptical views of the circle in all views.

20. Find the front view of the elliptical path of a circular section through the sphere in Fig. 21.37. The edge view of the section is shown in the top view.

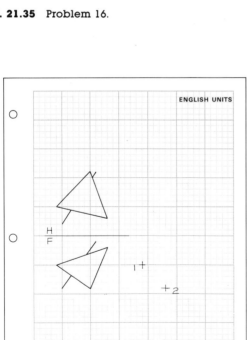

FIG. 21.36 Problems 17 and 18.

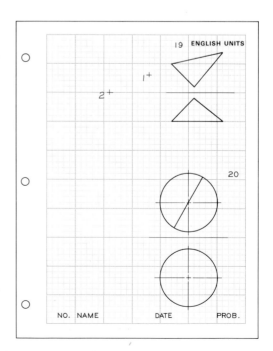

FIG. 21.37 Problems 19 and 20.

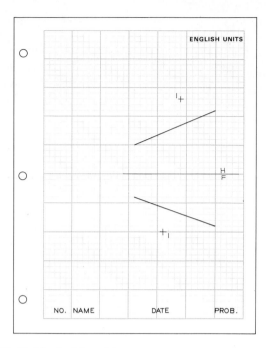

FIG. 21.38 Problem 21.

21. The line in Fig. 21.38 represents the center line of a right cylinder that has each circular end perpendicular to the axis. Complete the views of the cylinder and show the ends, which will appear as ellipses.

22

CHAPTER

Revolution

22.1
Introduction

The orthicon camera in Fig. 22.1 was designed to permit the camera to be revolved about three axes; thus it is possible to aim it in any direction to track space vehicles. This is just one of many designs that were based on the principles of revolution.

Revolution is another method of solving problems that can be solved by auxiliary views. In fact, revolution techniques were developed and used before auxiliary views came into use.

22.2
True length of a line in the front view

The simple object in Fig. 22.2 demonstrates how an inclined surface can be found true size by auxiliary view and by revolution. When the auxiliary view

FIG. 22.1 This orthicon camera is an advanced example of a design that utilizes principles of revolution. Its cradle was designed to permit the camera to be revolved into any position by revolving it about three axes. (Courtesy of ITT Industrial Laboratories.)

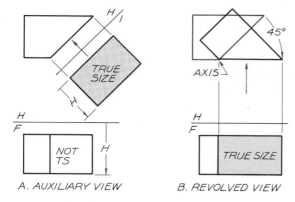

FIG. 22.2 The surface is found true size by an auxiliary view in Part A. At B, the surface is found true size by revolving the top view.

method is used, the observer changes position to an auxiliary vantage point, where he or she can look perpendicularly at the inclined surface.

When the revolution method is used, the top view of the object is revolved about an axis until the edge view of the inclined plane is perpendicular to the standard line of sight from the front view. In other words, the observer's line of sight does not change, but the conventional lines of sight between adjacent orthographic views are used.

A single line can be found true length in the front view by revolution, as shown in Fig. 22.3. By establishing the point view of an axis in the top view, Line *AB* is revolved into a position parallel to the frontal plane.

The top view represents the circular base of a right cone and the front view is the triangular view of a cone. Line *AB'* is the outside element of the cone and is therefore true length.

Figure 22.4 illustrates the technique of finding Line 1–2 true length in the front view. When in its first position, the observer's line of sight is not perpendicular to the triangular plane or Line 1–2. But when it is revolved to be perpendicular to the line of sight, the triangle appears true size and Line 1–2' is true length.

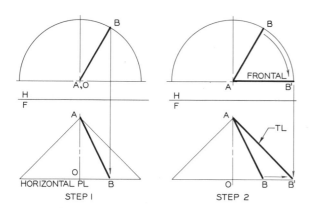

FIG. 22.3 True length in the front view

Step 1 The top view of Line *AB* is used as a radius to draw the base of a cone with Point *A* as the apex. The front view of the cone is drawn with a horizontal base through Point *B*. Line *AO* is the axis of the cone.

Step 2 The top view of Line *AB* is revolved to be parallel to the frontal plane. When projected to the front view, frontal Line *AB'* is the outside element of the cone and is true length.

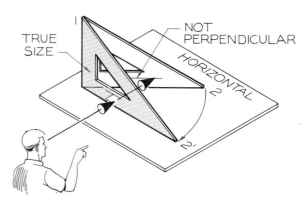

FIG. 22.4 Line 1–2 of the triangle does not appear true length in the front view because your line of sight is not perpendicular to it. When the triangle is revolved to a position where your line of sight is perpendicular to it, Line 1–2' can be seen true length.

22.3
True length of a line in the top view

A surface that appears as an edge in the front view can be found true size in the top view by a primary auxiliary view or by a single revolution (Fig. 22.5).

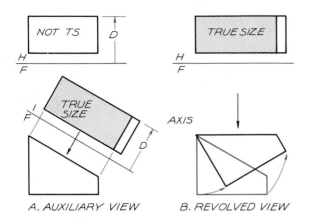

A. AUXILIARY VIEW   B. REVOLVED VIEW

FIG. 22.5 At A, the inclined plane is found true size by an auxiliary with a line of sight that is perpendicular to the surface. At B, the surface is found true size by revolving the front view until it is perpendicular to the standard line of sight from the top view.

The axis of revolution is located as a point in the front view, and is true length in the top view. The edge view of the plane is revolved until it is a horizontal edge in the front view (Fig. 22.5A). It is projected to the top view to find the surface true size. As in the auxiliary-view method, the depth dimension (D) does not change.

Line CD is found true length in the top view by revolving the line into a horizontal position in Step 2 of Fig. 22.6. The arc of revolution in the front view represents the base of a cone of revolution. Line CD' is true length in the top view since it is an outside element of the cone. Note that the depth dimension in the top view does not change.

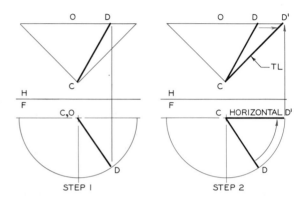

STEP 1   STEP 2

FIG. 22.6 True length of a line in the top view

Step 1 The front view of Line CD is used as a radius to draw the base of a cone with Point C as the apex. The top view of the cone is drawn with the base shown as a frontal plane.

Step 2 The front view of Line CD is revolved into position CD' where it is horizontal. When projected to the top view, CD' is the outside element of the cone and is true length.

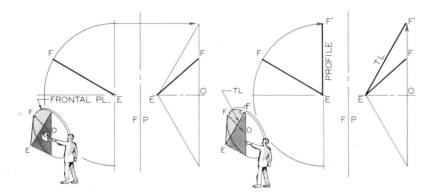

FIG. 22.7 True length of a line in the side view

Step 1 The front view of Line EF is used as a radius to draw the circular view of the base of a cone. The side view of the cone is drawn with its base through Point F that is a frontal edge.

Step 2 Line EF in the front view is revolved to position EF' where it is a profile line. Line EF' in the profile view is true length (TL), since it is a profile line and the outside element of the cone.

22.4
True length of a line in the profile view

Line *EF* in Fig. 22.7 is found true length by revolving it in the front view until it is parallel to the edge view of the profile plane (Step 1). The circular view of the cone is projected to the side view, where the triangular shape of the cone is seen. Since *EF'* is a profile in line in Step 2, *EF'* is true length in the side view, where it is the outside element of the cone.

In the previously covered examples, each line has been revolved about one of its ends. However, a line can be revolved about any point on its length. Line 5–6 in Fig. 22.8 is an example of a line that is found true length by revolving it about Point *O* near its midpoint.

22.5
Angles with a line and principal planes

The angle between a line and a plane will appear true size in the view where the plane is an edge and the line is true length. Two principal planes appear as edges in all principal views. Consequently, when a line appears true length in a principal view, the angle

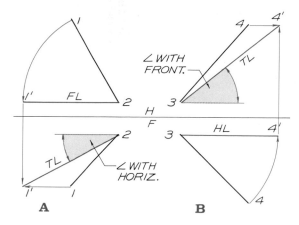

FIG. 22.9 Angles with principal planes

A. The angle with the horizontal plane can be measured in the front view if the line appears true length in this view.

B. The angle with the frontal plane can be measured in the top view if the line appears true length in this view.

between the line and two principal planes can be measured.

Two examples of finding lines true length by revolution are shown in Fig. 22.9

The angle between the horizontal and the profile planes can be measured in Part A in the front view. The angle with horizontal and profile planes can be measured in the top view in Part B.

22.6
True size of a plane

When a plane appears as an edge in a principal view (Fig. 22.10), it can be revolved to be parallel to the reference line (Step 1). The new front view is true size.

The plane in Fig. 22.11 is found true size by the combination of an auxiliary view and a single revolution. The plane is found as an edge projected from a true-length line in the plane. The edge view is revolved to be parallel to the F-1 reference line (Step 2). The true size of the plane is found by projecting the original points, 1, 2, and 3, in the front view parallel to the F-1 line to intersect the projectors from 1' and 2'. The true size of the plane could have been found by projecting from the top view to find the edge view as well.

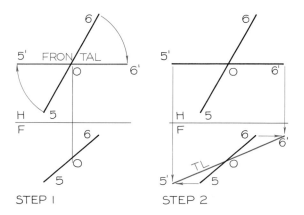

FIG. 22.8 In the preceding examples, the lines have been revolved about their ends, but they can be found true length by revolving them about any point on them. Line 5–6 is revolved into a frontal position in the top view and is found true length in the front view.

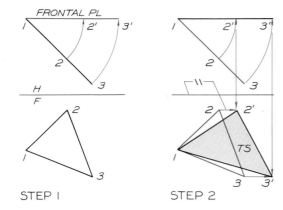

FRONTAL PL

STEP 1   STEP 2

FIG. 22.10 True size of a revolution

Step 1 The edge view of the plane is revolved to be parallel to the frontal plane.

Step 2 Points 2' and 3' are projected to the horizontal projectors from 2 and 3 in the front view.

22.7
True size of a plane by double revolution

The edge view of a plane can be found by revolution without the use of auxiliary views (Fig. 22.12). A frontal line is drawn on Plane 1–2–3 and the line appears true length in the front view. The plane is revolved until the true-length line is vertical in the front view (Step 1). The true-length line will project as a point in the top view, and therefore the plane will appear as an edge in this view (Step 2). Projectors from the top view of Points 2 and 3 are parallel to the H-F reference line.

A second revolution, called a **double revolution,** can be made to revolve this edge view of the plane until it is parallel to the frontal plane, as shown in Fig. 22.13. The top views of Points 1' and 2' are projected to the front view where Plane 1'–2'–3' is true size.

This second revolution could have been performed in Fig. 22.12, but this would have resulted in an overlapping of views that would have made it difficult to observe the separate steps.

Double revolution is used in Fig. 22.14 to find the oblique plane of the object, Plane 1–2–3, true size. In

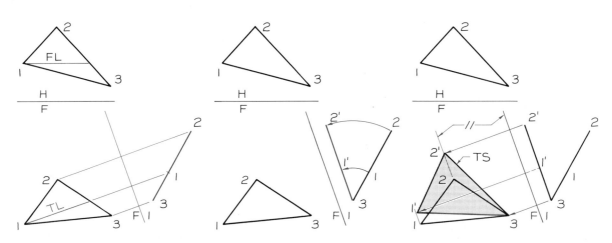

FIG. 22.11 True size of a plane by revolution

Step 1 Find the edge view of the plane by finding the point view of a true-length line on it in the front view.

Step 2 Revolve the edge view of the plane about one of its points (Point 3 in this example) until it is parallel to the F-1 line.

Step 3 Project Points 1' and 2' to the front view to the projectors from Points 1 and 2. These projectors must be parallel to the F-1 line.

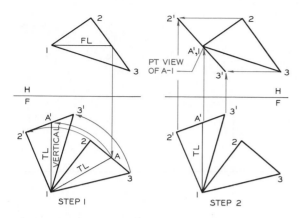

FIG. 22.12 Edge view of a plane

Step 1   A frontal line is found true length on the front view of the plane. The front view is revolved until the true-length line is vertical.

Step 2   Since the TL line (1–A') is vertical, it will appear as a point in the top view and the plane will appear as an edge, 1–2'–3'.

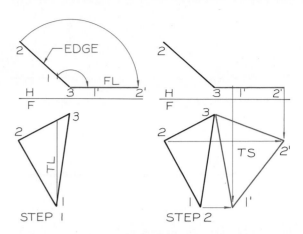

FIG. 22.13 True size of a plane

Step 1   When a plane appears as an edge in a principal view, it can be revolved to a position parallel to a reference line, the frontal plane in this case.

Step 2   Points 2' and 1' are projected to the front view to intersect with the horizontal projectors from the original Points 1 and 2. The plane is true size in this view.

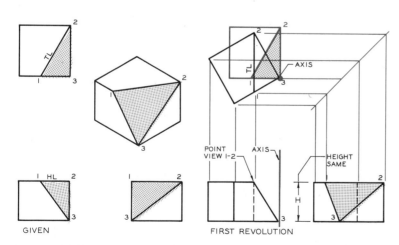

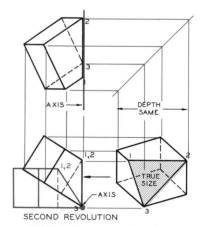

FIG. 22.14 True size by double revolution

Given   Three views of a block with an oblique plane across one corner.

Required   Find the plane true size by revolution.

Step 1   Since Line 1–2 is horizontal in the front view, it is true length in the top view. The top view is revolved into a position where Line 1–2 can be seen as a point in the front view.

Step 2   Plane 1–2–3 can be revolved into a vertical position in the front view, so that it will appear TS in the side view. The depth dimension does not change, since it is parallel to the axis of revolution.

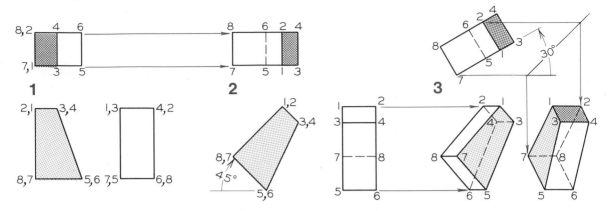

FIG. 22.15 Double revolution

Given Three views of a surface with an inclined plane.
Required Rotate the front view 45° and the top view 30°.

Step 1 The front view is revolved 45°, and the new width is projected to the top view where the depth is unchanged.

Step 2 The top view is revolved 30° to change the width and depth, but the height remains the same. The resulting front view is an axonometric pictorial.

Step 1, the true-length Line 1–2 on the plane is revolved in the top view until it is perpendicular to the frontal plane. Consequently, Line 1–2 appears as a point in the front view, and the plane appears as an edge. This changes the width and depth dimension, but the height dimension does not change.

In Step 2, the edge view of the plane is revolved

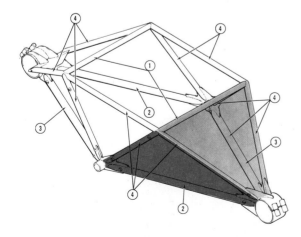

FIG. 22.16 The angle between any two planes of the helicopter engine mount can be found by using revolution principles. (Courtesy of Bell Helicopter Corp.)

into a vertical position parallel to the profile plane. The plane is found true size by projecting to the profile view where the depth dimension is unchanged and the height dimension has been increased.

A second example of double revolution of a solid is shown in Fig. 22.15 with dimensions transferred from view to view. The front view is revolved clockwise 45°, and new top and side views are drawn with the depth dimension remaining constant. The top view is revolved 30° counterclockwise in Step 2. A new front view and side view are constructed by using projectors from the side view found in Step 1 and the top view found in Step 2.

22.8
Angle between planes

The engine mount frame of a helicopter is an application where the angle between two intersecting planes must be found, in order to provide its design specifications (Fig. 22.16).

In Fig. 22.17, the angle between two planes is found by drawing the edge view of the dihedral angle (the angle between the planes) perpendicular to the line of intersection, and the plane of the angle is projected to front view (Step 1). The edge view of the

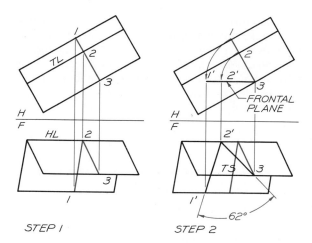

STEP 1 STEP 2

FIG. 22.17 Angle between planes

Step 1 A right section is drawn perpendicular to the TL line of intersection between the planes in the top view and is projected to the front view. The section is not true size in the front view.

Step 2 The edge view of the right section is revolved to position 1'–2'–3 in the top view to be parallel to the frontal plane. This section is projected to the front view, where it is true size.

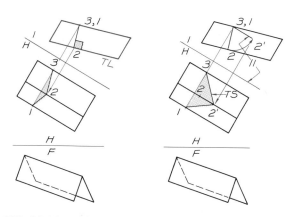

FIG. 22.18 Angle between oblique planes

Step 1 A TL view of the line of intersection is found in an auxiliary view projected from the top view. The right section is constructed perpendicular to the TL line of intersection and is projected to the top view.

Step 2 The edge view of the right section is revolved to be parallel to the H-1 reference line so the plane will appear TS in the top view revolved. The angle between the planes is Angle 1'–2'–3.

angle is revolved until it is a frontal plane in the top view, and it is then projected to the front view where its true-size view is found (Step 2).

A similar problem is solved in Fig. 22.18. In this example the line of intersection does not appear true length in the given views; consequently an auxiliary view is used in Step 1 to find it true length. The plane of the angle between the planes can be drawn as an edge perpendicular to the true-length line of intersection. The foreshortened view of Plane 1–2–3 is projected to the top view in Step 1. The edge view of Plane 1–2–3 is then revolved in the primary auxiliary view until it is parallel to the H-1 line (Step 2).

22.9
Location of directions

To solve more advanced problems of revolution, it is necessary that you be able to locate the basic directions of up, down, forward, and backward in any view that you are given. In Fig. 22.19A, the direction of backward and up are located by first drawing directional arrows in the given top and front views.

Line 4–5 is drawn pointing backward in the top view, and its front view appears as a point. Arrow 4–5 is projected to the auxiliary view as any other line to locate the direction of backward. By drawing the arrow on the other end of the line, you would find the direction of forward.

The direction of up is located in Fig. 22.19B by drawing Line 4–6 in the direction of up in the front view and in the top view as a point. The arrow is found in the primary auxiliary by the usual projection

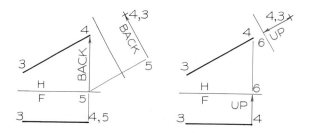

FIG. 22.19 To find the direction of backward, forward, up, and down in an auxiliary view, construct an arrow pointing in the desired direction in the given principal views and project this arrow to the auxiliary view. The directions of backward and up are shown here.

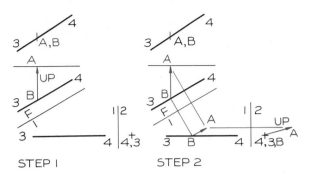

FIG. 22.20 Direction in a secondary auxiliary view

Step 1 To find the direction of up in the secondary auxiliary view, Arrow *AB* is drawn pointing upward in the front view. It appears as a point in the top view.

Step 2 Arrow *AB* is projected to the primary and secondary auxiliary views like any other line. The direction of up is located in the secondary auxiliary view.

method, and the direction of up is located. The direction of down would be in the opposite direction.

The location of directions in secondary auxiliary views are found in the same manner. The direction of up is found in Fig. 22.20 by beginning with an arrow

pointing upward in the front view and appearing as a point in the top view (Step 1). The arrow, *AB,* is projected from the front view to the primary, and then to a secondary auxiliary view to give the direction of up in all views. The other directions can be found in the same manner by beginning with the two given principal views of a known directional arrow.

22.10
Revolution of a point about an axis

In Fig. 22.21 it is required that Point *O* be revolved about Axis 3–4 to its most forward position. The circular path of revolution is drawn in the primary auxiliary view where the axis is a point (Step 1). The direction of forward is drawn in Step 2 and the new location of Point *O* is found at *O'.* By projecting back through the successive views, Point *O'* is found in each view. Note that *O'* lies on the line in the front view; this verifies that *O'* is in its most forward position.

The problem in Fig. 22.22 requires an additional auxiliary view since Axis 3–4 is not true length in the

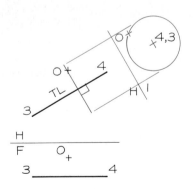
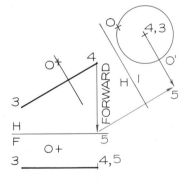
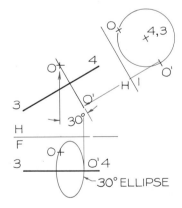

FIG. 22.21 Revolution about an axis

Step 1 To rotate Point *O* about Axis 3–4, it is necessary to find the point view of the axis in a primary auxiliary view. The circular path is drawn and the path of revolution is shown in the top view as an edge that is perpendicular to the axis.

Step 2 If it is required to rotate Point *O* to its most forward position, draw an arrow pointing forward in the top view. It will appear as a point in the front view. Arrow 4–5 is found in the auxiliary view to locate Point *O'.*

Step 3 Point *O'* is projected to the given views. The path of revolution appears as an ellipse in the front view since the axis is not TL. A 30° ellipse is used to construct the ellipse, since this is the angle your line of sight makes with the circular path.

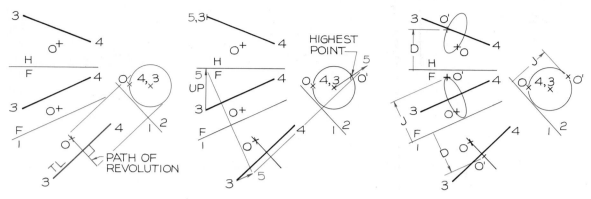

FIG. 22.22 Revolution of a point about an axis

Step 1 To rotate *O* about Axis 3–4, the axis is found as a point in a secondary auxiliary view where the circular path is drawn. The path appears as an edge in the primary auxiliary view where the axis is TL.

Step 2 If it is required to rotate *O* to its highest position, construct Arrow 3–5 in the front and top views that points upward. The direction of this arrow in the secondary auxiliary view locates the highest position, *O'*.

Step 3 Point *O'* is projected back to the given views by transferring the Dimensions *J* and *D* using your dividers. The highest point lies over the line in the top view to verify its position.

given views. Therefore the line must be found true length before it can be found as a point where the path of revolution can be drawn as a circle. Point *O* is revolved into its highest position, *O'*, where the "up" arrow, 3–5, is found in the secondary auxiliary view.

By projecting back to the given views, *O'* is located in each view. Its position in the top view is over the axis, which verifies that the point is located at its highest position.

The paths of revolution will appear as edges when their axes are true length, and as ellipses when their axes are not true length. The angle of the ellipse guide

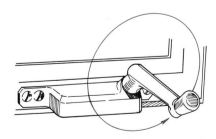

FIG. 22.23 This handcrank that is used on casement windows is an example of a problem that must be solved by using revolution principles. The handle must be properly positioned so it will not interfere with the window sill or wall.

for drawing the ellipse in the front view is the angle the projectors from the front view make with the edge view of the revolution in the primary auxiliary view. To find the ellipse in the top view, an auxiliary view must be used to find the path of revolution as an edge perpendicular to the true-length axis projected from the top view.

The handcrank of a casement window (Fig. 22.23) is an example of a problem that must be solved using revolution to determine the clearances between the sill and the window frame.

22.11
Revolution of a line about an axis

Line 3–4 is revolved about Axis 1–2 in Fig. 22.24. The point view of Axis 1–2 is found as a point in Step 1. A circle is drawn tangent to Line 3–4 with its center at the point view of 1–2, and arcs are drawn through each end of the line as well. The perpendicular is revolved into the desired position and the new endpoints are found, 3' and 4'.

The top view of Line 3'–4' is found by projecting parallel to the H-1 line from the original points of 3

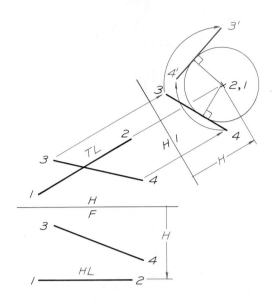

FIG. 22.24 Revolution of a line about an axis

Step 1 The axis, 1–2, is found as a point in the auxiliary view. Line 3–4 is revolved to its specified position in this view.

Step 2 The new position of Line 3–4 is projected back to the top view where projectors parallel to the top view of the H-1 reference line intersect those from the auxiliary view to find Line 3'–4'. Line 3'–4' is located in the front view.

FIG. 22.25 A conveyor chute must be properly installed so that two edges of its right section are vertical so the conveyors will function properly. This requires the application of the revolution of a prism about its axis. (Courtesy of Stephens-Adamson Manufacturing Co.)

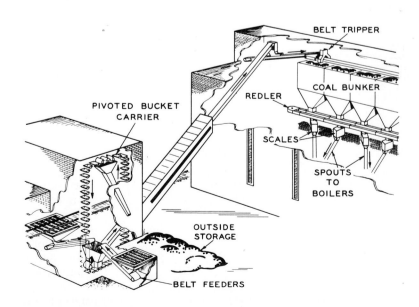

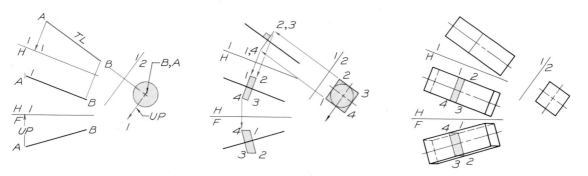

FIG. 22.26 Revolution of a prism about its axis

Step 1   Locate the point view of Center line *AB* in the secondary auxiliary view by drawing a circle about the axis with a diameter equal to one side of the square right section. Draw a vertical arrow in the front and top views and project it to the secondary auxiliary view to indicate the direction of vertical.

Step 2   Draw the right section, 1–2–3–4, in the secondary auxiliary view with two sides parallel to the vertical directional arrow. Project this section back to the successive views. The edge view of the section could have been located in any position along centerline *AB* in the primary auxiliary view, so long as it was perpendicular to the center line.

Step 3   Draw the lateral edges of the prism through the corners of the right section so that they are parallel to the center line in all views. Terminate the ends of the prism in the primary auxiliary view where they appear as edges that are perpendicular to the center line. Project the corner points of the ends to the top and front views.

and 4 in the top view, as shown in Step 2. These projectors intersect those from the auxiliary view. The front view is obtained by projecting from the top view and transferring the height dimensions from the primary auxiliary view (Step 2).

22.12
Revolution of a right prism about its axis

A coal chute between two buildings is shown in Fig. 22.25, which is used to convey coal at a continuous rate. It is necessary to have the sides of the enclosed chute vertical and the bottom of the chute's right section horizontal in order for the coal to be transported efficiently.

In Fig. 22.26, it is required that the right section be positioned about Centerline *AB* so that two of its sides will be vertical. This is done by finding the point view of the axis in Step 1, and the direction of up is projected to this view. In Step 2, the right section is

drawn about the axis so that two of its sides are parallel to the upward arrow. The right section is found in the other views in Step 2. In Step 3, the sides of the chute are constructed parallel to the axis.

The bottom of the chute's right section will be horizontal, and properly positioned for conveying coal.

22.13
Angle between a line and plane

The angle between a line and a plane is found by a combination of auxiliary views and revolution in Fig. 22.27. The plane is found true size in a secondary auxiliary view in Step 1.

The line is revolved in Step 2 until it is parallel to the 1–2 reference line. The line can then be found true length in the primary auxiliary view in Step 3. Since the line appears true length and the plane appears as an edge in this view, the true angle can be measured here.

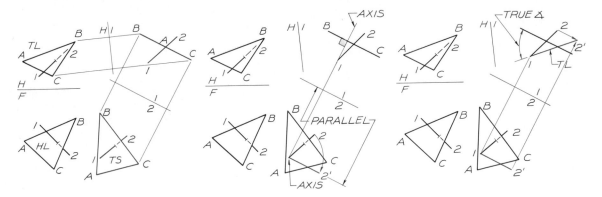

FIG. 22.27 Angle between a line and plane

Step 1 Construct Plane *ABC* as an edge in a primary auxiliary view, which can be projected from either view. Determine the true size of the plane in the secondary auxiliary view, and project Line 1–2 to each view.

Step 2 Revolve the secondary auxiliary view of the line until it is parallel to the 1–2 reference line. The axis of revolution appears as a point through point 1 in the secondary auxiliary view. The axis appears true length and is perpendicular to the 1–2 line and Plane *ABC* in the primary auxiliary view.

Step 3 Point 2' is projected to the primary auxiliary view where the true length of Line 1–2' is found by projecting the primary auxiliary view of Point 2 parallel to the 1–2 line. Since the plane appears as an edge and the line appears true length in this view, the true angle between them is found.

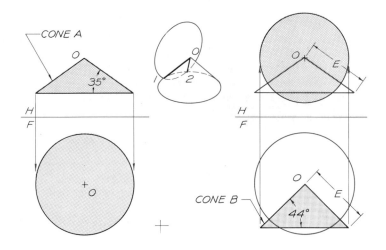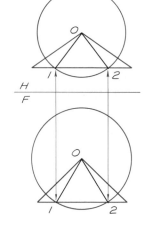

FIG. 22.28 A line at specified angles

Step 1 Draw a triangular view of a cone in the top view such that the outer elements make an angle of 35° with the frontal plane. Construct the circular view of the cone in the front view, using Point *O* as the apex. All elements of this cone make an angle of 35° with the frontal plane.

Step 2 Draw a triangular view of a cone in the front view such that the elements make an angle of 44° with the horizontal plane. Draw the elements of this cone equal in length to Element *E* of Cone *A*. All elements of Cone *B* make an angle of 44° with the horizontal plane.

Step 3 Since the elements of cones *A* and *B* are equal in length, there will be two common elements that lie on the surface of each cone, Elements *O*–1 and *O*–2. Locate Points 1 and 2 where the bases of the cone intersect in both views.

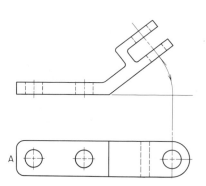

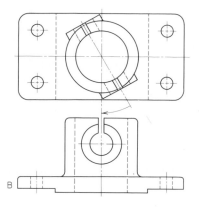

FIG. 22.29 It is conventional practice to revolve features of parts in order to show the features true size in the adjacent orthographic views. When this is done, it is unneccessary to show the arrows of rotation, since this is understood as standard practice.

22.14
A line at a specified angle with two principal planes

In Fig. 22.28 it is required that a line be drawn through Point O that will make angles of 35° with the frontal plane and 44° with the horizontal plane, and slopes forward and downward.

The cone containing elements making 35° with the frontal plane is drawn in Step 1. In Step 2, the cone with elements making 44° with the horizontal plane is drawn. The length of the elements of both cones must be equal, so they will intersect with equal elements.

Lines 0–1 and 0–2 are found in Step 3, where these lines are elements that lie on each cone and make the specified angles with the principal planes.

22.15
Revolution of parts on detail drawings

It is standard practice to revolve features such as those shown in Fig. 22.29 to make the views more descriptive. The front view of the part is true size because the top view has been revolved. This technique, called conventional practice, should be used to conserve effort while gaining additional clarity in the preparation of working drawings.

Problems

Use Size A (8½″ × 11″) sheets for the following problems and lay out the problems using instruments. Each square on the grid is equal to 0.20 in. or about 5 mm. The problems can be laid out on grid or plain paper. Label all reference planes and points in each problem with ⅛″ letters and/or numbers, using guidelines.

The crosses marked "1" and "2" are to be used for placing primary and secondary reference lines. The primary reference line should pass through "1" and the secondary through "2".

1–4. Find the true-length views of the lines in Fig. 22.30 by revolution.

5. Find the true-size view of the plane in Fig. 22.31 by an auxiliary view and a single revolution.

6. Find the true-size view of the plane in Fig. 22.31 by revolution.

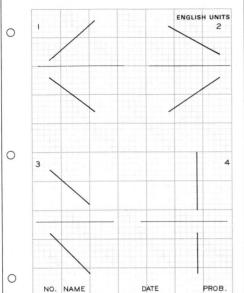

FIG. 22.30 Problems 1 through 4.

7–8. Find the true-size views of the planes in Fig. 22.32 by double revolution.

9–10. Find the dihedral angles between the planes in Fig. 22.33.

11–12. Revolve the points about the given axes in Fig. 22.34 and show the points in all views. In Problem 11, revolve the point into its most forward position, and in Problem 12 into its highest position.

13. The center line of a conveyor chute is given in the top and front views of Fig. 22.35. The chute has a 10-foot-square cross section. Construct the necessary views to revolve the 10-foot square into a position where two sides of the right section will be vertical planes. Show the chute in all views. Scale: $1'' = 10'$.

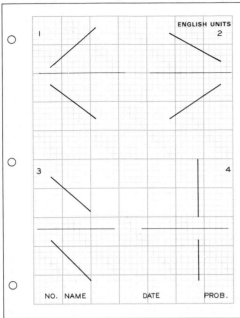

FIG. 22.31 Problems 5 and 6.

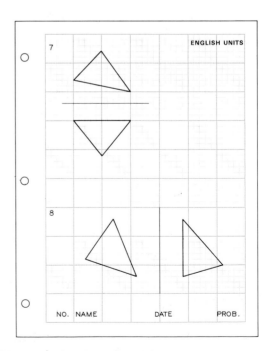

FIG. 22.32 Problems 7 and 8.

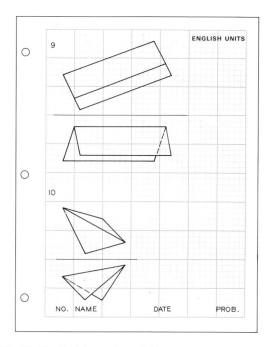

FIG. 22.33 Problems 9 and 10.

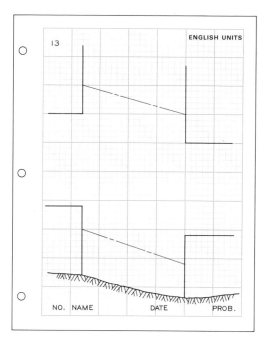

FIG. 22.35 Problem 13.

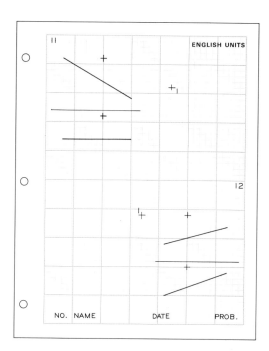

FIG. 22.34 Problems 11 and 12.

14–19. Lay out these problems on a Size B sheet (11″ × 17″) using the horizontal format. Position the crossing division lines at the center of the sheet. The grid is spaced at 0.25 in. or approximately 6 mm apart. Problem 14 requires that you lay out the prism in Area 1 (Fig. 22.36) and rotate it as specified about its corner Point 0 in the remaining areas of the sheet. Problems 15 through 19 require that you replace the object in Problem 14 with one of those given in Fig. 22.37, and rotate these objects through the angles specified in each step.

20. Draw the object in Fig. 22.38, but complete the top view by showing the inclined surface revolved into a true-size position. This will eliminate the need for an auxiliary view as presently shown.

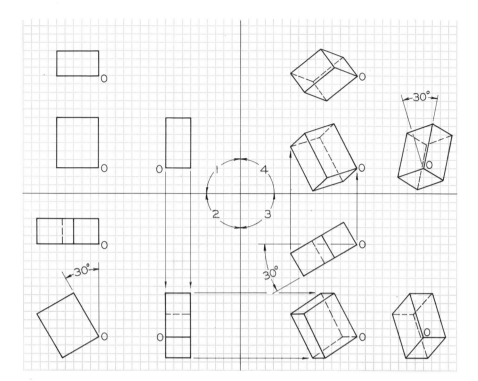

FIG. 22.36   Problem 14.

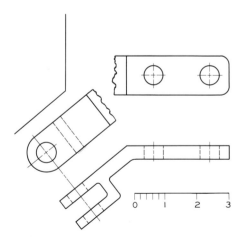

FIG. 22.37   Problems 15 through 19.

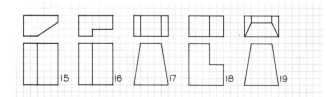

FIG. 22.38   Problem 20.

23

Vector Graphics

23.1
Introduction

When analyzing a system for strength, it is necessary to consider the forces of tension and compression within the system. These forces are represented by vectors. Vectors may also be used to represent other quantities such as distance, velocity, and electrical properties.

Graphical methods are useful in the solution of vector problems as alternative methods to conventional trigonometric and algebraic methods.

23.2
Basic definitions

A knowledge of the terminology of graphical vectors is necessary to understand the techniques of problem-solving with vectors.

FORCE A push or a pull that tends to produce motion. All forces have (1) magnitude, (2) direction, and (3) a point of application. A force is represented by the rope being pulled in Fig. 23.1A.

VECTOR A graphical representation of a quantity of force is drawn to scale to indicate magnitude, direction, and point of application. The vector shown in Fig. 23.1B represents the force of the rope pulling the weight, W.

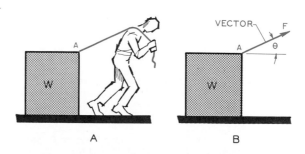

FIG. 23.1 Representation of a force by a vector

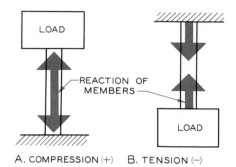

FIG. 23.2 Comparison of tension and compression in a member

MAGNITUDE The amount of push or pull is represented by the length of the vector line. Magnitude is usually measured in pounds or kilograms of force.

DIRECTION The inclination of a force (with respect to a reference coordinate system), indicated by a line with an arrow at one end.

POINT OF APPLICATION The point through which the force is applied on the object or member (Point A in Fig. 23.1A).

COMPRESSION The state created in a member by subjecting it to opposite pushing forces. A member tends to be shortened by the compression (Fig. 23.2A). Compression is represented by a plus sign ($+$) or the letter C.

TENSION The state created in a member by subjecting it to opposite pulling forces. A member tends to be stretched by tension, as shown in Fig. 23.2B. Tension is represented by a minus sign ($-$) or the letter T.

FORCE SYSTEM The combination of all forces acting on a given object. Figure 23.3 shows a force system.

RESULTANT A single force that can replace all the forces of a force system and have the same effect as the combined forces.

EQUILIBRANT The opposite of a resultant; it is the single force that can be used to counterbalance all forces of a force system.

COMPONENTS Any individual forces that, if combined, would result in a given single force. For example, Forces A and B are components of Resultant R_1 in Step 1 of Fig. 23.3.

SPACE DIAGRAM A diagram depicting the physical relationship between structural members. The force system in Fig. 23.3 is given as a space diagram.

VECTOR DIAGRAM A diagram composed of vectors that are scaled to their appropriate lengths to represent the forces within a given system. The vector diagram is used to solve for unknowns.

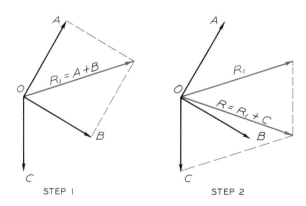

FIG. 23.3 Resultant by the parallelogram method

Step 1 Draw a parallelogram with its sides parallel to Vectors A and B. The Diagonal R_1, drawn from Point P to Point O, is the resultant of forces A and B.

Step 2 Draw a parallelogram using Vectors R_1 and C to find Diagonal R from P to Q. This is the resultant that can replace forces A, B, and C.

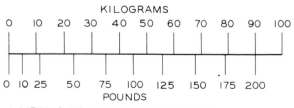

FIG. 23.4 The kilogram (kg) is the standard metric unit for measuring forces, which are represented by pounds in the English system: 1 kilogram = 2.2 pounds.

STATICS The study of forces and force systems that are in equilibrium.

METRIC UNITS The kilogram (kg) is the standard unit for indicating mass (loads). A comparison of kilograms with pounds is shown in Fig. 23.4. The metric ton is 1000 kilograms. One kilogram = 2.2 pounds.

23.3
Coplanar, concurrent force systems

When several forces, represented by vectors, act through a common point of application, the system is said to be **concurrent.** Vectors *A, B,* and *C* act through a single point in Fig. 23.3; therefore, this is a concurrent system. When only one view is necessary to show the true length of all vectors, as in Fig. 23.3, the system is **coplanar.**

The **resultant** represents the composite effect of all forces on the point of application. The resultant is found graphically by: (1) the parallelogram method and (2) the polygon method.

23.4
Resultant of a coplanar, concurrent system— parallelogram method

In Fig. 23.3, all the vectors lie in the same plane and act through a common point. The vectors are scaled to a known magnitude.

To apply the parallelogram method to determine the resultant, the vectors for a force system must be known and drawn to scale. Two vectors are used to find a parallelogram; the diagonal of the parallelogram is the resultant of these two vectors and has its point of origin at Point *P* (Fig. 23.3). Resultant R_1 can be called the **vector sum** of Vectors *A* and *B.*

Since Vectors *A* and *B* have been replaced by R_1, they can be disregarded in the next step of the solution. Again, Resultant R_1 and Vector *C* are resolved by completing a parallelogram, i.e., by drawing a line parallel to each vector. The diagonal of this parallelogram is the resultant of the entire system and is the

vector sum of R_1 and *C.* Resultant *R* can be analyzed as though it were the only force acting on the point; therefore, the analysis of a particular point-of-force application is simplified.

23.5
Resultant of a coplanar, concurrent system— polygon method

The system of forces shown in Fig. 23.3 is shown again in Fig. 23.5, but in this case the resultant is found by the polygon method. The forces are drawn to scale and in their true directions, with each force being drawn head-to-tail to form the polygon. In this example, the vectors are drawn in a clockwise sequence, beginning with Vector *A.* The polygon does not close; this means that the system is not in **equilibrium.** In other words, it would tend to be in motion, since the forces are not balanced. The Resultant *R* is drawn from the tail of Vector *A* to the head of Vector *C* to close the polygon.

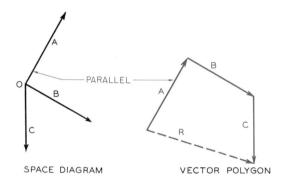

FIG. 23.5 Resultant of a coplanar, concurrent system as determined by the polygon method.

23.6
Resultant of a coplanar, concurrent system— analytical method

The analytical example in Fig. 23.6 is given to afford a comparison between graphical and analytical methods.

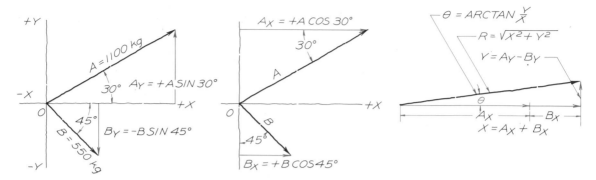

FIG. 23.6　Resultant by the analytical method

Required　An unscaled, freehand sketch of two forces is given. Find the resultant using the analytical method.

Step 1　The Y-components (vertical components) are found to be the sine functions of the angles the vectors make with the X-axis. The Y-component of A is positive and the Y-component of B is negative.

Step 2　The X-components (horizontal components) are the cosine functions of 30° and 45° in this case, both in the positive direction.

Step 3　The Y-components and X-components are summed to find the components of the resultant, X and Y. The Pythagorean theorem is applied to find the magnitude of the resultant. Its angle with the X-axis is the arctangent of Y/X.

In Step 1, the vertical components, which are parallel to the Y-axis, are drawn from the ends of both vectors to form right triangles, and their lengths are found through the use of sine functions of the angles the vectors make with the X-axis.

The horizontal component of each vector is drawn parallel to the X-axis through the end of each vector. The lengths of these components are found to be the cosine functions of the given vectors in Step 2.

The Y-components of each vector, A_y and B_y, can be added, since each lies in the same direction (Step 3). The resulting value is $Y = A_y - B_y$, since the components have opposite directions. The horizontal component is $X = A_x + B_x$, since both components have equal directions.

A right triangle is sketched using the X- and Y-components. The vertical and horizontal components are laid off head-to-tail to form a right triangle of forces. The magnitude of the resultant is found by the Pythagorean theorem,

$$R = \sqrt{X^2 + Y^2}.$$

The direction of the resultant measured from the X-axis is:

$$\text{angle } \theta = \arctan Y/X.$$

Law of sines

The law of sines is illustrated in Fig. 23.7A. When any three values are known, the remaining unknowns of a triangle can be computed. An example is given (Fig. 23.7B) where two sides of a triangle are vectors of known magnitude and direction.

> An **equilibrant** has the same magnitude, orientation, and point of application as the **resultant** in any system of forces.

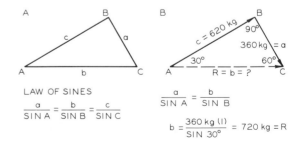

FIG. 23.7　The law of sines is illustrated in Part A. This principle is used in Part B to solve for Resultant R (B) when two angles and vectors are known.

The difference is their direction. The resultant of the system of forces shown in Fig. 23.8 is found by the parallelogram method. The equilibrant can be applied at Point O to balance the forces A and B and thereby cause the system to be in equilibrium.

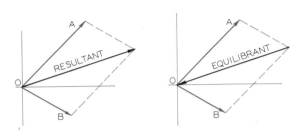

FIG. 23.8 The resultant and equilibrant are equal in all respects except the position of the arrowhead.

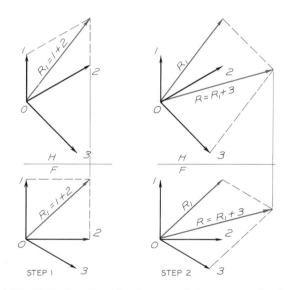

FIG. 23.9 Resultant by the parallelogram method

Step 1   Vectors 1 and 2 are used to construct a parallelogram in the top and front views. The diagonal, R_1, is the resultant of these two vectors.

Step 2   Vectors 3 and R_1 are used to construct a second parallelogram to find the views of the overall resultant, R_2.

23.7
Resultant of noncoplanar, concurrent forces— parallelogram method

When vectors lie in more than one plane of projection, they are said to be **noncoplanar;** therefore more than one view is necessary to analyze their spatial relationships. The resultant of a system of noncoplanar forces can be found by the parallelogram method, regardless of their number, if their true projections are given in two adjacent orthographic views.

Vectors 1 and 2 in Fig. 23.9 are used to construct the top and front views of a parallelogram. The diagonal of the parallelogram, R_1, is found in both views. As a check, the front view of R_1 must be an orthographic projection of its top view.

In Step 2, Resultant R_1 and Vector 3 are resolved to form Resultant R_2 in both views. The top and front views of R_2 must project orthographically. Resultant R_2 can be used to replace Vectors 1, 2, and 3. Since R_2 is an oblique line, its true length must be found by auxiliary view, as shown in Fig. 23.10, or by revolution.

23.8
Resultant of noncoplanar, concurrent forces— polygon method

The same system of forces that was given in Fig. 23.9 is solved in Fig. 23.10 for the resultant of the system by the polygon method.

In Step 1, each vector is laid head-to-tail in a clockwise direction, beginning with Vector 1. The vectors are drawn in each view to be orthographic projections at all times (Step 2). Since the vector polygon did not close, the system is not in equilibrium. The resultant, R, is constructed from the tail of Vector 1 to the head of Vector 3 in both views.

Resultant R is an oblique line and requires an auxiliary view to find its true length. The magnitude of the resultant can be measured in the true-length auxiliary view by using the same scale as was used to draw the original views.

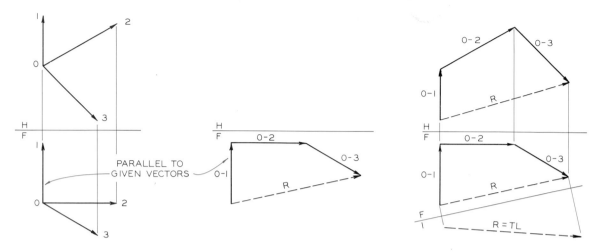

FIG. 23.10 Resultant by the polygon method

Step 1 Each vector is laid head-to-tail in the front view. The front view of the resultant is Vector *R*.

Step 2 The same vectors are laid head-to-tail in the top view to find the top view of *R*. The resultant is found true length by an auxiliary view.

FIG. 23.11 The cargo cranes on the cruise ship *Santa Rosa* are examples of coplanar, concurrent force systems that are designed to remain in equilibrium. (Courtesy of Exxon Corp.)

23.9
Forces in equilibrium

An example of a coplanar, concurrent structure in equilibrium can be seen in the loading cranes in Fig. 23.11, which are used for handling cargo.

The coplanar, concurrent structure given in Fig. 23.12 is designed to support a load of $W = 1000$ kg. The maximum loading in each is used to determine the type and size of structural members used in the structural design.

In Step 1, the only known force, $W = 1000$ kg, is laid off parallel to the given direction. Unknown forces *A* and *B* are drawn head-to-tail as vectors to close the force polygon.

In Step 2, Vectors *A* and *B* are analyzed to determine whether they are in tension or compression. Vector *B* points upward to the left, which is toward Point *O* when transferred to the structural diagram. A vector that acts toward a point of application is in compression. Vector *A* points away from Point *A* when transferred to the structural diagram and is therefore in tension.

A similar example of a force system involving a pulley is solved in Fig. 23.13 to determine the loads in the structural members caused by the weight of 100 lb. The only difference between this solution and the previous one is the construction of two equal vectors to represent the loads in the cable on both sides of the pulley.

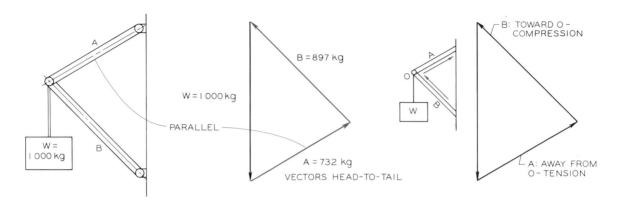

FIG. 23.12 Coplanar forces in equilibrium

Required Find the forces in the two structural members caused by the load of 1000 kg.

Step 1 Draw the load of 1000 kg as a vector. Draw Vectors A and B parallel to their directions from each end of W. Arrowheads are drawn head-to-tail.

Step 2 Vector A points away from Point O when transferred to the structural diagram. Therefore, Vector A is in tension. Vector B points toward Point O and is in compression.

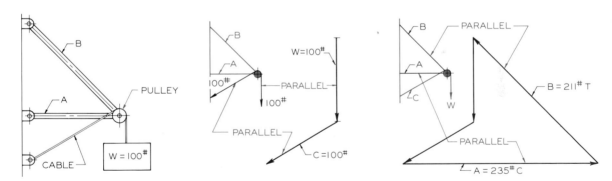

FIG. 23.13 Determination of forces in equilibrium

Required Find the forces in the members caused by the load of 100 lb supported by the pulley.

Step 1 The force of the cable is equal to 100 lb on both sides of the pulley. These two forces are drawn as vectors head-to-tail parallel to their directions in the space diagram.

Step 2 A and B are drawn to close the polygon, and arrowheads are placed to give a head-to-tail arrangement. The direction of A is toward the point of application, (compression); B is away from the point (tension).

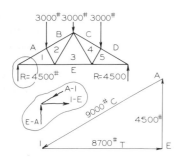

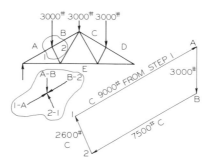

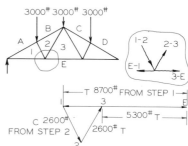

FIG. 23.14 Joint analysis of a truss

Step 1 The truss is labeled using Bow's notation, with letters between the exterior loads and numbers between interior members. The lower left joint can be analyzed, since it has only two unknowns, A–1 and 1–E. These vectors are found by drawing them parallel to the unknown vectors from both ends of the reaction of 4500 lb. The vectors are laid off in a head-to-tail order.

Step 2 Using the vector of 1–A found in Step 1 and Load AB, the two unknowns B–2 and 2–1 can be found. The known vectors are laid out beginning with Vector 1–A and moving clockwise about the joint. Vectors B–2 and 2–1 close the polygon. If a vector points toward the point of application, it is in compression; if away from the point, it is in tension.

Step 3 The third joint can be analyzed by laying out Vectors E–1 and 1–2 from the previous steps. Vectors 2–3 and 3–E close the polygon and are parallel to their directions in the space diagram. Vectors 2–3 and 3–E point away from the point of application and are in tension.

23.10

Truss analysis

Vector polygons can be used to analyze structural trusses to determine the loads in each member by two graphical methods: (1) joint-by-joint analysis, and (2) Maxwell diagrams.

Joint-by-joint analysis

The Fink truss shown in Fig. 23.14 is loaded with forces of 3000 lb that are concentrated at joints of the structural members. A special method of designating forces, called **Bow's notation,** is used. The exterior forces applied to the truss are labeled with letters placed between the forces. Numerals are placed between the interior members.

Each vector is referred to by the number on each of its sides by reading in a clockwise direction. For example, the first vertical load at the left is called AB, with A at the tail and B at the head of the vector.

We first analyze the joint at the left where the reaction of 4500 pounds (denoted by #) is known.

This force, reading in a clockwise direction about the joint, is called EA with an upward direction. The tail is labeled E and the head A. Continuing in a clockwise direction, the next force is A–1 and the next 1–E, which closes the polygon and ends with the beginning letter, E. The arrows are placed, beginning with the known vector EA, in a head-to-tail arrangement.

Tension and compression can be determined by relating the sense of each vector to the original joint. For example, A–1 points toward the joint and is in compression, while 1–E is away and in tension.

Since the truss is symmetrical and equally loaded, the loads in the members on the right will be equal to those on the left.

The other joints are analyzed in the same manner in Steps 2 and 3. The direction of the vectors is opposite at each end. Vector A–1 is toward the left in Step 1 and toward the right in Step 2.

Maxwell diagrams

The Maxwell diagram is exactly the same as the joint-by-joint analysis except that the polygons are positioned to overlap, with some vectors common to more than one polygon. Again, Bow's notation is used to good advantage.

The first step (Fig. 23.15) is to lay out the exterior loads beginning clockwise about the truss—*AB, BC, CD, DE,* and *EA*—head-to-tail. A letter is placed at each end of the vectors. Since they are parallel, this polygon will be a straight line.

The structural analysis begins at the joint through which Reaction *EA* acts. A free-body diagram is drawn to isolate this joint for easier analysis. The two unknowns are Members *A–1* and *1–E*. These vectors are drawn parallel to their direction in the truss in Step 1 of Fig. 23.15 with *A–1* beginning at Point *A* and *1–E* beginning at Point *E*. These directions are extended to locate Point 1. The vectors must be drawn head-to-tail. Because Resultant *EA* points upward, Vector *A–1* must have its tail at *A*, giving it a direction toward Point 1. By relating this to the free-body diagram, we can see that the direction is toward the point of appli-

cation, which means that *A–1* is in compression. Vector *1–E* points away from the joint, which means that it is in tension. The vectors are coplanar and can be scaled to determine their loads.

In Step 2, Vectors *1–A* and *AB* are known, while Vectors *B–2* and *2–1* are unknown, making it possible to solve for them. Vector *B–2* is drawn parallel to the structural member through Point *B* in the Maxwell diagram and the line of Vector *2–1* is extended from Point 1 until it intersects with *B–2*, where point 2 is located. The arrows of each vector are found by laying off each vector head-to-tail. Both Vectors *B–2* and *2–1* point toward the joint in the free-body diagram; therefore they are in compression.

The next joint is analyzed in sequence to find the stresses in *2–3* and *3–E* (Step 3). The truss will have equal forces on each side, since it is symmetrical and is loaded symmetrically. The total Maxwell diagram is drawn to illustrate the completed work in Step 3.

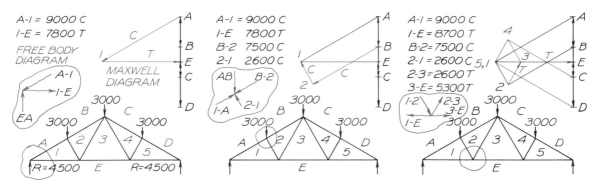

FIG. 23.15 Truss analysis

Step 1 Label the spaces between the outer forces of the truss with letters and the internal spaces with numbers, using Bow's notation. Add the given load vectors in a Maxwell diagram, and sketch a free-body diagram of the first joint. Using Vectors *EA, A–1* and *1–E* drawn head-to-tail, draw a vector diagram to find their magnitudes. Vector *A–1* is in compression (+) because its points toward the joint, and Vector *1–E* is in tension (−) because its points away from the joint.

Step 2 Draw a sketch of the next joint to be analyzed. Since *AB* and *A–1* are known, we have to determine only two unknowns, *2–1* and *B–2*. Draw these parallel to their direction, head-to-tail, in the Maxwell diagram using the existing vectors found in Step 1. Vectors *B–2* and *2–1* are in compression since each points toward the joint. Note that Vector *A–1* becomes *1–A* when read in a clockwise direction.

Step 3 Sketch a free-body diagram of the next joint to be analyzed. The unknowns in this case are *2–3* and *3–E*. Determine the true length of these members in the Maxwell diagram by drawing vectors parallel to given members to find Point 3. Vectors *2–3* and *3–E* are in tension because they act away from the joint. This same process is repeated to find the loads of the members on the opposite side.

FIG. 23.16 The structural members of this tripod support for a moon vehicle can be analyzed graphically to determine design loads. (Courtesy of NASA.)

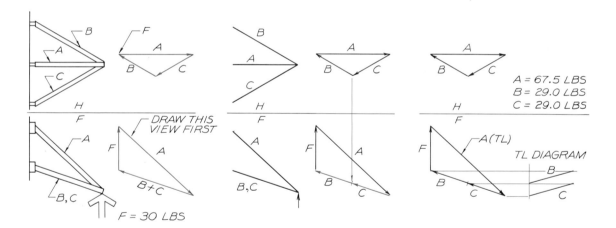

FIG. 23.17 Noncoplanar structural analysis—a special case

Step 1 Two forces, *B* and *C*, coincide in the front view, resulting in only two unknowns in this view. Vector *F* (30 lb) is drawn, and the other two unknowns are drawn parallel to their front view in the front view of the vector polygon. The top view of *A* can be found by projection, from which Vectors *B* and *C* can be found.

Step 2 The point of intersection of Vectors *B* and *C* in the top view is projected to the front view to separate these vectors. All vectors are drawn head-to-tail. Vectors *B* and *C* are in tension because their vectors are acting away from the point in the space diagram, while *A* is in compression.

Step 3 The top and front views found in Step 2 do not give the true lengths of Vectors *B* and *C*, since they are oblique. Their lengths are determined by a true-length diagram where they are scaled to find the forces in each member.

If all the polygons in the series do not close at every point with perfect symmetry, there is an error in construction. If the error of closure is very slight, it can be disregarded, since safety factors are generally applied in derivation of working stresses of structural systems to assure safe construction. Arrowheads are omitted on Maxwell diagrams, since each vector will have opposite directions when applied to different joints.

23.11
Noncoplanar structural analysis—special case

Three-dimensional structures systems require the use of descriptive geometry, since it is necessary to analyze the system in more than one plane. The manned flying system (MFS) in Fig. 23.16 can be analyzed to determine the forces in the support members (Fig. 23.17). Weight on the moon can be found by multiplying earth weight by a factor of 0.165. A tripod that must support 182 lb on earth has to support only 30 lb on the moon.

This example is a special case, since Members B and C lie in the same plane and appear as an edge in the front view.

A vector polygon is constructed in the front view in Step 1 of Fig. 23.17 by drawing Force F as a vector and using the other vectors as the other sides of the polygon. One of these vectors is actually a summation of Vectors B and C. The top view is drawn using Vectors B and C to close the polygon from each end of Vector A. In Step 2, the front view of Vectors B and C is found.

The true lengths of the vectors are found in a true-length diagram in Step 3. The vectors are measured to determine their loads. Vector A is found to be in compression because it points toward the point of concurrency. Vectors B and C are in tension.

23.12
Noncoplanar structural analysis—general case

The structural frame shown in Fig. 23.18 is attached to a vertical wall to support a load of $W = 600$ lb.

Since there are three unknowns in each of the views, we are required to construct an auxiliary view that will give the edge view of a plane containing two of the vectors, thereby reducing the number of unknowns to two (Step 1). We no longer need to refer to the front view.

A vector polygon is drawn by constructing vectors parallel to the members in the auxiliary view (Step 1). An adjacent orthographic view of the vector polygon is also drawn by constructing its vectors parallel to the members in the top view (Step 2). A true-length diagram is used Step 3 to find the true length of the vectors to determine their magnitudes.

A three-dimensional system vector is the side-boom tractors used for lowering pipe into a ditch during pipeline construction (Fig. 23.19).

23.13
Nonconcurrent, coplanar vectors

Forces *may* be applied in such a manner that they are **nonconcurrent,** as illustrated in Fig. 23.20. Bow's notation can be used to locate the resultant of this type of nonconcurrent system.

In Step 1, the vectors are laid off to form a vector polygon in which the closing vector is the resultant, $R = 68$ lb. Each vector is resolved into two components by randomly locating Point O on the interior or exterior of the polygon and connecting Point O with the end of each vector. The components, or strings, from Point O are equal and opposite components of adjacent vectors. For example, Component o–b is common to Vectors AB and BC. Since the strings from Point O are equal and opposite, the system has not changed statically.

In Step 2, each string is transferred to the space diagram of the vectors where it is drawn between the respective vectors to which it applies. (The figure thus produced is called a **funicular diagram.**) For instance, String o–b is drawn in the area between the Vectors AB and BC. String o–c is drawn in the C area to connect at the intersection of o–b and Vector BC. The point of intersection of the last two strings, o–a and o–d, locates a point through which the Resultant R will pass.

FIG. 23.18   Noncoplanar structural analysis—general case

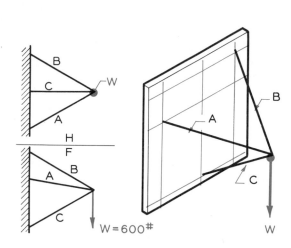

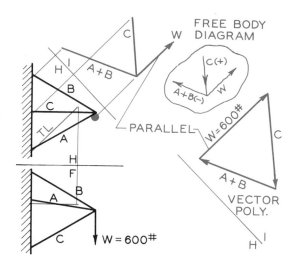

Given   The top and front views of a three-member frame that is attached to a vertical wall in such a way that it can support a maximum weight of 600 lb.

Required   Find the loads in the structural members.

Step 1   To limit the unknowns to two, construct an auxiliary view to find two vectors lying in the edge of a plane. Use the auxiliary view and top view in the remainder of the problem. Draw a vector polygon parallel to the members in the auxiliary view in which $W = 600$ lb is the only known vector. Sketch a free-body diagram for preliminary analysis.

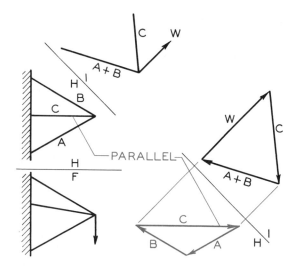

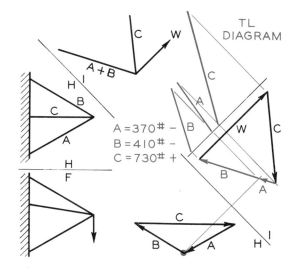

Step 2   Construct an orthographic projection of the view of the vector polygon found in Step 1 so that its vectors are parallel to the members in the top view. The reference plane between the two views is parallel to the H-1 plane. This portion of the problem is closely related to the problem in Fig. 23.17.

Step 3   Project the intersection of Vectors A and B in the horizontal view of the vector polygon to the auxiliary view polygon to establish the lengths of Vectors A and B. Determine the true lengths of all vectors in a true-length diagram and measure their magnitudes. Analyze for tension or compression, as covered in Section 23.10.

FIG. 23.19 Tractor sidebooms represent noncoplanar, concurrent systems of forces that can be solved graphically. (Courtesy of Trunkline Gas Co.)

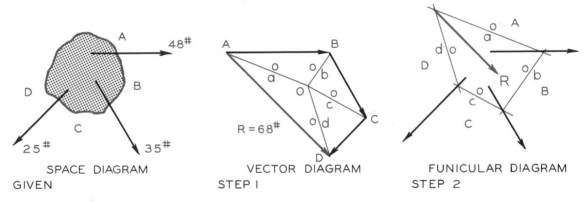

SPACE DIAGRAM	VECTOR DIAGRAM	FUNICULAR DIAGRAM
GIVEN	STEP 1	STEP 2

FIG. 23.20 Resultant of nonconcurrent forces

Required Find the resultant of the known forces applied to the above object. The forces are nonconcurrent.

Step 1 The vectors are drawn head-to-tail to find Resultant R. Point O is conveniently located for the construction of strings to the ends of each vector.

Step 2 Each string is drawn between the two vectors to which it applies. *Example: o–c* between *BC* and *CD*. These strings are connected in sequence until the strings *o–a* and *o–d* establish the position of R.

23.14
Nonconcurrent systems resulting in couples

A **couple** is the descriptive name given to two parallel, equal, and opposite forces which are separated by a distance and are applied to a member in such a manner to cause the member to rotate.

An important quantity associated with a couple is its **moment.** The moment of any force is a measure of its rotational effect. An example is shown in Fig. 23.21, in which two equal and opposite forces are applied to a wheel. The forces are separated by the Distance D. The moment of the couple is found by multiplying one of the forces by the perpendicular distance between it and the other: $F \times D$. If the force is 20 lb and the distance is 3 ft, the moment of the couple would be given as 60 ft-lb.

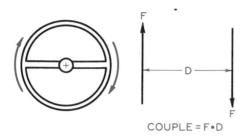

FIG. 23.21 Representation of a couple or moment

A series of parallel forces is applied to a beam in Fig. 23.22. The spaces between the vectors are labeled with letters which follow Bow's notation. We are required to determine the resultant.

After constructing a vector diagram, we have a straight line which is parallel to the direction of the forces and which closes at Point A. We then locate pole Point O and draw the strings of a funicular diagram.

The strings are transferred to the space diagram and are drawn in their respective spaces. For example, o–c is drawn in the C-space between Vectors BC and CD. The last two strings, o–d and o–a, do not close at

a common point, but are found to be parallel; the result is therefore a couple. The distance between the forces of the couple is the perpendicular distance, E, between Strings o–a and o–d in the space diagram, using the scale of the space diagram. The magnitude of the force is the scaled distance from Point O to A and D in the vector diagram, using the scale of the vector diagram. The moment of the couple is equal to 7.5 lb × E in a counterclockwise direction.

23.15
Resultant of parallel, nonconcurrent forces

Forces applied to beams, such as those shown in Fig. 23.23, are parallel and nonconcurrent in many instances, and they may have the effect of a couple, tending to cause a rotational motion.

The beam in Fig. 23.24 is on a rotational crane that is used to move building materials in a limited area. The magnitude of the weight W is unknown, but the counterbalance weight is known to be 2000 lb; Column R supports the beam as shown. Assuming

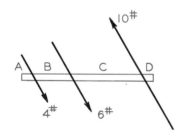

SPACE DIAGRAM

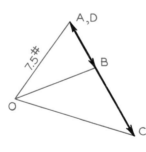

VECTOR DIAGRAM

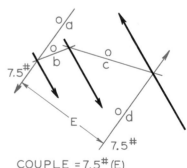

COUPLE = 7.5# (E)

FUNICULAR DIAGRAM

FIG. 23.22 Couple resultants

Required Find the resultant of these nonconcurrent forces applied to this beam.

Step 1 The spaces between each force are labeled in Bow's notation.

Step 2 The vectors are laid out head-to-tail; they will lie in a straight line since they are parallel. Pole Point O is located in a convenient location and the ends of each vector are connected with Point O.

Step 3 Strings o–a, o–b, o–c, and o–d are successively drawn between the vectors to which they apply. Since Strings o–a and o–d are parallel, the resultant will be a couple equal to 7.5 lb × E, where E is the distance between o–a and o–d.

FIG. 23.23 The boom of this crane can be analyzed for its resultant as a parallel, nonconcurrent system of forces when the cables have been disregarded.

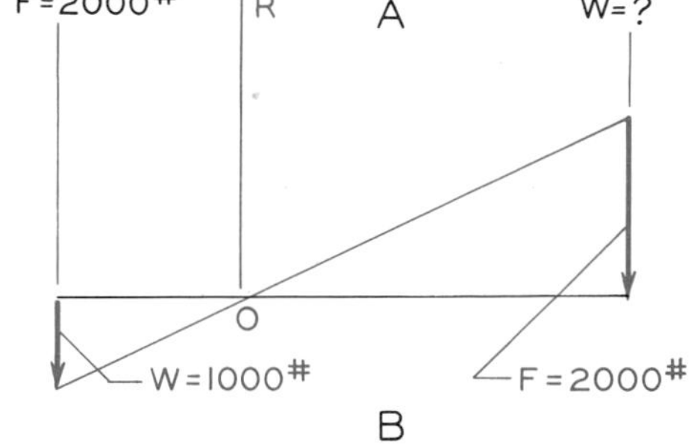

that the support cables have been omitted, we desire to find the weight W that would balance the beam.

The graphical solution (Fig. 23.24B) is found by constructing a line to represent the total distance between the Forces F and W. Point O is projected from the space diagram to this line. Point O is the point of balance where the summation of the moments will be equal to zero. Vectors F and W are drawn to scale at each end of the line by transposing them to the opposite ends of the beam. A line is drawn from the end of Vector F through Point O and extended to intersect the direction of Vector W. This point represents the end of Vector W, which can be scaled, resulting in a magnitude of 1000 lb.

23.16
Resultant of parallel, nonconcurrent forces on a beam

The beam given in Fig. 23.25 is supported at each end and must support three given loads. We are required to determine the magnitude of each support, R_1 and R_2, along with the resultant of the loads and its location. The spaces between all vectors are labeled in a clockwise direction with Bow's notation in Step 1, and a force diagram is drawn.

In Step 2 the lines of force in the space diagram are extended and the strings from the vector diagram are drawn in their respective spaces, parallel to their original direction. *Example:* String oa is drawn parallel to String OA in Space A between Forces EA and AB, and String ob is drawn in Space B beginning at the intersection of oa with Vector AB. The last string, oe, is drawn to close the funicular diagram. The direction of String oe is transferred to the force diagram, where it is laid off through Point O to intersect the load line at Point E. Vector DE represents Support R_2 (refer to Bow's notation as it was applied in Step 1). Vector EA represents Support R_1.

The magnitude of the resultant of the loads (Step 3) is the summation of the vertical downward forces, or the distance from A to D, or 500 lb. The location of the resultant is found by extending the extreme outside strings in the funicular diagram, oa and od, to their point of intersection. The resultant is found to have a magnitude of 500 lb, a vertical downward direction, and a point of application established by $\overline{X}$.

FIG. 23.24 Determining the resultant of parallel, nonconcurrent forces

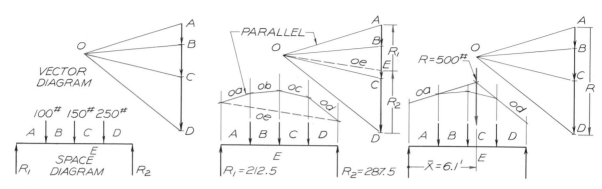

FIG. 23.25 Beam analysis with parallel loads

Step 1 Letter the spaces between the loads with Bow's notation. Find the graphical summation of the vectors by drawing them head-to-tail in a vector diagram at a convenient scale. Locate Pole Point O at a convenient location and draw strings from Point O to each end of the vectors.

Step 2 Extend the lines of force in the space diagram, and draw a funicular diagram with String o–a in the A-space, o–b in the B-space, o–c in the C-space, etc. The last string, which is drawn to close the diagram, is o–e. Transfer this string to the vector polygon and use it to locate Point E, thus establishing the lengths of R_1 and R_2, which are EA and DE, respectively.

Step 3 The resultant of the three downward forces will be equal to their graphical summation, Line AD. Locate the resultant by extending Strings o–a and o–d in the funicular diagram to a point of intersection. The Resultant $R = 500$ lb will act through this point in a downward direction. $\bar{X}$ is a locating dimension.

Problems

Problems should be presented in instrument drawings on Size A (8½" × 11") paper, grid or plain, using the format introduced in Section 2.33. Each grid square represents 0.20 in. All notes, sketches, drawings, and graphical work should be neatly prepared in keeping with good practices. Written matter should be legibly lettered using ⅛ in. guidelines.

1. In Fig. 23.26 (A) determine the resultant of the force system by the parallelogram method at the left of the sheet. Solve the same system using the vector polygon method at the right of the sheet. Scale 1" = 100 lb (note that each grid square equals 0.20 in.). (B) Determine the resultant of the concurrent, coplanar force system shown at the left of the sheet by the parallelogram method. Solve the same system using the polygon method at the right of the sheet. Scale 1" = 100 lb.

2. (A and B) In Fig. 23.27 solve for the resultant of each of the concurrent, noncoplanar force systems by the parallelogram method at the left of the sheet. Solve for the resultant of the same systems by the vector polygon method at the right of the sheet. Find the true length of the resultant in both problems. Letter all construction. Scale 1" = 600 lb.

3. (A and B) In Fig. 23.28, the concurrent, coplanar force systems are in equilibrium. Find the loads in each structural member. Use a scale of 1" = 300 lb in Part A and a scale of 1" = 200 lb in Part B. Show and label all construction.

4. In Fig. 23.29, solve for the loads in the structural members of the truss. Vector polygon scale 1" = 2000 lb. Label all construction.

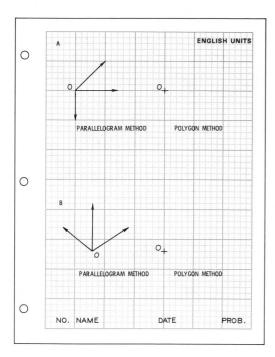

FIG. 23.26 Resultant of concurrent, coplanar vectors.

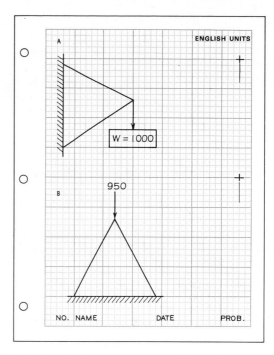

FIG. 23.28 Coplanar, concurrent forces in equilibrium.

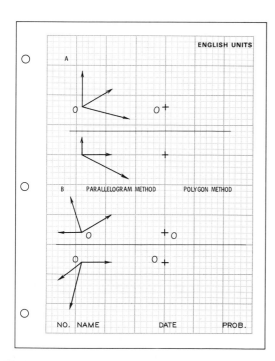

FIG. 23.27 Resultant of concurrent, noncoplanar vectors.

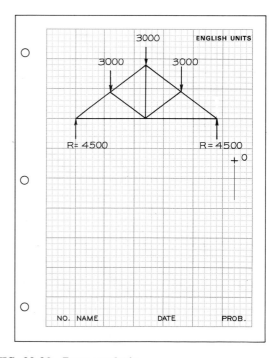

FIG. 23.29 Truss analysis.

518 CHAPTER 23 VECTOR GRAPHICS

5. In Fig. 23.30, solve for the loads in the structural members of the concurrent, noncoplanar force system. Find the true length of all vectors. Scale 1″ = 300 lb.

6. In Fig. 23.31, solve for the loads in the structural members of the concurrent, noncoplanar force system. Find the true length of all vectors. Scale 1″ = 400 lb.

7. (A) In Fig. 23.32, find the resultant of the coplanar, nonconcurrent force system. The vectors are drawn to a scale of 1″ = 100 lb. (B) In Part B of the figure, solve for the resultant of the coplanar, nonconcurrent force system. The vectors are given in their true positions and at the true distances from each other. The space diagram is drawn to a scale of 1″ = 1.0′. Draw the vectors to a scale of 1″ = 30 lb. Show all construction.

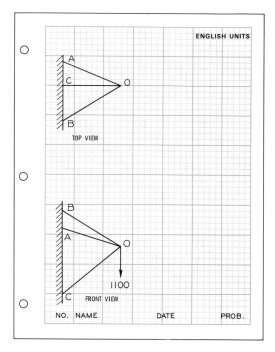

FIG. 23.31 Noncoplanar, concurrent forces in equilbrium.

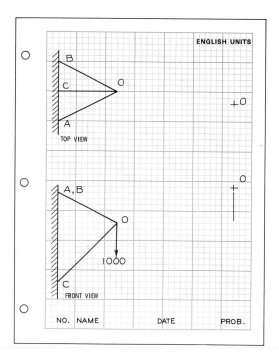

FIG. 23.30 Noncoplanar, concurrent forces in equilibrium.

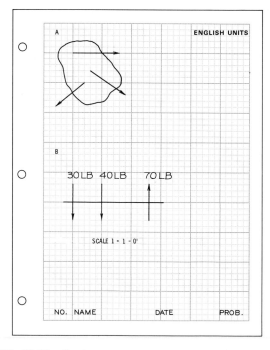

FIG. 23.32 Coplanar, nonconcurrent forces.

8. (A) In Fig. 23.33, determine the force that must be applied at *A* to balance the horizontal member supported at *B*. Scale 1″ = 100 lb. (B) In Part B of the figure, find the resultants at each end of the horizontal beam. Find the resultant of the downward loads and determine where it would be positioned. Scale 1″ = 600 lb.

9. Determine the forces in the three members of the tripod in Fig. 23.34. The tripod supports a load of *W* = 250 lb. Find the true lengths of all vectors.

10. The vectors in Fig. 23.35 each make an angle of 60° with the structural member on which they are applied. Find the resultant of this force system.

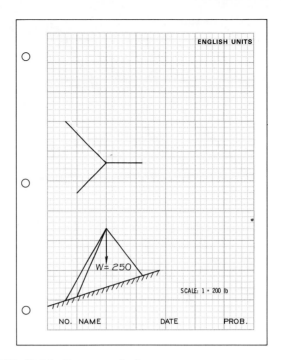

FIG. 23.34 Beam analysis.

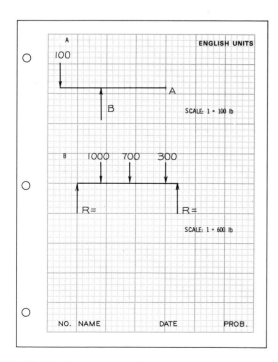

FIG. 23.33 Beam analysis.

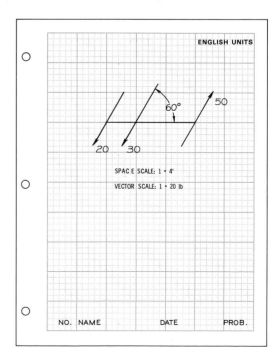

FIG. 23.35 Noncoplanar, concurrent forces in equilibrium.

CHAPTER

24

Intersections and Developments

24.1
Introduction

This chapter deals with the methods of finding lines of **intersections** between parts that join together. Usually these parts are made of sheet metal or plywood used to form concrete to a desired shape.

Once the intersections have been found, flat patterns called **developments** can be found. The patterns can then be laid out on the sheet metal and cut to conform to the desired shape.

You can see many examples of intersections and developments in Fig. 24.1, where a refinery is under construction. The principles covered in this chapter must be used to solve problems of this type.

24.2
Intersections of lines and planes

The basic steps of finding the intersection between a line and a plane are illustrated in Fig. 24.2. This is a special case since the plane appears as an edge, and

FIG. 24.1 This refinery installation illustrates many examples of the application of principles of intersections and developments.

520

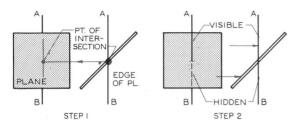

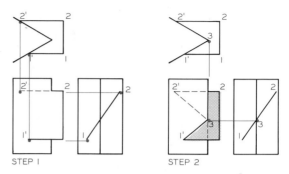

FIG. 24.2 Intersection of a line and a plane

Step 1 The point of intersection can be found in the view where the plane appears as an edge, the side view in this example.

Step 2 Visibility in the front view is determined by looking from the front view to the right side view.

FIG. 24.4 Intersection of a plane at a corner

Step 1 The intersecting plane appears as an edge in the side view. Intersection Points 1' and 2' are projected from the top and side views to the front view.

Step 2 The line of intersection from 1' to 2' must bend around the vertical corner at Point 3 in the top and side views. This point is projected to the front view to locate Line 1'–2'–3'.

the point of intersection can be easily seen in this view (Step 1). It is projected to the front view, and the visibility of the line is found in Step 2.

This principle is used in Fig. 24.3 to find the line of intersection between two planes. Since *EFGH* appears as an edge in the side view, points of intersection 1 and 2 can be found and projected to the front view and the visibility determined in Step 2. The intersection was found by finding the piercing points of Lines *AB* and *DC* and connecting these points.

The intersection of a plane at a corner of a prism results in a line of intersection that bends around the corner (Fig. 24.4). Piercing Points 2' and 1' are found in Step 1.

Corner Point 3 is seen in the side view of Step 2, where the vertical corner pierces the plane. Point 3 is projected to the corner in the front view. Point 2' is hidden in the front view since it is on the back side of the assembly.

The intersection of a plane and a prism is found in Fig. 24.5, where the plane appears as an edge. The points of intersection are found for each corner line and are connected; visibility is shown to complete the line of intersection.

An intersection between a plane and a prism is shown in Fig. 24.6, where the vertical corners of the prism are true length in the front view and the plane appears foreshortened in both views. Imaginary cutting planes are passed vertically through the planes of the prism in the top view to find the piercing points of the corners in the front view. The points are connected and the visibility is determined to complete the solution.

The intersection between a foreshortened plane and an oblique prism is found in Fig. 24.7. The plane is found as an edge in a primary auxiliary view. The piercing points of the corners of the prism are located in the auxiliary view and are projected back to the given views.

Points 1, 2, and 3 are projected from the auxiliary view to the given views as examples. Visibility is determined by analysis of crossing lines.

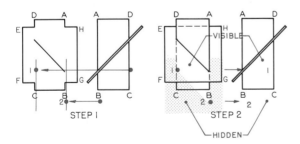

FIG. 24.3 Intersection between planes

Step 1 The points where Plane *EFGH* intersects Lines *AB* and *DC* are found in the view where the plane appears as an edge. These points are projected to the front view.

Step 2 Line 1–2 is the line of intersection. Visibility is determined by looking from the front view to the right-side view.

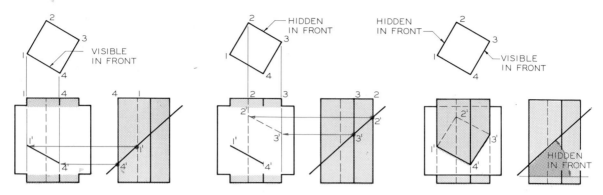

FIG. 24.5 Intersection of a plane and a prism

Step 1 Vertical Corners 1 and 4 intersect the edge view of the plane in the side view at 1' and 4'. These points are projected to the front view and are connected with a visible line.

Step 2 Vertical Corners 2 and 3 intersect the edge of the inclined plane at 2' and 3' in the side view. These points are connected in the front view with a hidden line.

Step 3 Lines 1'–2' and 3'–4' are drawn as hidden and visible lines, respectively. Visibility is determined by inspection of the top and side views and by projection to the front view.

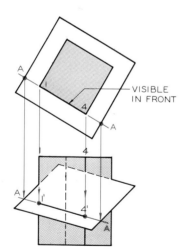

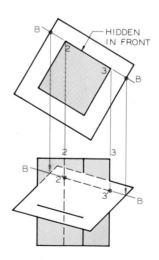

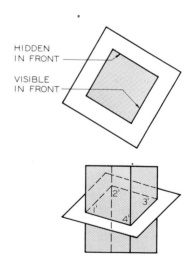

FIG. 24.6 Intersection of an oblique plane and a prism

Step 1 Vertical Cutting Plane A–A is passed through the vertical plane, 1–4, in the top view and is projected to the front view. Piercing Points 1' and 4' are found in this view.

Step 2 Vertical Plane B–B is passed through the top view of Plane 2–3 and is projected to the front view where Piercing Points 2' and 3' are found. Line 2'–3' is a hidden line.

Step 3 The line of intersection is completed by connecting the four points in the front view. Visibility in the front view is found by inspection of the top view.

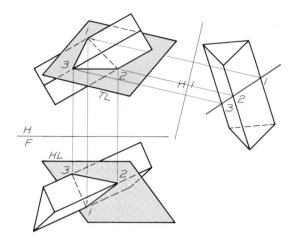

FIG. 24.7 The intersection between a plane and a prism can be found by constructing a view in which the plane appears as an edge.

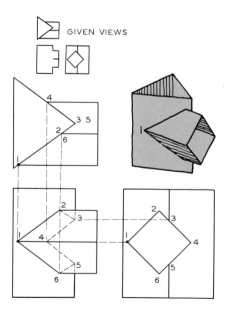

GIVEN VIEWS

FIG. 24.8 Three views of intersecting prisms. The points of intersection can be seen where intersecting planes appear as edges.

24.3
Intersections between prisms

The same principles used to find the intersection between a plane and a line are used to find the intersection between two prisms in Fig. 24.8. Piercing Points 1, 2, and 3 are found in the front view by projecting from the side and top views. Point X is located in the side view where line of intersection 1–2 bends around the vertical corner of the other prism. Points 1, X, and 2 are connected in sequence and the visibility is determined.

In Fig. 24.9, an inclined prism intersects a vertical prism. In Step 1, the end view of the inclined prism is found by an auxiliary view. In the auxiliary view, you can see where Plane 2–3 bends around Corner AB at Point X in Step 2.

Points of intersection 2' and 3' are projected from the top to the front view. The line of intersection 2'–X–3' can be drawn to complete this portion of the line of intersection. The remaining lines, 1'–3' and 2'–1', are connected to complete the solution (Step 3).

An alternative method of solving a problem of this type is shown in Fig. 24.10. In Step 1, piercing Points 1' and 2' are found in the front view by projecting from the top view. Point 5, the point where Line 1'–5–2' bends around vertical Corner AB, is found in Step 2. A cutting plane is passed through Corner AB in the top view. The front view of Point 5 is found and the lines of intersection are completed.

The conduit connector shown in Fig. 24.11 is an example of the application of intersecting planes and prisms.

24.4
Intersection of a plane and cylinder

The intersections of the components of the gas transmission system shown in Fig. 24.12 offer numerous examples of problems that were solved using the principles of intersections.

The intersection between a plane and a cylinder is found in Fig. 24.13, where the plane appears as an edge in one of the given views. Cutting planes are passed vertically through the top view of the cylinder

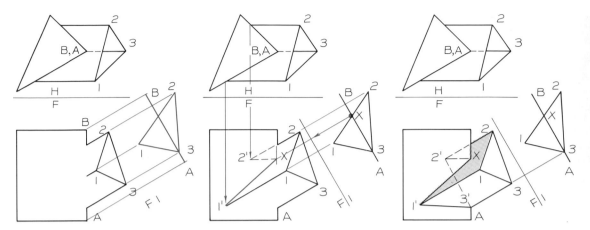

FIG. 24.9 Intersection between two prisms

Step 1 Construct the end view of the inclined prism by projecting an auxiliary view from the front view. Show only Line *AB* of the vertical prism in the auxiliary view.

Step 2 Locate Piercing Points 1′ and 2′ in the top and front views. Intersection Line 1′–2′ will not be a straight line, but it will bend around corner *AB* at Point *X*, which is projected from the auxiliary view.

Step 3 Intersection lines from 2′ and 1′ to 3′ do not bend around the corner. Therefore, these are drawn as straight lines. Line 1′–3′ is visible and Line 2′–3′ is invisible.

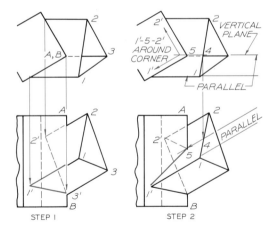

FIG. 24.10 Intersection between prisms

Step 1 The piercing points of Lines 1, 2, and 3 are found in the top view and are projected to the front view where Piercing Points 1′, 2′, and 3′ are found.

Step 2 A cutting plane is passed through Corner *AB* in the top view to locate Point 5 where line of intersection 1′–2′ bends around the vertical prism. Point 5 is found in the front view, and 1′–5–2′ is drawn.

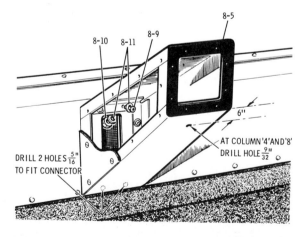

FIG. 24.11 This conduit connector was designed through the use of the principles of intersection of a plane and a prism. (Courtesy of the Federal Aviation Administration.)

FIG. 24.12 This complex of pipes and vessels contains many applications of intersections. (Courtesy of Trunkline Gas Co.)

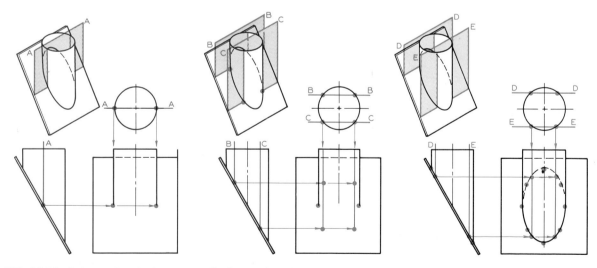

FIG. 24.13 Intersection between a cylinder and a plane

Step 1 A vertical cutting plane, A–A, is passed through the cylinder parallel to its axis to find two points of intersection.

Step 2 Two more cutting planes, B–B and C–C, are used to find four additional points in the top and left side views; these points are projected to the front view.

Step 3 Additional cutting planes are used to find more points. These points are connected to give an elliptical line of intersection.

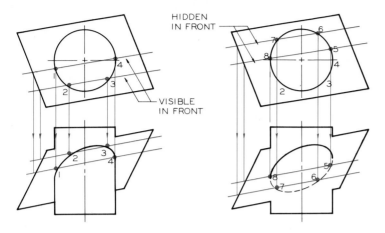

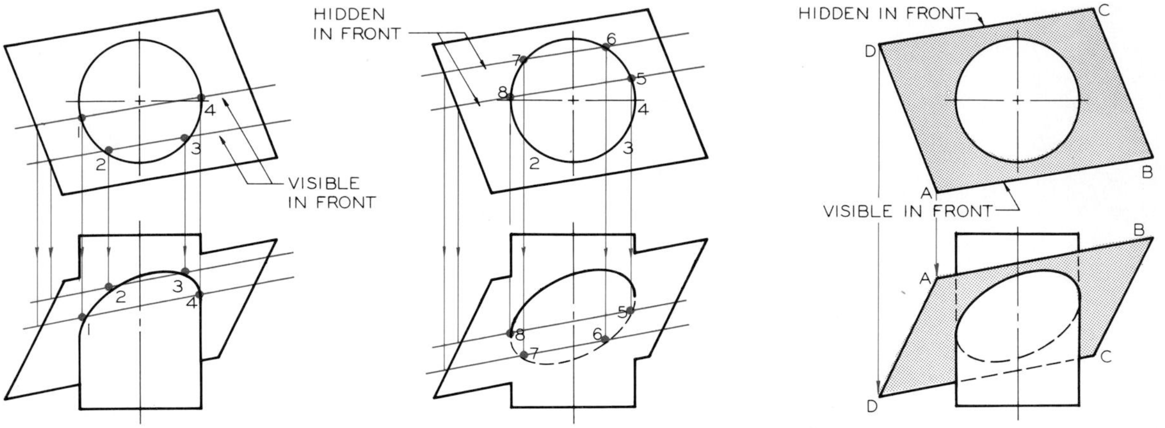

FIG. 24.14 Intersection of a cylinder and an oblique plane

Step 1 Vertical cutting planes are passed through the cylinder in the top view to establish elements on its surface and lines on the oblique plane. Piercing Points 1, 2, 3, and 4 are projected to the front view of their respective lines and are connected with a visible line.

Step 2 Additional cutting planes are used to find other piercing points—5, 6, 7, and 8—which are projected to the front view of their respective lines on the oblique plane. These are connected with a hidden line by inspection of the top view.

Step 3 Visibility of the plane and cylinder is completed in the front view. Line *AB* is found to be visible by inspection of the top view, and Line *CD* is found to be hidden.

to establish elements on the cylinder and their piercing points. The piercing points are projected to each view to find the line of intersection, which is an ellipse.

A more general problem is solved in Fig. 24.14, where the cylinder is vertical but the plane is oblique. Vertical cutting planes are passed through the cylinder and the plane in the top view to find piercing points

of the cylinder's elements on the plane. These points are projected to the front view to complete the line of intersection, an ellipse. The more cutting planes that are used, the more accurate will be the line of intersection.

The general case of the intersection between a plane and cylinder is solved in Fig. 24.15, where both are oblique in the given views. The edge view of the plane is found in an auxiliary view. Cutting planes are passed through the cylinder parallel to the cylinder's axis in the auxiliary view to find the piercing points.

The piercing points of the elements are connected to give elliptical lines of intersection in the given views. Visibility is determined by analysis.

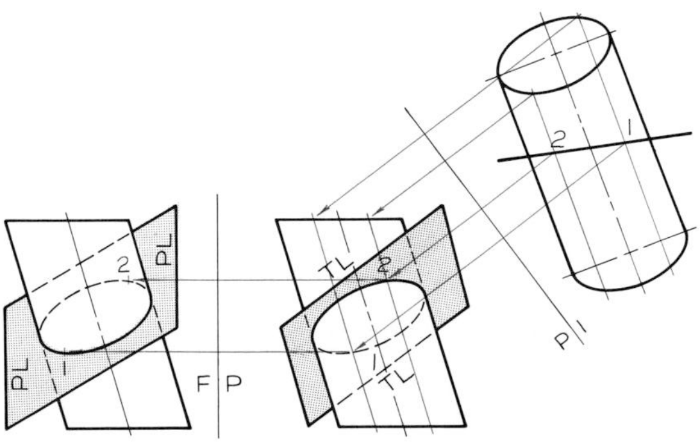

24.5
Intersections between cylinders and prisms

A series of vertical cutting planes is used in Fig. 24.16 to establish lines that lie on the surfaces of the cylinder and prism. A primary auxiliary view is drawn to show

FIG. 24.15 The intersection between an oblique cylinder and an oblique plane can be found by constructing a view that shows the plane as an edge.

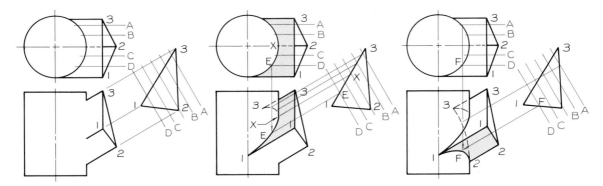

FIG. 24.16 Intersection between a cylinder and a prism

Step 1 Project an auxiliary view of the triangular prism from the front view to show three of its surfaces as edges. Pass frontal cutting planes through the top view of the cylinder and project them to the auxiliary view. The spacing between the planes is equal in both views.

Step 2 Locate points along the line of intersection of Plane 1–3 in the top view and project them to the front view. *Example:* Point E on cutting Plane D is found in the top and primary auxiliary views and projected to the front view where the projectors intersect. Point X is the point where visibility changes from visible to hidden in the front view.

Step 3 Determine the remaining points of intersection by using the same cutting planes. Point F is shown in the top and primary auxiliary views and is projected to the front view of 1–2. Connect the points and determine visibility. Judgment should be used in spacing the cutting planes so that they will produce the most accurate line of intersection.

the end view of the inclined prism. The vertical cutting planes are shown in this view also, spaced the same distance apart as in the top view (Step 1).

The line of intersection from 1 to 3 is projected from the auxiliary view to the front view, Step 2, where the intersection is an elliptical curve. The change of visibility of this line is found at Point X in the top and auxiliary views, and is projected to the front view. In Step 3, the process is continued to find the lines of intersection of the other two planes of the prism.

24.6
Intersections between two cylinders

The line of intersection between two perpendicular cylinders can be found by passing cutting planes through the cylinders parallel to the centerlines of each (Fig. 24.17). Each cutting plane locates the piercing points of two elements of one cylinder on an ele-

ment of the other cylinder. The points are connected and visibility is determined to complete the solution.

The intersection between nonperpendicular cylinders is found in Fig. 24.18 by a series of vertical cutting planes. Each cutting plane is passed through the cylinders parallel to the centerline of each. Points 1 and 2 are labeled on cutting Plane D as examples of points on the line of intersection. Other points are found in the same manner. The auxiliary view is an optional view that is not required for the solution of this problem, but it assists you in visualizing the problem. Points 1 and 2 are shown on cutting Plane D in the auxiliary view, where they can be projected to the front view as a check on the solution found when projecting from the top view.

24.7
Intersections between planes and cones

To find points of intersection on a cone, cutting planes can be used that are (1) perpendicular to the cone's

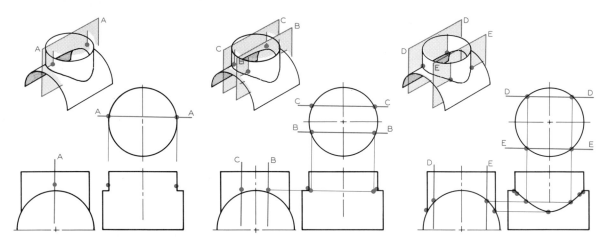

FIG. 24.17 Intersection between two cylinders

Step 1 A cutting plane, *A–A*, is passed through the cylinders parallel to the axes of both. Two points of intersection are found.

Step 2 Cutting Planes *C–C* and *B–B* are used to find four additional points of intersection.

Step 3 Cutting Planes *D–D* and *E–E* locate four more points. Points found in this manner give the line of intersection.

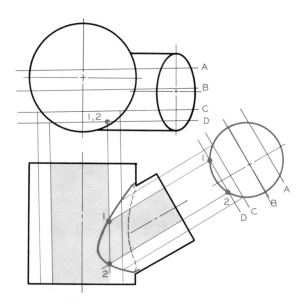

FIG. 24.18 The intersection between these cylinders is found by finding the end view of the inclined cylinder in an auxiliary view. Vertical cutting planes are used to find the piercing points of the elements of the cylinder and the line of intersection.

axis, or (2) parallel to the cone's axis. Horizontal cutting planes are shown in Fig. 24.19A where they are labeled as H_1 and H_2. The horizontal planes cut circular sections that appear true size in the top view.

The cutting planes in Fig. 24.19B are passed radially through the top view to establish elements on the surface of the cone that are projected to the front view. Points 1 and 2 are found on these elements in both views by projection.

A series of radial cutting planes is used to find elements on the cone in Fig. 24.20. These elements cross the edge view of the plane in the front view to locate piercing points that are projected to the top view of the same elements to form the line of intersection.

A cone and an oblique plane intersect in Fig. 24.21, and the line of intersection is found by using a series of horizontal cutting planes. The sections cut by these imaginary planes will be circles in the top view. In addition, the cutting planes locate lines on the oblique plane that intersect the same circular sections cut by each respective cutting plane. The points of intersection are found in the top view and are projected to the front view.

The horizontal cutting-plane method also could have been used to solve the example in Fig. 24.20.

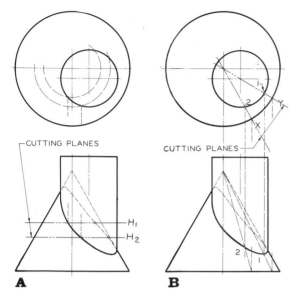

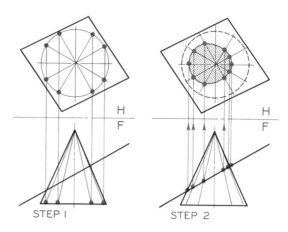

FIG. 24.20 Intersection of a plane and a cone

Step 1 Divide the base into even divisions in the top view and connect these points with the apex to establish elements on the cone. Project these elements to the front view.

Step 2 The piercing point of each element on the edge view of the plane is projected to the top view to the same elements, where they are connected to form the line of intersection.

FIG. 24.19 Intersections on conical surfaces can be found by cutting planes that pass through the cone parallel to its base (Part A). A second method at B shows radial cutting planes that pass through the cone's center line and perpendicular to its base.

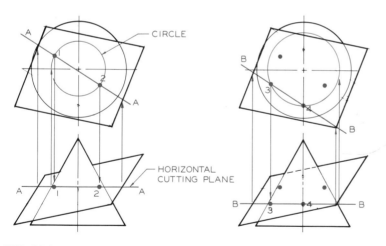

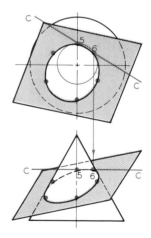

FIG. 24.21 Intersection of an oblique plane and a cone

Step 1 A horizontal cutting plane is passed through the front view to establish a circular section on the cone and a line on the oblique plane in the top view. The piercing point of this line must lie on the circular section. Piercing Points 1 and 2 are projected to the front view.

Step 2 Horizontal cutting Plane B–B is passed through the front view in the same manner to locate piercing Points 3 and 4 in the top view. These points are projected to the horizontal plane in the front view from the top view.

Step 3 Additional horizontal planes are used to find sufficient points to complete the line of intersection. Determination of the visibility completes the solution.

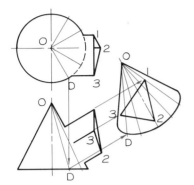
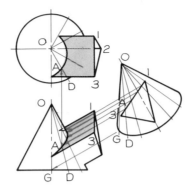
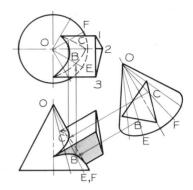

FIG. 24.22 Intersection between a cone and a prism

Step 1 Construct a primary auxiliary view to obtain the edge views of the lateral surfaces of the prism. In the auxiliary view, pass cutting planes through the cone which radiate from the apex to establish elements on the cone. Project the elements to the front and auxiliary views.

Step 2 Locate the piercing points of the cone's elements with the edge view of Plane 1–3 in the primary view and project them to the front and top views. *Example:* Point *A* lies on Element *OD* in the primary auxiliary view, so it is projected to the front and top views of Element *OD*.

Step 3 Locate the piercing points where the conical elements intersect the edge views of the planes of the prism in the auxiliary view. *Example:* Point *B* is found on *OE* in the primary auxiliary view and is projected to the front and top views of *OE*.

24.8
Intersections between cones and prisms

A primary auxiliary view is used to find the end view of the inclined prism that intersects the cone in Fig. 24.22, Step 1. Cutting planes that radiate from the apex of the cone in the top view are drawn in the auxiliary view to locate elements on the cone's surface that intersect the prism. These elements are drawn in the front view by projection.

Wherever the edge view of Plane 1–3 intersects an element in the auxiliary view, the piercing points are projected to the same element in the front and top views (Step 2). An extra cutting plane is passed through Point 3 in the auxiliary view to locate an element that is projected to the front and top views. Piercing Point 3 is projected to this element in sequence from the auxiliary view to the top view.

This same procedure is used to find the piercing points of the other two planes of the prism in Step 3. All projections of points of intersection originate in the auxiliary view, where the planes of the prism appear as edges.

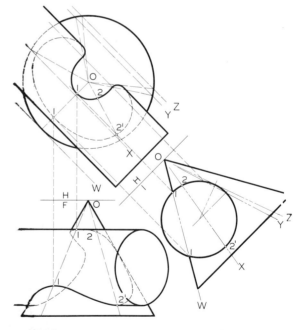

FIG. 24.23 The intersection between the cone and the cylinder is found by projecting an auxiliary view from the top view of the cone to find the circular view of the cylinder. Radial cutting planes are passed through the cone and the cylinder in the auxiliary view to locate piercing points of the cylinder's elements.

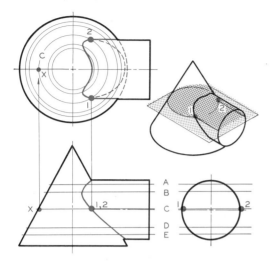

FIG. 24.24 Horizontal cutting planes are used to find the intersection between the cone and the cylinder. The cutting planes form circles in the top view.

FIG. 24.25 This electrically operated distributor illustrates intersections between a cone and a series of cylinders. (Courtesy of GATX.)

A similar problem is solved in Fig. 24.23, where a cylinder intersects a cone. The circular view of the cylinder is found in a primary auxiliary view. Points 2 and 2' are found in the auxiliary view on Element O–X where a radial cutting plane is passed through Apex O. Element O–X is found in the top and front views by projecting from the auxiliary view to locate Points 2 and 2'.

In Fig. 24.24, horizontal cutting planes are passed through the cone and the intersecting perpendicular cylinder to locate the line of intersection. A series of circular sections are found in the top view. Points 1 and 2 are found on cutting Plane C in the top view as examples, and are projected to the front view. Other points are found in this same manner.

This method is feasible only when the center line of the cylinder is perpendicular to the axis of the cone, so that circular sections can be found in the top view, rather than elliptical sections that would be difficult to draw.

The distributor housing in Fig. 24.25 is an example of an intersection between cylinders and a cone.

24.9
Intersections between pyramids and prisms

The intersection between an inclined prism and a pyramid is solved in Fig. 24.26. The end view of the inclined prism is found in primary auxiliary view; the pyramid is shown in this view also (Step 1). Radial Lines OB and OA are passed through Corners 1 and 3 in the auxiliary view (Step 2). The radial lines are projected from the auxiliary view to the front and top views. Intersecting Points 1 and 3 are located on OB and OA in each of these views by projection. Point 2 is the point where Line 1–3 bends around Corner OC. Lines of intersection 1–4 and 4–3 are found in Step 3; the visibility is determined; and the solution is completed.

A prism that is parallel to the base of a pyramid is shown in Fig. 24.27. Its lines of intersection are found by using a series of horizontal cutting planes that pass through the pyramid parallel to its base to form triangular sections in the top view.

The same cutting planes are passed through the corner lines of the prism in the front and auxiliary views. Each corner edge is extended in the top view to intersect the triangular section formed by the cut-

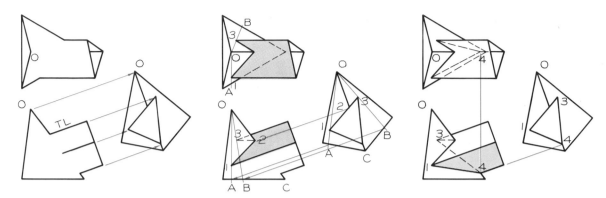

FIG. 24.26 Intersection between a prism and a pyramid

Step 1 Find the edge view of the surfaces of the prism by projecting an auxiliary view from the front view. Project the pyramid into this view also. Only the visible surfaces need be shown in this view.

Step 2 Pass Planes *A* and *B* through Apex *O* and Points 1 and 3 in the auxiliary view. Project Lines *OA* and *OB* to the front and top views. Project Points 1 and 3 to *OA* and *OB* in the principal views. Point 2 lies on Line *OC*. Connect Points 1, 2, and 3 to give the intersection of the upper plane.

Step 3 Point 4 lies on Line *OC* in the auxiliary view. Project this point to the principal views. Connect Point 4 to Points 3 and 1 to complete the intersections. Visibility is indicated. These geometric shapes are assumed to be hollow, as though constructed of sheet metal.

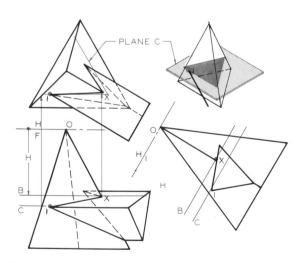

FIG. 24.27 The intersection between this pyramid and prism is found by finding the end view of the prism in an auxiliary view. Horizontal cutting planes are passed through the fold lines of the prism to find the piercing points and the line of intersection.

ting plane passing through it. Point 1 is given as an example.

Corner Point *X* is found by passing cutting Plane *B* through it in the auxiliary view where it crosses the corner line. This is where the line of intersection of this plane bends around the corner.

24.10
Intersections between spheres and planes

The line of intersection between a sphere and a plane is found in Fig. 24.28, where the plane appears as an edge in the front view. Horizontal cutting planes are passed through the front view to form circular sections in the top view. Two points are located on each cutting plane in the top view by projecting from the front view where the cutting plane crosses the edge view of the intersecting plane. The points are connected and the visibility is shown in the top view.

The elliptical intersection could have been drawn with an ellipse template that was selected by measuring the angle between the edge view of the plane in

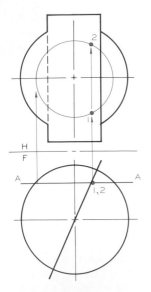 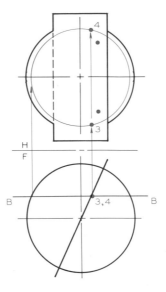 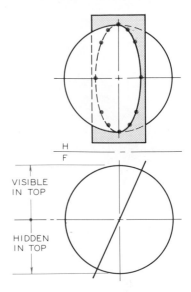

FIG. 24.28 Intersection of a sphere and a plane

Step 1 Horizontal cutting Plane A–A is passed through the front view of the sphere to establish a circular section in the top view. Piercing Points 1 and 2 are projected from the front view to the top view, where they lie on the circular section.

Step 2 Horizontal cutting Plane B–B is used to locate Piercing Points 3 and 4 in the top view by projecting to the circular section cut by the plane in the top view. Additional horizontal planes are used to find sufficient points in this manner.

Step 3 Visibility of the top view is found by inspection of the front view. The upper portion of the sphere will be visible in the top view and the lower portion will be hidden.

the front view and the projectors from the top view. The major diameter of the ellipse would be equal to the true diameter of the sphere, since the plane passes through the center of the sphere.

A general case of the intersection between a plane and a sphere is given in Fig. 24.29, where three points, 1, 2, and 3, are located on the sphere's surface. A circle is to be drawn through these three points and lie on the surface of the sphere.

The edge view of the plane is found by projecting from the top view to a primary auxiliary view. The circle on the sphere through 1, 2, and 3 will have an elliptical line of intersection in the top and front views. The ellipse template angle for the top view is the angle between the edge of the plane and the projector from the top view, 25°. The major diameter is drawn parallel to the true-length lines on Plane 1–2–3 in the top view.

The ellipse for the front view of the intersection is found in the same manner by finding the edge view of the plane in an auxiliary view projected from the

front view. The ellipse template angle is found to be 50°.

24.11
Intersections between spheres and prisms

The intersection between a sphere and a prism is found in Fig. 24.30 by drawing a series of vertical cutting planes in the top and side views. The planes form circular sections in the front view.

The intersections of the edges with the cutting planes in the side view are projected to their respective circles in the front view. Points 1 and 2 are located on cutting Plane A in the side view and on the circular path of A in the front view. Point X in the side view locates the point where the visibility changes in the front view.

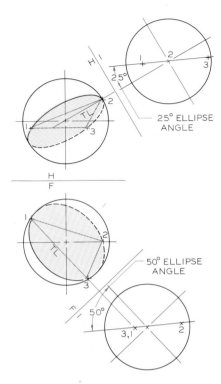

FIG. 24.29 Three points are given on the surface of the sphere through which a circle passes. This plane is found as an edge by projecting from the top and front views. Ellipse angles of 25° and 50° are found for drawing the top and front views of the elliptical intersections.

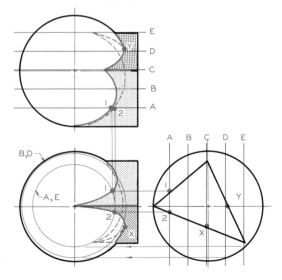

FIG. 24.30 Vertical cutting planes are used to find the intersection between a prism and a sphere.

Point Y in the side view is the point where the visibility of the intersection changes in the top view. Both of these points lie on the center lines of the sphere on the side view.

24.12
Miscellaneous intersections

A series of intersections of cutting planes passed through different figures is shown in Fig. 24.31. Radial lines are used to find a hyperbolic section on a cone at A. Horizontal cutting planes are used to locate a section in the top view of a **torus** (a donut-shaped object) at B.

Horizontal cutting planes are drawn to locate an elliptical section through a hemisphere at C with a supplementary auxiliary view. The section of a cutting plane through an oblique pyramid is shown at D.

The construction of runouts formed by fillets that intersect cylinders is shown in Fig. 24.32. Horizontal cutting planes are passed through the fillets in the front view and are projected to the top view to locate arcs formed by the cutting planes.

Points along the runout in the front view are found by projecting from the top view. Points 1 and 2 are shown as examples.

24.13
Principles of developments

The processing plant shown in Fig. 24.33 illustrates numerous examples of sheet metal shapes that were designed using the principles of developments. In other words, their patterns were laid out on flat stock and then formed to the proper shape by bending and seaming the joints.

Examples of standard hemmed edges and joints are shown in Fig. 24.34. The application of the sheet metal design will determine the best method of connecting the seams.

The development of the surfaces of three typical shapes into a flat pattern is shown in Fig. 24.35. The sides of a box are imagined to be unfolded into a common plane. The cylinder is rolled out for a distance that is equal to its circumference. The pattern of a right cone is developed using the length of the element as the radius.

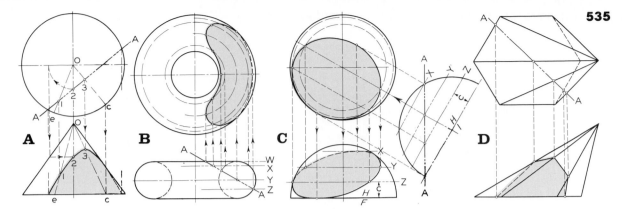

FIG. 24.31 Examples of cutting-plane intersections through four types of figures are shown here. (A) Radial lines are used to find a hyperbolic section through a cone. (B) An inclined plane is shown intersecting a torus. (C) The section cut by a plane through a hemisphere is shown. (D) A section cut by a vertical plane through a pyramid.

FIG. 24.33 Most of the surfaces shown in this refinery were made from flat stock that was fabricated to form these irregular shapes. These flat patterns are called developments.

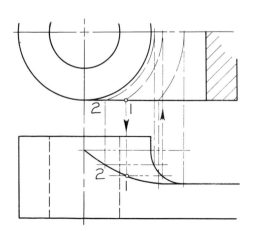

FIG. 24.32 Horizontal cutting planes are used to find the runout (intersection) where surfaces intersect the cylindrical ends of these parts.

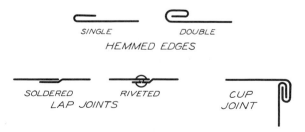

SINGLE   DOUBLE

HEMMED EDGES

SOLDERED   RIVETED   CUP JOINT

LAP JOINTS

FIG. 24.34 Examples of the types of seams that are used to join developments.

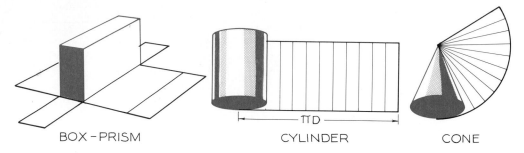

FIG. 24.35 Three standard types of developments: the box, cylinder, and cone

Patterns of shapes with parallel elements, such as the prisms and cylinders shown in Fig. 24.36A and B, are begun by constructing stretch-out lines that are parallel to the edge view of the right section of the parts. The distance around the right section is laid off along the stretch-out line. The prism and cylinder at C and D are inclined; consequently the right sections must be drawn perpendicular to their sides, not parallel to their bases.

An inside pattern (development) is more desirable than an outside pattern because most bending machines are designed to fold metal so that markings are folded inward, and because markings and scribings will be hidden when the pattern is assembled in final form. The method of denoting a pattern is by a series of lettered or numbered points about its layout. All lines on a development must be true length.

Seam lines (lines where the pattern is joined) should be the shortest lines. This results in the least expense of riveting or welding the pattern together to form the final shape, since it is the shortest seam possible.

24.14
Development of prisms

A flat pattern for a prism is developed in Fig. 24.37. Since the edges of the prism are vertical in the front view, its right section is perpendicular to these sides. The top view shows the right section true size. The stretch-out line is drawn parallel to the edge view of the right section, beginning with Point 1.

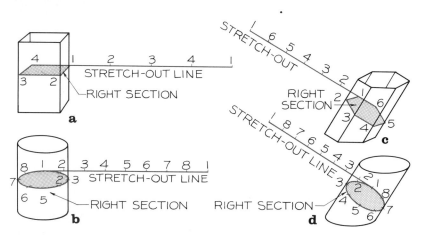

FIG. 24.36 The developments of right prisms and cylinders are found by rolling out the right sections along a stretch-out line (Parts A and B). When these figures are oblique, the right sections are found to be perpendicular to the sides of the prism and cylinder. The development is laid out along the stretch-out line that is parallel to the edge view of the right section (Parts C and D).

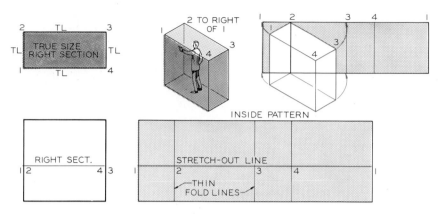

FIG. 24.37 The development of a rectangular prism to give an inside pattern. The stretch-out line is parallel to the edge view of the right section.

FIG. 24.38 Numerous examples of developments fabricated from sheet metal are shown in this plant. (Courtesy of Heat Engineering.)

lines are connected to form the limits of the developed surface. Fold lines are drawn as thin lines and the outside lines of a development are drawn as visible object lines.

The complex installation in Fig. 24.38 is composed of many developments ranging from simple prisms to more complicated shapes.

The development of the prism in Fig. 24.39 is similar to the example in Fig. 24.37 except that one end is beveled rather than square. The stretch-out line is drawn parallel to the edge view of the right section in the front view. The true-length distances around the right section are laid off along the stretch-out line and the fold lines are located. The lengths of the fold lines are found by projecting from the front view of these lines.

If an inside pattern is drawn and it is to be laid out to the right, Point 2 will be to the right of Point 1. This is determined by looking from the inside of the top view where 2 is seen to the right of 1.

To locate the fold lines on the pattern, Lines 2–3, 3–4, and 4–1 are transferred from the right section in the top view with your dividers to the stretch-out line. The length of each fold line is found by projecting its true length from the front view. The ends of the fold

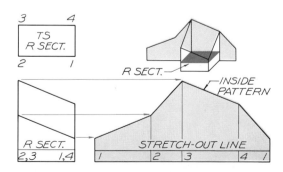

FIG. 24.39 The development of a rectangular prism with a beveled end to give an inside pattern. The stretch-out line is parallel to the right section.

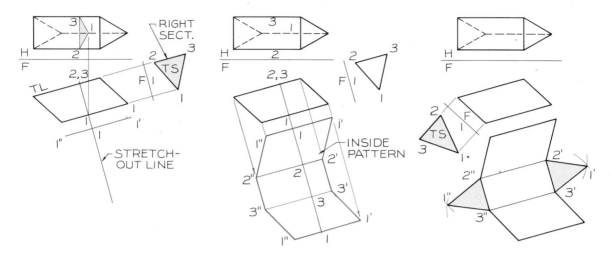

FIG. 24.40 Development of an oblique prism

Step 1 The edge view of the right section will be perpendicular to the true-length axis of the prism in the front view. Determine the true-size view of the right section by constructing an auxiliary view. Draw the stretch-out line parallel to the edge view of the right section. Project bend Line 1'–1" as the first line of the development.

Step 2 Since the pattern is developed toward the right, beginning with Line 1'–1", the next point is found to be Line 2'–2" by referring to the auxiliary view. Transfer true-length Lines 1–2, 2–3, and 3–1 from the right section to the stretch-out line to locate the elements. Determine the lengths of the bend lines by projection.

Step 3 Find the true-size views of the end pieces by projecting auxiliary views from the front view. Connect these surfaces to the development of the lateral sides to form the completed pattern. Fold lines are drawn with thin lines, while outside lines are drawn as regular object lines.

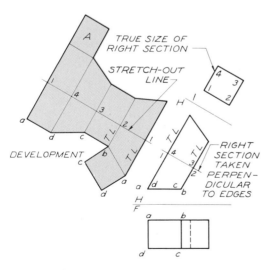

FIG. 24.41 The development of this oblique chute is found by finding the right section true-size in the auxiliary view. The stretch-out line is drawn parallel to the right section.

24.15
Development of oblique prisms

The prism in Fig. 24.40 is inclined to the horizontal plane, but its fold lines are true length in the front view. The right section is drawn as an edge perpendicular to these fold lines, and the stretch-out line is drawn as shown in Step 1. A true-size view of the right section is constructed in the auxiliary view.

In Step 2, the distances between the fold lines are transferred from the true-size right section to the stretch-out line. The lengths of the fold lines are found by projecting from the front view.

In Step 3, the ends of the prism are found and are attached to the pattern so that they can be folded into position.

In Fig. 24.41. The fold lines are true length in the top view; this enables you to draw the edge view of the right section perpendicular to the fold lines in the top view. The stretch-out line is drawn parallel to the edge view of the section, and the true size of the right section is found in an auxiliary view projected from the top view. The distances about the right section are transferred to the stretch-out line to locate the fold lines. The lengths of the fold lines are found by projecting from the top view. The end portions of the pattern are attached to the pattern to complete the construction.

A prism that does not project true length in either view can be developed as illustrated in Fig. 24.42. The fold lines are found true length in an auxiliary view projected from the front view. The right section will appear as an edge perpendicular to the fold lines in the auxiliary view. The true size of the right section is found in a secondary auxiliary view.

The stretch-out line is drawn parallel to the edge view on the right section. The fold lines are located on the stretch-out line by measuring around the right section in the secondary auxiliary view. The lengths of the fold lines are then projected to the development from the primary auxiliary view.

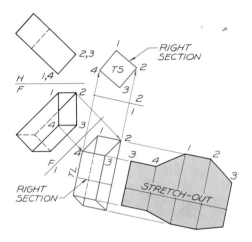

FIG. 24.42 The development of an oblique cylinder is found by finding an auxiliary view in which the fold lines are true length, and a secondary auxiliary view in which the right section appears true size. The stretch-out line is drawn parallel to the right section.

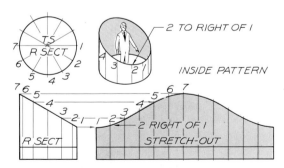

FIG. 24.43 The development of a right cylinder's inside pattern. The stretch-out line is parallel to the right section. Point 2 is to the right of Point 1 for an inside pattern.

24.16
Development of cylinders

The development of a cylinder is found in Fig. 24.43. Since the elements of the cylinder are true length in the front view, the right section will appear as an edge in this view, and true size in the top view. The stretch-out line is drawn parallel to the edge view of the right section, and Point 1 is chosen as the beginning point since it is the shortest element. To draw an inside pattern, assume that you are standing on the inside looking at Point 1 and you will see that Point 2 is to the right of 1. Therefore, the pattern is laid out with Point 2 to the right of Point 1.

The spacing between the elements in the top view can be conveniently done by drawing radial lines at 15° or 30° intervals. Using this technique, the elements will be equally spaced, making it convenient to lay them out along the stretch-out line. The lengths of the elements are found by projecting from the front view to complete the pattern.

An application of a developed cylinder with a beveled end is the air-conditioning duct from an automobile shown in Fig. 24.44. The development of a similar cylinder is shown in Fig. 24.45. The base of the front view is the edge view of the right section, which appears true size in the top view. Elements are located around the circumference of the right section in the top view. Two alternative methods are shown to illustrate how the elements are located at 30° intervals. One employs the 30°–60° triangle, and the other uses a compass with the radius equal to the radius of the right section.

The stretch-out line is drawn parallel to the right section in the front view. The total length is found mathematically by the formula $C = \pi D$. The stretch-out line is divided into the same number of divisions

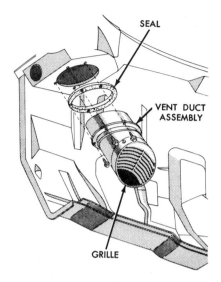

FIG. 24.44 This ventilator air duct was designed through the use of development principles. (Courtesy of Ford Motor Co.)

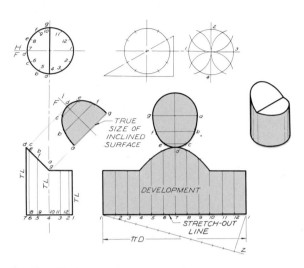

FIG. 24.45 The development of a right cylinder can be found by mathematically locating elements along the stretch-out line, which is equal in length to the circumference of the right section.

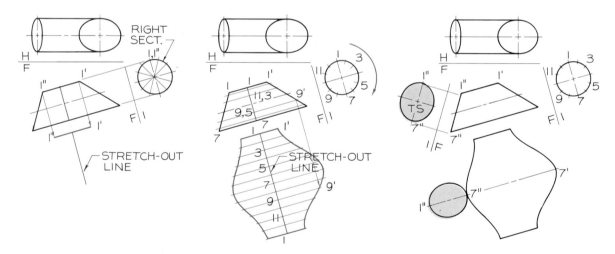

FIG. 24.46 Development of an oblique cylinder

Step 1 The right section appears as an edge in the front view, in which it is perpendicular to the true-length axis. Construct an auxiliary view to determine the true size of the right section. Draw a stretch-out line parallel to the edge view of the right section. Locate Element 1′–1″. Divide the right section into equal divisions.

Step 2 Project these elements to the front view from the right section. Transfer measurements between the points in the auxiliary view to the stretch-out line to locate the elements in the development. Determine the lengths of the elements by projection to complete the development.

Step 3 The development of the end pieces will require auxiliary views that project these surfaces as ellipses, as shown for the left end. Attach this true-size ellipse to the pattern. Note that the beginning line for the pattern was Line 1″–1′, the shortest element, for economy.

as there are elements, 12 in this case, to provide a high degree of accuracy in finding the circumference.

The pattern for the end of the cylinder is found by combining the partial top view and the auxiliary view. This end is connected to the overall pattern to complete the solution.

24.17
Development of oblique cylinders

The pattern for an oblique cylinder (Fig. 24.46) is found in the same manner as the previous examples, but with the addition of one preliminary step: The right section must be found true-size in an auxiliary view. A series of equally spaced elements is located around the right section in the auxiliary view and is projected back to the true-length view (Step 1). The stretch-out line is drawn parallel to the edge view of the right section in the front view.

In Step 2, the spacing between the elements is laid out along the stretch-out line, and the elements are drawn through these points perpendicular to the stretch-out line. The lengths of the elements are found by projecting from the front view.

The ends of the cylinder are found in Step 3 to complete the pattern. Only one end pattern is shown as an example.

The oblique cylinder in Fig. 24.47 is a more general case, where the elements are not true length in the given views. A primary auxiliary view is used to find a view where the elements are true length, and a secondary auxiliary view is drawn to find the true-size view of the right section. The stretch-out line is drawn parallel to the edge view of the right section in the primary auxiliary view, and the elements are located along this line by transferring their distances apart from the true-size right section.

The elements are drawn perpendicular to the stretch-out line. The length of each element is found by projecting from the primary auxiliary view. The endpoints are connected with a smooth curve to complete the pattern.

24.18
Development of pyramids

All lines used to draw a pattern must be true length. Pyramids have only a few lines that are true length in the given views; for this reason, the sloping corner lines must be found true-length at the outset.

The corner lines of a pyramid can be found by revolution, as shown in Fig. 24.48. Line 0–5 is revolved in the frontal position of 0–5' in the top view

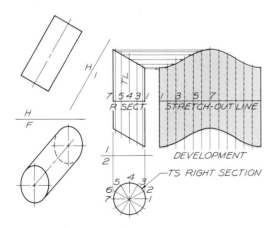

FIG. 24.47 The development of an oblique cylinder is found by constructing an auxiliary view in which the elements appear true length. The right section is found true size in a secondary auxiliary view.

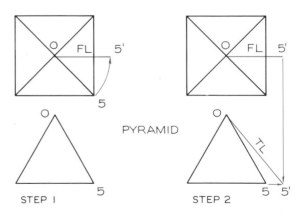

FIG. 24.48 True length by revolution

Step 1   Corner 0–5 of a pyramid is found true length by revolving it into the frontal plane in the top view, 0–5'.

Step 2   Point 5' is projected to the front view where 0–5' is true length.

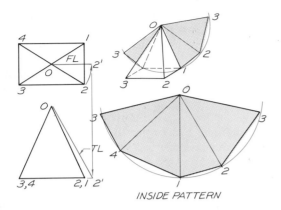

FIG. 24.49 Development of a right pyramid

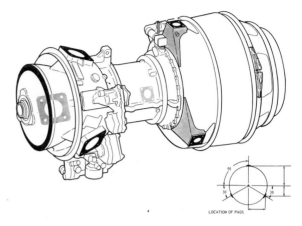

FIG. 24.51 Examples of pyramid shapes in the design of mounting pads for an engine. (Courtesy of Avco Lycoming.)

(Step 1). Since 0–5' is a frontal line, it will be true-length in the front view (Step 2).

The development of a pyramid is given in Fig. 24.49. Line 0–2 is revolved into the frontal plane in the top view to find its true length in the front view. All bend lines are equal in length since this is a right pyramid. Line 0–2' is used as a radius to construct the base circle for drawing the development. Distance 3–2 is transferred from the base in the top view to the development, where it forms a chord on the base circle. Lines 2–1, 1–4, and 4–3 are found in the same manner and in sequence. The bend lines are drawn as thin lines from the base to the apex, Point 0.

A variation of this problem is given in Fig. 24.50, in which the pyramid has been truncated or cut at an angle to its axis. The development of the inside pattern

is found in the same manner as the previous example; however, an additional step is required to establish the upper lines of the development. The true-length lines from the apex to Points 1', 2', 3', and 4' are found by revolution. These distances are laid on their respective lines of the pattern to locate the upper limits of the pattern.

The mounting pads in Fig. 24.51 are sections of pyramids that intersect an engine body.

The development of an oblique pyramid is shown in Fig. 24.52. In Step 1, the corner lines are found true length by revolution. Using these true-length lines and those that are given in the principal views, the pattern is drawn by triangulation using a compass (Step 2). In Step 3, the upper limits of the pattern are found by measuring along the fold lines from Point 0.

24.19
Development of cones

All elements of a right cone are equal in length, as illustrated in Fig. 24.53 where 0–6 is found true length by revolution. When revolved to 0–6' position, it is a frontal line and is therefore true length in the front view where it is an outside element of the cone. Point 7' is found by projecting horizontally to Element 0–6'.

The right cone in Fig. 24.54 is developed by dividing the base into equally spaced elements in the top view and by projecting them to the base in the front

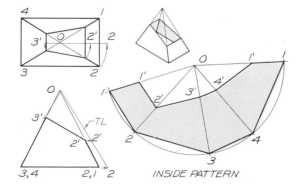

FIG. 24.50 The development of an inside pattern of a truncated pyramid. The corner lines are found true length by revolution.

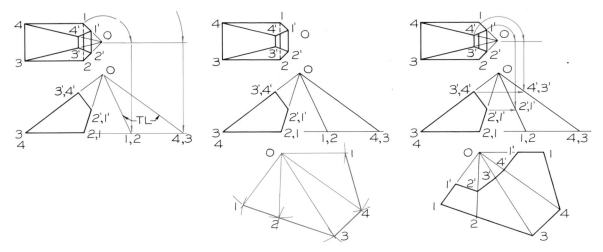

FIG. 24.52 Development of an oblique pyramid

Step 1 Revolve each of the bend lines in the top view until they are parallel to the frontal plane. Project to the front view where the true-length views of the revolved lines can be found. Let Point 0 remain stationary but project Points 1, 2, 3, and 4 horizontally in the front view to the projectors from the top view.

Step 2 The base lines appear true length in the top view. Using these true-length lines from the top view and the revolved lines in the front view, draw the development triangles. All triangles have one side and Point 0 in common. This gives a development of the surface, excluding the truncated section.

Step 3 The true lengths of the lines from Point 0 to Points 1', 2', 3', and 4' are found by revolving these lines. These distances are laid off from Point 0 along their respective lines to establish points along the upper edge of the pattern. The points are then connected by straight lines.

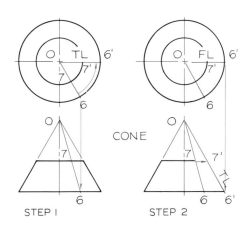

FIG. 24.53 True length by revolution

Step 1 An element of a cone, 0–6, is revolved into a frontal plane in the top view.

Step 2 Point 6' is projected to the front view where it is the outside element of the cone and is true length. Line 0–7' is found TL by projecting to the outside element in the front view.

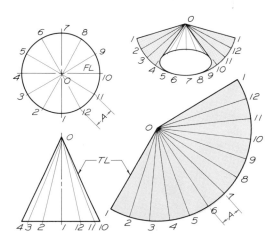

FIG. 24.54 The development of an inside pattern of a right cone. In this case, all elements are equal in length.

FIG. 24.55 An example of a conical shape that was formed from steel panels by applying principles of developments.

view. The outside elements in the front view, O–10 and O–4, are true length.

Using Element O–10 as a radius, draw the base circle of the development. The elements are located along the base circle equal to the chordal distances between them around the base in the top view. You can see that this is an inside pattern by inspecting the top view where Point 2 is to the right of Point 1 when viewed from the inside.

The sheet metal conical vessel in Fig. 24.55 is an example of a large vessel that was designed using principles of a development.

The development of a truncated cone is shown in Fig. 24.56. The pattern is found by laying out the total cone, ignoring the portion removed from it.

The upper portion that has been removed can be found by using true-length Line O–7' as the radius in the pattern view.

The hyperbolic sections through the front view of the cone can be found on their respective elements in the top and front views. Lines O–2' and O–3' are projected horizontally to the true length Element O–1 in the front view, where they will appear true length. These distances and others are measured off along their respective elements in the development to establish a smooth curve on the development.

24.20
Development of oblique cones

The development of an oblique cone is found in Fig. 24.57. The elements of this cone are of varying lengths, but the pattern will be symmetrical about an axis.

Elements are located in the top view by dividing the base into equal arcs and drawing the elements to Point O, the apex. These elements are projected to the frontal view. Each element is found true length by re-

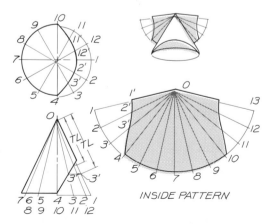

INSIDE PATTERN

FIG. 24.56 The development of a conical surface with a side opening

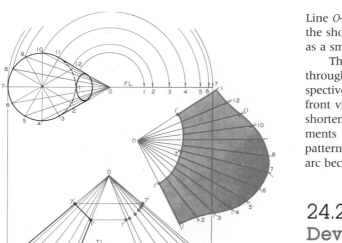

FIG. 24.57 The development of an oblique cone All elements must be found true length by a true-length diagram before the pattern can be constructed.

volving it into a frontal plane in the top view. When projected to the front view, the revolved lines are found true length in a true-length diagram.

The development is begun by constructing a series of adjacent triangles using the true-length elements and the chordal distances on the base in the top view.

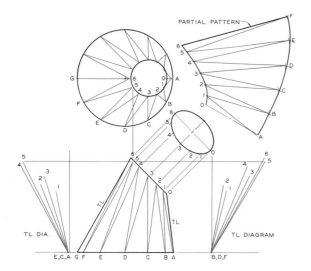

FIG. 24.58 The development of a partial pattern of a warped surface

Line O–1 is chosen for the line of separation since it is the shortest element. The base of the pattern is drawn as a smooth curve.

The distances from Point O to the upper cut through the cone are found by projecting to their respective elements in the true-length diagram from the front view. Line O–$7'$ is shown as an example. These shorter elements are located on their respective elements in the pattern to give the upper limits of the pattern. This will be an irregular curve rather than an arc because this is not a right cone.

24.21
Development of warped surfaces

The geometric shape in Fig. 24.58 is an approximate cone with a warped surface; it is similar to the oblique cone in Fig. 24.57. The development of this surface will be an approximation, since a warped surface cannot be laid out on a flat surface.

The surface is divided into a series of triangles in the top and front views by dividing the upper and lower views into equal sectors. The true lengths of all lines are found in the true-length diagrams drawn on both sides of the front and by projecting horizontally from the ends of the lines. If necessary, review Fig. 24.57 to see how the true-length diagrams were found.

The chordal distances between the points on the base appear true-length in the top view since the base is horizontal. However, an auxiliary view is necessary to find the true distance about the upper surface of the object.

The developed surface is found by triangulation using true-length lines from (1) the true-length diagram, (2) the horizontal base in the top view, and (3) the primary auxiliary view. Each point should be labeled as it is laid out to avoid confusion.

24.22
Development of transition pieces

A transition piece is a form that transforms the section at one end to a different shape at the other (Fig.

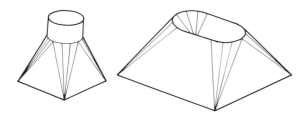

FIG. 24.59 Examples of transition pieces that join together parts that have different cross sections.

24.59). Huge transition pieces can be seen in the industrial installation in Fig. 24.60.

The development of a transition piece is shown in Fig. 24.61. In Step 1, radial elements are drawn from each corner to the equally spaced points on the circular end of the piece. Each of these lines is found true length by revolution.

In Step 2, the true-length lines are used with the true-length chordal distance in the top view to lay out

FIG. 24.60 Transition-piece developments are used to join a circular shape with a rectangular section. (Courtesy of Western Precipitation Group, Joy Manufacturing Co.)

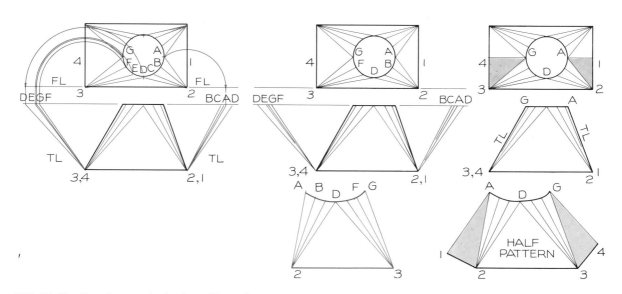

FIG. 24.61 Development of a transition piece

Step 1 Divide the circular edge of the surface into equal parts in the top view. Connect these points with bend lines to the corner points, 2 and 3. Find the true length (TL) of these lines by revolving them into a frontal plane and projecting them to the front view. These lines represent elements on the surface of an oblique cone.

Step 2 Using the TL lines found in the TL diagram and the lines on the circular edge in the top view, draw a series of triangles, which are joined together at common sides, to form the development. *Examples:* Arcs 2D and 2C are drawn from Point 2. Point C is found by drawing Arc DC from Point D. Line DC is TL in the top view.

Step 3 Construct the remaining planes, A–1–2 and G–3–4, by triangulation to complete the inside half-pattern of the transition piece. Draw the fold lines as thin lines at the places where the surface is to be bent slightly. The line of departure for the pattern is chosen along A–1, the shortest possible line.

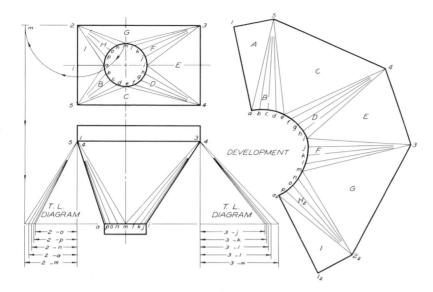

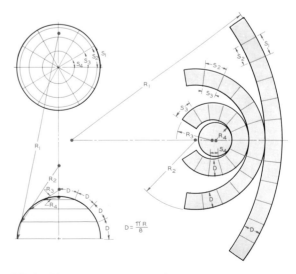

FIG. 24.62 The development of this transition piece is found by finding the conical elements true-length in a true-length diagram. Line 2–*m* is found true length as an example. The pattern is found by a series of triangulations.

a series of adjacent triangles to form the pattern beginning with element A–2.

The triangles A–1–2 and G–3–4 are added at each end of the pattern to complete the development of the half pattern.

A similar transition piece is developed in Fig. 24.62 using the same techniques of construction. Elements are established at the corners of the given views and are found true length in the true-length diagrams by revolution. By triangulation, using the true-length lines, the full pattern is drawn to complete the development.

24.23
Development of spheres— zone method

In Fig. 24.63, a development of a sphere is found by the zone method. A series of parallels, called **latitudes** in cartography, are drawn in the front view. Each is spaced an equal distance, *D*, apart along the surface in the front view. Distance *D* can be found mathematically to improve the accuracy of this step.

Cones are passed through the sphere's surface so that they pass through two parallels at the outer surface of the sphere. The largest cone with Element *R*1 is found by extending it through where the equator and the next parallel intersect on the sphere's surface

in the front view until *R*1 intersects the extended center line of the sphere. Elements *R*2, *R*3, and *R*4 are found by repeating this process.

The development is begun by laying out the largest zone, using *R*1 as the radius, on the arc that represents the base of an imaginary cone. The breadth of the zone is found by laying off Distance *D* from the front view to the development and drawing the upper portion of the zone with a radius equal to *R*1–*D*, using

FIG. 24.63 The zone method of finding the inside development of a spherical pattern

the same center. No regard is given to finding the arc lengths at this point.

The next zone is drawn using the Radius R2 with its center located on a line through the center of arc R1. The center of R2 is positioned along this line such that the arc to be drawn will be tangent to the preceding arc, which was drawn with Radius R1–D. The upper arc of this second zone is drawn with a radius of R2–D. The remaining zones are constructed successively in this manner. The last zone will appear as a circle with R4 as its radius.

The lengths of the arcs can be established by dividing the top view with vertical cutting planes that radiate through the poles. These lines, which lie on the surface of the sphere, are called **longitudes** in cartography. Arc Distances S1, S2, S3, and S4 are found on each parallel in the top view. These distances are measured off on the constructed arcs in the development. In this case there are 12 divisions, but smaller divisions would provide a more accurate measurement.

24.24
Development of spheres— gore method

Figure 24.64 illustrates an alternative method of developing a flat pattern for a sphere, using a series of spherical elements called **gores.** Equally spaced verti-

cal cutting planes are passed through the poles in the top view. Parallels are located in the front view by dividing the surface into equal zones of dimension D. The gores are projected to the front view.

A true-size view of one of the gores is projected from the top view. Dimensions can be checked mathematically for all points. A series of these gores is laid out in sequence to complete the pattern.

24.25
Development of elbows

An elbow is a cylindrical shape that turns a 90° angle. An elbow that is also a transition piece is shown in Fig. 24.65.

FIG. 24.65 The radial bend of the cylinder was developed using the technique covered in Fig. 20.66.

The method of constructing the pattern for an elbow is illustrated in Fig. 24.66. In Step 1, the 90° arc is drawn and the cylinder is drawn to size about its centerline. Divide this arc into one division less than the number of pieces that will be used in the elbow. Since three pieces are to be used in this example, the arc is divided into two along Line O–b. The two arcs are then bisected.

In Step 2, perpendiculars are drawn at the centerlines and O–b tangent to the arc. In Step 3, a half-circular view of the cylinder is drawn to locate equally spaced elements that are projected back to the given view parallel to the straight sides of the three pieces of the elbow. All elements will be true length in this view.

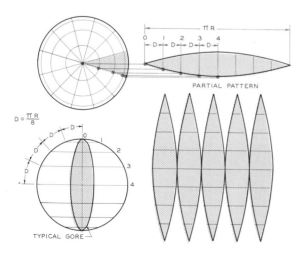

FIG. 24.64 The gore method of developing an inside pattern of a sphere

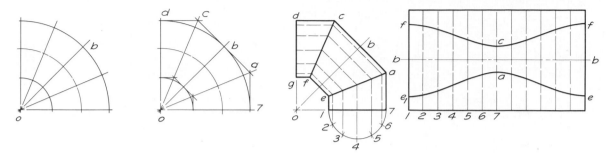

FIG. 24.66 Construction of an elbow

Step 1 Divide the arc into one division less than the number of pieces desired in the elbow (three in this case) along *Ob*. Bisect these arcs.

Step 2 Draw a tangent line through Point *b* to intersect Tangents *dc* and *a*–7.

Step 3 Draw a semicircle using 1–7 as the diameter. This is half of the right section that will be used to space the elements along the stretch-out line.

Step 4 The three pieces can be cut from a rectangular piece of material whose height is equal to the sum of *a*–7, *e*–*f*, and *cd*. The curves for each cut are found by transferring the lengths of the elements to parallel element lines in the pattern.

In Step 4, the flat pattern is developed, with the patterns of the two short pieces of the elbow located at the top and bottom of the rectangular piece. The middle segment is located between these two smaller patterns so that there will be no wasted material, and only two cuts will be necessary.

24.26
Development of straps

Figure 24.67 illustrates the steps of finding the development of a strap that has been bent to serve as a support bracket, between Point *A* on a vertical surface and Point *B* on a horizontal surface. In Step 1, the strap is drawn in the side view where the strap appears as an edge using the specified radius to show the bend. The bend is divided into equal arcs and is developed into a flat piece in the vertical plane through Point *A*. Point *B'* is located in this view.

In Step 2, the location of the hole at *B'* is found in the front view by projection. The true-size development of the strap is drawn in this view and it is projected back to the side view to complete that view of the strap.

In Step 3, the projected view of the strap in its bent position in the front view can be found by using projectors from the side view and the true pattern of the strap.

24.27
Intersections and developments in combination

Intersections between parts must be found before developments of each can be completed. An example of this type is shown in Fig. 24.68.

The intersection between the two prisms is found in the given top and front views. The pattern of the vertical prism is found to the left of the front view and the intersections are shown to indicate cuts that must be made in the pattern.

The pattern of the inclined prism cannot be found without the construction of an auxiliary view in which the fold lines appear true length. The true-size right section is found in a secondary auxiliary view. The fold lines of the pattern are laid off along a stretch-out line by transferring the distances around the right section to it.

The resulting two patterns can be cut out and folded to form shapes that will intersect as shown in the top and front views.

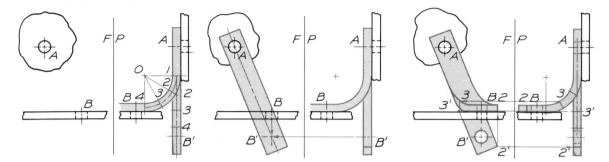

FIG. 24.67   Strap development

Step 1   Construct the edge view of the strap in the side view using the specified radius of bend. Locate Points 1, 2, 3, and 4 on the neutral axis at the bend. Revolve this portion of the strap into the vertical plane and measure the distances along this view of the neutral axis. Check the arc distances by mathematics. The hole is located at *B′* in this view.

Step 2   Construct the front view of *B′* by revolving Point *B* parallel to the profile plane until it intersects the projector from *B′* in the side view. Draw the center line of the true-size strap from *A* to *B′* in the front view. Add the outline of the strap around this center line and around the holes at each end, allowing enough material to provide sufficient strength.

Step 3   Determine the projection of the strap in the front view by projecting points from the given views. Points 3 and 2 are shown in the views to illustrate the system of projection used. The ends of the strap are drawn in each view to form true projections.

FIG. 24.68   Two prisms intersect in this example. Their intersections are found and the developments of each are found by using the principles covered in this chapter. The development of the inclined prism is found by using primary and secondary auxiliary views.

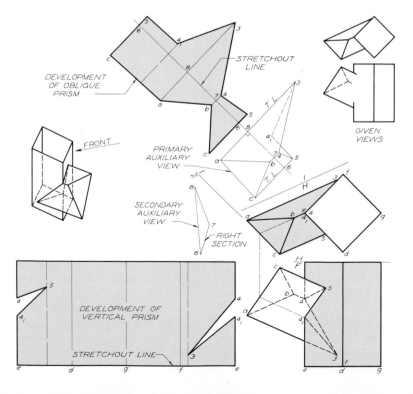

Problems

These problems are designed to fit on Size A and Size B sheets. Some problems can be grouped two or more on a single sheet. For example, two problems might be solved on a Size A sheet and four on a Size B sheet. The proper layout of a problem should be considered an important part of its solution.

Intersections

1–5. Lay out two problems on Size A sheets. Use the given scales to transfer the dimensions from the views to the drawing. In Problems 1–3, the labeled points represent points on cutting planes that pass through the objects. Find their lines of intersection and show visibility.

6–7. For Problem 6, lay out the top and front views twice on a Size A sheet, one above the other on the sheet. Using the auxiliary sections assigned, A, B, C, D, or E, find the intersection in the front and top views. Lay out the views of Problem 7 in the same manner and use right sections A through E as assigned to find the intersections.

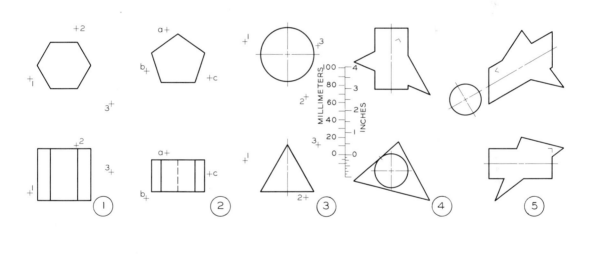

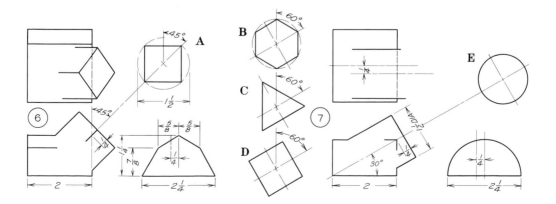

8–20. Lay out one problem per Size A sheet. Find the intersections and show the visibility for each problem.

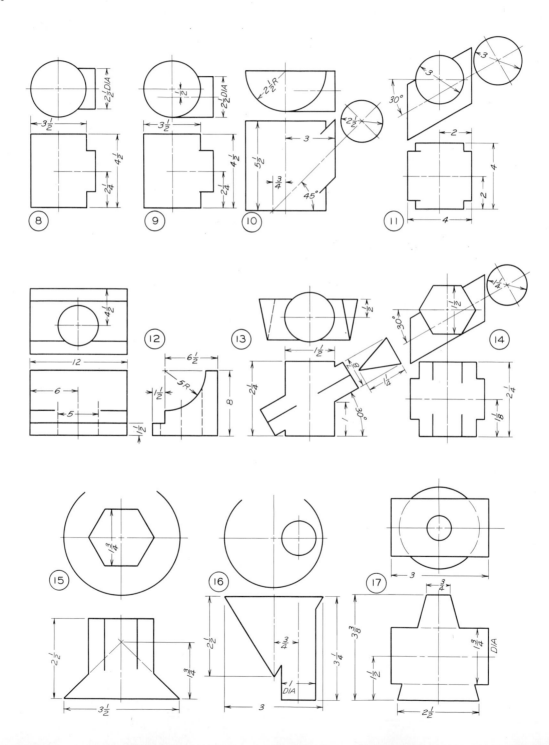

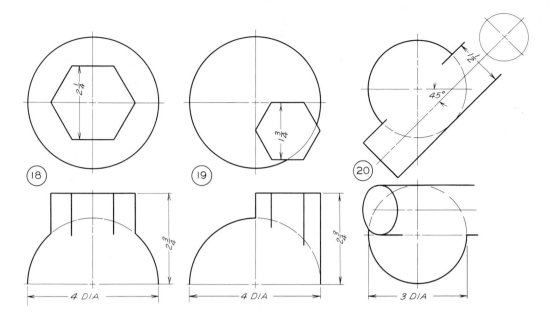

Developments

21–47. Lay out two problems per Size A sheet after
dividing the sheet in half by a division line. Find
the flat inside patterns of each and label the fold
lines in all views.

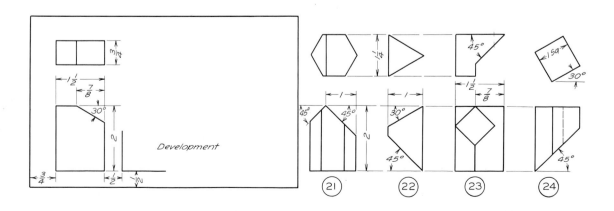

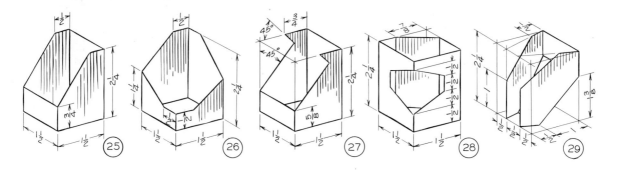

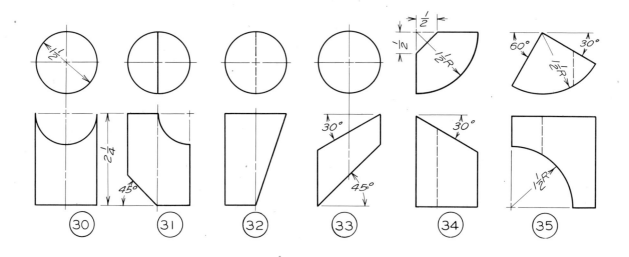

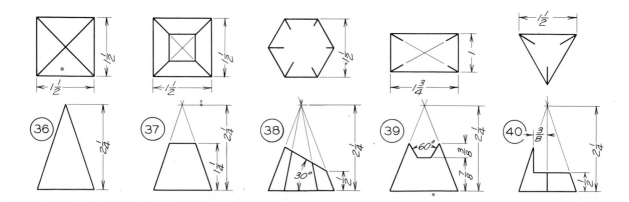

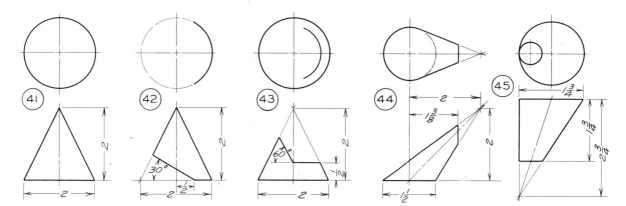

48–51. Lay out two problems per Size A sheet. Find the inside developments of each and remove the ends of the object that have been cut away by the cutting planes. Label the points in all views.

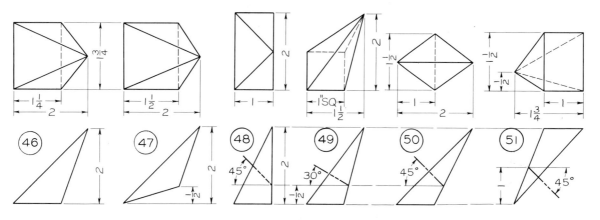

52–56. Lay out the problems with one per Size A sheet, using the scales given. Find the inside developments of each and label the points in all views.

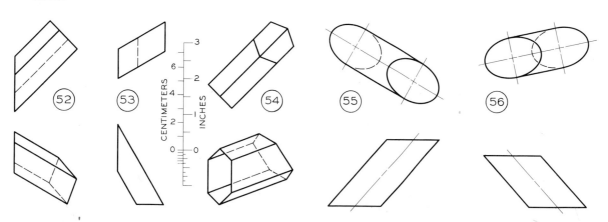

57–60. Lay out the problems with one per Size B sheet. Find the inside developments of each and label the points in all views.

Combination problems

61–72. Using Problems 6–17, lay them out with one per Size B sheet. Find the intersections and inside developments of each. Label the points in all views.

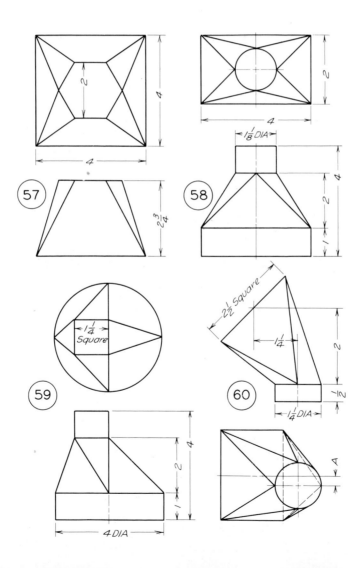

CHAPTER

25

Graphs

25.1

Introduction

Data and information expressed as numbers and words are difficult to analyze or evaluate unless they are transcribed into graphical form are called **graphs.** They are sometimes called **charts,** an acceptable term, but this term is more appropriate when referring to maps, which are specialized forms of graphs.

Graphs are helpful in the communication of data to others (Fig. 25.1); consequently, this is a popular means of briefing other people on trends that would otherwise be difficult to communicate. The trends of a plotted curve on a graph can be compared to the expressions on a person's face, which is a graph of sorts that reveals a person's feelings (Fig. 25.2). For example, a flat curve shows no change while an upwardly inclined curve indicates a positive increase. A downward curve, on the other hand, represents a downward trend and a negative result.

This chapter will deal with the more commonly used graphs. The basic types are:

1. Pie graphs,

2. Bar graphs,

FIG. 25.1 Graphs are helpful in presenting technical data to one's associates.

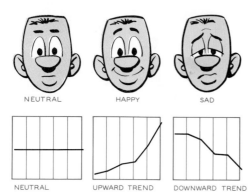

NEUTRAL HAPPY SAD

NEUTRAL UPWARD TREND DOWNWARD TREND

FIG. 25.2 Curves on a graph are similiar to expressions on a face.

3. Linear coordinate graphs,
4. Logarithmic coordinate graphs,
5. Semilogarithmic coordinate graphs, and
6. Schematics and diagrams.

25.2
Size proportions of graphs

Graphs may be used to illustrate technical reports that are reproduced in quantity, and they may be used for projection by slide or overhead projectors. In all cases, the proportion of the graph must be determined so that it will match the proportion of the space or the format of the visual aid.

If a graph is to be photographed by a 35-mm camera (Fig. 25.3), the graph must conform to the standard size of the 35-mm film that is used. This proportion is approximately 2 × 3, as shown in Fig. 25.4.

The proportions of the area in which the graph is to be drawn can be enlarged or reduced by using the diagonal-line method, as illustrated in Fig. 25.4.

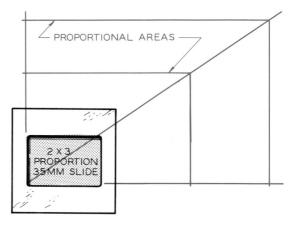

PROPORTIONAL AREAS

2 X 3 PROPORTION 35MM SLIDE

FIG. 25.4 This diagonal-line method can be used for construction areas that are proportional to those of a 35-mm slide.

FIG. 25.3 When graphs are drawn to be photographed, they must be laid out at a proportion that will best take advantage of the proportion of the film in the camera.

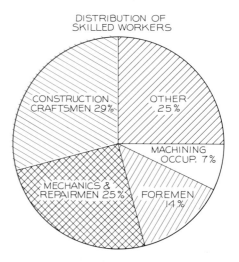

DISTRIBUTION OF SKILLED WORKERS

CONSTRUCTION CRAFTSMEN 29%

OTHER 25%

MACHINING OCCUP. 7%

MECHANICS & REPAIRMEN 25%

FOREMEN 14%

FIG. 25.5 A pie graph shows the relationship of the parts to a whole. It is effective only when there are a few parts.

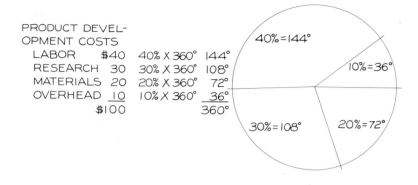 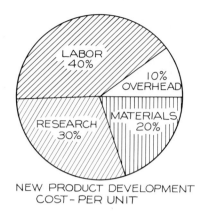

NEW PRODUCT DEVELOPMENT
COST – PER UNIT

FIG. 25.6 Drawing pie graphs

Step 1   The total sum of the parts, and the percentage of each is found. Each percentage is multiplied by 360° to find the angle of each sector.

Step 2   The circle is drawn and each sector is drawn using the degrees found in Step 1. The smaller sectors should be placed as nearly horizontal as possible.

Step 3   The sectors are labeled with names and percentages. In some cases it might be desirable to include the exact numbers in each sector as well.

25.3
Pie graphs

Pie graphs compare the relationship of parts to a whole when there are only a few parts. Figure 25.5 shows the distribution of skilled workers employed in industry.

The method of drawing a pie graph is shown in Fig. 25.6. The tabular data does not give as good an impression of the comparisons as does the pie graph, even though the data is quite simple.

To facilitate lettering within narrow spaces, the thin sectors should be placed as nearly horizontal as possible to provide more room for the label. The actual percentage should be given in all cases, and it may be desirable to give the actual numbers or values as well in each sector.

25.4
Bar graphs

Bar graphs are effective to compare values since they are well understood by the general public (Fig. 25.7). In this example, the bars not only show the overall production of timber (the total lengths of the bars),

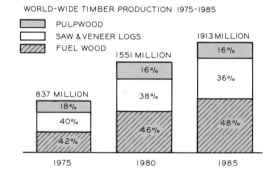

FIG. 25.7   In this example, each bar represents 100% of the total amount and each bar represents different totals.

but the portions of the total devoted to the three uses of the timber.

A bar graph can be composed of a single bar (Fig. 25.8). The total length of the bar is 100% and the bar is divided into lengths that are proportional to the percentages represented by each of the three parts of the bar.

The method of constructing a bar graph is given in Fig. 25.9. In this case the title of the graph is placed inside the graph where space is available. The title could have been placed under or over the graph.

The data should be sorted by arranging the bars in ascending or descending order, since it is desirable

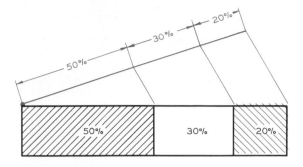

FIG. 25.8 The method of constructing a single bar graph where the sum of all of the parts will be 100%.

to know how the data represented by the bars rank from category to category (Fig. 25.10). An arbitrary arrangement of bars, alphabetically or numerically, results in a graph that is more difficult to evaluate than the descending arrangement at B.

If the data are sequential and involve time, such as sales per month, it would be less effective to rank the data in ascending order because it is more important to see variations in the data as related to periods of time. The determination of the method of arranging the bars depends upon the data and the judgment of the drafter.

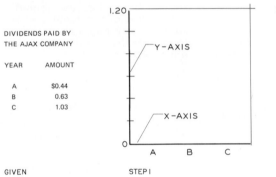
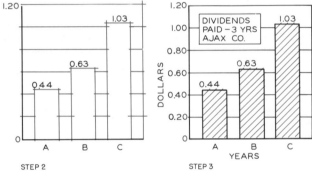

FIG. 25.9 Construction of a bar graph

Given These data are to be plotted as a bar graph.

Step 1 Lay off the vertical and horizontal axes so that the data will fit on the grid. Make the bars begin at zero.

Step 2 Construct and label the bars. The width of the bars should be different from the space between the bars. Horizontal grid lines should not pass through the bars.

Step 3 Strengthen lines, title the graph, label the axes, and crosshatch the bars.

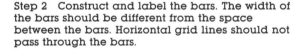

FIG. 25.10 The bars at A are arranged alphabetically. The resulting graph is not as easy to evaluate as the one at B, where the bars have been sorted and arranged in descending order.

Bars in a bar graph may be horizontal, as shown in Fig. 25.11, or vertical, as shown in Fig. 25.12. It is desirable for the bars of a graph to begin at zero to show a true comparison in the data.

25.5
Linear coordinate graphs

A typical coordinate graph is given in Fig. 25.13 with the accompanying notes that explain its important features. The axes are linear if the divisions along the axes are equally spaced.

EMPLOYMENT FUNCTIONS OF ENGINEERS

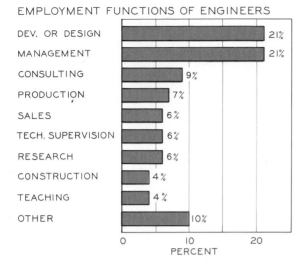

FIG. 25.11 A horizontal bar graph that is arranged in descending order to show where engineers are employed.

Points are plotted on the grid by using two measurements, called coordinates, made along each axis. The plotted points are indicated by using easy-to-draw symbols, such as circles, that can be drawn with a template.

The horizontal line of the graph is called the **abscissa** or *x*-axis, and the vertical scale at the left is called the **ordinate** or *y*-axis.

Once the points have been plotted, the curve is drawn from point to point. (The line that is drawn to represent the plotted points is called a curve whether it is a smooth or broken line.) The curve should not close up the plotted points, but they should be left as open circles or symbols.

The curve must be drawn as a heavy prominent line, since it is the most important part of the graph. In Fig. 25.13, there are two curves; therefore, it is helpful if they are drawn as different types of lines and labeled with a note and a leader. The title of the graph is placed inside the graph in a box to explain the graph.

Units are given along the *x*- and *y*-axes with labels that designate the units that the graph is comparing.

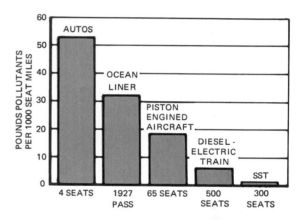

FIG. 25.12 This bar graph has the bars arranged in descending order to compare several sources of pollution. (Courtesy of Boeing Co.)

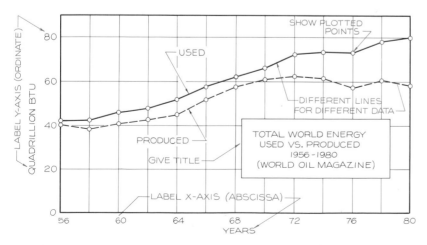

FIG. 25.13 The basic linear coordinate graph with the important features identified.

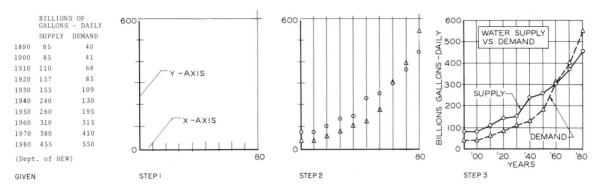

FIG. 25.14 Construction of a broken-line graph

Given A record of water supply and water demand since 1890 are to be plotted as a line graph.

Step 1 The vertical and horizontal axes are laid off to provide adequate space for the years and the largest values.

Step 2 Data points are plotted for the respective years. Different symbols are used for each curve.

Step 3 The data points are connected with straight lines, the axes are labeled, the graph is titled, and the lines are strengthened.

Broken-line graphs

The steps involved in drawing a linear coordinate graph are shown in Fig. 25.14. In this case, the points are connected with a broken-line curve since the data points are ten years apart on the *x*-axis. Thus it is impossible to assume that the change in the data is a smooth, continual progression from point to point.

For the best appearance the plotted points should not be crossed by the curve or the grid lines of the graph (Fig. 25.15). Each circle or symbol used to plot

FIG. 25.16 Any of these symbols or lines can be used effectively to represent different curves on a single graph. The symbols are about ⅛ in. (5 mm) in diameter.

points should be about ⅛ in. (5 mm) in diameter. Several approved symbols and lines are shown in Fig. 25.16.

The title of a graph can be placed in any of the positions shown in Fig. 25.17. The title should never be one as meaningless as "graph" or "coordinate graph." Instead, it should explain the graph by giving the important information such as company, date, source of the data, and the general comparisons being shown.

The proper calibration and labeling of axes is important to the appearance and usage of a graph. The example in Fig. 25.18A shows a properly executed axis. The axis at Part B has too many grid lines and too many units labeled along the axis. The units selected at C make it difficult to interpolate between the labeled values. For example, it is more difficult to lo-

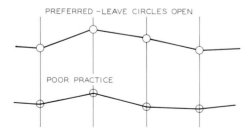

FIG. 25.15 The curve of a graph should be drawn from point to point, but it should not close up the symbols used to locate the plotted points.

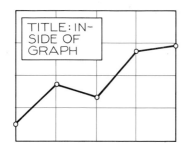

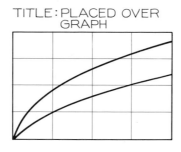

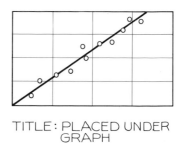

FIG. 25.17 Title placement on a graph

A. The title of a graph can be lettered inside a box placed within the area of the graph. The perimeter lines of the box should not coincide with grid lines.

B. The title can be placed above the graph. The title should be drawn in ⅛ in. letters, or slightly larger.

C. The title can be placed under a graph. It is good practice to be consistent when a series of graphs is used in the same report.

cate a value such as 22 by eye on this scale than on the one at A.

Smooth-line graphs

The strength of cement as related to curing time has been plotted in Fig. 25.19. You instinctively understand that the physical relationship between the strength of cement and its curing time will result in a smooth, continuous relationship that should be connected by a smooth curve. Even if the data points do not plot to lie on the curve, you know that deviation

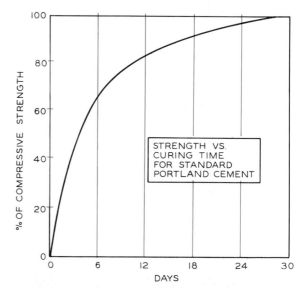

FIG. 25.19 When the process that is graphed involves gradual, continuous changes in relationships, the curve should be drawn as a smooth line.

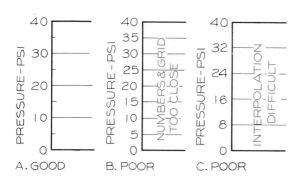

A. GOOD   B. POOR   C. POOR

FIG. 25.18 The scale at A is the best. It has about the right number of grid lines and divisions, and the numbers are given in well-spaced, easy-to-interpolate form. The numbers at B are too close and there are too many grid lines. The units at C are given in increments that make interpolation difficult by eye.

of the points from this curve is due to errors of measurement or the methods used in collecting the data for the points.

Similarly, the strength of clay tile, as related to its absorption characteristics, is an example of data that yields a smooth curve (Fig. 25.20). Note that the plotted data does not lie on the curve. Since you know

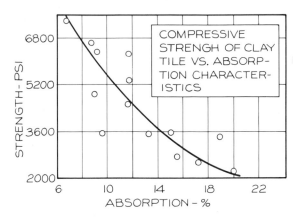

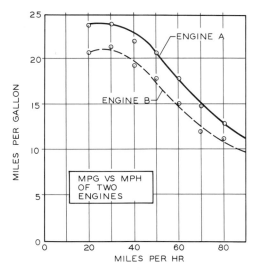

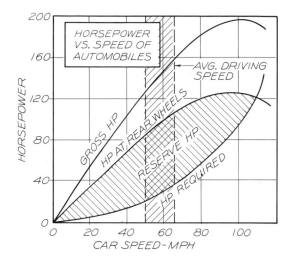

FIG. 25.20 If it is known that a relationship plotted on a graph should yield a smooth gradual curve, a smooth-line "best" curve is drawn to represent the average of the plotted points. You must use your judgment and knowledge of the data in cases of this type.

that the relationship is one that should be connected with a smooth curve, the **best curve** is drawn to interpret the data to give an average representation of the point.

There is a smooth-line curve relationship between miles per gallon and the speed at which a car is

FIG. 25.21 These are "best" curves that approximate the data without necessarily passing through each data point. Inspection of the data tells you that this curve should be a smooth-line curve rather than a broken-line curve.

driven. Two engines are compared in Fig. 25.21 with two smooth-line curves. The effect of speed on several automotive characteristics is compared in Fig. 25.22.

When a smooth-line curve is used to connect data points, there is the implication that you can **interpolate** between the plotted points to estimate other values. Points connected by a broken-line curve imply that you cannot interpolate between the plotted points.

FIG. 25.22 A linear coordinate graph is used here to analyze data affecting the design of an automobile's power system.

Straight-line graphs

Some graphs have neither broken-line curves nor smooth-line curves, but straight-line curves as shown in the example in Fig. 25.23. You can determine a third value from the two given values using this graph. For example, if you are driving 70 miles per hour and it takes 5 seconds to react to apply your brakes, you will have traveled 500 feet in this time.

Two-scale coordinate graphs

Graphs can be drawn with different scales in combination, such as the one shown in Fig. 25.24. The vertical scale at the left is in units of pounds, and the one

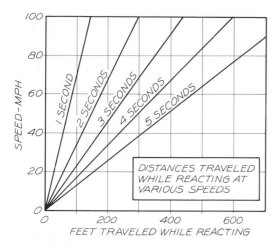

FIG. 25.23 A graph can be used to determine a third value when two variables are known. Taking this information from a graph is easier than computing each answer separately.

at the right is expressed in degrees of temperature. Both curves are drawn using their respective *y*-axes, and each curve is labeled.

Care must be taken to avoid confusing the reader of a graph of this type. These graphs are effective when comparing related variables such as the drag force and air temperature of a tire, as shown in this example.

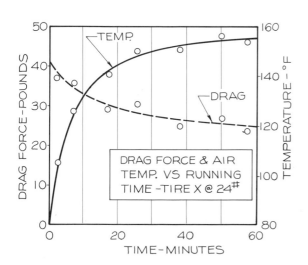

FIG. 25.24 This is a composite graph with different scales along each *y*-axis. The curves are labeled so that reference can be made to the applicable scale.

Optimization graphs

The optimization of the depreciation of an automobile and its increase in maintenance costs is shown in Fig. 25.25. These two sets of data are plotted and the curves cross at an *x*-axis value of four years. At this time, the cost of maintenance is equal to the value of the car, which indicates that this might be a desirable time to exchange it for a new one. ·

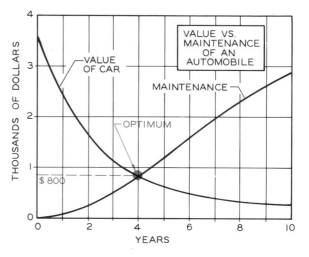

FIG. 25.25 This graph shows the optimum time to sell a car, based on the intersection of two curves that represent the depreciation of the car's value and its increasing maintenance costs.

Another optimization graph is constructed in Fig. 25.26. The manufacturing cost per unit is reduced as more units are made, but the warehousing cost increases. A third curve is found in Step 2, by adding the two curves. The "total" curve tells you that the optimum number to manufacture at a time is about 11,000 units. When more or fewer are manufactured, the expense per unit is greater.

Composite graphs

The graph in Fig. 25.27 is a composite between an area graph and a coordinate graph. The lower curve is plotted first. The upper curve is found by adding the

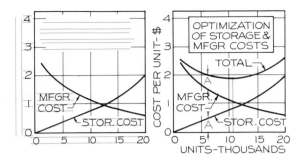

FIG. 25.26 Optimization graphs

Step 1   Lay out the graph and plot the given curves.

Step 2   Add the two curves to find a third curve. Distance A is shown transferred with dividers to locate a point on the third curve. The lowest point of the "total" curve is the optimum point of 11,000 units.

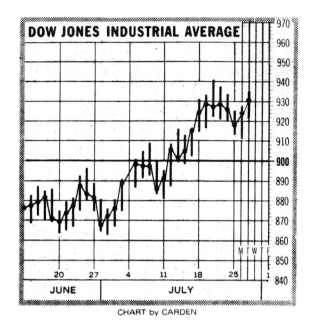

CHART by CARDEN

FIG. 25.28 This graph is a combination of a coordinate graph and a bar graph. The bars represent the ranges of selling during a day, and the broken-line curve connects the points at which the market closed each day.

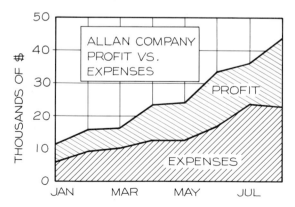

FIG. 25.27 This graph is a combination of a coordinate graph and an area graph. The upper curve represents the total of two values plotted, one above the other.

values to the lower curve so that the two areas represent the data. The upper curve is equal to the sum of the two *y*-values.

The graph in Fig. 25.28 is a combination of a coordinate graph and a bar graph that is used to show the Dow-Jones Industrial stock average. The bars represent the daily ranges in the index. The broken-line curve connects the points where the market closed for each day.

Break-even graphs

Break-even graphs are helpful in evaluating marketing and manufacturing costs that are used to determine the selling cost for a product. The break-even graph in Fig. 25.29 is drawn to reveal that 10,000 units must be sold at $3.50 each to cover the manufacturing and development costs. Sales in excess of 10,000 result in profit.

A second type of break-even graph (Fig. 25.30) uses the cost of manufacturing per unit versus the

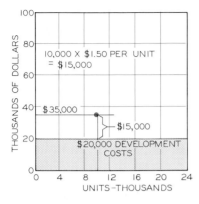

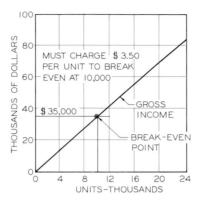

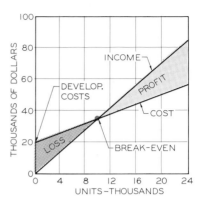

FIG. 25.29 Break-even graph

Step 1 This graph is drawn to shows the cost ($20,000 in this case) of developing the product. Each unit would cost $1.50 to manufacture if the total quantity were 10,000. This is a total investment of $35,000 for 10,000 units.

Step 2 In order for the manufacturer to break even at 10,000, the units must be sold for $3.50 each. Draw a line from zero through the break-even point of $35,000.

Step 3 The manufacturer's loss is $20,000 at zero units and becomes progressively less until the break-even point is reached. The profit is the difference between the cost and income to the right of the break-even point.

number of units produced. In this example, the development costs must be incorporated into the unit costs. The manufacturer can determine how many units must be sold to break even at a given price, or the price per unit if a given number is selected. In this example, a sales price of 80¢ requires that 8400 units be sold to break even.

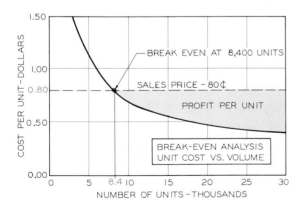

FIG. 25.30 The break-even point can be found on a graph that shows the relationship between the cost per unit, which includes the development cost, and the number of units produced. The sales price is a fixed price. The break-even point is reached when 8400 units have been sold at 80¢ each.

25.6
Logarithmic coordinate graphs

Both scales of a logarithmic grid are calibrated into divisions that are equal to the logarithms of the units represented. Commercially printed logarithmic grid paper is available in many variations for graphing data.

The graph in Fig. 25.31 has a logarithmic grid and shows the geometry of standard railroad cars as they relate to the tracks so that there will not be more than a maximum projection width of 12 feet around curves. Extremely large values can be shown on logarithmic grids since the lengths are considerably compressed.

25.7
Semilogarithmic coordinate graphs

Semilogarithmic graphs are referred to as **ratio graphs** because they give graphical representations of ratios. One scale, usually the vertical scale, is logarith-

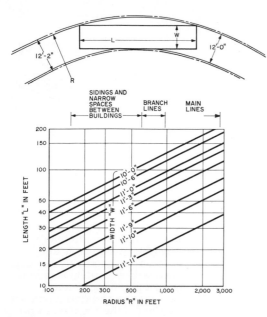

FIG. 25.31 This logarithmic graph shows the maximum load projection of 12 feet in relation to the length of a railroad car and the radius of the curve. (Courtesy of *Plant Engineering.*)

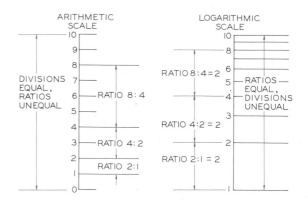

FIG. 25.33 The spacings on an arithmetic scale are equal, with unequal ratios between points. The spacings on logarithmic scales are unequal, but equal spaces represent equal ratios.

mic, and the other is linear (divided into equal divisions). Two curves that are parallel on a semilogarithmic graph have equal percentage increases.

Figure 25.32 shows the same data plotted on a linear grid and on a semilogarithmic grid. The semi-

logarithmic graph reveals that the percent of change from 0 to 5 is greater for Curve B than for Curve A, since Curve B is steeper. This comparison was not apparent in the plot on the linear grid.

The relationship between the linear scale and the logarithmic scale is shown in Fig. 25.33. Equal divisions along the linear scale have unequal ratios, and equal divisions along the log scale have equal ratios.

Log scales can be drawn to have one or many

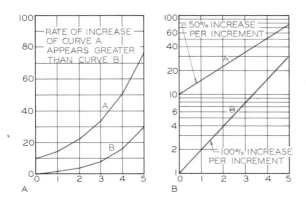

FIG. 25.32 When plotted on a standard grid, Curve A appears to be increasing at a greater rate than Curve B. However, the true rate of increase can be seen when the same data are plotted on a semilogarithmic graph in Part B.

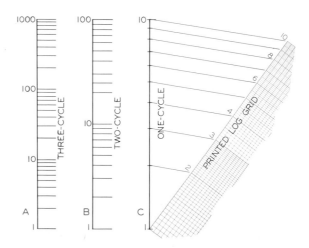

FIG. 25.34 Logarithmic paper can be purchased or drawn using several cycles. Three-, two-, and one-cycle scales are shown here. Calibrations can be drawn on a scale of any length by projecting from a printed scale, as shown in Part C.

cycles. Each cycle increases by a factor of 10. For example, the scale in Fig. 25.34A is a three-cycle scale, and the one at B is a two-cycle scale. When these must be drawn to a special length, commercially printed log scales can be used to graphically transfer the calibrations to the scale being drawn (Fig. 25.34C).

In Fig. 25.35, the calibrations along the log scale are separated by the difference in their logarithms. The logarithms are laid off using a convenient scale that is calibrated in decimal divisions.

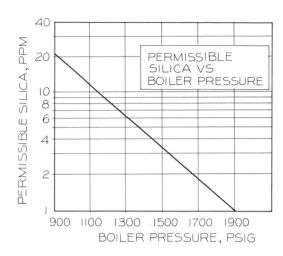

FIG. 25.36 A semilogarithmic graph is used to compare the permissible silica (parts per million) in relation to the boiler pressure.

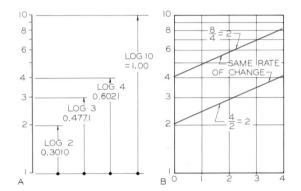

FIG. 25.35 A number's logarithm is used to locate its position on a log scale (A). This makes it possible to see the true rate of change at any location on a semilogarithmic graph (B).

It can be seen in Fig. 25.35B that parallel straight-line curves yield equal ratios of increase. Figure 25.36 is an example of a semilogarithmic graph used to present industrial data.

Semilog graphs maybe misunderstood by many people who do not recognize them as being different from linear coordinate graphs. Also, zero values cannot be shown on log scales.

Percentage graphs

The percent that one number is of another, or the percent increase of one number that is greater than the other, can be determined by using a semilogarithmic graph (Fig. 25.37).

Data plotted in Step 1 are used to find the percent that 30 is of 60, two points on the curve. The vertical

distance between them is equal to the difference of their logarithms. This distance is subtracted from the log of 100 at the right of the graph to give a value of 50% as a direct reading.

In Step 2, the percent of increase between two points is transferred from the grid to the lower end of the log scale and measured upward since the increase is greater than zero. These methods can be used to find percent increases or decreases of any sets of points on the grid.

25.8
Polar graphs

Polar graphs are drawn with a series of concentric circles with the origin at the center. Lines are drawn from the center toward the perimeter of the graph, where the data can be plotted through 360° by measuring values from the origin. The illumination of a lamp is shown in Fig. 25.38, where the maximum lighting of the lamp is 550 lumens at 35° from the vertical, for example.

This type of graph is used to plot the areas of illumination of all types of lighting fixtures and other applications. Polar graph paper is available commercially for drawing graphs of this type.

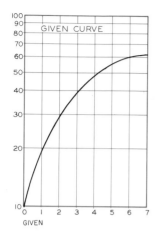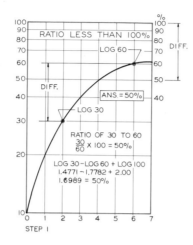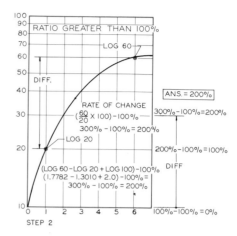

FIG. 25.37   Percentage graphs

Given   The data above are plotted on a semilogarithmic graph to enable you to determine percentages and ratios in same manner that you use a slide rule.

Step 1   The percent that a smaller number is of a larger number will be less than 100%. The log of 30 is subtracted from the log of 60 with dividers and is transferred to the percent scale at the right, where 30 is found to be 50% of 60.

Step 2   To find the percent of increase, a smaller number is divided into a larger number to give a value greater than 100%. The difference between the logs of 60 and 20 is found with dividers, and is measured upward from 100% at the right, to find a 200% increase.

25.9
Schematics

The **block diagram** in Fig. 25.39 shows the progressive steps of the completion of a construction project. Each step is blocked in and connected with arrows to explain the sequence of events.

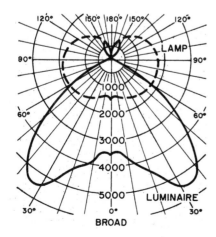

FIG. 25.38   A polar graph is used to show the illumination characteristics of luminaires.

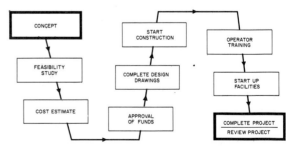

FIG. 25.39   This schematic shows a block diagram of the steps required to complete a project. (Courtesy of *Plant Engineering*.)

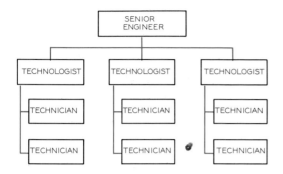

FIG. 25.40 This schematic shows the organization of a design team in a block diagram.

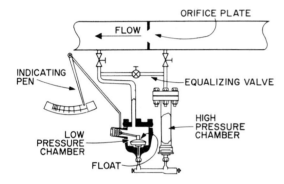

FIG. 25.41 A schematic showing the components of a gauge that measures the flow in a pipeline. (Courtesy of *Plant Engineering*.)

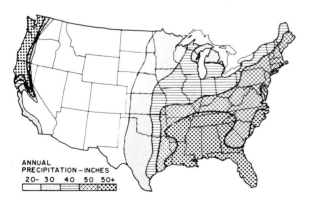

FIG. 25.42 A geographical graph that shows the weather characteristics of various areas. (Courtesy of the *Structural Clay Products Institute*.)

The organization of a company or a group of people can be depicted in an **organizational chart** of the type shown in Fig. 25.40. The offices represented by the blocks at the lower part of the graph are responsible to the blocks above them as they are connected by lines of authority. The lines connecting the blocks also suggest the routes for communication from one to another in an upward or downward direction.

The schematic in Fig. 25.41 is not a graph, nor is it a true view of the apparatus. Instead, it is a **schematic** that effectively shows how the parts and their functions relate to each other.

Geographical graphs are used to combine maps and other relationships such as weather (Fig. 25.42). Different symbols are used to represent the annual rainfall that various areas of the nation receive.

25.10
How to lie with graphs

Graphs can be used to distort data to the extent that the graph is actually lying. The three bar graphs in Fig. 25.43 present the same data, but a different impres-

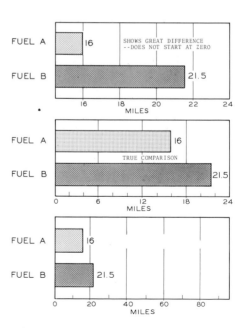

FIG. 25.43 All three graphs show the same data, but each gives a different impression of the data.

sion is given by each. The upper graph gives the impression that Fuel B is almost five times better than Fuel A because the bars do not begin at zero.

The center graph begins at zero, which gives a true comparison between the two bars. Fuel B is shown to be about 35% better than Fuel A.

The lower graph de-emphasizes the difference between the two fuels since the x-axis is much longer than is necessary. The implication of this horizontal scale is that a car might be expected to get over 90 miles per gallon. The bars appear insignificant on this graph, and even though they are drawn accurately, the difference in their lengths is de-emphasized.

The width of bars and the colors used can give a misleading impression in a bar graph. Beware of graphs in which the bars run off the graph. These graphs seldom give a true graphical picture.

Identical data are plotted on two graphs with different y-axes in Fig. 25.44. Graph A shows only a little variation in the data. Graph B gives a more dramatic effect because the expanded vertical scale emphasizes the difference in the data.

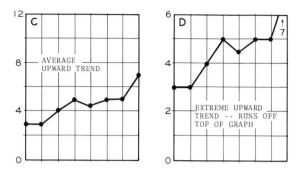

FIG. 25.45 The upward trend of the data at D appears greater than the curve at C. Both graphs show the same data.

The two graphs in Fig. 25.45 give a different appearance even though the same data are plotted on both. The data in Graph D appear to have a more significant rate of increase because of the selection of the vertical scale. The curve of Graph D is drawn to run off the top of the graph, giving the impression that the increase is too great to be contained on a graph.

Another method of misrepresenting data is the removal of a portion of the graph (Fig. 25.46). This distorts the visual comparison between the bars and negates the value of the graph.

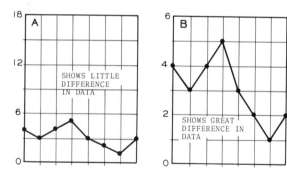

FIG. 25.44 Due to a variation in the scales, the change in the curve at B appears greater than in the curve at A, although the data is the same in each.

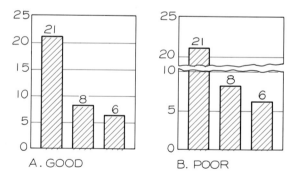

FIG. 25.46 This bar graph distorts the data at B because a portion of the graph has been removed, negating the pictorial value of the graph.

Problems

The problems below are to be drawn on Size A sheets (8½″ × 11″) in pencil or ink as specified. Follow the techniques that were covered in this chapter, and the examples that are given as the problems are being solved.

Pie graphs

1. Draw a pie graph that compares the employment of male youth between the ages of 16 and 21: Operators—25%; Craftsmen—9%; Professionals, technicians, and managers—6%; Clerical and sales—17%; Service—11%; Farm workers—13%; and Laborers—19%.

2. Draw a pie graph that shows the relationship between the following members of the technological team: Engineers—985,000; Technicians—932,000; and Scientists—410,000.

3. Construct a pie graph of the following percentages of the employment status of graduates of two-year technician programs one year after graduation; Employed—63%; Continuing full-time study—23%; Considering job offers—6%; Military—6%; and Other—2%.

4. Construct a pie graph that shows the relationship between the types of degrees held by engineers in aeronautical engineering: Bachelor—65%; Master—29%; and Ph.D.—6%.

5. Draw a pie graph for the following average annual expenditures of a state on public roads. The figures are: Construction—$13,600,000; Maintenance—$7,100,000; Purchase and upkeep of equipment—$2,900,000; Bonds—$5,600,000; Engineering and administration—$1,600,000.

6. Draw a pie graph that shows the data given in Problem 10.

7. Draw a pie graph that shows the data given in Problem 11.

Bar graphs

8. Draw a bar graph that depicts the unemployment rate of high-school graduates and dropouts in various age categories given in the following table:

Ages	Percent of labor force	
	Graduates	Dropouts
16–17	18	22
18–19	12.5	17.5
20–21	8	13
22–24	5	9

9. Draw a single bar that represents 100% of a die casting alloy. The proportional parts of the alloy are: Tin—16%; Lead—24%; Zinc—38.8%; Aluminum—16.4%; and Copper—4.8%.

10. Draw a bar graph that compares the number of skilled workers employed in various occupations. Arrange the graph for ease of interpretation and comparison of occupations. Use the following data: Carpenters—82,000; All-round machinists—310,000; Plumbers—350,000; Bricklayers—200,000; Appliance servicers—185,000; Automotive mechanics—760,000; Electricians—380,000; and Painters—400,000.

11. Draw a bar graph that represents the flow of a river in cubic feet per second (cfs) as shown in the following table. Show bars that represent the data for ten days in the first month only. Omit the second month.

Day of month	Rate of Flow in 1000 cfs	
	1st Month	2nd Month
1	19	19
2	130	70
3	228	79
4	129	33
5	65	19
6	32	14
7	17	15
8	13	11
9	22	19
10	32	27

12. Draw a bar graph that shows the airline distances in statute miles from New York to the cities listed in the table below. Arrange in ascending or descending order.

Berlin 3965

Buenos Aires 5300

Honolulu 4960

London 3465

Manila 8510

Mexico City 2090

Moscow 4665

Paris 3634

Tokyo 6740

13. Draw a bar graph that compares the corrosion resistance of the materials listed in the table below:

	Loss in Weight %	
	In Atmosphere	In Sea Water
Common Steel	100	100
10% Nickel Steel	70	80
25% Nickel Steel	20	55

14. Draw a bar graph using the data in Problem 1.

15. Draw a bar graph using the data in Problem 2.

16. Draw a bar graph using the data in Problem 3.

17. Construct a rectangular grid graph to show the accident experience of Company A. Plot the numbers of disabling accidents per million person-hours of work on the y-axis. Years will be plotted on the x-axis. Data: 1970–1.21; 1971–0.97; 1972–0.86; 1973–0.63; 1974–0.76; 1975–0.99; 1976–0.95; 1977–0.55; 1978–0.76; 1979–0.68; 1980–0.55; 1981–0.73; 1982–0.52; 1983–0.46.

Linear coordinate graphs

18. Using the data given in Table 25.1, draw a linear coordinate graph that compares the supply and demand of water in the United States from 1890 to 1980 in billions of gallons of water per day.

19. Present the data in Table 25.2 in a linear coordinate graph to decide which lamps should be selected to provide economical lighting for an industrial plant. The table gives the candlepower directly under the lamps (0°) and at various angles from the vertical when the lamps are mounted at a height of 25 feet.

20. Construct a linear coordinate graph that shows the relationship in energy costs (mills per kilowatt-hour) and the percent capacity of two types of power plants. Plot energy costs along the y-axis, and the capacity factor along the x-axis.

The plotted curve will compare the costs of a nuclear plant with a gas- or oil-fired plant. Data for a gas-fired plant: 17 mills, 10%; 12 mills, 20%; 8 mills, 40%; 7 mills, 60%; 6 mills, 80%; 5.8 mills, 100%. Nuclear-plant data: 24 mills, 10%; 14 mills, 20%; 7 mills, 40%; 5 mills, 60%; 4.2 mills, 80%; 3.7 mills, 100%.

21. Plot the data from Problem 17 as a linear coordinate graph.

22. Construct a linear coordinate graph to show the relationship between the transverse resilience in inch-pounds (y-axis) and the single-blow impact in foot-pounds (x-axis) of gray iron. Data: 21 fp, 375 ip; 22 fp, 350 ip; 23 fp, 380 ip; 30 fp, 400 ip; 32 fp, 420 ip; 33 fp, 410 ip; 38 fp, 510 ip; 45 fp, 615 ip; 50 fp, 585 ip; 60 fp, 785 ip; 70 fp, 900 ip; 75 fp, 920 ip.

23. Draw a linear coordinate graph to compare the two sets of data in the following table: capacity vs. diameter, and capacity vs. weight of a brine cooler. The horizontal scale is to be tons of capacity, and the vertical scales are to be outside diameter on the left, and weight (cwt) on the right.

Tons refrigerating capacity	Outside diameter, inches	Weight, cwt
15	22	25
30	28	46
50	34	73
85	42	116
100	46	136
130	50	164
160	58	215
210	60	263

Use 20 × 20 graph paper 8½ × 11. Horizontal scale of 1″ = 40 tons. Vertical scales of 1″ = 10″ of outside DIA, and 1″ = 40 cwt (hundred weight).

24. Draw a linear coordinate graph that shows the voltage characteristics for a generator as given in the following table of values: Abscissa—armature current in amperes (I_a); ordinate—terminal voltage in volts (E_t).

TABLE 25.1

	1890	1900	1910	1920	1930	1940	1950	1960	1970	1980
Supply	80	90	110	135	155	240	270	315	380	450
Demand	35	35	60	80	110	125	200	320	410	550

TABLE 25.2

Angle with vertical	0°	10°	20°	30°	40°	50°	60°	70°	80°	90°
Candlepower (thous.) 2-400W	37	34	25	12	5.5	2.5	2	0.5	0.5	0.5
Candlepower (thous.) 1-1000W	22	21	19	16	12.3	7	3	2	0.5	0.5

I_α	E_t	I_α	E_t	I_α	E_t
0	288	31.1	181.8	41.5	68
5.4	275	35.4	156	40.5	42.5
11.8	257	39.7	108	39.5	26.5
15.6	247	40.5	97	37.8	16
22.2	224.5	40.7	90	13.0	0
26.2	217	41.4	77.5		

25. Draw a linear coordinate graph for the centrifugal pump test data in the table below. The units along the x-axis are to be gallons per minute. There will be four curves to represent the variables given.

Gallons per min.	Discharge pressure	Water HP	Electric HP	Efficiency %
0	19.0	0.00	1.36	0.00
75	17.5	0.72	2.25	32.0
115	15.0	1.00	2.54	39.4
154	10.0	1.00	2.74	36.5
185	5.0	0.74	2.80	26.5
200	3.0	0.63	2.83	22.2

26. Draw a linear coordinate graph that compares two of the values shown in the table below— ultimate strength and elastic limit—with degrees of temperature labeled along the x-axis.

27. Draw linear coordinate graph that compares two of the values shown in the table in Problem 26— percent of elongation and percent of reduction of area of the cross section—with the degrees of temperature that will be represented along the x-axis.

Break-even graphs

28. Construct a break-even graph that shows the earnings for a new product that has a development cost of $12,000. The first 8000 will cost 50¢ each to manufacture, and you wish to break even at this quantity. What would be the profit at volumes of 20,000 and at 25,000?

29. Same as Problem 28 except that the development costs are $80,000, the manufacturing cost of the first 10,000 is $2.30 each, and the desired break-even point is 10,000. What would be the profit at a volume of 20,000 and at 30,000?

30. A manufacturer has incorporated the manufacturing and development costs into a cost-per-unit estimate. He wishes to sell the product at $1.50 each. On the y-axis plot cost per unit in dollars; on the x-axis, number of units in thousands. Data: 1000, $2.55; 2000, $2.01; 3000, $1.55; 4000, $1.20; 5000, $0.98; 6000, $0.81; 7000, $0.80;

°F	Ultimate strength	Elastic limit	Elongation %	Reduction of area %	Brinell hardness no.
400	257,500	208,000	10.8	31.3	500
500	247,000	224,500	12.5	39.5	483
600	232,500	214,000	13.3	42.0	453
700	207,500	193,500	15.0	47.5	410
800	180,500	169,000	17.0	52.5	358
900	159,500	146,500	18.5	56.5	313
1000	142,500	128,500	20.3	59.2	285
1100	126,500	114,000	23.0	60.8	255
1200	114,500	96,500	26.3	67.8	230
1300	108,000	85,500	25.8	58.3	235

TABLE 25.3

F	100	200	500	1000	2000	5000	10,000
$A(1)$	0.0028	0.002	0.0015	0.001	0.0006	0.0003	0.00013
$A(2)$	0.06	0.05	0.04	0.03	0.018	0.005	0.001

8000, $0.75; 9000, $0.73; 10,000, $0.70. How many must be sold to break even? What will be the total profit when 9000 are sold?

31. The cost per unit to produce a product by a manufacturing plant is given below. Construct a break-even graph with the cost per unit plotted on the y-axis and the number of units on the x-axis. Data: 1000, $5.90; 2000, $4.50; 3000, $3.80; 4000, $3.20; 5000, $2.85; 6000, $2.55; 7000, $2.30; 8000, $2.17; 9000, $2.00; 10,000, $0.95.

Logarithmic graphs

32. Using the data given in Table 25.3, construct a logarithmic graph where the vibration amplitude (A) is plotted as the ordinate and vibration frequency (F) as the abscissa. The data for Curve 1 represent the maximum limits of machinery in good condition with no danger from vibration. The data for Curve 2 are the lower limits of machinery that is being vibrated excessively to the danger point. The vertical scale should be three cycles and the horizontal scale two cycles.

33. Plot the data below on a two-cycle log graph to show the current in amperes (y-axis) versus the voltage in volts (x-axis) of precision temperature-sensing resistors. Data: 1 volt, 1.9 amps; 2 volts, 4 amps; 4 volts, 8 amps; 8 volts, 17 amps; 10 volts, 20 amps; 20 volts, 30 amps; 40 volts, 36 amps; 80 volts, 31 amps; 100 volts, 30 amps.

34. Plot the data from Problem 18 as a logarithmic graph.

35. Plot the data from Problem 24 as a logarithmic graph.

Semilogarithmic graphs

36. Construct a semilogarithmic graph with the y-axis a two-cycle log scale from 1 to 100 and the x-axis a linear scale from 1 to 7. Plot the data below to show the survivability of a shelter at varying distances from a one-megaton air burst. The data consists of overpressure in psi along the y-axis, and distance from ground zero in miles along the x-axis. The data points represent an 80% chance of survival of the shelter. Data: 1 mile, 55 psi; 2 miles, 11 psi; 3 miles, 4.5 psi; 4 miles, 2.5 psi; 5 miles, 2.0 psi; 6 miles, 1.3 psi.

37. The growth of two divisions of a company, Division A and Division B, is given in the data below. Plot the data on a rectilinear graph and on a semilog graph. The semilog graph should have a one-cycle log scale on the y-axis for sales in thousands of dollars, and a linear scale on the x-axis showing years for a six-year period. Data in dollars: 1 yr, A = $11,700 and B = $44,000; 2 yr, A = $19,500 and B = $50,000; 3 yr, A = $25,000 and B = $55,000; 4 yr, A = $32,000 and B = $64,000; 5 yr, A = $42,000 and B = $66,000; 6 yr, A = $48,000 and B = $75,000. Which division has the better growth rate?

38. Draw a semilog chart showing probable engineering progress. Use the following indices: 40,000 B.C. = 21; 30,000 B.C. = 21.5; 20,000 B.C. = 22; 16,000 B.C. = 23; 10,000 B.C. = 27; 6000 B.C. = 34; 4000 B.C. = 39; 2000 B.C. = 49; 500 B.C. = 60; A.D. 1900 = 100. Horizontal scale 1″ = 10,000 years. Height of cycle = about 5″. Two-cycle printed paper may be used if available.

39. Plot the data from Problem 24 as a semilogarithmic graph.

40. Plot the data from Problem 26 as a semilogarithmic graph.

Percentage graphs

41. Plot the data given in Problem 18 on a semilog graph to determine the percentages and ratios of the data. What is the percent of increase in the demand for water from 1890 to 1920? What percent of the demand is the supply for the following years: 1900, 1930, and 1970?

42. Using the graph plotted in Problem 37, determine the percent of increase of Division A and Division B from Year 1 to Year 4. What percent of sales of Division A are the sales of Division B at the end of Year 2? At the end of Year 6?

43. Plot two values from Problem 26—water horsepower and electric horsepower—on semilog paper compared with gallons per minute along the *x*-axis. What is the percent that water horsepower is of the electric horsepower when 75 gallons per minute are being pumped? What is the percent increase of the electric horsepower from 0 to 185 gallons per minute?

Organizational charts

44. Draw an organization chart for a city government organized as follows: The electorate elects school board, city council, and municipal court officers. The city council is responsible for the civil service commission, city manager, and city planning board. The city manager's duties cover finance, public safety, public works, health and welfare, and law.

45. Draw an organization chart for a manufacturing plant. The sales manager, chief engineer, treasurer, and general manager are responsible to the president. The general manager has three department heads: master mechanic, plant superintendent, and purchasing agent. The plant superintendent has charge of the shop foremen, under whom are the working forces, and also has direct charge of the shipping, tool and die, inspection, order, and stores and supplies departments.

Polar graphs

46. Construct a polar graph of the data given in Problem 19.

47. Construct a polar graph of the following illumination, in lumens at various angles, emitted from a luminaire. The zero-degrees position is vertically under the overhead lamp. Data: 0°, 12,000; 10°, 15,000; 20°, 10,000; 30°, 8000; 40°, 4200; 50°, 2500; 60°, 1000; 70°, 0. The illumination is symmetrical about the vertical.

26

Nomography

26.1
Nomography

An additional aid in analyzing data is a graphical computer called a **nomogram** or **nomograph.** A nomogram, or "number chart," is any graphical arrangement of calibrated scales and lines that may be used to facilitate calculations, usually those of a repetitive nature.

The term "nomogram" is frequently used to denote a specific type of scale arrangement called an alignment chart. Typical examples of alignment charts are shown in Fig. 26.1. Many other types are also used that have curved scales or other scale arrangements for more complex problems. The discussion of nomograms in this chapter will be limited to the simpler conversion, parallel-scale, and N-type graphs and their variations.

Using an alignment chart

An alignment graph is usually constructed to solve for one or more unknowns in a formula or empirical relationship between two or more quantities; for example, it can be used to convert degrees Celsius to degrees Fahrenheit, to find the size of a structural member to sustain a certain load, and so on. An alignment chart is read by placing a straightedge, or by drawing a line called an **isopleth,** across the scales of the chart and reading corresponding values from the scale on this line. The example in Fig. 26.2 shows readings for the formula $U + V = W$.

26.2
Alignment-graph scales

To construct any alignment graph, you must first determine the graduations of the scales that will be used to give the desired relationships. Alignment-graph scales are called **functional scales.** A functional scale is one that is graduated according to values of some *function* of a variable, but *calibrated* with values of the variable. A functional scale for $F(U) = U^2$ is illustrated in Fig. 26.3. It can be seen in this example that if a value of $U = 2$ was substituted into the equation, the position of U on the functional scale would be 4 units from zero, or $2^2 = 4$. This procedure can be repeated with all values of U by substitution.

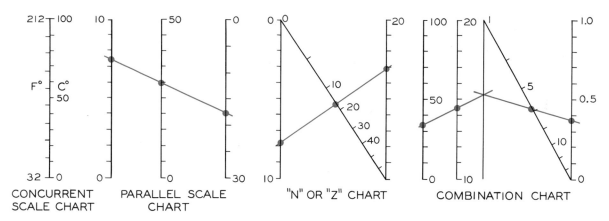

FIG. 26.1 Typical examples of types of alignment charts

The scale modulus

Since the graduations on a functional scale are spaced in proportion to values of the function, a proportionality, or scaling factor, is needed. This constant of proportionality is called the **scale modulus** and it is given by the equation

$$m = \frac{L}{F(U_2) - F(U_1)} \tag{1}$$

where

m = scale modulus, in inches per functional unit,

L = desired length of the scale, in inches,

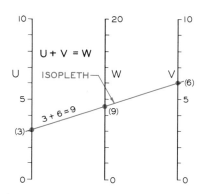

FIG. 26.2 Use of an isopleth to solve graphically for unknowns in the given equation.

VALUES OF U

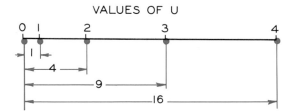

FIG. 26.3 Functional scale for units of measurement that are proportional to $F(U) = U^2$.

$F(U_2)$ = function value at the end of the scale,

$F(U_1)$ = function value at the start of the scale.

For example, suppose that we are to construct a functional scale for $F(U) = \sin U$, with $0° \leq U \leq 45°$ and a scale 6″ in length. Thus $L = 6″$, $F(U_2) = \sin 45° = 0.707$, $F(U_1) = \sin 0° = 0$. Therefore, Eq. (1) can be written in the following form by substitution:

$$m = \frac{6}{0.707 - 0} = 8.49 \text{ inches per (sine) unit}$$

The scale equation

Graduation and calibration of a functional scale are made possible by a **scale equation.** The general form of this equation may be written as a variation of Eq. (1) in the following form:

$$X = m[f(U) - F(U_1)] \tag{2}$$

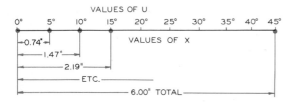

FIG. 26.4 Construction of a functional scale using values from Table 26.1, which were derived from the scale equation.

where

X = distance from the measuring point of the scale to any graduation point,

m = scale modulus,

$F(U)$ = functional value at the graduation point,

$F(U_1)$ = functional value at the measuring point of the scale.

For example, a functional scale is constructed for the previous equation, $F(U) = \sin U$ ($0° \leq U \leq 45°$). It has been determined that $m = 8.49$, $F(U) = \sin U$, and $F(U_1) = \sin 0° = 0$. Thus, by substitution the scale equation, (2), becomes

$$X = 8.49 (\sin U - 0) = 8.49 \sin U.$$

Using the equation, we can substitute values of U and construct a table of positions. In this case, the scale is calibrated at 5° intervals, as reflected in Table 26.1.

The values of X from the table give the positions, in inches, for the corresponding graduations, measured from the start of the scale ($U = 0°$); Fig. 26.4. It should be noted that the measuring point does *not* need to be at one end of the scale, but it is usually the most convenient point, especially if the functional value is zero at that point.

A graphical method of locating the functional values along a scale can be found as shown in Fig. 26.5 by the proportional-line method. The sine functions are measured off along a line at 5° intervals, with the

end of the line passing through the 0° end of the scale. The functions are transferred from the inclined line with parallel lines back to the scale where the functions are represented and labeled.

26.3
Concurrent scales

Concurrent scales are useful in the rapid conversion of one value into terms of a second system of measurement. Formulas of the type $F_1 = F_2$, which relate two variables, can be adapted to the concurrent-scale format. Typical examples might be the Fahrenheit–Celsius temperature relation,

$$°F = \tfrac{9}{5} °C + 32,$$

or the area of a circle,

$$A = \pi r^2.$$

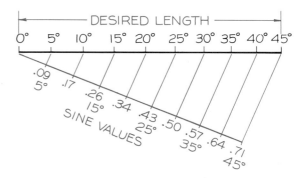

FIG. 26.5 A functional scale that shows the sine of the angles from 0° to 45° can be drawn graphically by the proportional-line method. The scale is drawn to a desired length and the sine values of angles at 5° intervals are laid off along a construction line that passes through the 0° end of the scale. These values are projected back to the scale.

TABLE 26.1

U	0°	5°	10°	15°	20°	25°	30°	35°	40°	45°
X	0	0.74	1.47	2.19	2.90	3.58	4.24	4.86	5.45	6.00

TABLE 26.2

r	1	2	3	4	5	6	7	8	9	10
X_r	0	0.15	0.40	0.76	1.21	1.77	2.42	3.18	4.04	5.00

Design of a concurrent-scale chart involves the construction of a functional scale for each side of the mathematical formula in such a manner that the **position** and **lengths** of each scale coincide. For example, to design a conversion chart 5 in. long that will give the areas of circles whose radii range from 1 to 10, we first write $F_1(A) = A$, $F_2(r) = \pi r^2$, and $r_1 = 1$, $r_2 = 10$. The scale modulus for r is

$$m_r = \frac{L}{F_2(r_2) - F_2(r_1)}$$

$$= \frac{5}{\pi(10)^2 - \pi(1)^2} = 0.0161.$$

Thus the scale equation for r becomes

$$X_r = m_r [F_2(r) - F_2(r_1)]$$
$$= 0.0161[\pi r^2 - \pi(1)^2]$$
$$= 0.0161\pi(r^2 - 1)$$
$$= 0.0505(r^2 - 1).$$

A table of values for X_r and r may now be completed as shown in Table 26.2. The r-scale can be drawn from this table, as shown in Fig. 26.6. From the original formula, $A = \pi r^2$, the limits of A are found to be $A_1 = \pi = 3.14$ and $A_2 = 100\pi = 314$. The scale modulus for concurrent scales is always the same for equal-length scales; therefore, $m_A = m_r = 0.0161$, and the scale equation for A becomes

$$X_A = m_A [F_1(A) - F_1(A)_1]$$
$$= 0.0161(A - 3.14).$$

The corresponding table of values is then computed for selected values of A, as shown in Table 26.3.

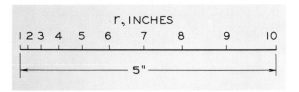

FIG. 26.6 Calibration of one scale of a concurrent scale chart using values from Table 26.2.

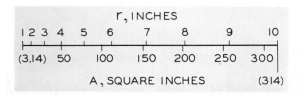

FIG. 26.7 The completed concurrent scale chart for the formula $A = \pi r^2$. Values for the A-scale are taken from Table 26.3.

The A-scale is now superimposed on the r-scale; its calibrations have been placed on the other side of the line to facilitate reading (Fig. 26.7). It may be desired to expand or contract one of the scales, in which case an alternative arrangement may be used, as shown in Fig. 26.8. The two scales are drawn parallel at any convenient distance, and calibrated in *opposite* directions. A different scale modulus and corresponding scale equation must be calculated for each scale if they are *not* the same length.

A graphical method can be used to construct concurrent scales, as shown in Fig. 26.9, by using the proportional-line method. Since there are 101.6 mm in 4 inches, the units of millimeters can be located on the

TABLE 26.3

A	(3.14)	50	100	150	200	250	300	(314)
X_A	0	0.76	1.56	2.36	3.16	3.96	4.76	5.00

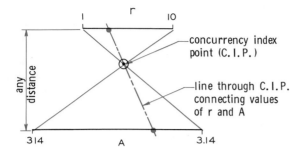

FIG. 26.8 Concurrent scale chart with unequal scales

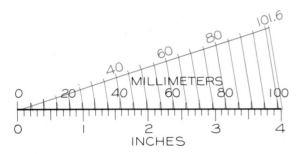

FIG. 26.9 The proportional-line method can be used to construct an alignment graph that converts inches to millimeters. This requires that the units at each end of the scales be known. For example, there are 101.6 millimeters in 4 inches.

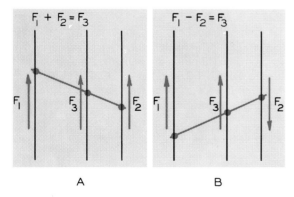

FIG. 26.10 Two common forms of parallel-scale alignment nomographs

upper side of the inch scale by projecting to the scale with a series of parallel projectors.

26.4
Construction of alignment graphs with three variables

For a formula of three functions (of one variable each), the general approach is to select the lengths and positions of *two* scales according to the range of variables and size of the chart desired. These are then calibrated by means of the scale equations, as shown in the preceding section. The position and calibration of the third scale will then depend upon these initial constructions. Although definite mathematical relationships exist that may be used to locate the third scale, graphical constructions are simpler and usually less subject to error. Examples of the various forms are presented in the following sections.

26.5
Parallel-scale nomographs

Many engineering relationships involve three variables that can be computed graphically on a repetitive basis. Any formula of the type $F_1 + F_2 = F_3$ may be represented as a parallel-scale alignment chart, as shown in Fig. 26.10A. Note that all scales increase (functionally) in the same direction and that the function of the middle scale represents the *sum* of the other two. Reversing the direction of any scale changes the sign of its function in the formula, as for $F_1 - F_2 = F_3$ in Fig. 26.10B.

To illustrate this type of alignment graph we will use the formula $Z = X + Y$, as illustrated in Fig. 26.11. The outer scales for X and Y are drawn and calibrated. They can be drawn to any length and positioned any distance apart, as shown in Fig. 26.12. Two sets of data that yield a Z of 8 in Step 1 are used to locate the parallel Z-scale. In Step 2, the Z-scale is drawn and divided into 16 equal units. The finished nomograph in Step 3 can be used to add various values of X and Y to find their sums along the Z-scale.

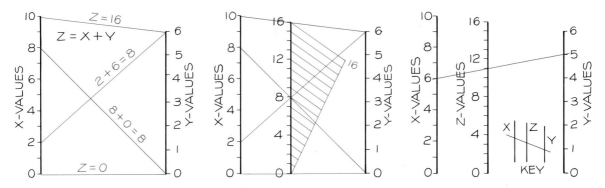

FIG. 26.11 Parallel-scale nomogram (linear)

Step 1 Two parallel scales are drawn at any length and calibrated. The location of the parallel Z-scale is found by selecting two sets of values that will give the same value of Z, 8 in this example. The ends of the Z-scale will have values of 0 and 16, the sum of the end values of X and Y.

Step 2 The Z-scale is drawn through the point located in Step 1 parallel to the other scales. The scale is calibrated from 0 to 16 by using the proportional-line method. Note that the two sets of X- and Y-values cross at 8, the sum of each set.

Step 3 The Z-scale is labeled and calibrated with easy-to-read units. A key is drawn to show how the nomograph is used. If the Y-scale were calibrated with 0 at the upper end instead of the bottom, a different Z-scale could be computed and the nomograph could be used for $Z = X - Y$.

A more complex alignment graph is illustrated in Fig. 26.13, where the formula $U + 2V = 3W$ is expressed in the form of a nomograph.

First it is necessary to determine and calibrate the two outer scales for U and V; we can make them any convenient length and position them any convenient distance apart, as shown in Fig. 26.12. These scales are used as the basis for the step-by-step construction shown in Fig. 26.13.

The limits of calibration for the middle scale are found by connecting the endpoints of the outer scales and substituting these values into the formula. Here, W is found to be 0 and 10 at the extreme ends (Step 1). Two pairs of corresponding values of U and V are selected that will give the same value of W. For example, values of $U = 0$ and $V = 7.5$ give a value of 5 for W. We also find that $W = 5$ when $U = 14$ and $V = 0.5$. We connect these corresponding pairs of values with isopleths to locate their intersection, which establishes the position of the W-scale.

Since the W-scale is linear ($3W$ is a linear function), it may be subdivided into uniform intervals (Step 2). For a nonlinear scale, the scale modulus (and the scale equation) may be found in Step 2 by substituting its length and its two end values into Eq. (1) of Section 26.2. The scales can be used to determine an infinite number of solutions when steps of two variables are known, (Step 3).

Parallel-scale graph with logarithmic scales

Problems involving formulas of the type $F_1 \times F_2 = F_3$ can be solved in a manner very similar to the example given in Fig. 26.1 when logarithmic scales are used.

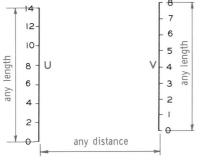

FIG. 26.12 Calibration of the outer scales for the formula $U + 2V = 3W$, where $0 \leq U \leq 14$ and $0 \leq V \leq 8$.

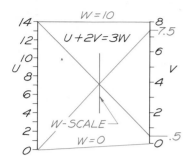
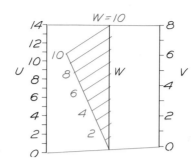
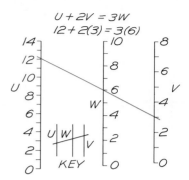

FIG. 26.13 Parallel-scale nomograph (linear)

Step 1 Substitute the end values of the U- and V-scales into the formula to find the end values of the W-scale: $W = 10$ and $W = 0$. Select two sets of U and V that will give the same value of W. *Example:* When $U = 0$ and $V = 7.5$, W will equal 5, and when $U = 14$ and $V = 0.5$, W will equal 5. Connect these sets of values; the intersection of their lines locates the position of the W-scale.

Step 2 Draw the W-scale parallel to the outer scales; its length is controlled by the previously established lines of $W = 10$ and $W = 0$. Since this scale is 10 linear divisions long, divide it graphically into ten units as shown. This will be a linear scale constructed as shown in Fig. 26.11.

Step 3 The nomogram can be used as illustrated by selecting any two known variables and connecting them with an isopleth to determine the third unknown. A key is always included to illustrate how the nomogram is to be used. An example of $U = 12$ and $V = 3$ is shown to verify the accuracy of the graph.

The first step in drawing a nomograph with logarithmic scales is learning how to transfer logarithmic functions to the scale. The graphical method is shown in Fig. 26.14, where units along the scale are found by projecting from a printed logarithmic scale with parallel lines. The mathematical method can also be used to locate these logarithmic spacings.

The formula $Z = XY$ is converted into a nomograph in Fig. 26.15. In Step 1, the X- and Y-scales are drawn within the desired limits from 1 to 10 on each. Sets of values of X and Y that yield the same value of Z, 10 in this case, are used to locate the z-axis. The limits of the Z-axis are 1 and 100.

In Step 2, the Z-axis is drawn and calibrated as a two-cycle log scale. In Step 3, a key is given to explain how an isopleth is used to add the logarithms of X and Y to give the log of Z. When logarithms are added the result is multiplication. Had the y-axis been calibrated in the opposite direction with 1 at the upper end and 10 at the lower end, a new z-axis could have been calibrated, and the nomograph used for the formula, $Z = Y/X$, since it would be subtracting logarithms.

A more advanced example of this type of problem is the formula $R = S\sqrt{T}$, for $0.1 \leq S \leq 1.0$ and $1 \leq T \leq 100$. Assume the scales to be 6 in. long. These scales need not be equal except for convenience. This formula may be converted into the required form by taking common logarithms of both sides, which gives

$$\log R = \log S + \tfrac{1}{2} \log T.$$

Thus, we have

$$F_1(S) + F_2(T) = F_3(R), \tag{1}$$

where $F_1(S) = \log S$, $F_2(T) = \frac{1}{2} \log T$, and $F_3(R) =$

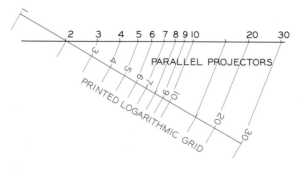

FIG. 26.14 Graphical calibration of a scale using logarithmic paper

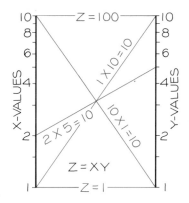
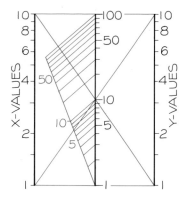
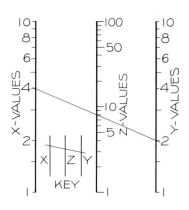

FIG. 26.15 Parallel-scale nomogram (logarithmic)

Step 1 For the equation $Z = XY$, parallel log scales are drawn that give the ranges of values for X and Y. Sets of X and Y points that yield the same value of Z, 10 in this example, are drawn. Their intersection locates the Z-scale with end values of 1 and 100.

Step 2 The Z-axis is graphically calibrated as a two-cycle logarithmic scale from 1 to 100. This scale is parallel to the X- and Y-scales.

Step 3 A key is drawn. An isopleth is drawn to show that $4 \times 2 = 8$. By reversing the Y-value scale from 1 down-ward to 10 and computing a different Z-scale, the nomogram could be used for the equation $Z = X/Y$.

log R. The scale modulus for $F_1(S)$ is, from Eq. (1),

$$m_S = \frac{6}{\log 1.0 - \log 0.1} = \frac{6}{0 - (-1)} = 6. \quad (2)$$

Choosing the scale measuring point from $S = 0.1$, we find from Eq. (2) that the scale equation for $F_1(S)$ is

$$X_S = 6(\log S - \log 0.1) = 6(\log S + 1). \quad (3)$$

Similarly, the scale modulus for $F_2(T)$ is

$$m_T = \frac{6}{\frac{1}{2}\log 100 - \frac{1}{2}\log 1} = \frac{6}{\frac{1}{2}(2) - \frac{1}{2}(0)} = 6. \quad (4)$$

Thus, the scale equation, measuring from $T = 1$, is:

$$X_T = 6(\tfrac{1}{2}\log T - \tfrac{1}{2}\log 1) = 3\log T. \quad (5)$$

The corresponding tables for the two scale equations may be computed as shown in Tables 26.4 and 26.5.

TABLE 26.4

S	0.1	0.2	0.3	0.4	0.5	0.6	0.7	0.8	0.9	1.0
X_S	0	1.80	2.88	3.61	4.19	4.67	5.07	5.42	5.72	6.00

TABLE 26.5

T	1	2	4	6	8	10	20	40	60	80	100
X_T	0	0.91	1.80	2.33	2.71	3.00	3.91	4.81	5.33	5.77	6.00

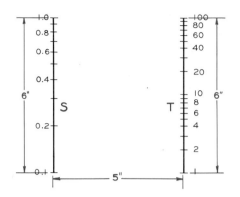

FIG. 26.16 Calibration of the outer scales for the formula $R = S\sqrt{T}$, where $0.1 \le S \le 1.0$ and $1 \le T \le 100$.

We shall position the two scales 5 in. apart, as shown in Fig. 26.16. The logarithmic scales are graduated using the values in Tables 26.4 and 26.5. The step-by-step procedure for constructing the remainder of the nomogram is given in Fig. 26.17 using the two outer scales determined here.

The end values of the middle (R) scale are found from the formula $R = S\sqrt{T}$ to be $R = 1.0\sqrt{100} = 10$ and $R = 0.1\sqrt{1} = 0.1$. Choosing a value of $R = 1.0$, we find that corresponding value pairs of S and T might be $S = 0.1$, $T = 100$, and $S = 1.0$, $T = 1.0$. We connect these pairs with isopleths in Step 1 and

position the middle scale at the intersection of the lines connecting the corresponding values. The R-scale is drawn parallel to the outer scales and is calibrated by deriving its scale modulus:

$$m_R = \frac{6}{\log 10 - \log 0.1} = \frac{6}{1 - (-1)} = 3.$$

Thus, its scale equation (measuring from $R = 0.1$) is

$$X_R = 3(\log R - \log 0.1) = 3(\log R + 1.0).$$

Table 26.6 is computed to give the values for the scale. These values are applied to the R-scale, as shown in Step 2. The finished nomogram can be used as illustrated in Step 3 to compute the unknown variables when two variables are given.

Note that this example illustrates a general method of creating a parallel-scale graph for all formulas of the type $F_1 + F_2 = F_3$ through the use of a table of values computed from the scale equation.

26.6
N- or Z-graphs

Whenever F_2 and F_3 are linear functions, we can partially avoid using logarithmic scales for formulas of the

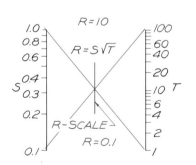
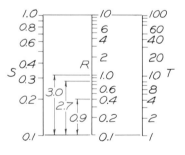
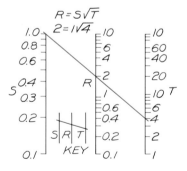

FIG. 26.17 Parallel-scale chart (logarithmic)

Step 1 Connect the end values of the outer scales to determine the extreme values of the R-scale, $R = 10$ and $R = 0.1$. Select corresponding values of S and T that will give the same value of R. Values of $S = 0.1$, $T = 100$ and $S = 1.0$, $T = 1.0$ give a value of $T = 1.0$. Connect the pairs to locate the position of the R-scale.

Step 2 Draw the R-scale to extend from 0.1 to 10. Calibrate it by substituting values determined from its scale equation. These values have been computed and tabulated in Table 26.6. The resulting tabulation is a logarithmic, two-cycle scale.

Step 3 Add labels to the finished nomogram and draw a key to indicate how it is to be used. An isopleth has been used to determine R when $S = 1.0$ and $T = 4$. The result of 2 is the same as that obtained mathematically, thus verifying the accuracy of the chart. Other combinations can be solved in this same manner.

TABLE 26.6

R	0.1	0.2	0.4	0.6	0.8	1.0	2.0	4.0	6.0	8.0	10.0
X_R	0	0.91	1.80	2.33	2.71	3.00	3.91	4.81	5.33	5.71	6.00

type

$$F_1 = \frac{F_2}{F_3}.$$

Instead, we use an N-graph, as shown in Fig. 26.18. The outer scales, or "legs," of the N are functional scales and will therefore be linear if F_2 and F_3 are linear, whereas if the same formula were drawn as a parallel-scale graph, all scales would have to be logarithmic.

Some main features of the N-chart are:

1. The outer scales are parallel functional scales of F_2 and F_3.

2. They increase (functionally) in *opposite* directions.

3. The diagonal scale connects the (functional) *zeros* of the outer scale.

4. In general, the diagonal scale is not a functional scale for the function F_1 and is generally nonlinear.

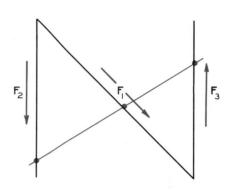

FIG. 26.18 An N-graph for solving an equation of the form $F_1 = F_2/F_3$.

Construction of an N-graph is simplified by the fact that locating the middle (diagonal) scale is usually less of a problem than it is for a parallel-scale graph. Calibration of the diagonal scale is most easily accomplished by graphical methods.

The steps in constructing a basic N-graph of the equation $Z = Y/X$ are shown in Fig. 26.19. In Step 1,

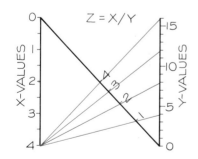
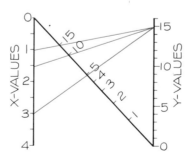
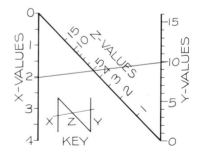

FIG. 26.19 An N-chart nomograph

Step 1 A Z-nomograph for the equation of $Z = Y/X$ can be drawn with two parallel scales. The zero ends of each scale are connected with a diagonal scale. Isopleths are drawn to locate whole units along the diagonal.

Step 2 Additional isopleths are drawn to locate other whole units along the diagonal. It is important that the units labeled on the diagonal be whole units that are easy to interpolate between when the nomogram is used.

Step 3 The diagonal scale is labeled and a key is drawn. An example isopleth is drawn to show that $10/2 = 5$. Note that the accuracy of the N-chart is greater at the 0 end of the diagonal. It approaches infinity at the other end.

the diagonal is drawn to connect the zero ends of the scales. Whole values are located along the diagonal by using combinations of X- and Y-values. It is important that the units located along the diagonal are whole values that are easy to interpolate between.

In Step 3, the diagonal is labeled and a key is given to explain how to use the nomograph. A sample isopleth is given that verifies the correctness of the graphical relationship between the scales.

A more advanced N-graph is constructed for the equation:

$$A = \frac{B + 2}{C + 5},$$

where $0 \le B \le 8$ and $0 \le C \le 15$. This equation follows the form of

$$F_1 = \frac{F_2}{F_3},$$

where $F_1(A) = A$, $F_2(B) = B + 2$, and $F_3(C) = C + 5$. Thus the outer scales will be for $B + 2$ and $C + 5$, and the diagonal scale will be for A.

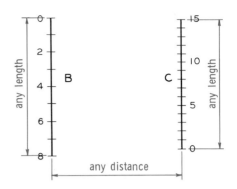

FIG. 26.20 Calibration of the outer scales of an N-graph for the equation $A = (B + 2)/(C + 5)$.

The construction is begun in the same manner as for a parallel-scale graph by selecting the layout of the outer scales (Fig. 26.20). As before, the limits of the diagonal scale are determined by connecting the endpoints on the outer scales, giving $A = 0.1$ for $B = 0$, $C = 15$ and $A = 2.0$ for $B = 8$, $C = 0$, as shown in the given portion of Fig. 26.21. The remainder of the construction is given in step form in the figure.

The diagonal scale is located by finding the **function zeros** of the outer scales, i.e., the points where

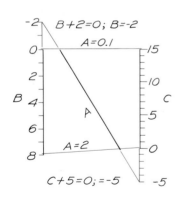 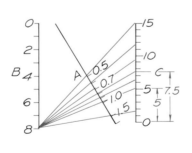 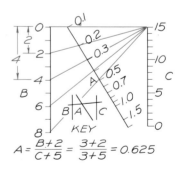

FIG. 26.21 Construction of an N-graph

Step 1 Locate the diagonal scale by finding the functional zeros of the outer scales. This is done by setting $B + 2 = 0$ and $C + 5 = 0$ which gives a zero value for A.

Step 2 Select the upper limit of one of the outer scales, $B = 8$ in this case, and substitute it into the given equation to find a series of values of C for the desired values of A, as shown in Table 26.7. Draw isopleths from $B = 8$ to the values of C to calibrate the A-scale.

Step 3 Calibrate the remainder of the A-scale in the same manner by substituting $C = 15$ into the equation to determine a series of values on the B-scale for desired values on the A-scale, as listed in Table 26.8. Draw isopleths from $C = 15$ to calibrate the A-scale as shown. Draw a key to indicate how the nomogram is used.

TABLE 26.7

A	2.0	1.5	1.0	0.9	0.8	0.7	0.6	0.5
C	0	1.67	5.0	6.11	7.50	9.28	11.7	15.0

$B + 2 = 0$ or $B = -2$, and $C + 5 = 0$ or $C = -5$. The diagonal scale may then be drawn by connecting these points as shown in Step 1. Calibration of the diagonal scale is most easily accomplished by substituting into the formula. Select the upper limit of an outer scale, for example, $B = 8$. This gives the formula

$$A = \frac{10}{C + 5}.$$

Solve this equation for the other outer scale variable,

$$C = \frac{10}{A} - 5.$$

Using this as a "scale equation," make a table of values for the desired values of A and corresponding values of C (up to the limit of C in the chart), as shown in Table 26.7. Connect isopleths from $B = 8$ to the tabulated values of C. Their intersections with the diagonal scale give the required calibrations for approximately half the diagonal scale, as shown in Step 2 of Fig. 26.21.

The remainder of the diagonal scale is calibrated by substituting the end value of the other outer scale ($C = 15$) into the formula, giving

$$A = \frac{B + 2}{20}.$$

Solving this for B yields

$$B = 20A - 2.$$

A table for the desired values of A can be constructed as shown in Table 26.8. Isopleths connecting $C = 15$

TABLE 26.8

A	0.5	0.4	0.3	0.2	0.1
B	8.0	6.0	4.0	2.0	0

with the tabulated values of B will locate the remaining calibrations on the A-scale, as shown in Step 3.

26.7
Combination forms of alignment graphs

The types of graphs discussed above may be used in combination to handle different types of formulas. For example, formulas of the type $F_1/F_2 = F_3/F_4$ (four variables) may be represented as *two* N-charts by the insertion of a "dummy" function. To do this, let

$$\frac{F_1}{F_2} = S \quad \text{and then} \quad S = \frac{F_3}{F_4}.$$

Each of these may be represented as shown in Part A of Fig. 26.22, where one N-graph is inverted and rotated 90°. In this way, the charts may be superimposed as shown in Part B if the S-scales are of equal length. The S-scale, being a "dummy" scale, does not need to be calibrated; it is merely a "turning" scale for intermediate values of S which do not actually enter into the formula itself. The chart is read with *two* isopleths which connect the four variable values and cross on the S-scale, as shown in Part C. Nomograms of this form are commonly called **ratio graphs.**

Formulas of the type $F_1 + F_2 = F_3F_4$ are handled similarly. As in the preceding example, a "dummy" function is used: $F_1 + F_2 = S$ and $S = F_3F_4$. In order to apply the superimposition principle, a more equitable arrangement is obtained by rewriting the equations as $F_2 = S - F_1$ and $F_3 = S/F_4$. These two equations then take the form of a parallel-scale nomogram and an N-graph, respectively, as shown in Part A of Fig. 26.23. Again the S-scales must be identical but need not be calibrated. The nomograms are superimposed in Part B. The S-scale is used as a "turning" scale for the two isopleths, as shown in Part C. Many other combinations are possible, limited only by the ingenuity of the nomographer in adapting formulas and scale arrangements to his needs.

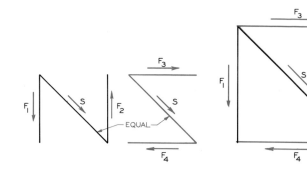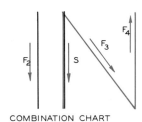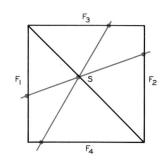

FIG. 26.22 Four-variable graph

Step 1 A combination chart can be developed to handle four variables in the form $F_1/F_2 = F_3/F_4$ by developing two N-graphs in the forms $F_1/F_2 = S$ and $F_3/F_4 = S$, where S is a dummy scale of equal length in both charts.

Step 2 If equal-length scales are used in each of the N-graphs and if the S-scales are equal, then the charts can be overlapped so that each is common to the S-scale.

Step 3 Two lines (isopleths) are drawn to cross at a common point on the S-scale. Numerous combinations of the four variables can be read on the surrounding scales. The S-scale need not be calibrated, since no values are read from it.

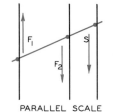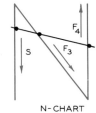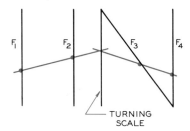

FIG. 26.23 Combination parallel-scale chart and N-graph

Step 1 Formulas of the type $F_1 + F_2 = F_3F_4$ can be combined into one nomogram by constructing a parallel-scale chart and an N-graph with an equal S-scale (a dummy scale).

Step 2 By superimposing the two equal S-scales, the two nomograms are combined into a combination chart. The S-scale need not be calibrated, since values are not read from it.

Step 3 The addition of the variables can be handled at the left of the chart. The S-scale is the turning scale from which the N-graph can be used to find the unknown variables.

Problems

The following problems are to be solved on Size A sheets (8½″ × 11″). Problems that involve geometric construction and mathematical calculations should show the construction and calculations as part of the solutions. If the mathematical calculations are extensive, these should be included on a separate sheet.

Concurrent scales

Construct concurrent scales for converting the following relationships of one type of unit to another. The range of units for the scales is given for each.

1. Kilometers and miles:

$$1.609 \text{ km} = 1 \text{ mile, from 10 to 100 miles.}$$

2. Liters and U.S. gallons:

$$1 \text{ liter} = 0.2692 \text{ U.S. gallons, from 1 to 10 liters.}$$

3. Knots and miles per hour:

$$1 \text{ knot} = 1.15 \text{ miles per hour, from 0 to 45 knots.}$$

4. Horsepower and British thermal units:

$$1 \text{ horsepower} = 42.4 \text{ btu, from 0 to 1200 hp.}$$

5. Centigrade and Fahrenheit:

$$°F = \tfrac{9}{5}°C + 32, \text{ from } 32°F \text{ to } 212°F.$$

6. Radius and area of a circle:

$$\text{Area} = \pi r^2, \text{ from } r = 0 \text{ to } 10.$$

7. Inches and millimeters:

$$1 \text{ inch} = 25.4 \text{ millimeters, from 0 to 5 inches.}$$

8. Numbers and their logarithms:

(use logarithm tables), numbers from 1 to 10.

Addition and subtraction nomographs

Construct parallel-scale nomographs to solve the following addition and subtraction problems.

9. $A = B + C$, where $B = 0$ to 10 and $C = 0$ to 5.

10. $Z = X + Y$, where $X = 0$ to 8 and $Y = 0$ to 12.

11. $Z = Y - X$, where $X = 0$ to 6 and $Y = 0$ to 24.

12. $A = C - B$, where $C = 0$ to 30 and $B = 0$ to 6.

13. $W = 2V + U$, where $U = 0$ to 12 and $V = 0$ to 9.

14. $W = 3U + V$, where $U = 0$ to 10 and $V = 0$ to 10.

15. Electrical current at a circuit junction: $I = I_1 + I_2$.
I = current entering the junction in amperes
I_1 = current leaving the junction, varying from 2 to 15 amps
I_2 = current leaving junction, varying from 7 to 36 amps.

16. Pressure change in fluid flowing in a pipe:

$$\Delta P = P_2 - P_1.$$

ΔP = pressure change between two points in pounds per square inch,
P_1 = pressure upstream, varying from 3 psi to 12 psi,
P_2 = pressure downstream, varying from 10 psi to 15 psi.

Multiplication and division: parallel scales

Construct parallel-scale nomographs with logarithmic scales that will perform the following multiplication and division operations:

17. Area of a rectangle: Area = Height × Width, where $H = 1$ to 10 and $W = 1$ to 12.

18. Area of a triangle: A = Base × Height/2, where $B = 1$ to 10 and $H = 1$ to 5.

19. Electrical potential between terminals of a conductor: $E = IR$.
E = electrical potential in volts,
I = current, varying from 1 to 10 amperes,
R = resistance, varying from 5 to 30 ohms.

20. Pythagorean theorem: $C^2 = A^2 + B^2$.
C = hypotenuse of a right triangle in centimeters,
A = one leg of the right triangle, varying from 5 to 50 cm,
B = second leg of the right triangle, varying from 20 to 80 cm.

21. Allowable pressure on a shaft bearing:

$$P = ZN/100.$$

P = pressure in pounds per square inch,
Z = viscosity of lubricant from 15 to 50 cp (centipoises),
N = angular velocity of shaft from 10 to 1000 rpm.

22. Miles per gallon an automobile travels: mpg = miles/gallon. Miles vary from 1 to 500 and gallons from 1 to 24.

23. Cost per mile (cpm) of an automobile: cpm = cost/miles. Miles vary from 1 to 500 and cost varies from $1 to $28.

24. $R = S\sqrt{T}$: where S varies from 1 to 10 and T from 1 to 10.

25. Angular velocity of a rotating body: $W = V/R$.
   W = angular velocity, in radians per second,
   V = Peripheral velocity, varying from 1 to 100 meters per second,
   R = radius, varying from 0.1 to 1 meter.

N-graphs

Construct N-graphs that will solve the following equations.

26. Stress = P/A: where P varies from 0 to 1000 psi and A varies from 0 to 15 square inches.

27. Volume of a cylinder: $V = \pi r^2 h$
   V = volume in cubic inches,
   r = radius, varying from 5 to 10 feet,
   h = height, varying from 2 to 20 inches.

28. Same as Problem 17.

29. Same as Problem 18.

30. Same as Problem 19.

31. Same as Problem 20.

32. Same as Problem 21.

33. Same as Problem 22.

34. Same as Problem 23.

35. Same as Problem 24.

36. Same as Problem 25.

Combination nomographs

37. Construct a combination nomograph to express the law of sines: a/sine $A = b$/sine B. Assume that a and b vary from 0 to 10, and that A and B vary from 0° to 90°.

38. Construct a combination nomograph to determine the velocity of sound in a solid, using the formula

$$C = \sqrt{\frac{E + 4\mu/3}{p}}$$

where E varies from 10^6 to 10^7 psi, μ varies from 1×10^6 to 2×10^6 psi, and C varies from 1000 to 1500 fps. (*Hint:* rewrite the formula as $C^2 p = E + 4/3\ \mu$.

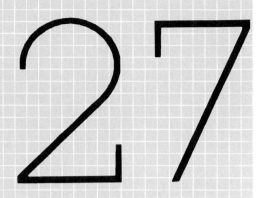

Empirical Equations and Calculus

27.1
Empirical data

Data gathered from laboratory experiments and tests of prototypes or from actual field tests are called **empirical data.** Often empirical data can be transformed to equation form by means of one of three types of equations to be covered here.

The analysis of empirical data begins with the plotting of the data on rectangular grids, logarithmic grids, and semilogarithmic grids. Curves are then sketched through each point to determine which of the grids renders a straight-line relationship (Fig. 27.1). When the data plots as a straight line, its equation may be determined.

27.2
Selection of points on a curve

Two methods of finding the equation of a curve are (1) the selected-points method and (2) the slope-intercept method. These are compared on a linear graph in Fig. 27.2.

Selected-points method

Two widely separated points, such as (1, 30) and 4, 60) can be selected on the curve and substituted in the equation below:

$$\frac{Y - 30}{X - 1} = \frac{60 - 30}{4 - 1}$$

The resulting data for the equation is

$$Y = 10X + 20.$$

Slope-intercept method

To apply the slope-intercept method, the intercept on the y-axis where $X = 0$ must be known. If the x-axis is logarithmic, then the log of $X = 1$ is 0 and the intercept must be found above the value of $X = 1$.

In Fig. 27.2B, the data do not intercept the y-axis; therefore the curve must be extended to find the intercept $B = 20$. The slope of the curve is found ($\Delta Y/\Delta X$) and substituted into the slope–intercept form to give the equation as $Y = 10X + 20$.

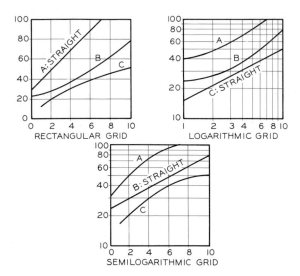

FIG. 27.1 Empirical data are plotted on each of these types of grids to determine which will render a straight-line plot. If the data can be plotted as a straight line on one of these grids, their equation can be found.

Other methods of converting data to equations are used, but the two methods illustrated here make the best use of the graphical process and are the most direct methods of introducing these concepts.

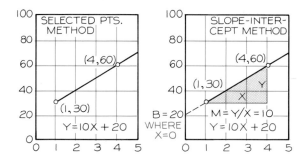

FIG. 27.2 The equation of a straight line on a grid can be determined by selecting any two points on the line. The slope-intercept method requires that the intercept be found where $X = 0$ on a semilog grid. This requires the extension of the curve to the y-axis.

27.3

The linear equation: $Y = MX + B$

The curve fitting the experimental data plotted in Fig. 27.3 is a straight line; therefore, we may assume that these data are linear, meaning that each measurement along the y-axis is directly proportional to x-axis units. We may use the slope-intercept method or the selected-points method to find the equation for the data.

In the slope-intercept method, two known points are selected along the curve. The vertical and horizontal differences between the coordinates of each of these points are determined to establish the right triangle shown in Part A of the figure. In the slope-intercept equation, $Y = MX + B$, M is the tangent of the angle between the curve and the horizontal, B is the intercept of the curve with the y-axis where $X = 0$, and X and Y are variables. In this example $M = {}^{30}/_5 = 6$ and the intercept is 20.

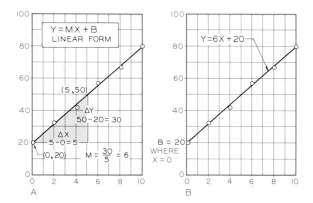

FIG. 27.3 (A) A straight line on an arithmetic grid will have an equation in the form $Y = MX + B$. The slope, M, is found to be 6. (B) The intercept, B, is found to be 20. The equation is written as $Y = 6X + 20$.

If the curve has sloped downward to the right, the slope would have been negative. By substituting this information into the slope-intercept equation, we obtain $Y = 6X + 20$, from which we can determine values of Y by substituting any value of X into the equation.

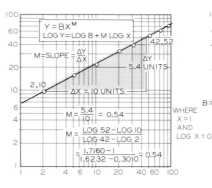
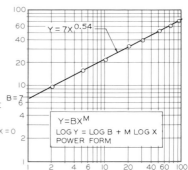

FIG. 27.4 The power equation, $Y = BX^M$

Given The data plotted on the rectangular grid give an approximation of a parabola. Since the data do not form a straight line on the rectangular grid, the equation will not be linear.

Step 1 The data forms a straight line on a logarithmic grid. The slope, M, can be found graphically with an engineer's scale, setting dX at 10 units and measuring the slope (dY) using the same scale.

Step 2 The intercept $B = 7$ is found where $X = 1$. The slope and intercept are substituted into the equation, which then becomes $Y = 7X^{0.54}$.

The selected-points method could also have been used to arrive at the same equation if the intercept were not known. By selecting two widely separated points such as (2, 32) and (10, 80), one can write the equation in this form:

$$\frac{Y - 32}{X - 2} = \frac{80 - 32}{10 - 2}, \qquad \therefore Y = 6X + 20$$

which results in the same equation as was found by the slope-intercept method $(Y = MX + B)$.

27.4
The power equation: $Y = BX^M$

When the data are plotted on a logarithmic grid (Fig. 27.4), they are found to form a straight line (Step 1). Therefore, we express the data in the form of a power equation in which Y is a function of X raised to a given power or $Y = BX^M$. The equation of the data is obtained by using the point where the curve intersects the y-axis where $X = 0$ to find B, and letting M equal the slope of the curve. Two known points are selected

on the curve to form the slope triangle. The engineers' scale can then be used, when the cycles along the X- and Y-axes are equal, to measure the slope between the coordinates of the two points.

If the horizontal distance of the right triangle is drawn to be 1 or 10 or a multiple of 10, the vertical distance can be read directly. In Step 2, the slope M (tangent of the angle) is found to be 0.54. The intercept B is 7; thus, the equation is $Y = 7X^{0.54}$, which can be evaluated for each value of Y by converting this power equation into the logarithmic form of log Y:

$$\log Y = \log B + M \log X,$$
$$\log Y = \log 7 + 0.54 \log X.$$

Note that when the slope-intercept method is used, the intercept is found on the y-axis where $X = 1$. In Fig. 27.5, the y-axis at the left of the graph has an X value of 0.1; consequently, the intercept is located midway across the graph where $X = 1$. This is analogous to the linear form of the equation, since the log of 1 is 0. The curve slopes downward to the right; thus the slope, M, is negative.

Base-10 logarithms are used in these examples, but natural logs could be used with e (2.718) as the base.

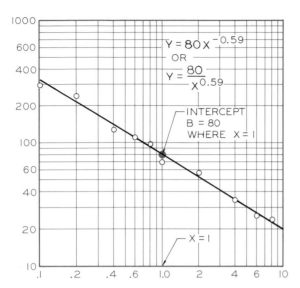

FIG. 27.5 When the slope-intercept equation is used, the intercept can be found only where $X = 1$. Therefore, in this example the intercept is found at the middle of the graph.

27.5

The exponential equation: $Y = BM^X$

When the data in Fig. 27.6 are plotted on a semilogarithmic grid (Step 1), they approximate a straight line for which we can write the equation $Y = BM^X$, where B is the Y-intercept of the curve and M is the slope of the curve. The procedure for deriving the equation is shown in Step 2, in which two points are selected along the curve so that a right triangle can be drawn to represent the differences between the coordinates of the points selected. The slope of the curve is found to be

$$\log M = \frac{\log 40 - \log 6}{8 - 3} = 0.1648,$$

or

$$M = (10)^{0.1648} = 1.46.$$

The value of M can be substituted in the equation in

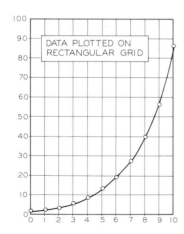

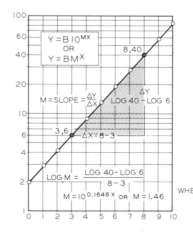

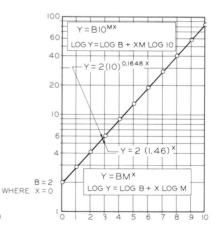

FIG. 27.6 The exponential equation: $Y = BM^X$

Given These data give a straight line on a semilogarithmic grid. Therefore, the data fit the equation form, $Y = BM^X$.

Step 1 The slope must be found by mathematical calculations; it cannot be found graphically. The slope may be written in either of the forms shown here.

Step 2 The intercept $B = 2$ is found where $X = 0$. The slopes, M, and B, are substituted into the equation to give $Y = 2(10)^{0.1648X}$ or $Y = 2(1.46)^X$.

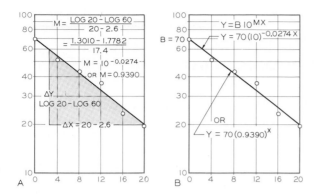

FIG. 27.7 When a curve slopes downward to the right, its slope is negative as calculated in Part A. Two forms of a final equation are shown in Part B by substitution.

the following manner:

$$Y = BM^X \qquad \text{or} \qquad Y = 2(1.46)^X,$$

$$Y = B(10)^{MX} \qquad \text{or} \qquad Y = 2(10)^{0.1648X},$$

where X is a variable that can be substituted into the equation to give an infinite number of values for Y. We can write this equation in its logarithmic form, which enables us to solve it for the unknown value of Y for any given value of X. The equation can be writ-

ten as

$$\log Y = \log B + X \log M$$

or

$$\log Y = \log 2 + X \log 1.46.$$

The same methods are used to find the slope of a curve with a negative slope. The curve of the data in Fig. 27.7 slopes downward to the right; therefore the slope is negative. Two points are selected in order to find which is the antilog of -0.0274. The intercept, 70, can be combined with the slope, M, to find the final equations, as illustrated in Step 2 of Fig. 27.7.

27.6
Applications of empirical graphs

Figure 27.8 is an example of how empirical data can be plotted to compare the transverse strength and impact resistance of gray iron. The data are somewhat scattered, but the best curve is drawn. Since the curve is a straight line on a linear graph, these data fit the equation form:

$$Y = MX + B.$$

Figure 27.9 is an example of how empirical data can be plotted to compare the specific weight (pounds per horsepower) of generators and hydraulic pumps versus horsepower. Note that the weight of these units decreases linearly as the horsepower increases. Therefore, these data can be written in the form of the power equation

$$Y = BX^M.$$

The half-life decay of radioactivity is plotted in Fig. 27.10 to show the relationship of decay to time. Since the half-life of different isotopes varies, different units would have to be assigned to time along the X-axis; however, the curve would be a straight line for all isotopes. The exponential form of the equation discussed in Section 27.5 can be applied to find the equation for these data in the form of

$$Y = BM^X.$$

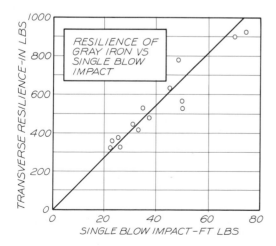

FIG. 27.8 The relationship between the transverse strength of gray iron and impact resistance results in a straight line with an equation of the form $Y = MX + B$.

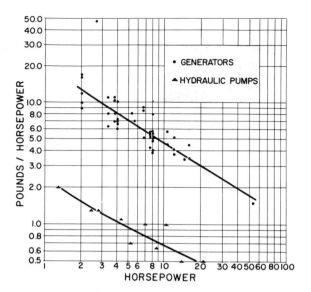

FIG. 27.9   Empirical data plotted on a logarithmic grid, showing the specific weight versus horsepower of electric generators and hydraulic pumps. The curve is the average of points plotted. (Courtesy of General Motors Corp., *Engineering Journal*.)

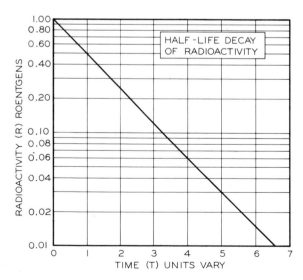

FIG. 27.10   The relative decay of radioactivity is plotted as a straight line on this semilog graph, making it possible for its equation to be found in the form $Y = BM^x$.

27.7
Introduction to graphical calculus

If the equation of the curve is known, traditional methods of calculus will solve the problem. However, many experimental data cannot be converted to standard equations. In these cases, it is desirable to use the graphical method of calculus which provides relatively accurate solutions to irregular problems.

The two basic forms of calculus are (1) **differential calculus,** and (2) **integral calculus.** Differential calculus is used to determine the rate of change of one variable with respect to another. For example, the curve plotted in Fig. 27.11A represents the relationship between two variables. Note that the Y-variable increases as the X-variable increases. The rate of change at any instant along the curve is the slope of a line that is tangent to the curve at that particular point. This exact slope is often difficult to determine graphically; consequently, it can be approximated by constructing a chord at a given interval, as shown in Fig. 27.11A. The slope of this chord can be measured by finding the tangent of $\Delta Y/\Delta X$.

This slope can represent miles per hour, weight versus length, or a number of other meaningful rates that are important to the analysis of data.

Integral calculus is the reverse of differential calculus. Integration is the process of finding the area under a given curve, which can be thought of generally

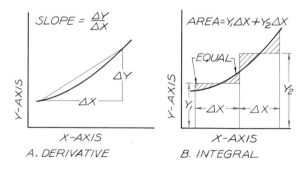

A.   The derivative of a curve is the change at any point that is the slope of the curve, *X/X*.

B.   The integral of a curve is the cumulative area enclosed by the curve, which is the summation of the products of the areas.

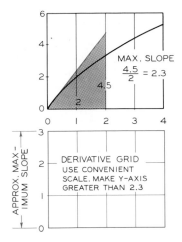 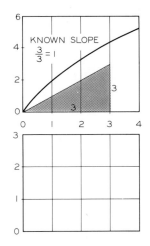 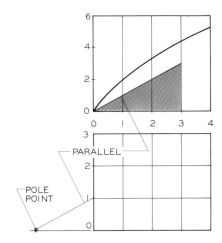

FIG. 27.12 Scales for graphical differentiation

Step 1 The maximum slope of a curve is found by constructing a line tangent to the curve where it is steepest. The maximum slope, 2.3, is found and the derivative grid is laid off with a maximum ordinate of 3.0 to accommodate the maximum value.

Step 2 A known slope is found on the given grid; this value is 1 in this example. The known slope has no relationship to the curve at this point.

Step 3 Construct a line from 1 on the Y-axis of the derivative grid that is parallel to the slope of the triangle constructed in the given grid. This line locates the pole point where it crosses the extension of the X-axis.

as the product of the two variables plotted on the x- and y-axes. The area under a curve is approximated by dividing one of the variables into a number of very small intervals, which become small rectangular areas at a particular zone under the curve, as shown in Fig. 27.11B. The bars are extended so that as much of the square end of the bar is under the curve as above the curve and the average height of the bar is, therefore, near its midpoint.

27.8
Graphical differentiation

Graphical differentiation is defined as the determination of the rate of change of two variables with respect to each other at any given point. Figure 27.12 illustrates the preliminary construction of the derivative scale.

STEP 1 The original data are plotted graphically and the axes are labeled with the proper units of measurement. A chord can be constructed to estimate the maximum slope by inspection. In the given curve, the maximum slope is estimated to be 2.3. A vertical scale is constructed in excess of this value to provide for the plotting of slopes that may exceed the estimate. The ordinate scale is drawn to a convenient scale to facilitate measurement.

STEP 2 A known slope is plotted on the given grid. This slope need not be related to the data curve in any way. In this case, the slope can be read directly as 1.

STEP 3 The pole can be found by drawing a line from the ordinate of 1 (the known slope) on the derivative scale parallel to the slope line. Similar triangles are used to obtain the pole.

The steps in completing the graphical differentiation are given in Fig. 27.13. The same horizontal intervals used in the given curve are projected directly beneath on the derivative scale. The maximum slope of the data curve is estimated to be slightly less than 12. A scale is selected that will provide an ordinate that will accommodate the maximum slope. A line is drawn from Point 12 on the ordinate axis of the derivative grid that is parallel to the known slope on the

FIG. 27.13 Graphical differentiation

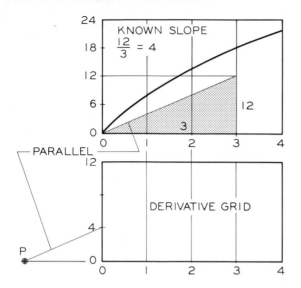

Required   Find the derivative curve of the given data.

Step 1   Find the derivative grid and the pole point using the construction illustrated in Fig. 27.12.

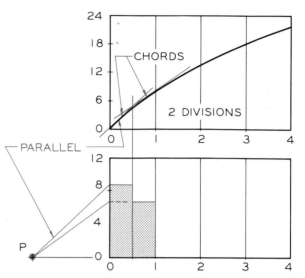

Step 2   Construct chords between intervals on the given curve and draw lines parallel to these chords through Point *P* on the derivative grid. These lines locate the heights of bars in their respective intervals. The first interval is divided in two since the curve is changing sharply in this interval.

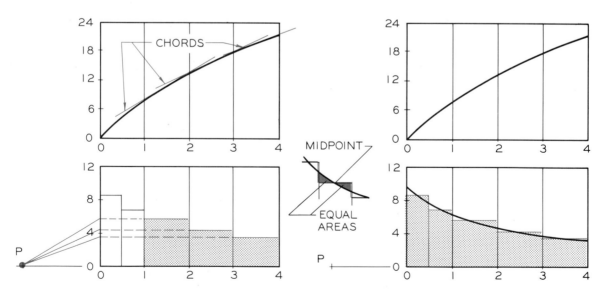

Step 3   Additional chords are drawn in the last two intervals. Lines parallel to these chords are drawn from the pole point to the *Y*-axis to find additional bars in their respective intervals.

Step 4   The vertical bars represent the different slopes of the curve at different intervals. The derivative curve is drawn through the midpoints of the bars so that the areas under and above the bars are approximately equal.

given curve grid. The point of intersection of this line and the extension of the *x*-axis is the pole point.

A series of chords is constructed on the given curve. Lines are constructed parallel to these chords through Point *P* and extended to the *y*-axis of the derivative grid to locate bars at each interval. The interval between 0 and 1 was divided in half to provide a more accurate plot. The curve is the sharpest in this interval. A smooth curve is constructed through the top of these bars in such a manner that the area above the horizontal top of the bar is the same as that below it. The rate of change, $\Delta Y/\Delta X$, can be found at any interval of the variable *X* by reading directly from the derivative graph at the value of *X* in question.

27.9
Applications of graphical differentiation

The mechanical handling shuttle shown in Fig. 27.14 is used to convert rotational motion into controlled linear motion. A scale drawing of the linkage components is given so that graphical analysis can be applied to determine the motion resulting from this system.

The linkage is drawn to show the end positions of Point *P*, which will be used as the zero point for plotting the travel versus the degrees of revolution. Since

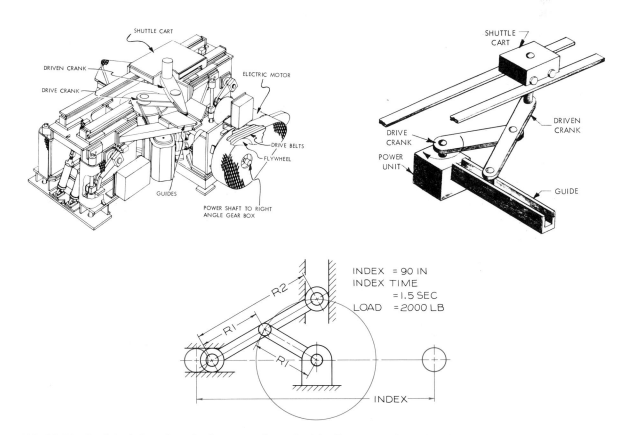

FIG. 27.14 A pictorial and scale drawing of an electrically powered mechanical handling shuttle used to move automobile parts on an assembly line. (Courtesy of General Motors Corp.)

rotation is constant at one revolution per three seconds, the degrees of revolution can be converted to time, as shown in the data curve given at the top of Fig. 27.15. The drive crank, R_1, is revolved at 30° intervals, and the distance that Point P travels from its end position is plotted on the graph to give the distance-versus-time relationship.

We determine the ordinate scale of the derivative grid by estimating the maximum slope of the data curve, which is found to be about 100 in/sec. A convenient scale is chosen that will be used for the derivative curve. A slope of 40 is drawn on the given data curve to be used in determining the location of Pole P in the derivative grid. From Point 40 on the derivative ordinate scale, we draw a line parallel to the known

slope, which is found on the given grid. Point P is the point where this line intersects the extension of the x-axis.

Chords are drawn on the given curve to approximate the slope at various points. Lines are constructed through Point P of the derivative scale parallel to the chord lines of the given curve and extended to the ordinate scale. The points thus obtained are then projected across to their respective intervals to form vertical bars. A smooth curve is drawn through the top of each of the bars to give an average of the bars. This derivative curve can be used to find the velocity of the shuttle in inches per second at any time interval.

The construction of the second derivative curve, the acceleration, is very similar to that of the first derivative. By inspecting the first derivative, we estimate the maximum slope to be 200 in/sec/sec. An easily measured scale is established for the ordinate scale of the second derivative curve. Point P is found in the same manner as the first derivative.

Chords are drawn at intervals on the first derivative curve. Lines are drawn parallel to these chords from Point P in the second derivative curve to the y-axis, where they are projected horizontally to their respective intervals to form a series of bars. A smooth curve is drawn through the tops of the bars to give a close approximation of the average areas of the bars. A minus scale is given to indicate deceleration.

The maximum acceleration is found to be at the extreme endpoints and the minimum acceleration is at 90°, where the velocity is the maximum. It can be seen from the velocity and acceleration plots that the parts being handled by the shuttle are accelerated at a rapid rate until the maximum velocity is attained at 90°, at which time deceleration begins and continues until the parts come to rest.

27.10

Graphical integration

Integration is the process of determining the area (product of two variables) under a given curve. For example, if the y-axis were pounds and the x-axis were feet, the integral curve would give the product of the variables, foot-pounds, at any interval of feet along the x-axis.

Figure 27.16 depicts the method of constructing scales for graphical integration.

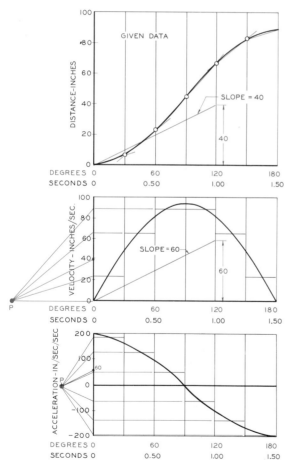

FIG. 27.15   Graphical determination of velocity and acceleration of the mechanical handling shuttle by differential calculus.

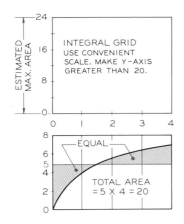
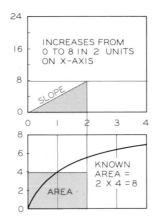
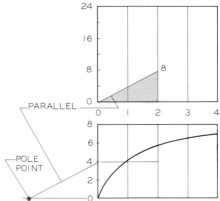

FIG. 27.16 Scales for graphical integration

Step 1 To determine the maximum value on the y-axis of the integral curve, a line is drawn to approximate the area under the given grid. This is found to be 20 and the y-axis of the grid is constructed with 24 as the maximum value.

Step 2 A known area, 8 in this case, is found in the given grid. A slope line from 0 to 8 is constructed in the integral grid directly above the known area, which establishes the integral for this model.

Step 3 A line is drawn from 4 on the y-axis of the given grid parallel to the slope line in the integral grid. This line from 4 crosses the extension of the x-axis to locate the pole point.

STEP 1 It is customary to locate the integral curve above the given data curve, since the integral will be an equation raised to a higher power. A line is drawn through the given data curve to approximate the total area under the curve, 4 × 5, or 20 square units of area. The ordinate scale is drawn on the integral curve in excess of 20 units to provide a margin for any overage. The horizontal scale intervals are projected from the given curve to the integral grid.

STEP 2 The ordinate at any point on the integral scale will represent the area under the curve as measured from the origin to that point on the given data grid. The ordinate at point 2 on the x-axis directly above the rectangle must be equal to its area of 8. A slope is drawn from the origin to the ordinate of 8.

STEP 3 Point P is found by drawing a line from Point 4 on the given grid parallel to the slope established in the integral grid. This line intersects the extension of the x-axis at Point P, the pole point.

This technique is used in Fig. 27.17 to integrate the equation of the given curve, $Y = 2X^2$.

From the given grid, the total area under the curve can be estimated to be less than 40 units, the maximum height of the y-axis on the integral curve. A convenient scale is selected for the ordinate, and the pole point, P, is found.

A series of vertical bars is constructed to approximate the areas under the curve at these intervals. The narrower the bars, the more accurate will be the resulting calculations. The interval between 1 and 2 was divided in half to provide a more accurate plot. The top lines of the bars are extended horizontally to the y-axis, where the points are then connected by lines to point P. Lines are drawn parallel to AP, BP, CP, DP, and EP in the integral grid to correspond to the respective intervals in the given grid. The intersection points of the chords are connected by a smooth curve—the integral curve to give the cumulative product of the X- and Y-variables along the x-axis. For example, the area under the curve at $X = 3$ can be read directly as 18.

Mathematical integration gives the following result for the area under the curve from 0 to 3:

$$\text{Area A} = \int_0^3 Y\, dX, \qquad \text{where } Y = 2X^2;$$

$$A = \int_0^3 2X^2\, dX = \frac{2}{3}X^3 \Big|_0^3 = 18.$$

FIG. 27.17 Graphical integration

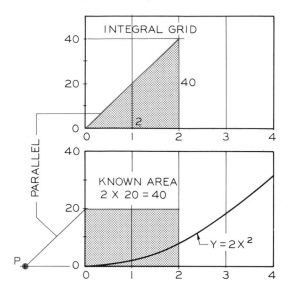

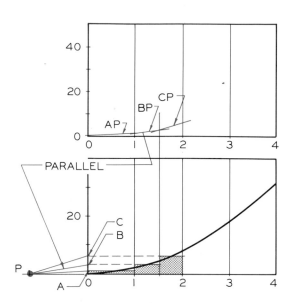

Required   Plot the integral curve of the given data.

Step 1   Find the pole point, *P*, using the technique illustrated in Fig. 27.16.

Step 2   Construct bars to approximate the areas under the curve. The interval from 1 to 2 was divided in half to improve the accuracy of the approximation. The heights of the bars are projected to the *Y*-axis and lines are drawn to the pole point. Sloping Lines *AP*, *BP*, and *CP* are drawn in their respective intervals parallel to the lines drawn to the pole, *P*.

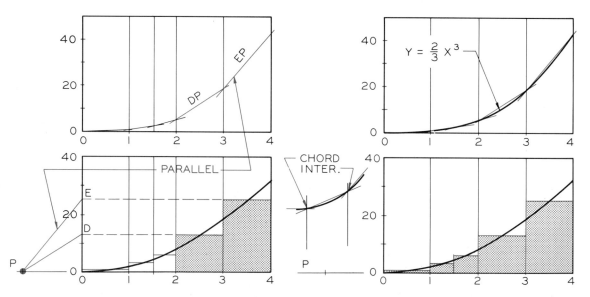

Step 3   Additional bars are drawn from 2 to 4 on the *X*-axis. The heights of the bars are projected to the *Y*-axis and rays are drawn to the pole point, *P*. Lines *DP* and *DE* are drawn in their respective intervals and parallel to their rays in the integral grid.

Step 4   The straight lines that were connected in the integral grid represent chords of the integral curve. Construct the integral curve to pass through the points where the chords intersect. Any ordinate value on the integral curve represents the cumulative area under the given curve from zero to that point on the *X*-axis.

27.11

Applications of graphical integration

Integration is commonly used in the study of the strength of materials to determine shear, moments, and deflections of beams. An example problem of this type is shown in Fig. 27.18, in which a truck exerts a total force of 36,000 lb on a beam of a bridge. The first step is to determine the resultants supporting each end of the beam.

A scale drawing of the beam is made with the loads concentrated at their respective positions. A force diagram is drawn, using Bow's notational system for laying out the vectors in sequence. Pole Point O is located and rays are drawn from the ends of each vector to Point O. The lines of force in the load diagram at the top of the figure are extended to the funicular diagram. Then lines are drawn parallel to the rays between the corresponding lines of force. For example, Ray OA is drawn in the A-interval in the funicular diagram. The closing ray of the funicular diagram, OE, is transferred to the vector diagram by drawing a parallel through Point O to locate Point E. Vector DE is the right-end resultant of 20.1 kips (one kip equals 1000 lb), and EA is the left-hand resultant of 15.9 kips. The

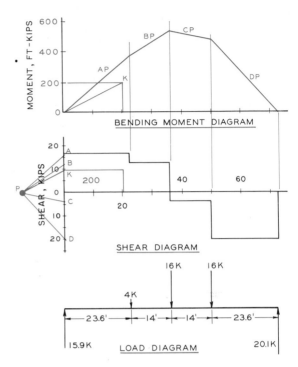

FIG. 27.19 Determination of shear and bending moment by graphical integration

origin of the resultant force of 36 kips is found by extending OA and OD in the funicular diagram to their point of intersection.

From the load diagram shown in Fig. 27.19 we can, by integration, find the shear diagram, which indicates the points in the beam where failure due to crushing is most critical. Since the applied loads are concentrated rather than uniformly applied, the shear diagram will be composed of straight-line segments. In the shear diagram, the left-end resultant of 15.9 kips is drawn to scale from the axis. The first load of 4 kips, acting in a downward direction, is subtracted from this value directly over its point of application, projected from the load diagram. The second load of 16 kips also exerts a downward force and so is subtracted from the 11.9 kips (15.9 − 4). The third load of 16 kips is also subtracted, and the right-end resultant will bring the shear diagram back to the x-axis. The beam must be designed to withstand the maximum shear at each support and minimum shear at the center.

The moment diagram is used to evaluate the bending characteristics of the applied loads in foot-pounds at any interval along the beam. The ordinate

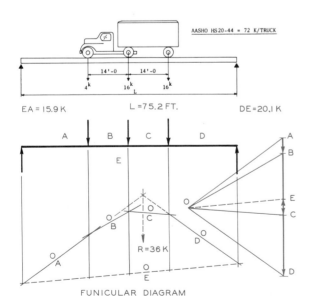

FIG. 27.18 Determination of the forces on a beam of a bridge and its total resultant.

of any X-value in the moment diagram must represent the cumulative foot-pounds in the shear diagram as measured from either end of the beam.

Pole Point P is located in the shear diagram by applying the method described in Fig. 27.18. A rectangular area of 200 ft-kips is found in the shear diagram. We estimate the total area to be less than 600 ft-kips, so we select a convenient scale that will allow an ordinate scale of 600 units for the moment diagram. We draw a known area of 200 (10 × 20) on the shear diagram. A diagonal line in the moment diagram is drawn that slopes upward from 0 to 200, where $X =$ 20. The diagonal, OK, is transferred to the shear dia-

gram, where it is drawn from the ordinate of the given rectangle to Point P on the extension of the x-axis. Rays AP, BP, CP, and DP are found in the shear diagram by projecting horizontally from the various values of shear. In the moment diagram, these rays are then drawn in their respective intervals to form a straight-line curve that represents the cumulative area of the shear diagram, which is in units of ft-kips. Maximum bending will occur at the center of the beam, where the shear is zero. The bending is scaled to be about 560 ft-kips. The beam selected for this span must be capable of withstanding a shear of 20.1 kips and a bending moment of 560 ft-kips.

Problems

The following problems are to be solved on Size A sheets ($8\frac{1}{2}'' \times 11''$). Solutions involving mathematical calculations should show these calculations on separate sheets if space is not available on the sheet where the graphical solution is drawn.

Empirical equations—logarithmic

1. Find the equation of the data shown in the following table. The empirical data compare input voltage (V) with input current in amperes (I) to a heat pump.

y-axis	V	0.8	1.3	1.75	1.85
x-axis	I	20	30	40	45

2. Find the equation of the data in the following table. The empirical data give the relationship between peak allowable current in amperes (I) with the overload operating time in cycles at 60 cycles per second (C). Place I on y-axis.

y-axis	I	2000	1840	1640	1480	1300	1200	1000
x-axis	C	1	2	5	10	20	50	100

3. Find the equation of the data in the following table. The empirical data for a low-voltage circuit breaker used on a welding machine give the maximum loading during weld in amperes (rms) for the percent of duty (pdc). Place rms along the y-axis and pdc along the x-axis.

y-axis	rms	7500	5200	4400	3400	2300	1700
x-axis	pdc	3	6	9	15	30	60

4. Construct a three-cycle × three-cycle logarithmic graph to find the equation of a machine's vibration during operation. Plot vibration displacement in mills along the y-axis and vibration frequency in cycles per minute (cpm) along the x-axis. Data: 100 cpm, 0.80 mills; 400 cpm, 0.22 mills; 1000

cpm, 0.09 mills; 10,000 cpm, 0.009 mills; 50,000 cpm, 0.0017 mills.

5. Find the equation of the data in the following table that compares the velocities of air moving over a plane surface in feet per second *(v)* at different heights in inches *(y)* above the surface. Plot *y*-values on the *y*-axis.

y	v
0.1	18.8
0.2	21.0
0.3	22.6
0.4	24.1
0.6	26.0
0.8	27.3
1.2	29.2
1.6	30.6
2.4	32.4
3.2	33.7

6. Find the equation of the data in the following table that shows the distance traveled in feet *(s)* at various times in seconds *(t)* of a test vehicle. Plot *s* on the *y*-axis and *t* on the *x*-axis.

t	s
1	15.8
2	63.3
3	146.0
4	264.0
5	420.0
6	580.0

Empirical equations—linear

7. Construct a linear graph to determine the equation for the yearly cost of a compressor in relationship to the compressor's size in horsepower. The yearly cost should be plotted on the *y*-axis and the compressor's size in horsepower on the *x*-axis. Data: 0 hp, $0; 50 hp, $2100; 100 hp, $4500; 150 hp, $6700; 200 hp, $9000; 250 hp, $11,400. What is the equation of these data?

8. Construct a linear graph to determine the equation for the cost of soil investigation by boring to determine the proper foundation design for vary-

ing sizes of buildings. Plot the cost of borings in dollars along the *y*-axis and the building area in sq ft along the *x*-axis. Data: 0 sq ft, $0; 25,000 sq ft, $35,000; 50,000 sq ft, $70,000; 750,000 sq ft, $100,000; 1,000,000 sq ft, $130,000.

9. Find the equation of the empirical data plotted in Fig. 27.8.

10. Plot the data in the table below on a linear graph and determine its equation. The empirical data show the deflection in centimeters of a spring *(d)* when it is loaded with different weights in kilograms *(W)*. Plot *W* along the *x*-axis and *d* along the *y*-axis

w	d
0	0.45
1	1.10
2	1.45
3	2.03
4	2.38
5	3.09

11. Plot the data in the table below on a linear graph and determine its equation. The empirical data show the temperatures that are read from a Fahrenheit thermometer at B and a centigrade thermometer at A. Plot the *A*-values along the *x*-axis and the *B*-values along the *y*-axis.

°A	°B
−6.8	20.0
6.0	43.0
16.0	60.8
32.2	90.0
52.0	125.8
76.0	169.0

Empirical equations— semilogarithmic

12. Construct a semilog graph of the following data to determine their equation. The *y*-axis should be a two-cycle log scale and the *x*-axis a 10-unit lin-

ear scale. Plot the voltage (E) along the y-axis and time (T) in sixteenths of a second along the x-axis to represent resistor voltage during capacitor charging. Data: 0, 10 volts; 2, 6 volts; 4, 3.6 volts; 6, 2.2 volts; 8, 1.4 volts; 10, 0.8 volts.

13. Find the equation of the data that is plotted in Fig. 27.10.

14. Construct a semilog graph of the following data to determine their equation. The y-axis should be a three-cycle log scale and the x-axis a linear scale from 0 to 250. These data give a comparison of the reduction factor, R (y-axis), with the mass thickness per square foot (x-axis) of a nuclear protection barrier. Data: 0, 1.0R; 100, 0.9R; 150, 0.028R; 200, 0.009R; 300, 0.0011R.

15. An engineering firm is considering its expansion by reviewing its past incomes that are shown in the table below. Their years of operation are represented by x, and N is their annual income in millions. Plot x along the x-axis and N along the y-axis and determine the equation of their progress.

x	N
1	0.05
2	0.08
3	0.12
4	0.20
5	0.32
6	0.51
7	0.80
8	1.30
9	2.05
10	3.25

Empirical equations—general types

16–21. Plot the experimental data shown in Table 27.1 on the grid where the data will appear as straight-line curves. Determine the equations of the data.

Calculus—differentiation

22. Plot the equation $Y = X^3/6$ as a rectangular graph. Graphically differentiate the curve to determine the first and second derivatives.

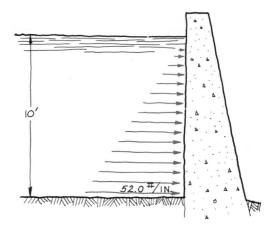

FIG. 27.20 Pressure on a 12 in.-wide section of a dam. (Problem 30.)

23. Plot the following data on a graph and find the derivative curve of the data on a graph placed below the first: $Y = 2X^2$.

24. Plot the following equation on a graph and find the derivative curve of the data on a graph placed below the first: $4Y = 8 - X^2$.

25. Plot the following data on a graph and find the derivative curve of the data on a graph placed below the first: $3Y = X^2 + 16$.

26. Plot the following data on a graph and find the derivative curve of the data on a graph placed below the first: $X = 3Y^2 - 5$.

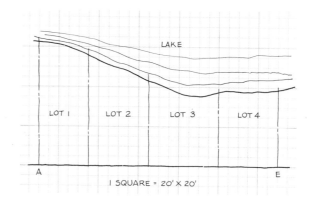

FIG. 27.21 Plot plan of a tract bounded by a lake front. (Problem 31.)

TABLE 27.1

A	x	0	40	80	120	160	200	240	280		
	y	4.0	7.0	9.8	12.5	15.3	17.2	21.0	24.0		
B	x	1	2	5	10	20	50	100	200	500	1000
	y	1.5	2.4	3.3	6.0	9.0	15.0	23.0	24.0	60	85
C	x	1	5	10	50	100	500	1000			
	y	3	10	19	70	110	400	700			
D	x	2	4	6	8	10	12	14			
	y	6.5	14.0	32.0	75.0	115.0	320	710			
E	x	0	2	4	6	8	10	12	14		
	y	20	34	53	96	115	270	430	730		
F	x	0	1	2	3	4	5	6	7	8	9	10
	y	1.8	2.1	2.2	2.5	2.7	3.0	3.4	3.7	4.1	4.5	5.0

Calculus—integration

27. Plot the following equation on a graph and find the integral curve of the data on a graph placed above the first: $Y = X^2$.

28. Plot the following equation on a graph and find the integral curve of the data on a graph placed above the first: $Y = 9 - X^2$.

29. Plot the following equation on a graph and find the integral curve of the data on a graph placed above the first: $Y = X$.

30. Using graphical calculus, analyze a vertical strip 12 in. wide on the inside face of the dam in Fig. 27.20. The force on this strip will be 52.0 lb/in. at the bottom of the dam. The first graph will be pounds per inch (ordinate) versus height in inches (abscissa). The second graph will be the integral of the first to give shear in pounds (ordinate) versus height in inches (abscissa). The third will be the integral of the second graph to give the moment in inch-pounds (ordinate) versus height in inches (abscissa). Convert these scales to give feet instead of inches.

31. A plot plan shows that a tract of land is bounded by a lake front (Fig. 27.21). By graphical integration, determine a graph that will represent the cumulative area of the land from Point A to E. What is the total area? What is the area of each lot?

CHAPTER

28

Pipe Drafting

28.1
Introduction

An understanding of pipe drafting begins with a familiarity with the types of pipe that are available. The commonly used types of pipe are (1) steel pipe, (2) cast-iron pipe, (3) copper, brass, and bronze pipe and tubing, and (4) plastic pipe.

The standards for the grades and weights for pipe and pipe fittings are specified by several organizations to ensure the uniformity of size and strength of interchangeable components. Several of these organizations are the American National Standards Institute (ANSI), the American Society for Test Materials (ASTM), the American Petroleum Institute (API), and the Manufacturers Standardization Society (MSS).

28.2
Welded and seamless steel pipe

Traditionally, steel pipe has been specified in three **weights standard** (STD), **extra strong** (XS), and **double extra strong** (XXS). These designations and

their specifications are listed in the ANSI B 36.10–1979 standards. However, additional designations for pipe, called **schedules,** have been introduced to provide the pipe designer with a wider selection of pipe to cover more applications.

The ten schedules are: Schedule 10, Schedule 20, Schedule 30, Schedule 40, Schedule 60, Schedule 80, Schedule 100, Schedule 120, Schedule 140, and Schedule 160. The wall thicknesses of the pipes vary from the thinnest, in Schedule 10, to the thickest, in Schedule 160. The outside diameters are of a constant size for pipes of the same nominal size in all schedules.

Schedule designations correspond to STD, XS, and XXS specifications in some cases, as is shown partially in Table 28.1. This table has been abbreviated from the ANSI B 36.10–1979 tables by omitting a number of the pipe sizes and schedules. The most often used schedules are 40, 80, and 120.

Pipes from the smallest size up to and including 12-in. pipes are specified by their inside diameter (ID), which means that the outside diameter (OD) is larger than the specified size. The inside diameters are the same size as the nominal sizes of the pipe for STD weight pipe. For XS and XXS pipe, the inside diameters are slightly different in size from the nominal size. Beginning with the 14-in. diameter pipes, the nominal sizes represent the outside diameters of the pipe.

The standard lengths for steel pipe are 20 ft and 40 ft. Seamless steel (SMLS STL) pipe is a smooth pipe with no weld seams along its length. Welded pipe is formed into a cylinder and is butt-welded (BW) at the seam, or it is joined with an electric resistance weld (ERW).

28.3
Cast-iron pipe

Cast-iron pipe is used for the transportation of liquids, water, gas, and sewerage. When used as a sewerage pipe, cast-iron pipe is referred to as "soil pipe." Cast-iron pipe is available in diameter sizes from 3 in. to 60 in.

The standard lengths of cast-iron pipe are 5 ft and 10 ft. Cast iron is more brittle and more subject to cracking when it is loaded than is steel pipe. Therefore, it should not be used where high pressures or weights will be applied to it.

28.4
Copper, brass, and bronze piping

Copper, brass, and bronze are used to manufacture piping and tubing for use in applications where there must be a high resistance to corrosive elements, such

TABLE 28.1
DIMENSIONS AND WEIGHTS OF WELDED AND SEAMLESS STEEL PIPE (ANSI B 36.10–1979)

Inch Units				Identification			SI Units		
Inch Nominal Size	O.D. (in.)	Wall Thk. (in.)	Weight lbs/ft	*STD XS XXS	Sch. no.		O.D. (mm)	Wall Thk. (mm)	Weight kg/m
½	0.84	0.11	0.85	STD	40		21.3	2.8	1.3
1	1.32	0.13	1.68	STD	40		33.4	3.4	2.5
1	1.3	0.18	2.17	XS	80		33.4	4.6	3.2
1	1.3	0.36	3.66	XXS			33.4	9.1	5.5
2	2.38	0.22	3.65	STD	40		60.3	3.9	5.4
2	2.38	0.22	5.02	XS	80		60.3	5.5	7.5
2	2.38	0.44	9.03	XXS			60.3	11.1	13.4
4	4.50	0.23	10.79	STD	40		114.3	6.0	16.1
4	4.50	0.34	14.98	XS	80		114.3	8.6	42.6
4	4.50	0.67	27.54	XXS			114.3	17.1	41.0
8	8.63	0.32	28.55	STD	40		219.1	8.2	42.6
8	8.63	0.50	43.39	XS	80		219.1	12.7	64.6
8	8.63	0.88	74.40	XXS			219.1	22.2	107.9
12	12.75	0.38	49.56	STD			323.0	9.5	67.9
12	12.75	0.50	65.42	XS			323.0	12.7	97.5
12	12.75	1.00	125.4	XXS	120		133.9	25.4	187.0
14	†14.00	0.38	54.57	STD	30		355.6	9.5	87.3
14	14.00	0.50	72.08	XS			355.6	12.7	107.4
18	18.00	0.38	70.59	STD			457	9.5	106.2
18	18.00	0.50	93.45	XS			457	12.7	139.2
24	24.00	0.38	94.62	STD	20		610	9.5	141.1
24	24.00	0.50	125.49	XS			610	12.7	187.1
30	30.00	0.38	118.65	STD			762	9.5	176.8
30	30.00	0.50	157.53	XS	20		762	12.7	234.7
40	40.00	0.38	158.70	STD			1016	9.5	236.5
40	40.00	0.50	210.90	XS			1016	12.7	314.2

*Standard (STD)
X-strong (XS)
XX-strong (XXS)

†Beginning with 14-in. DIA pipe, the nominal size represents the outside diameter (O.D.)

This table has been compressed by omitting many of the available pipe sizes. The nominal sizes of pipes that are listed in the complete table are: ⅛", ¼", ⅜", ½", ¾", 1", 1¼", 1½", 2", 2½", 3", 3½", 4", 5", 6", 8", 10", 12", 14", 16", 18", 20", 22", . . . (at 2" increments up to 60").

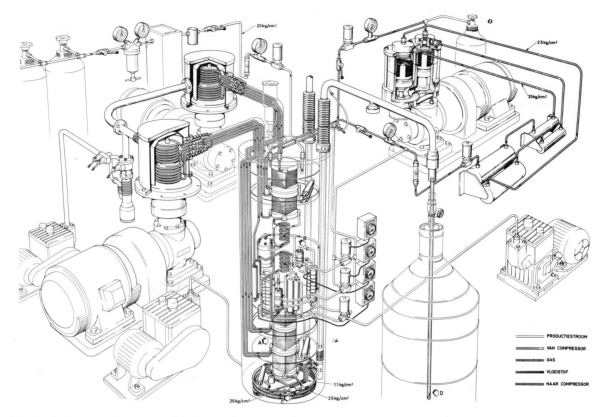

FIG. 28.1   The small pipes in this illustration are called tubes because they are less than 2 in. in diameter and can be bent to form angles.

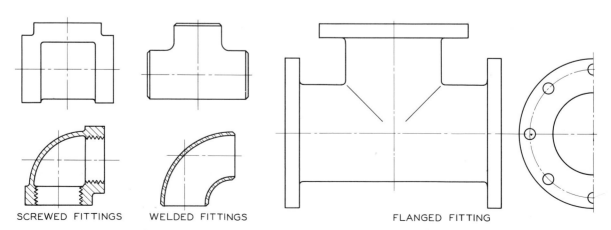

FIG. 28.2   The three basic types of joints are screwed, welded, and flanged joints.

as acidic soils and chemicals that are transmitted through the pipes. Copper pipe is used when the pipes are placed within or under concrete slab foundations of buildings. This ensures that the pipes will not have to be replaced because of corrosion. The standard length of pipes made of these nonferrous materials is 12 ft.

Tubing is a smaller-size pipe that can be easily bent when it is made of copper, brass, or bronze. An example of a system of tubing is shown in Fig. 28.1. The term piping is considered to apply to rigid pipes that are larger than tubes, usually in excess of 2 in. in diameter.

28.5
Miscellaneous pipes

Other materials that are used to manufacture pipes are aluminum, asbestos-cement, concrete, polyvinyl chloride (PVC), and various other plastics. Each of these materials has its special characteristics that make it desirable or economical for certain applications. The method of designing and detailing piping systems by the pipe drafter is essentially the same regardless of the piping material used.

28.6
Pipe joints

The basic connection in a pipe system is the joint where two straight sections of pipe fit together. Three types of joints are illustrated in Fig. 28.2: **screwed, welded,** and **flanged.**

SCREWED JOINTS are joined by pipe threads of the type covered in Chapter 9 and Appendix 10. Pipe threads are tapered at a ratio of 1 to 16 along the outside diameter (Fig. 28.3). As the pipes are screwed together, the threads bind to form a snug, locking fit. A cementing compound is applied to the threads before joining, to improve the seal.

FLANGED JOINTS, shown in Fig. 28.4, are welded to the straight sections of pipe, which are then bolted together around the perimeter of the flanges.

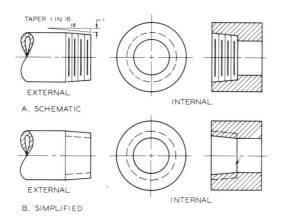

FIG. 28.3 A screwed pipe utilizes a pipe thread that binds tightly as they are screwed together.

WELDING NECK FLANGES LAP JOINT FLANGES SOCKET WELDING FLANGES SLIP-ON WELDING FLANGES THREADED FLANGES

FIG. 28.4 Types of flanged joints and the methods of attaching the flanges to the pipes. (Courtesy of Vogt Machine Co.)

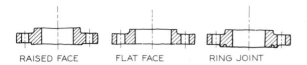

FIG. 28.5 Three types of flange faces are the raised face (RF), flat face (FF), and ring joint (RJ).

Flanged joints form strong rigid joints that can withstand high pressure and permit disassembly of the joints when needed. Several types of flange faces are shown in Fig. 28.5 and in Appendix 13.

WELDED JOINTS are joined by welded seams around the perimeter of the pipe to form butt welds. Welded joints are used extensively in ''big inch'' pipelines that are used for transporting petroleum products cross-country.

BELL AND SPIGOT (B&S) joints are used to join cast-iron pipes, (Fig. 28.6). The spigot is placed inside the bell and the two are sealed with molten lead or a sealing ring that snaps into position to form a sealed joint.

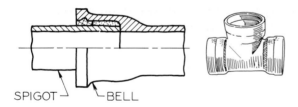

FIG. 28.6 A bell and spigot joint (B&S) is used to connect cast-iron pipes.

SOLDERING is used to connect smaller pipes and tubular connections. Soldering is usually limited to nonferrous tubing. However, screwed fittings are available to connect tubing, as shown in Fig. 28.7.

28.7
Screwed fittings

A number of standard fittings used to are shown in Fig. 28.8. The two types of graphical symbols that are used to represent fittings and pipe are **double-line**

symbols and **single-line symbols.**

Double-line symbols are more representative of the fittings and pipes since they are drawn to scale with double lines. Single-line symbols are more symbolic since the size of the pipe is drawn with a single line and the fittings are drawn as single-line schematic symbols.

Fittings are available in three weights: standard (STD), extra strong (XS), and double extra strong (XXS). These weights match the standard weights of the pipes with which they will be connnected. Other weights of fittings are available, but these three weights are stocked by practically all suppliers.

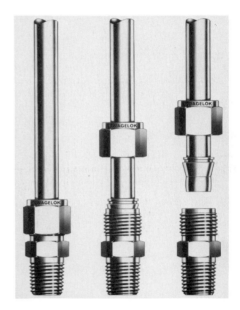

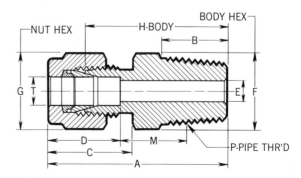

FIG. 28.7 This fitting is used to attach small tubing. (Courtesy of Crawford Fitting Co.)

REDUCER **HALF COUPLING** **PIPE CAP** **SQUARE HEAD PLUG** **HEX. HD. PLUG** **ROUND HEAD PLUG** **HEXAGON BUSHING** **FLUSH BUSHING**

DOUBLE-LINE SYMBOLS: SCREWED

SINGLE-LINE SYMBOLS: SCREWED

4 X 2 RED   3 X 3 HLF CPLG   1- CAP   2-SQ HD PLUG   3-HEX HD PLUG   4-RD HD PLUG   3 X 2 HEX BUSH.   2 X 1 FLUSH BUSH.

DESIGNATIONS

90° ELBOW   **TEE**   **45° ELBOW**   **CROSS**   **STREET ELBOW**   **LATERAL**   **COUPLING**

DOUBLE-LINE SYMBOLS: SCREWED

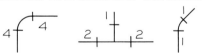  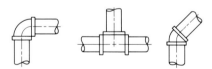 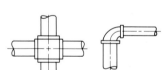 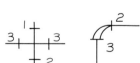 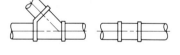 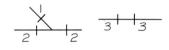

SINGLE-LINE SYMBOLS: SCREWED

4 X 90° ELL   2 X 2 X 1 TEE   1 X 45° ELL   3 X 3 X 2 X 1 CROSS   3 X 2 RED ELL   2 X 2 X 1 LAT   3 COUPL
DESIGNATIONS

FIG. 28.8 Examples of standard fittings for screwed connections along with the single-line and double-line symbols that are used to represent them. Nominal pipe sizes can be indicated by numbers placed near the joints. The major flow direction is labeled first, with the branches labeled second. The large openings are labeled to precede the smaller openings.

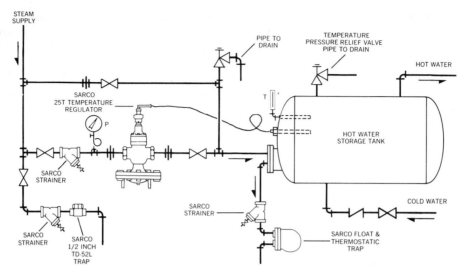

FIG. 28.9 This is a single-line piping system with the major valves represented as double-line symbols. (Courtesy of Sarco, Inc.)

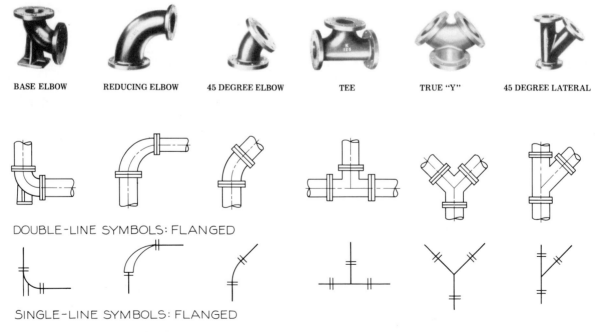

FIG. 28.10 Examples of standard flanged fittings along with the single-line and double-line symbols that are used to represent them.

A piping system of screwed fittings is shown in Fig. 28.9 with double-line symbols in a single-line system to call attention to them. These could just as well have been drawn using the single-line symbols.

The most common symbols for representing fittings are shown in Table 28.2 which have been extracted from ANSI Z 32.2.3 standards.

radii. The long-radius (LR) ells have radii that are approximately 1.5 times the nominal diameter of the large end of the ell. The radius of a short-radius ell is equal to the diameter of the larger end.

A table of dimensions for 125 LB and 250 LB cast-iron fittings is given in Appendixes 11–13.

28.8
Flanged fittings

Flanges are used to connect fittings into a piping system when heavy loads are supported in large pipes and where pressures are great. The use of flanges is expensive and their usage should be kept to a minimum if other joining methods can be used. Flanges are welded to straight pipe sections so that they can be bolted together.

Examples of several fittings are drawn as double-line and single-line symbols in Fig. 28.10. The elbow is commonly referred to as an "ell" and it is available in angles of turn of 90° and 45° in both long and short

28.9
Welded fittings

Welding is a method of joining pipes and fittings for permanent, pressure-resistant joints. Examples of double-line and single-line fittings connected by welding are shown in Fig. 28.11. Fittings are available with beveled edges that are prepared for welding.

A piping layout in Fig. 28.12 illustrates a series of welded joints with a double-line drawing. The location of the welded joints has been dimensioned. Several flanged fittings have been welded into the system in order for the flanges to be used. A comparison of flanged and welded joints is shown in Fig. 28.13.

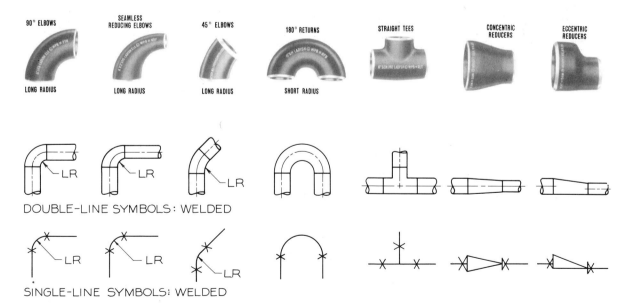

FIG. 28.11 Examples of standard fittings, welded along with the single-line and double-line symbols that are used to represent them.

TABLE 28.2

PIPING SYMBOLS

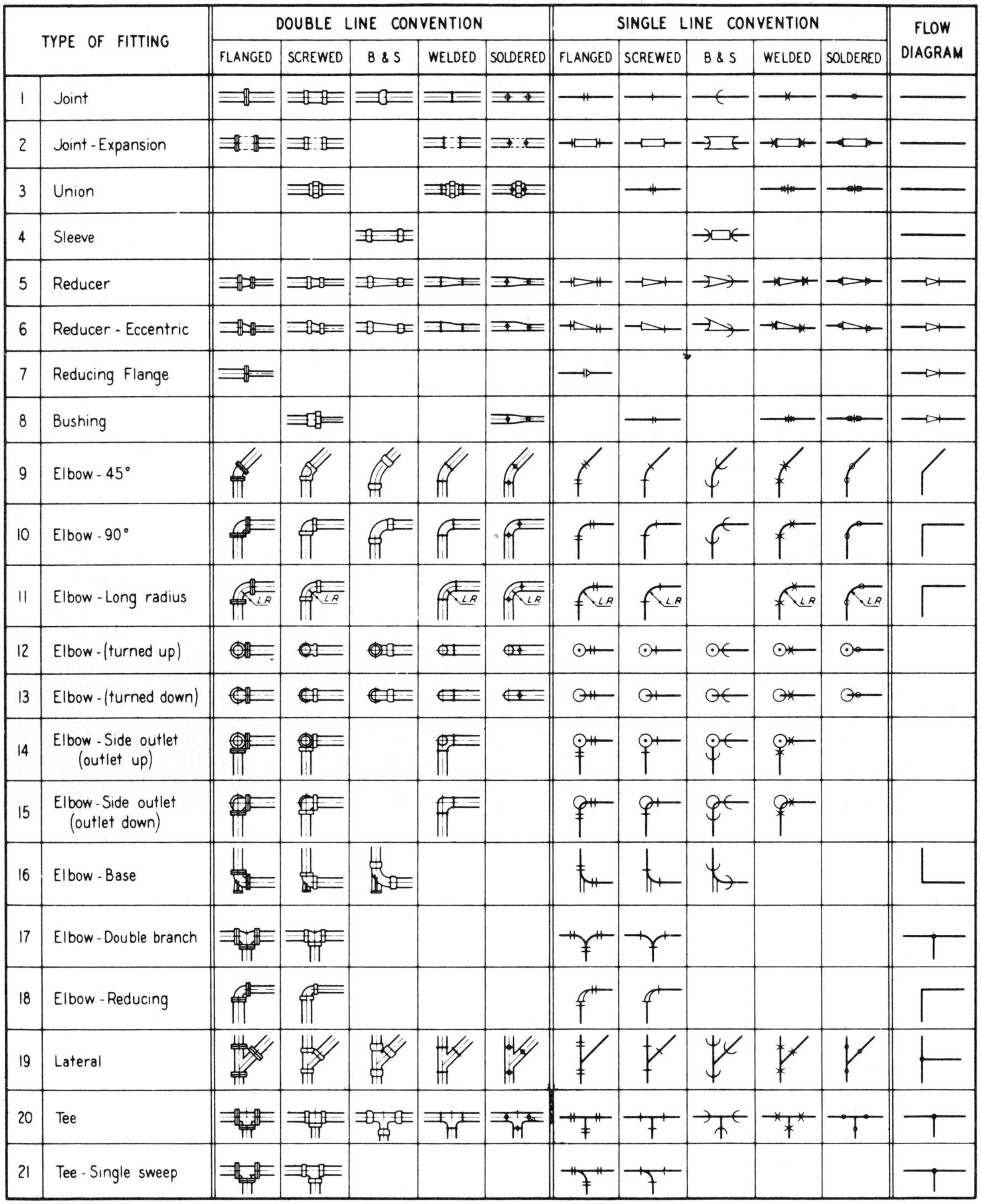

TYPE OF FITTING		DOUBLE LINE CONVENTION					SINGLE LINE CONVENTION					FLOW DIAGRAM
		FLANGED	SCREWED	B & S	WELDED	SOLDERED	FLANGED	SCREWED	B & S	WELDED	SOLDERED	
1	Joint											
2	Joint - Expansion											
3	Union											
4	Sleeve											
5	Reducer											
6	Reducer - Eccentric											
7	Reducing Flange											
8	Bushing											
9	Elbow - 45°											
10	Elbow - 90°											
11	Elbow - Long radius											
12	Elbow - (turned up)											
13	Elbow - (turned down)											
14	Elbow - Side outlet (outlet up)											
15	Elbow - Side outlet (outlet down)											
16	Elbow - Base											
17	Elbow - Double branch											
18	Elbow - Reducing											
19	Lateral											
20	Tee											
21	Tee - Single sweep											

Cont.

	TYPE OF FITTING	DOUBLE LINE CONVENTION					SINGLE LINE CONVENTION					FLOW DIAGRAM
		FLANGED	SCREWED	B & S	WELDED	SOLDERED	FLANGED	SCREWED	B & S	WELDED	SOLDERED	
22	Tee-Double sweep											
23	Tee-(outlet up)											
24	Tee-(outlet down)											
25	Tee-Side outlet (outlet up)											
26	Tee-Side outlet (outlet down)											
27	Cross											
28	Valve-Globe											
29	Valve-Angle											
30	Valve-Motor operated globe											
31	Valve-Gate											
32	Valve-Angle gate											
33	Valve-Motor operated gate											
34	Valve-Check											
35	Valve-Angle check											
36	Valve-Safety											
37	Valve-Angle safety											
38	Valve-Quick opening											
39	Valve-Float operating											
40	Stop Cock											

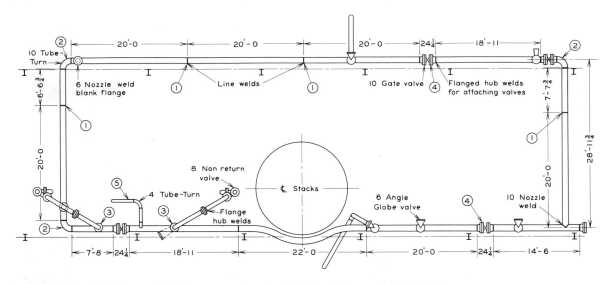

FIG. 28.12 A piping layout that uses a series of welded and flanged joints. This system is drawn using double-line symbols.

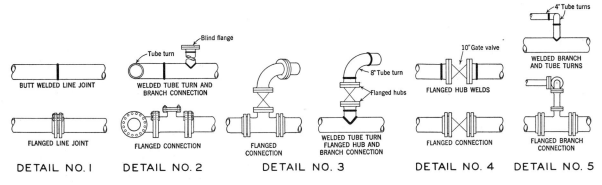

FIG. 28.13 A comparison of flanged and welded joints drawn using double-line symbols.

28.10
Valves

Valves are used to regulate the flow within a pipeline or to turn off the flow completely. Types of valves are gate, globe, angle, check, safety, diaphragm, float, and relief valves, to name a few. The three basic types—**gate, globe,** and **check valves**—are shown in Fig. 28.14 using single-line symbols.

GATE VALVES are used to turn the flow within a pipe on or off with the least restriction of flow through the valve. These valves are not meant to be used to regulate the degree of flow.

GLOBE VALVES are not only used to turn the flow on and off, but they are also used to regulate the flow to a desired level.

ANGLE VALVES are types of globe valves that turn at 90° angles at bends in the piping system. They

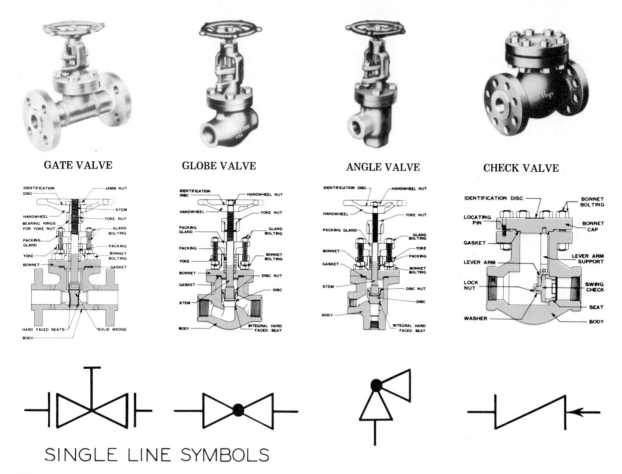

FIG. 28.14 The three basic types of valves are gate, globe, and check valves. (Photographs courtesy of Vogt Machine Co.)

have the same controlling features as the straight globe valves.

CHECK VALVES restrict the flow in the pipe to only one direction. A backward flow is checked by either a movable piston or a swinging washer that will be activated by a change of flow (Fig. 28.14).

The symbols for the other types of valves can be found in Fig. 28.15. Examples of double-line symbols depicting valves and fittings are shown in Fig. 28.16.

28.11
Fittings in orthographic views

Fittings and valves must be shown from any view in orthographic projection. Two and three views of typical fittings and valves are shown in Fig. 28.17. These fittings are drawn as single-line screwed fittings, but the same general principles can be used to represent other types of joints as double-line drawings. Several

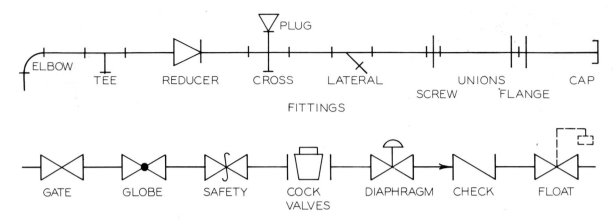

FIG. 28.15 Examples of types of valves and fittings.

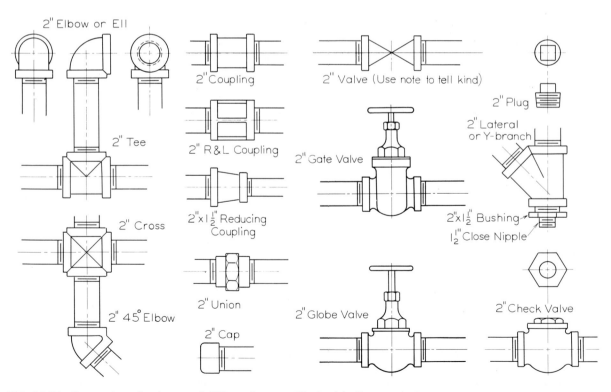

FIG. 28.16 Examples of valves and fittings drawn with double-line symbols.

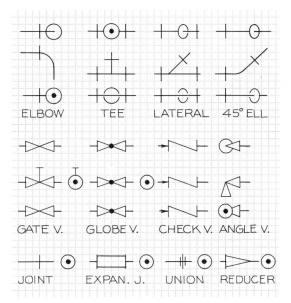

FIG. 28.17 Fittings and valves drawn with single-line symbols are drawn differently in the various orthographic views. Observe these differences and how the various views are drawn.

views of fittings are given using double-line symbols in Fig. 28.16.

Observe the various views of the fittings and notice how the direction of an elbow can be shown by a slight variation in the different views. The same techniques are used to represent tees and laterals.

A piping system is shown in a single orthographic view in Fig. 28.18, where a combination of double-line and single-line symbols are drawn. Note that arrows are used to give the direction of flow in the system. Different joints are screwed, welded, and flanged. Horizontal elevation lines are given to dimension the heights of each horizontal pipe. Station 5 + 12 − 0-¼″ represents a distance of 500 feet plus 12′ − 0-¼″, or 512′ − 0-¼″ from the beginning station point of 0 + 00.

The dimensions in Fig. 28.18 are measured from the centerlines of the pipes; this is indicated by the CL symbols. In some cases, the elevations of the pipes are dimensioned to the bottom of the pipe, which is abbreviated as BOP (Fig. 28.23).

28.12
Piping systems in pictorial

Isometric drawings of piping systems are very helpful in the representation of three-dimensional installations that would be difficult to interpret if drawn in orthographic projection. Isometric and axonometric drawings of piping systems are called **"spool drawings."** They can be drawn using either single-line or double-line symbols.

A three-dimensional piping system is drawn orthographically in Fig. 28.19 with top and front views. Although this is a relatively simple three-dimensional system, a thorough understanding of orthographic projection is required to read the drawing.

In Fig. 28.20, the piping system is drawn with all of the pipes revolved into the same horizontal plane. You will notice that the vertical pipes and their fittings are drawn true size in the top view. This is called a **developed** pipe drawing. The fittings and pipe sizes are noted on this preliminary sketch from which the finished drawing will be made in Fig. 28.21.

An axonometric schematic is drawn in Fig. 28.22 to explain the three-dimensional relationship of the parts of the system. The rounded bends in the elbows in an isometric drawing can be constructed with ellipses using the isometric ellipse template, or the corners can be drawn square to reduce the effort and time required.

A north arrow is drawn on the plan view of the piping system in Fig. 28.19; this can be used to orient the isometric pictorial. This north direction is not necessarily related to compass north, but it is a direction that is selected parallel to a major set of pipes within the system. In the isometric drawing, it is preferred that the north arrow point to the upper left or upper right corner of the pictorial.

28.13
Dimensioned isometrics

An isometric drawing can be drawn as a fully dimensioned and specified drawing from which a piping system can be constructed. The spool drawing in Fig. 28.23 is an example where the specifications for the pipe, fittings, flanges, and valves are noted on the drawing and are itemized in the bill of materials.

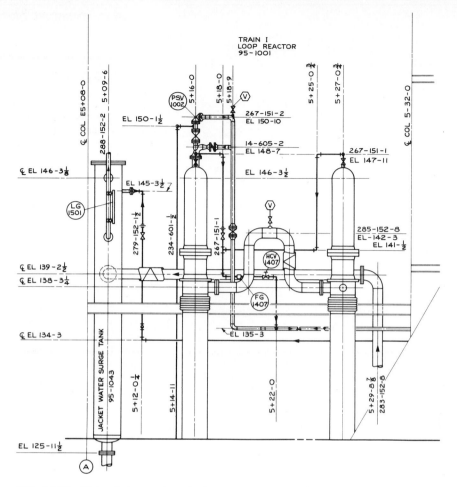

FIG. 28.18 This piping system is drawn using a combination of single-line and double-line symbols. The connections are shown as screwed, welded, and flanged. (Courtesy of Bechtel Corp.)

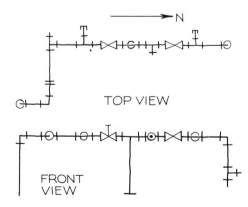

FIG. 28.19 Top and front views are used to represent this three-dimensional piping system using single-line symbols.

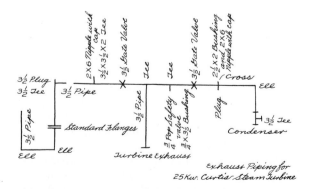

FIG. 28.20 The vertical pipes shown in Fig. 28.19 are revolved into the horizontal plane to form a developed drawing. The fittings and valves are noted on the sketch.

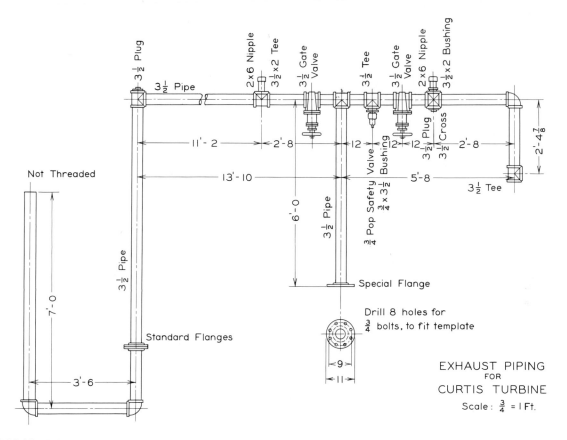

FIG 28.21 A finished developed drawing that shows all of the components in the system true size with double-line symbols.

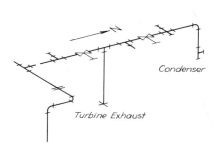

FIG. 28.22 An axonometric pictorial of the pipe system shown in Fig. 28.19 is drawn to give a three-dimensional picture of the system.

A number of abbreviations are used to specify piping components and fittings, as you can see by referring to the bill of materials. Many of the standard abbreviations associated with pipe drawings and specifications are given in Table 28.3. Part number 1, for example, is an 8-in. diameter pipe of a Schedule 40 weight that is made of seamless steel by the open hearth (OH) process. Instead of OH, the abbreviation EF may be used, which is the abbreviation for electric furnace.

Under the column "materials," you will notice a code that begins with the letter A, such as A–53. The letter A is used to represent a grade of carbon steel that is listed in Table A of the ANSI B31.3: *Petroleum Refinery Piping* Standards. The codes for fittings, flanges, and valves are taken from the manufacturers' catalogs of these products.

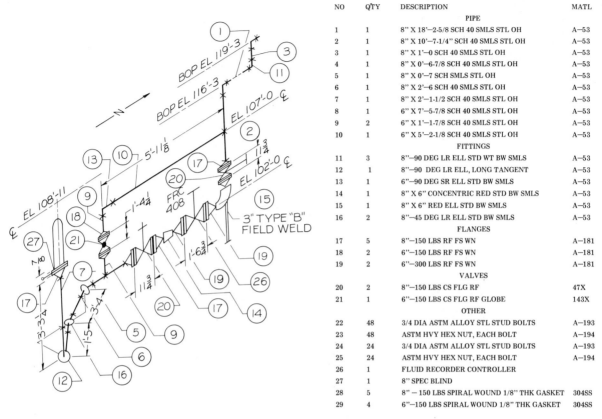

NO	QTY	DESCRIPTION	MATL
		PIPE	
1	1	8" X 18'—2-5/8 SCH 40 SMLS STL OH	A—53
2	1	8" X 10'—7.1/4" SCH 40 SMLS STL OH	A—53
3	1	8" X 1'—0 SCH 40 SMLS STL OH	A—53
4	1	8" X 0'—6.7/8 SCH 40 SMLS STL OH	A—53
5	1	8" X 0'—7 SCH SMLS STL OH	A—53
6	1	8" X 2'—6 SCH 40 SMLS STL OH	A—53
7	1	8" X 2'—1.1/2 SCH 40 SMLS STL OH	A—53
8	1	6" X 7'—5.7/8 SCH 40 SMLS STL OH	A—53
9	2	6" X 1'—1.7/8 SCH 40 SMLS STL OH	A—53
10	1	6" X 5'—2.1/8 SCH 40 SMLS STL OH	A—53
		FITTINGS	
11	3	8"—90 DEG LR ELL STD WT BW SMLS	A—53
12	1	8"—90 DEG LR ELL, LONG TANGENT	A—53
13	1	6"—90 DEG SR ELL STD BW SMLS	A—53
14	1	8" X 6" CONCENTRIC RED STD BW SMLS	A—53
15	1	8" X 6" RED ELL STD BW SMLS	A—53
16	2	8"—45 DEG LR ELL STD BW SMLS	A—53
		FLANGES	
17	5	8"—150 LBS RF FS WN	A—181
18	2	6"—150 LBS RF FS WN	A—181
19	2	6"—300 LBS RF FS WN	A—181
		VALVES	
20	2	8"—150 LBS CS FLG RF	47X
21	1	6"—150 LBS CS FLG RF GLOBE	143X
		OTHER	
22	48	3/4 DIA ASTM ALLOY STL STUD BOLTS	A—193
23	48	ASTM HVY HEX NUT, EACH BOLT	A—194
24	24	3/4 DIA ASTM ALLOY STL STUD BOLTS	A—193
25	24	ASTM HVY HEX NUT, EACH BOLT	A—194
26	1	FLUID RECORDER CONTROLLER	
27	1	8" SPEC BLIND	
28	5	8" — 150 LBS SPIRAL WOUND 1/8" THK GASKET	304SS
29	4	6"—150 LBS SPIRAL WOUND 1/8" THK GASKET	304SS

FIG. 28.23 This dimensioned isometric pictorial is called a "spool drawing." It is sufficiently complete that it can serve as a working drawing when used with the bill of materials.

A suggested format for spool drawings is given in Fig. 28.24. This format is used by the Bechtel Corporation, a major construction company, in designing and constructing pipelines and refineries.

are given by following the general rules of working drawings. In addition, the welded joints are specified to ensure that the vessel is properly fabricated to withstand the pressures and weights that it will be subjected to.

28.14
Vessel detailing

Vessels are containers, usually cylindrical in shape, that are used to contain petroleum products and other chemicals. The cylinders can be installed in vertical or horizontal positions. Vessels can also be spherical or ellipsoidal.

A detailed drawing of a cylindrical vessel is shown in Fig. 28.25. Its dimensions and specifications

28.15
Summary

The area of pipe drafting is a complex study in graphics and technology worthy of a sizeable textbook on this field alone. The coverage here has been presented as an introduction to familiarize you with the basics of the field. The standards of pipe drafting vary to a notable degree from company to company.

TABLE 28.3
STANDARD ABBREVIATIONS ASSOCIATED WITH PIPE SPECIFICATIONS

AVG	average	FS	forged steel	SPEC	specification
BC	bolt circle	FSS	forged stainless steel	SR	short radius
BE	beveled ends	FW	field weld	SS	stainless steel
BF	blind flange	GALV	galvanized	STD	standard
BM	bill of materials	GR	grade	STL	steel
BOP	bottom of pipe	ID	inside diameter	STM	steam
B&S	bell & spigot	INS	insulate	SW	socketweld
BWG	Birmingham wire gauge	IPS	iron pipe size	SWP	standard working pressure
CAS	cast alloy steel	LR	long radius	TC	test connection
CI	cast iron	LW	lap weld	TE	threaded end
CO	clean out	MI	malleable iron	TEMP	temperature
CONC	concentric	MFG	manufacture	T&G	tongue & groove
CPLG	coupling	OD	outside diameter	TOS	top of steel
CS	carbon steel, cast steel	OH	open hearth	TYP	typical
DWG	drawing	PE	plain end—not beveled	VC	vitrified clay
ECC	eccentric	PR	pair	WE	weld end
EF	electric furnace	RED	reducer	WN	weld neck
EFW	electric fusion weld	RF	raised face	WB	welded bonnet
ELEV	elevation	RTG or RJ	ring type joint	WT	weight
ERW	electric resistance weld	SCH	schedule	XS	extra strong
FF	flat face	SCRD	screwed	XXS	double extra strong
FLG	flange	SMLS	seamless		
FOB	flat on bottom	SO	slip-on		

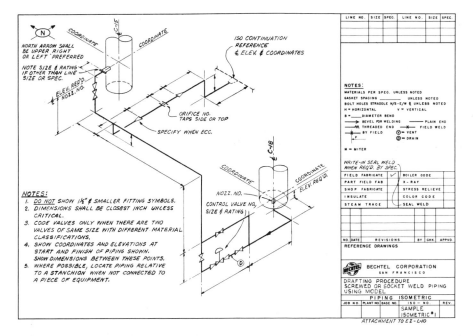

FIG. 28.24 A suggested format for preparing spool drawings by the Bechtel Corp. is shown here. (Courtesy of the Bechtel Corp.)

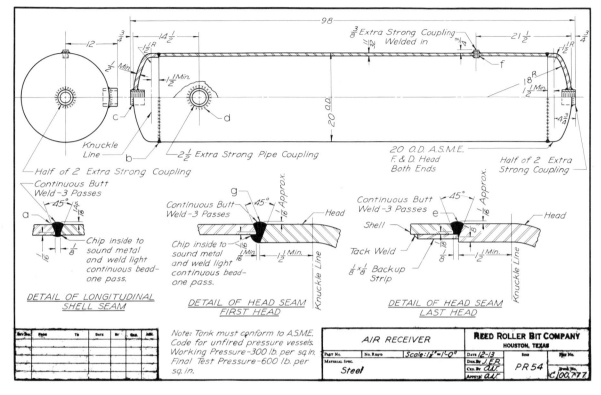

FIG. 28.25   A detail drawing of a cylindrical vessel that is connected into a pipe system.

Problems

1. On a Size A sheet, draw five orthographic views of the fittings listed below. The views should include the front view, the top view, the bottom view, and the left and right views. Draw two fittings per page. Refer to Fig. 28.17 and Table 28.2 as guides in making these drawings. Use single-line symbols to draw the following screwed fittings: 90° ell, 45° ell, tee, lateral, cap reducing ell, cross, concentric reducer, check valve, union, globe valve, gate valve, and bushing.

2. Same as Problem 1, but draw the fittings as flanged fittings.

3. Same as Problem 1, but draw the fittings as welded fittings.

4. Same as Problem 1, but draw the fittings as double-line screwed fittings.

5. Same as Problem 1, but draw the fittings as double-line flanged fittings.

6. Same as Problem 1, but draw the fittings as double-line welded fittings.

7. Convert the single-line sketch in Fig. 28.26 into a double-line system that will fit on a Size A sheet.

8. Convert the single-line pipe system in Fig. 28.27 into a double-line drawing that will fit on a Size A sheet, using the graphical scale given in the drawing to select the best scale for the system.

9. Convert the pipe system shown in isometric in

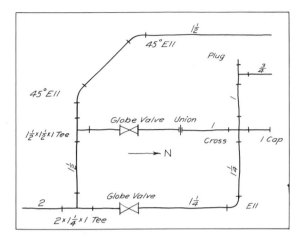

FIG. 28.26 Problem 7.

surements from the given drawing, and select a convenient scale.

12. Convert the orthograhic drawing of the pipe system in Fig. 28.19 into a double-line isometric drawing that will fit on a Size B sheet.

13. Convert the orthographic pipe system in Fig. 28.9 into a single-line isometric drawing that will fit on a Size B sheet. Estimate the dimensions.

14. Convert the orthographic pipe system in Fig. 28.9 into a double-line orthographic view that will fit on a Size B sheet.

Fig. 28.20 into a double-line isometric drawing that will fit on a Size B sheet.

10. Convert the pipe system given in Fig. 28.20 into a two-view orthographic drawing using single-line symbols.

11. Convert the isometric drawing of the pipe system in Fig. 28.24 into a two-view orthographic draw-ing that will fit on a Size B sheet. Take the mea-

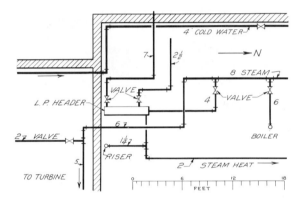

FIG. 28.27 Problem 8.

29

Electric/Electronics Drafting

29.1
Introduction

Electric/electronics drafting is a specialty area of the field of drafting technology. Electrical drafting is related to the transmission of electrical power that is used in large quantities in homes and industry for lighting, heating, and equipment operation. Electronics drafting deals with circuits in which electronic tubes or transistors are used, where power is used in smaller quantities. Examples of electronic equipment are radios, televisions, computers, and similar products.

Electronics drafters are responsible for the preparation of drawings that will be used in fabricating the circuit, and thereby bringing the product into being. They will work from sketches and specifications developed by the engineer or electronics technologist. This chapter will review the drafting practices that are necessary for the preparation of electronic diagrams. Most of the principles covered are shown applied in the Schematic Diagram in Fig. 29.5.

A major portion of the text in this chapter has been extracted from ANSI Y14.15, *Electrical and Electronics Diagrams,* the standards that regulate the drafting techniques used in this area. The symbols used were taken from ANSI Y32.2, *Graphic Symbols for Electrical and Electronics Diagrams.*

29.2
Types of diagrams

Electronic circuits are classified and drawn in the format of one of the following types of diagrams:

1. **Single-line diagrams,**
2. **Schematic diagrams,** or
3. **Connection diagrams.**

The suggested line weights for drawing these diagrams are shown in Fig. 29.1.

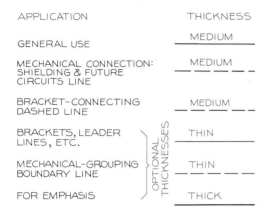

APPLICATION	THICKNESS
GENERAL USE	MEDIUM
MECHANICAL CONNECTION: SHIELDING & FUTURE CIRCUITS LINE	MEDIUM
BRACKET–CONNECTING DASHED LINE	MEDIUM
BRACKETS, LEADER LINES, ETC.	THIN
MECHANICAL-GROUPING BOUNDARY LINE	THIN
FOR EMPHASIS	THICK

OPTIONAL THICKNESSES

FIG. 29.1 The recommended line weights for drawing electronics diagrams.

Single-line diagrams

Single-line diagrams use single lines and graphic symbols to show an electric circuit or system of circuits and the parts and devices within it. A single line is used to represent both AC and DC systems, as illustrated in Fig. 29.2. An example of a single-line dia-

gram of an audio system is shown in Fig. 29.3. Primary circuits are indicated by thick connecting lines, and medium lines are used to represent connections to the current and potential sources.

Single-line diagrams show the connections of meters, major equipment, and instruments. In addition, ratings are often given to supplement the graphic symbols to provide such information as kilowatt ratings, voltages, cycles and revolutions per minute, and generator ratings (Fig. 29.4).

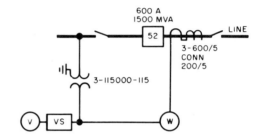

FIG. 29.2 A portion of a single-line diagram where heavy lines represent the primary circuits, and medium lines represent the connections to the current and potential sources. (Courtesy of ANSI.)

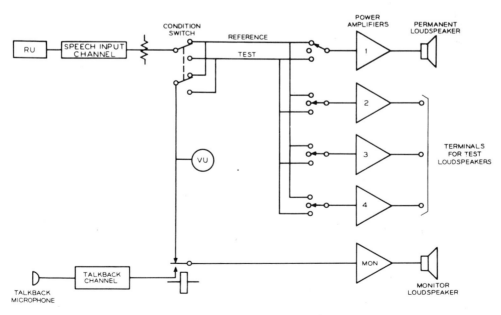

FIG. 29.3 A typical single-line diagram for illustrating electronics and communications circuits. (Courtesy of ANSI.)

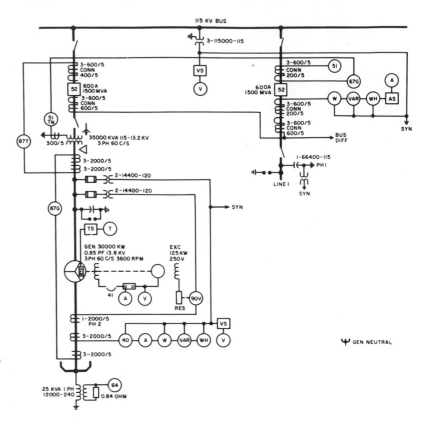

FIG. 29.4 A single-line diagram that is used to illustrate a power switchgear complete with the device designations noted on the diagram. (Courtesy of ANSI.)

Schematic diagrams

Schematic diagrams use graphic symbols to show the electrical connections and functions of a specific circuit arrangement. Although the schematic diagram enables one to trace the circuit and its functions, the physical size, shapes, and locations at various components are not given. A schematic diagram is illustrated in Fig. 29.5 (see pages 634–635); this diagram should be referred to for applications of the principles that are covered in this chapter.

Connection diagrams

Connection diagrams show the connections and installations of the parts and devices of the system. In addition to showing the internal connections, external connections, or both, they show the physical arrangement of the parts. It can be described as an installation diagram such as the one shown in Fig. 29.6.

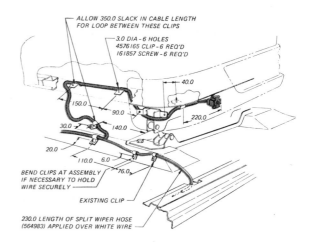

FIG. 29.6 This is a three-dimensional connection diagram that shows the circuit and its components with the necessary dimensions to explain how it is connected or installed. (Courtesy of the General Motors Corp.)

29.3
Schematic diagram connecting symbols

The most basic symbols of a circuit are those that are used to represent connections of parts within the circuit. The use of dots to show connections is optional; it is preferable to omit them if clarity is not sacrificed by so doing. Connections, or junctions, are indicated by using small black dots, as shown in Fig. 29.7A. The dots distinguish between connecting lines and those that simply pass over each other (Part B). It is preferred that connecting wires have single junctions wherever possible.

When the layout of a circuit does not permit the use of single junctions, and lines within the circuit must cross, then dots must be used to distinguish between crossing and connecting lines (Parts C and D).

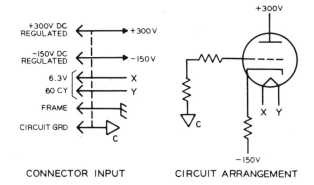

CONNECTOR INPUT   CIRCUIT ARRANGEMENT

FIG. 29.8 Circuits may be interrupted and connections not shown by lines if they are properly labeled to clarify their relationship to the part of the circuit that has been removed. The connections above are labeled to match those on the left and right sides of the illustration. (Courtesy of ANSI.)

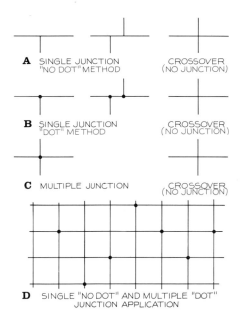

FIG. 29.7 Connections should be shown with single-point junctions as shown at (A). Dots may be used to call attention to connections as shown at (B); and dots *must* be used when there are multiples of the type shown at (C).

INTERRUPTED PATHS are breaks in lines within a schematic diagram that are interrupted to conserve space when this can be done without confusion. For example, the circuit in Fig. 29.8 has been interrupted

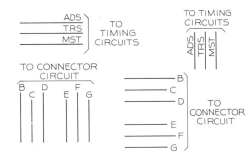

FIG. 29.9 Brackets and notes may be used to specify the destinations of interrupted circuits, as shown in this illustration.

since the lines do not connect the left and the right sides of the circuit. Instead, the ends of the lines are labeled to correspond to the matching notes at the other side of the interrupted circuit.

There will be occasions where sets of lines in a horizontal or vertical direction will be interrupted (Fig. 29.9). Brackets will be used to interrupt the circuit and notes will be placed outside the brackets to indicate the destinations of the wires or their connections.

In some cases, a dashed line is used to connect brackets that interrupt circuits (Fig. 29.10). The dashed line should be drawn so that it will not be mistaken as a continuation of one of the lines within the bracket.

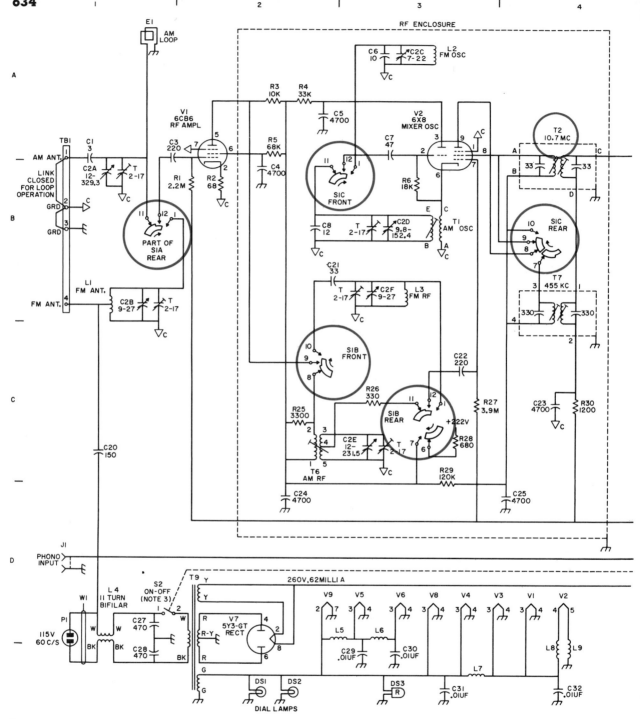

FIG. 29.5 A typical schematic diagram of an AM/FM radio circuit with all parts of the system labeled. The title of a drawing of this type should specify that it is a schematic diagram in addition to the type of circuit being presented. (Courtesy of ANSI.)

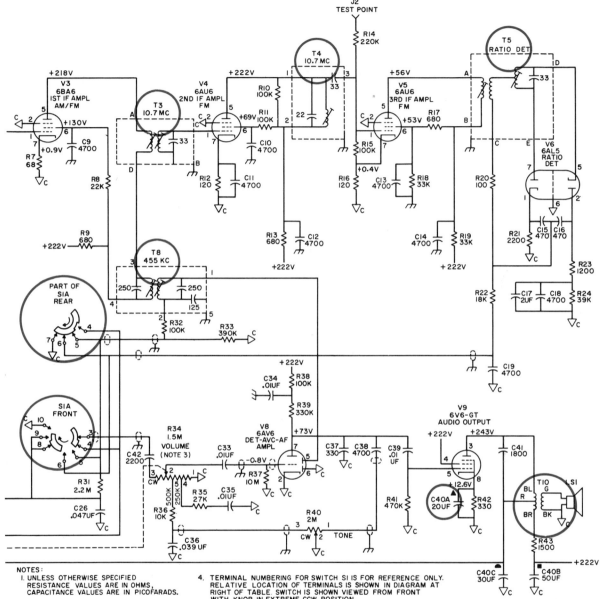

NOTES:
1. UNLESS OTHERWISE SPECIFIED RESISTANCE VALUES ARE IN OHMS, CAPACITANCE VALUES ARE IN PICOFARADS.

2. VOLTAGES MEASURED TO COMMON WIRING WITH VTVM.

3. SWITCH S2 IS PART OF POTENTIOMETER R34. TERMINAL NUMBERING IS FOR REFERENCE ONLY. RELATIVE LOCATION OF TERMINALS IS SHOWN IN DIAGRAM AT RIGHT.

4. TERMINAL NUMBERING FOR SWITCH SI IS FOR REFERENCE ONLY. RELATIVE LOCATION OF TERMINALS IS SHOWN IN DIAGRAM AT RIGHT OF TABLE. SWITCH IS SHOWN VIEWED FROM FRONT WITH KNOB IN EXTREME CCW POSITION.

R34 TERMINAL SIDE

POS	FUNCTION	SWITCH SECTIONS AND TERMINALS CONNECTED					
		SIA		SIB		SIC	
		FRONT	REAR	FRONT	REAR	FRONT	REAR
I	PHONO (SHOWN)	3-4	5-6-7	—	—	—	8-9
2	AM RADIO	3-5, 8-9	4-6-7, 11-12	8-9	6-7, 11-12	11-12	7-8, 9-10
3	FM RADIO	3-6, 9-10	4-5-7, 12-1	9-10	6-7, 12-1	12-1	8-9

HIGHEST REFERENCE DESIGNATIONS		
C42	L9	R43
REFERENCE DESIGNATIONS NOT USED		

SD-802151

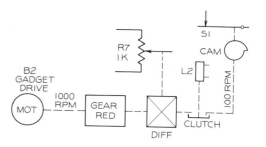

FIG. 29.10 The connections of interrupted circuits can be indicated by using brackets and a dashed line in addition to labeling the lines. The dashed line should not be drawn to appear as an extension of one of the lines in the circuit.

FIG. 29.11 If mechanical functions are closely related to electrical functions, it may be desirable to link the mechanical components within the schematic diagram.

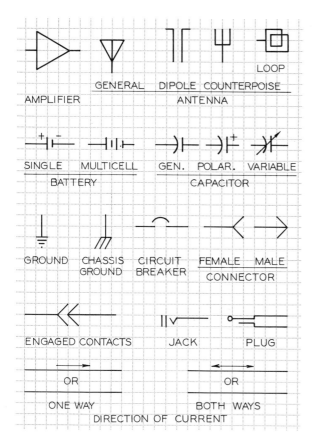

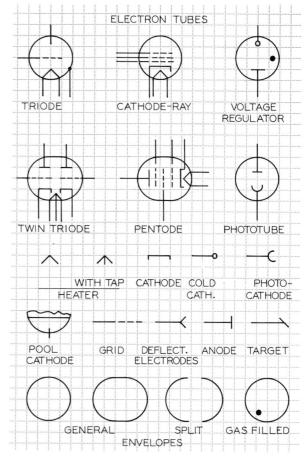

FIG. 29.12 These graphic symbols parts within a schematic diagram are drawn on a 5 mm (0.20 in.) grid. The suggested sizes of the symbols can be found by taking the dimensions from the grid to draw the symbols full size.

FIG. 29.13 The upper six symbols are used to represent often-used types of electron tubes. The interpretation of the parts that make up each symbol is given in the lower half of the figure.

MECHANICAL LINKAGES that are closely related to electronic functions may be shown as part of a schematic diagram (Fig. 29.11). An arrangement of this type helps clarify the relationship of the electronics circuit with the mechanical components.

29.4
Graphic symbols

The electronics drafter must be familiar with the basic graphic symbols that are used to represent the parts and devices within electrical and electronics circuits.

The symbols covered in this article are extracted from the ANSI Y32.2 standard, are adequate for practically all diagrams. However, when a highly specialized part needs to be shown and a symbol for it is not provided in these standards, it is permissible for the drafter to develop his or her own symbol provided it is properly labeled and its meaning clearly conveyed.

The symbols that are presented in Figs. 29.12–29.16 are drawn on a grid of 5 mm (0.20 in.) squares that have been reduced. The actual dimensions of the symbols can be approximated by using the grid and equating each square to its full-size measurement of 5

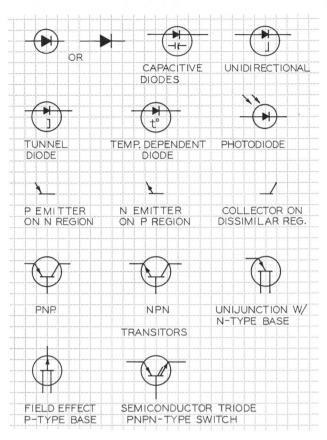

FIG. 29.15 Graphics symbols for semiconductor devices and transistors. The arrows in the middle of the figure illustrate the meanings of the arrows used in the transistor symbols shown below them.

mm. Symbols may be drawn larger or smaller to fit the size of your layout provided the relative proportions of the symbols are kept about the same.

The symbols in Figs. 29.12–29.16 are but a few of the more commonly used symbols. There are between five and six hundred different symbols in the ANSI standards for variations of the basic electrical/electronics symbols.

Electronics symbols can be drawn with conventional drawing equipment and instruments by approximating the proportions of the symbols; drafting templates of electronics symbols are available if you wish to save time and effort (Fig. 29.17).

Some of the symbols have been noted to provide designations of their sizes or ratings. The need for this additional information depends on the requirements and the usage of the schematic diagrams.

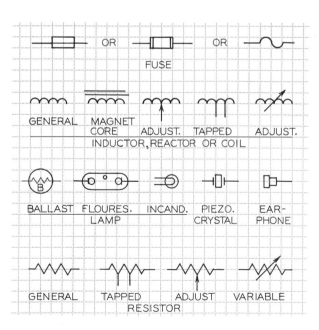

FIG. 29.14 Graphics symbols of standard circuit components

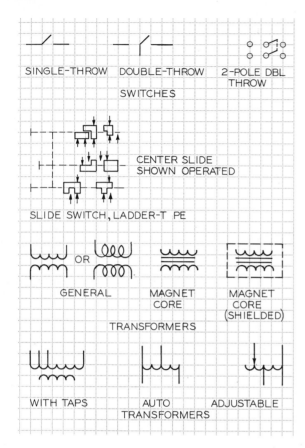

FIG. 29.16 Graphics symbols for representing switches and transformers

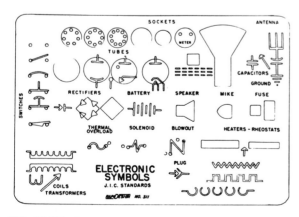

FIG. 29.17 A number of various types of template are available for drawing the graphics symbols in schematic diagrams. (Courtesy of Frederick Post Co.)

29.5
Terminals

Terminals are the ends of devices that are attached in a circuit with connecting wires. Examples of devices with terminals that are specified in circuit diagrams are switches, relays, and transformers. The graphic symbol for a terminal is an open circle that is the same size as the solid circle used to indicate a connection.

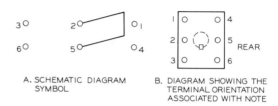

FIG. 29.18 An example of a method of labeling the terminals of a toggle switch on a schematic diagram (left), and a diagram that illustrates the toggle switch when it is viewed from its rear (right).

SWITCHES are used to turn a circuit on or off, or else to actuate a certain part of it while turning another off. Examples of labeling switches are shown in the schematic diagram in Fig. 29.5. In this case, a table is used with notes to clarify the switching connections of the terminals.

When a group of parts is enclosed or shielded (drawn enclosed with dashed lines) and the terminal circles have been omitted, the terminal markings

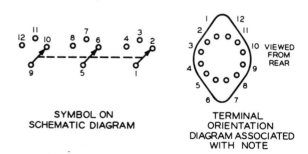

FIG. 29.19 An example of a rotary switch as it would appear on a schematic diagram (left), and a diagram that shows the numbered terminals of the switch when viewed from its rear. (Courtesy of ANSI.)

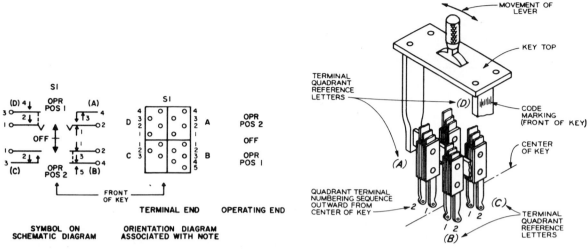

FIG. 29.20 An example of a typical lever switch as it would appear on a schematic diagram (left, Part A) and an orientation diagram that shows the numbered terminals of the switch when viewed from its operating end (right, Part A). A pictorial of the lever switch and its four quadrants is given in Part B of the figure. (Courtesy of ANSI.)

should be placed immediately outside the enclosure, as shown in Fig. 29.5 at T2, T3, T4, T5, and T8. The terminal identifications should be added to the graphic symbols that correspond to the actual physical markings that appear on or near the terminals of the part, such as 10.7 MC for transformer T3 in Fig. 29.5. Several examples of notes and symbols that explain the parts of a diagram are shown in Figs. 29.18, 29.19, and 29.20.

Colored wires or symbols are often used to identify the various leads that connect to terminals. When colored wires need to be identified on a diagram, the colors are lettered on the drawing, as was done for the transformer T10 of Fig. 29.5. An example of the identification of a capacitor, C40, that is marked with geometric symbols is shown in Fig. 29.5.

ROTARY TERMINALS are used to regulate the resistance in some circuits, and the direction of rotation of the dial is indicated on the schematic diagram. The abbreviations CW (clockwise) or CCW (counterclockwise) are placed adjacent to the movable contact when it is in its extreme clockwise position, as shown in Fig. 29.21A. The movable contact has an arrow at its end.

If the device terminals are not marked, numbers may be used with the resistor symbol and the number 2 assigned to the adjustable contact (Fig. 29.21B).

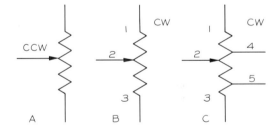

FIG. 29.21 To indicate the direction of rotation of rotary switches on a schematic diagram, the abbreviations CW (clockwise) and CCW (counterclockwise) are placed near the movable contact (Part A). If the device terminals are not marked, numbers may be used with the resistor symbols and the number 2 assigned to the adjustable contact (Part B). Additional contacts may be labeled as shown at C.

Other fixed taps may be sequentially numbered and added as shown in Part C.

The position of a switch as it relates to the function of a circuit should be indicated on a schematic diagram. A method of showing functions of a variable switch is shown in Fig. 29.22. The arrow represents the movable end of the switch that can be positioned to connect with several circuits. The different func-

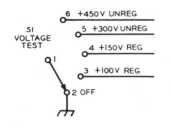

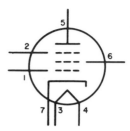

FUNCTIONS SHOWN AT SYMBOL

SI VOLTAGE TEST	
FUNCTION	TERM.
OFF	1-2
+ 100V REG	1-3
+ 150V REG	1-4
+300V UNREG	1-5
+450V UNREG	1-6

FUNCTIONS SHOWN IN TABULAR FORM

FIG. 29.22 For more complex switches, position-to-position function relations may be shown using symbols on the schematic diagram, or by a table of values located elsewhere on the diagram. (Courtesy of ANSI.)

tional positions of the rotary switch are shown both by symbol and by table.

Another method of representing a rotary switch is shown in Fig. 29.23 by symbol and by table. The tabular form is preferred due to the complexity of this particular switch. The dashes between the numbers in the table indicate that the numbers have been connected. For example, when the switch is in Position 2 the following terminals are connected: 1 and 3, 5 and 7, and 9 and 11. A table of this type is used at the bottom of Fig. 29.5.

Electron tubes have pins that fit into sockets that have terminals connecting into circuits. Pins are labeled with numbers placed outside the symbol used to represent the tube, as shown in Fig. 29.24, and are numbered in a clockwise direction with the tube viewed from its bottom.

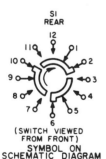

SYMBOL ON SCHEMATIC DIAGRAM

SI (REAR)		
POS	FUNCTION	TERM.
1	OFF(SHOWN)	1-2,5-6,9-10
2	STANDBY	1-3,5-7,9-11
3	OPERATE	1-4,5-8,9-12

FUNCTIONS SHOWN IN TABULAR FORM

FIG. 29.23 A rotary switch may be shown on a schematic diagram with its terminals labeled as shown at the left, or its functions can be given in a table placed elsewhere on the drawing as shown at the right. Dashes are used to indicate the linkage of the numbered terminals. For example, 1–2 means that terminals 1 and 2 are connected in the "off" position. (Courtesy of ANSI.)

FIG. 29.24 Tube pin numbers should be placed outside the tube envelope and adjacent to the connecting lines. (Courtesy of ANSI.)

29.6 Separation of parts

In complex circuits, it is often advantageous to separate elements of a multi-element part with portions of the graphic symbols drawn in different locations on the drawing. An example of this method of separation is the switch labeled S1A, S1B, S1C, etc., in Fig. 29.5. The switch is labeled S1 and the letters that follow, called suffixes, are used to designate different parts of the same switch. Suffix letters may also be used to label subdivisions of an enclosed unit that is made up of a series of internal parts, such as the crystal unit shown in Fig. 29.25. These crystals are referred to as Y1A and Y1B.

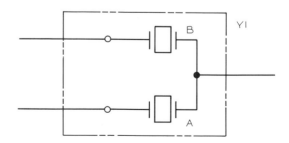

FIG. 29.25 As subdivisions within the complete part, Crystals A and B are referred to as Y1A and Y1B.

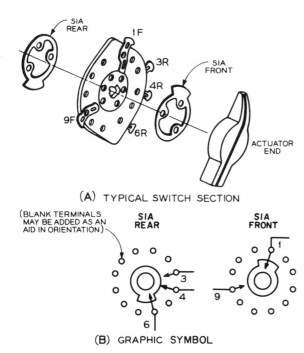

(A) TYPICAL SWITCH SECTION

(BLANK TERMINALS
MAY BE ADDED AS AN
AID IN ORIENTATION)

S1A
REAR

S1A
FRONT

(B) GRAPHIC SYMBOL

FIG. 29.26 Parts of rotary switches are designated with suffix letters A, B, C, etc., and are referred to as S1A, S1B, S1C, etc. The words FRONT and REAR are added to these designations when both sides of the switch are used. (Courtesy of ANSI.)

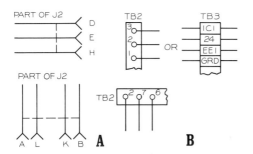

FIG. 29.27 The portions of connectors or terminal boards are functionally separated on a diagram; the words PART OF may precede the reference designation of the entire portion. Or conventional breaks can be used to indicate graphically that the part drawn is only a portion of the whole.

Rotary switches of the type shown in Fig. 29.26 are designated as S1A, S1B, etc. The suffix letters A, B, etc., are labeled in sequence beginning with the knob and working away from it. Each end of the various sections of the switch should be viewed from the same end. When the rear and front of the switches need to be used, the words FRONT and REAR are added to the designations.

Portions of items such as terminal boards, connectors, or rotary switches may be separated on a diagram. The words PART OF may precede the identification of the portion of the circuit of which it is a part, as shown in Fig. 29.27A. A second method of showing a part of a system is by using conventional break lines that make the note, PART OF, unnecessary.

29.7
Reference designations

A combination of letters and numbers that identify items on a schematic diagram are called reference designations. These designations are used to identify the components not only on the drawing, but in the related documents that refer to them. Reference designations should be placed close to the symbols that represent the replaceable items of a circuit on a drawing. Items that are not separately replaceable may be identified if this is considered necessary. Mounting devices for electron tubes, lamps, fuses, etc., are seldom identified on schematic diagrams.

It is standard practice to begin each reference designation with an uppercase letter that may be followed by a numeral with no hyphen between them. The number usually represents a portion of the part being represented. The lowest number of a designation should be assigned to begin at the upper left of the schematic diagram and proceed consecutively from left to right and top to bottom throughout the drawing.

Some of the standard abbreviations used to designate parts of an assembly are: amplifier–A, battery–BT, capacitor–C, connector–J, piezoelectric crystal–Y, fuse–F, electron tube–V, generator–G, rectifier–CR, resistor–R, transformer–T, and transistor–Q.

As the circuit is being designed, some of the numbered elements may be deleted from the circuit drawing. The numbered elements that remain should not be renumbered even though there is a missing element within the sequence of numbers used to label the

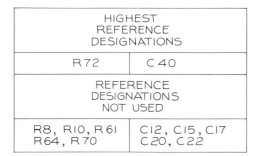

HIGHEST REFERENCE DESIGNATIONS	
R 72	C 40
REFERENCE DESIGNATIONS NOT USED	
R8, RIO, R 6I R64, R 70	C I2, CI5, CI7 C 20, C 22

FIG. 29.28 Reference designations are used to identify parts of a circuit. They are labeled in a numerical sequence from left to right beginning at the upper left of the diagram. If parts are later deleted from the system, the ones deleted should be listed in a table, along with the highest reference number designations.

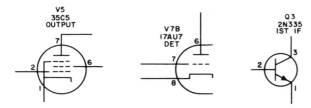

FIG. 29.29 Three lines of notes can be used with electron tubes to specify reference designations, type designation, and function. This information should be located adjacent to the symbol, and preferably above it. (Courtesy of ANSI.)

parts. Instead, a table of the type shown in Fig. 29.28 can be used to list the parts that have been omitted from the circuit. The highest designations are also given in the table as a check to be sure that all parts were considered.

Electron tubes are labeled not only with reference designations but with type designation and circuit function, as shown in Fig. 29.29. This information is labeled in three lines, such as V5/35C5/OUTPUT, which are located adjacent to the symbol.

29.8
Numerical units of function

Functional units such as the values of resistance, capacitance, inductance, and voltage should be specified with the fewest number of zeros by using the multipliers in Fig. 29.30A as prefixes. Examples using this method of expression are shown in the B and C parts of Fig. 29.30, where units of resistance and capacitance are given. When four-digit numbers are given, the commas should be omitted. One thousand should be written as 1000, not as 1,000. You should recognize and use the lowercase or uppercase prefixes as indicated in the table of Fig. 29.30.

A general note can be used where certain units are repeated on a drawing to reduce time and effort:

UNLESS OTHERWISE SPECIFIED: RESISTANCE VALUES ARE IN OHMS. CAPACITANCE VALUES ARE IN MICROFARADS.

FIG. 29.30 Multipliers should be used to reduce the number of zeros in a number (Part A). Examples of expressing units of capacitance and resistance are shown at B and C. (Courtesy of ANSI.)

Multiplier	Prefix	Symbol	
		Method 1	Method 2
10^{12}	tera	T	T
10^{9}	giga	G	G
10^{6} (1,000,000)	mega	M	M
10^{3} (1000)	kilo	k	K
10^{-3} (.001)	milli	m	MILLI
10^{-6} (.000001)	micro	μ	U
10^{-9}	nano	n	N
10^{-12}	pico	p	P
10^{-15}	femto	f	F
10^{-18}	atto	a	A

A. MULTIPLIERS

Range in Ohms	Express as	Example
Less than 1000	ohms	.031 470
1000 to 99,999	ohms or kilohms	1800 15,853 10k 82k
100,000 to 999,999	kilohms or megohms	220k .22M
1,000,000 or more	megohms	3.3M

B. RESISTANCE

Range in Picofarads	Express as	Example
Less than 10,000	picofarads	152.4 pF 4700 pF
10,000 or more	microfarads	.015 μF 30 μF

C. CAPACITANCE

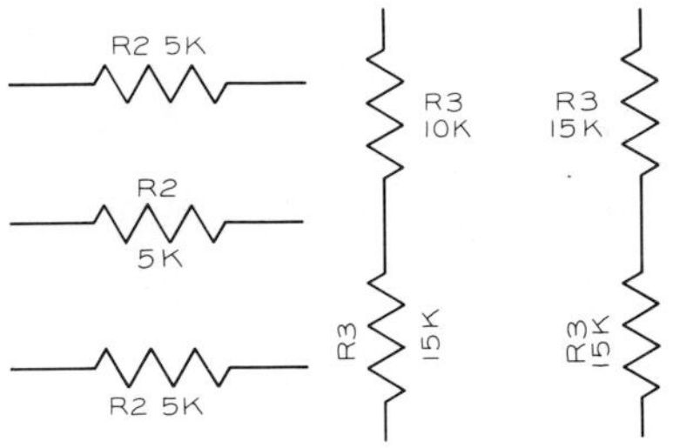

FIG. 29.31 Methods of labeling the units of resistance on a schematic diagram

or

CAPACITANCE VALUES ARE IN PICOFARADS.

A note for specifying capacitance values is:

CAPACITANCE VALUES SHOWN AS NUMBERS EQUAL TO OR GREATER THAN UNITS ARE IN pF AND NUMBERS LESS THAN UNITY ARE IN μF.

Examples of the placement of the reference designations and the numerical values of resistors are shown in Fig. 29.31.

29.9
Functional identification of parts

The readability of a circuit is improved if parts are labeled to indicate their functions. Test points are labeled on drawings with the letters ''Tp'' and their suffix numbers. The sequence of the suffix numbers should be the same as the sequence of troubleshooting the circuit when it is defective. As an alternative, the test function can be indicated on the diagram below the reference designation.

Additional information may be included on a schematic diagram to aid in the maintenance of the system:

DC resistance of windings and coils.

Critical input and output impedance values.

Wave shapes (voltage or current) at significant points.

Wiring requirements for critical ground points, shielding, pairing, etc.

Power or voltage ratings of parts.

Caution notation for electrical hazards at maintenance points.

Circuit voltage values at significant points (tube pins, test points, terminal boards, etc.).

Zones (grid system) on complex schematics.

Signal flow direction in main signal paths shall be emphasized.

29.10
Printed circuits

Printed circuits are universally used for miniature electronic components and computer systems. The degree of miniaturization that has occurred in the electronics industry can be seen in Fig. 29.32, where the function of a vacuum tube has been replaced by a chip transistor the size of a dot.

The drawings of printed circuits are drawn up to four times or more the size that the circuit will ultimately be printed. The drawings are usually drawn in black India ink on acetate film and are then photographically reduced to the desired size. The circuit is ''printed'' onto an insulated board made of plastic or

FIG. 29.32 The result of the miniaturization of electronic devices can be seen here where a solid state logic technology (SLT) chip the size of a dot gives the same function as its predecessors, the transistor and the vacuum tube. (Courtesy of IBM.)

FIG. 29.33 This is a magnified view of a printed circuit where the circuit has actually been printed and etched on a circuit board, and the devices are then soldered into position. (Courtesy of Bishop Industries Corp.)

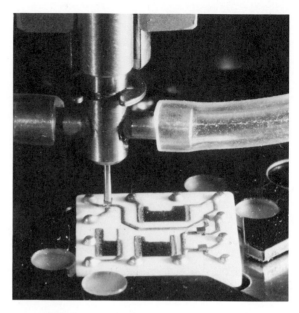

FIG. 29.35 A chip is shown being attached to the one-half inch square SLT module at a rate of better than one per second. (Courtesy of IBM.)

ceramics, and the devices within the circuit are connected and soldered (Fig. 29.33).

The steps of fabricating a solid state logic technology (SLT) module that was used in the IBM 360 computer is shown in Fig. 29.34. The tiny module (about

one-half inch square) had its chip transistor positioned and attached in the last step of assembly before it was encased in its protective metal shell (Fig. 29.35).

Some printed circuits are printed on both sides of the circuit board; this requires two photographic negatives, as shown in Fig. 29.36, that were made from positive drawings (black lines on a white background). Each drawing for each side can be made on separate sheets of acetate that are laid over each other when the second diagram is drawn. However, a more efficient method uses red and blue tape that can be

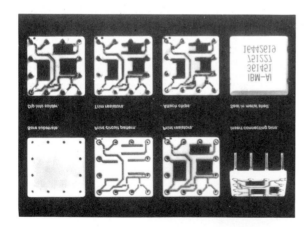

FIG. 29.34 The steps of fabricating a printed circuit are shown here where the circuit and its parts are printed on a ceramic substrate base, and the pins are connected. Next, it is dipped into solder to connect the parts of the circuit and to build up the printed circuit, the resistors are trimmed and the chips are attached to prepare the module for sealing in a metal shield. (Courtesy of IBM.)

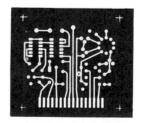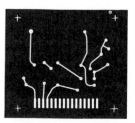

FIG. 29.36 A printed circuit that is attached to both sides of the circuit board requires two circuit drawings, one for each side, that are photographically converted to reduced negatives for printing. (Courtesy of Bishop Industries Corp.)

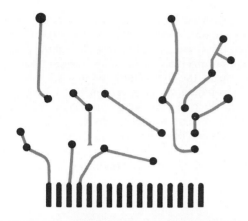

FIG. 29.37 By using two colors, such as blue and red, one circuit drawing can be made, and two negatives made from the same drawing by using camera filters that screen out one of the colors with each shot. The circuits are then printed on each side of the board. (Courtesy of Bishop Industries Corp.)

used for making a single drawing (Fig. 29.37) from which two negatives are photographically made. Filters are used on the process camera to drop out the red for one negative and a different filter for dropping out the blue for the second negative.

Printed circuits are usually coated with silicone varnish to prevent malfunction because of the collection of moisture or dust on the surface. They may also be enclosed in protective shells.

29.11
Shortcut symbols

Several manufacturers produce preprinted symbols that can be used for "drawing" high quality electronic circuits and printed circuits. The symbols are available on sheets or on tapes that can be burnished onto the surface of the drawing to form a permanent schematic diagram (Fig. 29.38).

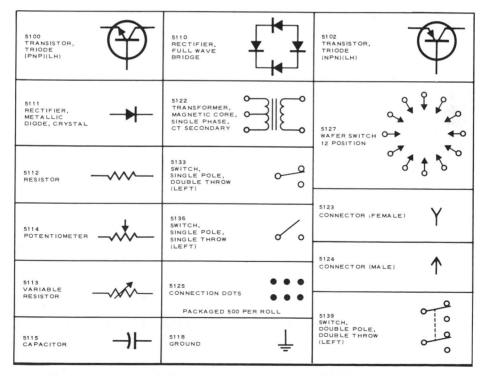

FIG. 29.38 Stick-on symbols are available for laying out schematic diagrams rather than drawing them. The resulting layouts have a higher contrast and sharpness that improves their reproducibility and reduction for publication. (Courtesy of Bishop Industries Corp.)

The symbols can be connected with matching tape to represent wires between them instead of drawing the lines. Schematic diagrams made with these materials are of a very high quality that reproduces well when reduced in size for publication in specifications or in technical journals.

29.12
Installation drawings

Many types of electric/electronics drawings are used to produce the finished installation, from the designer who visualized the system at the outset of the project to the contractor who builds it. Drawings are used to design the circuit, detail its parts for fabrication, specify the arrangement of the devices within the system, and instruct the contractor how to install the project.

A combination arrangement and wiring diagram drawing is shown in Fig. 29.39, where the system is shown in a front and right-side view. The wiring diagram explains how the wires and components within the system are connected for the metal-encased switchgear. Bus bars are conductors for the primary circuits.

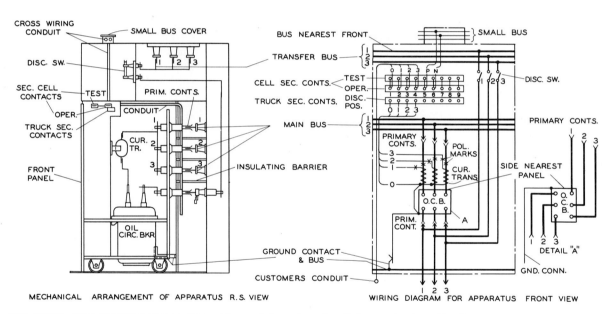

FIG. 29.39 This drawing shows views of a metal-enclosed switchgear to describe the arrangement of the apparatus; it also gives the wiring diagram for the unit.

Problems

1. On a Size A sheet, make a schematic diagram of the circuit shown in Fig. 29.40.

2. On a Size A sheet, make a schematic diagram of the circuit shown in Fig. 29.41.

3. On a Size A sheet, make a schematic diagram of the circuit shown in Fig. 29.42.

4. On a Size A sheet, make a schematic diagram of the circuit shown in Fig. 29.43.

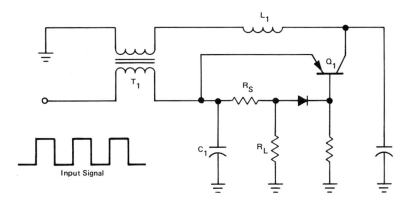

FIG. 29.40 Problem 1: A low-pass inductive-input filter. (Courtesy of NASA.)

FIG. 29.41 Problem 2: A quadruple-sampling processor. (Courtesy of NASA.)

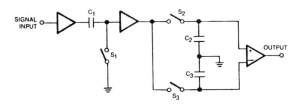

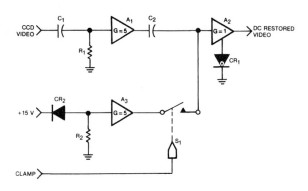

FIG. 29.42 Problem 3: A temperature-compensating DC restorer circuit designed to condition the video circuit of a Fairchild area-image sensor. (Courtesy of NASA.)

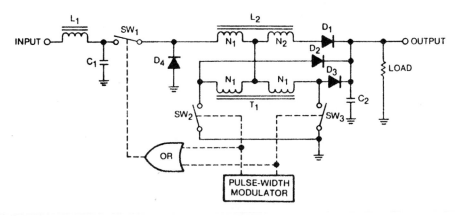

FIG. 29.43 Problem 4: A "buck/boost" voltage regulator. (Courtesy of NASA.)

5. On a Size A sheet, make a schematic diagram of the circuit shown in Fig. 29.44.

6. On a Size B sheet, make a schematic diagram of the circuit shown in Fig. 29.45.

7. On a Size C sheet, make a schematic diagram of the circuit shown in Fig. 29.46.

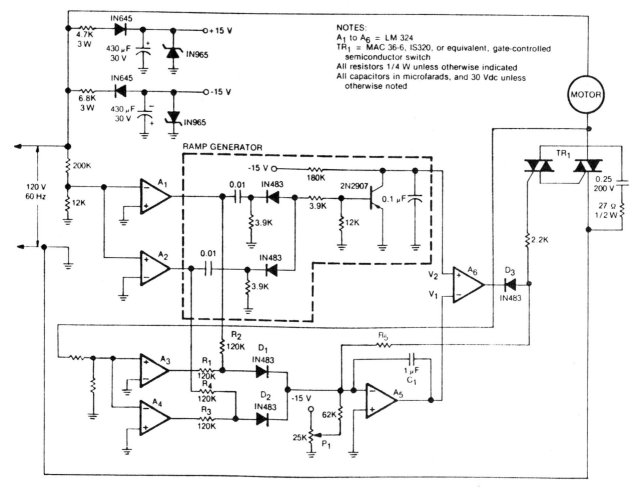

FIG. 29.44 Problem 5: An improved power-factor controller. (Courtesy of NASA.)

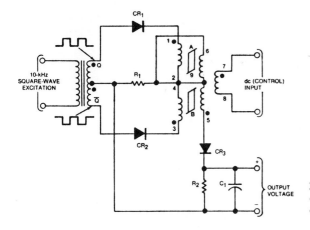

FIG. 29.45 Problem 6: A magnetic-amplifier DC transducer. (Courtesy of NASA.)

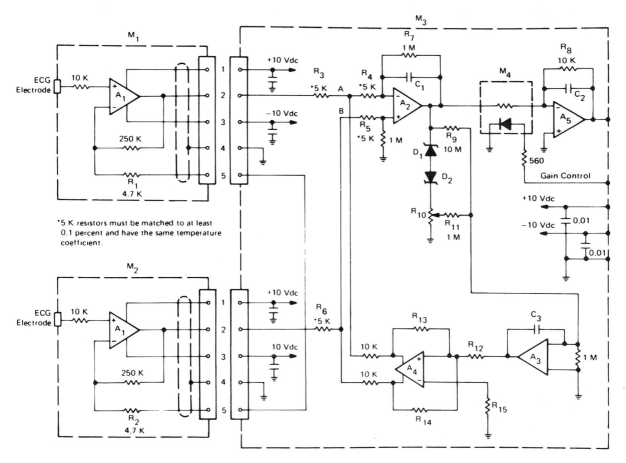

FIG. 29.46 Problem 7: An electrocardiograph (EKG) signal conditioner. (Courtesy of NASA.)

8. On a Size C sheet, make a schematic diagram of the circuit shown in Fig. 29.47, and give the parts list.

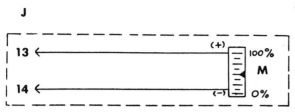

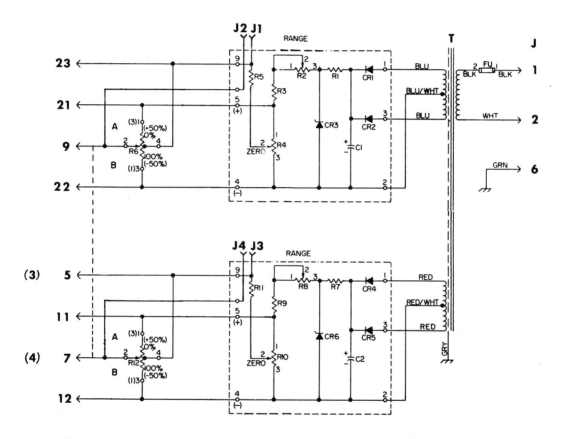

COMPONENT	DESCRIPTION	
CR1, CR2, CR3, CR4	DIODE	1N3193
CR3 & CR6	ZENER DIODE	24V 3W ± .5V TOL
R1 & R7	RESISTOR	560 Ω 5W 5%
R3 & R9	RESISTOR	56 Ω
R5 & R11	RESISTOR	2.7K 1/2W 5%
R2 & R8	TRIM POT	50 Ω
R4 & R10	TRIM POT	5K
R6 & R12	POT	500 Ω 2W DUAL
C1 & C2	CAPACITOR	60 MFD 60V
T	TRANSFORMER	
FU	FUSE 1 AMP	
M	EDGEWISE METER	
J	CONNECTOR	
J1, J2, J3, J4	TEST JACK	

TYPE	R6 SCALE		M SCALE	
	A	B	BOTTOM	TOP
RS1100C	0%	100%	0%	100%
RS1110C	0%	100%	PER ENGINEERING DATA	
RS3100C	+50%	-50%	-50%	+50%
RS4100C	0%	100%	0%	100%
RS2120C	100%	0%	OMIT	
RS2100C	100%	0%	0%	100%

NOTE: OUTPUT WIRED TO TERMINALS 3 AND 4 ON
TYPES RS1100C, RS1110C, AND RS3100C.

OUTPUT WIRED TO TERMINALS 5 AND 7 ON
TYPES RS2100C, RS2120C, AND RS4100C.

FIG. 29.47 Problem 8: A schematic diagram of a dual output manual station and its parts list. (Courtesy of NASA.)

CHAPTER

30

Computer Graphics

30.1
Introduction

Computer graphics, a method of making drawings with a computer and a plotter, has come into broad usage with more growth expected. Computer graphic systems that once cost several hundred thousand dollars are now available at much more affordable prices.

Computer-aided manufacturing (CAM) is the use of computerized production machinery, including metal-cutting machines such as lathes, drills, and milling machines. Modern CAM systems control complicated robots used to assemble everything from delicate printed circuit boards to heavy metal automobile parts. These robots can have as many as ten arms that may weld, drill holes, and tighten bolts in a sequence of steps.

Computer-aided design (CAD) is the computer-aided process of solving design problems in all areas of technology. The specifications of the design can be stored and recalled for further modification and evaluation. The graphic display of the CAD system aids designers in viewing and studying their designs as they are being developed.

Computer-aided design drafting (CADD) is used to produce the final working drawings once the design has been finalized. The designer can display the drawings on a screen, called a **cathode-ray tube (CRT),** before the final copies of the drawings are drawn as "hard copies" by the plotter. Some industries use programs similar to those written for making the drawings to drive the automated machinery in the shop and make the finished part, thereby eliminating the need for drawings made on paper.

30.2
Applications of computer graphics

One application of computer graphics is the design and drawing of printed circuits of the type shown in Fig. 30.1. Printed circuits are drawn as much as five times larger than their true size and then reduced photographically. A computer-driven plotter will draw the circuit to an accuracy of within approximately 0.001 in.

FIG. 30.1 An often-used application of computer graphics is the drawing of printed circuits for electronic systems. They can be drawn very accurately, photographically reduced, and then fabricated. (Courtesy of Prime Computer.)

Another application is the design and representation in both orthographic and pictorial views of piping systems. Once the system has been completed, the design and its specifications can be stored for rapid recall when the system needs to be reexamined (Fig. 30.2).

Computer graphics is used as an aid in finite element analysis, where a series of elements is used to represent an irregular three-dimensional shape. In Fig. 30.3, the designer is digitizing (assigning numerical coordinates to) the elements of a gun mount by working from a multiview drawing at his right. The mount is then displayed by the computer system as an isometric on the screen of the CRT.

Clothing patterns for a wide selection of graduated sizes (Fig. 30.4) can be drawn and cut using computer graphics. The automatic cutter follows the computer-generated path to cut the most economical patterns from a piece of material.

Pictorials drawn by the computer allow both technical and nontechnical people to observe a three-dimensional design. Most 3-D computer graphics systems permit the observer to view a design from any angle as it revolves on the screen.

30.3
Hardware systems

A basic computer graphics system consists of a **computer,** a **terminal,** a **plotter,** and a **printer.** Additional devices such as **digitizers** and **light pens** may be used for direct input of graphics information (Fig. 30.5).

Computer

Computers are the devices that receive the input of the programmer, execute the programs, and produce some form of output. The largest computers, **mainframes,** are big, fast, powerful, and expensive. The smallest computers, **microcomputers,** are in widespread use for personal and business applications. Microcomputers are excellent for graphic applications where massive data storage and high speed are not essential and low price and small size are important.

The **minicomputer,** a medium-sized computer, lies between the mainframe and the microcomputer in both performance and cost.

FIG. 30.2 The designer is using computer graphics to plot a piping system in color on the screen of a video display. (Courtesy of Applicon.)

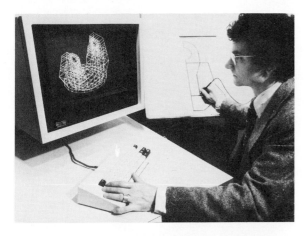

FIG. 30.3 This engineer has constructed a 3-D finite element model of a gun mount on the Applicon display by digitizing a multiview engineering drawing. (Courtesy of Applicon.)

FIG. 30.5 This Producer drafting system is composed of a keyboard, CRT, digitizer, computer, and plotter. (Courtesy of Bausch & Lomb.)

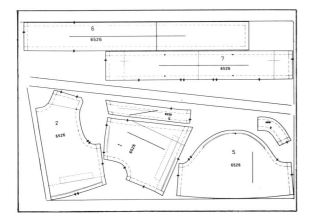

FIG. 30.4 This clothing pattern was designed, plotted, and cut out by utilizing computer graphics on a Versatec plotter. Programs have been written that will plot this same pattern for graduated sizes of clothing. (Courtesy of Versatec.)

FIG. 30.6 An engineer is shown working at a computer graphics system's terminal where he has access to a keyboard and a menu tablet. (Courtesy of CADD-23, Brodhead Garrett Co.)

Terminal

The terminal is the device by which the user communicates with the computer. It usually consists of a keyboard and some type of output device such as a typewriter or a television screen (Fig. 30.6). Most graphics terminals use one of the three types of television screens or CRTs (cathode-ray tubes). One type is **raster scanned,** which means that the picture display is being refreshed or scanned from left to right and top to bottom at a rate of about 60 times per second.

A second type is the **storage tube,** where the image is drawn on the screen like a drawing on an erasable blackboard.

The third and most powerful is the **vector-refreshed tube,** which means that each line in the picture is being redrawn continuously by the computer. The vector-refreshed tube requires more computer power than is available from the minicomputers and the microcomputers.

Plotters

The plotter is the machine that is directed by the computer and the program to make a drawing. The two basic types of plotters are **flatbed** and **drum.** The flatbed plotter has a flat bed to which the drawing paper is attached. A pen is moved about over the paper in a raised or lowered position to complete the drawing (Fig. 30.7). The drum plotter uses a special type of paper held on a spool and rolled over a rotating drum (Fig. 30.8). The drum rotates in two directions as the pen suspended above the surface of the paper moves left or right along the drum.

FIG. 30.8 A drum plotter used for plotting the output from the computer. (Courtesy of Hewlett-Packard.)

FIG. 30.7 A close-up of the flat-bed plotter that uses four pens of different colors to plot the output from the computer. (Courtesy of Computervision.)

Printers

The printer can be thought of as a typewriter operated by the program and computer. The speed and type quality offered by the printer are features that determine the purchase price. One of the most important applications of the printer in a computer system is the printing of a copy of a computer program in "hard copy" form that can be used for review and modification by the programmer.

Digitizers

The digitizer is a device that is used to input the coordinates of points of a drawing into the computer by tracing the drawing located on a digitizer board. A pen-like stylus connected to the computer is often used for this purpose. Some systems using the refreshed or scanned types of CRT terminals use a **light pen** that can establish or change points on the screen. These points can be either graphic data or commands used by the computer.

30.4
Computer graphics for the microcomputer

The microcomputer is being used widely as an economical machine for performing computer graphic functions because it is adequate for the needs of most applications. Many new graphics software programs are being created for use with the microcomputer.

Versions of software programs are AutoCAD, CADD-23, CADplan, Personal Designer, and VersaCAD. Each program works best if used with a computer, a hard disk, and 512 Kb of RAM. A typical IBM microcomputer graphics system with a tablet digitizer is shown in Fig. 30.9.

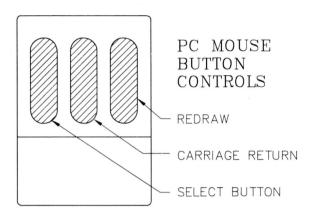

FIG. 30.10 A three-button Mouse that is used with a pad to provide three functions.

30.5
AutoCAD computer graphics

The remainder of Chapter 30 will describe the use of **AutoCAD 2.15** (AutoCAD is a trademark of Autodesk Inc.) with an **IBM XT Personal Computer** (512 RAM, 10 Mb hard disk), a **Mouse Systems mouse,** and a **Hewlett-Packard 7475A plotter.**

FIG. 30.9 New CAD software designed for the new generation of personal computers, VERSACAD ADVANCED, is a general-purpose, computer-aided drafting program from. (Courtesy of T&W Systems.)

Although computer graphics software differs among the firms producing it, a number of similarities exist, making it possible to transfer knowledge of one system to the next. The serious user must refer to the AutoCAD manual to broaden her or his understanding of this software since only introductory AutoCAD topics are discussed.

AutoCAD supports several types of tablets and digitizing devices for inputting data. Although data can be input by the keyboard, the drafter will find a tablet or a mouse much faster.

The mouse (Fig. 30.10) has three buttons that can be used to select points, give a carriage return, and redraw the screen. Like a large ball bearing, mice work by a mechanical principle and others use an optical light transmitted to a light-sensitive grid.

30.6
Drawing layers

The beginner, who has not used computer graphics or AutoCAD, can benefit from turning the machine on, loading the system by typing ACAD, and responding to the screen prompts to see how much can be understood from them. For example, if you want to draw a line, select **DRAW** from the menu and **LINE** from the submenu. By experimentation, you become familiar with the system, its prompts, and the menu. But soon, you will need specific instructions before you will be able to use effectively this powerful tool.

AutoCAD provides an infinite number of layers on which to make a drawing. Each has its own name, color, and line type. For example, you may have a layer named **HIDDEN** that appears on the screen in yellow with dashed lines drawn on it to represent hidden lines. No more than one name, color, and line type can be assigned to a single layer.

Layers can be used efficiently to reduce the duplication of effort that several applications of a basic layer would require. For example, it is common practice for an architect to use the same floor plan for several different applications: one each for dimensions, furniture arrangement, floor finishes, electrical details, etc. The basic plan can be used for all these applications by turning on the needed layers and turning off others.

SETTING LAYERS For basic applications the following layers are sufficient for working drawings:

Layer Name	State	Color	Line Type
Visible	on/off	1 Red	Continuous
Hidden	on/off	2 Yellow	Hidden
Center	on/off	3 Green	Center
Hatch	on/off	4 Cyan	Continuous
Dimen	on/off	5 Blue	Continuous
Cut	on/off	6 Magenta	Phantom
0	on/off	7 White	Continuous

Layers have names that correspond to their line types with a different color to distinguish them on a color monitor. The 0 layer is the system default layer that cannot be deleted but it can be turned off.

By selecting **LAYERS** from the Root Menu, you will recieve a sub-command of **LAYER,** which you select from the keyboard, from the screen with your mouse or by typing LAYER at the keyboard. The word LAYER is a command and it is followed with a carriage return (CR). The following is AutoCAD's dialogue with you while you are setting layers:

- Command: **LAYER**
- ?/Set/New/On/Off/Color/Ltype: **NEW** (or **N**) (CR)
- Layer name(s) ⟨0⟩: **VISIBLE,HIDDEN, CENTER,HATCH,DIMEN,CUT** (CR).

This series of responses creates six new layers by name and each will have a default color of white and a de-

fault line type of continuous. To set the color of each layer, you must respond in the following manner:

- ?/Set/New/On/Off/Color/Ltype: **COLOR** (or **C**) (CR)
- Color: **RED** (CR)*
- Layer name(s) ⟨0⟩: **VISIBLE** (CR).

The Visible layer has now been assigned the color of red, which could have been assigned with the number 1 instead of the word, RED. The colors of the other layers must be assigned in the same manner to their respective layers one at a time. Line types are assigned to layers in the following manner:

- ?/Set/New/On/Off/Color/Ltype: **Ltype** (or **L**) (CR)
- Linetype (or ?) ⟨CONTINUOUS⟩: **HIDDEN** (CR)
- Layer name(s) ⟨0⟩: **HIDDEN** (CR).

The lines on the Hidden layer will now be drawn with dashed (Hidden) lines.

You must **Set** a layer to make it the current one in order to draw on it. It is set by responding in the following manner:

- ?/Set/New/On/Off/Ltype: **SET** (or **S**) (CR)
- Layer name(s) ⟨0⟩: **VISIBLE** (CR) (CR).†

Any line you draw now will appear on the Visible layer in red with continuous lines. Once you have defined a number of layers and drawn on each, they can be turned on or off by the On/Off command in the following manner:

- ?/Set/New/On/Off/Ltype: **On** or **OFF**
- Layer name(s) ⟨0⟩: **VISIBLE** or ***** (for all layers)
- (or, **VISIBLE,HIDDEN,CENTER,** to turn these layers on or off).

Even though a number of layers are on, you can draw on only one—**the current layer.** By selecting the question mark (?), the screen will display a current listing of the layers, their line types, their colors, and their on/off status. Press function key F1 to flip the screen back to the graphics editor.

(VCR) will be used for carriage return.
†two returns

To save the file of layers and their specifications, use the utility command **END,** which returns you to the main menu. But first, let's set additional parameters as shown in the next section before we **END.**

30.7
Setting screen parameters

To set screen parameters, you must become familiar with the terms LIMITS, GRID, SNAP, UNITS, ORTHO, and the function keys. **LIMITS** are used to give the size of the screen area that represents the size of the paper on which your drawing will be plotted. If you are making a full-size drawing that will fit on an A-size sheet (11″ × 8.5″), your drawing area will be about 10.5″ × 8″ for most plotters. If you are using millimeters, the limits will be 254 × 198. Limits are set as follows:

- Command: **LIMITS** (CR)
- Lower left corner: ⟨0.00,0.00⟩: (CR) (Accept default value)
- Upper right corner: ⟨24.00,36.00⟩: **10,7.8** (CR).

A grid on the screen that can be used as a drawing aid can be set in the following manner:

- Command: **GRID** On/Off/Value (X)/Aspect ⟨0.00, 0.00⟩: **.2** (CR).

This command paints the screen with a dot pattern that is 0.2 in. apart (Fig. 30.11). To make the newly assigned limits fill the screen's working area, use the

ZOOM ALL command from the **DISPLAY** command on the Root Menu.

The units to be used in your drawing must be assigned before setting the limits. By typing or selecting the **UNITS** command, AutoCAD will give you the following selection:

1. Scientific (2.67E + 02)
2. Decimal (267.00 inches or millimeters)
3. Engineering (12′–4.50″)
4. Architectural (16′–3½″).

Select the type of desired units and respond to the prompts to specify the desired details, such as the number of decimal places you want. Decimal units are used for the metric system (millimeters) and the English system (inches). (To return to the drawing editor mode, press the F1 function key.)

The **SNAP** command can be used to make the cursor on the screen snap to a visible or invisible grid. If your grid is set at spacings of 0.2 in., the SNAP command makes the cursor snap to the grid or an invisible 0.1 grid between the visible grid points (Fig. 30.11).

- Command: **SNAP** On/Off/Value/Aspect/Rotate/ Style: 0.2 (CR).

The **ORTHO** command forces all lines drawn in this mode to be either vertical or horizontal to aid in making orthographic views, where most lines meet at right angles.

- Command: **ORTHO** and select On or Off from the screen menu.

ORTHO can be turned off by using **CTRL O** or by pressing function key F8 (Fig. 30.12). Function keys can be used to turn screen parameters on and off once they have been set. **Running Coordinates** can be obtained to give the coordinates of the cursor's position by pressing F6.

The status line (Fig. 30.13) at the top of the screen tells you the current layer by name; SNAP is on when an S appears, ORTHO is on when an O appears, and FILL is on when F appears (coordinates appear when this feature is on).

If these parameters were inserted while you were setting layers, they would be saved when you **END**ed the file. This file should be saved with its set parameters as an ''empty'' file named **FORMAT.** This file can be called up for editing each new drawing rather than

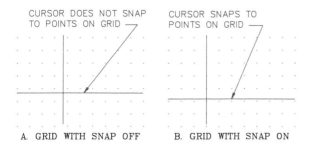

A. GRID WITH SNAP OFF B. GRID WITH SNAP ON

FIG. 30.11 The SNAP command can be used to make the cursor stop on a visible (or invisible) grid on the screen.

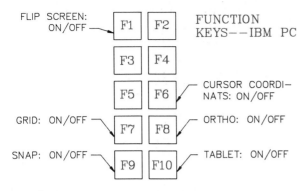

FIG. 30.12 The function keys on the IBM microcomputer can be used to activate six functions.

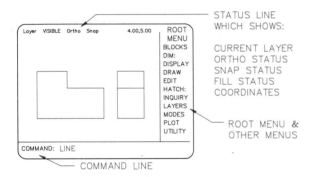

FIG. 30.13 The line at the top of the screen, the status line, gives the modes you are operating with. The command line appears at the bottom of screen.

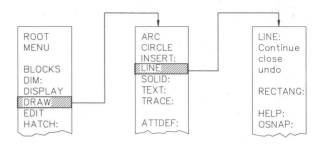

FIG. 30.14 You may progress from the root menu to the levels of submenus by selecting commands with the keyboard or mouse.

assigned parameters will appear on the screen ready for a new drawing. To draw lines select **DRAW** from the root menu and **LINE** from the submenu that follows (Fig. 30.14). The command line on the screen prompts you as follows:

- Command: **Line** (CR)
- From point: (Select a point or give coordinates)
- To point: (Select second point or give coordinates)
- To point: (ETC. You can continue in this manner with a series of connected lines.)

As the lines are drawn by moving the cursor about the screen, the current line will "rubber band" from its last point (Fig. 30.15). Points can be entered at the keyboard as absolute or relative coordinates

beginning a new drawing. Once the drawing is completed in this file, **SAVE** the drawing, give it a different name, and then **QUIT,** responding **YES** to the question, "Do you wish to discard all changes in the drawing?" In this manner, you have emptied your original file, FORMAT, without disturbing its parameters, which can be used over again, and you have saved a drawing that was made from it.

30.8
Basics of drawing lines

From the main menu, enter 2 to edit a drawing and respond with **FORMAT** when prompted for the name of the drawing to edit. The empty drawing file with its

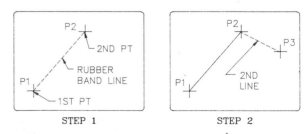

FIG. 30.15 Drawing a line

Step 1 Command: **LINE**
From point: Select P1
To point: Select P2 (The line is drawn.)

Step 2 To point: Select P3 (The line is drawn)
Press carriage return to disengage the rubber band.

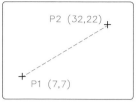

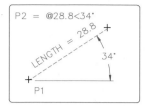

A. ABSOLUTE COORDINATES   B. RELATIVE COORDINATES

FIG. 30.16 A. Absolute coordinates can be entered at the keyboard to establish the ends of a line.

B. Relative coordinates, which are relative to the previous point on the screen, can be entered at the keyboard.

(Fig. 30.16). **Absolute coordinates** are located with respect to (0,0), the lower left corner of the drawing area and **relative coordinates** are measured from the last point. Relative points are written with the symbol @ followed by the distance from the last point plus < followed by the angle the line makes with the horizontal measured counterclockwise. Example: @28.8<34. The status line gives the length and angle from the last point while the current line rubber bands from point to point.

When drawing polygons, the last side can be made to close by using the **CLOSE** command as shown in Fig. 30.17.

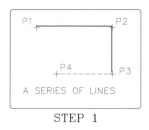

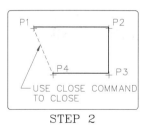

STEP 1   STEP 2

FIG. 30.17 The CLOSE command

Step 1   A series of lines are drawn using the LINE command from P1 through P4. (The last line is a rubber band line until the next point is selected.)

Step 2   After P4 has been selected, select CLOSE, and the line will be drawn to the first point of the series, P1.

30.9
Drawing circles

The command for drawing circles can be found under the **DRAW** submenu where you may give the center and radius, the center and diameter, or three points. By using **DRAG,** you can move the cursor and see the size of the circle change until it is the desired size (Fig. 30.18). DRAG can be turned on or off by inserting the DRAG command, followed by On or Off.

- Command: **CIRCLE** (CR)
- 3P/2P/⟨Center point⟩:
- (Locate center, C, with the cursor.)
- Diameter/⟨Radius⟩: Select radius with cursor, and the circle is drawn.

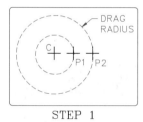

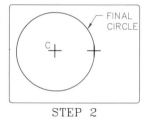

STEP 1   STEP 2

FIG. 30.18 Drawing a circle

Step 1   Command: **CIRCLE**
3P/2P ⟨Center point⟩: Locate center, C, with cursor. Move cursor to positions P1 and P2, and the circle will change as the radius is changed.

Step 2   The final radius is selected with the select button and the circle is drawn.

30.10
Drawing arcs

The **ARC** subcommand is found under the DRAW command. Arcs can be drawn by using nine combinations of variables such as starting point, center, angle, ending point, length of arc, and radius. For example, the S,C,E version requires that you locate the starting point (S), the center (C), and the ending point (E) (Fig. 30.19). The arc begins at Pt. S, but Pt. E need not lie on the arc. Arcs are drawn counterclockwise by

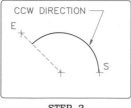

FIG. 30.19 The ARC command

Step 1 Command: **ARC**
Command: (Select S, C, E)
Command: Arc Start Point or C: Select Pt. S Center/
End ⟨Second point⟩: C Center: Select **C**.

Step 2 Angle/Length of chord ⟨End point⟩: Select Pt.
E. (The arc is drawn to an imaginary line from the
center to Pt. E counterclockwise.)

default.

- Command: ARC
- Center/⟨Start point⟩: C (CR)
- Center: Select Center, C
- Start point: Select Pt. S
- Angle/Length of chord/⟨End point⟩: select end
  point.

30.11
Enlarging and reducing drawings

Portions of a drawing or the entire drawing can be
enlarged by the **ZOOM** subcommand, a submenu under the **DISPLAY** command. In Fig. 30.20, a part of
the drawing is too small to read, but with a ZOOM
Window, you can select the part you want enlarged
to fill the screen.

- Command: **ZOOM**
- Magnification or type (ACELPW): **WINDOW** or
  W (CR)
- First point: Select P1
- Second Point: Select P2
  (These responses establish the window of the
  new picture and the area is enlarged to fill the
  screen.)

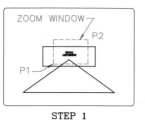
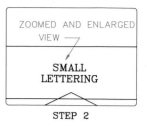

FIG. 30.20 The ZOOM command

Step 1 Command: ZOOM (CR)
Magnification or type (ACELPW): **W** (CR)
First point: Select P1

Step 2 Second point: Select P2. The window will be
enlarged to fill the screen.

Instead of responding with **Window,** you can respond with **All, Center, Extents, Left, Previous,** or
a number to indicate the factor by which you want the
size of the drawing to be changed to. **ALL** makes the
drawing limits as large as possible to fit within the display screen. **CENTER** allows you to pick the center of
the drawing and the desired degree of magnification
or reduction. **EXTENTS** enlarges the drawing to fill
the screen beyond the screen's limits. By responding
with **L,** you can pick the lower left corner and the
height of the drawing you wish to enlarge. **PRE-
VIOUS** displays the last view that was displayed on
the screen. Views can be zoomed repetitively an infinite number of times.

30.12
Erasing lines

The **ERASE** command, a subcommand under **EDIT,**
enables you to remove portions of a drawing. The
LAST command erases single entities (lines, text, circles, arcs, and blocks) removing one at a time, working backward from the last one drawn. The **WIN-
DOW** is the second option for erasing. However, only
entities directly within the window will be erased (Fig.
30.21).

- Command: **ERASE** (CR)
- Select objects or Window or Last: **WINDOW** or
  W (CR)
- Lower left corner: Select P1
- Upper right corner: Select P2, 4 found

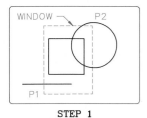

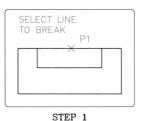

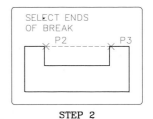

FIG. 30.21 The ERASE command

Step 1 Command: Erase
Select Objects or Window or Last: **W**
Lower left corner: Select P1
Upper right corner: Select P2.

Step 2 Press Return (CR) and the rectangle within the window is erased. Entities that are partially within the window are not erased.

FIG. 30.23 The BREAK command

Step 1 Command: BREAK (CR)
Select object: Select P1 on line to be broken
Enter second point or F: F. (CR)

Step 2 Enter first point: Select P2
Enter second point: Select P3 (CR) (The line is broken from P2 to P3.)

- Select objects or Window or Last: **RETURN**
     (The entities are erased.)

The default of the ERASE command is **Select Objects** which requires that you use the cursor to indicate those entities on the screen to be erased (Fig. 30.22). The **RETURN** erases the entities. If you erase something by mistake, the **OOPS** command restores only the last erasure.

   The **BREAK** command removes a portion of an entity, such as a line (Fig. 30.23).

- Command: **BREAK**
- Select object: Select Point P1 on the line to be broken

- Enter second point or F: **F** (F says that you now wish to select the first point) (CR)
- Enter first point: Select P2
- Enter second point: Select P3
     (The line is broken from P2 to P3.)

You could have omitted the step where the F response was given if the break hadn't begun and ended at the intersections of lines. But this extra step ensures that the computer understands which line is to be broken.

30.13
Fillets

FILLETS can be drawn to any desired radius between two non-parallel lines whether they intersect or not. The fillet is drawn and the lines are trimmed (Fig. 30.24).

- Command: **FILLET** (CR)
- Polyline Radius/⟨Select two lines⟩: **R** (CR)
- Enter fillet radius: **1.5** (CR)
- Command: (CR)
- FILLET Polyline Radius/⟨Select two lines⟩: Select P1 and P2 (CR)
     (The fillet is drawn and lines are trimmed.)

By entering a fillet radius of 0, nonintersecting lines will be extended automatically to form a perfect inter-

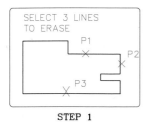

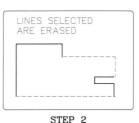

FIG. 30.22 The ERASE command—entities

Step 1 Command: ERASE (CR)
Select Objects or Window or Last: (O is default)
Select entities (lines) with P1, P2, and P3.

Step 2 Press RETURN (CR) and the selected lines are erased.

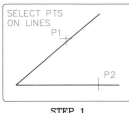

FIG. 30.24 The FILLET command

Step 1   Command: FILLET
Radius/(Select two lines): R (CR)
Enter fillet radius: 1.5 (CR)
Command: (CR)
FILLET Radius/(Select two lines):

Step 2   Select with P1 and P2. The fillet is drawn and the lines are trimmed.

section. Once the radius is assigned, it remains in memory as the default radius for drawn additional fillets.

30.14
Text and numerals

The **TEXT** command, a subcommand under the DRAW command, inserts plotted text in a drawing from the keyboard at any size and angle (Fig. 30.25).

- Command: **TEXT** (CR)
- Starting point or (ARCS): Select point with cursor

FIG. 30.25 The TEXT command defaults to left justified text and the point indicated represents the lower left corner of capital letters. Part A.

Text can also be centered, right justified, or aligned so that the first and last letters of a word or line span the distance between two selected points, Part B.

- Height ⟨0.18⟩: **0.5** (CR)
- Rotation angle ⟨0⟩: (CR)
- Text: **TEXT** (CR)
  (The word, TEXT, is written beginning with the point selected and is left justified.)

Other responses are **A** (aligned), changes the text to fit exactly between two endpoints (Part A); **C** (centered), centers the text about a central point; **R** (right justified), justifies the text to the point located at the right; and **S** (style), selects a text font style of lettering from available ones.

30.15
Moving and
copying drawings

A drawing can be moved to a new position by the **MOVE** command or duplicated by the **COPY** command. When copied, the original drawing is left in its original position and a copy of it is located where specified. The figure drawing in Fig. 30.26 is moved to a new position by the following commands:

- Command: **MOVE** (CR)
- Select Objects or Window or Last: **WINDOW** or **W** (CR)

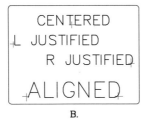

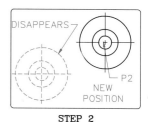

FIG. 30.26 The MOVE command

Step 1   Command: MOVE
Select Objects or Window or Last: **W** (CR)
First point: Select Pt.
Second point: Select Pt.
Select Objects of Window or Last: (CR)
Base point or displacement: Select 1st Pt.

Step 2   Second point of displacement: Select 2nd Pt. The object is drawn at its new position, P2, and the original drawing disappears.

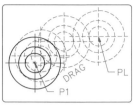

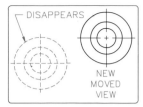

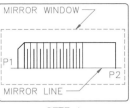

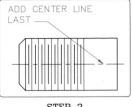

| STEP 1 | STEP 2 | STEP 1 | STEP 2 |

FIG. 30.27 The DRAG command

FIG. 30.28 The MIRROR command

Step 1   Command: MOVE (CR)

When prompted for "Second point or displacement." respond with DRAG. As the cursor is moved, the drawing is DRAGGED about the screen.

Step 2   When moved to the desired location, press the select button to draw the object in its final position. The original drawing disappears.

- First point: Select point
- Second point: Select point, 5 found
- Select objects or Window or Last: (CR)
- Base point or displacement: Select 1st Point
- Second point of displacement: Select 2nd Point
  (The drawing is moved.)

The **COPY** procedure is the same as **MOVE** except the beginning command is **COPY.**

A drawing can be moved or copied by dragging it into position by the **DRAG** command (Fig. 30.27). **DRAG** is activated by responding

- Second point of displacement: **DRAG.**

As the cursor moves about the screen, the drawing to be moved or copied moves dynamically until it is set by pressing the select button of the mouse.

Besides WINDOW, you may use the default to **select the entities** to be moved one at a time with the cursor. Also, you may select **LAST,** which will move the last entity such as a line, arc, circle, text, or block.

30.16
Mirroring drawings

Symmetrical objects can be drawn by drawing a portion of the figure and mirroring the drawing about one or more axes. The schematic threads in Fig. 30.28 are

Step 1   Command: MIRROR (CR)
Select Objects or Window or Last: Window drawing to be mirrored
First point on mirror line: Select P1
Second point: Select P2.

Step 2   Delete old objects? ⟨N⟩ N **(CR)**
The object is mirrored about the mirror line. Draw the center line.

drawn in the following manner:

- Command: **MIRROR** (CR)
- Select objects or Window or Last: **WINDOW** (CR)
- First point: Select window
- Second point: Select window
- Select objects or Window or Last: (CR)
- First point or mirror line: Select P1
- Second point: Select P2
- Delete old objects? ⟨N⟩ Respond with **Y** or **N.**

30.17
Snapping to objects (OSNAP)

You will wish to draw lines that SNAP to features of other objects within drawings rather than snapping to a grid. This type of snap is called an **OSNAP,** short for object snap. For example, to draw a line from an intersection of lines to the endpoint of a line (Fig. 30.29), the following procedure is used:

- Command: **LINE** (CR)
- From Point: (Select **OSNAP** on the screen menu)
- Select **INTERSEC,** which is short for intersection
- LINE from Point: **intersec of**

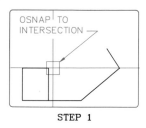
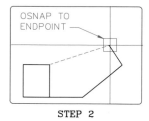

STEP 1   STEP 2

FIG. 30.29 The OSNAP command

Step 1   Command: LINE
From point: Select OSNAP on the screen menu
Select INTERSEC (intersection)
Select intersection point.

Step 2   To point: Select OSNAP from the screen
menu
Select ENDPOINT
To point: Endpoint of (Move cursor to endpoint of
line and press select button.) The line is drawn.

Move cursor with target to the intersection
point where the line is to begin and press the
select button.

- To point: Select **OSNAP** from the screen menu
  Select **ENDPOINT**

- To point: **Endpoint of**
  Move cursor with target to the end of the line
  and press the select button.
  (The line will be drawn.)

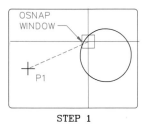
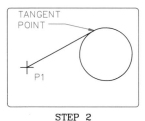

STEP 1   STEP 2

FIG. 30.30 The OSNAP tangent command

Step 1   Command: LINE
From point: Select P1
To point: Select OSNAP from the screen menu and
select TANGENT

Step 2   To point: Move cursor to point on circle and
press select. The line is drawn from P1 tangent to
the circle on the side where the tangent point was
selected.

In Fig. 30.30 a line is drawn from P1, a tangent to the
circle. The same procedure as shown above is used,
but instead of specifying **ENDPOINT** for the second
point, **TANGENT** is selected on the side of the circle
where the tangent point is to be located. The tangent
point is found automatically and the line is drawn.

Other OSNAP options permit you to snap to the
following: centers of arcs and circles, insertion points
of blocks, midpoints of lines, nearest points of an object, points (nodes) perpendicular to lines, tangent to
arcs and circles, and quadrant points of a circle.

30.18
Blocks

A powerful feature of computer graphics is the option
to build a file of drawings or symbols used repetitively
on drawings. In AutoCAD, these filed drawings are
called **BLOCKS.** They are drawn in the conventional
manner and are blocked as follows (Fig. 30.31):

- Command: **BLOCK** (CR)
- BLOCK name (or ?): **SI** (CR)
- Insertion base point: Select insert point
- Select objects or Window or Last: **WINDOW** or
  W (CR)
- First point: Select point
- Second point: Select point
- Select objects or Window or Last: (CR)
  (The block is filed into memory and disappears
  from the screen.)

The block is inserted into the drawing by the following
steps:

- Command: **INSERT** (CR)
- Block name (or ?): **SI** (CR)
- Insertion point: Select point with cursor
- X scale factor ⟨1⟩ or Corner: **0.5** (CR)
- Y scale factor ⟨default = X⟩: (CR)
- Rotation angle ⟨0⟩: (CR)
  (The Block, SI, is redrawn at 50% of its original
  size.)

BLOCKS inserted as entities cannot be edited by erasing parts of them or breaking lines within them. When

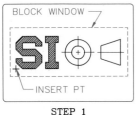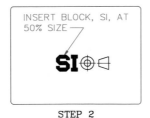

STEP 1   STEP 2

FIG. 30.31 The BLOCK command

Step 1   Make a drawing that you wish to Block and respond to the Command: BLOCK as follows:
Block name (or ?): SI
Insertion point: Center or Select insert point
Select objects or Window or Last: W (Window the drawing)
(The object disappears into memory.)

Step 2   To Insert, respond to the Command: INSERT as follows:
Block name (or ?): SI (CR)
Insertion point: Select point with cursor
x scale factor ⟨1⟩ /or Corner/XYZ: **0.5** (CR)
y scale factor ⟨default = x⟩ (CR)
Rotation angle ⟨0⟩: (CR)
(Block SI is inserted at 50% scale.)

you attempt to erase a portion of a block, the whole entity is erased. However, blocks can be edited if they are inserted with a star in front of their name. Example: Block name (or ?): ***SI**

Blocks can be used only on the current drawing file unless they are converted to **WBLOCKs,** a Write Block. WBLOCKs converts a BLOCK to a permanent BLOCK. A WBLOCK conversion is performed as follows:

- Command: **WBLOCK** (CR)

- File Name: **SI** (CR) (This assigns the name of the WBLOCK.)

- Block Name: **SI** (CR) (This is the name of the BLOCK that is being changed to a WBLOCK.)

A library of WBLOCKs can relieve the drafter of making repetitive drawings and will improve productivity.

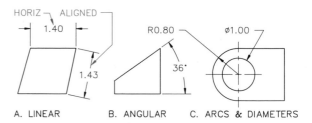

A. LINEAR   B. ANGULAR   C. ARCS & DIAMETERS

Fig. 30.32 The types of dimensions that appear on a drawing.

30.19
Dimensioning

Figure 30.32 shows the types of dimensions usable with AutoCAD (these options are in the submenu of the **DIM** command). When you dimension a line, a submenu prompts you to specify if the dimension line is to be horizontal, vertical, aligned, or rotated. In Fig. 30.33, a horizontal line is dimensioned by selecting the two endpoints of the line (P1 and P2) and locating the dimension line with P3. The dimension of 2.40", which is measured by the program, is accepted by a RETURN and the dimension is shown on the drawing (Step 2). All drawings are drawn full size when using computer graphics. The scaling process occurs during the plotting of the drawings.

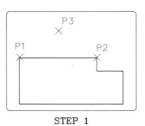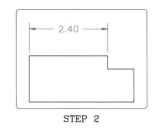

STEP 1   STEP 2

FIG. 30.33 Dimensioning a line

Step 1   Command: DIM as follows:
DIM: LINEAR/ HORIZONTAL
First extension line origin or RETURN to select: P1
Second extension line origin: P2
Dimension line location: P3.

Step 2   Press carriage return (CR) and the measurement appears at the command line, 2.40. Press return to accept this value and the dimension line is drawn.

DIM VARS	DEFAULT	
DIMSCALE	1.0000	Overall scale factor
DIMASZ	0.1800	Arrow size
DIMCEN	0.0900	Center mark size
DIMEXO	0.0645	Extension line offset
DIMDLI	0.3800	Dimension line increment
DIMEXE	0.1800	Extension beyond dim. line
DIMTP	0.0000	Plus tolerance
DIMTM	0.0000	Minus tolerance
DIMTXT	0.1800	Text height
DIMTSZ	0.0000	Tick size
DIMTOL	*OFF	Add +/− to dimension text
DIMLIM	*OFF	Generate dimension limits
DIMTIH	ON	Text inside extension is horizontal
DIMTOH	ON	Text outside extension is horizontal
DIMSE1	OFF	Suppress first extension line
DIMSE2	OFF	Suppress second extension line
DIMTAD	OFF	Place text above dimension line

FIG. 30.34 A listing of the DIM VARS that can be selected and changed when dimensioning.

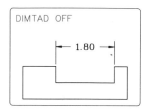
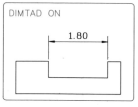

A. TEXT IN DIMEN. LINE   B. TEXT ABOVE DIMEN. LINE

FIG. 30.36 The command DIMTAD places the text ABOVE the dimension line when **ON**, Part B. When **OFF**, the text is inserted within the dimension line, Part A

Prior to any dimensioning attempts, you should check the dimensioning variables, **Dim Vars,** which can be called up on the screen and displayed one at a time. Figure 30.34 gives a complete listing of the dimensioning variables. Figure 30.35 shows all ratios of these variables based on the letter height of the dimension numerals and letters. Once the variables are inserted, they become default variables. Also, the **UNIT** command must be used to assign the number of decimals you want when dimension units are displayed. Two decimal places are used for inches and none are used for millimeters. Architectural units of feet and inches can be used as an option for units.

Dimensions can be placed over or within the dimension lines by using the command **DIMTAD** turned on or off (Fig. 30.36).

Once a series of horizontal dimensions are linked end to end, the **CONTINUE** command can be used to attach successive dimension lines as shown in Fig. 30.37. The CONTINUE command suppresses the first extension line since one is left from the previous dimension.

BASELINE dimensioning can be used to give a series of dimensions originating from the same baseline. The **DIMDLI** variable (dimension line increment for continuation) separates automatically the parallel dimension lines (Fig 30.38).

30.20
Dimensioning arcs and circles

Dimensions of circles will be given (Fig. 30.39) based on the size of the circle, unless you override the defaults in the system. The computer follows the same decision process in selecting the style of dimensions as you would when using a pencil. Figure 30.40 shows the process of dimensioning a circle. A point on the circumference is selected with the cursor and the diameter is computed automatically and placed across the circle.

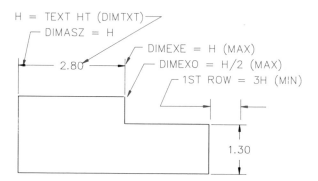

FIG. 30.35 The dimensioning variables are based on the height of the lettering used on a drawing, usually about one-eighth of an inch high.

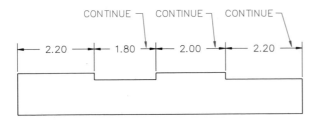

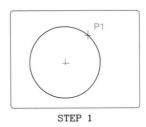 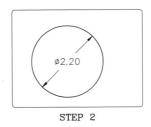

FIG. 30.37 When dimensions are placed end to end, the command CONTINUE can be used to specify the "first extension line origin" after the first dimension line has been drawn.

FIG. 30.40 Dimensioning a circle

Step 1 Command: DIAMETER
Select arc or circle: Select P1.

Step 2 Dimension text ⟨2.20⟩: Return to accept this dimension. The diametric dimension is drawn from P1 through the center.

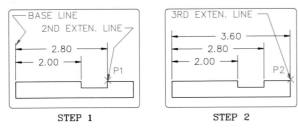

FIG. 30.38 The BASELINE command

Step 1 Command: BASELINE
Dimension the first line as shown in the previous figure. When prompted for the "first extension line" for the next dimension, respond with BASELINE, and you will be asked for the "second extension line." Select P1.

Step 2 The second dimension line will be drawn. Respond to "Dim:" with BASELINE again and when asked for "second extension line," select P2. Continue in this manner for any number of dimensions using the same base line.

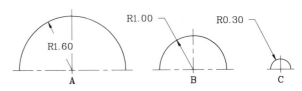

FIG. 30.41 Arcs will be dimensioned by one of the formats given here, depending upon the size of the radius.

Arcs are dimensioned with an R placed in front of the dimension of the radius (Fig. 30.41). The commands for dimensioning arcs are almost identical to those for dimensioning circles.

LEADERS can be used to position diametric and radial dimensions at desired locations while still giving the radius and diameter symbols in front of the measurements. Figure 30.42 gives the steps of dimensioning an arc.

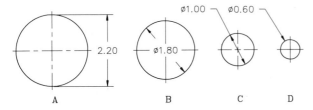

FIG. 30.39 Types of dimensions that are available for dimensioning circles with AutoCAD.

30.21
Dimensioning angles

Dimensioning an angle begins by selecting the two lines of the angle and the location for the dimension line arc. The dimension value will be centered in the arc between the arrows, where room permits. Varia-

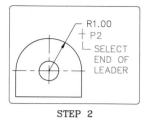

STEP 1 STEP 2

FIG. 30.42 Dimensioning with a LEADER

Step 1 Command: LEADER
Leader start: P1.

Step 2 To point: Select P2)(CR)
Dimension text ⟨1.00⟩: (CR) to accept
(The leader and dimension are drawn.)

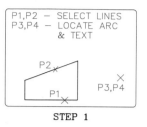
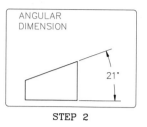

STEP 1 STEP 2

FIG. 30.44 The ANGULAR command

Step 1 Command: ANGULAR
Select first line: P1
Select second line: P2
Enter dimension line arc location: P3 (CR)

Step 2 Enter text location: P4. The angular
dimension is drawn.

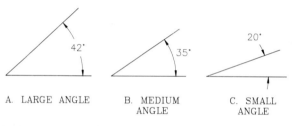

A. LARGE ANGLE   B. MEDIUM   C. SMALL
                    ANGLE   ANGLE

FIG. 30.43 Angles will be dimensioned in one of
the following formats using AutoCAD.

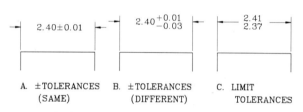

A. ±TOLERANCES   B. ±TOLERANCES   C. LIMIT
   (SAME)   (DIFFERENT)   TOLERANCES

FIG. 30.45 Dimensions can be toleranced in any
of these three formats.

tions in dimensional angles (Fig. 30.43) occur from
insufficient space for a particular angular dimension.
Figure 30.44 shows the commands for dimensioning
an angle.

30.22
Toleranced dimensions

Dimensions can be toleranced automatically with any
of the forms in Fig. 30.45 by setting the **Dim Vars,
DIMTP** (plus tolerance), **DIMTM** (minus tolerance),
and the **DIMTOL** to ON. As long as DIMTOL is on,
all dimensions will have tolerance applied to them. By
turning on the **Dim Vars, DIMLIM,** AutoCAD will
compute the upper and lower limits of the dimension
and apply them to the dimension line. Figure 30.46
gives the steps of giving tolerance limits on a dimension.

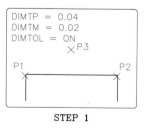
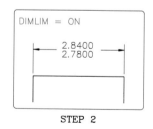

STEP 1 STEP 2

FIG. 30.46 Toleranced dimensions

Step 1 To tolerance a dimension, set Dim Vars,
DIMTOL, to on. Set DIMTP (plus tolerance) and
DIMTM (minus tolerance) to the desired values. Set
DIMLIM to on to convert the tolerances into the limit
form.

Step 2 Dimension the line by using P1, P2, and P3
to specify the first extension line, second extension
line, and the dimension locations. The toleranced
dimension is drawn.

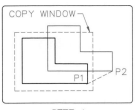
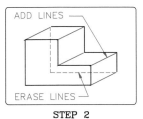

STEP 1   STEP 2

FIG. 30.47 An oblique pictorial

Step 1   Draw the frontal surface of the oblique and copy this view at P2, which is the desired angle and distance from P1.

Step 2   Connect the corner points and erase the invisible lines to complete the oblique.

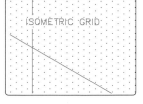

A. ORTHOGRAPHIC GRID   B. ISOMETRIC GRID

FIG. 30.48 The orthographic grid is called the STANDARD style of the SNAP mode, Part A. The ISOMETRIC style of the SNAP mode is shown at Part B.

30.23
Oblique pictorials

An oblique pictorial can be constructed as shown in Fig. 30.47. The front view, an orthographic view, is constructed and copied behind the first view at the angle desired for the receding axis. With OSNAP, the endpoints are connected for the visible edges and invisible lines are erased. Circles can be drawn as true circles on the true-size front surface, but circular features should be avoided on the receding planes.

STEP 1   STEP 2

FIG. 30.49 The isometric pictorial

Step 1   Draw the front view of the isometric pictorial. Copy this front view at its proper location.

Step 2   Connect the corner points and erase the invisible lines. The cursor lines can be moved into three positions using Control E or ISOPLANE.

30.24
Isometric pictorials

The **STYLE** command which is found under the SNAP command can be used to change the rectangular GRID (called **STANDARD, S**) to **ISOMETRIC (I),** where the dots are aligned vertically and at 30 degrees with the horizontal axis (Fig. 30.48). In addition, the lines of the cursor on the screen align with two of the isometric axes. They can be rotated by **Control E** or by activating **ISOPLANE** from the screen menu. When **ORTHO** is turned on, lines are forced to be drawn parallel to the isometric axes. The steps for constructing an isometric using the grid are shown in Fig. 30.49.

To depict circular features in isometric, construct a four-center ellipse that is one unit across its diameter (Fig. 30.50). This ellipse is made into a BLOCK and inserted where needed. It can be scaled to fit different drawings by giving it an x-value factor during insertion. Also, it can be rotated to fit each isometric plane.

An example of an isometric pictorial with elliptical features is shown in Fig. 30.51. The ellipse was inserted using the four-center ellipse BLOCK (see Fig. 30.50). The size of the block is changed by entering the size factor when you are prompted for the x-value. It is helpful if the block is designed to be one unit across (a UNIT block), so its size can be changed for factor specification. For example, a BLOCK 2.48″ in size will be multiplied by an x-value of 2.48.

Remember to use a star in front of the block, *BLOCK, if you wish to edit it after insertion.

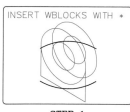

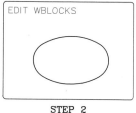

STEP 1   STEP 2

FIG. 30.50 The four-center ellipse block

Step 1   Construct a rhombus that is one unit along both axes. Using the four-center ellipse method, construct an ellipse using four arcs. (The rhombus should be drawn on a construction layer that can be turned off so the lines will not show.)

Step 2   Make a WBLOCK of this ellipse using the center the insertion point. This block can be INSERTED as a *BLOCK so it can be broken when needed.

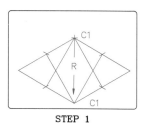

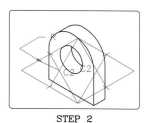

STEP 1   STEP 2

FIG. 30.51 An isometric drawing with elliptical features

Step 1   The Unit Block of the ellipse developed in the previous figure is inserted and sized (x scale factor) to match the drawing's dimensions. The ellipse can be rotated as a BLOCK to fit any of the three isometric surfaces.

Step 2   The back side of the isometric is drawn. The hidden lines are removed with the BREAK command.

30.25

Applications

Any type of conceivable drawing can be made with practical usage of computer graphics.

FIG. 30.52 A circle graph is an example of drawing plotted on a flat bed plotter. (Courtesy of Houston Instrument, Inc.)

GRAPHS   The circle graph (Fig. 30.52) was generated from drawing commands that have been covered in this chapter.

TWO-D DRAWING   The Lufkin oil well pump (Fig. 30.53) was drawn using VersaCAD(tm) software. Notice that this drawing is a combination of points, lines, arcs, and circles.

THREE-D DRAWINGS   A three-dimensional drawing of a pulley was drawn in Fig. 30.54 by using elements of the pulley's geometry.

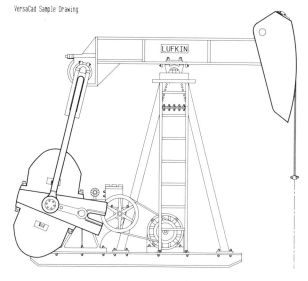

VersaCad Sample Drawing

FIG. 30.53 A two-dimensional drawing of an oil pump. (Courtesy of T&W Systems, Inc.)

Figure name:4-step pulley

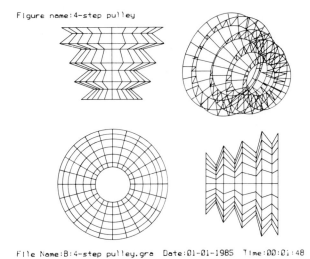

File Name:B:4-step pulley.gro   Date:01-01-1985   Time:00:01:48

FIG. 30.54 A two-dimensional and a three-dimensional plot of a pulley. (Courtesy of Brodhead-Garrett CO., CADD-23.)

30.26
Summary of autoCAD

The previous examples have given the more basic applications of AutoCAD. Many commands and options within commands were omitted because of space limitations. AutoCAD and similar software packages offer significant assistance to the drafter, designer, and engineer in efficiently preparing production drawings.

Suppliers of microcomputer software

Sources of microcomputer software programs that can be used for the production of graphics are

AutoCAD
Autodesk Inc.
2320 Marinship Way
Sausalito, CA 94965
(415) 331-0356

Personal Designer
Computervision Inc.
50 Mall Road
Burlington, MA 01803
(617) 275-1800

CADD-23
Brodhead Garrett Co.
4560 East 71st Street
Cleveland, OH 44105
1-800-321-6730

VersaCAD T&W Systems
7372 Prince Drive Suite 106
Huntington Beach, CA 92647
(714) 847-9960

CADplan
Personal CAD Systems, Inc.
981 University Avenue
Los Gatos, CA 95030
(408) 354-7193

APPENDIXES

Contents

APPENDIX 1
ABBREVIATIONS (ANSI Z 32.13)

Word	Abbreviation	Word	Abbreviation	Word	Abbreviation
Abbreviate	ABBR	Bench mark	BM	Centigrade	C
Absolute	ABS	Between	BET.	Centigram	CG
Account	ACCT	Between centers	BC	Centimeter	cm
Actual	ACT.	Between		Chain	CH
Adapter	ADPT	perpendiculars	BP	Chamfer	CHAM
Addendum	ADD.	Bevel	BEV	Change notice	CN
Adjust	ADJ	Bill of material	B/M	Change order	CO
Advance	ADV	Birmingham wire gage	BWG	Channel	CHAN
After	AFT.	Blueprint	BP	Check	CHK
Aggregate	AGGR	Board	BD	Check valve	CV
Air condition	AIR COND	Boiler	BLR	Chemical	CHEM
Airport	AP	Bolt circle	BC	Chord	CHD
Airplane	APL	Both sides	BS	Circle	CIR
Allowance	ALLOW	Bottom	BOT	Circuit	CKT
Alloy	ALY	Bottom chord	BC	Circular	CIR
Alteration	ALT	Boundary	BDY	Circular pitch	CP
Alternate	ALT	Bracket	BRKT	Circumference	CIRC
Alternating current	AC	Brake horsepower	BHP	Clockwise	CW
Altitude	ALT	Brass	BRS	Coated	CTD
Aluminum	AL	Brazing	BRZG	Cold drawn	CD
American National		Break	BRK	Cold drawn steel	CDS
Standard	AMER NATL STD	Breaker	BKR	Cold finish	CF
American wire gage	AWG	Bridge	BRDG	Cold punched	CP
Ammeter	AM	Brinnell hardness	BH	Cold rolled	CR
Amount	AMT	British Standard	BR STD	Cold rolled steel	CRS
Ampere	AMP	British Thermal Units	BTU	Column	COL
Anneal	ANL	Broach	BRO	Combination	COMB.
Antenna	ANT.	Bronze	BRZ	Combustion	COMB
Apparatus	APP	Brown & Sharp (Wire gage,		Commutator	COMM
Appendix	APPX	same as AWG)	B&S	Company	CO
Approved	APPD	Building	BLDG	Concentric	CONC
Approximate	APPROX	Bulkhead	BHD	Concrete	CONC
Arc weld	ARC/W	Bureau	BU	Condition	COND
Area	A	Bureau of Standards	BU STD	Connect	CONN
Armature	ARM.	Bushing	BUSH.	Contact	CONT
Asbestos	ASB	Button	BUT.	Cord	CD
Asphalt	ASPH	Buzzer	BUZ	Corporation	CORP
Assembly	ASSY	By-pass	BYP	Corrugate	CORR
Association	ASSN			Cotter	COT
Atomic	AT	Cabinet	CAB.	Counterclockwise	CCW
Authorized	AUTH	Cadmium plate	CD PL	Counterbore	CBORE
Auxiliary	AUX	Calculate	CALC	Counterdrill	CDRILL
Avenue	AVE	Calibrate	CAL	Counterpunch	CPUNCH
Average	AVG	Calorie	CAL	Countersink	CSK
Avoirdupois	AVDP	Capacitor	CAP	Coupling	CPLG
Azimuth	AZ	Cap screw	CAP SCR	Crank	CRK
		Case harden	CH	Cross section	XSECT
Babbitt	BAB	Cast iron	CI	Cubic	CU
Back pressure	BP	Cast steel	CS	Cubic centimeter	cc
Balance	BAL	Casting	CSTG	Cubic feet per minute	CFM
Ball bearing	BB	Castle nut	CAS NUT	Cubic feet per second	CFS
Barometer	BAR	Catalogue	CAT.	Cubic foot	CU FT
Barrel	BBL	Cement	CEM	Cubic inch	CU IN.
Base line	BL	Center	CTR	Cubic meter	CU M
Base plate	BP	Centerline	CL	Cubic yard	CU YD
Battery	BAT.	Center of gravity	CG	Current	CUR
Bearing	BRG	Center to center	C to C	Cylinder	CYL

Cont.

APPENDIX 1
ABBREVIATIONS (ANSI Z 32.13) (Cont.)

Word	Abbreviation	Word	Abbreviation	Word	Abbreviation
Decimal	DEC	Fillister	FIL	Illustrate	ILLUS
Dedendum	DED	Filter	FLT	Inch	(") IN.
Degree	(°) DEG	Finish	FIN.	Inches per second	IPS
Department	DEPT	Finish all over	FAO	Include	INCL
Design	DSGN	Flange	FLG	Industrial	IND
Detail	DET	Flat head	FH	Information	INFO
Develop	DEV	Fluid	FL	Inside diameter	ID
Diagonal	DIAG	Focus	FOC	Instrument	INST
Diagram	DIAG	Foot	(') FT	Insulate	INS
Diameter	DIA	Forging	FORG	Interior	INT
Diametrical pitch	DP	Forward	FWD	Internal	INT
Dimension	DIM.	Foundation	FDN	Intersect	INT
Direct current	DC	Foundry	FDRY	Iron	I
Discharge	DISCH	Frequency	FREQ	Irregular	IRREG
Distance	DIST	Front	FR		
District	DIST			Jack	J
Ditto	DO.			Joint	JT
Dovetail	DVTL	Gage	GA	Junction	JCT
Dowel	DWL	Gallon	GAL	Junction box	JB
Down	DN	Galvanize	GALV		
Dozen	DOZ	Galvanized iron	GI	Key	K
Drafting	DFTG	Galvanized steel	GS	Keyseat	KST
Draftsman	DFTSMN	Gasket	GSKT	Keyway	KWY
Drawing	DWG	General	GEN	Kiln-dried	KD
Drill	DR	Government	GOVT	Kip (1000 lb)	K
Drive	DR	Governor	GOV	Knots	KN
Drop forge	DF	Grade	GR		
		Grade line	GL	Laboratory	LAB
		Gram	G	Lateral	LAT
Each	EA	Gravity	G	Latitude	LAT
East	E	Grind	GRD	Left	L
Eccentric	ECC	Groove	GRV	Left hand	LH
Effective	EFF	Ground	GRD	Length	LG
Elbow	ELL	Gypsum	GYP	Letter	LTR
Electric	ELEC			Light	LT
Elevation	ELEV			Line	L
Engineer	ENGR	Half-round	½ RD	Logarithm	LOG.
Equal	EQ	Handle	HDL	Lubricate	LUB
Equipment	EQUIP.	Hanger	HGR	Lumber	LBR
Equivalent	EQUIV	Hard	H		
Estimate	EST	Hard-drawn	HD		
Exterior	EXT	Harden	HDN	Machine	MACH
Extra heavy	X HVY	Hardware	HDW	Malleable	MALL
Extra strong	X STR	Head	HD	Malleable iron	MI
		Headless	HDLS	Manhole	MH
		Headquarters	HQ	Manual	MAN.
Fabricate	FAB	Heat	HT	Manufacture	MFR
Face to face	F to F	Heat treat	HT TR	Material	MATL
Fahrenheit	F	Hexagon	HEX	Maximum	MAX
Fairing	FAIR.	High-pressure	HP	Mechanical	MECH
Far side	FS	High-speed	HS	Mechanism	MECH
Federal	FED.	Horizontal	HOR	Median	MED
Feet	(') FT	Horsepower	HP	Metal	MET.
Feet per minute	FPM	Hot rolled	HR	Meter (Instrument or	
Feet per second	FPS	Hot rolled steel	HRS	measure of length)	M
Field	FLD	Hour	HR	Miles	MI
Figure	FIG.	Hundredweight	CWT	Miles per gallon	MPG
Fillet	FIL	Hydraulic	HYD	Miles per hour	MPH

APPENDIX 1
ABBREVIATIONS (ANSI Z 32.13) (Cont.)

Word	Abbreviation	Word	Abbreviation	Word	Abbreviation
Millimeter	MM	Precast	PRCST	Shaft	SFT
Minimum	MIN	Prefabricated	PREFAB	Sketch	SK
Minute	(') MIN	Preferred	PFD	Sleeve	SLV
Miscellaneous	MISC	Prepare	PREP	Slide	SL
Mixture	MIX.	Pressure	PRESS.	Slotted	SLOT.
Model	MOD	Pressure angle	PA.	Socket	SOC
Month	MO	Process	PROC	Solder	SLD
Morse taper	MOR T	Production	PROD	South	S
Multiple	MULT	Profile	PF	Space	SP
		Project	PROJ	Special	SPL
National	NATL	Proof	PRF	Specific gravity	SP GR
Near side	NS			Spherical	SPHER
Negative	NEG	Quadrant	QUAD	Spot faced	SF
Neutral	NEUT	Quart	QT	Spring	SPG
Nipple	NIP.	Quarter	QTR	Square	SQ
Nominal	NOM	Quarter-round	¼ RD	Stainless	STN
Normal	NOR			Stainless steel	SST
North	N	Radial	RAD	Standard	STD
Not to scale	NTS	Radius	R	Station	STA
Number	NO.	Railroad	RR	Steel	STL
		Ream	RM	Stock	STK
Obsolete	OBS	Received	RECD	Straight	STR
Octagon	OCT	Record	REC	Structural	STR
Ohm	Ω	Rectangle	RECT	Substitute	SUB
On center	OC	Reference	REF	Summary	SUM.
Opposite	OPP	Reference line	REF L	Supply	SUP
Optical	OPT	Relief	REL	Surface	SUR
Original	ORIG	Remove	REM	Symbol	SYM
Ounce	OZ	Require	REQ	Symmetrical	SYM
Outlet	OUT.	Required	REQD	System	SYS
Outside diameter	OD	Return	RET.		
Outside face	OF	Revise	REV	Tangent	TAN.
Outside radius	OR	Revolution	REV	Taper	TPR
Overall	OA	Revolutions per minute	RPM	Technical	TECH
		Rheostat	RHEO	Temperature	TEMP
Pack	PK	Right	R	Template	TEMP
Packing	PKG	Right hand	RH	Tensile strength	TS
Parallel	PAR.	Rivet	RIV	Tension	TENS.
Part	PT	Rockwell hardness	RH	Thick	THK
Patent	PAT.	Roller bearing	RB	Thousand	M
Permanent	PERM	Room	RM	Thousand pound	KIP
Perpendicular	PERP	Root diameter	RD	Thread	THD
Photograph	PHOTO	Root mean square	RMS	Tolerance	TOL
Piece	PC	Rough	RGH	Tongue & groove	T&G
Pint	PT	Round	RD	Tool steel	TS
Pitch	P	Rubber	RUB.	Tooth	T
Pitch circle	PC			Total	TOT
Pitch diameter	PD	Safety	SAF	Transfer	TRANS
Plastic	PLSTC	Sand blast	SD BL	Typical	TYP
Plate	PL	Schedule	SCH		
Point	PT	Screen	SCRN	Ultimate	ULT
Polish	POL	Screw	SCR	Unit	U
Position	POS	Sea level	SL	Universal	UNIV
Positive	POS	Second	SEC		
Pound	LB	Section	SECT	Vacuum	VAC
Pounds per square inch	PSI	Separate	SEP	Valve	V
Power	PWR	Set screw	SS	Variable	VAR

Cont.

APPENDIX 1
ABBREVIATIONS (ANSI Z 32.13) (Cont.)

Word	Abbreviation	Word	Abbreviation	ABBREVIATIONS FOR COLORS	
Vertical	VERT	West	W	Amber	AMB
Volt	V	Width	W	Black	BLK
Voltmeter	VM	Wood	WD	Blue	BLU
Volume	VOL	Woodruff	WDF	Brown	BRN
		Wrought iron	WI	Green	GRN
Washer	WASH.			Orange	ORN
Watt	W	Yard	YD	White	WHT
Weight	WT	Year	YR	Yellow	YEL

APPENDIX 2
CONVERSION TABLES

Length conversions

Angstrom units	$\times\ 1 \times 10^{-10}$	= meters
	$\times\ 1 \times 10^{-4}$	= microns
	$\times\ 1.650\ 763\ 73 \times 10^{-4}$	= wavelengths of orange-red line of krypton 86
Cables	$\times\ 120$	= fathoms
	$\times\ 720$	= feet
	$\times\ 219.456$	= meters
Fathoms	$\times\ 6$	= feet
	$\times\ 1.828\ 8$	= meters
Feet	$\times\ 12$	= inches
	$\times\ 0.3048$	= meters
Furlongs	$\times\ 660$	= feet
	$\times\ 201.168$	= meters
	$\times\ 220$	= yards
Inches	$\times\ 2.54 \times 10^{8}$	= Angstroms
	$\times\ 25.4$	= millimeters
	$\times\ 8.333\ 33 \times 10^{-2}$	= feet
Kilometers	$\times\ 3.280\ 839 \times 10^{3}$	= feet
	$\times\ 0.62$	= miles
	$\times\ 0.539\ 956$	= nautical miles
	$\times\ 0.621\ 371$	= statute miles
	$\times\ 1.093\ 613 \times 10^{3}$	= yards
Light-years	$\times\ 9.460\ 55 \times 10^{12}$	= kilometers
	$\times\ 5.878\ 51 \times 10^{12}$	= statute miles
Meters	$\times\ 1 \times 10^{10}$	= Angstroms
	$\times\ 3.280\ 839\ 9$	= feet
	$\times\ 39.370\ 079$	= inches
	$\times\ 1.093\ 61$	= yards
Microns	$\times\ 10^{4}$	= Angstroms
	$\times\ 10^{-4}$	= centimeters
	$\times\ 10^{-6}$	= meters
Nautical Miles (International)	$\times\ 8.439\ 049$	= cables
	$\times\ 6.076\ 115\ 49 \times 10^{3}$	= feet
	$\times\ 1.852 \times 10^{3}$	= meters
	$\times\ 1.150\ 77$	= statute miles

APPENDIX 2
CONVERSION TABLES (Cont.)

Length conversions		
Statute Miles	$\times$ 5.280 $\times$ 10^3	= feet
	$\times$ 8	= furlongs
	$\times$ 6.336 0 $\times$ 10^4	= inches
	$\times$ 1.609 34	= kilometers
	$\times$ 8.689 7 $\times$ 10^{-1}	= nautical miles
Miles	$\times$ 10^{-3}	= inches
	$\times$ 2.54 $\times$ 10^{-2}	= millimeters
	$\times$ 25.4	= micrometers
	$\times$ 0.61	= kilometers
Yards	$\times$ 3	= feet
	$\times$ 9.144 $\times$ 10^{-1}	= meters
Feet/hour	$\times$ 3.048 $\times$ 10^{-4}	= kilometers/hour
	$\times$ 1.645 788 $\times$ 10^{-4}	= knots
Feet/minute	$\times$ 0.3048	= meters/minute
	$\times$ 5.08 $\times$ 10^{-3}	= meters/second
Feet/second	$\times$ 1.097 28	= kilometers/hour
	$\times$ 18.288	= meters/minute
Kilometers/hour	$\times$ 3.280 839 $\times$ 10^3	= feet/hour
	$\times$ 54.680 66	= feet/minute
	$\times$ 0.277 777	= meters/second
	$\times$ 0.621 371	= miles/hour
Kilometers/minute	$\times$ 3.280 839 $\times$ 10^3	= feet/minute
	$\times$ 37.282 27	= miles/hour
Knots	$\times$ 6.076 115 $\times$ 10^3	= feet/hour
	$\times$ 101.268 5	= feet/minute
	$\times$ 1.687 809	= feet/second
	$\times$ 1.852	= kilometers/hour
	$\times$ 30.866	= meters/minute
	$\times$ 0.514 4	= meters/second
	$\times$ 1.150 77	= statute miles/hour
Meters/hour	$\times$ 3.280 839	= feet/hour
	$\times$ 88	= feet/minute
	$\times$ 1.466	= feet/second
	$\times$ 1 $\times$ 10^{-3}	= kilometers/hour
	$\times$ 1.667 $\times$ 10^{-2}	= meters/minute
Feet/second^2	$\times$ 1.097 28	= kilometers/hour/second
	$\times$ 0.304 8	= meters/second^2

Area conversions		
Acres	$\times$ 4.046 85 $\times$ 10^{-3}	= square kilometers
	$\times$ 4.046 856 $\times$ 10^3	= square meters
	$\times$ 4.356 0 $\times$ 10^4	= square feet
Ares	$\times$ 2.471 053 8 $\times$ 10^{-2}	= acres
	$\times$ 1	= square dekameters
	$\times$ 10^2	= square meters
Barns	$\times$ 1 $\times$ 10^{-28}	= square meters
Circular mils	$\times$ 1 $\times$ 10^{-6}	= circular inches
	$\times$ 5.067 074 8 $\times$ 10^{-4}	= square millimeters
	$\times$ 0.785 398 1	= square mils

Cont.

APPENDIX 2
CONVERSION TABLES (Cont.)

Area conversions

Hectares	$\times$ 2.471 05	= acres
	$\times$ 10^2	= ares
	$\times$ 10^4	= square meters
Square feet	$\times$ 2.295 684 $\times$ 10^{-5}	= acres
	$\times$ 9.290 3 $\times$ 10^{-4}	= ares
	$\times$ 144	= square inches
	$\times$ 9.290 304 $\times$ 10^{-2}	= square meters
Square inches	$\times$ 1.273 239 5 $\times$ 10^6	= circular mils
	$\times$ 6.944 4 $\times$ 10^{-3}	= square feet
	$\times$ 6.451 6 $\times$ 10^{-4}	= square meters
Square kilometers	$\times$ 247.105 38	= acres
	$\times$ 1.076 391 0 $\times$ 10^7	= square feet
	$\times$ 1.000	= cubic meters
	$\times$ 1.307 950 6	= cubic yards
	$\times$ 219.969	= imperial gallons
Liters	$\times$ 10^3	= cubic centimeters
	$\times$ 1.000 $\times$ 10^6	= cubic millimeters
	$\times$ 1.000 $\times$ 10^{-3}	= cubic meters
	$\times$ 61.023 74	= cubic inches
	$\times$ 3.531 5 $\times$ 10^{-2}	= cubic feet
	$\times$ 1.307 95 $\times$ 10^{-3}	= cubic yards
	$\times$ 0.22	= gallons
	$\times$ 0.219 969	= imperial gallons
	$\times$ 0.879 877	= imperial quarts
Imperial pints	$\times$ 0.125	= imperial gallons
	$\times$ 0.568 261	= liters
	$\times$ 20	= imperial fluid ounces
	$\times$ 0.5	= imperial quarts
	$\times$ 568.260 9	= cubic centimeters
Imperial quarts	$\times$ 1.136 52 $\times$ 10^3	= cubic centimeters
	$\times$ 69.354 8	= cubic inches
	$\times$ 1.136 522 8	= liters

Power conversions

British Thermal Units/hour	$\times$ 2.928 7 $\times$ 10^{-4}	= kilowatts
	$\times$ 0.292 875	= watts
BTU/minute	$\times$ 1.757 25 $\times$ 10^{-2}	= kilowatts
BTU/pound	$\times$ 2.324 4	= joules/gram
BTU/second	$\times$ 1.413 91	= horsepower
	$\times$ 107.514	= kilogrammeters/second
	$\times$ 1.054 35	= kilowatts
	$\times$ 1.054 35 $\times$ 10^3	= watts
Foot-pound-force/hour	$\times$ 5.050 $\times$ 10^{-7}	= horsepower
	$\times$ 3.766 16 $\times$ 10^{-7}	= kilowatts
Foot-pound-force/ minute	$\times$ 3.030 303 $\times$ 10^{-5}	= horsepower
	$\times$ 2.259 70 $\times$ 10^{-2}	= joules/second
	$\times$ 2.259 70 $\times$ 10^{-5}	= kilowatts
Horsepower	$\times$ 42.435 6	= BTU/minute
	$\times$ 550	= footpounds/second
	$\times$ 0.746	= kilowatts
	$\times$ 746	= joules/second

APPENDIX 2
CONVERSION TABLES (Cont.)

Power conversions

Kilogrammeters/second	$\times$ 9.806 65	= watts
Kilowatts	$\times$ 3.414 43 $\times 10^3$	= BTU/hour
	$\times$ 2.655 22 $\times 10^6$	= footpounds/hour
	$\times$ 4.425 37 $\times 10^4$	= footpounds/minute
	$\times$ 737.562	= footpounds/second
	$\times$ 1.019 726 $\times 10^7$	= gramcentimeters/second
	$\times$ 1.341 02	= horsepower
	$\times$ 3.6 $\times 10^6$	= joules/hour
	$\times 10^3$	= joules/second
	$\times$ 3.671 01 $\times 10^5$	= kilogrammeters/hour
	$\times$ 999.835	= international watt
Watts	$\times$ 44.253 7	= footpounds/minute
	$\times$ 1.341 02 $\times 10^{-3}$	= horsepower
	$\times$ 1	= joules/second

Time conversions

(No attempt has been made in this brief treatment to correlate solar, mean solar, sidereal, and mean sidereal days.)

Mean solar days	$\times$ 24	= mean solar hours
Mean solar hours	$\times$ 3.600 $\times 10^3$	= mean solar seconds
	$\times$ 60	= mean solar minutes

Angle conversions

Degrees	$\times$ 60	= minutes
	$\times$ 1.745 329 3 $\times 10^{-2}$	= radians
Degrees/foot	$\times$ 5.726 145 $\times 10^{-4}$	= radians/centimeter
Degrees/minute	$\times$ 2.908 8 $\times 10^{-4}$	= radians/second
	$\times$ 4.629 629 $\times 10^{-5}$	= revolutions/second
Degrees/second	$\times$ 1.745 329 3 $\times 10^{-2}$	= radians/second
	$\times$ 0.166	= revolutions/minute
	$\times$ 2.77 $\times 10^{-3}$	= revolutions/second
Minutes	$\times$ 1.667 $\times 10^{-2}$	= degrees
	$\times$ 2.908 8 $\times 10^{-4}$	= radians
	$\times$ 60	= seconds
Radians	$\times$ 0.159 154	= circumferences
	$\times$ 57.295 77	= degrees
	$\times$ 3.437 746 $\times 10^3$	= minutes
Seconds	$\times$ 2.777 $\times 10^{-4}$	= degrees
	$\times$ 1.667 $\times 10^{-2}$	= minutes
	$\times$ 4.848 136 8 $\times 10^{-6}$	= radians
Steradians	$\times$ 0.159 154 9	= hemispheres
	$\times$ 7.957 74 $\times 10^{-2}$	= spheres
	$\times$ 0.636 619 7	= spherical right angles

Mass conversions

Grains	$\times$ 6.479 8 $\times 10^{-2}$	= grams
	$\times$ 2.285 71 $\times 10^{-3}$	= avoirdupois ounces

Cont.

APPENDIX 2
CONVERSION TABLES (Cont.)

Mass conversions

Grams	$\times$ 15.432 358	= grains
	$\times$ 3.527 396 $\times 10^{-2}$	= avoirdupois ounces
	$\times$ 2.204 62 $\times 10^{-3}$	= avoirdupois pounds
Kilograms	$\times$ 564.383 4	= avoirdupois drams
	$\times$ 2.204 622 6	= avoirdupois pounds
	$\times$ 2.2	= pounds
	$\times$ 9.842 065 $\times 10^{-4}$	= long tons
	$\times$ 10^{-3}	= metric tons
	$\times$ 1.102 31 $\times 10^{-3}$	= short tons
Avoirdupois ounces	$\times$ 28.349 5	= grams
	$\times$ 6.25 $\times 10^{-2}$	= avoirdupois pounds
	$\times$ 0.911 458	= troy ounces
Avoirdupois pounds	$\times$ 256	= drams
	$\times$ 4.535 923 7 $\times 10^2$	= grams
	$\times$ 0.453 592 4	= kilograms
	$\times$ 16	= ounces
Long tons	$\times$ 2.24 $\times 10^3$	= avoirdupois pounds
	$\times$ 1.106 046 9	= metric tons
	$\times$ 1.12	= short tons
Metric tons	$\times$ 10^3	= kilograms
	$\times$ 2.204 622 $\times 10^3$	= avoirdupois pounds
Short tons	$\times$ 2 $\times 10^3$	= avoirdupois pounds
	$\times$ 907.184 74	= kilograms

Force conversions

Dynes	$\times$ 10^{-5}	= newtons
Newtons	$\times$ 10^5	= dynes
	$\times$ 0.224 808	= pounds-force
Pounds	$\times$ 4.448 22	= newtons

Energy conversions

British Thermal Units (thermochemical)	$\times$ 1.054 35 $\times 10^3$	= joules
	$\times$ 2.928 27 $\times 10^{-4}$	= kilowatthours
	$\times$ 1.054 35 $\times 10^3$	= wattseconds
Foot-pound-force	$\times$ 1.355 818 0	= joules
	$\times$ 0.138 255	= kilogramforce-meters
	$\times$ 3.766 16 $\times 10^{-7}$	= kilowatthours
	$\times$ 1.355 818 0	= newtonmeters
Joules	$\times$ 9.484 5 $\times 10^{-4}$	= British Thermal Units
	$\times$ 0.737 562	= foot-pounds-force
	$\times$ 0.101 971 6	= kilogramforce-meters
	$\times$ 2.777 7 $\times 10^{-7}$	= kilowatthours
	$\times$ 1	= wattseconds
Kilogramforce-meters	$\times$ 9.287 7 $\times 10^{-3}$	= British Thermal Units
	$\times$ 7.233 01	= foot-pounds-force
	$\times$ 9.806 65	= joules
	$\times$ 9.806 65	= newtonmeters
	$\times$ 2.724 0 $\times 10^{-3}$	= watthours

APPENDIX 2
CONVERSION TABLES (Cont.)

Energy conversions

Kilowatthours	$\times$ 3.409 52 $\times$ 10^3	= British Thermal Units
	$\times$ 2.655 22 $\times$ 10^6	= foot-pounds-force
	$\times$ 1.341 02	= horsepowerhours
	$\times$ 3.6 $\times$ 10^6	= joules
	$\times$ 3.670 98 $\times$ 10^5	= kilogramforce-meters
Newtonmeters	$\times$ 0.101 971	= kilogramforce-meters
	$\times$ 0.737 562	= poundforce-feet
Watthours	$\times$ 3.414 43	= British Thermal Units
	$\times$ 2.655 22 $\times$ 10^3	= foot-pounds-force
	$\times$ 3.6 $\times$ 10^3	= joules
	$\times$ 3.670 98 $\times$ 10^2	= kilogramforce-meters

Pressure conversions

Atmospheres	$\times$ 1.013 25	= bars
	$\times$ 1.033 23 $\times$ 10^3	= grams/square centimeter
	$\times$ 1.033 23 $\times$ 10^7	= grams/square meter
	$\times$ 14.696 0	= pounds/square inch
	$\times$ 760	= torrs
	$\times$ 101	= kilopascals
Bars	$\times$ 0.986 923	= atmospheres
	$\times$ 10^6	= baryes
	$\times$ 1.019 716 $\times$ 10^7	= grams/square meter
	$\times$ 1.019 716 $\times$ 10^4	= kilogramsforce/square meter
	$\times$ 14.503 8	= poundsforce/square inch
Baryes	$\times$ 10^{-6}	= bars
Inches of mercury	$\times$ 3.386 4 $\times$ 10^{-2}	= bars
	$\times$ 345.316	= kilogramsforce/square meter
	$\times$ 70.726 2	= poundsforce/square foot
Pascal	$\times$ 1	= newton/square meter

APPENDIX 3
LOGARITHMS OF NUMBERS

N	0	1	2	3	4	5	6	7	8	9
1.0	.0000	.0043	.0086	.0128	.0170	.0212	.0253	.0294	.0334	.0374
1.1	.0414	.0453	.0492	.0531	.0569	.0607	.0645	.0682	.0719	.0755
1.2	.0792	.0828	.0864	.0899	.0934	.0969	.1004	.1038	.1072	.1106
1.3	.1139	.1173	.1206	.1239	.1271	.1303	.1335	.1367	.1399	.1430
1.4	.1461	.1492	.1523	.1553	.1584	.1614	.1644	.1673	.1703	.1732
1.5	.1761	.1790	.1818	.1847	.1875	.1903	.1931	.1959	.1987	.2014
1.6	.2041	.2068	.2095	.2122	.2148	.2175	.2201	.2227	.2253	.2279
1.7	.2304	.2330	.2355	.2380	.2405	.2430	.2455	.2480	.2504	.2529
1.8	.2553	.2577	.2601	.2625	.2648	.2672	.2695	.2718	.2742	.2765
1.9	.2788	.2810	.2833	.2856	.2878	.2900	.2923	.2945	.2967	.2989
2.0	.3010	.3032	.3054	.3075	.3096	.3118	.3139	.3160	.3181	.3201
2.1	.3222	.3243	.3263	.3284	.3304	.3324	.3345	.3365	.3385	.3404
2.2	.3424	.3444	.3464	.3483	.3502	.3522	.3541	.3560	.3579	.3598
2.3	.3617	.3636	.3655	.3674	.3692	.3711	.3729	.3747	.3766	.3784
2.4	.3802	.3820	.3838	.3856	.3874	.3892	.3909	.3927	.3945	.3962
2.5	.3979	.3997	.4014	.4031	.4048	.4065	.4082	.4099	.4116	.4133
2.6	.4150	.4166	.4183	.4200	.4216	.4232	.4249	.4265	.4281	.4298
2.7	.4314	.4330	.4346	.4362	.4378	.4393	.4409	.4425	.4440	.4456
2.8	.4472	.4487	.4502	.4518	.4533	.4548	.4564	.4579	.4594	.4609
2.9	.4624	.4639	.4654	.4669	.4683	.4698	.4713	.4728	.4742	.4757
3.0	.4771	.4786	.4800	.4814	.4829	.4843	.4857	.4871	.4886	.4900
3.1	.4914	.4928	.4942	.4955	.4969	.4983	.4997	.5011	.5024	.5038
3.2	.5051	.5065	.5079	.5092	.5105	.5119	.5132	.5145	.5159	.5172
3.3	.5185	.5198	.5211	.5224	.5237	.5250	.5263	.5276	.5289	.5302
3.4	.5315	.5328	.5340	.5353	.5366	.5378	.5391	.5403	.5416	.5428
3.5	.5441	.5453	.5465	.5478	.5490	.5502	.5514	.5527	.5539	.5551
3.6	.5563	.5575	.5587	.5599	.5611	.5623	.5635	.5647	.5658	.5670
3.7	.5682	.5694	.5705	.5717	.5729	.5740	.5752	.5763	.5775	.5786
3.8	.5798	.5809	.5821	.5832	.5843	.5855	.5866	.5877	.5888	.5899
3.9	.5911	.5922	.5933	.5944	.5955	.5966	.5977	.5988	.5999	.6010
4.0	.6021	.6031	.6042	.6053	.6064	.6075	.6085	.6096	.6107	.6117
4.1	.6128	.6138	.6149	.6160	.6170	.6180	.6191	.6201	.6212	.6222
4.2	.6232	.6243	.6253	.6263	.6274	.6284	.6294	.6304	.6314	.6325
4.3	.6335	.6345	.6355	.6365	.6375	.6385	.6395	.6405	.6415	.6425
4.4	.6435	.6444	.6454	.6464	.6474	.6484	.6493	.6503	.6513	.6522
4.5	.6532	.6542	.6551	.6561	.6571	.6580	.6590	.6599	.6609	.6618
4.6	.6628	.6637	.6646	.6656	.6665	.6675	.6684	.6693	.6702	.6712
4.7	.6721	.6730	.6739	.6749	.6758	.6767	.6776	.6785	.6794	.6803
4.8	.6812	.6821	.6830	.6839	.6848	.6857	.6866	.6875	.6884	.6893
4.9	.6902	.6911	.6920	.6928	.6937	.6946	.6955	.6964	.6972	.6981
5.0	.6990	.6998	.7007	.7016	.7024	.7033	.7042	.7050	.7059	.7067
5.1	.7076	.7084	.7093	.7101	.7110	.7118	.7126	.7135	.7143	.7152
5.2	.7160	.7168	.7177	.7185	.7193	.7202	.7210	.7218	.7226	.7235
5.3	.7243	.7251	.7259	.7267	.7275	.7284	.7292	.7300	.7308	7316
5.4	.7324	.7332	.7340	.7348	.7356	.7364	.7372	.7380	.7388	.7396
N	0	1	2	3	4	5	6	7	8	9

APPENDIX 3
LOGARITHMS OF NUMBERS (Cont.)

N	0	1	2	3	4	5	6	7	8	9
5.5	.7404	.7412	.7419	.7427	.7435	.7443	.7451	.7459	.7466	.7474
5.6	.7482	.7490	.7497	.7505	.7513	.7520	.7528	.7536	.7543	.7551
5.7	.7559	.7566	.7574	.7582	.7589	.7597	.7604	.7612	.7619	.7627
5.8	.7634	.7642	.7649	.7657	.7664	.7672	.7679	.7686	.7694	.7701
5.9	.7709	.7716	.7723	.7731	.7738	.7745	.7752	.7760	.7767	.7774
6.0	.7782	.7789	.7796	.7803	.7810	.7818	.7825	.7832	.7839	.7846
6.1	.7853	.7860	.7868	.7875	.7882	.7889	.7896	.7903	.7910	.7917
6.2	.7924	.7931	.7938	.7945	.7952	.7959	.7966	.7973	.7980	.7987
6.3	.7993	.8000	.8007	.8014	.8021	.8028	.8035	.8041	.8048	.8055
6.4	.8062	.8069	.8075	.8082	.8089	.8096	.8102	.8109	.8116	.8122
6.5	.8129	.8136	.8142	.8149	.8156	.8162	.8169	.8176	.8182	.8189
6.6	.8195	.8202	.8209	.8215	.8222	.8228	.8235	.8241	.8248	.8254
6.7	.8261	.8267	.8274	.8280	.8287	.8293	.8299	.8306	.8312	.8319
6.8	.8325	.8331	.8338	.8344	.8351	.8357	.8363	.8370	.8376	.8382
6.9	.8388	.8395	.8401	.8407	.8414	.8420	.8426	.8432	.8439	.8445
7.0	.8451	.8457	.8463	.8470	.8476	.8482	.8488	.8494	.8500	.8506
7.1	.8513	.8519	.8525	.8531	.8537	.8543	.8549	.8555	.8561	.8567
7.2	.8573	.8579	.8585	.8591	.8597	.8603	.8609	.8615	.8621	.8627
7.3	.8633	.8639	.8645	.8651	.8657	.8663	.8669	.8675	.8681	.8686
7.4	.8692	.8698	.8704	.8710	.8716	.8722	.8727	.8733	.8739	.8745
7.5	.8751	.8756	.8762	.8768	.8774	.8779	.8785	.8791	.8797	.8802
7.6	.8808	.8814	.8820	.8825	.8831	.8837	.8842	.8848	.8854	.8859
7.7	.8865	.8871	.8876	.8882	.8887	.8893	.8899	.8904	.8910	.8915
7.8	.8921	.8927	.8932	.8938	.8943	.8949	.8954	.8960	.8965	.8971
7.9	.8976	.8982	.8987	.8993	.8998	.9004	.9009	.9015	.9020	.9025
8.0	.9031	.9036	.9042	.9047	.9053	.9058	.9063	.9069	.9074	.9079
8.1	.9085	.9090	.9096	.9101	.9106	.9112	.9117	.9122	.9128	.9133
8.2	.9138	.9143	.9149	.9154	.9159	.9165	.9170	.9175	.9180	.9186
8.3	.9191	.9196	.9201	.9206	.9212	.9217	.9222	.9227	.9232	.9238
8.4	.9243	.9248	.9253	.9258	.9263	.9269	.9274	.9279	.9284	.9289
8.5	.9294	.9299	.9304	.9309	.9315	.9320	.9325	.9330	.9335	.9340
8.6	.9345	.9350	.9355	.9360	.9365	.9370	.9375	.9380	.9385	.9390
8.7	.9395	.9400	.9405	.9410	.9415	.9420	.9425	.9430	.9435	.9440
8.8	.9445	.9450	.9455	.9460	.9465	.9469	.9474	.9479	.9484	.9489
8.9	.9494	.9499	.9504	.9509	.9513	.9518	.9523	.9528	.9533	.9538
9.0	.9542	.9547	.9552	.9557	.9562	.9566	.9571	.9576	.9581	.9586
9.1	.9590	.9595	.9600	.9605	.9609	.9614	.9619	.9624	.9628	.9633
9.2	.9638	.9643	.9647	.9652	.9657	.9661	.9666	.9671	.9675	.9680
9.3	.9685	.9689	.9694	.9699	.9703	.9708	.9713	.9717	.9722	.9727
9.4	.9731	.9736	.9741	.9745	.9750	.9754	.9759	.9763	.9768	.9773
9.5	.9777	.9782	.9786	.9791	.9795	.9800	.9805	.9809	.9814	.9818
9.6	.9823	.9827	.9832	.9836	.9841	.9845	.9850	.9854	.9859	.9863
9.7	.9868	.9872	.9877	.9881	.9886	.9890	.9894	.9899	.9903	.9908
9.8	.9912	.9917	.9921	.9926	.9930	.9934	.9939	.9943	.9948	.9952
9.9	.9956	.9961	.9965	.9969	.9974	.9978	.9983	.9987	.9991	.9996
N	0	1	2	3	4	5	6	7	8	9

APPENDIX 4
VALUES OF TRIGONOMETRIC FUNCTIONS

Degrees	Radians	Sine	Tangent	Cotangent	Cosine		
0° 00′	.0000	.0000	.0000		1.0000	1.5708	90° 00′
10′	.0029	.0029	.0029	343.77	1.0000	1.5679	50′
20′	.0058	.0058	.0058	171.89	1.0000	1.5650	40′
30′	.0087	.0087	.0087	114.59	1.0000	1.5621	30′
40′	.0116	.0116	.0116	85.940	.9999	1.5592	20′
50′	.0145	.0145	.0145	68.750	.9999	1.5563	10′
1° 00′	.0175	.0175	.0175	57.290	.9998	1.5533	89° 00′
10′	.0204	.0204	.0204	49.104	.9998	1.5504	50′
20′	.0233	.0233	.0233	42.964	.9997	1.5475	40′
30′	.0262	.0262	.0262	38.188	.9997	1.5446	30′
40′	.0291	.0291	.0291	34.368	.9996	1.5417	20′
50′	.0320	.0320	.0320	31.242	.9995	1.5388	10′
2° 00′	.0349	.0349	.0349	28.636	.9994	1.5359	88° 00′
10′	.0378	.0378	.0378	26.432	.9993	1.5330	50′
20′	.0407	.0407	.0407	24.542	.9992	1.5301	40′
30′	.0436	.0436	.0437	22.904	.9990	1.5272	30′
40′	.0465	.0465	.0466	21.470	.9989	1.5243	20′
50′	.0495	.0494	.0495	20.206	.9988	1.5213	10′
3° 00′	.0524	.0523	.0524	19.081	.9986	1.5184	87° 00′
10′	.0553	.0552	.0553	18.075	.9985	1.5155	50′
20′	.0582	.0581	.0582	17.169	.9983	1.5126	40′
30′	.0611	.0610	.0612	16.350	.9981	1.5097	30′
40′	.0640	.0640	.0641	15.605	.9980	1.5068	20′
50′	.0669	.0669	.0670	14.924	.9978	1.5039	10′
4° 00′	.0698	.0698	.0699	14.301	.9976	1.5010	86° 00′
10′	.0727	.0727	.0729	13.727	.9974	1.4981	50′
20′	.0756	.0756	.0758	13.197	.9971	1.4952	40′
30′	.0785	.0785	.0787	12.706	.9969	1.4923	30′
40′	.0814	.0814	.0816	12.251	.9967	1.4893	20′
50′	.0844	.0843	.0846	11.826	.9964	1.4864	10′
5° 00′	.0873	.0872	.0875	11.430	.9962	1.4835	85° 00′
10′	.0902	.0901	.0904	11.059	.9959	1.4806	50′
20′	.0931	.0929	.0934	10.712	.9957	1.4777	40′
30′	.0960	.0958	.0963	10.385	.9954	1.4748	30′
40′	.0989	.0987	.0992	10.078	.9951	1.4719	20′
50′	.1018	.1016	.1022	9.7882	.9948	1.4690	10′
6° 00′	.1047	.1045	.1051	9.5144	.9945	1.4661	84° 00′
10′	.1076	.1074	.1080	9.2553	.9942	1.4632	50′
20′	.1105	.1103	.1110	9.0098	.9939	1.4603	40′
30′	.1134	.1132	.1139	8.7769	.9936	1.4573	30′
40′	.1164	.1161	.1169	8.5555	.9932	1.4544	20′
50′	.1193	.1190	.1198	8.3450	.9929	1.4515	10′
7° 00′	.1222	.1219	.1228	8.1443	.9925	1.4486	83° 00′
10′	.1251	.1248	.1257	7.9530	.9922	1.4457	50′
20′	.1280	.1276	.1287	7.7704	.9918	1.4428	40′
30′	.1309	.1305	.1317	7.5958	.9914	1.4399	30′
40′	.1338	.1334	.1346	7.4287	.9911	1.4370	20′
50′	.1367	.1363	.1376	7.2687	.9907	1.4341	10′
8° 00′	.1396	.1392	.1405	7.1154	.9903	1.4312	82° 00′
10′	.1425	.1421	.1435	6.9682	.9899	1.4283	50′
20′	.1454	.1449	.1465	6.8269	.9894	1.4254	40′
30′	.1484	.1478	.1495	6.6912	.9890	1.4224	30′
40′	.1513	.1507	.1524	6.5606	.9886	1.4195	20′
50′	.1542	.1536	.1554	6.4348	.9881	1.4166	10′
9° 00′	.1571	.1564	.1584	6.3138	.9877	1.4137	81° 00′
		Cosine	Cotangent	Tangent	Sine	Radians	Degrees

APPENDIX 4
VALUES OF TRIGONOMETRIC FUNCTIONS (Cont.)

Degrees	Radians	Sine	Tangent	Cotangent	Cosine		
9° 00'	.1571	.1564	.1584	6.3138	.9877	1.4137	81° 00'
10'	.1600	.1593	.1614	6.1970	.9872	1.4108	50'
20'	.1629	.1622	.1644	6.0844	.9868	1.4079	40'
30'	.1658	.1650	.1673	5.9758	.9863	1.4050	30'
40'	.1687	.1679	.1703	5.8708	.9858	1.4021	20'
50'	.1716	.1708	.1733	5.7694	.9853	1.3992	10'
10° 00'	.1745	.1736	.1763	5.6713	.9848	1.3963	80° 00'
10'	.1774	.1765	.1793	5.5764	.9843	1.3934	50'
20'	.1804	.1794	.1823	5.4845	.9838	1.3904	40'
30'	.1833	.1822	.1853	5.3955	.9833	1.3875	30'
40'	.1862	.1851	.1883	5.3093	.9827	1.3846	20'
50'	.1891	.1880	.1914	5.2257	.9822	1.3817	10'
11° 00'	.1920	.1908	.1944	5.1446	.9816	1.3788	79° 00'
10'	.1949	.1937	.1974	5.0658	.9811	1.3759	50'
20'	.1978	.1965	.2004	4.9894	.9805	1.3730	40'
30'	.2007	.1994	.2035	4.9152	.9799	1.3701	30'
40'	.2036	.2022	.2065	4.8430	.9793	1.3672	20'
50'	.2065	.2051	.2095	4.7729	.9787	1.3643	10'
12° 00'	.2094	.2079	.2126	4.7046	.9781	1.3614	78° 00'
10'	.2123	.2108	.2156	4.6382	.9775	1.3584	50'
20'	.2153	.2136	.2186	4.5736	.9769	1.3555	40'
30'	.2182	.2164	.2217	4.5107	.9763	1.3526	30'
40'	.2211	.2193	.2247	4.4494	.9757	1.3497	20'
50'	.2240	.2221	.2278	4.3897	.9750	1.3468	10'
13° 00'	.2269	.2250	.2309	4.3315	.9744	1.3439	77° 00'
10'	.2298	.2278	.2339	4.2747	.9737	1.3410	50'
20'	.2327	.2306	.2370	4.2193	.9730	1.3381	40'
30'	.2356	.2334	.2401	4.1653	.9724	1.3352	30'
40'	.2385	.2363	.2432	4.1126	.9717	1.3323	20'
50'	.2414	.2391	.2462	4.0611	.9710	1.3294	10'
14° 00'	.2443	.2419	.2493	4.0108	.9703	1.3265	76° 00'
10'	.2473	.2447	.2524	3.9617	.9696	1.3235	50'
20'	.2502	.2476	.2555	3.9136	.9689	1.3206	40'
30'	.2531	.2504	.2586	3.8667	.9681	1.3177	30'
40'	.2560	.2532	.2617	3.8208	.9674	1.3148	20'
50'	.2589	.2560	.2648	3.7760	.9667	1.3119	10'
15° 00'	.2618	.2588	.2679	3.7321	.9659	1.3090	75° 00'
10'	.2647	.2616	.2711	3.6891	.9652	1.3061	50'
20'	.2676	.2644	.2742	3.6470	.9644	1.3032	40'
30'	.2705	.2672	.2773	3.6059	.9636	1.3003	30'
40'	.2734	.2700	.2805	3.5656	.9628	1.2974	20'
50'	.2763	.2728	.2836	3.5261	.9621	1.2945	10'
16° 00'	.2793	.2756	.2867	3.4874	.9613	1.2915	74° 00'
10'	.2822	.2784	.2899	3.4495	.9605	1.2886	50'
20'	.2851	.2812	.2931	3.4124	.9596	1.2857	40'
30'	.2880	.2840	.2962	3.3759	.9588	1.2828	30'
40'	.2909	.2868	.2994	3.3402	.9580	1.2799	20'
50'	.2938	.2896	.3026	3.3052	.9572	1.2770	10'
17° 00'	.2967	.2924	.3057	3.2709	.9563	1.2741	73° 00'
10'	.2996	.2952	.3089	3.2371	.9555	1.2712	50'
20'	.3025	.2979	.3121	3.2041	.9546	1.2683	40'
30'	.3054	.3007	.3153	3.1716	.9537	1.2654	30'
40'	.3083	.3035	.3185	3.1397	.9528	1.2625	20'
50'	.3113	.3062	.3217	3.1084	.9520	1.2595	10'
18° 00'	.3142	.3090	.3249	3.0777	.9511	1.2566	72° 00'
		Cosine	Cotangent	Tangent	Sine	Radians	Degrees

Cont.

APPENDIX 4
VALUES OF TRIGONOMETRIC FUNCTIONS (Cont.)

Degrees	Radians	Sine	Tangent	Cotangent	Cosine		
18° 00′	.3142	.3090	.3249	3.0777	.9511	1.2566	72° 00′
10′	.3171	.3118	.3281	3.0475	.9502	1.2537	50′
20′	.3200	.3145	.3314	3.0178	.9492	1.2508	40′
30′	.3229	.3173	.3346	2.9887	.9483	1.2479	30′
40′	.3258	.3201	.3378	2.9600	.9474	1.2450	20′
50′	.3287	.3228	.3411	2.9319	.9465	1.2421	10′
19° 00′	.3316	.3256	.3443	2.9042	.9455	1.2392	71° 00′
10′	.3345	.3283	.3476	2.8770	.9446	1.2363	50′
20′	.3374	.3311	.3508	2.8502	.9436	1.2334	40′
30′	.3403	.3338	.3541	2.8239	.9426	1.2305	30′
40′	.3432	.3365	.3574	2.7980	.9417	1.2275	20′
50′	.3462	.3393	.3607	2.7725	.9407	1.2246	10′
20° 00′	.3491	.3420	.3640	2.7475	.9397	1.2217	70° 00′
10′	.3520	.3448	.3673	2.7228	.9387	1.2188	50′
20′	.3549	.3475	.3706	2.6985	.9377	1.2159	40′
30′	.3578	.3502	.3739	2.6746	.9367	1.2130	30′
40′	.3607	.3529	.3772	2.6511	.9356	1.2101	20′
50′	.3636	.3557	.3805	2.6279	.9346	1.2072	10′
21° 00′	.3665	.3584	.3839	2.6051	.9336	1.2043	69° 00′
10′	.3694	.3611	.3872	2.5826	.9325	1.2014	50′
20′	.3723	.3638	.3906	2.5605	.9315	1.1985	40′
30′	.3752	.3665	.3939	2.5386	.9304	1.1956	30′
40′	.3782	.3692	.3973	2.5172	.9293	1.1926	20′
50′	.3811	.3719	.4006	2.4960	.9283	1.1897	10′
22° 00′	.3840	.3746	.4040	2.4751	.9272	1.1868	68° 00′
10′	.3869	.3773	.4074	2.4545	.9261	1.1839	50′
20′	.3898	.3800	.4108	2.4342	.9250	1.1810	40′
30′	.3927	.3827	.4142	2.4142	.9239	1.1781	30′
40′	.3956	.3854	.4176	2.3945	.9228	1.1752	20′
50′	.3985	.3881	.4210	2.3750	.9216	1.1723	10′
23° 00′	.4014	.3907	.4245	2.3559	.9205	1.1694	67° 00′
10′	.4043	.3934	.4279	2.3369	.9194	1.1665	50′
20′	.4072	.3961	.4314	2.3183	.9182	1.1636	40′
30′	.4102	.3987	.4348	2.2998	.9171	1.1606	30′
40′	.4131	.4014	.4383	2.2817	.9159	1.1577	20′
50′	.4160	.4041	.4417	2.2637	.9147	1.1548	10′
24° 00′	.4189	.4067	.4452	2.2460	.9135	1.1519	66° 00′
10′	.4218	.4094	.4487	2.2286	.9124	1.1490	50′
20′	.4247	.4120	.4522	2.2113	.9112	1.1461	40′
30′	.4276	.4147	.4557	2.1943	.9100	1.1432	30′
40′	.4305	.4173	.4592	2.1775	.9088	1.1403	20′
50′	.4334	.4200	.4628	2.1609	.9075	1.1374	10′
25° 00′	.4363	.4226	.4663	2.1445	.9063	1.1345	65° 00′
10′	.4392	.4253	.4699	2.1283	.9051	1.1316	50′
20′	.4422	.4279	.4734	2.1123	.9038	1.1286	40′
30′	.4451	.4305	.4770	2.0965	.9026	1.1257	30′
40′	.4480	.4331	.4806	2.0809	.9013	1.1228	20′
50′	.4509	.4358	.4841	2.0655	.9001	1.1199	10′
26° 00′	.4538	.4384	.4877	2.0503	.8988	1.1170	64° 00′
10′	.4567	.4410	.4913	2.0353	.8975	1.1141	50′
20′	.4596	.4436	.4950	2.0204	.8962	1.1112	40′
30′	.4625	.4462	.4986	2.0057	.8949	1.1083	30′
40′	.4654	.4488	.5022	1.9912	.8936	1.1054	20′
50′	.4683	.4514	.5059	1.9768	.8923	1.1025	10′
27° 00′	.4712	.4540	.5095	1.9626	.8910	1.0996	63° 00′
		Cosine	Cotangent	Tangent	Sine	Radians	Degrees

APPENDIX 4
VALUES OF TRIGONOMETRIC FUNCTIONS (Cont.)

Degrees	Radians	Sine	Tangent	Cotangent	Cosine		
27° 00′	.4712	.4540	.5095	1.9626	.8910	1.0996	63° 00′
10′	.4741	.4566	.5132	1.9486	.8897	1.0966	50′
20′	.4771	.4592	.5169	1.9347	.8884	1.0937	40′
30′	.4800	.4617	.5206	1.9210	.8870	1.0908	30′
40′	.4829	.4643	.5243	1.9074	.8857	1.0879	20′
50′	.4858	.4669	.5280	1.8940	.8843	1.0850	10′
28° 00′	.4887	.4695	.5317	1.8807	.8829	1.0821	62° 00′
10′	.4916	.4720	.5354	1.8676	.8816	1.0792	50′
20′	.4945	.4746	.5392	1.8546	.8802	1.0763	40′
30′	.4974	.4772	.5430	1.8418	.8788	1.0734	30′
40′	.5003	.4797	.5467	1.8291	.8774	1.0705	20′
50′	.5032	.4823	.5505	1.8165	.8760	1.0676	10′
29° 00′	.5061	.4848	.5543	1.8040	.8746	1.0647	61° 00′
10′	.5091	.4874	.5581	1.7917	.8732	1.0617	50′
20′	.5120	.4899	.5619	1.7796	.8718	1.0588	40′
30′	.5149	.4924	.5658	1.7675	.8704	1.0559	30′
40′	.5178	.4950	.5696	1.7556	.8689	1.0530	20′
50′	.5207	.4975	.5735	1.7437	.8675	1.0501	10′
30° 00′	.5236	.5000	.5774	1.7321	.8660	1.0472	60° 00′
10′	.5265	.5025	.5812	1.7205	.8646	1.0443	50′
20′	.5294	.5050	.5851	1.7090	.8631	1.0414	40′
30′	.5323	.5075	.5890	1.6977	.8616	1.0385	30′
40′	.5352	.5100	.5930	1.6864	.8601	1.0356	20′
50′	.5381	.5125	.5969	1.6753	.8587	1.0327	10′
31° 00′	.5411	.5150	.6009	1.6643	.8572	1.0297	59° 00′
10′	.5440	.5175	.6048	1.6534	.8557	1.0268	50′
20′	.5469	.5200	.6088	1.6426	.8542	1.0239	40′
30′	.5498	.5225	.6128	1.6319	.8526	1.0210	30′
40′	.5527	.5250	.6168	1.6212	.8511	1.0181	20′
50′	.5556	.5275	.6208	1.6107	.8496	1.0152	10′
32° 00′	.5585	.5299	.6249	1.6003	.8480	1.0123	58° 00′
10′	.5614	.5324	.6289	1.5900	.8465	1.0094	50′
20′	.5643	.5348	.6330	1.5798	.8450	1.0065	40′
30′	.5672	.5373	.6371	1.5697	.8434	1.0036	30′
40′	.5701	.5398	.6412	1.5597	.8418	1.0007	20′
50′	.5730	.5422	.6453	1.5497	.8403	.9977	10′
33° 00′	.5760	.5446	.6494	1.5399	.8387	.9948	57° 00′
10′	.5789	.5471	.6536	1.5301	.8371	.9919	50′
20′	.5818	.5495	.6577	1.5204	.8355	.9890	40′
30′	.5847	.5519	.6619	1.5108	.8339	.9861	30′
40′	.5876	.5544	.6661	1.5013	.8323	.9832	20′
50′	.5905	.5568	.6703	1.4919	.8307	.9803	10′
34° 00′	.5934	.5592	.6745	1.4826	.8290	.9774	56° 00′
10′	.5963	.5616	.6787	1.4733	.8274	.9745	50′
20′	.5992	.5640	.6830	1.4641	.8258	.9716	40′
30′	.6021	.5664	.6873	1.4550	.8241	.9687	30′
40′	.6050	.5688	.6916	1.4460	.8225	.9657	20′
50′	.6080	.5712	.6959	1.4370	.8208	.9628	10′
35° 00′	.6109	.5736	.7002	1.4281	.8192	.9599	55° 00′
10′	.6138	.5760	.7046	1.4193	.8175	.9570	50′
20′	.6167	.5783	.7089	1.4106	.8158	.9541	40′
30′	.6196	.5807	.7133	1.4019	.8141	.9512	30′
40′	.6225	.5831	.7177	1.3934	.8124	.9483	20′
50′	.6254	.5854	.7221	1.3848	.8107	.9454	10′
36° 00′	.6283	.5878	.7265	1.3764	.8090	.9425	54° 00′
		Cosine	Cotangent	Tangent	Sine	Radians	Degrees

Cont.

APPENDIX 4
VALUES OF TRIGONOMETRIC FUNCTIONS (Cont.)

Degrees	Radians	Sine	Tangent	Cotangent	Cosine		
36° 00′	.6283	.5878	.7265	1.3764	.8090	.9425	54° 00′
10′	.6312	.5901	.7310	1.3680	.8073	.9396	50′
20′	.6341	.5925	.7355	1.3597	.8056	.9367	40′
30′	.6370	.5948	.7400	1.3514	.8039	.9338	30′
40′	.6400	.5972	.7445	1.3432	.8021	.9308	20′
50′	.6429	.5995	.7490	1.3351	.8004	.9279	10′
37° 00′	.6458	.6018	.7536	1.3270	.7986	.9250	53° 00′
10′	.6487	.6041	.7581	1.3190	.7969	.9221	50′
20′	.6516	.6065	.7627	1.3111	.7951	.9192	40′
30′	.6545	.6088	.7673	1.3032	.7934	.9163	30′
40′	.6574	.6111	.7720	1.2954	.7916	.9134	20′
50′	.6603	.6134	.7766	1.2876	.7898	.9105	10′
38° 00′	.6632	.6157	.7813	1.2799	.7880	.9076	52° 00′
10′	.6661	.6180	.7860	1.2723	.7862	.9047	50′
20′	.6690	.6202	.7907	1.2647	.7844	.9018	40′
30′	.6720	.6225	.7954	1.2572	.7826	.8988	30′
40′	.6749	.6248	.8002	1.2497	.7808	.8959	20′
50′	.6778	.6271	.8050	1.2423	.7790	.8930	10′
39° 00′	.6807	.6293	.8098	1.2349	.7771	.8901	51° 00′
10′	.6836	.6316	.8146	1.2276	.7753	.8872	50′
20′	.6865	.6338	.8195	1.2203	.7735	.8843	40′
30′	.6894	.6361	.8243	1.2131	.7716	.8814	30′
40′	.6923	.6383	.8292	1.2059	.7698	.8785	20′
50′	.6952	.6406	.8342	1.1988	.7679	.8756	10′
40° 00′	.6981	.6428	.8391	1.1918	.7660	.8727	50° 00′
10′	.7010	.6450	.8441	1.1847	.7642	.8698	50′
20′	.7039	.6472	.8491	1.1778	.7623	.8668	40′
30′	.7069	.6494	.8541	1.1708	.7604	.8639	30′
40′	.7098	.6517	.8591	1.1640	.7585	.8610	20′
50′	.7127	.6539	.8642	1.1571	.7566	.8581	10′
41° 00′	.7156	.6561	.8693	1.1504	.7547	.8552	49° 00′
10′	.7185	.6583	.8744	1.1436	.7528	.8523	50′
20′	.7214	.6604	.8796	1.1369	.7509	.8494	40′
30′	.7243	.6626	.8847	1.1303	.7490	.8465	30′
40′	.7272	.6648	.8899	1.1237	.7470	.8436	20′
50′	.7301	.6670	.8952	1.1171	.7451	.8407	10′
42° 00′	.7330	.6691	.9004	1.1106	.7431	.8378	48° 00′
10′	.7359	.6713	.9057	1.1041	.7412	.8348	50′
20′	.7389	.6734	.9110	1.0977	.7392	.8319	40′
30′	.7418	.6756	.9163	1.0913	.7373	.8290	30′
40′	.7447	.6777	.9217	1.0850	.7353	.8261	20′
50′	.7476	.6799	.9271	1.0786	.7333	.8232	10′
43° 00′	.7505	.6820	.9325	1.0724	.7314	.8203	47° 00′
10′	.7534	.6841	.9380	1.0661	.7294	.8174	50′
20′	.7563	.6862	.9435	1.0599	.7274	.8145	40′
30′	.7592	.6884	.9490	1.0538	.7254	.8116	30′
40′	.7621	.6905	.9545	1.0477	.7234	.8087	20′
50′	.7650	.6926	.9601	1.0416	.7214	.8058	10′
44° 00′	.7679	.6947	.9657	1.0355	.7193	.8029	46° 00′
10′	.7709	.6967	.9713	1.0295	.7173	.7999	50′
20′	.7738	.6988	.9770	1.0235	.7153	.7970	40′
30′	.7767	.7009	.9827	1.0176	.7133	.7941	30′
40′	.7796	.7030	.9884	1.0117	.7112	.7912	20′
50′	.7825	.7050	.9942	1.0058	.7092	.7883	10′
45° 00′	.7854	.7071	1.0000	1.0000	.7071	.7854	45° 00′
		Cosine	Cotangent	Tangent	Sine	Radians	Degrees

APPENDIX 5
WEIGHTS AND MEASURES

UNITED STATES SYSTEM

LINEAR MEASURE

Inches	Feet	Yards	Rods	Furlongs	Miles
1.0 =	.08333 =	.02778 =	.0050505 =	.00012626 =	.00001578
12.0 =	1.0 =	.33333 =	.0606061 =	.00151515 =	.00018939
36.0 =	3.0 =	1.0 =	.1818182 =	.00454545 =	.00056818
198.0 =	16.5 =	5.5 =	1.0 =	.025 =	.003125
7920.0 =	660.0 =	220.0 =	40.0 =	1.0 =	.125
63360.0 =	5280.0 =	1760.0 =	320.0 =	8.0 =	1.0

SQUARE AND LAND MEASURE

Sq. Inches	Square Feet	Square Yards	Sq. Rods	Acres	Sq. Miles
1.0 =	.006944 =	.000772			
144.0 =	1.0 =	.111111			
1296.0 =	9.0 =	1.0 =	.03306 =	.000207	
39204.0 =	272.25 =	30.25 =	1.0 =	.00625 =	.0000098
	43560.0 =	4840.0 =	160.0 =	1.0 =	.0015625
		3097600.0 =	102400.0 =	640.0 =	1.0

AVOIRDUPOIS WEIGHTS

Grains	Drams	Ounces	Pounds	Tons
1.0 =	.03657 =	.002286 =	.00143 =	.0000000714
27.34375 =	1.0 =	.0625 =	.003906 =	.00000195
437.5 =	16.0 =	1.0 =	.0625 =	.00003125
7000.0 =	256.0 =	16.0 =	1.0 =	.0005
14000000.0 =	512000.0 =	32000.0 =	2000.0 =	1.0

DRY MEASURE

Pints	Quarts	Pecks	Cubic Feet	Bushels
1.0 =	.5 =	.0625 =	.01945 =	.01563
2.0 =	1.0 =	.125 =	.03891 =	.03125
16.0 =	8.0 =	1.0 =	.31112 =	.25
51.42627 =	25.71314 =	3.21414 =	1.0 =	.80354
64.0 =	32.0 =	4.0 =	1.2445 =	1.0

LIQUID MEASURE

Gills	Pints	Quarts	U. S. Gallons	Cubic Feet
1.0 =	.25 =	.125 =	.03125 =	.00418
4.0 =	1.0 =	.5 =	.125 =	.01671
8.0 =	2.0 =	1.0 =	.250 =	.03342
32.0 =	8.0 =	4.0 =	1.0 =	.1337
			7.48052 =	1.0

METRIC SYSTEM

UNITS

Length—Meter : Mass—Gram : Capacity—Liter
for pure water at 4°C. (39.2°F.)
1 cubic decimeter or 1 liter = 1 kilogram

$$1000 \text{ Milli} \begin{cases} meters \text{ (mm)} \\ grams \text{ (mg)} \\ liters \text{ (ml)} \end{cases} = 100 \text{ Centi} \begin{cases} meters \text{ (cm)} \\ grams \text{ (cg)} \\ liters \text{ (cl)} \end{cases} = 10 \text{ Deci} \begin{cases} meters \text{ (dm)} \\ grams \text{ (dg)} \\ liters \text{ (dl)} \end{cases} = 1 \begin{cases} meter \\ gram \\ liter \end{cases}$$

$$1000 \begin{cases} meters \\ grams \\ liters \end{cases} = 100 \text{ Deka} \begin{cases} meters \text{ (dkm)} \\ grams \text{ (dkg)} \\ liters \text{ (dkl)} \end{cases} = 10 \text{ Hecto} \begin{cases} meters \text{ (hm)} \\ grams \text{ (hg)} \\ liters \text{ (hl)} \end{cases} = 1 \text{ Kilo} \begin{cases} meter \text{ (km)} \\ gram \text{ (kg)} \\ liter \text{ (kl)} \end{cases}$$

1 Metric Ton	= 1000 Kilograms
100 Square Meters	= 1 Are
100 Ares	= 1 Hectare
100 Hectares	= 1 Square Kilometer

APPENDIX 6
DECIMAL EQUIVALENTS AND TEMPERATURE CONVERSION

DECIMAL EQUIVALENTS — INCH-MILLIMETER CONVERSION TABLE

1/2	1/4	1/8	1/16	1/32	1/64	Decimals	Millimeters
			1	1	1	.015625	.396875
					3	.031250	.793750
				3	5	.046875	1.190625
					7	.062500	1.587500
		1	3	5	9	.078125	1.984375
					11	.093750	2.381250
				7	13	.109375	2.778125
					15	.125000	3.175000
	1		5	9	17	.140625	3.571875
					19	.156250	3.968750
				11	21	.171875	4.365625
					23	.187500	4.762500
		3	7	13	25	.203125	5.159375
					27	.218750	5.556250
				15	29	.234375	5.953125
					31	.250000	6.350000
1						.265625	6.746875

1/2	1/4	1/8	1/16	1/32	1/64	Decimals	Millimeters
	1			9	17	.265625	6.746875
					19	.281250	7.143750
			5			.296875	7.540625
						.312500	7.937500
		3		11	21	.328125	8.334375
					23	.343750	8.731250
			7			.359375	9.128125
						.375000	9.525000
				13	25	.390625	9.921875
					27	.406250	10.318750
				15	29	.421875	10.715625
					31	.437500	11.112500
1						.453125	11.509375

1/2	1/4	1/8	1/16	1/32	1/64	Decimals	Millimeters
				17	33	.515625	13.096875
					35	.531250	13.493750
			9			.546875	13.890625
						.562500	14.287500
		5		19	37	.578125	14.684375
					39	.593750	15.081250
				21	41	.609375	15.478125
			11		43	.625000	15.875000
						.640625	16.271875
				23	45	.656250	16.668750
	3				47	.671875	17.065625
						.687500	17.462500
			13	25	49	.703125	17.859375
					51	.718750	18.256250
		7		27	53	.734375	18.653125
					55	.750000	19.050000
			15	29	57	.765625	19.446875
					59	.781250	19.843750
				31	61	.796875	20.240625
2	4	8	16	32	63	.812500	20.637500
					64		

Decimals	Millimeters
.828125	21.034375
.843750	21.431250
.859375	21.828125
.875000	22.225000
.890625	22.621875
.906250	23.018750
.921875	23.415625
.937500	23.812500
.953125	24.209375
.968750	24.606250
.984375	25.003125
1.000000	25.400000

APPENDIX 6

DECIMAL EQUIVALENTS AND TEMPERATURE CONVERSION (Cont.)

TEMPERATURE CONVERSION

−210 to 0

C.	C. or F.	F.
−134	−210	−346
−129	−200	−328
−123	−190	−310
−118	−180	−292
−112	−170	−274
−107	−160	−256
−101	−150	−238
−95.6	−140	−220
−90.0	−130	−202
−84.4	−120	−184
−78.9	−110	−166
−73.3	−100	−148
−67.8	−90	−130
−62.2	−80	−112
−56.7	−70	−94
−51.1	−60	−76
−45.6	−50	−58
−40.0	−40	−40
−34.4	−30	−22
−28.9	−20	−4
−23.3	−10	14
−17.8	0	32

1 to 25

C.	C. or F.	F.
−17.2	1	33.8
−16.7	2	35.6
−16.1	3	37.4
−15.6	4	39.2
−15.0	5	41.0
−14.4	6	42.8
−13.9	7	44.6
−13.3	8	46.4
−12.8	9	48.2
−12.2	10	50.0
−11.7	11	51.8
−11.1	12	53.6
−10.6	13	55.4
−10.0	14	57.2
−9.44	15	59.0
−8.89	16	60.8
−8.33	17	62.6
−7.78	18	64.4
−7.22	19	66.2
−6.67	20	68.0
−6.11	21	69.8
−5.56	22	71.6
−5.00	23	73.4
−4.44	24	75.2
−3.89	25	77.0

26 to 50

C.	C. or F.	F.
−3.33	26	78.8
−2.78	27	80.6
−2.22	28	82.4
−1.67	29	84.2
−1.11	30	86.0
−0.56	31	87.8
0	32	89.6
0.56	33	91.4
1.11	34	93.2
1.67	35	95.0
2.22	36	96.8
2.78	37	98.6
3.33	38	100.4
3.89	39	102.2
4.44	40	104.0
5.00	41	105.8
5.56	42	107.6
6.11	43	109.4
6.67	44	111.2
7.22	45	113.0
7.78	46	114.8
8.33	47	116.6
8.89	48	118.4
9.44	49	120.2
10.0	50	122.0

51 to 75

C.	C. or F.	F.
10.6	51	123.8
11.1	52	125.6
11.7	53	127.4
12.2	54	129.2
12.8	55	131.0
13.3	56	132.8
13.9	57	134.6
14.4	58	136.4
15.0	59	138.2
15.6	60	140.0
16.1	61	141.8
16.7	62	143.6
17.2	63	145.4
17.8	64	147.2
18.3	65	149.0
18.9	66	150.8
19.4	67	152.6
20.0	68	154.4
20.6	69	156.2
21.1	70	158.0
21.7	71	159.8
22.2	72	161.6
22.8	73	163.4
23.3	74	165.2
23.9	75	167.0

76 to 100

C.	C. or F.	F.
24.4	76	168.8
25.0	77	170.6
25.6	78	172.4
26.1	79	174.2
26.7	80	176.0
27.2	81	177.8
27.8	82	179.6
28.3	83	181.4
28.9	84	183.2
29.4	85	185.0
30.0	86	186.8
30.6	87	188.6
31.1	88	190.4
31.7	89	192.2
32.2	90	194.0
32.8	91	195.8
33.3	92	197.6
33.9	93	199.4
34.4	94	201.2
35.0	95	203.0
35.6	96	204.8
36.1	97	206.6
36.7	98	208.4
37.2	99	210.2
37.8	100	212.0

101 to 340

C.	C. or F.	F.
43	110	230
49	120	248
54	130	266
60	140	284
66	150	302
71	160	320
77	170	338
82	180	356
88	190	374
93	200	392
99	210	410
100	212	413
104	220	428
110	230	446
116	240	464
121	250	482
127	260	500
132	270	518
138	280	536
143	290	554
149	300	572
154	310	590
160	320	608
166	330	626
171	340	644

341 to 490

C.	C. or F.	F.
177	350	662
182	360	680
188	370	698
193	380	716
199	390	734
204	400	752
210	410	770
216	420	788
221	430	806
227	440	824
232	450	842
238	460	860
243	470	878
249	480	896
254	490	914

491 to 750

C.	C. or F.	F.
260	500	932
266	510	950
271	520	968
277	530	986
282	540	1004
288	550	1022
293	560	1040
299	570	1058
304	580	1076
310	590	1094
316	600	1112
321	610	1130
327	620	1148
332	630	1166
338	640	1184
343	650	1202
349	660	1220
354	670	1238
360	680	1256
366	690	1274
371	700	1292
377	710	1310
382	720	1328
388	730	1346
393	740	1364
399	750	1382

INTERPOLATION FACTORS

C.		F.	C.		F.
0.56	1	1.8	3.33	6	10.8
1.11	2	3.6	3.89	7	12.6
1.67	3	5.4	4.44	8	14.4
2.22	4	7.2	5.00	9	16.2
2.78	5	9.0	5.56	10	18.0

$$°F = \frac{9}{5}(°C) + 32$$

$$°C = \frac{5}{9}(°F - 32)$$

NOTE:—The numbers in bold face type refer to the temperature either in degrees Centigrade or Fahrenheit which it is desired to convert into the other scale. If converting from Fahrenheit degrees to Centigrade degrees the equivalent temperature will be found in the left column, while if converting from degrees Centigrade to degrees Fahrenheit, the answer will be found in the column on the right.

APPENDIX 7
WEIGHTS AND SPECIFIC GRAVITIES

Substance	Weight Lb. per Cu. Ft.	Specific Gravity	Substance	Weight Lb. per Cu. Ft.	Specific Gravity
METALS, ALLOYS, ORES			**TIMBER, U. S. SEASONED**		
Aluminum, cast, hammered	165	2.55-2.75	Moisture Content by Weight:		
Brass, cast, rolled	534	8.4-8.7	Seasoned timber 15 to 20%		
Bronze, 7.9 to 14% Sn	509	7.4-8.9	Green timber up to 50%		
Bronze, aluminum	481	7.7	Ash, white, red	40	0.62-0.65
Copper, cast, rolled	556	8.8-9.0	Cedar, white, red	22	0.32-0.38
Copper ore, pyrites	262	4.1-4.3	Chestnut	41	0.66
Gold, cast, hammered	1205	19.25-19.3	Cypress	30	0.48
Iron, cast, pig	450	7.2	Fir, Douglas spruce	32	0.51
Iron, wrought	485	7.6-7.9	Fir, eastern	25	0.40
Iron, spiegel-eisen	468	7.5	Elm, white	45	0.72
Iron, ferro-silicon	437	6.7-7.3	Hemlock	29	0.42-0.52
Iron ore, hematite	325	5.2	Hickory	49	0.74-0.84
Iron ore, hematite in bank	160-180	....	Locust	46	0.73
Iron ore, hematite loose	130-160	....	Maple, hard	43	0.68
Iron ore, limonite	237	3.6-4.0	Maple, white	33	0.53
Iron ore, magnetite	315	4.9-5.2	Oak, chestnut	54	0.86
Iron slag	172	2.5-3.0	Oak, live	59	0.95
Lead	710	11.37	Oak, red, black	41	0.65
Lead ore, galena	465	7.3-7.6	Oak, white	46	0.74
Magnesium, alloys	112	1.74-1.83	Pine, Oregon	32	0.51
Manganese	475	7.2-8.0	Pine, red	30	0.48
Manganese ore, pyrolusite	259	3.7-4.6	Pine, white	26	0.41
Mercury	849	13.6	Pine, yellow, long-leaf	44	0.70
Monel Metal	556	8.8-9.0	Pine, yellow, short-leaf	38	0.61
Nickel	565	8.9-9.2	Poplar	30	0.48
Platinum, cast, hammered	1330	21.1-21.5	Redwood, California	26	0.42
Silver, cast, hammered	656	10.4-10.6	Spruce, white, black	27	0.40-0.46
Steel, rolled	490	7.85	Walnut, black	38	0.61
Tin, cast, hammered	459	7.2-7.5			
Tin ore, cassiterite	418	6.4-7.0			
Zinc, cast, rolled	440	6.9-7.2			
Zinc ore, blende	253	3.9-4.2	**VARIOUS LIQUIDS**		
			Alcohol, 100%	49	0.79
			Acids, muriatic 40%	75	1.20
			Acids, nitric 91%	94	1.50
VARIOUS SOLIDS			Acids, sulphuric 87%	112	1.80
Cereals, oats .... bulk	32	....	Lye, soda 66%	106	1.70
Cereals, barley ....bulk	39	....	Oils, vegetable	58	0.91-0.94
Cereals, corn, rye ....bulk	48	....	Oils, mineral, lubricants	57	0.90-0.93
Cereals, wheat ....bulk	48	....	Water, 4°C. max. density	62.428	1.0
Hay and Straw ....bales	20		Water, 100°C.	59.830	0.9584
Cotton, Flax, Hemp	93	1.47-1.50	Water, ice	56	0.88-0.92
Fats	58	0.90-0.97	Water, snow, fresh fallen	8	.125
Flour, loose	28	0.40-0.50	Water, sea water	64	1.02-1.03
Flour, pressed	47	0.70-0.80			
Glass, common	156	2.40-2.60			
Glass, plate or crown	161	2.45-2.72	**GASES**		
Glass, crystal	184	2.90-3.00			
Leather	59	0.86-1.02	Air, 0°C. 760 mm.	.08071	1.0
Paper	58	0.70-1.15	Ammonia	.0478	0.5920
Potatoes, piled	42		Carbon dioxide	.1234	1.5291
Rubber, caoutchouc	59	0.92-0.96	Carbon monoxide	.0781	0.9673
Rubber goods	94	1.0-2.0	Gas, illuminating	.028-.036	0.35-0.45
Salt, granulated, piled	48	....	Gas, natural	.038-.039	0.47-0.48
Saltpeter	67	....	Hydrogen	.00559	0.0693
Starch	96	1.53	Nitrogen	.0784	0.9714
Sulphur	125	1.93-2.07	Oxygen	.0892	1.1056
Wool	82	1.32			

The specific gravities of solids and liquids refer to water at 4°C., those of gases to air at 0°C. and 760 mm. pressure. The weights per cubic foot are derived from average specific gravities, except where stated that weights are for bulk, heaped or loose material, etc.

(Courtesy of the American Institute of Steel Construction.)

APPENDIX 7
WEIGHTS AND SPECIFIC GRAVITIES (Cont.)

Substance	Weight Lb. per Cu. Ft.	Specific Gravity	Substance	Weight Lb. per Cu. Ft.	Specific Gravity
ASHLAR MASONRY			**MINERALS**		
Granite, syenite, gneiss......	165	2.3-3.0	Asbestos....................	153	2.1-2.8
Limestone, marble............	160	2.3-2.8	Barytes....................	281	4.50
Sandstone, bluestone........	140	2.1-2.4	Basalt....................	184	2.7-3.2
			Bauxite....................	159	2.55
MORTAR RUBBLE MASONRY			Borax....................	109	1.7-1.8
			Chalk....................	137	1.8-2.6
Granite, syenite, gneiss......	155	2.2-2.8	Clay, marl................	137	1.8-2.6
Limestone, marble............	150	2.2-2.6	Dolomite................	181	2.9
Sandstone, bluestone........	130	2.0-2.2	Feldspar, orthoclase...........	159	2.5-2.6
			Gneiss, serpentine............	159	2.4-2.7
DRY RUBBLE MASONRY			Granite, syenite............	175	2.5-3.1
Granite, syenite, gneiss......	130	1.9-2.3	Greenstone, trap............	187	2.8-3.2
Limestone, marble............	125	1.9-2.1	Gypsum, alabaster...........	159	2.3-2.8
Sandstone, bluestone........	110	1.8-1.9	Hornblende................	187	3.0
			Limestone, marble..........	165	2.5-2.8
BRICK MASONRY			Magnesite................	187	3.0
Pressed brick	140	2.2-2.3	Phosphate rock, apatite......	200	3.2
Common brick................	120	1.8-2.0	Porphyry....................	172	2.6-2.9
Soft brick....................	100	1.5-1.7	Pumice, natural............	40	0.37-0.90
			Quartz, flint............	165	2.5-2.8
CONCRETE MASONRY			Sandstone, bluestone........	147	2.2-2.5
Cement, stone, sand............	144	2.2-2.4	Shale, slate............	175	2.7-2.9
Cement, slag, etc..............	130	1.9-2.3	Soapstone, talc............	169	2.6-2.8
Cement, cinder, etc............	100	1.5-1.7			
			STONE, QUARRIED, PILED		
VARIOUS BUILDING MATERIALS			Basalt, granite, gneiss........	96	....
Ashes, cinders..................	40-45	....	Limestone, marble, quartz	95	....
Cement, portland, loose......	90	....	Sandstone................	82	....
Cement, portland, set........	183	2.7-3.2	Shale....................	92	....
Lime, gypsum, loose........	53-64	....	Greenstone, hornblende......	107	....
Mortar, set....................	103	1.4-1.9			
Slags, bank slag............	67-72	....			
Slags, bank screenings........	98-117	....	**BITUMINOUS SUBSTANCES**		
Slags, machine slag............	96	....	Asphaltum....................	81	1.1-1.5
Slags, slag sand............	49-55	....	Coal, anthracite..............	97	1.4-1.7
			Coal, bituminous............	84	1.2-1.5
EARTH, ETC., EXCAVATED			Coal, lignite............	78	1.1-1.4
Clay, dry......................	63	....	Coal, peat, turf, dry........	47	0.65-0.85
Clay, damp, plastic............	110	....	Coal, charcoal, pine........	23	0.28-0.44
Clay and gravel, dry............	100	....	Coal, charcoal, oak........	33	0.47-0.57
Earth, dry, loose..............	76	....	Coal, coke....................	75	1.0-1.4
Earth, dry, packed..............	95	....	Graphite....................	131	1.9-2.3
Earth, moist, loose............	78	....	Paraffine....................	56	0.87-0.91
Earth, moist, packed............	96	....	Petroleum....................	54	0.87
Earth, mud, flowing............	108	....	Petroleum, refined........	50	0.79-0.82
Earth, mud, packed............	115	....	Petroleum, benzine........	46	0.73-0.75
Riprap, limestone............	80-85	....	Petroleum, gasoline........	42	0.66-0.69
Riprap, sandstone............	90	....	Pitch....................	69	1.07-1.15
Riprap, shale................	105	....	Tar, bituminous..............	75	1.20
Sand, gravel, dry, loose......	90-105	....			
Sand, gravel, dry, packed....	100-120	....			
Sand, gravel, dry, wet......	118-120	....			
			COAL AND COKE, PILED		
EXCAVATIONS IN WATER			Coal, anthracite..............	47-58	....
Sand or gravel..................	60	....	Coal, bituminous, lignite..	40-54	....
Sand or gravel and clay......	65	....	Coal, peat, turf............	20-26	....
Clay..........................	80	....	Coal, charcoal............	10-14	....
River mud......................	90	....	Coal, coke................	23-32	....
Soil..........................	70	....			
Stone riprap....................	65	....			

The specific gravities of solids and liquids refer to water at 4°C., those of gases to air at 0°C. and 760 mm. pressure. The weights per cubic foot are derived from average specific gravities, except where stated that weights are for bulk, heaped or loose material, etc.

APPENDIX 8

WIRE AND SHEET METAL GAGES

WIRE AND SHEET METAL GAGES
IN DECIMALS OF AN INCH

Name of Gage	United States Standard Gage*		The United States Steel Wire Gage	American or Brown & Sharpe Wire Gage	New Birmingham Standard Sheet & Hoop Gage	British Imperial or English Legal Standard Wire Gage	Birmingham or Stubs Iron Wire Gage	Name of Gage
Principal Use	Uncoated Steel Sheets and Light Plates		Steel Wire except Music Wire	Non-Ferrous Sheets and Wire	Iron and Steel Sheets and Hoops	Wire	Strips, Bands, Hoops and Wire	Principal Use
Gage No.	Weight Oz. per Sq. Ft.	Approx. Thickness Inches	Thickness, Inches					Gage No.
7/0's			.4900		.6666	.500		7/0's
6/0's			.4615	.5800	.625	.464		6/0's
5/0's			.4305	.5165	.5883	.432	.500	5/0's
4/0's			.3938	.4600	.5416	.400	.454	4/0's
3/0's			.3625	.4096	.500	.372	.425	3/0's
2/0's			.3310	.3648	.4452	.348	.380	2/0's
0			.3065	.3249	.3964	.324	.340	0
1			.2830	.2893	.3532	.300	.300	1
2			.2625	.2576	.3147	.276	.284	2
3	160	.2391	.2437	.2294	.2804	.252	.259	3
4	150	.2242	.2253	.2043	.250	.232	.238	4
5	140	.2092	.2070	.1819	.2225	.212	.220	5
6	130	.1943	.1920	.1620	.1981	.192	.203	6
7	120	.1793	.1770	.1443	.1764	.176	.180	7
8	110	.1644	.1620	.1285	.1570	.160	.165	8
9	100	.1495	.1483	.1144	.1398	.144	.148	9
10	90	.1345	.1350	.1019	.1250	.128	.134	10
11	80	.1196	.1205	.0907	.1113	.116	.120	11
12	70	.1046	.1055	.0808	.0991	.104	.109	12
13	60	.0897	.0915	.0720	.0882	.092	.095	13
14	50	0747	.0800	.0641	.0785	.080	.083	14
15	45	.0673	.0720	.0571	.0699	.072	.072	15
16	40	.0598	.0625	.0508	.0625	.064	.065	16
17	36	.0538	.0540	.0453	.0556	.056	.058	17
18	32	.0478	.0475	.0403	.0495	.048	.049	18
19	28	.0418	.0410	.0359	.0440	.040	.042	19
20	24	.0359	.0348	.0320	.0392	.036	.035	20
21	22	.0329	.0318	.0285	.0349	.032	.032	21
22	20	.0299	.0286	.0253	.0313	.028	.028	22
23	18	.0269	.0258	.0226	.0278	.024	.025	23
24	16	.0239	.0230	.0201	.0248	.022	.022	24
25	14	.0209	.0204	.0179	.0220	.020	.020	25
26	12	.0179	.0181	.0159	.0196	.018	.018	26
27	11	.0164	.0173	.0142	.0175	.0164	.016	27
28	10	.0149	.0162	.0126	.0156	.0148	.014	28
29	9	.0135	.0150	.0113	.0139	.0136	.013	29
30	8	.0120	.0140	.0100	.0123	.0124	.012	30
31	7	.0105	.0132	.0089	.0110	.0116	.010	31
32	6.5	.0097	.0128	.0080	.0098	.0108	.009	32
33	6	.0090	.0118	.0071	.0087	.0100	.008	33
34	5.5	.0082	.0104	.0063	.0077	.0092	.007	34
35	5	.0075	.0095	.0056	.0069	.0084	.005	35
36	4.5	.0067	.0090	.0050	.0061	.0076	.004	36
37	4.25	.0064	.0085	.0045	.0054	.0068		37
38	4	.0060	.0080	.0040	.0048	.0060		38
39			.0075	.0035	.0043	.0052		39
40			.0070	.0031	.0039	.0048		40

* U. S. Standard Gage is officially a weight gage, in oz. per sq. ft. as tabulated. The Approx. Thickness shown is the "Manufacturers' Standard" of the American Iron and Steel Institute, based on steel as weighing 501.81 lbs. per cu. ft. (489.6 true weight plus 2.5 percent for average over-run in area and thickness). The A.I.S.I. standard nomenclature for flat rolled carbon steel is as follows:

Widths, Inches	Thicknesses, Inch						
	0.2500 and thicker	0.2499 to 0.2031	0.2030 to 0.1875	0.1874 to 0.0568	0.0567 to 0.0344	0.0343 to 0.0255	0.0254 to 0.0142	0.0141 and thinner
To 3½ incl.	Bar	Bar	Strip	Strip	Strip	Strip	Sheet	Sheet
Over 3½ to 6 incl.	Bar	Bar	Strip	Strip	Strip	Sheet	Sheet	Sheet
" 6 to 12 "	Plate	Strip	Strip	Strip	Sheet	Sheet	Sheet	Sheet
" 12 to 32 "	Plate	Sheet	Sheet	Sheet	Sheet	Sheet	Sheet	Black Plate
" 32 to 48 "	Plate	Sheet	Sheet	Sheet	Sheet	Sheet	Sheet	Sheet
" 48	Plate	Plate	Plate	Sheet	Sheet	Sheet	Sheet	——

APPENDIX 9
PIPING SYMBOLS

TYPE OF FITTING		DOUBLE LINE CONVENTION					SINGLE LINE CONVENTION					FLOW DIAGRAM
		FLANGED	SCREWED	B & S	WELDED	SOLDERED	FLANGED	SCREWED	B & S	WELDED	SOLDERED	
1	Joint											
2	Joint - Expansion											
3	Union											
4	Sleeve											
5	Reducer											
6	Reducer - Eccentric											
7	Reducing Flange											
8	Bushing											
9	Elbow - 45°											
10	Elbow - 90°											
11	Elbow - Long radius											
12	Elbow - (turned up)											
13	Elbow - (turned down)											
14	Elbow - Side outlet (outlet up)											
15	Elbow - Side outlet (outlet down)											
16	Elbow - Base											
17	Elbow - Double branch											
18	Elbow - Reducing											
19	Lateral											
20	Tee											
21	Tee - Single sweep											

Cont.

APPENDIX 9
PIPING SYMBOLS (Cont.)

TYPE OF FITTING		DOUBLE LINE CONVENTION					SINGLE LINE CONVENTION					FLOW DIAGRAM
		FLANGED	SCREWED	B & S	WELDED	SOLDERED	FLANGED	SCREWED	B & S	WELDED	SOLDERED	
22	Tee-Double sweep											
23	Tee-(outlet up)											
24	Tee-(outlet down)											
25	Tee-Side outlet (outlet up)											
26	Tee-Side outlet (outlet down)											
27	Cross											
28	Valve-Globe											
29	Valve-Angle											
30	Valve-Motor operated globe											Motor operated
31	Valve-Gate											
32	Valve-Angle gate											
33	Valve-Motor operated gate											Motor operated
34	Valve-Check											
35	Valve-Angle check											
36	Valve-Safety											
37	Valve-Angle safety											
38	Valve-Quick opening											
39	Valve-Float operating											
40	Stop Cock											

APPENDIX 10

AMERICAN STANDARD TAPER PIPE THREADS, NPT[1]

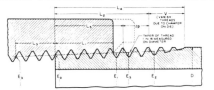

1	2	3	4	5	6	7	8	9	10	11
				Pitch Di-ameter at	Hand-Tight Engagement			Effective Thread, External		
Nominal Pipe Size	Outside Diameter of Pipe D	Threads per Inch n	Pitch of Thread p	Beginning of External Thread E_0	Length[2] L_1		Dia E_1	Length L_2		Dia E_2
					In.	Thds		In.	Thds	In.
$\frac{1}{16}$	0.3125	27	0.03704	0.27118	0.160	4.32	0.28118	0.2611	7.05	0.28750
$\frac{1}{8}$	0.405	27	0.03704	0.36351	0.180	4.86	0.37476	0.2639	7.12	0.38000
$\frac{1}{4}$	0.540	18	0.05556	0.47739	0.200	3.60	0.48989	0.4018	7.23	0.50250
$\frac{3}{8}$	0.675	18	0.05556	0.61201	0.240	4.32	0.62701	0.4078	7.34	0.63750
$\frac{1}{2}$	0.840	14	0.07143	0.75843	0.320	4.48	0.77843	0.5337	7.47	0.79179
$\frac{3}{4}$	1.050	14	0.07143	0.96768	0.339	4.75	0.98887	0.5457	7.64	1.00179
1	1.315	$11\frac{1}{2}$	0.08696	1.21363	0.400	4.60	1.23863	0.6828	7.85	1.25630
$1\frac{1}{4}$	1.660	$11\frac{1}{2}$	0.08696	1.55713	0.420	4.83	1.58338	0.7068	8.13	1.60130
$1\frac{1}{2}$	1.900	$11\frac{1}{2}$	0.08696	1.79609	0.420	4.83	1.82234	0.7235	8.32	1.84130
2	2.375	$11\frac{1}{2}$	0.08696	2.26902	0.436	5.01	2.29627	0.7565	8.70	2.31630
$2\frac{1}{2}$	2.875	8	0.12500	2.71953	0.682	5.46	2.76216	1.1375	9.10	2.79062
3	3.500	8	0.12500	3.34062	0.766	6.13	3.38850	1.2000	9.60	3.41562
$3\frac{1}{2}$	4.000	8	0.12500	3.83750	0.821	6.57	3.88881	1.2500	10.00	3.91562
4	4.500	8	0.12500	4.33438	0.844	6.75	4.38712	1.3000	10.40	4.41562
5	5.563	8	0.12500	5.39073	0.937	7.50	5.44929	1.4063	11.25	5.47862
6	6.625	8	0.12500	6.44609	0.958	7.66	6.50597	1.5125	12.10	6.54062
8	8.625	8	0.12500	8.43359	1.063	8.50	8.50003	1.7125	13.70	8.54062
10	10.750	8	0.12500	10.54531	1.210	9.68	10.62094	1.9250	15.40	10.66562
12	12.750	8	0.12500	12.53281	1.360	10.88	12.61781	2.1250	17.00	12.66562
14 OD	14.000	8	0.12500	13.77500	1.562	12.50	13.87262	2.2500	18.90	13.91562
16 OD	16.000	8	0.12500	15.76250	1.812	14.50	15.87575	2.4500	19.60	15.91562
18 OD	18.000	8	0.12500	17.75000	2.000	16.00	17.87500	2.6500	21.20	17.91562
20 OD	20.000	8	0.12500	19.73750	2.125	17.00	19.87031	2.8500	22.80	19.91562
24 OD	24.000	8	0.12500	23.71250	2.375	19.00	23.86094	3.2500	26.00	23.91562

All dimensions are given in inches.

[1] The basic dimensions of the American Standard Taper Pipe Thread are given in inches to four or five decimal places. While this implies a greater degree of precision than is ordinarily attained, these dimensions are the basis of gage dimensions and are so expressed for the purpose of eliminating errors in computations.

[2] Also length of thin ring gage and length from gaging notch to small end of plug gage.

(Courtesy of ANSI; B2.1–1960.)

APPENDIX 11
AMERICAN STANDARD 250-LB CAST IRON FLANGED FITTINGS

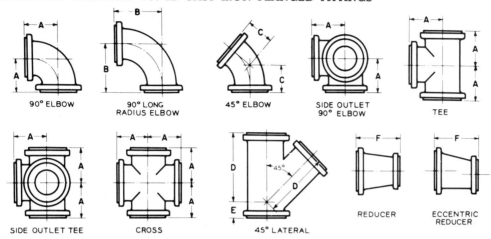

90° ELBOW 90° LONG RADIUS ELBOW 45° ELBOW SIDE OUTLET 90° ELBOW TEE

SIDE OUTLET TEE CROSS 45° LATERAL REDUCER ECCENTRIC REDUCER

Dimensions of 250-lb Cast Iron Flanged Fittings

Nominal Pipe Size	Flanges			Fittings		Straight					
	Dia of Flange	Thickness of Flange (Min)	Dia of Raised Face	Inside Dia of Fittings (Min)	Wall Thickness	Center to Face 90 Deg Elbow Tees, Crosses and True "Y"	Center to Face 90 Deg Long Radius Elbow	Center to Face 45 Deg Elbow	Center to Face Lateral	Short Center to Face True "Y" and Lateral	Face to Face Reducer
						A	B	C	D	E	F
1	4 7/8	11/16	2 11/16	1	7/16	4	5	2	6 1/2	2	....
1 1/4	5 1/4	3/4	3 1/16	1 1/4	7/16	4 1/4	5 1/2	2 1/2	7 1/4	2 1/4	....
1 1/2	6 1/8	13/16	3 9/16	1 1/2	7/16	4 1/2	6	2 3/4	8 1/2	2 1/2	....
2	6 1/2	7/8	4 3/16	2	7/16	5	6 1/2	3	9	2 1/2	5
2 1/2	7 1/2	1	4 15/16	2 1/2	1/2	5 1/2	7	3 1/2	10 1/2	2 1/2	5 1/2
3	8 1/4	1 1/8	5 11/16	3	9/16	6	7 3/4	3 1/2	11	3	6
3 1/2	9	1 3/16	6 5/16	3 1/2	9/16	6 1/2	8 1/2	4	12 1/2	3	6 1/2
4	10	1 1/4	6 15/16	4	5/8	7	9	4 1/2	13 1/2	3	7
5	11	1 3/8	8 5/16	5	11/16	8	10 1/4	5	15	3 1/2	8
6	12 1/2	1 7/16	9 11/16	6	3/4	8 1/2	11 1/2	5 1/2	17 1/2	4	9
8	15	1 5/8	11 15/16	8	13/16	10	14	6	20 1/2	5	11
10	17 1/2	1 7/8	14 1/16	10	15/16	11 1/2	16 1/2	7	24	5 1/2	12
12	20 1/2	2	16 7/16	12	1	13	19	8	27 1/2	6	14
14	23	2 1/8	18 15/16	13 1/4	1 1/8	15	21 1/2	8 1/2	31	6 1/2	16
16	25 1/2	2 1/4	21 1/16	15 1/4	1 1/4	16 1/2	24	9 1/2	34 1/2	7 1/2	18
18	28	2 3/8	23 5/16	17	1 3/8	18	26 1/2	10	37 1/2	8	19
20	30 1/2	2 1/2	25 9/16	19	1 1/2	19 1/2	29	10 1/2	40 1/2	8 1/2	20
24	36	2 3/4	30 5/16	23	1 5/8	22 1/2	34	12	47 1/2	10	24
30	43	3	37 3/16	29	2	27 1/2	41 1/2	15	....	....	30

All dimensions are given in inches.
(Courtesy of ANSI; B16.1–1967.)

APPENDIX 12
AMERICAN STANDARD 125-LB CAST IRON FLANGED FITTINGS

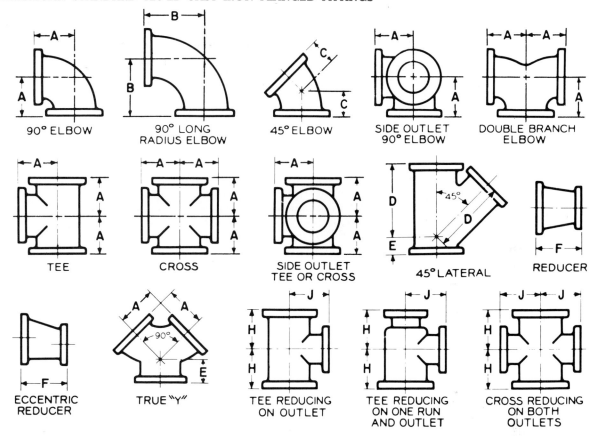

90° ELBOW

90° LONG RADIUS ELBOW

45° ELBOW

SIDE OUTLET 90° ELBOW

DOUBLE BRANCH ELBOW

TEE

CROSS

SIDE OUTLET TEE OR CROSS

45° LATERAL

REDUCER

ECCENTRIC REDUCER

TRUE "Y"

TEE REDUCING ON OUTLET

TEE REDUCING ON ONE RUN AND OUTLET

CROSS REDUCING ON BOTH OUTLETS

Cont.

APPENDIX 12
AMERICAN STANDARD 125-LB CAST IRON FLANGED FITTINGS (Cont.)

	Flanges		General		Straight Fittings						Reducing Fittings (Short Body Patterns)		
											Tees and Crosses		
Nominal Pipe Size	Dia of Flange	Thickness of Flange (Min)	Inside Dia of Flange Fittings	Wall Thickness	Center to Face 90 deg Elbow Tees, Crosses True "Y" and Double Branch Elbow A	Center to Face 90 deg Long Radius El-bow B	Center to Face 45 deg Elbow C	Center to Face Lateral D	Short Center to Face True "Y" and Lateral E	Face to Face Reducer F	Size of Outlet and Smaller	Center to Face Run H	Center to Face Outlet or Side Outlet J
1	4¼	7/16	1	5/16	3½	5	1¾	5¾	1¾	. . .			
1¼	4⅝	½	1¼	5/16	3¾	5½	2	6¼	1¾	. . .			
1½	5	9/16	1½	5/16	4	6	2¼	7	2				
2	6	⅝	2	5/16	4½	6½	2½	8	2½	5			
2½	7	11/16	2½	5/16	5	7	3	9½	2½	5½			
3	7½	¾	3	⅜	5½	7¾	3	10	3	6			
3½	8½	13/16	3½	7/16	6	8½	3½	11½	3	6½	All reducing tees and crosses, sizes 16 in. and smaller, shall have same center to face dimensions as straight size fittings, corresponding to the size of the largest opening.		
4	9	15/16	4	½	6½	9	4	12	3	7			
5	10	15/16	5	½	7½	10¼	4½	13½	3½	8			
6	11	1	6	9/16	8	11½	5	14½	3½	9			
8	13½	1⅛	8	⅝	9	14	5½	17½	4½	11			
10	16	1 3/16	10	¾	11	16½	6½	20½	5	12			
12	19	1¼	12	13/16	12	19	7½	24½	5½	14			
14	21	1⅜	14	⅞	14	21½	7½	27	6	16			
16	23½	1 7/16	16	1	15	24	8	30	6½	18			
18	25	1 9/16	18	1 1/16	16½	26½	8½	32	7	19	12	13	15½
20	27½	1 11/16	20	1⅛	18	29	9½	35	8	20	14	14	17
24	32	1⅞	24	1¼	22	34	11	40½	9	24	16	15	19
30	38¾	2⅛	30	1 7/16	25	41½	15	49	10	30	20	18	23
36	46	2⅜	36	1⅝	28*	49	18	. . .	. . .	36	24	20	26
42	53	2⅝	42	1 13/16	31*	56½	21	. . .	. . .	42	24	23	30
48	59½	2¾	48	2	34*	64	24	. . .	. . .	48	30	26	34

All dimensions are given in inches.
(Courtesy of ANSI; B16.1–1967.)

APPENDIX 13

AMERICAN STANDARD 125-LB CAST IRON FLANGES*

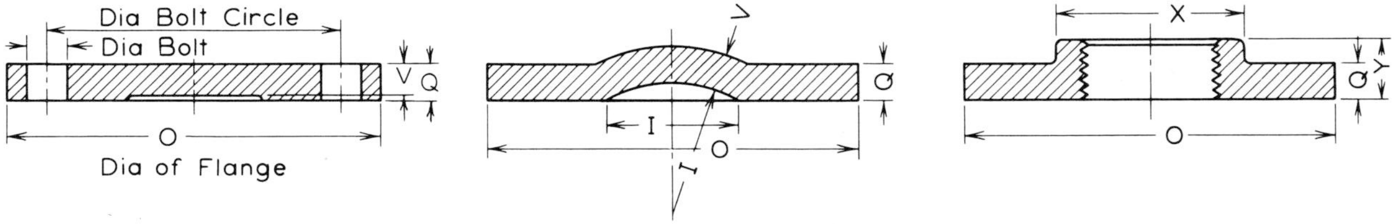

Size *I*	*O*	*Q*	*V*	*X*	*Y*	Dia. Bolt Circle	No. of Bolts	Dia. Bolts	Dia. Bolt Holes	Length of Bolts
1	4 1/4	7/16	—	1 15/16	0.68	3 1/8	4	1/2	5/8	1 3/4
1 1/4	4 5/8	1/2	—	2 5/16	0.76	3 1/2	4	1/2	5/8	2
1 1/2	5	9/16	—	2 9/16	0.87	3 7/8	4	1/2	5/8	2
2	6	5/8	—	3 1/16	1.00	4 3/4	4	5/8	3/4	2 1/4
2 1/2	7	11/16	—	3 9/16	1.14	5 1/2	4	5/8	3/4	2 1/2
3	7 1/2	3/4	—	4 1/4	1.20	6	4	5/8	3/4	2 1/2
3 1/2	8 1/2	13/16	—	4 13/16	1.25	7	8	5/8	3/4	2 3/4
4	9	15/16	—	5 5/16	1.30	7 1/2	8	5/8	3/4	3
5	10	15/16	—	6 7/16	1.41	8 1/2	8	3/4	7/8	3 1/4
6	11	1	—	7 9/16	1.51	9 1/2	8	3/4	7/8	3 1/4
8	13 1/2	1 1/8	—	9 11/16	1.71	11 3/4	8	3/4	7/8	3 1/2
10	16	1 3/16	—	11 15/16	1.93	14 1/4	12	7/8	1	3 3/4
12	19	1 1/4	12/16	14 1/16	2.13	17	12	7/8	1	3 3/4
14 O.D.	21	1 3/8	7/8	15 3/8	2.25	18 3/4	12	1	1 1/8	4 1/4
16 O.D.	23 1/2	1 7/16	1	17 1/2	2.45	21 1/4	16	1 1/8	1 1/8	4 1/2
18 O.D.	25	1 9/16	1 1/16	19 5/8	2.65	22 3/4	16	1 1/8	1 1/4	4 3/4

All dimensions in inches.

* Extracted from American Standards, "Cast-Iron Pipe Flanges and Flanged Fittings" (ANSI B16.1), with the permission of the publisher, The American Society of Mechanical Engineers.

APPENDIX 14

AMERICAN NATIONAL STANDARD 125-LB CAST IRON SCREWED FITTINGS*

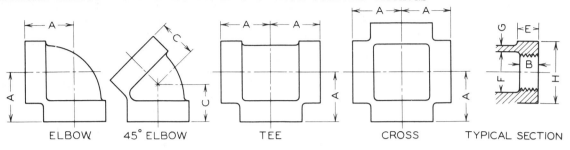

ELBOW 45° ELBOW TEE CROSS TYPICAL SECTION

Nominal Pipe Size	A	C	B Min	E Min	F Min	F Max	G Min	H Min
¼	0.81	0.73	0.32	0.38	0.540	0.584	0.110	0.93
⅜	0.95	0.80	0.36	0.44	0.675	0.719	0.120	1.12
½	1.12	0.88	0.43	0.50	0.840	0.897	0.130	1.34
¾	1.31	0.98	0.50	0.56	1.050	1.107	0.155	1.63
1	1.50	1.12	0.58	0.62	1.315	1.385	0.170	1.95
1¼	1.75	1.29	0.67	0.69	1.660	1.730	0.185	2.39
1½	1.94	1.43	0.70	0.75	1.900	1.970	0.200	2.68
2	2.25	1.68	0.75	0.84	2.375	2.445	0.220	3.28
2½	2.70	1.95	0.92	0.94	2.875	2.975	0.240	3.86
3	3.08	2.17	0.98	1.00	3.500	3.600	0.260	4.62
3½	3.42	2.39	1.03	1.06	4.000	4.100	0.280	5.20
4	3.79	2.61	1.08	1.12	4.500	4.600	0.310	5.79
5	4.50	3.05	1.18	1.18	5.563	5.663	0.380	7.05
6	5.13	3.46	1.28	1.28	6.625	6.725	0.430	8.28
8	6.56	4.28	1.47	1.47	8.625	8.725	0.550	10.63
10	8.08	5.16	1.68	1.68	10.750	10.850	0.690	13.12
12	9.50	5.97	1.88	1.88	12.750	12.850	0.800	15.47
14 O.D.	10.40	—	2.00	2.00	14.000	14.100	0.880	16.94
16 O.D.	11.82	—	2.20	2.20	16.000	16.100	1.000	19.30

All dimensions in inches.

* Extracted from American National Standards, "Cast-Iron Screwed Fittings, 125- and 250-lb" (ANSI B16.4), with the permission of the publisher, The American Society of Mechanical Engineers.

APPENDIX 15
AMERICAN NATIONAL STANDARD UNIFIED INCH SCREW THREADS
(UN AND UNR THREAD FORM)*

Sizes			Threads per Inch											Sizes
			Series with Graded Pitches			Series with Constant Pitches								
Primary	Secondary	Basic Major Diameter	Coarse UNC	Fine UNF	Extra Fine UNEF	4UN	6UN	8UN	12UN	16UN	20UN	28UN	32UN	
0		0.0600	—	80	—	—	—	—	—	—	—	—	—	0
	1	0.0730	64	72	—	—	—	—	—	—	—	—	—	1
2		0.0860	56	64	—	—	—	—	—	—	—	—	—	2
	3	0.0990	48	56	—	—	—	—	—	—	—	—	—	3
4		0.1120	40	48	—	—	—	—	—	—	—	—	—	4
5		0.1250	40	44	—	—	—	—	—	—	—	—	—	5
6		0.1380	32	40	—	—	—	—	—	—	—	—	UNC	6
8		0.1640	32	36	—	—	—	—	—	—	—	—	UNC	8
10		0.1900	24	32	—	—	—	—	—	—	—	—	UNF	10
	12	0.2160	24	28	32	—	—	—	—	—	—	UNF	UNEF	12
$\frac{1}{4}$		0.2500	20	28	32	—	—	—	—	—	UNC	UNF	UNEF	$\frac{1}{4}$
$\frac{5}{16}$		0.3125	18	24	32	—	—	—	—	—	20	28	UNEF	$\frac{5}{16}$
$\frac{3}{8}$		0.3750	16	24	32	—	—	—	—	UNC	20	28	UNEF	$\frac{3}{8}$
$\frac{7}{16}$		0.4375	14	20	28	—	—	—	—	16	UNF	UNEF	32	$\frac{7}{16}$
$\frac{1}{2}$		0.5000	13	20	28	—	—	—	—	16	UNF	UNEF	32	$\frac{1}{2}$
$\frac{9}{16}$		0.5625	12	18	24	—	—	—	UNC	16	20	28	32	$\frac{9}{16}$
$\frac{5}{8}$		0.6250	11	18	24	—	—	—	12	16	20	28	32	$\frac{5}{8}$
	$\frac{11}{16}$	0.6875	—	—	24	—	—	—	12	16	20	28	32	$\frac{11}{16}$
$\frac{3}{4}$		0.7500	10	16	20	—	—	—	12	UNF	UNEF	28	32	$\frac{3}{4}$
	$\frac{13}{16}$	0.8125	—	—	20	—	—	—	12	16	UNEF	28	32	$\frac{13}{16}$
$\frac{7}{8}$		0.8750	9	14	20	—	—	—	12	16	UNEF	28	32	$\frac{7}{8}$
	$\frac{15}{16}$	0.9375	—	—	20	—	—	—	12	16	UNEF	28	32	$\frac{15}{16}$
1		1.0000	8	12	20	—	—	UNC	UNF	16	UNEF	28	32	1
	$1\frac{1}{16}$	1.0625	—	—	18	—	—	8	12	16	20	28	—	$1\frac{1}{16}$
$1\frac{1}{8}$		1.1250	7	12	18	—	—	8	UNF	16	20	28	—	$1\frac{1}{8}$
	$1\frac{3}{16}$	1.1875	—	—	18	—	—	8	12	16	20	28	—	$1\frac{3}{16}$
$1\frac{1}{4}$		1.2500	7	12	18	—	—	8	UNF	16	20	28	—	$1\frac{1}{4}$
	$1\frac{5}{16}$	1.3125	—	—	18	—	—	8	12	16	20	28	—	$1\frac{5}{16}$
$1\frac{3}{8}$		1.3750	6	12	18	—	UNC	8	UNF	16	20	28	—	$1\frac{3}{8}$
	$1\frac{7}{16}$	1.4375	—	—	18	—	6	8	12	16	20	28	—	$1\frac{7}{16}$
$1\frac{1}{2}$		1.5000	6	12	18	—	UNC	8	UNF	16	20	28	—	$1\frac{1}{2}$
	$1\frac{9}{16}$	1.5625	—	—	18	—	6	8	12	16	20	—	—	$1\frac{9}{16}$
$1\frac{5}{8}$		1.6250	—	—	18	—	6	8	12	16	20	—	—	$1\frac{5}{8}$
	$1\frac{11}{16}$	1.6875	—	—	18	—	6	8	12	16	20	—	—	$1\frac{11}{16}$
$1\frac{3}{4}$		1.7500	5	—	—	—	6	8	12	16	20	—	—	$1\frac{3}{4}$
	$1\frac{13}{16}$	1.8125	—	—	—	—	6	8	12	16	20	—	—	$1\frac{13}{16}$
$1\frac{7}{8}$		1.8750	—	—	—	—	6	8	12	16	20	—	—	$1\frac{7}{8}$
	$1\frac{15}{16}$	1.9375	—	—	—	—	6	8	12	16	20	—	—	$1\frac{15}{16}$
2		2.0000	$4\frac{1}{2}$	—	—	—	6	8	12	16	20	—	—	2
	$2\frac{1}{8}$	2.1250	—	—	—	—	6	8	12	16	20	—	—	$2\frac{1}{8}$
$2\frac{1}{4}$		2.2500	$4\frac{1}{2}$	—	—	—	6	8	12	16	20	—	—	$2\frac{1}{4}$
	$2\frac{3}{8}$	2.3750	—	—	—	—	6	8	12	16	20	—	—	$2\frac{3}{8}$
$2\frac{1}{2}$		2.5000	4	—	—	UNC	6	8	12	16	20	—	—	$2\frac{1}{2}$
	$2\frac{5}{8}$	2.6250	—	—	—	4	6	8	12	16	20	—	—	$2\frac{5}{8}$
$2\frac{3}{4}$		2.7500	4	—	—	UNC	6	8	12	16	20	—	—	$2\frac{3}{4}$
	$2\frac{7}{8}$	2.8750	—	—	—	4	6	8	12	16	20	—	—	$2\frac{7}{8}$

* Series designation shown indicates the UN thread form; however, the UNR thread form may be specified by substituting UNR in place of UN in all designations for external use only.

Cont.

APPENDIX 15
AMERICAN NATIONAL STANDARD UNIFIED INCH SCREW THREADS
(UN AND UNR THREAD FORM)* (Cont.)

Sizes		Basic Major Diameter	Series with Graded Pitches			Threads per Inch								Sizes
						Series with Constant Pitches								
Primary	Secondary		Coarse UNC	Fine UNF	Extra Fine UNEF	4UN	6UN	8UN	12UN	16UN	20UN	28UN	32UN	Sizes
3		3.0000	4	—	—	UNC	6	8	12	16	20	—	—	3
	$3\frac{1}{8}$	3.1250	—	—	—	4	6	8	12	16	—	—	—	$3\frac{1}{8}$
$3\frac{1}{4}$		3.2500	4	—	—	UNC	6	8	12	16	—	—	—	$3\frac{1}{4}$
	$3\frac{3}{8}$	3.3750	—	—	—	4	6	8	12	16	—	—	—	$3\frac{3}{8}$
$3\frac{1}{2}$		3.5000	4	—	—	UNC	6	8	12	16	—	—	—	$3\frac{1}{2}$
	$3\frac{5}{8}$	3.6250	—	—	—	4	6	8	12	16	—	—	—	$3\frac{5}{8}$
$3\frac{3}{4}$		3.7500	4	—	—	UNC	6	8	12	16	—	—	—	$3\frac{3}{4}$
	$3\frac{7}{8}$	3.8750	—	—	—	4	6	8	12	16	—	—	—	$3\frac{7}{8}$
4		4.0000	4	—	—	UNC	6	8	12	16	—	—	—	4
	$4\frac{1}{8}$	4.1250	—	—	—	4	6	8	12	16	—	—	—	$4\frac{1}{8}$
$4\frac{1}{4}$		4.2500	—	—	—	4	6	8	12	16	—	—	—	$4\frac{1}{4}$
	$4\frac{3}{8}$	4.3750	—	—	—	4	6	8	12	16	—	—	—	$4\frac{3}{8}$
$4\frac{1}{2}$		4.5000	—	—	—	4	6	8	12	16	—	—	—	$4\frac{1}{2}$
	$4\frac{5}{8}$	4.6250	—	—	—	4	6	8	12	16	—	—	—	$4\frac{5}{8}$
$4\frac{3}{4}$		4.7500	—	—	—	4	6	8	12	16	—	—	—	$4\frac{3}{4}$
	$4\frac{7}{8}$	4.8750	—	—	—	4	6	8	12	16	—	—	—	$4\frac{7}{8}$
5		5.0000	—	—	—	4	6	8	12	16	—	—	—	5
	$5\frac{1}{8}$	5.1250	—	—	—	4	6	8	12	16	—	—	—	$5\frac{1}{8}$
$5\frac{1}{4}$		5.2500	—	—	—	4	6	8	12	16	—	—	—	$5\frac{1}{4}$
	$5\frac{3}{8}$	5.3750	—	—	—	4	6	8	12	16	—	—	—	$5\frac{3}{8}$
$5\frac{1}{2}$		5.5000	—	—	—	4	6	8	12	16	—	—	—	$5\frac{1}{2}$
	$5\frac{5}{8}$	5.6250	—	—	—	4	6	8	12	16	—	—	—	$5\frac{5}{8}$
$5\frac{3}{4}$		5.7500	—	—	—	4	6	8	12	16	—	—	—	$5\frac{3}{4}$
	$5\frac{7}{8}$	5.8750	—	—	—	4	6	8	12	16	—	—	—	$5\frac{7}{8}$
6		6.0000	—	—	—	4	6	8	12	16	—	—	—	6

(Courtesy of ANSI; B1.1–1974.)

APPENDIX 16

TAP DRILL SIZES FOR AMERICAN NATIONAL AND UNIFIED COARSE AND FINE THREADS

$p = \text{pitch} = \dfrac{1}{\text{No. thrd. per in.}}$

$d = \text{depth} = p \times .649519$

$f = \text{flat} = \dfrac{p}{8}$

$\text{pitch diameter} = d - \dfrac{.6495}{N}$

For Nos. 575 and 585 Screw Thread Micrometers

Size	Threads per inch NC UNC	Threads per inch NF UNF	Outside Diameter Inches	Pitch Diameter Inches	Root Diameter Inches	Tap Drill Approx. 75% Full Thread	Decimal Equiv. of Tap Drill
0	..	80	.0600	.0519	.0438	3/64	.0469
1	64	..	.0730	.0629	.0527	53	.0595
1	..	72	.0730	.0640	.0550	53	.0595
2	56	..	.0860	.0744	.0628	50	.0700
2	..	64	.0860	.0759	.0657	50	.0700
3	48	..	.0990	.0855	.0719	47	.0785
3	..	56	.0990	.0874	.0758	46	.0810
4	40	..	.1120	.0958	.0795	43	.0890
4	..	48	.1120	.0985	.0849	42	.0935
5	40	..	.1250	.1088	.0925	38	.1015
5	..	44	.1250	.1102	.0955	37	.1040
6	32	..	.1380	.1177	.0974	36	.1065
6	..	40	.1380	.1218	.1055	33	.1130
8	32	..	.1640	.1437	.1234	29	.1360
8	..	36	.1640	.1460	.1279	29	.1360
10	24	..	.1900	.1629	.1359	26	.1470
10	..	32	.1900	.1697	.1494	21	.1590
12	24	..	.2160	.1889	.1619	16	.1770
12	..	28	.2160	.1928	.1696	15	.1800
1/4	20	..	.2500	.2175	.1850	7	.2010
1/4	..	28	.2500	.2268	.2036	3	.2130
5/16	18	..	.3125	.2764	.2403	F	.2570
5/16	..	24	.3125	.2854	.2584	I	.2720
3/8	16	..	.3750	.3344	.2938	5/16	.3125
3/8	..	24	.3750	.3479	.3209	Q	.3320
7/16	14	..	.4375	.3911	.3447	U	.3680
7/16	..	20	.4375	.4050	.3726	25/64	.3906
1/2	13	..	.5000	.4500	.4001	27/64	.4219
1/2	..	20	.5000	.4675	.4351	29/64	.4531
9/16	12	..	.5625	.5084	.4542	31/64	.4844
9/16	..	18	.5625	.5264	.4903	33/64	.5156
5/8	11	..	.6250	.5660	.5069	17/32	.5312
5/8	..	18	.6250	.5889	.5528	37/64	.5781
3/4	10	..	.7500	.6850	.6201	21/32	.6562
3/4	..	16	.7500	.7094	.6688	11/16	.6875
7/8	9	..	.8750	.8028	.7307	49/64	.7656
7/8	..	14	.8750	.8286	.7822	13/16	.8125

Cont.

APPENDIX 16

TAP DRILL SIZES FOR AMERICAN NATIONAL AND UNIFIED COARSE AND FINE THREADS (Cont.)

Size	Threads per inch NC UNC	NF UNF	Outside Diameter Inches	Pitch Diameter Inches	Root Diameter Inches	Tap Drill Approx. 75% Full Thread	Decimal Equiv. of Tap Drill
1	8	..	1.0000	.9188	.8376	$\frac{7}{8}$	.8750
1	..	12	1.0000	.9459	.8917	$\frac{59}{64}$	.9219
1⅛	7	..	1.1250	1.0322	.9394	$\frac{63}{64}$	.9844
1⅛	..	12	1.1250	1.0709	1.0168	$1\frac{3}{64}$	1.0469
1¼	7	..	1.2500	1.1572	1.0644	$1\frac{7}{64}$	1.1094
1¼	..	12	1.2500	1.1959	1.1418	$1\frac{11}{64}$	1.1719
1⅜	6	..	1.3750	1.2667	1.1585	$1\frac{7}{32}$	1.2187
1⅜	..	12	1.3750	1.3209	1.2668	$1\frac{19}{64}$	1.2969
1½	6	..	1.5000	1.3917	1.2835	$1\frac{11}{32}$	1.3437
1½	..	12	1.5000	1.4459	1.3918	$1\frac{27}{64}$	1.4219
1¾	5	..	1.7500	1.6201	1.4902	$1\frac{9}{16}$	1.5625
2	4½	..	2.0000	1.8557	1.7113	$1\frac{25}{32}$	1.7812
2¼	4½	..	2.2500	2.1057	1.9613	$2\frac{1}{32}$	2.0313
2½	4	..	2.5000	2.3376	2.1752	2¼	2.2500
2¾	4	..	2.7500	2.5876	2.4252	2½	2.5000
3	4	..	3.0000	3.8376	2.6752	2¾	2.7500
3¼	4	..	3.2500	3.0876	2.9252	3	3.0000
3½	4	..	3.5000	3.3376	3.1752	3¼	3.2500
3¾	4	..	3.7500	3.5876	3.4252	3½	3.5000
4	4	..	4.0000	3.3786	3.6752	3¾	3.7500

(Courtesy of the L. S. Starrett Company.)

APPENDIX 17

LENGTH OF THREAD ENGAGEMENT GROUPS

Nominal Size Diam. Over	Nominal Size Diam. To and Incl	Pitch P	Length of Thread Engagement Group S To and Incl	Length of Thread Engagement Group N Over	Length of Thread Engagement Group N To and Incl	Length of Thread Engagement Group L Over
1.5	2.8	0.2	0.5	0.5	1.5	1.5
		0.25	0.6	0.6	1.9	1.9
		0.35	0.8	0.8	2.6	2.6
		0.4	1	1	3	3
		0.45	1.3	1.3	3.8	3.8
2.8	5.6	0.35	1	1	3	3
		0.5	1.5	1.5	4.5	4.5
		0.6	1.7	1.7	5	5
		0.7	2	2	6	6
		0.75	2.2	2.2	6.7	6.7
		0.8	2.5	2.5	7.5	7.5
5.6	11.2	0.75	2.4	2.4	7.1	7.1
		1	3	3	9	9
		1.25	4	4	12	12
		1.5	5	5	15	15
11.2	22.4	1	3.8	3.8	11	11
		1.25	4.5	4.5	13	13
		1.5	5.6	5.6	16	16
		1.75	6	6	18	18
		2	8	8	24	24
		2.5	10	10	30	30

Nominal Size Diam. Over	Nominal Size Diam. To and Incl	Pitch P	Length of Thread Engagement Group S To and Incl	Length of Thread Engagement Group N Over	Length of Thread Engagement Group N To and Incl	Length of Thread Engagement Group L Over
22.4	45	1	4	4	12	12
		1.5	6.3	6.3	19	19
		2	8.5	8.5	25	25
		3	12	12	36	36
		3.5	15	15	45	45
		4	18	18	53	53
		4.5	21	21	63	63
45	90	1.5	7.5	7.5	22	22
		2	9.5	9.5	28	28
		3	15	15	45	45
		4	19	19	56	56
		5	24	24	71	71
		5.5	28	28	85	85
		6	32	32	95	95
90	180	2	12	12	36	36
		3	18	18	53	53
		4	24	24	71	71
		6	36	36	106	106
180	355	3	20	20	60	60
		4	26	26	80	80
		6	40	40	118	118

All dimensions are given in millimeters. (Courtesy of ISO Standards.)

APPENDIX 18
ISO METRIC SCREW THREAD STANDARD SERIES

Nominal Size Dia. (mm) Column[a]			Series with Graded Pitches		Pitches (mm) Series with Constant Pitches												Nominal Size Dia. (mm)
1	2	3	Coarse	Fine	6	4	3	2	1.5	1.25	1	0.75	0.5	0.35	0.25	0.2	
0.25			0.075	—	—	—	—	—	—	—	—	—	—	—	—	—	0.25
0.3			0.08	—	—	—	—	—	—	—	—	—	—	—	—	—	0.3
	0.35		0.09	—	—	—	—	—	—	—	—	—	—	—	—	—	0.35
0.4			0.1	—	—	—	—	—	—	—	—	—	—	—	—	—	0.4
	0.45		0.1	—	—	—	—	—	—	—	—	—	—	—	—	—	0.45
0.5			0.125	—	—	—	—	—	—	—	—	—	—	—	—	—	0.5
	0.55		0.125	—	—	—	—	—	—	—	—	—	—	—	—	—	0.55
0.6			0.15	—	—	—	—	—	—	—	—	—	—	—	—	—	0.6
	0.7		0.175	—	—	—	—	—	—	—	—	—	—	—	—	—	0.7
0.8			0.2	—	—	—	—	—	—	—	—	—	—	—	—	—	0.8
	0.9		0.225	—	—	—	—	—	—	—	—	—	—	—	—	—	0.9
			0.25	—	—	—	—	—	—	—	—	—	—	—	—	0.2	1
	1.1		0.25	—	—	—	—	—	—	—	—	—	—	—	—	0.2	1.1
1.2			0.25	—	—	—	—	—	—	—	—	—	—	—	—	0.2	1.2
	1.4		0.3	—	—	—	—	—	—	—	—	—	—	—	—	0.2	1.4
1.6			0.35	—	—	—	—	—	—	—	—	—	—	—	—	0.2	1.6
	1.8		0.35	—	—	—	—	—	—	—	—	—	—	—	—	0.2	1.8
2			0.4	—	—	—	—	—	—	—	—	—	—	—	0.25	—	2
	2.2		0.45	—	—	—	—	—	—	—	—	—	—	—	0.25	—	2.2
2.5			0.45	—	—	—	—	—	—	—	—	—	—	0.35	—	—	2.5
3			0.5	—	—	—	—	—	—	—	—	—	—	0.35	—	—	3
	3.5		0.6	—	—	—	—	—	—	—	—	—	—	0.35	—	—	3.5
4			0.7	—	—	—	—	—	—	—	—	—	0.5	—	—	—	4
	4.5		0.75	—	—	—	—	—	—	—	—	—	0.5	—	—	—	4.5
5			0.8	—	—	—	—	—	—	—	—	—	0.5	—	—	—	5
		5.5	—	—	—	—	—	—	—	—	—	—	0.5	—	—	—	5.5
6			1	—	—	—	—	—	—	—	—	0.75	—	—	—	—	6
		7	1	—	—	—	—	—	—	—	—	0.75	—	—	—	—	7
8			1.25	1	—	—	—	—	—	—	1	0.75	—	—	—	—	8
		9	1.25	—	—	—	—	—	—	—	1	0.75	—	—	—	—	9
10			1.5	1.25	—	—	—	—	—	1.25	1	0.75	—	—	—	—	10
	11		1.5	1.25	—	—	—	—	—	—	1	0.75	—	—	—	—	11
12			1.75	1.25	—	—	—	—	1.5	1.25	1	—	—	—	—	—	12
	14		2	1.5	—	—	—	—	1.5	1.25[b]	1	—	—	—	—	—	14
		15	—	—	—	—	—	—	1.5	—	1	—	—	—	—	—	15
16			2	1.5	—	—	—	—	1.5	—	1	—	—	—	—	—	16
		17	—	—	—	—	—	—	1.5	—	1	—	—	—	—	—	17
	18		2.5	1.5	—	—	—	2	1.5	—	1	—	—	—	—	—	18
20			2.5	1.5	—	—	—	2	1.5	—	1	—	—	—	—	—	20
	22		2.5	1.5	—	—	—	2	1.5	—	1	—	—	—	—	—	22

[a] Thread diameter should be selected from columns 1, 2 or 3, with preference being in that order.
[b] Pitch 1.25 mm in combination with diameter 14 mm has been included for sparkplug applications.
[c] Diameter 35 mm has been included for bearing locknut applications.

The use of pitches shown in parentheses should be avoided wherever possible.

The pitches enclosed in the bold frame, together with the corresponding nominal diameters in columns 1 and 2, are those combinations which have been established by ISO Recommendations as a selected "coarse" and "fine" series for commercial fasteners.

APPENDIX 18
ISO METRIC SCREW THREAD STANDARD SERIES (Cont.)

Nominal Size Dia. (mm) Column[a]			Series with Graded Pitches		Pitches (mm) Series with Constant Pitches												Nominal Size Dia. (mm)
1	2	3	Coarse	Fine	6	4	3	2	1.5	1.25	1	0.75	0.5	0.35	0.25	0.2	
24			3	2	—	—	—	2	1.5	—	1	—	—	—	—	—	24
		25	—	—	—	—	—	2	1.5	—	1	—	—	—	—	—	25
		26	—	—	—	—	—	—	1.5	—	1	—	—	—	—	—	26
	27		3	2	—	—	—	2	1.5	—	1	—	—	—	—	—	27
		28	—	—	—	—	—	2	1.5	—	1	—	—	—	—	—	28
30			3.5	2	—	—	(3)	2	1.5	—	1	—	—	—	—	—	30
		32	—	—	—	—	—	2	1.5	—	—	—	—	—	—	—	32
	33		3.5	2	—	—	(3)	2	1.5	—	—	—	—	—	—	—	33
		35[c]	—	—	—	—	—	—	1.5	—	—	—	—	—	—	—	35[c]
36			4	3	—	—	—	2	1.5	—	—	—	—	—	—	—	36
		38	—	—	—	—	—	—	1.5	—	—	—	—	—	—	—	38
	39		4	3	—	—	—	2	1.5	—	—	—	—	—	—	—	39
		40	—	—	—	—	3	2	1.5	—	—	—	—	—	—	—	40
42			4.5	3	—	4	3	2	1.5	—	—	—	—	—	—	—	42
	45		4.5	3	—	4	3	2	1.5	—	—	—	—	—	—	—	45
48			5	3	—	4	3	2	1.5	—	—	—	—	—	—	—	48
		50	—	—	—	—	3	2	1.5	—	—	—	—	—	—	—	50
	52		5	3	—	4	3	2	1.5	—	—	—	—	—	—	—	52
		55	—	—	—	4	3	2	1.5	—	—	—	—	—	—	—	55
56			5.5	4	—	4	3	2	1.5	—	—	—	—	—	—	—	56
		58	—	—	—	4	3	2	1.5	—	—	—	—	—	—	—	58
	60		5.5	4	—	4	3	2	1.5	—	—	—	—	—	—	—	60
		62	—	—	—	4	3	2	1.5	—	—	—	—	—	—	—	62
64			6	4	—	4	3	2	1.5	—	—	—	—	—	—	—	64
		65	—	—	—	4	3	2	1.5	—	—	—	—	—	—	—	65
	68		6	4	—	4	3	2	1.5	—	—	—	—	—	—	—	68
		70	—	—	6	4	3	2	1.5	—	—	—	—	—	—	—	70
72			—	—	6	4	3	2	1.5	—	—	—	—	—	—	—	72
		75	—	—	6	4	3	2	1.5	—	—	—	—	—	—	—	75
	76		—	—	6	4	3	2	1.5	—	—	—	—	—	—	—	76
		78	—	—	—	—	—	2	—	—	—	—	—	—	—	—	78
80			—	—	6	4	3	2	1.5	—	—	—	—	—	—	—	80
		82	—	—	—	—	—	2	—	—	—	—	—	—	—	—	82
	85		—	—	6	4	3	2	—	—	—	—	—	—	—	—	85
90			—	—	6	4	3	2	—	—	—	—	—	—	—	—	90

Cont.

APPENDIX 18
ISO METRIC SCREW THREAD STANDARD SERIES (Cont.)

Nominal Size Dia. (mm)			Pitches (mm)													Nominal Size Dia. (mm)
Column[a]			Series with Graded Pitches		Series with Constant Pitches											
1	2	3	Coarse	Fine	6	4	3	2	1.5	1.25	1	0.75	0.5	0.35	0.25	0.2	
	95		—	—	6	4	3	2	—	—	—	—	—	—	—	—	95
100			—	—	6	4	3	2	—	—	—	—	—	—	—	—	100
	105		—	—	6	4	3	2	—	—	—	—	—	—	—	—	105
110			—	—	6	4	3	2	—	—	—	—	—	—	—	—	110
	115		—	—	6	4	3	2	—	—	—	—	—	—	—	—	115
	120		—	—	6	4	3	2	—	—	—	—	—	—	—	—	120
125			—	—	6	4	3	2	—	—	—	—	—	—	—	—	125
	130		—	—	6	4	3	2	—	—	—	—	—	—	—	—	130
		135	—	—	6	4	3	2	—	—	—	—	—	—	—	—	135
140			—	—	6	4	3	2	—	—	—	—	—	—	—	—	140
		145	—	—	6	4	3	2	—	—	—	—	—	—	—	—	145
	150		—	—	6	4	3	2	—	—	—	—	—	—	—	—	150
		155	—	—	6	4	3	—	—	—	—	—	—	—	—	—	155
160			—	—	6	4	3	—	—	—	—	—	—	—	—	—	160
		165	—	—	6	4	3	—	—	—	—	—	—	—	—	—	165
	170		—	—	6	4	3	—	—	—	—	—	—	—	—	—	170
		175	—	—	6	4	3	—	—	—	—	—	—	—	—	—	175
180			—	—	6	4	3	—	—	—	—	—	—	—	—	—	180
		185	—	—	6	4	3	—	—	—	—	—	—	—	—	—	185
	190		—	—	6	4	3	—	—	—	—	—	—	—	—	—	190
		195	—	—	6	4	3	—	—	—	—	—	—	—	—	—	195
200			—	—	6	4	3	—	—	—	—	—	—	—	—	—	200
		205	—	—	6	4	3	—	—	—	—	—	—	—	—	—	205
	210		—	—	6	4	3	—	—	—	—	—	—	—	—	—	210
220			—	—	6	4	3	—	—	—	—	—	—	—	—	—	220
		225	—	—	6	4	3	—	—	—	—	—	—	—	—	—	225
		230	—	—	6	4	3	—	—	—	—	—	—	—	—	—	230
		235	—	—	6	4	3	—	—	—	—	—	—	—	—	—	235
	240		—	—	6	4	3	—	—	—	—	—	—	—	—	—	240
		245	—	—	6	4	3	—	—	—	—	—	—	—	—	—	245
250			—	—	6	4	3	—	—	—	—	—	—	—	—	—	250
		255	—	—	6	4	—	—	—	—	—	—	—	—	—	—	255
	260		—	—	6	4	—	—	—	—	—	—	—	—	—	—	260
		265	—	—	6	4	—	—	—	—	—	—	—	—	—	—	265
		270	—	—	6	4	—	—	—	—	—	—	—	—	—	—	270
		275	—	—	6	4	—	—	—	—	—	—	—	—	—	—	275
280			—	—	6	4	—	—	—	—	—	—	—	—	—	—	280
		285	—	—	6	4	—	—	—	—	—	—	—	—	—	—	285
		290	—	—	6	4	—	—	—	—	—	—	—	—	—	—	290
		295	—	—	6	4	—	—	—	—	—	—	—	—	—	—	295
	300		—	—	6	4	—	—	—	—	—	—	—	—	—	—	300

[1] Thread diameter should be selected from columns 1, 2, or 3; with preference being in that order.

APPENDIX 19

SQUARE AND ACME THREADS

Size	Threads per Inch	Size	Threads per Inch
$\frac{3}{8}$	12	2	$2\frac{1}{2}$
$\frac{7}{16}$	10	$2\frac{1}{4}$	2
$\frac{1}{2}$	10	$2\frac{1}{2}$	2
$\frac{9}{16}$	8	$2\frac{3}{4}$	2
$\frac{5}{8}$	8	3	$1\frac{1}{2}$
$\frac{3}{4}$	6	$3\frac{1}{4}$	$1\frac{1}{2}$
$\frac{7}{8}$	5	$3\frac{1}{2}$	$1\frac{1}{3}$
1	5	$3\frac{3}{4}$	$1\frac{1}{3}$
$1\frac{1}{8}$	4	4	$1\frac{1}{3}$
$1\frac{1}{4}$	4	$4\frac{1}{4}$	$1\frac{1}{3}$
$1\frac{1}{2}$	3	$4\frac{1}{2}$	1
$1\frac{3}{4}$	$2\frac{1}{2}$	over $4\frac{1}{2}$	1

APPENDIX 20
AMERICAN STANDARD SQUARE BOLTS AND NUTS

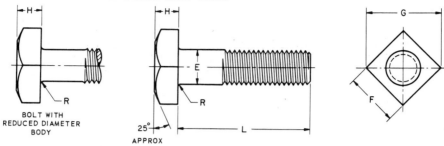

BOLT WITH
REDUCED DIAMETER
BODY

25° APPROX

Dimensions of Square Bolts

Nominal Size or Basic Product Dia		Body Dia E	Width Across Flats F			Width Across Corners G		Height H			Radius of Fillet R
		Max	Basic	Max	Min	Max	Min	Basic	Max	Min	Max
1/4	0.2500	0.260	3/8	0.3750	0.362	0.530	0.498	11/64	0.188	0.156	0.031
5/16	0.3125	0.324	1/2	0.5000	0.484	0.707	0.665	13/64	0.220	0.186	0.031
3/8	0.3750	0.388	9/16	0.5625	0.544	0.795	0.747	1/4	0.268	0.232	0.031
7/16	0.4375	0.452	5/8	0.6250	0.603	0.884	0.828	19/64	0.316	0.278	0.031
1/2	0.5000	0.515	3/4	0.7500	0.725	1.061	0.995	21/64	0.348	0.308	0.031
5/8	0.6250	0.642	15/16	0.9375	0.906	1.326	1.244	27/64	0.444	0.400	0.062
3/4	0.7500	0.768	1 1/8	1.1250	1.088	1.591	1.494	1/2	0.524	0.476	0.062
7/8	0.8750	0.895	1 5/16	1.3125	1.269	1.856	1.742	19/32	0.620	0.568	0.062
1	1.0000	1.022	1 1/2	1.5000	1.450	2.121	1.991	21/32	0.684	0.628	0.093
1 1/8	1.1250	1.149	1 11/16	1.6875	1.631	2.386	2.239	3/4	0.780	0.720	0.093
1 1/4	1.2500	1.277	1 7/8	1.8750	1.812	2.652	2.489	27/32	0.876	0.812	0.093
1 3/8	1.3750	1.404	2 1/16	2.0625	1.994	2.917	2.738	29/32	0.940	0.872	0.093
1 1/2	1.5000	1.531	2 1/4	2.2500	2.175	3.182	2.986	1	1.036	0.964	0.093

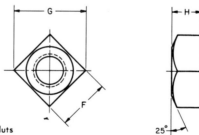

Dimensions of Square Nuts

Nominal Size or Basic Major Dia of Thread		Width Across Flats F			Width Across Corners G		Thickness H		
		Basic	Max	Min	Max	Min	Basic	Max	Min
1/4	0.2500	7/16	0.4375	0.425	0.619	0.584	7/32	0.235	0.203
5/16	0.3125	9/16	0.5625	0.547	0.795	0.751	17/64	0.283	0.249
3/8	0.3750	5/8	0.6250	0.606	0.884	0.832	21/64	0.346	0.310
7/16	0.4375	3/4	0.7500	0.728	1.061	1.000	3/8	0.394	0.356
1/2	0.5000	13/16	0.8125	0.788	1.149	1.082	7/16	0.458	0.418
5/8	0.6250	1	1.0000	0.969	1.414	1.330	35/64	0.569	0.525
3/4	0.7500	1 1/8	1.1250	1.088	1.591	1.494	21/32	0.680	0.632
7/8	0.8750	1 5/16	1.3125	1.269	1.856	1.742	49/64	0.792	0.740
1	1.0000	1 1/2	1.5000	1.450	2.121	1.991	7/8	0.903	0.847
1 1/8	1.1250	1 11/16	1.6875	1.631	2.386	2.239	1	1.030	0.970
1 1/4	1.2500	1 7/8	1.8750	1.812	2.652	2.489	1 3/32	1.126	1.062
1 3/8	1.3750	2 1/16	2.0625	1.994	2.917	2.738	1 13/64	1.237	1.169
1 1/2	1.5000	2 1/4	2.2500	2.175	3.182	2.986	1 5/16	1.348	1.276

(Courtesy of ANSI; B18.2.1–1965 and ANSI; B18.2.2–1965.)

APPENDIX 21
AMERICAN STANDARD HEXAGON HEAD BOLTS AND NUTS

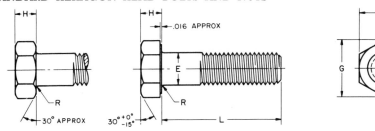

Dimensions of Hex Cap Screws (Finished Hex Bolts)

Nominal Size or Basic Product Dia		Body Dia E		Width Across Flats F			Width Across Corners G		Height H			Radius of Fillet R	
		Max	Min	Basic	Max	Min	Max	Min	Basic	Max	Min	Max	Min
1/4	0.2500	0.2500	0.2450	7/16	0.4375	0.428	0.505	0.488	5/32	0.163	0.150	0.025	0.015
5/16	0.3125	0.3125	0.3065	1/2	0.5000	0.489	0.577	0.557	13/64	0.211	0.195	0.025	0.015
3/8	0.3750	0.3750	0.3690	9/16	0.5625	0.551	0.650	0.628	15/64	0.243	0.226	0.025	0.015
7/16	0.4375	0.4375	0.4305	5/8	0.6250	0.612	0.722	0.698	9/32	0.291	0.272	0.025	0.015
1/2	0.5000	0.5000	0.4930	3/4	0.7500	0.736	0.866	0.840	5/16	0.323	0.302	0.025	0.015
9/16	0.5625	0.5625	0.5545	13/16	0.8125	0.798	0.938	0.910	23/64	0.371	0.348	0.045	0.020
5/8	0.6250	0.6250	0.6170	15/16	0.9375	0.922	1.083	1.051	25/64	0.403	0.378	0.045	0.020
3/4	0.7500	0.7500	0.7410	1 1/8	1.1250	1.100	1.299	1.254	15/32	0.483	0.455	0.045	0.020
7/8	0.8750	0.8750	0.8660	1 5/16	1.3125	1.285	1.516	1.465	35/64	0.563	0.531	0.065	0.040
1	1.0000	1.0000	0.9900	1 1/2	1.5000	1.469	1.732	1.675	39/64	0.627	0.591	0.095	0.060
1 1/8	1.1250	1.1250	1.1140	1 11/16	1.6875	1.631	1.949	1.859	11/16	0.718	0.658	0.095	0.060
1 1/4	1.2500	1.2500	1.2390	1 7/8	1.8750	1.812	2.165	2.066	25/32	0.813	0.749	0.095	0.060
1 3/8	1.3750	1.3750	1.3630	2 1/16	2.0625	1.994	2.382	2.273	27/32	0.878	0.810	0.095	0.060
1 1/2	1.5000	1.5000	1.4880	2 1/4	2.2500	2.175	2.598	2.480	15/16	0.974	0.902	0.095	0.060
1 3/4	1.7500	1.7500	1.7380	2 5/8	2.6250	2.538	3.031	2.893	1 3/32	1.134	1.054	0.095	0.060
2	2.0000	2.0000	1.9880	3	3.0000	2.900	3.464	3.306	1 7/32	1.263	1.175	0.095	0.060
2 1/4	2.2500	2.2500	2.2380	3 3/8	3.3750	3.262	3.897	3.719	1 3/8	1.423	1.327	0.095	0.060
2 1/2	2.5000	2.5000	2.4880	3 3/4	3.7500	3.625	4.330	4.133	1 17/32	1.583	1.479	0.095	0.060
2 3/4	2.7500	2.7500	2.7380	4 1/8	4.1250	3.988	4.763	4.546	1 11/16	1.744	1.632	0.095	0.060
3	3.0000	3.0000	2.9880	4 1/2	4.5000	4.350	5.196	4.959	1 7/8	1.935	1.815	0.095	0.060

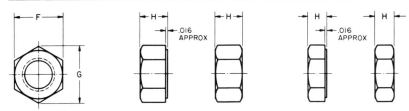

Dimensions of Hex Nuts and Hex Jam Nuts

Nominal Size or Basic Major Dia of Thread		Width Across Flats F			Width Across Corners G		Thickness Hex Nuts H			Thickness Hex Jam Nuts H		
		Basic	Max	Min	Max	Min	Basic	Max	Min	Basic	Max	Min
1/4	0.2500	7/16	0.4375	0.428	0.505	0.488	7/32	0.226	0.212	5/32	0.163	0.150
5/16	0.3125	1/2	0.5000	0.489	0.577	0.557	17/64	0.273	0.258	3/16	0.195	0.180
3/8	0.3750	9/16	0.5625	0.551	0.650	0.628	21/64	0.337	0.320	7/32	0.227	0.210
7/16	0.4375	11/16	0.6875	0.675	0.794	0.768	3/8	0.385	0.365	1/4	0.260	0.240
1/2	0.5000	3/4	0.7500	0.736	0.866	0.840	7/16	0.448	0.427	5/16	0.323	0.302
9/16	0.5625	7/8	0.8750	0.861	1.010	0.982	31/64	0.496	0.473	5/16	0.324	0.301
5/8	0.6250	15/16	0.9375	0.922	1.083	1.051	35/64	0.559	0.535	3/8	0.387	0.363
3/4	0.7500	1 1/8	1.1250	1.088	1.299	1.240	41/64	0.665	0.617	27/64	0.446	0.398
7/8	0.8750	1 5/16	1.3125	1.269	1.516	1.447	3/4	0.776	0.724	31/64	0.510	0.458
1	1.0000	1 1/2	1.5000	1.450	1.732	1.653	55/64	0.887	0.831	35/64	0.575	0.519
1 1/8	1.1250	1 11/16	1.6875	1.631	1.949	1.859	31/32	0.999	0.939	39/64	0.639	0.579
1 1/4	1.2500	1 7/8	1.8750	1.812	2.165	2.066	1 1/16	1.094	1.030	23/32	0.751	0.687
1 3/8	1.3750	2 1/16	2.0625	1.994	2.382	2.273	1 11/64	1.206	1.138	25/32	0.815	0.747
1 1/2	1.5000	2 1/4	2.2500	2.175	2.598	2.480	1 9/32	1.317	1.245	27/32	0.880	0.808

(Courtesy of ANSI; B18.2.1–1965 and ANSI; B18.2.2–1965.)

APPENDIX 22

FILLISTER HEAD AND ROUND HEAD CAP SCREWS

Fillister Head Cap Screws

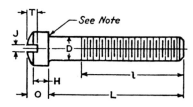

Nom-inal Size	D Body Diameter		A Head Diameter		H Height of Head		O Total Height of Head		J Width of Slot		T Depth of Slot	
	Max	Min	Max	Min	Max	Min	Max	Min	Max	Min	Max	Min
1/4	0.250	0.245	0.375	0.363	0.172	0.157	0.216	0.194	0.075	0.064	0.097	0.077
5/16	0.3125	0.307	0.437	0.424	0.203	0.186	0.253	0.230	0.084	0.072	0.115	0.090
3/8	0.375	0.369	0.562	0.547	0.250	0.229	0.314	0.284	0.094	0.081	0.142	0.112
7/16	0.4375	0.431	0.625	0.608	0.297	0.274	0.368	0.336	0.094	0.081	0.168	0.133
1/2	0.500	0.493	0.750	0.731	0.328	0.301	0.413	0.376	0.106	0.091	0.193	0.153
9/16	0.5625	0.555	0.812	0.792	0.375	0.346	0.467	0.427	0.118	0.102	0.213	0.168
5/8	0.625	0.617	0.875	0.853	0.422	0.391	0.521	0.478	0.133	0.116	0.239	0.189
3/4	0.750	0.742	1.000	0.976	0.500	0.466	0.612	0.566	0.149	0.131	0.283	0.223
7/8	0.875	0.866	1.125	1.098	0.594	0.556	0.720	0.668	0.167	0.147	0.334	0.264
1	1.000	0.990	1.312	1.282	0.656	0.612	0.803	0.743	0.188	0.166	0.371	0.291

All dimensions are given in inches.

The radius of the fillet at the base of the head:
   For sizes 1/4 to 3/8 in. incl. is 0.016 min and 0.031 max,
        7/16 to 9/16 in. incl. is 0.016 min and 0.047 max,
        5/8 to 1 in. incl. is 0.031 min and 0.062 max.

Round Head Cap Screws

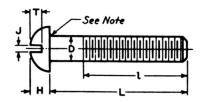

Nom-inal Size	D Body Diameter		A Head Diameter		H Height of Head		J Width of Slot		T Depth of Slot	
	Max	Min	Max	Min	Max	Min	Max	Min	Max	Min
1/4	0.250	0.245	0.437	0.418	0.191	0.175	0.075	0.064	0.117	0.097
5/16	0.3125	0.307	0.562	0.540	0.245	0.226	0.084	0.072	0.151	0.126
3/8	0.375	0.369	0.625	0.603	0.273	0.252	0.094	0.081	0.168	0.138
7/16	0.4375	0.431	0.750	0.725	0.328	0.302	0.094	0.081	0.202	0.167
1/2	0.500	0.493	0.812	0.786	0.354	0.327	0.106	0.091	0.218	0.178
9/16	0.5625	0.555	0.937	0.909	0.409	0.378	0.118	0.102	0.252	0.207
5/8	0.625	0.617	1.000	0.970	0.437	0.405	0.133	0.116	0.270	0.220
3/4	0.750	0.742	1.250	1.215	0.546	0.507	0.149	0.131	0.338	0.278

All dimensions are given in inches.

Radius of the fillet at the base of the head:
   For sizes 1/4 to 3/8 in. incl. is 0.016 min and 0.031 max,
        7/16 to 9/16 in. incl..is 0.016 min and 0.047 max,
        5/8 to 1 in..incl. is 0.031 min and 0.062 max.

(Courtesy of ANSI; B18.6.2–1956.)

APPENDIX 23
FLAT HEAD CAP SCREWS

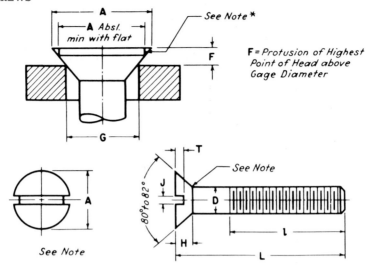

Nominal Size	D Body Diameter		A Head Diameter			G Gaging Diameter	H Height of Head	J Width of Slot		T Depth of Slot		F Protrusion Above Gaging Diameter	
	Max	Min	Max	Min	Absolute Min with Flat		Average	Max	Min	Max	Min	Max	Min
1/4	0.250	0.245	0.500	0.477	0.452	0.4245	0.140	0.075	0.064	0.068	0.045	0.0452	0.0307
5/16	0.3125	0.307	0.625	0.598	0.567	0.5376	0.177	0.084	0.072	0.086	0.057	0.0523	0.0354
3/8	0.375	0.369	0.750	0.720	0.682	0.6507	0.210	0.094	0.081	0.103	0.068	0.0594	0.0401
7/16	0.4375	0.431	0.8125	0.780	0.736	0.7229	0.210	0.094	0.081	0.103	0.068	0.0649	0.0448
1/2	0.500	0.493	0.875	0.841	0.791	0.7560	0.210	0.106	0.091	0.103	0.068	0.0705	0.0495
9/16	0.5625	0.555	1.000	0.962	0.906	0.8691	0.244	0.118	0.102	0.120	0.080	0.0775	0.0542
5/8	0.625	0.617	1.125	1.083	1.020	0.9822	0.281	0.133	0.116	0.137	0.091	0.0846	0.0588
3/4	0.750	0.742	1.375	1.326	1.251	1.2085	0.352	0.149	0.131	0.171	0.115	0.0987	0.0682
7/8	0.875	0.866	1.625	1.568	1.480	1.4347	0.423	0.167	0.147	0.206	0.138	0.1128	0.0776
1	1.000	0.990	1.875	1.811	1.711	1.6610	0.494	0.188	0.166	0.240	0.162	0.1270	0.0870
1 1/8	1.125	1.114	2.062	1.992	1.880	1.8262	0.529	0.196	0.178	0.257	0.173	0.1401	0.0964
1 1/4	1.250	1.239	2.312	2.235	2.110	2.0525	0.600	0.211	0.193	0.291	0.197	0.1542	0.1056
1 3/8	1.375	1.363	2.562	2.477	2.340	2.2787	0.665	0.226	0.208	0.326	0.220	0.1684	0.1151
1 1/2	1.500	1.488	2.812	2.720	2.570	2.5050	0.742	0.258	0.240	0.360	0.244	0.1825	0.1245

All dimensions are given in inches.

The maximum and minimum head diameters, A, are extended to the theoretical sharp corners.

The radius of the fillet at the base of the head shall not exceed 0.4 Max. D.

*Edge of head may be flat as shown or slightly rounded.

(Courtesy of ANSI; B18.6.2–1956.)

APPENDIX 24
MACHINE SCREWS

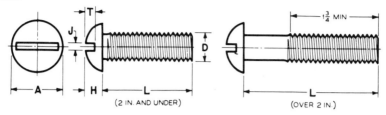

(2 IN. AND UNDER)

(OVER 2 IN.)

Dimensions of Slotted Round Head Machine Screws

Nominal Size	D Diameter of Screw	A Head Diameter		H Head Height		J Width of Slot		T Depth of Slot	
	Basic	Max	Min	Max	Min	Max	Min	Max	Min
0	0.0600	0.113	0.099	0.053	0.043	0.023	0.016	0.039	0.029
1	0.0730	0.138	0.122	0.061	0.051	0.026	0.019	0.044	0.033
2	0.0860	0.162	0.146	0.069	0.059	0.031	0.023	0.048	0.037
3	0.0990	0.187	0.169	0.078	0.067	0.035	0.027	0.053	0.040
4	0.1120	0.211	0.193	0.086	0.075	0.039	0.031	0.058	0.044
5	0.1250	0.236	0.217	0.095	0.083	0.043	0.035	0.063	0.047
6	0.1380	0.260	0.240	0.103	0.091	0.048	0.039	0.068	0.051
8	0.1640	0.309	0.287	0.120	0.107	0.054	0.045	0.077	0.058
10	0.1900	0.359	0.334	0.137	0.123	0.060	0.050	0.087	0.065
12	0.2160	0.408	0.382	0.153	0.139	0.067	0.056	0.096	0.073
1/4	0.2500	0.472	0.443	0.175	0.160	0.075	0.064	0.109	0.082
5/16	0.3125	0.590	0.557	0.216	0.198	0.084	0.072	0.132	0.099
3/8	0.3750	0.708	0.670	0.256	0.237	0.094	0.081	0.155	0.117
7/16	0.4375	0.750	0.707	0.328	0.307	0.094	0.081	0.196	0.148
1/2	0.5000	0.813	0.766	0.355	0.332	0.106	0.091	0.211	0.159
9/16	0.5625	0.938	0.887	0.410	0.385	0.118	0.102	0.242	0.183
5/8	0.6250	1.000	0.944	0.438	0.411	0.133	0.116	0.258	0.195
3/4	0.7500	1.250	1.185	0.547	0.516	0.149	0.131	0.320	0.242

All dimensions are given in inches.

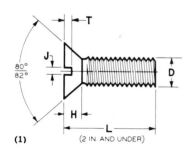

(1)

(2 IN. AND UNDER)

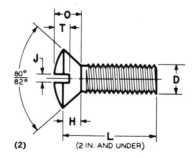

(2)

(2 IN. AND UNDER)

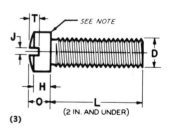

SEE NOTE

(3)

(2 IN. AND UNDER)

Three other common forms of machine screws are shown above: (1) flat head, (2) oval head, and (3) fillister head. Although dimension tables are not given for these three types of machine screws in this text, their general dimensions are closely related to those shown in the table above. Additional information on these screws can be obtained from ANSI; B18.6.3–1962.

(Courtesy of ANSI; B18.6.3–1962.)

APPENDIX 25

AMERICAN STANDARD MACHINE SCREWS

(The proportions of the screws can be found by multiplying the major diameter, D, by the factors given below.)

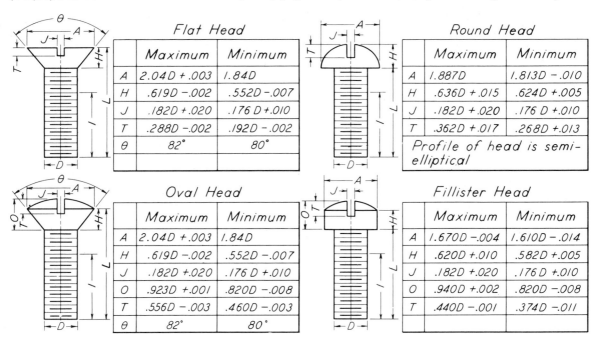

Flat Head

	Maximum	Minimum
A	2.04D + .003	1.84D
H	.619D −.002	.552D −.007
J	.182D +.020	.176D +.010
T	.288D −.002	.192D −.002
θ	82°	80°

Round Head

	Maximum	Minimum
A	1.887D	1.813D − .010
H	.636D +.015	.624D +.005
J	.182D +.020	.176D +.010
T	.362D +.017	.268D +.013

Profile of head is semi-elliptical

Oval Head

	Maximum	Minimum
A	2.04D +.003	1.84D
H	.619D −.002	.552D −.007
J	.182D +.020	.176D +.010
O	.923D +.001	.820D −.008
T	.556D −.003	.460D −.003
θ	82°	80°

Fillister Head

	Maximum	Minimum
A	1.670D −.004	1.610D − .014
H	.620D +.010	.582D +.005
J	.182D +.020	.176D +.010
O	.940D +.002	.820D −.008
T	.440D −.001	.374D −.011

APPENDIX 26

AMERICAN STANDARD MACHINE TAPERS*

No. of Taper	Taper per Foot (Basic)	Origin of Series	No. of Taper	Taper per Foot (Basic)	Origin of Series	No. of Taper	Taper per Foot (Basic)	Origin of Series	No. of Taper	Taper per Foot (Basic)	Origin of Series
0.239	0.50200	Brown & Sharpe	*	0.62326	Morse	250	0.750	¾ in. per ft.	600	0.750	¾ in. per ft.
.299	.50200	Brown & Sharpe	4½	.62400	Morse	300	.750	¾ in. per ft.	800	0.750	¾ in. per ft.
.375	.50200	Brown & Sharpe	5	.63151	Morse	350	.750	¾ in. per ft.	1000	0.750	¾ in. per ft.
1	.59858	Morse	6	.62565	Morse	400	.750	¾ in. per ft.	1200	0.750	¾ in. per ft.
2	.59941	Morse	7	.62400	Morse	450	.750	¾ in. per ft.			
3	.60235	Morse	200	.750	¾ in. per ft.	500	.750	¾ in. per ft.			

All dimensions in inches.

* Extracted from American Standards, "Machine Tapers, Self-Holding and Steep Taper Series" (ASA B5,10-1960), with the permission of the publisher, The American Society of Mechanical Engineers.

APPENDIX 27

AMERICAN NATIONAL STANDARD SQUARE HEAD SET SCREWS (ANSI B18.6.2)

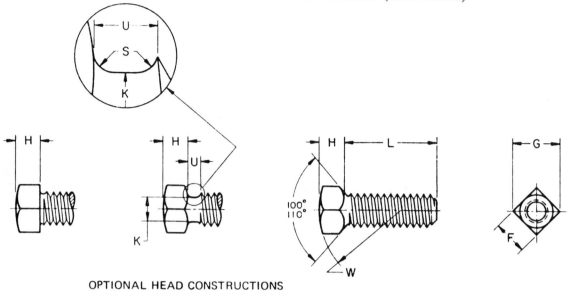

OPTIONAL HEAD CONSTRUCTIONS

Nominal Size[1] or Basic Screw Diameter		F Width Across Flats		G Width Across Corners		H Head Height		K Neck Relief Diameter		S Neck Relief Fillet Radius	U Neck Relief Width	W Head Radius
		Max	Min	Max	Min	Max	Min	Max	Min	Max	Min	Min
10	0.1900	0.188	0.180	0.265	0.247	0.148	0.134	0.145	0.140	0.027	0.083	0.48
1/4	0.2500	0.250	0.241	0.354	0.331	0.196	0.178	0.185	0.170	0.032	0.100	0.62
5/16	0.3125	0.312	0.302	0.442	0.415	0.245	0.224	0.240	0.225	0.036	0.111	0.78
3/8	0.3750	0.375	0.362	0.530	0.497	0.293	0.270	0.294	0.279	0.041	0.125	0.94
7/16	0.4375	0.438	0.423	0.619	0.581	0.341	0.315	0.345	0.330	0.046	0.143	1.09
1/2	0.5000	0.500	0.484	0.707	0.665	0.389	0.361	0.400	0.385	0.050	0.154	1.25
9/16	0.5625	0.562	0.545	0.795	0.748	0.437	0.407	0.454	0.439	0.054	0.167	1.41
5/8	0.6250	0.625	0.606	0.884	0.833	0.485	0.452	0.507	0.492	0.059	0.182	1.56
3/4	0.7500	0.750	0.729	1.060	1.001	0.582	0.544	0.620	0.605	0.065	0.200	1.88
7/8	0.8750	0.875	0.852	1.237	1.170	0.678	0.635	0.731	0.716	0.072	0.222	2.19
1	1.0000	1.000	0.974	1.414	1.337	0.774	0.726	0.838	0.823	0.081	0.250	2.50
1 1/8	1.1250	1.125	1.096	1.591	1.505	0.870	0.817	0.939	0.914	0.092	0.283	2.81
1 1/4	1.2500	1.250	1.219	1.768	1.674	0.966	0.908	1.064	1.039	0.092	0.283	3.12
1 3/8	1.3750	1.375	1.342	1.945	1.843	1.063	1.000	1.159	1.134	0.109	0.333	3.44
1 1/2	1.5000	1.500	1.464	2.121	2.010	1.159	1.091	1.284	1.259	0.109	0.333	3.75

[1] Where specifying nominal size in decimals, zeros preceding decimal and in the fourth decimal place shall be omitted.

APPENDIX 28

AMERICAN NATIONAL STANDARD POINTS FOR SQUARE HEAD SET SCREWS (ANSI B18.6.2)

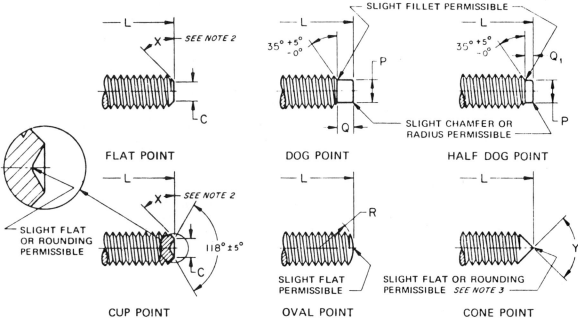

FLAT POINT   DOG POINT   HALF DOG POINT

CUP POINT   OVAL POINT   CONE POINT

Nominal Size[1] or Basic Screw Diameter		C Cup and Flat Point Diameters		P Dog and Half Dog Point Diameters		Q Point Length Dog		Q₁ Point Length Half Dog		R Oval Point Radius +0.031 -0.000	Y Cone Point Angle 90° ±2° For These Nominal Lengths or Longer; 118° ±2° For Shorter Screws
		Max	Min	Max	Min	Max	Min	Max	Min		
10	0.1900	0.102	0.088	0.127	0.120	0.095	0.085	0.050	0.040	0.142	1/4
1/4	0.2500	0.132	0.118	0.156	0.149	0.130	0.120	0.068	0.058	0.188	5/16
5/16	0.3125	0.172	0.156	0.203	0.195	0.161	0.151	0.083	0.073	0.234	3/8
3/8	0.3750	0.212	0.194	0.250	0.241	0.193	0.183	0.099	0.089	0.281	7/16
7/16	0.4375	0.252	0.232	0.297	0.287	0.224	0.214	0.114	0.104	0.328	1/2
1/2	0.5000	0.291	0.270	0.344	0.334	0.255	0.245	0.130	0.120	0.375	9/16
9/16	0.5625	0.332	0.309	0.391	0.379	0.287	0.275	0.146	0.134	0.422	5/8
5/8	0.6250	0.371	0.347	0.469	0.456	0.321	0.305	0.164	0.148	0.469	3/4
3/4	0.7500	0.450	0.425	0.562	0.549	0.383	0.367	0.196	0.180	0.562	7/8
7/8	0.8750	0.530	0.502	0.656	0.642	0.446	0.430	0.227	0.211	0.656	1
1	1.0000	0.609	0.579	0.750	0.734	0.510	0.490	0.260	0.240	0.750	1 1/8
1 1/8	1.1250	0.689	0.655	0.844	0.826	0.572	0.552	0.291	0.271	0.844	1 1/4
1 1/4	1.2500	0.767	0.733	0.938	0.920	0.635	0.615	0.323	0.303	0.938	1 1/2
1 3/8	1.3750	0.848	0.808	1.031	1.011	0.698	0.678	0.354	0.334	1.031	1 5/8
1 1/2	1.5000	0.926	0.886	1.125	1.105	0.760	0.740	0.385	0.365	1.125	1 3/4

[1] Where specifying nominal size in decimals, zeros preceding decimal and in the fourth decimal place shall be omitted.

[2] Point angle X shall be 45° plus 5°, minus 0°, for screws of nominal lengths equal to or longer than those listed in Column Y, and 30° minimum for screws of shorter nominal lengths.

[3] The extent of rounding or flat at apex of cone point shall not exceed an amount equivalent to 10 per cent of the basic screw diameter.

APPENDIX 29
AMERICAN NATIONAL STANDARD SLOTTED HEADLESS SET SCREWS (ANSI B18.6.2)

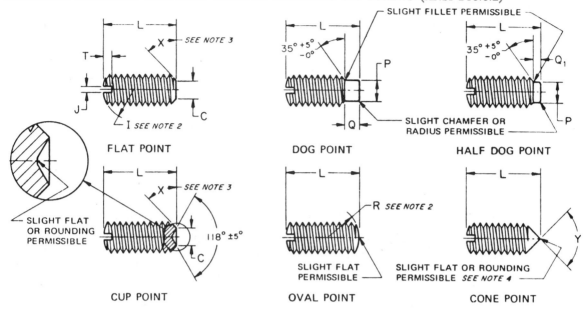

FLAT POINT DOG POINT HALF DOG POINT

CUP POINT OVAL POINT CONE POINT

Nominal Size[1] or Basic Screw Diameter		I[2] Crown Radius	J Slot Width		T Slot Depth		C Cup and Flat Point Diameters		P Dog Point Diameters		Q Point Length Dog		Q₁ Point Length Half Dog		R[2] Oval Point Radius	Y Cone Point Angle 90° ±2° For These Nominal Lengths or Longer; 118° ±2° For Shorter Screws
		Basic	Max	Min	Max	Min	Max	Min	Max	Min	Max	Min	Max	Min	Basic	
0	0.0600	0.060	0.014	0.010	0.020	0.016	0.033	0.027	0.040	0.037	0.032	0.028	0.017	0.013	0.045	5/64
1	0.0730	0.073	0.016	0.012	0.020	0.016	0.040	0.033	0.049	0.045	0.040	0.036	0.021	0.017	0.055	3/32
2	0.0860	0.086	0.018	0.014	0.025	0.019	0.047	0.039	0.057	0.053	0.046	0.042	0.024	0.020	0.064	7/64
3	0.0990	0.099	0.020	0.016	0.028	0.022	0.054	0.045	0.066	0.062	0.052	0.048	0.027	0.023	0.074	1/8
4	0.1120	0.112	0.024	0.018	0.031	0.025	0.061	0.051	0.075	0.070	0.058	0.054	0.030	0.026	0.084	5/32
5	0.1250	0.125	0.026	0.020	0.036	0.026	0.067	0.057	0.083	0.078	0.063	0.057	0.033	0.027	0.094	3/16
6	0.1380	0.138	0.028	0.022	0.040	0.030	0.074	0.064	0.092	0.087	0.073	0.067	0.038	0.032	0.104	3/16
8	0.1640	0.164	0.032	0.026	0.046	0.036	0.087	0.076	0.109	0.103	0.083	0.077	0.043	0.037	0.123	1/4
10	0.1900	0.190	0.035	0.029	0.053	0.043	0.102	0.088	0.127	0.120	0.095	0.085	0.050	0.040	0.142	1/4
12	0.2160	0.216	0.042	0.035	0.061	0.051	0.115	0.101	0.144	0.137	0.115	0.105	0.060	0.050	0.162	5/16
1/4	0.2500	0.250	0.049	0.041	0.068	0.058	0.132	0.118	0.156	0.149	0.130	0.120	0.068	0.058	0.188	5/16
5/16	0.3125	0.312	0.055	0.047	0.083	0.073	0.172	0.156	0.203	0.195	0.161	0.151	0.083	0.073	0.234	3/8
3/8	0.3750	0.375	0.068	0.060	0.099	0.089	0.212	0.194	0.250	0.241	0.193	0.183	0.099	0.089	0.281	7/16
7/16	0.4375	0.438	0.076	0.068	0.114	0.104	0.252	0.232	0.297	0.287	0.224	0.214	0.114	0.104	0.328	1/2
1/2	0.5000	0.500	0.086	0.076	0.130	0.120	0.291	0.270	0.344	0.334	0.255	0.245	0.130	0.120	0.375	9/16
9/16	0.5625	0.562	0.096	0.086	0.146	0.136	0.332	0.309	0.391	0.379	0.287	0.275	0.146	0.134	0.422	5/8
5/8	0.6250	0.625	0.107	0.097	0.161	0.151	0.371	0.347	0.469	0.456	0.321	0.305	0.164	0.148	0.469	3/4
3/4	0.7500	0.750	0.134	0.124	0.193	0.183	0.450	0.425	0.562	0.549	0.383	0.367	0.196	0.180	0.562	7/8

[1] Where specifying nominal size in decimals, zeros preceding decimal and in the fourth decimal place shall be omitted.

[2] Tolerance on radius for nominal sizes up to and including 5 (0.125 in.) shall be plus 0.015 in. and minus 0.000, and for larger sizes, plus 0.031 in. and minus 0.000. Slotted ends on screws may be flat at option of manufacturer.

[3] Point angle X shall be 45° plus 5°, minus 0°, for screws of nominal lengths equal to or longer than those listed in Column Y, and 30° minimum for screws of shorter nominal lengths.

[4] The extent of rounding or flat at apex of cone point shall not exceed an amount equivalent to 10 per cent of the basic screw diameter.

APPENDIX 30

TWIST DRILL SIZES

Number Size Drills

Size	Drill Diameter		Size	Drill Diameter		Size	Drill Diameter		Size	Drill Diameter	
	Inches	mm		Inches	mm		Inches	mm		Inches	mm
1	0.2280	5.7912	21	0.1590	4.0386	41	0.0960	2.4384	61	0.0390	0.9906
2	0.2210	5.6134	22	0.1570	3.9878	42	0.0935	2.3622	62	0.0380	0.9652
3	0.2130	5.4102	23	0.1540	3.9116	43	0.0890	2.2606	63	0.0370	0.9398
4	0.2090	5.3086	24	0.1520	3.8608	44	0.0860	2.1844	64	0.0360	0.9144
5	0.2055	5.2197	25	0.1495	3.7973	45	0.0820	2.0828	65	0.0350	0.8890
6	0.2040	5.1816	26	0.1470	3.7338	46	0.0810	2.0574	66	0.0330	0.8382
7	0.2010	5.1054	27	0.1440	3.6576	47	0.0785	1.9812	67	0.0320	0.8128
8	0.1990	5.0800	28	0.1405	3.5560	48	0.0760	1.9304	68	0.0310	0.7874
9	0.1960	4.9784	29	0.1360	3.4544	49	0.0730	1.8542	69	0.0292	0.7417
10	0.1935	4.9149	30	0.1285	3.2639	50	0.0700	1.7780	70	0.0280	0.7112
11	0.1910	4.8514	31	0.1200	3.0480	51	0.0670	1.7018	71	0.0260	0.6604
12	0.1890	4.8006	32	0.1160	2.9464	52	0.0635	1.6129	72	0.0250	0.6350
13	0.1850	4.6990	33	0.1130	2.8702	53	0.0595	1.5113	73	0.0240	0.6096
14	0.1820	4.6228	34	0.1110	2.8194	54	0.0550	1.3970	74	0.0225	0.5715
15	0.1800	4.5720	35	0.1100	2.7940	55	0.0520	1.3208	75	0.0210	0.5334
16	0.1770	4.4958	36	0.1065	2.7051	56	0.0465	1.1684	76	0.0200	0.5080
17	0.1730	4.3942	37	0.1040	2.6416	57	0.0430	1.0922	77	0.0180	0.4572
18	0.1695	4.3053	38	0.1015	2.5781	58	0.0420	1.0668	78	0.0160	0.4064
19	0.1660	4.2164	39	0.0995	2.5273	59	0.0410	1.0414	79	0.0145	0.3638
20	0.1610	4.0894	40	0.0980	2.4892	60	0.0400	1.0160	80	0.0135	0.3429

Metric Drill Sizes   Preferred sizes are in color type. Decimal-inch equivalents are for reference only.

Drill Diameter		Drill Diameter		Drill Diameter		Drill Diameter		Drill Diameter		Drill Diameter		Drill Diameter	
mm	in.	mm	in.	mm	in.	mm	in.	mm	in.	mm	in.	mm	in.
.40	.0157	1.03	.0406	2.20	.0866	5.00	.1969	10.00	.3937	21.50	.8465	48.00	1.8898
.42	.0165	1.05	.0413	2.30	.0906	5.20	.2047	10.30	.4055	22.00	.8661	50.00	1.9685
.45	.0177	1.08	.0425	2.40	.0945	5.30	.2087	10.50	.4134	23.00	.9055	51.50	2.0276
.48	.0189	1.10	.0433	2.50	.0984	5.40	.2126	10.80	.4252	24.00	.9449	53.00	2.0866
.50	.0197	1.15	.0453	2.60	.1024	5.60	.2205	11.00	.4331	25.00	.9843	54.00	2.1260
.52	.0205	1.20	.0472	2.70	.1063	5.80	.2283	11.50	.4528	26.00	1.0236	56.00	2.2047
.55	.0217	1.25	.0492	2.80	.1102	6.00	.2362	12.00	.4724	27.00	1.0630	58.00	2.2835
.58	.0228	1.30	.0512	2.90	.1142	6.20	.2441	12.50	.4921	28.00	1.1024	60.00	2.3622
.60	.0236	1.35	.0531	3.00	.1181	6.30	.2480	13.00	.5118	29.00	1.1417		
.62	.0244	1.40	.0551	3.10	.1220	6.50	.2559	13.50	.5315	30.00	1.1811		
.65	.0256	1.45	.0571	3.20	.1260	6.70	.2638	14.00	.5512	31.00	1.2205		
.68	.0268	1.50	.0591	3.30	.1299	6.80	.2677	14.50	.5709	32.00	1.2598		
.70	.0276	1.55	.0610	3.40	.1339	6.90	.2717	15.00	.5906	33.00	1.2992		
.72	.0283	1.60	.0630	3.50	.1378	7.10	.2795	15.50	.6102	34.00	1.3386		
.75	.0295	1.65	.0650	3.60	.1417	7.30	.2874	16.00	.6299	35.00	1.3780		
.78	.0307	1.70	.0669	3.70	.1457	7.50	.2953	16.50	.6496	36.00	1.4173		
.80	.0315	1.75	.0689	3.80	.1496	7.80	.3071	17.00	.6693	37.00	1.4567		
.82	.0323	1.80	.0709	3.90	.1535	8.00	.3150	17.50	.6890	38.00	1.4961		
.85	.0335	1.85	.0728	4.00	.1575	8.20	.3228	18.00	.7087	39.00	1.5354		
.88	.0346	1.90	.0748	4.10	.1614	8.50	.3346	18.50	.7283	40.00	1.5748		
.90	.0354	1.95	.0768	4.20	.1654	8.80	.3465	19.00	.7480	41.00	1.6142		
.92	.0362	2.00	.0787	4.40	.1732	9.00	.3543	19.50	.7677	42.00	1.6535		
.95	.0374	2.05	.0807	4.50	.1772	9.20	.3622	20.00	.7874	43.50	1.7126		
.98	.0386	2.10	.0827	4.60	.1811	9.50	.3740	20.50	.8071	45.00	1.7717		
1.00	.0394	2.15	.0846	4.80	.1890	9.80	.3858	21.00	.8268	46.50	1.8307		

Cont.

APPENDIX 30

TWIST DRILL SIZES (Cont.)

Letter Size Drills

Size	Drill Diameter		Size	Drill Diameter		Size	Drill Diameter		Size	Drill Diameter	
	Inches	mm		Inches	mm		Inches	mm		Inches	mm
A	0.234	5.944	H	0.266	6.756	O	0.316	8.026	V	0.377	9.576
B	0.238	6.045	I	0.272	6.909	P	0.323	8.204	W	0.386	9.804
C	0.242	6.147	J	0.277	7.036	Q	0.332	8.433	X	0.397	10.084
D	0.246	6.248	K	0.281	7.137	R	0.339	8.611	Y	0.404	10.262
E	0.250	6.350	L	0.290	7.366	S	0.348	8.839	Z	0.413	10.490
F	0.257	6.528	M	0.295	7.493	T	0.358	9.093			
G	0.261	6.629	N	0.302	7.601	U	0.368	9.347			

(Courtesy of General Motors Corporation.)

APPENDIX 31

STRAIGHT PINS

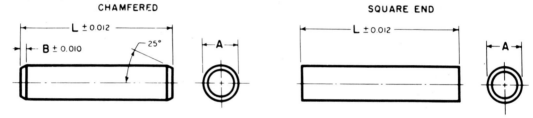

CHAMFERED   SQUARE END

Nominal Diameter	Diameter A		Chamfer B
	Max	Min	
0.062	0.0625	0.0605	0.015
0.094	0.0937	0.0917	0.015
0.109	0.1094	0.1074	0.015
0.125	0.1250	0.1230	0.015
0.156	0.1562	0.1542	0.015
0.188	0.1875	0.1855	0.015
0.219	0.2187	0.2167	0.015
0.250	0.2500	0.2480	0.015
0.312	0.3125	0.3095	0.030
0.375	0.3750	0.3720	0.030
0.438	0.4375	0.4345	0.030
0.500	0.500	0.4970	0.030

All dimensions are given in inches.

These pins must be straight and free from burrs or any other defects that will affect their serviceability.

(Courtesy of ANSI; B5.20–1958.)

APPENDIX 32
STANDARD KEYS AND KEYWAYS

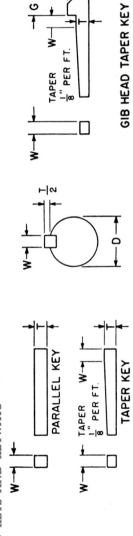

PARALLEL KEY

TAPER KEY
TAPER 1/8" PER FT.

GIB HEAD TAPER KEY
TAPER 1/8" PER FT.

SPROCKET BORE (= SHAFT DIAM.) INCHES D	KEYWAY DIMENSIONS — INCHES				KEY DIMENSIONS — INCHES					GIB HEAD DIMENSIONS — INCHES				KEY TOLERANCES TAPER AND GIB HEAD	
	For Square Key		For Flat Key		Square		Flat		TOLERANCE ON W AND T (−)	Square Key		Flat Key		W (−)	T (+)
	Width W	Depth T/2	Width W	Depth T/2	Width W	Height T	Width W	Height T		H	G	H	G		
1/2 — 9/16	1/8	1/16	1/8	3/64	1/8 × 1/8		1/8 × 3/32		0.002	1/4	7/32	3/16	1/8	0.002	0.002
5/8 — 7/8	3/16	3/32	3/16	1/16	3/16 × 3/16		3/16 × 1/8		0.002	5/16	9/32	1/4	3/16	0.002	0.002
15/16 — 1 1/4	1/4	1/8	1/4	3/32	1/4 × 1/4		1/4 × 3/16		0.002	7/16	11/32	5/16	1/4	0.002	0.002
1 3/16 — 1 3/8	5/16	5/32	5/16	1/8	5/16 × 5/16		5/16 × 1/4		0.002	9/16	13/32	3/8	5/16	0.002	0.002
1 7/16 — 1 3/4	3/8	3/16	3/8	1/8	3/8 × 3/8		3/8 × 1/4		0.002	11/16	15/32	7/16	3/8	0.002	0.002
1 13/16 — 2 1/4	1/2	1/4	1/2	3/16	1/2 × 1/2		1/2 × 3/8		0.0025	7/8	19/32	5/8	1/2	0.0025	0.0025
2 5/16 — 2 3/4	5/8	5/16	5/8	7/32	5/8 × 5/8		5/8 × 7/16		0.0025	1 1/16	23/32	3/4	5/8	0.0025	0.0025
2 7/8 — 3 1/4	3/4	3/8	3/4	1/4	3/4 × 3/4		3/4 × 1/2		0.0025	1 1/4	7/8	7/8	3/4	0.0025	0.0025
3 3/8 — 3 3/4	7/8	7/16	7/8	5/16	7/8 × 7/8		7/8 × 5/8		0.003	1 1/2	1	1 1/16	7/8	0.003	0.003
3 7/8 — 4 1/2	1	1/2	1	3/8	1 × 1		1 × 3/4		0.003	1 3/4	1 3/16	1 1/4	1	0.003	0.003
4 3/4 — 5 1/2	1 1/4	5/8	1 1/4	7/16	1 1/4 × 1 1/4		1 1/4 × 7/8		0.003	2	1 7/16	1 1/2	1 1/4	0.003	0.003
5 3/4 — 7 3/8	1 1/2	3/4	1 1/2	1/2	1 1/2 × 1 1/2		1 1/2 × 1		0.003	2 1/2	1 3/4	1 3/4	1 1/2	0.003	0.003
7 1/2 — 9 7/8	1 3/4	7/8	..	..	1 3/4 × 1 3/4		.. × ..		0.004	3	2	..	..	0.004	0.004
10 — 12 1/2	2	1	..	..	2 × 2		.. × ..		0.004	3 1/2	2 3/8	..	..	0.004	0.004

Standard Keyway Tolerances: Straight Keyway — Width (W) + .005 / − .000   Depth (T/2) + .010 / − .000

Taper Keyway — Width (W) + .005 / − .000   Depth (T/2) + .000 / − .010

APPENDIX 33
WOODRUFF KEYS

USA STANDARD

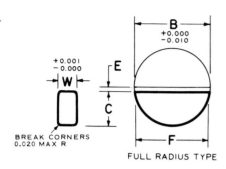

FULL RADIUS TYPE

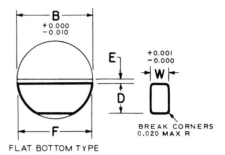

FLAT BOTTOM TYPE

Woodruff Keys

Key No.	Nominal Key Size W × B	Actual Length F +0.000-0.010	Height of Key				Distance Below Center E
			C		D		
			Max	Min	Max	Min	
202	$\frac{1}{16} \times \frac{1}{4}$	0.248	0.109	0.104	0.109	0.104	$\frac{1}{64}$
202.5	$\frac{1}{16} \times \frac{5}{16}$	0.311	0.140	0.135	0.140	0.135	$\frac{1}{64}$
302.5	$\frac{3}{32} \times \frac{5}{16}$	0.311	0.140	0.135	0.140	0.135	$\frac{1}{64}$
203	$\frac{1}{16} \times \frac{3}{8}$	0.374	0.172	0.167	0.172	0.167	$\frac{1}{64}$
303	$\frac{3}{32} \times \frac{3}{8}$	0.374	0.172	0.167	0.172	0.167	$\frac{1}{64}$
403	$\frac{1}{8} \times \frac{3}{8}$	0.374	0.172	0.167	0.172	0.167	$\frac{1}{64}$
204	$\frac{1}{16} \times \frac{1}{2}$	0.491	0.203	0.198	0.194	0.188	$\frac{3}{64}$
304	$\frac{3}{32} \times \frac{1}{2}$	0.491	0.203	0.198	0.194	0.188	$\frac{3}{64}$
404	$\frac{1}{8} \times \frac{1}{2}$	0.491	0.203	0.198	0.194	0.188	$\frac{3}{64}$
305	$\frac{3}{32} \times \frac{5}{8}$	0.612	0.250	0.245	0.240	0.234	$\frac{1}{16}$
405	$\frac{1}{8} \times \frac{5}{8}$	0.612	0.250	0.245	0.240	0.234	$\frac{1}{16}$
505	$\frac{5}{32} \times \frac{5}{8}$	0.612	0.250	0.245	0.240	0.234	$\frac{1}{16}$
605	$\frac{3}{16} \times \frac{5}{8}$	0.612	0.250	0.245	0.240	0.234	$\frac{1}{16}$
406	$\frac{1}{8} \times \frac{3}{4}$	0.740	0.313	0.308	0.303	0.297	$\frac{1}{16}$
506	$\frac{5}{32} \times \frac{3}{4}$	0.740	0.313	0.308	0.303	0.297	$\frac{1}{16}$
606	$\frac{3}{16} \times \frac{3}{4}$	0.740	0.313	0.308	0.303	0.297	$\frac{1}{16}$
806	$\frac{1}{4} \times \frac{3}{4}$	0.740	0.313	0.308	0.303	0.297	$\frac{1}{16}$
507	$\frac{5}{32} \times \frac{7}{8}$	0.866	0.375	0.370	0.365	0.359	$\frac{1}{16}$
607	$\frac{3}{16} \times \frac{7}{8}$	0.866	0.375	0.370	0.365	0.359	$\frac{1}{16}$
707	$\frac{7}{32} \times \frac{7}{8}$	0.866	0.375	0.370	0.365	0.359	$\frac{1}{16}$
807	$\frac{1}{4} \times \frac{7}{8}$	0.866	0.375	0.370	0.365	0.359	$\frac{1}{16}$
608	$\frac{3}{16} \times 1$	0.992	0.438	0.433	0.428	0.422	$\frac{1}{16}$
708	$\frac{7}{32} \times 1$	0.992	0.438	0.433	0.428	0.422	$\frac{1}{16}$
808	$\frac{1}{4} \times 1$	0.992	0.438	0.433	0.428	0.422	$\frac{1}{16}$
1008	$\frac{5}{16} \times 1$	0.992	0.438	0.433	0.428	0.422	$\frac{1}{16}$
1208	$\frac{3}{8} \times 1$	0.992	0.438	0.433	0.428	0.422	$\frac{1}{16}$
609	$\frac{3}{16} \times 1\frac{1}{8}$	1.114	0.484	0.479	0.475	0.469	$\frac{5}{64}$
709	$\frac{7}{32} \times 1\frac{1}{8}$	1.114	0.484	0.479	0.475	0.469	$\frac{5}{64}$
809	$\frac{1}{4} \times 1\frac{1}{8}$	1.114	0.484	0.479	0.475	0.469	$\frac{5}{64}$
1009	$\frac{5}{16} \times 1\frac{1}{8}$	1.114	0.484	0.479	0.475	0.469	$\frac{5}{64}$

(Courtesy of ANSI; B17.2-1967.)

APPENDIX 34
WOODRUFF KEYSEATS

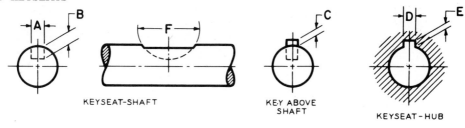

KEYSEAT-SHAFT   KEY ABOVE SHAFT   KEYSEAT-HUB

Keyseat Dimensions

Key Number	Nominal Size Key	Keyseat — Shaft					Key Above Shaft	Keyseat — Hub	
		Width A*		Depth B	Diameter F		Height C	Width D	Depth E
		Min	Max	+0.005 -0.000	Min	Max	+0.005 -0.005	+0.002 -0.000	+0.005 -0.000
202	1/16 × 1/4	0.0615	0.0630	0.0728	0.250	0.268	0.0312	0.0635	0.0372
202.5	1/16 × 5/16	0.0615	0.0630	0.1038	0.312	0.330	0.0312	0.0635	0.0372
302.5	3/32 × 5/16	0.0928	0.0943	0.0882	0.312	0.330	0.0469	0.0948	0.0529
203	1/16 × 3/8	0.0615	0.0630	0.1358	0.375	0.393	0.0312	0.0635	0.0372
303	3/32 × 3/8	0.0928	0.0943	0.1202	0.375	0.393	0.0469	0.0948	0.0529
403	1/8 × 3/8	0.1240	0.1255	0.1045	0.375	0.393	0.0625	0.1260	0.0685
204	1/16 × 1/2	0.0615	0.0630	0.1668	0.500	0.518	0.0312	0.0635	0.0372
304	3/32 × 1/2	0.0928	0.0943	0.1511	0.500	0.518	0.0469	0.0948	0.0529
404	1/8 × 1/2	0.1240	0.1255	0.1355	0.500	0.518	0.0625	0.1260	0.0685
305	3/32 × 5/8	0.0928	0.0943	0.1981	0.625	0.643	0.0469	0.0948	0.0529
405	1/8 × 5/8	0.1240	0.1255	0.1825	0.625	0.643	0.0625	0.1260	0.0685
505	5/32 × 5/8	0.1553	0.1568	0.1669	0.625	0.643	0.0781	0.1573	0.0841
605	3/16 × 5/8	0.1863	0.1880	0.1513	0.625	0.643	0.0937	0.1885	0.0997
406	1/8 × 3/4	0.1240	0.1255	0.2455	0.750	0.768	0.0625	0.1260	0.0685
506	5/32 × 3/4	0.1553	0.1568	0.2299	0.750	0.768	0.0781	0.1573	0.0841
606	3/16 × 3/4	0.1863	0.1880	0.2143	0.750	0.768	0.0937	0.1885	0.0997
806	1/4 × 3/4	0.2487	0.2505	0.1830	0.750	0.768	0.1250	0.2510	0.1310
507	5/32 × 7/8	0.1553	0.1568	0.2919	0.875	0.895	0.0781	0.1573	0.0841
607	3/16 × 7/8	0.1863	0.1880	0.2763	0.875	0.895	0.0937	0.1885	0.0997
707	7/32 × 7/8	0.2175	0.2193	0.2607	0.875	0.895	0.1093	0.2198	0.1153
807	1/4 × 7/8	0.2487	0.2505	0.2450	0.875	0.895	0.1250	0.2510	0.1310
608	3/16 × 1	0.1863	0.1880	0.3393	1.000	1.020	0.0937	0.1885	0.0997
708	7/32 × 1	0.2175	0.2193	0.3237	1.000	1.020	0.1093	0.2198	0.1153
808	1/4 × 1	0.2487	0.2505	0.3080	1.000	1.020	0.1250	0.2510	0.1310
1008	5/16 × 1	0.3111	0.3130	0.2768	1.000	1.020	0.1562	0.3135	0.1622
1208	3/8 × 1	0.3735	0.3755	0.2455	1.000	1.020	0.1875	0.3760	0.1935
609	3/16 × 1 1/8	0.1863	0.1880	0.3853	1.125	1.145	0.0937	0.1885	0.0997
709	7/32 × 1 1/8	0.2175	0.2193	0.3697	1.125	1.145	0.1093	0.2198	0.1153
809	1/4 × 1 1/8	0.2487	0.2505	0.3540	1.125	1.145	0.1250	0.2510	0.1310
1009	5/16 × 1 1/8	0.3111	0.3130	0.3228	1.125	1.145	0.1562	0.3135	0.1622

(Courtesy of ANSI; B17.2–1967.)

APPENDIX 35
TAPER PINS

Number	7/0	6/0	5/0	4/0	3/0	2/0	0	1	2	3	4	5	6	7	8	9	10
Size (large end)	0.0625	0.0780	0.0940	0.1090	0.1250	0.1410	0.1560	0.1720	0.1930	0.2190	0.2500	0.2890	0.3410	0.4090	0.4920	0.5910	0.7060
Length, L																	
0.375	X	X															
0.500	X	X	X														
0.625	X	X	X	X													
0.750		X	X	X	X												
0.875			X	X	X	X											
1.000			X	X	X	X	X										
1.250				X	X	X	X	X									
1.500					X	X	X	X	X								
1.750						X	X	X	X	X							
2.000							X	X	X	X	X						
2.250								X	X	X	X	X					
2.500									X	X	X	X	X				
2.750									X	X	X	X	X	X			
3.000										X	X	X	X	X	X		
3.250										X	X	X	X	X	X	X	
3.500											X	X	X	X	X	X	X
3.750												X	X	X	X	X	X
4.000													X	X	X	X	X
4.250														X	X	X	X
4.500															X	X	X
4.750															X	X	X
5.000															X	X	X
5.250																X	X
5.500																X	X
5.750																X	X
6.000																X	X

All dimensions are given in inches.

Standard reamers are available for pins given above the line.

Pins Nos. 11 (size 0.8600), 12 (size 1.032), 13 (size 1.241), and 14 (1.523) are special sizes—hence their lengths are special.

To find small diameter of pin, multiply the length by 0.2083 and subtract the result from the large diameter.

(Courtesy of ANSI; B5.20–1958.)

APPENDIX 36
PLAIN WASHERS

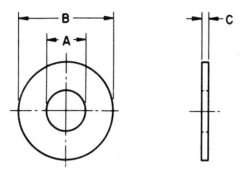

Dimensions of Preferred Sizes of Type A Plain Washers[a]

When specifying washers on drawings or in notes, give the inside diameter, outside diameter, and the thickness.
Example: 0.938 × 1.750 × 0.134 TYPE A PLAIN WASHER.

Nominal Washer Size[b]			Inside Diameter A			Outside Diameter B			Thickness C		
				Tolerance			Tolerance				
			Basic	Plus	Minus	Basic	Plus	Minus	Basic	Max	Min
—	—		0.078	0.000	0.005	0.188	0.000	0.005	0.020	0.025	0.016
—	—		0.094	0.000	0.005	0.250	0.000	0.005	0.020	0.025	0.016
—	—		0.125	0.008	0.005	0.312	0.008	0.005	0.032	0.040	0.025
No. 6	0.138		0.156	0.008	0.005	0.375	0.015	0.005	0.049	0.065	0.036
No. 8	0.164		0.188	0.008	0.005	0.438	0.015	0.005	0.049	0.065	0.036
No. 10	0.190		0.219	0.008	0.005	0.500	0.015	0.005	0.049	0.065	0.036
$\frac{3}{16}$	0.188		0.250	0.015	0.005	0.562	0.015	0.005	0.049	0.065	0.036
No. 12	0.216		0.250	0.015	0.005	0.562	0.015	0.005	0.065	0.080	0.051
$\frac{1}{4}$	0.250	N	0.281	0.015	0.005	0.625	0.015	0.005	0.065	0.080	0.051
$\frac{1}{4}$	0.250	W	0.312	0.015	0.005	0.734[c]	0.015	0.007	0.065	0.080	0.051
$\frac{5}{16}$	0.312	N	0.344	0.015	0.005	0.688	0.015	0.007	0.065	0.080	0.051
$\frac{5}{16}$	0.312	W	0.375	0.015	0.005	0.875	0.030	0.007	0.083	0.104	0.064
$\frac{3}{8}$	0.375	N	0.406	0.015	0.005	0.812	0.015	0.007	0.065	0.080	0.051
$\frac{3}{8}$	0.375	W	0.438	0.015	0.005	1.000	0.030	0.007	0.083	0.104	0.064
$\frac{7}{16}$	0.438	N	0.469	0.015	0.005	0.922	0.015	0.007	0.065	0.080	0.051
$\frac{7}{16}$	0.438	W	0.500	0.015	0.005	1.250	0.030	0.007	0.083	0.104	0.064
$\frac{1}{2}$	0.500	N	0.531	0.015	0.005	1.062	0.030	0.007	0.095	0.121	0.074
$\frac{1}{2}$	0.500	W	0.562	0.015	0.005	1.375	0.030	0.007	0.109	0.132	0.086

[a] Preferred sizes are for the most part from series previously designated "Standard Plate" and "SAE." Where common sizes existed in the two series, the SAE size is designated "N" (narrow) and the Standard Plate "W" (wide). These sizes as well as all other sizes of Type A Plain Washers are to be ordered by ID, OD, and thickness dimensions.

[b] Nominal washer sizes are intended for use with comparable nominal screw or bolt sizes.

[c] The 0.734 in., 1.156 in., and 1.469 in. outside diameters avoid washers which could be used in coin-operated devices.

Cont.

APPENDIX 36
PLAIN WASHERS (Cont.)

Nominal Washer Size[b]			Inside Diameter A Basic	Tolerance Plus	Tolerance Minus	Outside Diameter B Basic	Tolerance Plus	Tolerance Minus	Thickness C Basic	Max	Min
$\frac{9}{16}$	0.562	N	0.594	0.015	0.005	1.156[c]	0.030	0.007	0.095	0.121	0.074
$\frac{9}{16}$	0.562	W	0.625	0.015	0.005	1.469[c]	0.030	0.007	0.109	0.132	0.086
$\frac{5}{8}$	0.625	N	0.656	0.030	0.007	1.312	0.030	0.007	0.095	0.121	0.074
$\frac{5}{8}$	0.625	W	0.688	0.030	0.007	1.750	0.030	0.007	0.134	0.160	0.108
$\frac{3}{4}$	0.750	N	0.812	0.030	0.007	1.469	0.030	0.007	0.134	0.160	0.108
$\frac{3}{4}$	0.750	W	0.812	0.030	0.007	2.000	0.030	0.007	0.148	0.177	0.122
$\frac{7}{8}$	0.875	N	0.938	0.030	0.007	1.750	0.030	0.007	0.134	0.160	0.108
$\frac{7}{8}$	0.875	W	0.938	0.030	0.007	2.250	0.030	0.007	0.165	0.192	0.136
1	1.000	N	1.062	0.030	0.007	2.000	0.030	0.007	0.134	0.160	0.108
1	1.000	W	1.062	0.030	0.007	2.500	0.030	0.007	0.165	0.192	0.136
$1\frac{1}{8}$	1.125	N	1.250	0.030	0.007	2.250	0.030	0.007	0.134	0.160	0.108
$1\frac{1}{8}$	1.125	W	1.250	0.030	0.007	2.750	0.030	0.007	0.165	0.192	0.136
$1\frac{1}{4}$	1.250	N	1.375	0.030	0.007	2.500	0.030	0.007	0.165	0.192	0.136
$1\frac{1}{4}$	1.250	W	1.375	0.030	0.007	3.000	0.030	0.007	0.165	0.192	0.136
$1\frac{3}{8}$	1.375	N	1.500	0.030	0.007	2.750	0.030	0.007	0.165	0.192	0.136
$1\frac{3}{8}$	1.375	W	1.500	0.045	0.010	3.250	0.045	0.010	0.180	0.213	0.153
$1\frac{1}{2}$	1.500	N	1.625	0.030	0.007	3.000	0.030	0.007	0.165	0.192	0.136
$1\frac{1}{2}$	1.500	W	1.625	0.045	0.010	3.500	0.045	0.010	0.180	0.213	0.153
$1\frac{5}{8}$	1.625		1.750	0.045	0.010	3.750	0.045	0.010	0.180	0.213	0.153
$1\frac{3}{4}$	1.750		1.875	0.045	0.010	4.000	0.045	0.010	0.180	0.213	0.153
$1\frac{7}{8}$	1.875		2.000	0.045	0.010	4.250	0.045	0.010	0.180	0.213	0.153
2	2.000		2.125	0.045	0.010	4.500	0.045	0.010	0.180	0.213	0.153
$2\frac{1}{4}$	2.250		2.375	0.045	0.010	4.750	0.045	0.010	0.220	0.248	0.193
$2\frac{1}{2}$	2.500		2.625	0.045	0.010	5.000	0.045	0.010	0.238	0.280	0.210
$2\frac{3}{4}$	2.750		2.875	0.065	0.010	5.250	0.065	0.010	0.259	0.310	0.228
3	3.000		3.125	0.065	0.010	5.500	0.065	0.010	0.284	0.327	0.249

(Courtesy of ANSI; B27.2–1965.)

APPENDIX 37

LOCK WASHERS (ANSI B27.1)

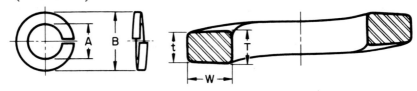

Dimensions of Regular* Helical Spring Lock Washers

Nominal Washer Size		Inside Diameter A		Outside Diameter B	Washer Section	
					Width W	Thickness $\frac{T+t}{2}$
		Min	Max	Max**	Min	Min
No. 2	0.086	0.088	0.094	0.172	0.035	0.020
No. 3	0.099	0.101	0.107	0.195	0.040	0.025
No. 4	0.112	0.115	0.121	0.209	0.040	0.025
No. 5	0.125	0.128	0.134	0.236	0.047	0.031
No. 6	0.138	0.141	0.148	0.250	0.047	0.031
No. 8	0.164	0.168	0.175	0.293	0.055	0.040
No. 10	0.190	0.194	0.202	0.334	0.062	0.047
No. 12	0.216	0.221	0.229	0.377	0.070	0.056
¼	0.250	0.255	0.263	0.489	0.109	0.062
⁵⁄₁₆	0.312	0.318	0.328	0.586	0.125	0.078
³⁄₈	0.375	0.382	0.393	0.683	0.141	0.094
⁷⁄₁₆	0.438	0.446	0.459	0.779	0.156	0.109
½	0.500	0.509	0.523	0.873	0.171	0.125
⁹⁄₁₆	0.562	0.572	0.587	0.971	0.188	0.141
⅝	0.625	0.636	0.653	1.079	0.203	0.156
¹¹⁄₁₆	0.688	0.700	0.718	1.176	0.219	0.172
¾	0.750	0.763	0.783	1.271	0.234	0.188
¹³⁄₁₆	0.812	0.826	0.847	1.367	0.250	0.203
⅞	0.875	0.890	0.912	1.464	0.266	0.219
¹⁵⁄₁₆	0.938	0.954	0.978	1.560	0.281	0.234
1	1.000	1.017	1.042	1.661	0.297	0.250
1¹⁄₁₆	1.062	1.080	1.107	1.756	0.312	0.266
1⅛	1.125	1.144	1.172	1.853	0.328	0.281
1³⁄₁₆	1.188	1.208	1.237	1.950	0.344	0.297
1¼	1.250	1.271	1.302	2.045	0.359	0.312
1⁵⁄₁₆	1.312	1.334	1.366	2.141	0.375	0.328
1⅜	1.375	1.398	1.432	2.239	0.391	0.344
1⁷⁄₁₆	1.438	1.462	1.497	2.334	0.406	0.359
1½	1.500	1.525	1.561	2.430	0.422	0.375

*Formerly designated Medium Helical Spring Lock Washers.

**The maximum outside diameters specified allow for the commercial tolerances on cold drawn wire.

APPENDIX 38
COTTER PINS

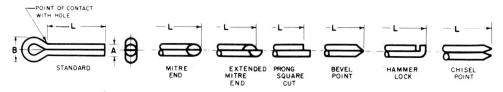

Nominal Diameter	Diameter A		Outside Eye Diameter B Min	Hole Sizes Recommended
	Max	Min		
0.031	0.032	0.028	$\frac{1}{16}$	$\frac{3}{64}$
0.047	0.048	0.044	$\frac{3}{32}$	$\frac{1}{16}$
0.062	0.060	0.056	$\frac{1}{8}$	$\frac{5}{64}$
0.078	0.076	0.072	$\frac{5}{32}$	$\frac{3}{32}$
0.094	0.090	0.086	$\frac{3}{16}$	$\frac{7}{64}$
0.109	0.104	0.100	$\frac{7}{32}$	$\frac{1}{8}$
0.125	0.120	0.116	$\frac{1}{4}$	$\frac{9}{64}$
0.141	0.134	0.130	$\frac{9}{32}$	$\frac{5}{32}$
0.156	0.150	0.146	$\frac{5}{16}$	$\frac{11}{64}$
0.188	0.176	0.172	$\frac{3}{8}$	$\frac{13}{64}$
0.219	0.207	0.202	$\frac{7}{16}$	$\frac{15}{64}$
0.250	0.225	0.220	$\frac{1}{2}$	$\frac{17}{64}$
0.312	0.280	0.275	$\frac{5}{8}$	$\frac{5}{16}$
0.375	0.335	0.329	$\frac{3}{4}$	$\frac{3}{8}$
0.438	0.406	0.400	$\frac{7}{8}$	$\frac{7}{16}$
0.500	0.473	0.467	1	$\frac{1}{2}$
0.625	0.598	0.590	$1\frac{1}{4}$	$\frac{5}{8}$
0.750	0.723	0.715	$1\frac{1}{2}$	$\frac{3}{4}$

All dimensions are given in inches.

A certain amount of leeway is permitted in the design of the head; however, the outside diameters given should be adhered to.

Prongs are to be parallel; ends shall not be open.

Points may be blunt, bevel, extended prong, mitre, etc., and purchaser may specify type required.

Lengths shall be measured as shown on the above illustration (L-dimension).

Cotter pins shall be free from burrs or any defects that will affect their serviceability.

(Courtesy of ANSI; B5.20–1958.)

APPENDIX 39
AMERICAN STANDARD RUNNING AND SLIDING FITS

Limits are in thousandths of an inch.
Limits for hole and shaft are applied algebraically to the basic size to obtain the limits of size for the parts.
Data in bold face are in accordance with ABC agreements.
Symbols H5, g5, etc., are Hole and Shaft designations used in ABC System.

Nominal Size Range Inches Over	To	Class RC 1 Limits of Clearance	RC 1 Hole H5	RC 1 Shaft g4	Class RC 2 Limits of Clearance	RC 2 Hole H6	RC 2 Shaft g5	Class RC 3 Limits of Clearance	RC 3 Hole H7	RC 3 Shaft f6	Class RC 4 Limits of Clearance	RC 4 Hole H8	RC 4 Shaft f7
0	− 0.12	0.1 / 0.45	+ 0.2 / 0	− 0.1 / − 0.25	0.1 / 0.55	+ 0.25 / 0	− 0.1 / − 0.3	0.3 / 0.95	+ 0.4 / 0	− 0.3 / − 0.55	0.3 / 1.3	+ 0.6 / 0	− 0.3 / − 0.7
0.12	− 0.24	0.15 / 0.5	+ 0.2 / 0	− 0.15 / − 0.3	0.15 / 0.65	+ 0.3 / 0	− 0.15 / − 0.35	0.4 / 1.12	+ 0.5 / 0	− 0.4 / − 0.7	0.4 / 1.6	+ 0.7 / 0	− 0.4 / − 0.9
0.24	− 0.40	0.2 / 0.6	0.25 / 0	− 0.2 / − 0.35	0.2 / 0.85	+ 0.4 / 0	− 0.2 / − 0.45	0.5 / 1.5	+ 0.6 / 0	− 0.5 / − 0.9	0.5 / 2.0	+ 0.9 / 0	− 0.5 / − 1.1
0.40	− 0.71	0.25 / 0.75	+ 0.3 / 0	− 0.25 / − 0.45	0.25 / 0.95	+ 0.4 / 0	− 0.25 / − 0.55	0.6 / 1.7	+ 0.7 / 0	− 0.6 / − 1.0	0.6 / 2.3	+ 1.0 / 0	− 0.6 / − 1.3
0.71	− 1.19	0.3 / 0.95	+ 0.4 / 0	− 0.3 / − 0.55	0.3 / 1.2	+ 0.5 / 0	− 0.3 / − 0.7	0.8 / 2.1	+ 0.8 / 0	− 0.8 / − 1.3	0.8 / 2.8	+ 1.2 / 0	− 0.8 / − 1.6
1.19	− 1.97	0.4 / 1.1	+ 0.4 / 0	− 0.4 / − 0.7	0.4 / 1.4	+ 0.6 / 0	− 0.4 / − 0.8	1.0 / 2.6	+ 1.0 / 0	− 1.0 / − 1.6	1.0 / 3.6	+ 1.6 / 0	− 1.0 / − 2.0
1.97	− 3.15	0.4 / 1.2	+ 0.5 / 0	− 0.4 / − 0.7	0.4 / 1.6	+ 0.7 / 0	− 0.4 / − 0.9	1.2 / 3.1	+ 1.2 / 0	− 1.2 / − 1.9	1.2 / 4.2	+ 1.8 / 0	− 1.2 / − 2.4
3.15	− 4.73	0.5 / 1.5	+ 0.6 / 0	− 0.5 / − 0.9	0.5 / 2.0	+ 0.9 / 0	− 0.5 / − 1.1	1.4 / 3.7	+ 1.4 / 0	− 1.4 / − 2.3	1.4 / 5.0	+ 2.2 / 0	− 1.4 / − 2.8
4.73	− 7.09	0.6 / 1.8	+ 0.7 / 0	− 0.6 / − 1.1	0.6 / 2.3	+ 1.0 / 0	− 0.6 / − 1.3	1.6 / 4.2	+ 1.6 / 0	− 1.6 / − 2.6	1.6 / 5.7	+ 2.5 / 0	− 1.6 / − 3.2
7.09	− 9.85	0.6 / 2.0	+ 0.8 / 0	− 0.6 / − 1.2	0.6 / 2.6	+ 1.2 / 0	− 0.6 / − 1.4	2.0 / 5.0	+ 1.8 / 0	− 2.0 / − 3.2	2.0 / 6.6	+ 2.8 / 0	− 2.0 / − 3.8
9.85	− 12.41	0.8 / 2.3	+ 0.9 / 0	− 0.8 / − 1.4	0.8 / 2.9	+ 1.2 / 0	− 0.8 / − 1.7	2.5 / 5.7	+ 2.0 / 0	− 2.5 / − 3.7	2.5 / 7.5	+ 3.0 / 0	− 2.5 / − 4.5
12.41	− 15.75	1.0 / 2.7	+ 1.0 / 0	− 1.0 / − 1.7	1.0 / 3.4	+ 1.4 / 0	− 1.0 / − 2.0	3.0 / 6.6	+ / 0	− 3.0 / − 4.4	3.0 / 8.7	+ 3.5 / 0	− 3.0 / − 5.2
15.75	− 19.69	1.2 / 3.0	+ 1.0 / 0	− 1.2 / − 2.0	1.2 / 3.8	+ 1.6 / 0	− 1.2 / − 2.2	4.0 / 8.1	+ 1.6 / 0	− 4.0 / − 5.6	4.0 / 10.5	+ 4.0 / 0	− 4.0 / − 6.5
19.69	− 30.09	1.6 / 3.7	+ 1.2 / 0	− 1.6 / − 2.5	1.6 / 4.8	+ 2.0 / 0	− 1.6 / − 2.8	5.0 / 10.0	+ 3.0 / 0	− 5.0 / − 7.0	5.0 / 13.0	+ 5.0 / 0	− 5.0 / − 8.0
30.09	− 41.49	2.0 / 4.6	+ 1.6 / 0	− 2.0 / − 3.0	2.0 / 6.1	+ 2.5 / 0	− 2.0 / − 3.6	6.0 / 12.5	+ 4.0 / 0	− 6.0 / − 8.5	6.0 / 16.0	+ 6.0 / 0	− 6.0 / −10.0
41.49	− 56.19	2.5 / 5.7	+ 2.0 / 0	− 2.5 / − 3.7	2.5 / 7.5	+ 3.0 / 0	− 2.5 / − 4.5	8.0 / 16.0	+ 5.0 / 0	− 8.0 / −11.0	8.0 / 21.0	+ 8.0 / 0	− 8.0 / −13.0
56.19	− 76.39	3.0 / 7.1	+ 2.5 / 0	− 3.0 / − 4.6	3.0 / 9.5	+ 4.0 / 0	− 3.0 / − 5.5	10.0 / 20.0	+ 6.0 / 0	−10.0 / −14.0	10.0 / 26.0	+10.0 / 0	−10.0 / −16.0
76.39	− 100.9	4.0 / 9.0	+ 3.0 / 0	− 4.0 / − 6.0	4.0 / 12.0	+ 5.0 / 0	− 4.0 / − 7.0	12.0 / 25.0	+ 8.0 / 0	−12.0 / −17.0	12.0 / 32.0	+12.0 / 0	−12.0 / −20.0
100.9	− 131.9	5.0 / 11.5	+ 4.0 / 0	− 5.0 / − 7.5	5.0 / 15.0	+ 6.0 / 0	− 5.0 / − 9.0	16.0 / 32.0	+10.0 / 0	−16.0 / −22.0	16.0 / 36.0	+16.0 / 0	−16.0 / −26.0
131.9	− 171.9	6.0 / 14.0	+ 5.0 / 0	− 6.0 / − 9.0	6.0 / 19.0	+ 8.0 / 0	− 6.0 / −11.0	18.0 / 38.0	+ 8.0 / 0	−18.0 / −26.0	18.0 / 50.0	+20.0 / 0	−18.0 / −30.0
171.9	− 200	8.0 / 18.0	+ 6.0 / 0	− 8.0 / −12.0	8.0 / 22.0	+10.0 / 0	− 8.0 / −12.0	22.0 / 48.0	+16.0 / 0	−22.0 / −32.0	22.0 / 63.0	+25.0 / 0	−22.0 / −38.0

(Courtesy of USASI; B4.1–1955.)

Cont.

APPENDIX 39

AMERICAN STANDARD RUNNING AND SLIDING FITS (Cont.)

Class RC 5			Class RC 6			Class RC 7			Class RC 8			Class RC 9			Nominal Size Range Inches	
Limits of Clearance	Standard Limits Hole H8	Standard Limits Shaft e7	Limits of Clearance	Standard Limits Hole H9	Standard Limits Shaft e8	Limits of Clearance	Standard Limits Hole H9	Standard Limits Shaft d8	Limits of Clearance	Standard Limits Hole H10	Standard Limits Shaft c9	Limits of Clearance	Standard Limits Hole H11	Standard Limits Shaft	Over	To
0.6 1.6	+ 0.6 − 0	− 0.6 − 1.0	0.6 2.2	+ 1.0 − 0	− 0.6 − 1.2	1.0 2.6	+ 1.0 0	− 1.0 − 1.6	2.5 5.1	+ 1.6 0	− 2.5 − 3.5	4.0 8.1	+ 2.5 0	− 4.0 − 5.6	0	0.12
0.8 2.0	+ 0.7 − 0	− 0.8 − 1.3	0.8 2.7	+ 1.2 − 0	− 0.8 − 1.5	1.2 3.1	+ 1.2 0	− 1.2 − 1.9	2.8 5.8	+ 1.8 0	− 2.8 − 4.0	4.5 9.0	+ 3.0 0	− 4.5 − 6.0	0.12	0.24
1.0 2.5	+ 0.9 − 0	− 1.0 − 1.6	1.0 3.3	+ 1.4 − 0	− 1.0 − 1.9	1.6 3.9	+ 1.4 0	− 1.6 − 2.5	3.0 6.6	+ 2.2 0	− 3.0 − 4.4	5.0 10.7	+ 3.5 0	− 5.0 − 7.2	0.24	0.40
1.2 2.9	+ 1.0 − 0	− 1.2 − 1.9	1.2 3.8	+ 1.6 − 0	− 1.2 − 2.2	2.0 4.6	+ 1.6 0	− 2.0 − 3.0	3.5 7.9	+ 2.8 0	− 3.5 − 5.1	6.0 12.8	+ 4.0 − 0	− 6.0 − 8.8	0.40	0.71
1.6 3.6	+ 1.2 − 0	− 1.6 − 2.4	1.6 4.8	+ 2.0 − 0	− 1.6 − 2.8	2.5 5.7	+ 2.0 0	− 2.5 − 3.7	4.5 10.0	+ 3.5 0	− 4.5 − 6.5	7.0 15.5	+ 5.0 0	− 7.0 − 10.5	0.71	1.19
2.0 4.6	+ 1.6 − 0	− 2.0 − 3.0	2.0 6.1	+ 2.5 − 0	− 2.0 − 3.6	3.0 7.1	+ 2.5 0	− 3.0 − 4.6	5.0 11.5	+ 4.0 0	− 5.0 − 7.5	8.0 18.0	+ 6.0 0	− 8.0 − 12.0	1.19	1.97
2.5 5.5	+ 1.8 − 0	− 2.5 − 3.7	2.5 7.3	+ 3.0 − 0	− 2.5 − 4.3	4.0 8.8	+ 3.0 0	− 4.0 − 5.8	6.0 13.5	+ 4.5 0	− 6.0 − 9.0	9.0 20.5	+ 7.0 0	− 9.0 − 13.5	1.97	3.15
3.0 6.6	+ 2.2 − 0	− 3.0 − 4.4	3.0 8.7	+ 3.5 − 0	− 3.0 − 5.2	5.0 10.7	+ 3.5 0	− 5.0 − 7.2	7.0 15.5	+ 5.0 0	− 7.0 − 10.5	10.0 24.0	+ 9.0 0	− 10.0 − 15.0	3.15	4.73
3.5 7.6	+ 2.5 − 0	− 3.5 − 5.1	3.5 10.0	+ 4.0 − 0	− 3.5 − 6.0	6.0 12.5	+ 4.0 0	− 6.0 − 8.5	8.0 18.0	+ 6.0 0	− 8.0 − 12.0	12.0 28.0	+ 10.0 0	− 12.0 − 18.0	4.73	7.09
4.0 8.6	+ 2.8 − 0	− 4.0 − 5.8	4.0 11.3	+ 4.5 0	− 4.0 − 6.8	7.0 14.3	+ 4.5 0	− 7.0 − 9.8	10.0 21.5	+ 7.0 0	− 10.0 − 14.5	15.0 34.0	+ 12.0 0	− 15.0 − 22.0	7.09	9.85
5.0 10.0	+ 3.0 0	− 5.0 − 7.0	5.0 13.0	+ 5.0 0	− 5.0 − 8.0	8.0 16.0	+ 5.0 0	− 8.0 − 11.0	12.0 25.0	+ 8.0 0	− 12.0 − 17.0	18.0 38.0	+ 12.0 0	− 18.0 − 26.0	9.85	12.41
6.0 11.7	+ 3.5 0	− 6.0 − 8.2	6.0 15.5	+ 6.0 0	− 6.0 − 9.5	10.0 19.5	+ 6.0 0	− 10.0 13.5	14.0 29.0	+ 9.0 0	− 14.0 − 20.0	22.0 45.0	+ 14.0 0	− 22.0 − 31.0	12.41	15.75
8.0 14.5	+ 4.0 0	− 8.0 −10.5	8.0 18.0	+ 6.0 0	− 8.0 −12.0	12.0 22.0	+ 6.0 0	− 12.0 − 16.0	16.0 32.0	+10.0 0	− 16.0 − 22.0	25.0 51.0	+ 16.0 0	− 25.0 − 35.0	15.75	19.69
10.0 18.0	+ 5.0 0	−10.0 −13.0	10.0 23.0	+ 8.0 0	−10.0 −15.0	16.0 29.0	+ 8.0 0	− 16.0 − 21.0	20.0 40.0	+12.0 0	− 20.0 − 28.0	30.0 62.0	+ 20.0 0	− 30.0 − 42.0	19.69	30.09
12.0 22.0	+ 6.0 0	−12.0 −16.0	12.0 28.0	+10.0 0	−12.0 −18.0	20.0 36.0	+10.0 0	− 20.0 − 26.0	25.0 51.0	+16.0 0	− 25.0 − 35.0	40.0 81.0	+ 25.0 0	− 40.0 − 56.0	30.09	41.49
16.0 29.0	+ 8.0 0	−16.0 −21.0	16.0 36.0	+12.0 0	−16.0 −24.0	25.0 45.0	+12.0 0	− 25.0 − 33.0	30.0 62.0	+20.0 0	− 30.0 − 42.0	50.0 100	+ 30.0 0	− 50.0 − 70.0	41.49	56.19
20.0 36.0	+10.0 0	−20.0 −26.0	20.0 46.0	+16.0 0	−20.0 −30.0	30.0 56.0	+16.0 0	− 30.0 − 40.0	40.0 81.0	+25.0 0	− 40.0 − 56.0	60.0 125	+ 40.0 0	− 60.0 − 85.0	56.19	76.39
25.0 45.0	+12.0 0	−25.0 −33.0	25.0 57.0	+20.0 0	−25.0 −37.0	40.0 72.0	+20.0 0	− 40.0 − 52.0	50.0 100	+30.0 0	− 50.0 − 70.0	80.0 160	+ 50.0 0	− 80.0 −110	76.39	100.9
30.0 56.0	+16.0 0	−30.0 −40.0	30.0 71.0	+25.0 0	−30.0 −46.0	50.0 91.0	+25.0 0	− 50.0 − 66.0	60.0 125	+40.0 0	− 60.0 − 85.0	100 200	+ 60.0 0	−100 −140	100.9	131.9
35.0 67.0	+20.0 0	−35.0 −47.0	35.0 85.0	+30.0 0	−35.0 −55.0	60.0 110.0	+30.0 0	− 60.0 − 80.0	80.0 160	+50.0 0	− 80.0 −110	130 260	+ 80.0 0	−130 −180	131.9	171.9
45.0 86.0	+25.0 0	−45.0 −61.0	45.0 110.0	+40.0 0	−45.0 −70.0	80.0 145.0	+40.0 0	− 80.0 −105.0	100 200	+60.0 0	−100 −140	150 310	+100 0	−150 −210	171.9	200

(Courtesy of ANSI; B4.1–1955.)

APPENDIX 40
AMERICAN STANDARD CLEARANCE LOCATIONAL FITS

Limits are in thousandths of an inch.

Limits for hole and shaft are applied algebraically to the basic size to obtain the limits of size for the parts.

Data in bold face are in accordance with ABC agreements.

Symbols H9, f8, etc., are Hole and Shaft designations used in ABC System.

Nominal Size Range Inches Over	To	Class LC 1 Limits of Clearance	Hole H6	Shaft h5	Class LC 2 Limits of Clearance	Hole H7	Shaft h6	Class LC 3 Limits of Clearance	Hole H8	Shaft h7	Class LC 4 Limits of Clearance	Hole H10	Shaft h9	Class LC 5 Limits of Clearance	Hole H7	Shaft g6
0 —	0.12	0 / 0.45	+0.25 / −0	+0 / −0.2	0 / 0.65	+0.4 / −0	+0 / −0.25	0 / 1	+0.6 / −0	+0 / −0.4	0 / 2.6	+1.6 / −0	+0 / −1.0	0.1 / 0.75	+0.4 / −0	−0.1 / −0.35
0.12—	0.24	0 / 0.5	+0.3 / −0	+0 / −0.2	0 / 0.8	+0.5 / −0	+0 / −0.3	0 / 1.2	+0.7 / −0	+0 / −0.5	0 / 3.0	+1.8 / −0	+0 / −1.2	0.15 / 0.95	+0.5 / −0	−0.15 / −0.45
0.24—	0.40	0 / 0.65	+0.4 / −0	+0 / −0.25	0 / 1.0	+0.6 / −0	+0 / −0.4	0 / 1.5	+0.9 / −0	+0 / −0.6	0 / 3.6	+2.2 / −0	+0 / −1.4	0.2 / 1.2	+0.6 / −0	−0.2 / −0.6
0.40—	0.71	0 / 0.7	+0.4 / −0	+0 / −0.3	0 / 1.1	+0.7 / −0	+0 / −0.4	0 / 1.7	+1.0 / −0	+0 / −0.7	0 / 4.4	+2.8 / −0	+0 / −1.6	0.25 / 1.35	+0.7 / −0	−0.25 / −0.65
0.71—	1.19	0 / 0.9	+0.5 / −0	+0 / −0.4	0 / 1.3	+0.8 / −0	+0 / −0.5	0 / 2	+1.2 / −0	+0 / −0.8	0 / 5.5	+3.5 / −0	+0 / −2.0	0.3 / 1.6	+0.8 / −0	−0.3 / −0.8
1.19—	1.97	0 / 1.0	+0.6 / −0	+0 / −0.4	0 / 1.6	+1.0 / −0	+0 / −0.6	0 / 2.6	+1.6 / −0	+0 / −1	0 / 6.5	+4.0 / −0	+0 / −2.5	0.4 / 2.0	+1.0 / −0	−0.4 / −1.0
1.97—	3.15	0 / 1.2	+0.7 / −0	+0 / −0.5	0 / 1.9	+1.2 / −0	+0 / −0.7	0 / 3	+1.8 / −0	+0 / −1.2	0 / 7.5	+4.5 / −0	+0 / −3	0.4 / 2.3	+1.2 / −0	−0.4 / −1.1
3.15—	4.73	0 / 1.5	+0.9 / −0	+0 / −0.6	0 / 2.3	+1.4 / −0	+0 / −0.9	0 / 3.6	+2.2 / −0	+0 / −1.4	0 / 8.5	+5.0 / −0	+0 / −3.5	0.5 / 2.8	+1.4 / −0	−0.5 / −1.4
4.73—	7.09	0 / 1.7	+1.0 / −0	+0 / −0.7	0 / 2.6	+1.6 / −0	+0 / −1.0	0 / 4.1	+2.5 / −0	+0 / −1.6	0 / 10	+6.0 / −0	+0 / −4	0.6 / 3.2	+1.6 / −0	−0.6 / −1.6
7.09—	9.85	0 / 2.0	+1.2 / −0	+0 / −0.8	0 / 3.0	+1.8 / −0	+0 / −1.2	0 / 4.6	+2.8 / −0	+0 / −1.8	0 / 11.5	+7.0 / −0	+0 / −4.5	0.6 / 3.6	+1.8 / −0	−0.6 / −1.8
9.85—	12.41	0 / 2.1	+1.2 / −0	+0 / −0.9	0 / 3.2	+2.0 / −0	+0 / −1.2	0 / 5	+3.0 / −0	+0 / −2.0	0 / 13	+8.0 / −0	+0 / −5	0.7 / 3.9	+2.0 / −0	−0.7 / −1.9
12.41—	15.75	0 / 2.4	+1.4 / −0	+0 / −1.0	0 / 3.6	+2.2 / −0	+0 / −1.4	0 / 5.7	+3.5 / −0	+0 / −2.2	0 / 15	+9.0 / −0	+0 / −6	0.7 / 4.3	+2.2 / −0	−0.7 / −2.1
15.75—	19.69	0 / 2.6	+1.6 / −0	+0 / −1.0	0 / 4.1	+2.5 / −0	+0 / −1.6	0 / 6.5	+4 / −0	+0 / −2.5	0 / 16	+10.0 / −0	+0 / −6	0.8 / 4.9	+2.5 / −0	−0.8 / −2.4
19.69—	30.09	0 / 3.2	+2.0 / −0	+0 / −1.2	0 / 5.0	+3 / −0	+0 / −2	0 / 8	+5 / −0	+0 / −3	0 / 20	+12.0 / −0	+0 / −8	0.9 / 5.9	+3.0 / −0	−0.9 / −2.9
30.09—	41.49	0 / 4.1	+2.5 / −0	+0 / −1.6	0 / 6.5	+4 / −0	+0 / −2.5	0 / 10	+6 / −0	+0 / −4	0 / 26	+16.0 / −0	+0 / −10	1.0 / 7.5	+4.0 / −0	−1.0 / −3.5
41.49—	56.19	0 / 5.0	+3.0 / −0	+0 / −2.0	0 / 8.0	+5 / −0	+0 / −3	0 / 13	+8 / −0	+0 / −5	0 / 32	+20.0 / −0	+0 / −12	1.2 / 9.2	+5.0 / −0	−1.2 / −4.2
56.19—	76.39	0 / 6.5	+4.0 / −0	+0 / −2.5	0 / 10	+6 / −0	+0 / −4	0 / 16	+10 / −0	+0 / −6	0 / 41	+25.0 / −0	+0 / −16	1.2 / 11.2	+6.0 / −0	−1.2 / −5.2
76.39—	100.9	0 / 8.0	+5.0 / −0	+0 / −3.0	0 / 13	+8 / −0	+0 / −5	0 / 20	+12 / −0	+0 / −8	0 / 50	+30.0 / −0	+0 / −20	1.4 / 14.4	+8.0 / −0	−1.4 / −6.4
100.9 —	131.9	0 / 10.0	+6.0 / −0	+0 / −4.0	0 / 16	+10 / −0	+0 / −6	0 / 26	+16 / −0	+0 / −10	0 / 65	+40.0 / −0	+0 / −25	1.6 / 17.6	+10.0 / −0	−1.6 / −7.6
131.9 —	171.9	0 / 13.0	+8.0 / −0	+0 / −5.0	0 / 20	+12 / −0	+0 / −8	0 / 32	+20 / −0	+0 / −12	0 / 8	+50.0 / −0	+0 / −30	1.8 / 21.8	+12.0 / −0	−1.8 / −9.8
171.9 —	200	0 / 16.0	+10.0 / −0	+0 / −6.0	0 / 26	+16 / −0	+0 / −10	0 / 41	+25 / −0	+0 / −16	0 / 100	+60.0 / −0	+0 / −40	1.8 / 27.8	+16.0 / −0	−1.8 / −11.8

(Courtesy of USASI; B4.1–1955.)

Cont.

A-64

APPENDIX 40
AMERICAN STANDARD CLEARANCE LOCATIONAL FITS (Cont.)

Class LC 6			Class LC 7			Class LC 8			Class LC 9			Class LC 10			Class LC 11			Nominal Size Range Inches	
Limits of Clearance	Standard Limits Hole H9	Standard Limits Shaft f8	Limits of Clearance	Standard Limits Hole H10	Standard Limits Shaft e9	Limits of Clearance	Standard Limits Hole H10	Standard Limits Shaft d9	Limits of Clearance	Standard Limits Hole H11	Standard Limits Shaft c10	Limits of Clearance	Standard Limits Hole H12	Standard Limits Shaft	Limits of Clearance	Standard Limits Hole H13	Standard Limits Shaft	Over	To
0.3 / 1.9	+1.0 / 0	−0.3 / −0.9	0.6 / 3.2	+1.6 / 0	−0.6 / −1.6	1.0 / 3.6	+1.6 / −0	−1.0 / −2.0	2.5 / 6.6	+2.5 / −0	−2.5 / −4.1	4 / 12	+4 / −0	−4 / −8	5 / 17	+6 / −0	−5 / −11	0	0.12
0.4 / 2.3	+1.2 / 0	−0.4 / −1.1	0.8 / 3.8	+1.8 / 0	−0.8 / −2.0	1.2 / 4.2	+1.8 / −0	−1.2 / −2.4	2.8 / 7.6	+3.0 / −0	−2.8 / −4.6	4.5 / 14.5	+5 / −0	−4.5 / −9.5	6 / 20	+7 / −0	−6 / −13	0.12	0.24
0.5 / 2.8	+1.4 / 0	−0.5 / −1.4	1.0 / 4.6	+2.2 / 0	−1.0 / −2.4	1.6 / 5.2	+2.2 / −0	−1.6 / −3.0	3.0 / 8.7	+3.5 / −0	−3.0 / −5.2	5 / 17	+6 / −0	−5 / −11	7 / 25	+9 / −0	−7 / −16	0.24	0.40
0.6 / 3.2	+1.6 / 0	−0.6 / −1.6	1.2 / 5.6	+2.8 / 0	−1.2 / −2.8	2.0 / 6.4	+2.8 / −0	−2.0 / −3.6	3.5 / 10.3	+4.0 / −0	−3.5 / −6.3	6 / 20	+7 / −0	−6 / −13	8 / 28	+10 / −0	−8 / −18	0.40	0.71
0.8 / 4.0	+2.0 / 0	−0.8 / −2.0	1.6 / 7.1	+3.5 / 0	−1.6 / −3.6	2.5 / 8.0	+3.5 / −0	−2.5 / −4.5	4.5 / 13.0	+5.0 / −0	−4.5 / −8.0	7 / 23	+8 / −0	−7 / −15	10 / 34	+12 / −0	−10 / −22	0.71	1.19
1.0 / 5.1	+2.5 / 0	−1.0 / −2.6	2.0 / 8.5	+4.0 / 0	−2.0 / −4.5	3.0 / 9.5	+4.0 / −0	−3.0 / −5.5	5 / 15	+6 / −0	−5 / −9	8 / 28	+10 / −0	−8 / −18	12 / 44	+16 / −0	−12 / −28	1.19	1.97
1.2 / 6.0	+3.0 / 0	−1.2 / −3.0	2.5 / 10.0	+4.5 / 0	−2.5 / −5.5	4.0 / 11.5	+4.5 / −0	−4.0 / −7.0	6 / 17.5	+7 / −0	−6 / −10.5	10 / 34	+12 / −0	−10 / −22	14 / 50	+18 / −0	−14 / −32	1.97	3.15
1.4 / 7.1	+3.5 / 0	−1.4 / −3.6	3.0 / 11.5	+5.0 / 0	−3.0 / −6.5	5.0 / 13.5	+5.0 / −0	−5.0 / −8.5	7 / 21	+9 / −0	−7 / −12	11 / 39	+14 / −0	−11 / −25	16 / 60	+22 / −0	−16 / −38	3.15	4.73
1.6 / 8.1	+4.0 / 0	−1.6 / −4.1	3.5 / 13.5	+6.0 / 0	−3.5 / −7.5	6 / 16	+6 / −0	−6 / −10	8 / 24	+10 / −0	−8 / −14	12 / 44	+16 / −0	−12 / −28	18 / 68	+25 / −0	−18 / −43	4.73	7.09
2.0 / 9.3	+4.5 / 0	−2.0 / −4.8	4.0 / 15.5	+7.0 / 0	−4.0 / −8.5	7 / 18.5	+7 / −0	−7 / −11.5	10 / 29	+12 / −0	−10 / −17	16 / 52	+18 / −0	−16 / −34	22 / 78	+28 / −0	−22 / −50	7.09	9.85
2.2 / 10.2	+5.0 / 0	−2.2 / −5.2	4.5 / 17.5	+8.0 / 0	−4.5 / −9.5	7 / 20	+8 / −0	−7 / −12	12 / 32	+12 / −0	−12 / −20	20 / 60	+20 / −0	−20 / −40	28 / 88	+30 / −0	−28 / −58	9.85	12.41
2.5 / 12.0	+6.0 / 0	−2.5 / −6.0	5.0 / 20.0	+9.0 / 0	−5 / −11	8 / 23	+9 / −0	−8 / −14	14 / 37	+14 / −0	−14 / −23	22 / 66	+22 / −0	−22 / −44	30 / 100	+35 / −0	−30 / −65	12.41	15.75
2.8 / 12.8	+6.0 / 0	−2.8 / −6.8	5.0 / 21.0	+10.0 / 0	−5 / −11	9 / 25	+10 / −0	−9 / −15	16 / 42	+16 / −0	−16 / −26	25 / 75	+25 / −0	−25 / −50	35 / 115	+40 / −0	−35 / −75	15.75	19.69
3.0 / 16.0	+8.0 / 0	−3.0 / −8.0	6.0 / 26.0	+12.0 / −0	−6 / −14	10 / 30	+12 / −0	−10 / −18	18 / 50	+20 / −0	−18 / −30	28 / 88	+30 / −0	−28 / −58	40 / 140	+50 / −0	−40 / −90	19.69	30.09
3.5 / 19.5	+10.0 / 0	−3.5 / −9.5	7.0 / 33.0	+16.0 / −0	−7 / −17	12 / 38	+16 / −0	−12 / −22	20 / 61	+25 / −0	−20 / −36	30 / 110	+40 / −0	−30 / −70	45 / 165	+60 / −0	−45 / −105	30.09	41.49
4.0 / 24.0	+12.0 / 0	−4.0 / −12.0	8.0 / 40.0	+20.0 / −0	−8 / −20	14 / 46	+20 / −0	−14 / −26	25 / 75	+30 / −0	−25 / −45	40 / 140	+50 / −0	−40 / −90	60 / 220	+80 / −0	−60 / −140	41.49	56.19
4.5 / 30.5	+16.0 / 0	−4.5 / −14.5	9.0 / 50.0	+25.0 / −0	−9 / −25	16 / 57	+25 / −0	−16 / −32	30 / 95	+40 / −0	−30 / −55	50 / 170	+60 / −0	−50 / 110	70 / 270	+100 / −0	−70 / −170	56.19	76.39
5.0 / 37.0	+20.0 / 0	−5 / −17	10.0 / 60.0	+30.0 / −0	−10 / −30	18 / 68	+30 / −0	−18 / −38	35 / 115	+50 / −0	−35 / −65	50 / 210	+80 / −0	−50 / −130	80 / 330	+125 / −0	−80 / −205	76.39	100.9
6.0 / 47.0	+25.0 / 0	−6 / −22	12.0 / 67.0	+40.0 / −0	−12 / −27	20 / 85	+40 / −0	−20 / −45	40 / 140	+60 / −0	−40 / −80	60 / 260	+100 / −0	−60 / −160	90 / 410	+160 / −0	−90 / −250	100.9	131.9
7.0 / 57.0	+30.0 / 0	−7 / −27	14.0 / 94.0	+50.0 / −0	−14 / −44	25 / 105	+50 / −0	−25 / −55	50 / 180	+80 / −0	−50 / −100	80 / 330	+125 / −0	−80 / −205	100 / 500	+200 / −0	−100 / −300	131.9	171.9
7.0 / 72.0	+40.0 / 0	−7 / −32	14.0 / 114.0	+60.0 / −0	−14 / −54	25 / 125	+60 / −0	−25 / −65	50 / 210	+100 / −0	−50 / −110	90 / 410	+160 / −0	−90 / −250	125 / 625	+250 / −0	−125 / −375	171.9	200

(Courtesy of ANSI; B4.1–1955.)

APPENDIX 41
AMERICAN STANDARD TRANSITION LOCATIONAL FITS

Limits are in thousandths of an inch.

Limits for hole and shaft are applied algebraically to the basic size to obtain the limits of size for the mating parts.

Data in bold face are in accordance with ABC agreements.

"Fit" represents the maximum interference (minus values) and the maximum clearance (plus values).

Symbols H7, js6, etc., are Hole and Shaft designations used in ABC System.

Nominal Size Range Inches Over	To	Class LT 1 Fit	Hole H7	Shaft js6	Class LT 2 Fit	Hole H8	Shaft js7	Class LT 3 Fit	Hole H7	Shaft k6	Class LT 4 Fit	Hole H8	Shaft k7	Class LT 5 Fit	Hole H7	Shaft n6	Class LT 6 Fit	Hole H7	Shaft n7
0	0.12	−0.10 / +0.50	+0.4 / −0	+0.10 / −0.10	−0.2 / +0.8	+0.6 / −0	+0.2 / −0.2							−0.5 / +0.15	+0.4 / −0	+0.5 / +0.25	−0.65 / +0.15	+0.4 / −0	+0.65 / +0.25
0.12	0.24	−0.15 / +0.65	+0.5 / −0	+0.15 / −0.15	−0.25 / +0.95	+0.7 / −0	+0.25 / −0.25							−0.6 / +0.2	+0.5 / −0	+0.6 / +0.3	−0.8 / +0.2	+0.5 / −0	+0.8 / +0.3
0.24	0.40	−0.2 / +0.8	+0.6 / −0	+0.2 / −0.2	−0.3 / +1.2	+0.9 / −0	+0.3 / −0.3	−0.5 / +0.5	+0.6 / −0	+0.5 / +0.1	−0.7 / +0.8	+0.9 / −0	+0.7 / +0.1	−0.8 / +0.2	+0.6 / −0	+0.8 / +0.4	−1.0 / +0.2	+0.6 / −0	+1.0 / +0.4
0.40	0.71	−0.2 / +0.9	+0.7 / −0	+0.2 / −0.2	−0.35 / +1.35	+1.0 / −0	+0.35 / −0.35	−0.5 / +0.6	+0.7 / −0	+0.5 / +0.1	−0.8 / +0.9	+1.0 / −0	+0.8 / +0.1	−0.9 / +0.2	+0.7 / −0	+0.9 / +0.5	−1.2 / +0.2	+0.7 / −0	+1.2 / +0.5
0.71	1.19	−0.25 / +1.05	+0.8 / −0	+0.25 / −0.25	−0.4 / +1.6	+1.2 / −0	+0.4 / −0.4	−0.6 / +0.7	+0.8 / −0	+0.6 / +0.1	−0.9 / +1.1	+1.2 / −0	+0.9 / +0.1	−1.1 / +0.2	+0.8 / −0	+1.1 / +0.6	−1.4 / +0.2	+0.8 / −0	+1.4 / +0.6
1.19	1.97	−0.3 / +1.3	+1.0 / −0	+0.3 / −0.3	−0.5 / +2.1	+1.6 / −0	+0.5 / −0.5	−0.7 / +0.9	+1.0 / −0	+0.7 / +0.1	−1.1 / +1.5	+1.6 / −0	+1.1 / +0.1	−1.3 / +0.3	+1.0 / −0	+1.3 / +0.7	−1.7 / +0.3	+1.0 / −0	+1.7 / +0.7
1.97	3.15	−0.3 / +1.5	+1.2 / −0	+0.3 / −0.3	−0.6 / +2.4	+1.8 / −0	+0.6 / −0.6	−0.8 / +1.1	+1.2 / −0	+0.8 / +0.1	−1.3 / +1.7	+1.8 / −0	+1.3 / +0.1	−1.5 / +0.4	+1.2 / −0	+1.5 / +0.8	−2.0 / +0.4	+1.2 / −0	+2.0 / +0.8
3.15	4.73	−0.4 / +1.8	+1.4 / −0	+0.4 / −0.4	−0.7 / +2.9	+2.2 / −0	+0.7 / −0.7	−1.0 / +1.3	+1.4 / −0	+1.0 / +0.1	−1.5 / +2.1	+2.2 / −0	+1.5 / +0.1	−1.9 / +0.4	+1.4 / −0	+1.9 / +1.0	−2.4 / +0.4	+1.4 / −0	+2.4 / +1.0
4.73	7.09	−0.5 / +2.1	+1.6 / −0	+0.5 / −0.5	−0.8 / +3.3	+2.5 / −0	+0.8 / −0.8	−1.1 / +1.5	+1.6 / −0	+1.1 / +0.1	−1.7 / +2.4	+2.5 / −0	+1.7 / +0.1	−2.2 / +0.4	+1.6 / −0	+2.2 / +1.2	−2.8 / +0.4	+1.6 / −0	+2.8 / +1.2
7.09	9.85	−0.6 / +2.4	+1.8 / −0	+0.6 / −0.6	−0.9 / +3.7	+2.8 / −0	+0.9 / −0.9	−1.4 / +1.6	+1.8 / −0	+1.4 / +0.2	−2.0 / +2.6	+2.8 / −0	+2.0 / +0.2	−2.6 / +0.4	+1.8 / −0	+2.6 / +1.4	−3.2 / +0.4	+1.8 / −0	+3.2 / +1.4
9.85	12.41	−0.6 / +2.6	+2.0 / −0	+0.6 / −0.6	−1.0 / +4.0	+3.0 / −0	+1.0 / −1.0	−1.4 / +1.8	+2.0 / −0	+1.4 / +0.2	−2.2 / +2.8	+3.0 / −0	+2.2 / +0.2	−2.6 / +0.6	+2.0 / −0	+2.6 / +1.4	−3.4 / +0.6	+2.0 / −0	+3.4 / +1.4
12.41	15.75	−0.7 / +2.9	+2.2 / −0	+0.7 / −0.7	−1.0 / +4.5	+3.5 / −0	+1.0 / −1.0	−1.6 / +2.0	+2.2 / −0	+1.6 / +0.2	−2.4 / +3.3	+3.5 / −0	+2.4 / +0.2	−3.0 / +0.6	+2.2 / −0	+3.0 / +1.6	−3.8 / +0.6	+2.2 / −0	+3.8 / +1.6
15.75	19.69	−0.8 / +3.3	+2.5 / −0	+0.8 / −0.8	−1.2 / +5.2	+4.0 / −0	+1.2 / −1.2	−1.8 / +2.3	+2.5 / −0	+1.8 / +0.2	−2.7 / +3.8	+4.0 / −0	+2.7 / +0.2	−3.4 / +0.7	+2.5 / −0	+3.4 / +1.8	−4.3 / +0.7	+2.5 / −0	+4.3 / +1.8

(Courtesy of ANSI; B4.1–1955.)

APPENDIX 42
AMERICAN STANDARD INTERFERENCE LOCATIONAL FITS

Limits are in thousandths of an inch.
Limits for hole and shaft are applied algebraically to the
basic size to obtain the limits of size for the parts.
Data in bold face are in accordance with ABC agreements,
Symbols H7, p6, etc., are Hole and Shaft designations
used in ABC System.

Nominal Size Range Inches Over	To	Class LN 1 Limits of Interference	Class LN 1 Hole H6	Class LN 1 Shaft n5	Class LN 2 Limits of Interference	Class LN 2 Hole H7	Class LN 2 Shaft p6	Class LN 3 Limits of Interference	Class LN 3 Hole H7	Class LN 3 Shaft r6
0	0.12	0 / 0.45	+ 0.25 / − 0	+0.45 / +0.25	0 / 0.65	+ 0.4 / − 0	+ 0.65 / + 0.4	0.1 / 0.75	+ 0.4 / − 0	+ 0.75 / + 0.5
0.12	0.24	0 / 0.5	+ 0.3 / − 0	+0.5 / +0.3	0 / 0.8	+ 0.5 / − 0	+ 0.8 / + 0.5	0.1 / 0.9	+ 0.5 / − 0	+ 0.9 / + 0.6
0.24	0.40	0 / 0.65	+ 0.4 / − 0	+0.65 / +0.4	0 / 1.0	+ 0.6 / − 0	+ 1.0 / + 0.6	0.2 / 1.2	+ 0.6 / − 0	+ 1.2 / + 0.8
0.40	0.71	0 / 0.8	+ 0.4 / − 0	+0.8 / +0.4	0 / 1.1	+ 0.7 / − 0	+ 1.1 / + 0.7	0.3 / 1.4	+ 0.7 / − 0	+ 1.4 / + 1.0
0.71	1.19	0 / 1.0	+ 0.5 / − 0	+1.0 / +0.5	0 / 1.3	+ 0.8 / − 0	+ 1.3 / + 0.8	0.4 / 1.7	+ 0.8 / − 0	+ 1.7 / + 1.2
1.19	1.97	0 / 1.1	+ 0.6 / − 0	+1.1 / +0.6	0 / 1.6	+ 1.0 / − 0	+ 1.6 / + 1.0	0.4 / 2.0	+ 1.0 / − 0	+ 2.0 / + 1.4
1.97	3.15	0.1 / 1.3	+ 0.7 / − 0	+1.3 / +0.7	0.2 / 2.1	+ 1.2 / − 0	+ 2.1 / + 1.4	0.4 / 2.3	+ 1.2 / − 0	+ 2.3 / + 1.6
3.15	4.73	0.1 / 1.6	+ 0.9 / − 0	+1.6 / +1.0	0.2 / 2.5	+ 1.4 / − 0	+ 2.5 / + 1.6	0.6 / 2.9	+ 1.4 / − 0	+ 2.9 / + 2.0
4.73	7.09	0.2 / 1.9	+ 1.0 / − 0	+1.9 / +1.2	0.2 / 2.8	+ 1.6 / − 0	+ 2.8 / + 1.8	0.9 / 3.5	+ 1.6 / − 0	+ 3.5 / + 2.5
7.09	9.85	0.2 / 2.2	+ 1.2 / − 0	+2.2 / +1.4	0.2 / 3.2	+ 1.8 / − 0	+ 3.2 / + 2.0	1.2 / 4.2	+ 1.8 / − 0	+ 4.2 / + 3.0
9.85	12.41	0.2 / 2.3	+ 1.2 / − 0	+2.3 / +1.4	0.2 / 3.4	+ 2.0 / − 0	+ 3.4 / + 2.2	1.5 / 4.7	+ 2.0 / − 0	+ 4.7 / + 3.5
12.41	15.75	0.2 / 2.6	+ 1.4 / − 0	+2.6 / +1.6	0.3 / 3.9	+ 2.2 / − 0	+ 3.9 / + 2.5	2.3 / 5.9	+ 2.2 / − 0	+ 5.9 / + 4.5
15.75	19.69	0.2 / 2.8	+ 1.6 / − 0	+2.8 / +1.8	0.3 / 4.4	+ 2.5 / − 0	+ 4.4 / + 2.8	2.5 / 6.6	+ 2.5 / − 0	+ 6.6 / + 5.0
19.69	30.09		+ 2.0 / − 0		0.5 / 5.5	+ 3 / − 0	+ 5.5 / + 3.5	4 / 9	+ 3 / − 0	+ 9 / + 7
30.09	41.49		+ 2.5 / − 0		0.5 / 7.0	+ 4 / − 0	+ 7.0 / + 4.5	5 / 11.5	+ 4 / − 0	+11.5 / + 9
41.49	56.19		+ 3.0 / − 0		1 / 9	+ 5 / − 0	+ 9 / + 6	7 / 15	+ 5 / − 0	+15 / +12
56.19	76.39		+ 4.0 / − 0		1 / 11	+ 6 / − 0	+11 / + 7	10 / 20	+ 6 / − 0	+20 / +16
76.39	100.9		+ 5.0 / − 0		1 / 14	+ 8 / − 0	+14 / + 9	12 / 25	+ 8 / − 0	+25 / +20
100.9	131.9		+ 6.0 / − 0		2 / 18	+10 / − 0	+18 / +12	15 / 31	+10 / − 0	+31 / +25
131.9	171.9		+ 8.0 / − 0		4 / 24	+12 / − 0	+24 / +16	18 / 38	+12 / − 0	+38 / +30
171.9	200		+10.0 / − 0		4 / 30	+16 / − 0	+30 / +20	24 / 50	+16 / − 0	+50 / +40

(Courtesy of ANSI; B4.1–1955.)

APPENDIX 43
AMERICAN STANDARD FORCE AND SHRINK FITS

Limits are in thousandths of an inch.
Limits for hole and shaft are applied algebraically to the basic size to obtain the limits of size for the parts.
Data in bold face are in accordance with ABC agreements.
Symbols H7, s6, etc., are Hole and Shaft designations used in ABC System.

Nominal Size Range Inches Over — To	Class FN 1 Limits of Interference	Class FN 1 Std Limits Hole H6	Class FN 1 Std Limits Shaft	Class FN 2 Limits of Interference	Class FN 2 Std Limits Hole H7	Class FN 2 Std Limits Shaft s6	Class FN 3 Limits of Interference	Class FN 3 Std Limits Hole H7	Class FN 3 Std Limits Shaft t6	Class FN 4 Limits of Interference	Class FN 4 Std Limits Hole H7	Class FN 4 Std Limits Shaft u6	Class FN 5 Limits of Interference	Class FN 5 Std Limits Hole H8	Class FN 5 Std Limits Shaft x7
0 — 0.12	0.05	+0.25	+0.5	0.2	+0.4	+0.85				0.3	+0.4	+0.95	0.3	+0.6	+1.3
	0.5	−0	+0.3	0.85	−0	+0.6				0.95	−0	+0.7	1.3	−0	+0.9
0.12 — 0.24	0.1	+0.3	+0.6	0.2	+0.5	+1.0				0.4	+0.5	+1.2	0.5	+0.7	+1.7
	0.6	−0	+0.4	1.0	−0	+0.7				1.2	−0	+0.9	1.7	−0	+1.2
0.24 — 0.40	0.1	+0.4	+0.75	0.4	+0.6	+1.4				0.6	+0.6	+1.6	0.5	+0.9	+2.0
	0.75	−0	+0.5	1.4	−0	+1.0				1.6	−0	+1.2	2.0	−0	+1.4
0.40 — 0.56	0.1	−0.4	+0.8	0.5	+0.7	+1.6				0.7	+0.7	+1.8	0.6	+1.0	+2.3
	0.8	−0	+0.5	1.6	−0	+1.2				1.8	−0	+1.4	2.3	−0	+1.6
0.56 — 0.71	0.2	+0.4	+0.9	0.5	+0.7	+1.6				0.7	+0.7	+1.8	0.8	+1.0	+2.5
	0.9	−0	+0.6	1.6	−0	+1.2				1.8	−0	+1.4	2.5	−0	+1.8
0.71 — 0.95	0.2	+0.5	+1.1	0.6	+0.8	+1.9				0.8	+0.8	+2.1	1.0	+1.2	+3.0
	1.1	−0	+0.7	1.9	−0	+1.4				2.1	−0	+1.6	3.0	−0	+2.2
0.95 — 1.19	0.3	+0.5	+1.2	0.6	+0.8	+1.9	0.8	+0.8	+2.1	1.0	+0.8	+2.3	1.3	+1.2	+3.3
	1.2	−0	+0.8	1.9	−0	+1.4	2.1	−0	+1.6	2.3	−0	+1.8	3.3	−0	+2.5
1.19 — 1.58	0.3	+0.6	+1.3	0.8	+1.0	+2.4	1.0	+1.0	+2.6	1.5	+1.0	+3.1	1.4	+1.6	+4.0
	1.3	−0	+0.9	2.4	−0	+1.8	2.6	−0	+2.0	3.1	−0	+2.5	4.0	−0	+3.0
1.58 — 1.97	0.4	+0.6	+1.4	0.8	+1.0	+2.4	1.2	+1.0	+2.8	1.8	+1.0	+3.4	2.4	+1.6	+5.0
	1.4	−0	+1.0	2.4	−0	+1.8	2.8	−0	+2.2	3.4	−0	+2.8	5.0	−0	+4.0
1.97 — 2.56	0.6	+0.7	+1.8	0.8	+1.2	+2.7	1.3	+1.2	+3.2	2.3	+1.2	+4.2	3.2	+1.8	+6.2
	1.8	−0	+1.3	2.7	−0	+2.0	3.2	−0	+2.5	4.2	−0	+3.5	6.2	−0	+5.0
2.56 — 3.15	0.7	+0.7	+1.9	1.0	+1.2	+2.9	1.8	+1.2	+3.7	2.8	+1.2	+4.7	4.2	+1.8	+7.2
	1.9	−0	+1.4	2.9	−0	+2.2	3.7	−0	+3.0	4.7	−0	+4.0	7.2	−0	+6.0
3.15 — 3.94	0.9	+0.9	+2.4	1.4	+1.4	+3.7	2.1	+1.4	+4.4	3.6	+1.4	+5.9	4.8	+2.2	+8.4
	2.4	−0	+1.8	3.7	−0	+2.8	4.4	−0	+3.5	5.9	−0	+5.0	8.4	−0	+7.0
3.94 — 4.73	1.1	+0.9	+2.6	1.6	+1.4	+3.9	2.6	+1.4	+4.9	4.6	+1.4	+6.9	5.8	+2.2	+9.4
	2.6	−0	+2.0	3.9	−0	+3.0	4.9	−0	+4.0	6.9	−0	+6.0	9.4	−0	+8.0
4.73 — 5.52	1.2	+1.0	+2.9	1.9	+1.6	+4.5	3.4	+1.6	+6.0	5.4	+1.6	+8.0	7.5	+2.5	+11.6
	2.9	−0	+2.2	4.5	−0	+3.5	6.0	−0	+5.0	8.0	−0	+7.0	11.6	−0	+10.0
5.52 — 6.30	1.5	+1.0	+3.2	2.4	+1.6	+5.0	3.4	+1.6	+6.0	5.4	+1.6	+8.0	9.5	+2.5	+13.6
	3.2	−0	+2.5	5.0	−0	+4.0	6.0	−0	+5.0	8.0	−0	+7.0	13.6	−0	+12.0
6.30 — 7.09	1.8	+1.0	+3.5	2.9	+1.6	+5.5	4.4	+1.6	+7.0	6.4	+1.6	+9.0	9.5	+2.5	+13.6
	3.5	−0	+2.8	5.5	−0	+4.5	7.0	−0	+6.0	9.0	−0	+8.0	13.6	−0	+12.0
7.09 — 7.88	1.8	+1.2	+3.8	3.2	+1.8	+6.2	5.2	+1.8	+8.2	7.2	+1.8	+10.2	11.2	+2.8	+15.8
	3.8	−0	+3.0	6.2	−0	+5.0	8.2	−0	+7.0	10.2	−0	+9.0	15.8	−0	+14.0
7.88 — 8.86	2.3	+1.2	+4.3	3.2	+1.8	+6.2	5.2	+1.8	+8.2	8.2	+1.8	+11.2	13.2	+2.8	+17.8
	4.3	−0	+3.5	6.2	−0	+5.0	8.2	−0	+7.0	11.2	−0	+10.0	17.8	−0	+16.0
8.86 — 9.85	2.3	+1.2	+4.3	4.2	+1.8	+7.2	6.2	+1.8	+9.2	10.2	+1.8	+13.2	13.2	+2.8	+17.8
	4.3	−0	+3.5	7.2	−0	+6.0	9.2	−0	+8.0	13.2	−0	+12.0	17.8	−0	+16.0
9.85 — 11.03	2.8	+1.2	+4.9	4.0	+2.0	+7.2	7.0	+2.0	+10.2	10.0	+2.0	+13.2	15.0	+3.0	+20.0
	4.9	−0	+4.0	7.2	−0	+6.0	10.2	−0	+9.0	13.2	−0	+12.0	20.0	−0	+18.0
11.03 — 12.41	2.8	+1.2	+4.9	5.0	+2.0	+8.2	7.0	+2.0	+10.2	12.0	+2.0	+15.2	17.0	+3.0	+22.0
	4.9	−0	+4.0	8.2	−0	+7.0	10.2	−0	+9.0	15.2	−0	+14.0	22.0	−0	+20.0
12.41 — 13.98	3.1	+1.4	+5.5	5.8	+2.2	+9.4	7.8	+2.2	+11.4	13.8	+2.2	+17.4	18.5	+3.5	+24.2
	5.5	−0	+4.5	9.4	−0	+8.0	11.4	−0	+10.0	17.4	−0	+16.0	24.2	+0	+22.0
13.98 — 15.75	3.6	+1.4	+6.1	5.8	+2.2	+9.4	9.8	+2.2	+13.4	15.8	+2.2	+19.4	21.5	+3.5	+27.2
	6.1	−0	+5.0	9.4	−0	+8.0	13.4	−0	+12.0	19.4	−0	+18.0	27.2	−0	+25.0
15.75 — 17.72	4.4	+1.6	+7.0	6.5	+2.5	+10.6	9.5	+2.5	+13.6	17.5	+2.5	+21.6	24.0	+4.0	+30.5
	7.0	−0	+6.0	10.6	−0	+9.0	13.6	−0	+12.0	21.6	−0	+20.0	30.5	−0	+28.0
17.72 — 19.69	4.4	+1.6	+7.0	7.5	+2.5	+11.6	11.5	+2.5	+15.6	19.5	+2.5	+23.6	26.0	+4.0	+32.5
	7.0	−0	+6.0	11.6	−0	+10.0	15.6	−0	+14.0	23.6	−0	+22.0	32.5	−0	+30.0

(Courtesy of ANSI; B4.1–1955.)

A-68

APPENDIX 44
THE INTERNATIONAL TOLERANCE GRADES (ANSI B4.2)

Dimensions are in mm.

Over	Up to and including	IT01	IT0	IT1	IT2	IT3	IT4	IT5	IT6	IT7	IT8	IT9	IT10	IT11	IT12	IT13	IT14	IT15	IT16
0	3	0.0003	0.0005	0.0008	0.0012	0.002	0.003	0.004	0.006	0.010	0.014	0.025	0.040	0.060	0.100	0.140	0.250	0.400	0.600
3	6	0.0004	0.0006	0.001	0.0015	0.0025	0.004	0.005	0.008	0.012	0.018	0.030	0.048	0.075	0.120	0.180	0.300	0.480	0.750
6	10	0.0004	0.0006	0.001	0.0015	0.0025	0.004	0.006	0.009	0.015	0.022	0.036	0.058	0.090	0.150	0.220	0.360	0.580	0.900
10	18	0.0005	0.0008	0.0012	0.002	0.003	0.005	0.008	0.011	0.018	0.027	0.043	0.070	0.110	0.180	0.270	0.430	0.700	1.100
18	30	0.0006	0.001	0.0015	0.0025	0.004	0.006	0.009	0.013	0.021	0.033	0.052	0.084	0.130	0.210	0.330	0.520	0.840	1.300
30	50	0.0006	0.001	0.0015	0.0025	0.004	0.007	0.011	0.016	0.025	0.039	0.062	0.100	0.160	0.250	0.390	0.620	1.000	1.600
50	80	0.0008	0.0012	0.002	0.003	0.005	0.008	0.013	0.019	0.030	0.046	0.074	0.120	0.190	0.300	0.460	0.740	1.200	1.900
80	120	0.001	0.0015	0.0025	0.004	0.006	0.010	0.015	0.022	0.035	0.054	0.087	0.140	0.220	0.350	0.540	0.870	1.400	2.200
120	180	0.0012	0.002	0.0035	0.005	0.008	0.012	0.018	0.025	0.040	0.063	0.100	0.160	0.250	0.400	0.630	1.000	1.600	2.500
180	250	0.002	0.003	0.0045	0.007	0.010	0.014	0.020	0.029	0.046	0.072	0.115	0.185	0.290	0.460	0.720	1.150	1.850	2.900
250	315	0.0025	0.004	0.006	0.008	0.012	0.016	0.023	0.032	0.052	0.081	0.130	0.210	0.320	0.520	0.810	1.300	2.100	3.200
315	400	0.003	0.005	0.007	0.009	0.013	0.018	0.025	0.036	0.057	0.089	0.140	0.230	0.360	0.570	0.890	1.400	2.300	3.600
400	500	0.004	0.006	0.008	0.010	0.015	0.020	0.027	0.040	0.063	0.097	0.155	0.250	0.400	0.630	0.970	1.550	2.500	4.000
500	630	0.0045	0.006	0.009	0.011	0.016	0.022	0.030	0.044	0.070	0.110	0.175	0.280	0.440	0.700	1.100	1.750	2.800	4.400
630	800	0.005	0.007	0.010	0.013	0.018	0.025	0.035	0.050	0.080	0.125	0.200	0.320	0.500	0.800	1.250	2.000	3.200	5.000
800	1000	0.0055	0.008	0.011	0.015	0.021	0.029	0.040	0.056	0.090	0.140	0.230	0.360	0.560	0.900	1.400	2.300	3.600	5.600
1000	1250	0.0065	0.009	0.013	0.018	0.024	0.034	0.046	0.066	0.105	0.165	0.260	0.420	0.660	1.050	1.650	2.600	4.200	6.600
1250	1600	0.008	0.011	0.015	0.021	0.029	0.040	0.054	0.078	0.125	0.195	0.310	0.500	0.780	1.250	1.950	3.100	5.000	7.800
1600	2000	0.009	0.013	0.018	0.025	0.035	0.048	0.065	0.092	0.150	0.230	0.370	0.600	0.920	1.500	2.300	3.700	6.000	9.200
2000	2500	0.011	0.015	0.022	0.030	0.041	0.057	0.077	0.110	0.175	0.280	0.440	0.700	1.100	1.750	2.800	4.400	7.000	11.000
2500	3150	0.013	0.018	0.026	0.036	0.050	0.069	0.093	0.135	0.210	0.330	0.540	0.860	1.350	2.100	3.300	5.400	8.600	13.500

Basic sizes — Tolerance grades[3]

[3] IT Values for tolerance grades larger than IT16 can be calculated by using the following formulas:
IT17 = IT12 × 10; IT18 = IT13 × 10; etc.

APPENDIX 45
PREFERRED HOLE BASIS CLEARANCE FITS—CYLINDRICAL FITS (ANSI B4.2)

AMERICAN NATIONAL STANDARD
PREFERRED METRIC LIMITS AND FITS

ANSI B4.2-1978

Dimensions in mm.

BASIC SIZE		LOOSE RUNNING Hole H11	Shaft c11	Fit	FREE RUNNING Hole H9	Shaft d9	Fit	CLOSE RUNNING Hole H8	Shaft f7	Fit	SLIDING Hole H7	Shaft g6	Fit	LOCATIONAL CLEARANCE Hole H7	Shaft h6	Fit
1	MAX	1.060	0.940	0.180	1.025	0.980	0.070	1.014	0.994	0.030	1.010	0.998	0.018	1.010	1.000	0.016
	MIN	1.000	0.880	0.060	1.000	0.955	0.020	1.000	0.984	0.006	1.000	0.992	0.002	1.000	0.994	0.000
1.2	MAX	1.260	1.140	0.180	1.225	1.180	0.070	1.214	1.194	0.030	1.210	1.198	0.018	1.210	1.200	0.016
	MIN	1.200	1.080	0.060	1.200	1.155	0.020	1.200	1.184	0.006	1.200	1.192	0.002	1.200	1.194	0.000
1.6	MAX	1.660	1.540	0.180	1.625	1.580	0.070	1.614	1.594	0.030	1.610	1.598	0.018	1.610	1.600	0.016
	MIN	1.600	1.480	0.060	1.600	1.555	0.020	1.600	1.584	0.006	1.600	1.592	0.002	1.600	1.594	0.000
2	MAX	2.060	1.940	0.180	2.025	1.980	0.070	2.014	1.994	0.030	2.010	1.998	0.018	2.010	2.000	0.016
	MIN	2.000	1.880	0.060	2.000	1.955	0.020	2.000	1.984	0.006	2.000	1.992	0.002	2.000	1.994	0.000
2.5	MAX	2.560	2.440	0.180	2.525	2.480	0.070	2.514	2.494	0.030	2.510	2.498	0.018	2.510	2.500	0.016
	MIN	2.500	2.380	0.060	2.500	2.455	0.020	2.500	2.484	0.006	2.500	2.492	0.002	2.500	2.494	0.000
3	MAX	3.060	2.940	0.180	3.025	2.980	0.070	3.014	2.994	0.030	3.010	2.998	0.018	3.010	3.000	0.016
	MIN	3.000	2.880	0.060	3.000	2.955	0.020	3.000	2.984	0.006	3.000	2.992	0.002	3.000	2.994	0.000
4	MAX	4.075	3.930	0.220	4.030	3.970	0.090	4.018	3.990	0.040	4.012	3.996	0.024	4.012	4.000	0.020
	MIN	4.000	3.855	0.070	4.000	3.940	0.030	4.000	3.978	0.010	4.000	3.988	0.004	4.000	3.992	0.000
5	MAX	5.075	4.930	0.220	5.030	4.970	0.090	5.018	4.990	0.040	5.012	4.996	0.024	5.012	5.000	0.020
	MIN	5.000	4.855	0.070	5.000	4.940	0.030	5.000	4.978	0.010	5.000	4.988	0.004	5.000	4.992	0.000
6	MAX	6.075	5.930	0.220	6.030	5.970	0.090	6.018	5.990	0.040	6.012	5.996	0.024	6.012	6.000	0.020
	MIN	6.000	5.855	0.070	6.000	5.940	0.030	6.000	5.978	0.010	6.000	5.988	0.004	6.000	5.992	0.000
8	MAX	8.090	7.920	0.260	8.036	7.960	0.112	8.022	7.987	0.050	8.015	7.995	0.029	8.015	8.000	0.024
	MIN	8.000	7.830	0.080	8.000	7.924	0.040	8.000	7.972	0.013	8.000	7.986	0.005	8.000	7.991	0.000
10	MAX	10.090	9.920	0.260	10.036	9.960	0.112	10.022	9.987	0.050	10.015	9.995	0.029	10.015	10.000	0.024
	MIN	10.000	9.830	0.080	10.000	9.924	0.040	10.000	9.972	0.013	10.000	9.986	0.005	10.000	9.991	0.000
12	MAX	12.110	11.905	0.315	12.043	11.950	0.136	12.027	11.984	0.061	12.018	11.994	0.035	12.018	12.000	0.029
	MIN	12.000	11.795	0.095	12.000	11.907	0.050	12.000	11.966	0.016	12.000	11.983	0.006	12.000	11.989	0.000
16	MAX	16.110	15.905	0.315	16.043	15.950	0.136	16.027	15.984	0.061	16.018	15.994	0.035	16.018	16.000	0.029
	MIN	16.000	15.795	0.095	16.000	15.907	0.050	16.000	15.966	0.016	16.000	15.983	0.006	16.000	15.989	0.000
20	MAX	20.130	19.890	0.370	20.052	19.935	0.169	20.033	19.980	0.074	20.021	19.993	0.041	20.021	20.000	0.034
	MIN	20.000	19.760	0.110	20.000	19.883	0.065	20.000	19.959	0.020	20.000	19.980	0.007	20.000	19.987	0.000
25	MAX	25.130	24.890	0.370	25.052	24.935	0.169	25.033	24.980	0.074	25.021	24.993	0.041	25.021	25.000	0.034
	MIN	25.000	24.760	0.110	25.000	24.883	0.065	25.000	24.959	0.020	25.000	24.980	0.007	25.000	24.987	0.000
30	MAX	30.130	29.890	0.370	30.052	29.935	0.169	30.033	29.980	0.074	30.021	29.993	0.041	30.021	30.000	0.034
	MIN	30.000	29.760	0.110	30.000	29.883	0.065	30.000	29.959	0.020	30.000	29.980	0.007	30.000	29.987	0.000

Cont.

APPENDIX 45

PREFERRED HOLE BASIS CLEARANCE FITS—CYLINDRICAL FITS (Cont.)

AMERICAN NATIONAL STANDARD
PREFERRED METRIC LIMITS AND FITS

ANSI B4.2-1978

Dimensions in mm.

BASIC SIZE		LOOSE RUNNING Hole H11	Shaft c11	Fit	FREE RUNNING Hole H9	Shaft d9	Fit	CLOSE RUNNING Hole H8	Shaft f7	Fit	SLIDING Hole H7	Shaft g6	Fit	LOCATIONAL CLEARANCE Hole H7	Shaft h6	Fit
40	MAX	40.160	39.880	0.440	40.062	39.920	0.204	40.039	39.975	0.089	40.025	39.991	0.050	40.025	40.000	0.041
	MIN	40.000	39.720	0.120	40.000	39.858	0.080	40.000	39.950	0.025	40.000	39.975	0.009	40.000	39.984	0.000
50	MAX	50.160	49.870	0.450	50.062	49.920	0.204	50.039	49.975	0.089	50.025	49.991	0.050	50.025	50.000	0.041
	MIN	50.000	49.710	0.130	50.000	49.858	0.080	50.000	49.950	0.025	50.000	49.975	0.009	50.000	49.984	0.000
60	MAX	60.190	59.860	0.520	60.074	59.900	0.248	60.046	59.970	0.106	60.030	59.990	0.059	60.030	60.000	0.049
	MIN	60.000	59.670	0.140	60.000	59.826	0.100	60.000	59.940	0.030	60.000	59.971	0.010	60.000	59.981	0.000
80	MAX	80.190	79.850	0.530	80.074	79.900	0.248	80.046	79.970	0.106	80.030	79.990	0.059	80.030	80.000	0.049
	MIN	80.000	79.660	0.150	80.000	79.826	0.100	80.000	79.940	0.030	80.000	79.971	0.010	80.000	79.981	0.000
100	MAX	100.220	99.830	0.610	100.087	99.880	0.294	100.054	99.964	0.125	100.035	99.988	0.069	100.035	100.000	0.057
	MIN	100.000	99.610	0.170	100.000	99.793	0.120	100.000	99.929	0.036	100.000	99.966	0.012	100.000	99.978	0.000
120	MAX	120.220	119.820	0.620	120.087	119.880	0.294	120.054	119.964	0.125	120.035	119.988	0.069	120.035	120.000	0.057
	MIN	120.000	119.600	0.180	120.000	119.793	0.120	120.000	119.929	0.036	120.000	119.966	0.012	120.000	119.978	0.000
160	MAX	160.250	159.790	0.710	160.100	159.855	0.345	160.063	159.957	0.146	160.040	159.986	0.079	160.040	160.000	0.065
	MIN	160.000	159.540	0.210	160.000	159.755	0.145	160.000	159.917	0.043	160.000	159.961	0.014	160.000	159.975	0.000
200	MAX	200.290	199.760	0.820	200.115	199.830	0.400	200.072	199.950	0.168	200.046	199.985	0.090	200.046	200.000	0.075
	MIN	200.000	199.470	0.240	200.000	199.715	0.170	200.000	199.904	0.050	200.000	199.956	0.015	200.000	199.971	0.000
250	MAX	250.290	249.720	0.860	250.115	249.830	0.400	250.072	249.950	0.168	250.046	249.985	0.090	250.046	250.000	0.075
	MIN	250.000	249.430	0.280	250.000	249.715	0.170	250.000	249.904	0.050	250.000	249.956	0.015	250.000	249.971	0.000
300	MAX	300.320	299.670	0.970	300.130	299.810	0.450	300.081	299.944	0.189	300.052	299.983	0.101	300.052	300.000	0.084
	MIN	300.000	299.350	0.330	300.000	299.680	0.190	300.000	299.892	0.056	300.000	299.951	0.017	300.000	299.968	0.000
400	MAX	400.360	399.600	1.120	400.140	399.790	0.490	400.089	399.938	0.208	400.057	399.982	0.111	400.057	400.000	0.093
	MIN	400.000	399.240	0.400	400.000	399.650	0.210	400.000	399.881	0.062	400.000	399.946	0.018	400.000	399.964	0.000
500	MAX	500.400	499.520	1.280	500.155	499.770	0.540	500.097	499.932	0.228	500.063	499.980	0.123	500.063	500.000	0.103
	MIN	500.000	499.120	0.480	500.000	499.615	0.230	500.000	499.869	0.068	500.000	499.940	0.020	500.000	499.960	0.000

APPENDIX 46

PREFERRED HOLE BASIS TRANSITION AND INTERFERENCE FITS—CYLINDRICAL FITS (ANSI B4.2)

AMERICAN NATIONAL STANDARD
PREFERRED METRIC LIMITS AND FITS

ANSI B4.2-1978

Dimensions in mm.

BASIC SIZE		LOCATIONAL TRANSN. Hole H7	Shaft k6	Fit	LOCATIONAL TRANSN. Hole H7	Shaft n6	Fit	LOCATIONAL INTERF. Hole H7	Shaft p6	Fit	MEDIUM DRIVE Hole H7	Shaft s6	Fit	FORCE Hole H7	Shaft u6	Fit
1	MAX	1.010	1.006	0.010	1.010	1.010	0.006	1.010	1.012	0.004	1.010	1.020	-0.004	1.010	1.024	-0.008
	MIN	1.000	1.000	-0.006	1.000	1.004	-0.010	1.000	1.006	-0.012	1.000	1.014	-0.020	1.000	1.018	-0.024
1.2	MAX	1.210	1.206	0.010	1.210	1.210	0.006	1.210	1.212	0.004	1.210	1.220	-0.004	1.210	1.224	-0.008
	MIN	1.200	1.200	-0.006	1.200	1.204	-0.010	1.200	1.206	-0.012	1.200	1.214	-0.020	1.200	1.218	-0.024
1.6	MAX	1.610	1.606	0.010	1.610	1.610	0.006	1.610	1.612	0.004	1.610	1.620	-0.004	1.610	1.624	-0.008
	MIN	1.600	1.600	-0.006	1.600	1.604	-0.010	1.600	1.606	-0.012	1.600	1.614	-0.020	1.600	1.618	-0.024
2	MAX	2.010	2.006	0.010	2.010	2.010	0.006	2.010	2.012	0.004	2.010	2.020	-0.004	2.010	2.024	-0.008
	MIN	2.000	2.000	-0.006	2.000	2.004	-0.010	2.000	2.006	-0.012	2.000	2.014	-0.020	2.000	2.018	-0.024
2.5	MAX	2.510	2.506	0.010	2.510	2.510	0.006	2.510	2.512	0.004	2.510	2.520	-0.004	2.510	2.524	-0.008
	MIN	2.500	2.500	-0.006	2.500	2.504	-0.010	2.500	2.506	-0.012	2.500	2.514	-0.020	2.500	2.518	-0.024
3	MAX	3.010	3.006	0.010	3.010	3.010	0.006	3.010	3.012	0.004	3.010	3.020	-0.004	3.010	3.024	-0.008
	MIN	3.000	3.000	-0.006	3.000	3.004	-0.010	3.000	3.006	-0.012	3.000	3.014	-0.020	3.000	3.018	-0.024
4	MAX	4.012	4.009	0.011	4.012	4.016	0.004	4.012	4.020	0.000	4.012	4.027	-0.007	4.012	4.031	-0.011
	MIN	4.000	4.001	-0.009	4.000	4.008	-0.016	4.000	4.012	-0.020	4.000	4.019	-0.027	4.000	4.023	-0.031
5	MAX	5.012	5.009	0.011	5.012	5.016	0.004	5.012	5.020	0.000	5.012	5.027	-0.007	5.012	5.031	-0.011
	MIN	5.000	5.001	-0.009	5.000	5.008	-0.016	5.000	5.012	-0.020	5.000	5.019	-0.027	5.000	5.023	-0.031
6	MAX	6.012	6.009	0.011	6.012	6.016	0.004	6.012	6.020	0.000	6.012	6.027	-0.007	6.012	6.031	-0.011
	MIN	6.000	6.001	-0.009	6.000	6.008	-0.016	6.000	6.012	-0.020	6.000	6.019	-0.027	6.000	6.023	-0.031
8	MAX	8.015	8.010	0.014	8.015	8.019	0.005	8.015	8.024	0.000	8.015	8.032	-0.008	8.015	8.037	-0.013
	MIN	8.000	8.001	-0.010	8.000	8.010	-0.019	8.000	8.015	-0.024	8.000	8.023	-0.032	8.000	8.028	-0.037
10	MAX	10.015	10.010	0.014	10.015	10.019	0.005	10.015	10.024	0.000	10.015	10.032	-0.008	10.015	10.037	-0.013
	MIN	10.000	10.001	-0.010	10.000	10.010	-0.019	10.000	10.015	-0.024	10.000	10.023	-0.032	10.000	10.028	-0.037
12	MAX	12.018	12.012	0.017	12.018	12.023	0.006	12.018	12.029	0.000	12.018	12.039	-0.010	12.018	12.044	-0.015
	MIN	12.000	12.001	-0.012	12.000	12.012	-0.023	12.000	12.018	-0.029	12.000	12.028	-0.039	12.000	12.033	-0.044
16	MAX	16.018	16.012	0.017	16.018	16.023	0.006	16.018	16.029	0.000	16.018	16.039	-0.010	16.018	16.044	-0.015
	MIN	16.000	16.001	-0.012	16.000	16.012	-0.023	16.000	16.018	-0.029	16.000	16.028	-0.039	16.000	16.033	-0.044
20	MAX	20.021	20.015	0.019	20.021	20.028	0.006	20.021	20.035	-0.001	20.021	20.048	-0.014	20.021	20.054	-0.020
	MIN	20.000	20.002	-0.015	20.000	20.015	-0.028	20.000	20.022	-0.035	20.000	20.035	-0.048	20.000	20.041	-0.054
25	MAX	25.021	25.015	0.019	25.021	25.028	0.006	25.021	25.035	-0.001	25.021	25.048	-0.014	25.021	25.061	-0.027
	MIN	25.000	25.002	-0.015	25.000	25.015	-0.028	25.000	25.022	-0.035	25.000	25.035	-0.048	25.000	25.048	-0.061
30	MAX	30.021	30.015	0.019	30.021	30.028	0.006	30.021	30.035	-0.001	30.021	30.048	-0.014	30.021	30.061	-0.027
	MIN	30.000	30.002	-0.015	30.000	30.015	-0.028	30.000	30.022	-0.035	30.000	30.035	-0.048	30.000	30.048	-0.061

Cont.

APPENDIX 46
PREFERRED HOLE BASIS TRANSITION AND INTERFERENCE FITS—CYLINDRICAL FITS (ANSI B4.2) (Cont.)

AMERICAN NATIONAL STANDARD
PREFERRED METRIC LIMITS AND FITS

ANSI B4.2–1978

Dimensions in mm.

BASIC SIZE		LOCATIONAL TRANSN. Hole H7	Shaft k6	Fit	LOCATIONAL TRANSN. Hole H7	Shaft n6	Fit	LOCATIONAL INTERF. Hole H7	Shaft p6	Fit	MEDIUM DRIVE Hole H7	Shaft s6	Fit	FORCE Hole H7	Shaft u6	Fit
40	MAX	40.025	40.018	0.023	40.025	40.033	0.008	40.025	40.042	-0.001	40.025	40.059	-0.018	40.025	40.076	-0.035
	MIN	40.000	40.002	-0.018	40.000	40.017	-0.033	40.000	40.026	-0.042	40.000	40.043	-0.059	40.000	40.060	-0.076
50	MAX	50.025	50.018	0.023	50.025	50.033	0.008	50.025	50.042	-0.001	50.025	50.059	-0.018	50.025	50.086	-0.045
	MIN	50.000	50.002	-0.018	50.000	50.017	-0.033	50.000	50.026	-0.042	50.000	50.043	-0.059	50.000	50.070	-0.086
60	MAX	60.030	60.021	0.028	60.030	60.039	0.010	60.030	60.051	-0.002	60.030	60.072	-0.023	60.030	60.106	-0.057
	MIN	60.000	60.002	-0.021	60.000	60.020	-0.039	60.000	60.032	-0.051	60.000	60.053	-0.072	60.000	60.087	-0.106
80	MAX	80.030	80.021	0.028	80.030	80.039	0.010	80.030	80.051	-0.002	80.030	80.078	-0.029	80.030	80.121	-0.072
	MIN	80.000	80.002	-0.021	80.000	80.020	-0.039	80.000	80.032	-0.051	80.000	80.059	-0.078	80.000	80.102	-0.121
100	MAX	100.035	100.025	0.032	100.035	100.045	0.012	100.035	100.059	-0.002	100.035	100.093	-0.036	100.035	100.146	-0.089
	MIN	100.000	100.003	-0.025	100.000	100.023	-0.045	100.000	100.037	-0.059	100.000	100.071	-0.093	100.000	100.124	-0.146
120	MAX	120.035	120.025	0.032	120.035	120.045	0.012	120.035	120.059	-0.002	120.035	120.101	-0.044	120.035	120.166	-0.109
	MIN	120.000	120.003	-0.025	120.000	120.023	-0.045	120.000	120.037	-0.059	120.000	120.079	-0.101	120.000	120.144	-0.166
160	MAX	160.040	160.028	0.037	160.040	160.052	0.013	160.040	160.068	-0.003	160.040	160.125	-0.060	160.040	160.215	-0.150
	MIN	160.000	160.003	-0.028	160.000	160.027	-0.052	160.000	160.043	-0.068	160.000	160.100	-0.125	160.000	160.190	-0.215
200	MAX	200.046	200.033	0.042	200.046	200.060	0.015	200.046	200.079	-0.004	200.046	200.151	-0.076	200.046	200.265	-0.190
	MIN	200.000	200.004	-0.033	200.000	200.031	-0.060	200.000	200.050	-0.079	200.000	200.122	-0.151	200.000	200.236	-0.265
250	MAX	250.046	250.033	0.042	250.046	250.060	0.015	250.046	250.079	-0.004	250.046	250.169	-0.094	250.046	250.313	-0.238
	MIN	250.000	250.004	-0.033	250.000	250.031	-0.060	250.000	250.050	-0.079	250.000	250.140	-0.169	250.000	250.284	-0.313
300	MAX	300.052	300.036	0.048	300.052	300.066	0.018	300.052	300.088	-0.004	300.052	300.202	-0.118	300.052	300.382	-0.298
	MIN	300.000	300.004	-0.036	300.000	300.034	-0.066	300.000	300.056	-0.088	300.000	300.170	-0.202	300.000	300.350	-0.382
400	MAX	400.057	400.040	0.053	400.057	400.073	0.020	400.057	400.098	-0.005	400.057	400.244	-0.151	400.057	400.471	-0.378
	MIN	400.000	400.004	-0.040	400.000	400.037	-0.073	400.000	400.062	-0.098	400.000	400.208	-0.244	400.000	400.435	-0.471
500	MAX	500.063	500.045	0.058	500.063	500.080	0.023	500.063	500.108	-0.005	500.063	500.292	-0.189	500.063	500.580	-0.477
	MIN	500.000	500.005	-0.045	500.000	500.040	-0.080	500.000	500.068	-0.108	500.000	500.252	-0.292	500.000	500.540	-0.580

APPENDIX 47

PREFERRED SHAFT BASIS CLEARANCE FITS—CYLINDRICAL FITS (ANSI B4.2)

AMERICAN NATIONAL STANDARD
PREFERRED METRIC LIMITS AND FITS

ANSI B4.2-1978

Dimensions in mm.

BASIC SIZE		LOOSE RUNNING Hole C11	Shaft h11	Fit	FREE RUNNING Hole D9	Shaft h9	Fit	CLOSE RUNNING Hole F8	Shaft h7	Fit	SLIDING Hole G7	Shaft h6	Fit	LOCATIONAL CLEARANCE Hole H7	Shaft h6	Fit
1	MAX	1.120	1.000	0.180	1.045	1.000	0.070	1.020	1.000	0.030	1.012	1.000	0.018	1.010	1.000	0.016
	MIN	1.060	0.940	0.060	1.020	0.975	0.020	1.006	0.990	0.006	1.002	0.994	0.002	1.000	0.994	0.000
1.2	MAX	1.320	1.200	0.180	1.245	1.200	0.070	1.220	1.200	0.030	1.212	1.200	0.018	1.210	1.200	0.016
	MIN	1.260	1.140	0.060	1.220	1.175	0.020	1.206	1.190	0.006	1.202	1.194	0.002	1.200	1.194	0.000
1.6	MAX	1.720	1.600	0.180	1.645	1.600	0.070	1.620	1.600	0.030	1.612	1.600	0.018	1.610	1.600	0.016
	MIN	1.660	1.540	0.060	1.620	1.575	0.020	1.606	1.590	0.006	1.602	1.594	0.002	1.600	1.594	0.000
2	MAX	2.120	2.000	0.180	2.045	2.000	0.070	2.020	2.000	0.030	2.012	2.000	0.018	2.010	2.000	0.016
	MIN	2.060	1.940	0.060	2.020	1.975	0.020	2.006	1.990	0.006	2.002	1.994	0.002	2.000	1.994	0.000
2.5	MAX	2.620	2.500	0.180	2.545	2.500	0.070	2.520	2.500	0.030	2.512	2.500	0.018	2.510	2.500	0.016
	MIN	2.560	2.440	0.060	2.520	2.475	0.020	2.506	2.490	0.006	2.502	2.494	0.002	2.500	2.494	0.000
3	MAX	3.120	3.000	0.180	3.045	3.000	0.070	3.020	3.000	0.030	3.012	3.000	0.018	3.010	3.000	0.016
	MIN	3.060	2.940	0.060	3.020	2.975	0.020	3.006	2.990	0.006	3.002	2.994	0.002	3.000	2.994	0.000
4	MAX	4.145	4.000	0.220	4.060	4.000	0.090	4.028	4.000	0.040	4.016	4.000	0.024	4.012	4.000	0.020
	MIN	4.070	3.925	0.070	4.030	3.970	0.030	4.010	3.988	0.010	4.004	3.992	0.004	4.000	3.992	0.000
5	MAX	5.145	5.000	0.220	5.060	5.000	0.090	5.028	5.000	0.040	5.016	5.000	0.024	5.012	5.000	0.020
	MIN	5.070	4.925	0.070	5.030	4.970	0.030	5.010	4.988	0.010	5.004	4.992	0.004	5.000	4.992	0.000
6	MAX	6.145	6.000	0.220	6.060	6.000	0.090	6.028	6.000	0.040	6.016	6.000	0.024	6.012	6.000	0.020
	MIN	6.070	5.925	0.070	6.030	5.970	0.030	6.010	5.988	0.010	6.004	5.992	0.004	6.000	5.992	0.000
8	MAX	8.170	8.000	0.260	8.076	8.000	0.112	8.035	8.000	0.050	8.020	8.000	0.029	8.015	8.000	0.024
	MIN	8.080	7.910	0.080	8.040	7.964	0.040	8.013	7.985	0.013	8.005	7.991	0.005	8.000	7.991	0.000
10	MAX	10.170	10.000	0.260	10.076	10.000	0.112	10.035	10.000	0.050	10.020	10.000	0.029	10.015	10.000	0.024
	MIN	10.080	9.910	0.080	10.040	9.964	0.040	10.013	9.985	0.013	10.005	9.991	0.005	10.000	9.991	0.000
12	MAX	12.205	12.000	0.315	12.093	12.000	0.136	12.043	12.000	0.061	12.024	12.000	0.035	12.018	12.000	0.029
	MIN	12.095	11.890	0.095	12.050	11.957	0.050	12.016	11.982	0.016	12.006	11.989	0.006	12.000	11.989	0.000
16	MAX	16.205	16.000	0.315	16.093	16.000	0.136	16.043	16.000	0.061	16.024	16.000	0.035	16.018	16.000	0.029
	MIN	16.095	15.890	0.095	16.050	15.957	0.050	16.016	15.982	0.016	16.006	15.989	0.006	16.000	15.989	0.000
20	MAX	20.240	20.000	0.370	20.117	20.000	0.169	20.053	20.000	0.074	20.028	20.000	0.041	20.021	20.000	0.034
	MIN	20.110	19.870	0.110	20.065	19.948	0.065	20.020	19.979	0.020	20.007	19.987	0.007	20.000	19.987	0.000
25	MAX	25.240	25.000	0.370	25.117	25.000	0.169	25.053	25.000	0.074	25.028	25.000	0.041	25.021	25.000	0.034
	MIN	25.110	24.870	0.110	25.065	24.948	0.065	25.020	24.979	0.020	25.007	24.987	0.007	25.000	24.987	0.000
30	MAX	30.240	30.000	0.370	30.117	30.000	0.169	30.053	30.000	0.074	30.028	30.000	0.041	30.021	30.000	0.034
	MIN	30.110	29.870	0.110	30.065	29.948	0.065	30.020	29.979	0.020	30.007	29.987	0.007	30.000	29.987	0.000

Cont.

APPENDIX 47
PREFERRED SHAFT BASIS CLEARANCE FITS—CYLINDRICAL FITS (Cont.)

AMERICAN NATIONAL STANDARD
PREFERRED METRIC LIMITS AND FITS

ANSI B4.2-1978

Dimensions in mm.

BASIC SIZE		LOOSE RUNNING Hole C11	Shaft h11	Fit	FREE RUNNING Hole D9	Shaft h9	Fit	CLOSE RUNNING Hole F8	Shaft h7	Fit	SLIDING Hole G7	Shaft h6	Fit	LOCATIONAL CLEARANCE Hole H7	Shaft h6	Fit
40	MAX	40.280	40.000	0.440	40.142	40.000	0.204	40.064	40.000	0.089	40.034	40.000	0.050	40.025	40.000	0.041
	MIN	40.120	39.840	0.120	40.080	39.938	0.080	40.025	39.975	0.025	40.009	39.984	0.009	40.000	39.984	0.000
50	MAX	50.290	50.000	0.450	50.142	50.000	0.204	50.064	50.000	0.089	50.034	50.000	0.050	50.025	50.000	0.041
	MIN	50.130	49.840	0.130	50.080	49.938	0.080	50.025	49.975	0.025	50.009	49.984	0.009	50.000	49.984	0.000
60	MAX	60.330	60.000	0.520	60.174	60.000	0.248	60.076	60.000	0.106	60.040	60.000	0.059	60.030	60.000	0.049
	MIN	60.140	59.810	0.140	60.100	59.926	0.100	60.030	59.970	0.030	60.010	59.981	0.010	60.000	59.981	0.000
80	MAX	80.340	80.000	0.530	80.174	80.000	0.248	80.076	80.000	0.106	80.040	80.000	0.059	80.030	80.000	0.049
	MIN	80.150	79.810	0.150	80.100	79.926	0.100	80.030	79.970	0.030	80.010	79.981	0.010	80.000	79.981	0.000
100	MAX	100.390	100.000	0.610	100.207	100.000	0.294	100.090	100.000	0.125	100.047	100.000	0.069	100.035	100.000	0.057
	MIN	100.170	99.780	0.170	100.120	99.913	0.120	100.036	99.965	0.036	100.012	99.978	0.012	100.000	99.978	0.000
120	MAX	120.400	120.000	0.620	120.207	120.000	0.294	120.090	120.000	0.125	120.047	120.000	0.069	120.035	120.000	0.057
	MIN	120.180	119.780	0.180	120.120	119.913	0.120	120.036	119.965	0.036	120.012	119.978	0.012	120.000	119.978	0.000
160	MAX	160.460	160.000	0.710	160.245	160.000	0.345	160.106	160.000	0.146	160.054	160.000	0.079	160.040	160.000	0.065
	MIN	160.210	159.750	0.210	160.145	159.900	0.145	160.043	159.960	0.043	160.014	159.975	0.014	160.000	159.975	0.000
200	MAX	200.530	200.000	0.820	200.285	200.000	0.400	200.122	200.000	0.168	200.061	200.000	0.090	200.046	200.000	0.075
	MIN	200.240	199.710	0.240	200.170	199.885	0.170	200.050	199.954	0.050	200.015	199.971	0.015	200.000	199.971	0.000
250	MAX	250.570	250.000	0.860	250.285	250.000	0.400	250.122	250.000	0.168	250.061	250.000	0.090	250.046	250.000	0.075
	MIN	250.280	249.710	0.280	250.170	249.885	0.170	250.050	249.954	0.050	250.015	249.971	0.015	250.000	249.971	0.000
300	MAX	300.650	300.000	0.970	300.320	300.000	0.450	300.137	300.000	0.189	300.069	300.000	0.101	300.052	300.000	0.084
	MIN	300.330	299.680	0.330	300.190	299.870	0.190	300.056	299.948	0.056	300.017	299.968	0.017	300.000	299.968	0.000
400	MAX	400.760	400.000	1.120	400.350	400.000	0.490	400.151	400.000	0.208	400.075	400.000	0.111	400.057	400.000	0.093
	MIN	400.400	399.640	0.400	400.210	399.860	0.210	400.062	399.943	0.062	400.018	399.964	0.018	400.000	399.964	0.000
500	MAX	500.880	500.000	1.280	500.385	500.000	0.540	500.165	500.000	0.228	500.083	500.000	0.123	500.063	500.000	0.103
	MIN	500.480	499.600	0.480	500.230	499.845	0.230	500.068	499.937	0.068	500.020	499.960	0.020	500.000	499.960	0.000

APPENDIX 48

PREFERRED SHAFT BASIS TRANSITION AND INTERFERENCE FITS—CYLINDRICAL FITS

AMERICAN NATIONAL STANDARD
PREFERRED METRIC LIMITS AND FITS

ANSI B4.2-1978

Dimensions in mm.

BASIC SIZE		LOCATIONAL TRANSN. Hole K7	Shaft h6	Fit	LOCATIONAL TRANSN. Hole N7	Shaft h6	Fit	LOCATIONAL INTERF. Hole P7	Shaft h6	Fit	MEDIUM DRIVE Hole S7	Shaft h6	Fit	FORCE Hole U7	Shaft h6	Fit
1	MAX	1.000	1.000	0.006	0.996	1.000	0.002	0.994	1.000	0.000	0.986	1.000	-0.008	0.982	1.000	-0.012
	MIN	0.990	0.994	-0.010	0.986	0.994	-0.014	0.984	0.994	-0.016	0.976	0.994	-0.024	0.972	0.994	-0.028
1.2	MAX	1.200	1.200	0.006	1.196	1.200	0.002	1.194	1.200	0.000	1.186	1.200	-0.008	1.182	1.200	-0.012
	MIN	1.190	1.194	-0.010	1.186	1.194	-0.014	1.184	1.194	-0.016	1.176	1.194	-0.024	1.172	1.194	-0.028
1.6	MAX	1.600	1.600	0.006	1.596	1.600	0.002	1.594	1.600	0.000	1.586	1.600	-0.008	1.582	1.600	-0.012
	MIN	1.590	1.594	-0.010	1.586	1.594	-0.014	1.584	1.594	-0.016	1.576	1.594	-0.024	1.572	1.594	-0.028
2	MAX	2.000	2.000	0.006	1.996	2.000	0.002	1.994	2.000	0.000	1.986	2.000	-0.008	1.982	2.000	-0.012
	MIN	1.990	1.994	-0.010	1.986	1.994	-0.014	1.984	1.994	-0.016	1.976	1.994	-0.024	1.972	1.994	-0.028
2.5	MAX	2.500	2.500	0.006	2.496	2.500	0.002	2.494	2.500	0.000	2.486	2.500	-0.008	2.482	2.500	-0.012
	MIN	2.490	2.494	-0.010	2.486	2.494	-0.014	2.484	2.494	-0.016	2.476	2.494	-0.024	2.472	2.494	-0.028
3	MAX	3.000	3.000	0.006	2.996	3.000	0.002	2.994	3.000	0.000	2.986	3.000	-0.008	2.982	3.000	-0.012
	MIN	2.990	2.994	-0.010	2.986	2.994	-0.014	2.984	2.994	-0.016	2.976	2.994	-0.024	2.972	2.994	-0.028
4	MAX	4.003	4.000	0.011	3.996	4.000	0.004	3.992	4.000	0.000	3.985	4.000	-0.007	3.981	4.000	-0.011
	MIN	3.991	3.992	-0.009	3.984	3.992	-0.016	3.980	3.992	-0.020	3.973	3.992	-0.027	3.969	3.992	-0.031
5	MAX	5.003	5.000	0.011	4.996	5.000	0.004	4.992	5.000	0.000	4.985	5.000	-0.007	4.981	5.000	-0.011
	MIN	4.991	4.992	-0.009	4.984	4.992	-0.016	4.980	4.992	-0.020	4.973	4.992	-0.027	4.969	4.992	-0.031
6	MAX	6.003	6.000	0.011	5.996	6.000	0.004	5.992	6.000	0.000	5.985	6.000	-0.007	5.981	6.000	-0.011
	MIN	5.991	5.992	-0.009	5.984	5.992	-0.016	5.980	5.992	-0.020	5.973	5.992	-0.027	5.969	5.992	-0.031
8	MAX	8.005	8.000	0.014	7.996	8.000	0.005	7.991	8.000	0.000	7.983	8.000	-0.008	7.978	8.000	-0.013
	MIN	7.990	7.991	-0.010	7.981	7.991	-0.019	7.976	7.991	-0.024	7.968	7.991	-0.032	7.963	7.991	-0.037
10	MAX	10.005	10.000	0.014	9.996	10.000	0.005	9.991	10.000	0.000	9.983	10.000	-0.008	9.978	10.000	-0.013
	MIN	9.990	9.991	-0.010	9.981	9.991	-0.019	9.976	9.991	-0.024	9.968	9.991	-0.032	9.963	9.991	-0.037
12	MAX	12.006	12.000	0.017	11.995	12.000	0.006	11.989	12.000	0.000	11.979	12.000	-0.010	11.974	12.000	-0.015
	MIN	11.988	11.989	-0.012	11.977	11.989	-0.023	11.971	11.989	-0.029	11.961	11.989	-0.039	11.956	11.989	-0.044
16	MAX	16.006	16.000	0.017	15.995	16.000	0.006	15.989	16.000	0.000	15.979	16.000	-0.010	16.000	16.000	-0.015
	MIN	15.988	15.989	-0.012	15.977	15.989	-0.023	15.971	15.989	-0.029	15.961	15.989	-0.039	15.956	15.989	-0.044
20	MAX	20.006	20.000	0.019	19.993	20.000	0.006	19.986	20.000	0.000	19.973	20.000	-0.014	19.967	20.000	-0.020
	MIN	19.985	19.987	-0.015	19.972	19.987	-0.028	19.965	19.987	-0.035	19.952	19.987	-0.048	19.946	19.987	-0.054
25	MAX	25.006	25.000	0.019	24.993	25.000	0.006	24.986	25.000	0.000	24.973	25.000	-0.014	24.960	25.000	-0.027
	MIN	24.985	24.987	-0.015	24.972	24.987	-0.028	24.965	24.987	-0.035	24.952	24.987	-0.048	24.939	24.987	-0.061
30	MAX	30.006	30.000	0.019	29.993	30.000	0.006	29.986	30.000	0.000	29.973	30.000	-0.014	29.960	30.000	-0.027
	MIN	29.985	29.987	-0.015	29.972	29.987	-0.028	29.965	29.987	-0.035	29.952	29.987	-0.048	29.939	29.987	-0.061

Cont.

APPENDIX 48

PREFERRED SHAFT BASIS TRANSITION AND INTERFERENCE FITS—CYLINDRICAL FITS (Cont.)

AMERICAN NATIONAL STANDARD
PREFERRED METRIC LIMITS AND FITS

ANSI B4.2-1978

Dimensions in mm.

BASIC SIZE		LOCATIONAL TRANSN. Hole K7	Shaft h6	Fit	LOCATIONAL TRANSN. Hole N7	Shaft h6	Fit	LOCATIONAL INTERF. Hole P7	Shaft h6	Fit	MEDIUM DRIVE Hole S7	Shaft h6	Fit	FORCE Hole U7	Shaft h6	Fit
40	MAX	40.007	40.000	0.023	39.992	40.000	0.008	39.983	40.000	-0.001	39.966	40.000	-0.018	39.949	40.000	-0.035
	MIN	39.982	39.984	-0.018	39.967	39.984	-0.033	39.958	39.984	-0.042	39.941	39.984	-0.059	39.924	39.984	-0.076
50	MAX	50.007	50.000	0.023	49.992	50.000	0.008	49.983	50.000	-0.001	49.966	50.000	-0.018	49.935	50.000	-0.045
	MIN	49.982	49.984	-0.018	49.967	49.984	-0.033	49.958	49.984	-0.042	49.941	49.984	-0.059	49.914	49.984	-0.086
60	MAX	60.009	60.000	0.028	59.991	60.000	0.010	59.979	60.000	-0.002	59.958	60.000	-0.023	59.924	60.000	-0.057
	MIN	59.975	59.981	-0.021	59.961	59.981	-0.039	59.949	59.981	-0.051	59.928	59.981	-0.072	59.894	59.981	-0.106
80	MAX	80.009	80.000	0.028	79.991	80.000	0.010	79.979	80.000	-0.002	79.952	80.000	-0.029	79.909	80.000	-0.072
	MIN	79.979	79.981	-0.021	79.961	79.981	-0.039	79.949	79.981	-0.051	79.922	79.981	-0.078	79.879	79.981	-0.121
100	MAX	100.010	100.000	0.032	99.990	100.000	0.012	99.976	100.000	-0.002	99.942	100.000	-0.036	99.889	100.000	-0.089
	MIN	99.975	99.978	-0.025	99.955	99.978	-0.045	99.941	99.978	-0.059	99.907	99.978	-0.093	99.854	99.978	-0.146
120	MAX	120.010	120.000	0.032	119.990	120.000	0.012	119.976	120.000	-0.002	119.934	120.000	-0.044	119.869	120.000	-0.109
	MIN	119.975	119.978	-0.025	119.955	119.978	-0.045	119.941	119.978	-0.059	119.899	119.978	-0.101	119.834	119.978	-0.166
160	MAX	160.012	160.000	0.037	159.988	160.000	0.013	159.972	160.000	-0.003	159.915	160.000	-0.060	159.825	160.000	-0.150
	MIN	159.972	159.975	-0.028	159.948	159.975	-0.052	159.932	159.975	-0.068	159.875	159.975	-0.125	159.785	159.975	-0.215
200	MAX	200.013	200.000	0.042	199.986	200.000	0.015	199.967	200.000	-0.004	199.895	200.000	-0.076	199.781	200.000	-0.190
	MIN	199.967	199.971	-0.033	199.940	199.971	-0.060	199.921	199.971	-0.079	199.849	199.971	-0.151	199.735	199.971	-0.265
250	MAX	250.013	250.000	0.042	249.986	250.000	0.015	249.967	250.000	-0.004	249.877	250.000	-0.094	249.733	250.000	-0.238
	MIN	249.967	249.971	-0.033	249.940	249.971	-0.060	249.921	249.971	-0.079	249.831	249.971	-0.169	249.687	249.971	-0.313
300	MAX	300.016	300.000	0.048	299.986	300.000	0.018	299.964	300.000	-0.004	299.850	300.000	-0.118	299.670	300.000	-0.298
	MIN	299.964	299.968	-0.036	299.934	299.968	-0.066	299.912	299.968	-0.088	299.798	299.968	-0.202	299.618	299.968	-0.382
400	MAX	400.017	400.000	0.053	399.984	400.000	0.020	399.959	400.000	-0.005	399.813	400.000	-0.151	399.586	400.000	-0.378
	MIN	399.960	399.964	-0.040	399.927	399.964	-0.073	399.902	399.964	-0.098	399.756	399.964	-0.244	399.529	399.964	-0.471
500	MAX	500.018	500.000	0.058	499.983	500.000	0.023	499.955	500.000	-0.005	499.771	500.000	-0.189	499.483	500.000	-0.477
	MIN	499.955	499.960	-0.045	499.920	499.960	-0.080	499.892	499.960	-0.108	499.708	499.960	-0.292	499.420	499.960	-0.580

APPENDIX 49
HOLE SIZES FOR NON-PREFERRED DIAMETERS

Basic Size	C11	D9	F8	G7	H7	H8	H9	H11	K7	N7	P7	S7	U7
OVER 0 TO 3	+0.120 +0.060	+0.045 +0.020	+0.020 +0.006	+0.012 +0.002	+0.010 0.000	+0.014 0.000	+0.025 0.000	+0.060 0.000	0.000 −0.010	−0.004 −0.014	−0.006 −0.016	−0.014 −0.024	−0.018 −0.028
OVER 3 TO 6	+0.145 +0.070	+0.060 +0.030	+0.028 +0.010	+0.016 +0.004	+0.012 0.000	+0.018 0.000	+0.030 0.000	+0.075 0.000	+0.003 −0.009	−0.004 −0.016	−0.008 −0.020	−0.015 −0.027	−0.019 −0.031
OVER 6 TO 10	+0.170 +0.080	+0.076 +0.040	+0.035 +0.013	+0.020 +0.005	+0.015 0.000	+0.022 0.000	+0.036 0.000	+0.090 0.000	+0.005 −0.010	−0.004 −0.019	−0.009 −0.024	−0.017 −0.032	−0.022 −0.037
OVER 10 TO 14	+0.205 +0.095	+0.093 +0.050	+0.043 +0.016	+0.024 +0.006	+0.018 0.000	+0.027 0.000	+0.043 0.000	+0.110 0.000	+0.006 −0.012	−0.005 −0.023	−0.011 −0.029	−0.021 −0.039	−0.026 −0.044
OVER 14 TO 18	+0.205 +0.095	+0.093 +0.050	+0.043 +0.016	+0.024 +0.006	+0.018 0.000	+0.027 0.000	+0.043 0.000	+0.110 0.000	+0.006 −0.012	−0.005 −0.023	−0.011 −0.029	−0.021 −0.039	−0.026 −0.044
OVER 18 TO 24	+0.240 +0.110	+0.117 +0.065	+0.053 +0.020	+0.028 +0.007	+0.021 0.000	+0.033 0.000	+0.052 0.000	+0.130 0.000	+0.006 −0.015	−0.007 −0.028	−0.014 −0.035	−0.027 −0.048	−0.033 −0.054
OVER 24 TO 30	+0.240 +0.110	+0.117 +0.065	+0.053 +0.020	+0.028 +0.007	+0.021 0.000	+0.033 0.000	+0.052 0.000	+0.130 0.000	+0.006 −0.015	−0.007 −0.028	−0.014 −0.035	−0.027 −0.048	−0.040 −0.061
OVER 30 TO 40	+0.280 +0.120	+0.142 +0.080	+0.064 +0.025	+0.034 +0.009	+0.025 0.000	+0.039 0.000	+0.062 0.000	+0.160 0.000	+0.007 −0.018	−0.008 −0.033	−0.017 −0.042	−0.034 −0.059	−0.051 −0.076
OVER 40 TO 50	+0.290 +0.130	+0.142 +0.080	+0.064 +0.025	+0.034 +0.009	+0.025 0.000	+0.039 0.000	+0.062 0.000	+0.160 0.000	+0.007 −0.018	−0.008 −0.033	−0.017 −0.042	−0.034 −0.059	−0.061 −0.086
OVER 50 TO 65	+0.330 +0.140	+0.174 +0.100	+0.076 +0.030	+0.040 +0.010	+0.030 0.000	+0.046 0.000	+0.074 0.000	+0.190 0.000	+0.009 −0.021	−0.009 −0.039	−0.021 −0.051	−0.042 −0.072	−0.076 −0.106
OVER 65 TO 80	+0.340 +0.150	+0.174 +0.100	+0.076 +0.030	+0.040 +0.010	+0.030 0.000	+0.046 0.000	+0.074 0.000	+0.190 0.000	+0.009 −0.021	−0.009 −0.039	−0.021 −0.051	−0.048 −0.078	−0.091 −0.121
OVER 80 TO 100	+0.390 +0.170	+0.207 +0.120	+0.090 +0.036	+0.047 +0.012	+0.035 0.000	+0.054 0.000	+0.087 0.000	+0.220 0.000	+0.010 −0.025	−0.010 −0.045	−0.024 −0.059	−0.058 −0.093	−0.111 −0.146

Cont.

APPENDIX 49
HOLE SIZES FOR NON-PREFERRED DIAMETERS (Cont.)

Basic Size	c11	d9	f8	g7	h7	h8	h9	h11	k7	n7	p7	s7	u7
OVER 100 TO 120	+0.400 / +0.180	+0.207 / +0.120	+0.090 / +0.036	+0.047 / +0.012	+0.035 / 0.000	+0.054 / 0.000	+0.087 / 0.000	+0.220 / 0.000	+0.010 / −0.025	−0.010 / −0.045	−0.024 / −0.059	−0.066 / −0.101	−0.131 / −0.166
OVER 120 TO 140	+0.450 / +0.200	+0.245 / +0.145	+0.106 / +0.043	+0.054 / +0.014	+0.040 / 0.000	+0.063 / 0.000	+0.100 / 0.000	+0.250 / 0.000	+0.012 / −0.028	−0.012 / −0.052	−0.028 / −0.068	−0.077 / −0.117	−0.155 / −0.195
OVER 140 TO 160	+0.460 / +0.210	+0.245 / +0.145	+0.106 / +0.043	+0.054 / +0.014	+0.040 / 0.000	+0.063 / 0.000	+0.100 / 0.000	+0.250 / 0.000	+0.012 / −0.028	−0.012 / −0.052	−0.028 / −0.068	−0.085 / −0.125	−0.175 / −0.215
OVER 160 TO 180	+0.480 / +0.230	+0.245 / +0.145	+0.106 / +0.043	+0.054 / +0.014	+0.040 / 0.000	+0.063 / 0.000	+0.100 / 0.000	+0.250 / 0.000	+0.012 / −0.028	−0.012 / −0.052	−0.028 / −0.068	−0.093 / −0.133	−0.195 / −0.235
OVER 180 TO 200	+0.530 / +0.240	+0.285 / +0.170	+0.122 / +0.050	+0.061 / +0.015	+0.046 / 0.000	+0.072 / 0.000	+0.115 / 0.000	+0.290 / 0.000	+0.013 / −0.033	−0.014 / −0.060	−0.033 / −0.079	−0.105 / −0.151	−0.219 / −0.265
OVER 200 TO 225	+0.550 / +0.260	+0.285 / +0.170	+0.122 / +0.050	+0.061 / +0.015	+0.046 / 0.000	+0.072 / 0.000	+0.115 / 0.000	+0.290 / 0.000	+0.013 / −0.033	−0.014 / −0.060	−0.033 / −0.079	−0.113 / −0.159	−0.241 / −0.287
OVER 225 TO 250	+0.570 / +0.280	+0.285 / +0.170	+0.122 / +0.050	+0.061 / +0.015	+0.046 / 0.000	+0.072 / 0.000	+0.115 / 0.000	+0.290 / 0.000	+0.013 / −0.033	−0.014 / −0.060	−0.033 / −0.079	−0.123 / −0.169	−0.267 / −0.313
OVER 250 TO 280	+0.620 / +0.300	+0.320 / +0.190	+0.137 / +0.056	+0.069 / +0.017	+0.052 / 0.000	+0.081 / 0.000	+0.130 / 0.000	+0.320 / 0.000	+0.016 / −0.036	−0.014 / −0.066	−0.036 / −0.088	−0.138 / −0.190	−0.295 / −0.347
OVER 280 TO 315	+0.650 / +0.330	+0.320 / +0.190	+0.137 / +0.056	+0.069 / 0.017	+0.052 / 0.000	+0.081 / 0.000	+0.130 / 0.000	+0.320 / 0.000	+0.016 / −0.036	−0.014 / −0.066	−0.036 / −0.088	−0.150 / −0.202	−0.330 / −0.382
OVER 315 TO 355	+0.720 / +0.360	+0.350 / +0.210	+0.151 / +0.062	+0.075 / +0.018	+0.057 / 0.000	+0.089 / 0.000	+0.140 / 0.000	+0.360 / 0.000	+0.017 / −0.040	−0.016 / −0.073	−0.041 / −0.098	−0.169 / −0.226	−0.369 / −0.426
OVER 355 TO 400	+0.760 / +0.400	+0.350 / +0.210	+0.151 / +0.062	+0.075 / +0.018	+0.057 / 0.000	+0.089 / 0.000	+0.140 / 0.000	+0.360 / 0.000	+0.017 / −0.040	−0.016 / −0.073	−0.041 / −0.098	−0.187 / −0.244	−0.414 / −0.471
OVER 400 TO 450	+0.840 / +0.440	+0.385 / +0.230	+0.165 / +0.068	+0.083 / +0.020	+0.063 / 0.000	+0.097 / 0.000	+0.155 / 0.000	+0.400 / 0.000	+0.018 / −0.045	−0.017 / −0.080	−0.045 / −0.108	−0.209 / −0.272	−0.467 / −0.530
OVER 450 TO 500	+0.880 / +0.480	+0.385 / +0.230	+0.165 / +0.068	+0.083 / +0.020	+0.063 / 0.000	+0.097 / 0.000	+0.155 / 0.000	+0.400 / 0.000	+0.018 / −0.045	−0.017 / −0.080	−0.045 / −0.108	−0.229 / −0.292	−0.517 / −0.580

APPENDIX 50
SHAFT SIZES FOR NON-PREFERRED DIAMETERS

Basic Size	c11	d9	f7	g6	h6	h7	h9	h11	k6	n6	p6	s6	u6
OVER 0 TO 3	−0.060 / −0.120	−0.020 / −0.045	−0.006 / −0.016	−0.002 / −0.008	0.000 / −0.006	0.000 / −0.010	0.000 / −0.025	0.000 / −0.060	+0.006 / 0.000	+0.010 / +0.004	+0.012 / +0.006	+0.020 / +0.014	+0.024 / +0.018
OVER 3 TO 6	−0.070 / −0.145	−0.030 / −0.060	−0.010 / −0.022	−0.004 / −0.012	0.000 / −0.008	0.000 / −0.012	0.000 / −0.030	0.000 / −0.075	+0.009 / +0.001	+0.016 / +0.008	+0.020 / +0.012	+0.027 / +0.019	+0.031 / +0.023
OVER 6 TO 10	−0.080 / −0.170	−0.040 / −0.076	−0.013 / −0.028	−0.005 / −0.014	0.000 / −0.009	0.000 / −0.015	0.000 / −0.036	0.000 / −0.090	+0.010 / +0.001	+0.019 / +0.010	+0.024 / +0.015	+0.032 / +0.023	+0.037 / +0.028
OVER 10 TO 14	−0.095 / −0.205	−0.050 / −0.093	−0.016 / −0.034	−0.006 / −0.017	0.000 / −0.011	0.000 / −0.018	0.000 / −0.043	0.000 / −0.110	+0.012 / +0.001	+0.023 / +0.012	+0.029 / +0.018	+0.039 / +0.028	+0.044 / +0.033
OVER 14 TO 18	−0.095 / −0.205	−0.050 / −0.093	−0.016 / −0.034	−0.006 / −0.017	0.000 / −0.011	0.000 / −0.018	0.000 / −0.043	0.000 / −0.110	+0.012 / +0.001	+0.023 / +0.012	+0.029 / +0.018	+0.039 / +0.028	+0.044 / +0.033
OVER 18 TO 24	−0.110 / −0.240	−0.065 / −0.117	−0.020 / −0.041	−0.007 / −0.020	0.000 / −0.013	0.000 / −0.021	0.000 / −0.052	0.000 / −0.130	+0.015 / +0.002	+0.028 / +0.015	+0.035 / +0.022	+0.048 / +0.035	+0.054 / +0.041
OVER 24 TO 30	−0.110 / −0.240	−0.065 / −0.117	−0.020 / −0.041	−0.007 / −0.020	0.000 / −0.013	0.000 / −0.021	0.000 / −0.052	0.000 / −0.130	+0.015 / +0.002	+0.028 / +0.015	+0.035 / +0.022	+0.048 / +0.035	+0.061 / +0.048
OVER 30 TO 40	−0.120 / −0.280	−0.080 / −0.142	−0.025 / −0.050	−0.009 / −0.025	0.000 / −0.016	0.000 / −0.025	0.000 / −0.062	0.000 / −0.160	+0.018 / +0.002	+0.033 / +0.017	+0.042 / +0.026	+0.059 / +0.043	+0.076 / +0.060
OVER 40 TO 50	−0.130 / −0.290	−0.080 / −0.142	−0.025 / −0.050	−0.009 / −0.025	0.000 / −0.016	0.000 / −0.025	0.000 / −0.062	0.000 / −0.160	+0.018 / +0.002	+0.033 / +0.017	+0.042 / +0.026	+0.059 / +0.043	+0.086 / +0.070
OVER 50 TO 65	−0.140 / −0.330	−0.100 / −0.174	−0.030 / −0.060	−0.010 / −0.029	0.000 / −0.019	0.000 / −0.030	0.000 / −0.074	0.000 / −0.190	+0.021 / +0.002	+0.039 / +0.020	+0.051 / −0.032	+0.072 / +0.053	+0.106 / +0.087
OVER 65 TO 80	−0.150 / −0.340	−0.100 / −0.174	−0.030 / −0.060	−0.010 / −0.029	0.000 / −0.019	0.000 / −0.030	0.000 / −0.074	0.000 / −0.190	+0.021 / +0.002	+0.039 / +0.020	+0.051 / +0.032	+0.078 / +0.059	+0.121 / +0.102
OVER 80 TO 100	−0.170 / −0.390	−0.120 / −0.207	−0.036 / −0.071	−0.012 / −0.034	0.000 / −0.022	0.000 / −0.035	0.000 / −0.087	0.000 / −0.220	+0.025 / +0.003	+0.045 / +0.023	+0.059 / +0.037	+0.093 / +0.071	+0.146 / +0.124

Cont.

APPENDIX 50
SHAFT SIZES FOR NON-PREFERRED DIAMETERS (Cont.)

Basic Size	c11	d9	f7	g6	h6	h7	h9	h11	k6	n6	p6	s6	u6
OVER 100 TO 120	−0.180 −0.400	−0.120 −0.207	−0.036 −0.071	−0.012 −0.034	0.000 −0.022	0.000 −0.035	0.000 −0.087	0.000 −0.220	+0.025 +0.003	+0.045 +0.023	+0.059 +0.037	+0.101 +0.079	+0.166 +0.144
OVER 120 TO 140	−0.200 −0.450	−0.145 −0.245	−0.043 −0.083	−0.014 −0.039	0.000 −0.025	0.000 −0.040	0.000 −0.100	0.000 −0.250	+0.028 +0.003	+0.052 +0.027	+0.068 +0.043	+0.117 +0.092	+0.195 +0.170
OVER 140 TO 160	−0.210 −0.460	−0.145 −0.245	−0.043 −0.083	−0.014 −0.039	0.000 −0.025	0.000 −0.040	0.000 −0.100	0.000 −0.250	+0.028 +0.003	+0.052 +0.027	+0.068 +0.043	+0.125 +0.100	+0.215 +0.190
OVER 160 TO 180	−0.230 −0.480	−0.145 −0.245	−0.043 −0.083	−0.014 −0.039	0.000 −0.025	0.000 −0.040	0.000 −0.100	0.000 −0.250	+0.028 +0.003	+0.052 +0.027	+0.068 +0.043	+0.133 +0.108	+0.235 +0.210
OVER 180 TO 200	−0.240 −0.530	−0.170 −0.285	−0.050 −0.096	−0.015 −0.044	0.000 −0.029	0.000 −0.046	0.000 −0.115	0.000 −0.290	+0.033 +0.004	+0.060 +0.031	+0.079 +0.050	+0.151 +0.122	+0.265 +0.236
OVER 200 TO 225	−0.260 −0.550	−0.170 −0.285	−0.050 −0.096	−0.015 −0.044	0.000 −0.029	0.000 −0.046	0.000 −0.115	0.000 −0.290	+0.033 +0.004	+0.060 +0.031	+0.079 +0.050	+0.159 +0.130	+0.287 +0.258
OVER 225 TO 250	−0.280 −0.570	−0.170 −0.285	−0.050 −0.096	−0.015 −0.044	0.000 −0.029	0.000 −0.046	0.000 −0.115	0.000 −0.290	+0.033 +0.004	+0.060 +0.031	+0.079 +0.050	+0.169 +0.140	+0.313 +0.284
OVER 250 TO 280	−0.300 −0.620	−0.190 −0.320	−0.056 −0.108	−0.017 −0.049	0.000 −0.032	0.000 −0.052	0.000 −0.130	0.000 −0.320	+0.036 +0.004	+0.066 +0.034	+0.088 +0.056	+0.190 +0.158	+0.347 +0.315
OVER 280 TO 315	−0.330 −0.650	−0.190 −0.320	−0.056 −0.108	−0.017 −0.049	0.000 −0.032	0.000 −0.052	0.000 −0.130	0.000 −0.320	+0.036 +0.004	+0.066 +0.034	+0.088 +0.056	+0.202 +0.170	+0.382 +0.350
OVER 315 TO 355	−0.360 −0.720	−0.210 −0.350	−0.062 −0.119	−0.018 −0.054	0.000 −0.036	0.000 −0.057	0.000 −0.140	0.000 −0.360	+0.040 +0.004	+0.073 +0.037	+0.098 +0.062	+0.226 +0.190	+0.426 +0.390
OVER 355 TO 400	−0.400 −0.760	−0.210 −0.350	−0.062 −0.119	−0.018 −0.054	0.000 −0.036	0.000 −0.057	0.000 −0.140	0.000 −0.360	+0.040 +0.004	+0.073 +0.037	+0.098 +0.062	+0.244 +0.208	+0.471 +0.435
OVER 400 TO 450	−0.440 −0.840	−0.230 −0.385	−0.068 −0.131	−0.020 −0.060	0.000 −0.040	0.000 0.063	0.000 −0.155	0.000 0.400	+0.045 +0.005	+0.080 +0.040	+0.108 +0.068	+0.272 +0.232	+0.530 +0.490
OVER 450 TO 500	−0.480 −0.880	−0.230 −0.385	−0.068 −0.131	−0.020 −0.060	0.000 −0.040	0.000 −0.063	0.000 −0.155	0.000 −0.400	+0.045 +0.005	+0.080 +0.040	+0.108 +0.068	+0.292 +0.252	+0.580 +0.540

APPENDIX 51
ENGINEERING FORMULAS

Motion

S = distance (inches, feet, miles)
t = time (seconds, minutes, hours)
v = average velocity (feet per second, miles per hour, etc.
v_1 = initial velocity
v_2 = final velocity
a = acceleration (feet per second per second)

(1) $S = vt$

(2) $V(\text{avg.}) = \dfrac{v_2 + v_1}{2}$

(3) $S = \left(\dfrac{v_1 + v_2}{2}\right) \dfrac{\text{ft}}{\text{sec}} (t \text{ sec})$

(4) $a = \dfrac{v_2 - v_1}{t}$

(5) $S = v_1 t + \frac{1}{2} a t^2$

Angular Motion

V = linear velocity
N = number of revolutions per min
θ = angular distance in radians
1 radian = $360°/2\pi = 57.3°$
ω(omega) = average angular velocity
$\quad = \theta/t$ (rad per sec, rev per min)
ω_1 = initial velocity
ω_2 = final velocity
α(alpha) = angular acceleration = rad per sec^2
S = length of arc
r = radius of arc
D = diameter

(6) $\theta = (\text{avg. } \omega)t$

(7) $\omega(\text{avg.}) = \dfrac{\omega_2 + \omega_1}{2}$

(8) $\alpha = \dfrac{\omega_2 - \omega_1}{t}$

(9) $\omega_2 = \omega_1 + \alpha t$

(10) $\theta = \omega_1 t + \dfrac{\alpha t^2}{2}$

(11) $V = \pi D N$ or $V = r\omega$ (ft per sec, ft per min, etc.)

(12) $\omega = \dfrac{2\pi r n}{r} = 2\pi N$

Force and Acceleration

F = force (pounds)
M = mass
a = acceleration (ft per sec^2)
g = gravitational acceleration = 32.2 ft/sec^2
W = weight (pounds)
$M = F/a = W/g$ (units of mass in slugs)

Work

W = work (ft · lb)
F = force (lb)
d = distance
$W = Fd$

Power

W = work
t = time

1 horsepower = $550 \dfrac{\text{ft} \cdot \text{lb}}{\text{sec}}$

Avg. power = $\dfrac{W}{t} = \dfrac{\text{ft} \cdot \text{lb}}{\text{sec}}$ or $\dfrac{\text{ft} \cdot \text{lb}}{\text{min}}$ etc.

Kinetic Energy

W = weight
V = velocity
g = 32.2 ft per sec^2
K.E. = $WV^2/2g$

APPENDIX 52
GRADING GRAPH

This graph can be used to determine the individual grades of members of a team and to compute grade averages for those who do extra assignments.

The percent participation of each team member should be determined by the team as a whole (see Chapter 2 problems).

Example: written or oral report grades

Overall team grade: 82

Team members $N = 5$	Contribution $C = \%$	$F = CN$	Grade (graph)
J. Doe	20%	100	82.0
H. Brown	16%	80	76.4
L. Smith	24%	120	86.0
R. Black	20%	100	82.0
T. Jones	20%	100	82.0
	100%		

Example: quiz or problem sheet grades

Number assigned: 30
Number extra: 6

Total 36

Average grade for total (36): 82

$$F = \frac{\text{No. completed} \times 100}{\text{No. assigned}} = \frac{36 \times 100}{30} = 120$$

Final grade (from graph): 86.0

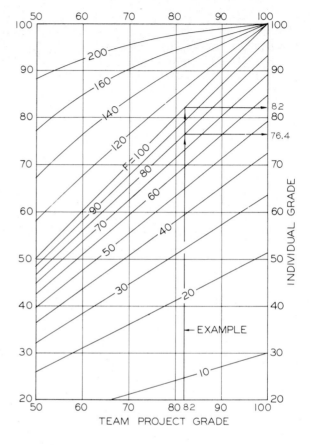

Fig. A50.—1. Grading graph.

INDEX

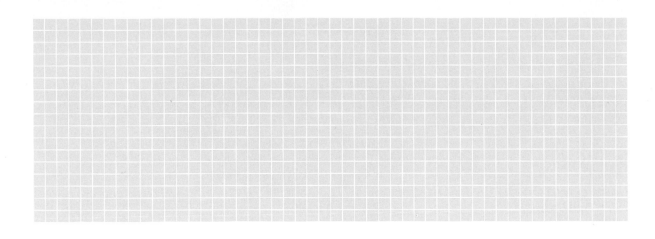